D0926515

Atlas of
the Breeding Birds
of Ontario

Atlas of
the Breeding Birds
of Ontario

Compiled by

Michael D. Cadman, Paul F. J. Eagles, Frederick M. Helleiner

Federation of Ontario Naturalists
and the
Long Point Bird Observatory

University of Waterloo Press

ISBN 0-88898-074-4

University of Waterloo Press
Dana Porter Library
University of Waterloo
Waterloo, Ontario, Canada
N2L 3G1

Printed at T. H. Best Printing Company
Don Mills, Ontario

Canadian Cataloguing in Publication Data
Main entry under title:
Atlas of the breeding birds of Ontario

Bibliography: p.
Includes index.
ISBN 0-88898-074-4

1. Birds - Ontario - Geographical distribution.
2. Birds - Ontario - Habitat. 3. Bird population -
Ontario. I. Cadman, Michael D. (Michael Derrick),
1955- . II. Eagles, Paul F. J. (Paul Franklin
John). III. Helleiner, Frederick M., 1933- .

QL685.5.06A75 1987   598.29713   C87-093832-0

Federation of Ontario Naturalists

Long Point Bird Observatory

The Federation of Ontario Naturalists was formed in 1931 when 7 local nature clubs and 28 individuals joined forces to "express the opinions and concerns of naturalists" in Ontario. The organization has since grown to 70 clubs and over 20,000 individuals. FON members have a common interest in the environment and wish to see it properly managed and protected. Ongoing efforts include wetlands conservation, environmental education, parks protection, nature-oriented trips and tours, curtailment of acid deposition, and the establishment and management of FON Nature Reserves. The FON served as the administrative home for the atlas project.

Federation of Ontario Naturalists
355 Lesmill Road
Don Mills, Ontario
M3B 2W8

The Long Point Bird Observatory, established in 1960 as the first of its kind in North America, is a membership-supported research organization dedicated to utilizing volunteer talents, energy and enthusiasm in collecting long-term population data on birds. Although centred on Long Point, the observatory conducts province-wide surveys on heronries, birds at feeders, loons on lakes and similar studies, of which the *Atlas of the Breeding Birds of Ontario* is the most ambitious to date. The observatory also extends its encouragement to amateur involvement in bird studies through its administration of an arms-length granting body, the James L. Baillie Memorial Fund for Bird Research and Preservation, a major sponsor of field work for the atlas project.

Long Point Bird Observatory
P.O. Box 160
Port Rowan, Ontario
N0E 1M0

Contents

Species and hybrids recorded in southern Ontario

Species recorded only in northern Ontario

Accepted breeders not reported in 1981-85 and miscellaneous species

Foreword

My dictionary describes an atlas as a book of maps, but that definition does not do justice to this volume. There are plenty of books that show maps, even maps of bird distribution: what distinguishes this atlas is the way in which these maps were produced. While most maps of plant and animal distribution are based on collections of more or less randomly gathered records, accumulated over a considerable period of time, the maps in this book represent the results of a great cooperative effort to document the breeding bird species in Ontario by covering the area thoroughly in a short period of time. It is a snapshot of bird distribution in the province, albeit a five year time exposure from 1981 to 1985.

When a group of ornithologists met in Waterloo in 1979, the concept of a Breeding Bird Atlas of Ontario was an idea whose time had come. Not that the concept was totally new. It had been pioneered by the *Atlas of Breeding Birds in Britain and Ireland*, published in 1976. The idea had already spread to continental Europe, Africa, and Australia, and several atlases were already under way in the United States. Nor was there a dearth of information about the distribution of breeding birds in Ontario. As early as 1886, T. McIlwraith had published *The Birds of Ontario*. J.L. Baillie and P. Harrington had summarized what was then known in *The Distribution of Breeding Birds in Ontario* in 1936 and 1937. Several regional studies detailing local distribution had been published, including H.R. Quilliam's *History of the Birds of Kingston, Ontario* (1965). W.E. Godfrey's *Birds of Canada* (1966) contained range maps and a revision was in preparation (1986). More detailed maps were shortly to appear in the two-volume work *Breeding Birds of Ontario: Nidiology and Distribution* (Vol. 1, 1983), by G.K. Peck and R.D. James, based on nest record cards. J.M. Speirs was preparing his *Birds of Ontario* (1985) which would contain maps based on breeding bird surveys. Yet none of these or other valuable sources that were or have since become available provided up-to-date, detailed data on the distribution of Ontario birds comparable to that in the British Atlas. For the steering committee that continued to meet, such a goal seemed both desirable and attainable.

It was an exciting challenge. Ontario spans over 1,500 km from the remnant Carolinian forests bordering Lake Erie in the south to the arctic tundra on the shores of Hudson Bay in the north. It covers an enormous area. While southern Ontario is settled and mostly accessible, much of northern Ontario is wild and remote. The avifauna of the province is rich and varied, including 75% of the bird species in Canada. Fortunately, there has always been a substantial number of naturalists interested in birds and that number was growing rapidly. Moreover, there were organizations through which birdwatchers could be reached and cooperative efforts mobilized. As it turned out, more people took part than we could have guessed - over 1,300 by the time the field work for the Atlas was completed.

Conducting this type of survey is a very different proposition from the gradual accumulation of records as they occur in the normal course of events. Because of the need for a huge amount of field work in a short period of time, a great deal of organization and coordination is required. In some ways it resembles a military exercise. A number of different elements are absolutely essential to make a project like this work and the Ontario Breeding Bird Atlas was exceptionally well served in this respect.

We received much encouragement from our two sponsoring organizations, the Federation of Ontario Naturalists (FON) and the Long Point Bird Observatory (LPBO). The FON provided a corporate identity and a roof over our heads, and backed the publication of the Atlas. The LPBO provided much needed expertise as well as tangible help to atlassers from the James L. Baillie Memorial Fund.

A great deal of financial and other assistance was provided from outside sources, enabling us to do an organized and thorough job. The Canadian Wildlife Service together with Supply and Services Canada got us off the ground and supported us all the way. Steve Curtis and Dan Welsh were especially helpful. The Ontario Ministry of Natural Resources provided both financial and logistic help. Tim Millard and his staff were very supportive - Irene Bowman deserves special mention - and many field staff of the Ministry took part in atlas surveys. We have the expertise of the Canada Lands Directorate to thank for the excellent maps that grace this book: Nicole Chartrand and Mike Comeau played key roles in the development and implementation of the computer mapping system. The University of Waterloo was very generous with space and facilities for our staff and served as our publisher. Additional staff were hired and trained through funding from the Department of Employment and Immigration. We received grants from the Ontario Heritage Foundation and from a number of private organizations: the Richard Ivey Foundation, the Westwind Charitable Foundation and the Sir Joseph Flavelle Foundation. World Wildlife Fund Canada provided both grants and the assistance of their staff. To all these must go our profound appreciation.

Many distinguished naturalists and professionals served on committees and met frequently to develop policies and methods, find support, scrutinize the records, design the Atlas, and plan its publication and sale. Authors, reviewers, editors, and artists laboured to produce an accurate, informative and attractive volume. A few individuals were crucial. George Francis, who guided the steering commitee, served as our first general chairman. Paul Eagles provided leadership in many ways and chaired the publication committee. Fred Helleiner chaired the committee overseeing the surveys of northern Ontario. These last two, together with Mike Cadman, took on the daunting task of editing a third of the species accounts each, and their names appear as the compilers of this volume. Three others, whose labours behind the scenes were indispensible, deserve special mention. David Hussell chaired the important technical advisory and data review committees. As part of these demanding tasks, his critical advice on many aspects of the project undoubtedly improved the quality of the atlas. Ross James was a member of several important committees, where his expertise on Ontario's avifauna contributed to the accuracy of the final product. In addition, he assembled the artwork, which complements the species accounts. Dan Welsh not only served as our direct link with Environment Canada, but also took an active role in data management. We are, of course, greatly indebted to Robert Bateman for his evocative cover illustration. Other individuals who helped in special ways are mentioned in the Acknowledgements.

We were ably served by our staff: David Balser, Mike Cadman, Karen Hall, Judith Kennedy, Anne Nash and Peter Tallon. Others worked for shorter periods. Dave was our computer expert, and everyone helped to dig us out from under mounds of data. Mike, our coordinator, did almost everything and kept us organized, busy and remarkably happy! He was also an author and our principal editor. We have been incredibly fortunate to have all these excellent people working with us.

Regional Coordinators, all volunteers, played a crucial part. They were called upon to do a staggering amount of paper work: for what use are countless bird records if they are disorganized, transcribed incorrectly or unverified? One can only explain their efforts as a labour of love.

Finally, the most important element in this cooperative endeavour was provided by the atlassers who conducted the field surveys. Of course they (I should say we, for nearly all of us were involved) were doing what comes naturally - birding and making lists for a constructive purpose - but that scarcely does justice to the knowledge of birds, long hours and travels so freely donated. Without this 'grass-roots'

support, no one could contemplate, much less afford a bird atlas on this scale. The atlassers are listed in the Acknowledgements.

This Atlas represents the fruits of our labour. It is a mine of information for the reader. Many intriguing patterns of distribution and abundance are depicted in these pages. One has only to compare the distributions of different species or to view them against the background maps in the Introduction to get an inkling of the possibilities. Scientists will study these patterns as illustrated by examples in the end papers. Resource agencies, conservationists and planners will find the atlas data an important component of the background information required for the designation of rare species, the protection of natural areas and rational land-use. For the naturalist this book will predict what is to be found in a given area and, by omission, what is a new and exciting find.

At the outset I described the Atlas as a snapshot through a window in time. This means that it is as up-to-date as we could make it when the survey was completed. It also means that it can serve as a baseline against which to identify changes which may take place in the future. Eventually naturalists may wish to take another snapshot to compare with this one. No doubt some of the readers of this volume will take part in such an effort. In the meantime there will be much new information about birds to add to the record. Moreover, other groups of plants and animals may be treated in the same way. This first atlas of its kind in Canada is an example of what can be done and a tribute to all who took part in its production.

J. Bruce Falls
Chairman, Atlas Management Committee

Acknowledgements

A project of the magnitude of the Ontario Breeding Bird Atlas can work only with the cooperation of a large number of organizations and individuals. We are truly fortunate to have been the benefactors of the time, effort, funding, commitment and energy expended on behalf of the atlas project by such a large number of talented people.

The following is a listing of organizations and people who helped make the project work, classified by their contribution:

Sponsoring Organizations

- Federation of Ontario Naturalists
- Long Point Bird Observatory

Atlas Supporters

- Austin Airways (transportation)
- Bell Canada (transportation)
- Canada Department of Employment and Immigration (funds, staff)
- Canada Department of Indian Affairs and Northern Development (funds)
- Canada Department of Supply and Services (funds)
- Digital Equipment Corporation of Canada (computer equipment, software)
- Environment Canada - Canadian Wildlife Service (funds, computer advice, lodging, logistics)
- Environment Canada - Lands Directorate (computer mapping)
- Environment Canada - Parks Service (funds)
- Federation of Ontario Naturalists (administration)
- Gulf Canada (funds)
- The James L. Baillie Memorial Fund for Bird Research and Preservation (funds)
- Long Point Bird Observatory (technical advice, fund-raising)
- Ontario Geological Survey (transportation and lodging)
- Ontario Heritage Foundation (funds)
- Ontario Ministry of Natural Resources (funds, transportation, lodging, logistical support)
- The Richard Ivey Foundation (funds)
- The Sir Joseph Flavelle Foundation (funds)
- University of Waterloo, Department of Recreation and Leisure Studies (office space, computer equipment and software, accounting, office supplies)
- The Westwind Charitable Foundation (funds)
- World Wildlife Fund Canada (funds)

Committee Members

Management Committee

Chairman: (1980-1982, Committee member 1983-1987), G.R. Francis, Department of Environment and Resource Studies, University of Waterloo

Chairman: (1983-1987), J.B. Falls, Department of Zoology, University of Toronto

I. Bowman (1980-1987), Wildlife Branch, Ministry of Natural Resources

P.F.J. Eagles (1980-1987), Department of Recreation and Leisure Studies, University of Waterloo

C.D. Fowle (1980-1987), Department of Biology, York University

C.E. Goodwin (1980-1981), Ontario Regional Editor, *American Birds*

D.J.T. Hussell (1980-1987), Long Point Bird Observatory, and later, Wildlife Research Section, Ministry of Natural Resources

R. Knapton (1980-1986), Department of Biology, Brock University

H.G. Lumsden (1981-1987), Wildlife Research Section, Ministry of Natural Resources

M.K. McNicholl (1986-1987), Long Point Bird Observatory

D.A. Sutherland (1983-1987), Ontario Field Ornithologists

D.A. Welsh (1980-1987), Ontario Region, Canadian Wildlife Service

Technical Advisory Committee

Chairman: D.J.T. Hussell (1981-1987)
P.F.J. Eagles (1981-1987)
F.M. Helleiner (1981-1987), Department of Geography, Trent University
R.D. James (1981-1987), Department of Ornithology, Royal Ontario Museum
H.G. Lumsden (1981-1987)
A.L.A. Middleton (1981-1987), Department of Zoology, University of Guelph
D. Sadler (1981-1987), Regional Coordinator, Peterborough
D.A. Sutherland (1983-1987)
D.A. Welsh (1981-1987)

Publication Committee

Chairman: R. Knapton (1982-1985)
Chairman: P.F.J. Eagles (1985-1987) (Committee member 1982-1987)
I. Bowman (1985-1987)
J.B. Falls (1985-1987)
G.R. Francis (1985-1987)
F.M. Helleiner (1985-1987)
D.J.T. Hussell (1985-1987)
R.D. James (1985-1987)

Data Access Committee

Chairman: G.R. Francis (1983-1987)
C.D. Fowle (1983-1987)
D. Sadler (1983-1987)

Data Review Committee

Chairman: D.J.T. Hussell (1984-1987)
R. Curry (1984-1987), Regional Coordinator, Hamilton
R.D. James (1984-1987)
R.B.H. Smith (1984-1987), Regional Coordinator (1983), Long Point
D.A. Sutherland (1984-1987)

Northern Committee

Chairman: F.M. Helleiner (1982-1985)
R.D. James (1982-1985)
H.G. Lumsden (1982-1985)
J.P. Prevett (1982-1985), Southwestern Region, Ministry of Natural Resources
D.A. Welsh (1982-1985)

Data Management Committee

Chairman: D.A. Welsh (1982-1987)
P.F.J. Eagles (1982-1987)
D.J.T. Hussell (1982-1987)

Staff

Coordinator: Michael D. Cadman (1981-1987)
Assistant Coordinator:
Deborah Mojzes (1981)
Claudia Thompson (1981)
Anne Nash (1982-1983)
Judith Kennedy (1984-1985)
Karen Hall (1986-1987)
Database Manager: David A. Balser (1984-1987)
Editorial Assistants:
Leslie Beattie
Karen Jeffrey
Liza Ordubegian
David Prescott
Peter Tallon

Regional Coordinators

- listed in Table 1

Authors

- credited at the end of each species account

Editors of species accounts

M.D. Cadman, P.F.J. Eagles, G.R. Francis, F.M. Helleiner, D.J.T. Hussell, R.D. James, D.R.C. Prescott

Reviewers of species accounts

K. Abraham, E.R. Armstrong, M. Austen, M. Biro, H. Blokpoel, C. Blomme, G.R. Bortolotti, I. Bowman, A. Bryant, D. Bucknell, M.D. Cadman, J. Carley, G. Carpentier, T. Cheskey, J. Coey, F. Cooke, D.P. Coulson, H.G. Currie, R. Curry, K. De Smet, D. Dennis, P. Dent, B. Duncan, E.H. Dunn, P.F.J. Eagles, A.J. Erskine, J.B. Falls, N. Flood, D.M. Fraser, M.E. Gartshore, M. Gawn, C.E. Goodwin, A.G. Gordon, J. Harcus, R. Harris, R. Hawker, F.M. Helleiner, D.J.T. Hussell, R.D. James, R. John, J. Kamstra, R. Knapton, A.B. Lambert, H.G. Lumsden, V. Macins, J.D. McCracken, W.D. McIlveen, M. McLaren, P.L. McLaren, D. McNicol, M.K. McNicholl, A. Mills, R.D. Montgomerie, R.I.G. Morrison, D.J. Mountjoy, B.J. Naylor, E. Nol, T.D. Nudds, M. Oldham, M. Parker, G.K. Peck, B. Penak, S. Peruniak, D.R.C. Prescott, J.P. Prevett, T.E. Quinney, J.D. Rising, C.J. Risley, S.S. Rosa, R.K. Ross, C.J. Sanders, D.M. Scott, R.B.H. Smith, D.H. Speirs, J.M. Speirs, D. Strickland, D.A. Sutherland, K.J. Szuba, R. Tozer, C.A. Walker, P.J. Weatherhead, R.D. Weir, D.A. Welsh, D.V. Weseloh, W. Wilson, P.A. Woodliffe

Text reviewers

M.D. Cadman, E.H. Dunn, P.F.J. Eagles, J.B. Falls, G.R. Francis, F.M. Helleiner, D.J.T. Hussell, A.B. Lambert, R.B.H. Smith, D.A. Welsh

Illustrators

C. Blomme, I. Bowman, D. Fewster, J. Hawke, S. House, R. James, I. Jones, C. McManiman, T. Norris, J. Schmelefske, M. Tasker, R. Tuckerman

Graphic Design

Carol Sabean, David Bartholomew

Base Map Production

Barry Levely

Cover Illustration

Robert Bateman

Atlassers

Ken Abraham, Rita Adams, Christine Adderson, William Addison, David Agro, Vel and Jamshed Ahmed, Bruce Aikins, J. Ainger, P. Aird, Michael Akey, Value Alatalu, K. Albo, Hugh Aldis, Algonquin Park Museum, Christie Allan, Tom Allaway, Margaret Allen, Esther C. Allin, Ken Allison, Max Alton, David Ambridge, Alice Anderson, Kathleen Anderson, Norval T. Anderson, Robert F. Andrle, B.T. Aniskowicz, C.D. Ankney, Maris Apse, Robert Ardiel, Agnes Armstrong, Fred Armstrong, J. Mac Armstrong, Jean Armstrong, John B. Armstrong, Ted Armstrong, Rob Arnup, K. Elizabeth Arthur, Don Arthurs, Robert Ashall, Jerry Asling, Doug Asquith, Ian Attridge, Allan Aubin, Yves Aubrey, Jeff Aufmann, Madeline Austen, Merv Austin, Sandy Austin, F. Avis, Harold H. Axtell

Adolf Baier, Ed Bailey, Phil Bailey, Barbara Bain, Margaret Bain, Heather Baines, Penny Baines, Mary H. Baird, Pam Baird, Sue Bajcor, Roy Baker, Bob Baldrey, Donald H. Baldwin, P.W. Ball, David A. Balser, Peter E. Bannon, Diana Banville, D. Banwell, Gerry Barbe, Mrs. Barber, Rosalind E. Barden, Diana Barker, Ray Barker, B.G. Barnard, Dean Barner, Fred Barrett, Harry B. Barrett, Dennis Barry, Tim Barry, Jo Barten, J.E. Bartl, Howard Battae, Ron Bauman, John Bax, Don Baxter, Tom Baxter,

D.J. Bazinet, R. Beach, Ron Beacock, Leslie Beattie, Moira Beauchesne, Gregor Beck, O. Bedard, Agnes Beemer, Robert Behrstock, Ron Belanger, Don Belfry, Arthur E.S. Bell, Chris Bell, Frank H. Bell, John Bell, John K. Bell, Karen Bellamy, Gordon Bellerby, Anne Belton, J. Bendell, Craig Benkman, Carol Bennett, Gerry Bennett, Ken Bennison, Roger Bernard, L. Berry, Godfrey and Marjorie Bethune, Greg Betteridge, Lois and Gordon Bews, Tony Bigg, F.J. Billinghurst, C.B. Birchard, Gerry and Hazel Bird, Michael Biro, Steve Bissell, Paul Bisson, John E. Black, Keith C. Black, R. Black, Ron Black, Walter Blackburn, Martin Blagdurn, Cathy Blake, Richard P. Blake, Peter Blancher, John Blaney, Sean Blaney, J. Blayden, Nancy Blogge, Hans Blokpoel, Chris and Gloria Blomme, Steve Blomme, F. Bloomer, Fred Bodsworth, Neville Bodsworth, Arthur Boissonneau, Margarette Bon, Peter Bondy, E. Boote, Gilian Booth, J.S. Booth, Judy Borer, Gary Bortolotti, Marc Bosc, Mark Bosch, Michael Bostock, Wilf Botham, C. Bouchard, D. Bouvier, Jacques M. Bouvier, Bob Bowles, R. Bowles, Irene Bowman, Brian Boyce, Alan and Muriel Boyd, Sylvia and Peter Boyle, Robert A. Bracken, Sidney Bradbury, Keith Bradle, Fred Bradley, G.A. Bradley, J. Bradley, W.L. Bradley, Michael Bradstreet, Robert O. Braley, Marianne Braun, Neil Bredin, Gary Breedyk, Kay Brennan, R.J. Brenner, Dave Brewer, Kathy Bricker, Ernie Bridger, Paul J. Bridges, Art Briggs-Jude, Monty Brigham, R. Bringeman, H. and R. Brintnell, John Brisbane, Bill Briscoe, Gail Brodie, Marie Brooks, Joan Brooksbank, Dave Broughton, Richard Brouillet, Vivian Browmwell, Chris and Debbie Brown, Dale Brown, Harry Brown, James and Margaret Brown, Kathleen Brown, Quentin Brown, Tom Brown, Jo-ann Brown Besner, Alyson Browne, Isobel Browne, Mark H.R. Browning, Ghislain Brunet, Dan Brunton, Mark Brunton, Mike and Sue Bryan, Andrew Bryant, Mary A. Bryant, Bruce Brydon, Don Bucknell, Jackie Budd, Richard Buell, Steve Buneta, Brian Bunn, D. and B. Burcher, Bill Burgess, Jane Burgess, Joe Burgess, K. Burk, Anna Burke, Peter Burke, Kirsten Burling, Robert Burton, Len Butcher, Caroline Butson, M. and T. Button, C. Buxton, Marshall Byle

Mike Cadman, John and Kay Calder, Bill Calvert, Gordon Cameron, Joy and Howard Cameron, Craig Campbell, Danny Campbell, Drew Campbell, John Campbell, Kaye Campbell, Lloyd Campbell, Malcolm Campbell, Marie Campbell, R.M. Campbell, Terry Campbell, Theresa Campbell, Jil Campling, Peter Cannell, Jacques Cantin, Rob Capell, Alexander Cappell, Lorraine Capstick, Ernie Carhart, John Carley, Bob Carlisle, Barbara Carlson, Margaret Carney, Ann Carpenter, Donna Gibbs Carpenter, A. Geoffrey Carpentier, Brian Carson, Betty Carter, Eric Carter, Glen Carter, Jonathan Carter, Ron Cartmell, John Cartwright, Jim Cashmore, Pam Cathrae, Rick Cathrae, Annabel Cathrall, Don and Barb Cavin, Pierre Cerenelli, Jacques Chabot, Pierre Chagnon, Scott

Chalmers, Joan Chamberlain, K. Chamberlain, Dave Chantler, Paul Chapman, Joan Chapple, J.P. Charlebois, Barbara Charlton, R. Charlton, Allen Chartier, Sharon Chatfield, Rosalind Chaundy, B. Cherriere, Robert Cheskey, Ted Cheskey, Norm Chesterfield, Alice Chisholm, Roger Chittenden, Mrs. Paddy Chitty, Mark Chojnacki, Patsy Christgau, David Christie, Peter Christie, W.B. Christie, Charles A. Christy, Philip Chu, John Church, David E. Clapp, Bill Clark, Caryl Clark, Jo-Ann Clark, Michael Clark, Roy Clark, Sherri Clark, Tom Clark, Will Clark, Dale Clarke, J.C. Clarke, Gary Clay, Mike and Pat Cleary, Mrs. Alec Clements, Brad Clements, Gerry Clements, V. Clements, Joyce Cloughly, Larry Cluchey, Glenn Coady, W.M. Coburn, Connie Cochrane, Madge Cochrane, Bill Cockburn, Jim Coey, Dave Coffey, David Coffman, Robert D. Cohen, Brian Coleman, Sheila Collett, Bruce Collins, Beverly Combe, Lyla Commandant, Scott Connop, Laurie Consaul, Frank Cook, Glen Cook, Margaret Cook, Fred Cooke, James V. Cookson, D. Copeland, Robert Copeland, John Coppell, Peggy and Al Corbould, John Corveth, David Cotterell, John Cotterell, Daryl P. Coulson, Ken Coulthart, Fred Cowell, Beth Cowieson, Bob Cox, Ellie T. Cox, Ernie Cox, Gerry Cox, Irma Coyne, Terry Crabe, Lisa Craig, Jack Cranmer-Byng, Janice Craske, Dan Crawford, Eleanor Crawford, George B. Crawford, John R. Cree, Jan Crider, W.J. Crins, H. Croal, Joel and Mary Crookham, B. Cruickshank, Mary Cryer, Dorothy Crysler, David Cuddy, Don Cuddy, Catherine Culbert, Walter Cullingworth, David Cunning, Dave Cunningham, Robin Cunningham, Hugh Currie, Bob Curry, Mrs. Curry, Grant A. Curtis, Steve Curtis, Lucille Cusson, W.A. Cutfield, Eddie Czerwinski

Joan Dale, G. Dales, J.H. Dales, Dennis and Darlene Dalke, C.A. Dalker, Kenneth Dance, Leslie Dandy, Jim Daniel, Julie Daoust, Ian Davidson, Larry Davies, Manuela G. Davilla, Glen Davison, Arn Dawe, Kenneth Dawe, Les and Pam Dawson, Rick Dawson, Dave De Boer, Ken De Smet, Pat Deacon, Rod Deakin, Joanne Dean, Joyce DeBoer, Mary Deer, Henrik Deichmann, Roy DeJong, Michael Delorey, Norman E. Denis, Pam Dennett, Allan Dennis, Darrell Dennis, Russ Dennison, Peter Dent, Carol Dersch, Francois Dessureault, K. Dewar, Terry Dickie, I.W. Dickson, Ron Diebolt, Bruce DiLabio, Lorne J. Dittmer, Kevin Dixon, Grant Dobson, Rick Dodd, Robert Donald, Murray Doney, Joan Doolittle, P. Dorais, Bob Dorney, Michel Dostaler, Tom Doubleday, Roger Doucet, C. Douglas, Jean Douwes, J. Dowall, Rick Dowson, Erv Drabek, Wayne Drabick, Rasa Draugelis, Laura Dubblestein, Terry Duguid, M. Duke, Faye Dunaway, David Dunbar, Bruce Duncan, G.J. Duncan, Les and Norrie Dunks, Erica Dunn, Peter Dunn, Pat and Jim Dunsire, Mervin Dupuis, Michael Dyer, Tom Dyke, Tim Dyson

Paul F.J. Eagles, Russell Eagles, Ryan Eagles, Gord Eason, Brian Eaton, Christopher Eckert, Bill Edmunds, Colin

Edwards, Joan Edwards, R.K.F. Edwards, Dave Elder, Michelle Elder, Carol Elion, Chris Ellingwood, David Ellingwood, Herb Elliot, Margaret Elliot, Joel Ellis, Marg Engberg, Jack Enns, Jill Entwhistle, A.J. Erskine, Nick Escott, William Este, Dave Euler, Chris Evans, Garfield Evans, Jack Evans, Jayne Evans, L. Verne Evans, Mike Evans, Ross Evans, Brent Evered, Duncan Evered, Mary Ewert

Steven D. Faccio, Earl Fairbanks, Jim Fairchild, George and Jean Fairfield, Joan Fairhead, John Fallis, J. Bruce Falls, Stephen Falls, Blayne Farnan, Terry Farquarson, Martin Farrell, K. Fawthrop, Luc and Xavier Fazio, Jane Feltz, Blair Ferguson, Don Ferguson, Jim Ferguson, Mark Ferguson, Wayne Ferguson, Carl Ferren, David Fewster, Marie Fick, David Fidler, Norm Fidler, Pat Field, Bill Fielding, C. and I. Fillion, Don Fillman, Tom Filose, Terry Findley, Barb Finlay, Mrs. D. Finley, Charles Finnigan, Owen Fisher, Bill Fitch, John L. Fitchett, Manson K. Fleguel, Deb Fletcher, Hazel Fletcher, R.M. Fletcher, Michael Flieg, Norman Flindall, Joachim W. Floegel, Nancy Flood, Ethel Fogarty, W.W. Fogelman, Dale Fogle, M. Foley, Kay Forbes, Bruce Ford, David Ford, W.H. Forman, Roy Forrester, Richard Forster, Alec Foster, Ivan and Fran Foster, John F. Foster, Murray Foster, Marc Fournier, C. David Fowle, Don Fowler, Helen Fowler, Roger Foxall, George Francis, John Francis, Jon Francis, Velma Franklin, Peter M. Franks, Al Fraser, Donald M. Fraser, Greg Fraser, J. Keith Fraser, John G. Frazer, Teresa Freschette, Victor Froese, R.D. Fry, Mike Furber

Sandy Gage, Denis Gagne, Allison Gagnon, Ruthann Gairdner, Peggy Gale, T.D. Galloway, Denys Gardiner, R.E. Gardiner, Irv Gardner, R. Gardner, Grant Garette, Tom Garner, Jim Garratt, Mary E. Gartshore, Chris Gates, A. Gauthier, M. Gauthier, M. Gawn, Stephen Gawn, Rosemary D. Gaymer, John Geale, Bob Geddes, Martin Geleynse, Allie Gemmell, Greg Gemmell, Connie and Roy Genge, Alan German, Carol German, Elsie German, Richard Gerson, Hal Gibbard, Brian L. Gibson, David Gibson, Jeff Gibson, Susan M. Gibson, Taimi Gibson, Gregor Gilbert, Terry Gill, F. Ed Gillan, Charles Gillard, Michael Gillespie, G. Girdwood, William G. Girling, E. Glanert, Jim M. Glenday, John D. Gliddon, Bev Glover, Dale Goble, J.H. Godden, Corey Goldman, Jim Goltz, Peter Good, Clive Goodwin, Alan G. Gordon, Donny Gordon, Ruth Gordon, Bob Gorman, Alyce Gorter, Annette L. Gosnell, Cathy Gouchie, Joyce Gould, Pete Goulding, Duncan Gow, Mary Gow, Marcus and Clare Grace, Richard Graefe, Roger Graf, Don Graham, Gordon Graham, H. Graham, Jim Graham, David Grainger, E.E. Grainger, Joan Grainger, Janet Grand, John Grandfield, Mrs. Alex Grandy, Victor Granholm, Don Grant, Josephine Grawbarger, Alex Gray, Ann and Brian Gray, Bob Gray, Ken Gray, Peter Greathead, A. Green, Allan Greenbaum, Andy Greenwood, Craig Greenwood, Eileen Greenwood, Alec Grey, Sheila Gribble, Ann Griffin, Gail and Barry Griffith, Barry Griffiths, Marion and George Grisdale, M. Groenewegen, Judy Grossmam, Jill Guertin, Donald H. Gunn, W.W.H. Gunn, Don Gunter

Sylvia Haavisto, John Haddock, Mrs. D. Hadwin, Erik Hagberg, John G. Haggeman, Jim Haggert, Roy Hakala, Anne Hall, Don and Vivian Hall, Karen Hall, Jean Hall-Armstrong, Larry Halyk, Harry Hamblin, Jack Hamer, Stewart Hamill, Doug Hamilton, Emily Hamilton, Heather Hamilton, J.G. Hamilton, Ron Hamilton, Fred Hammel, G. Hamse, Ron and Sharon Hancock, Dorice Hanes, G.R. Hanes, Alan Haney, Gary Hanna, Jack Hanna, Violet Hanna, Sandra Hannah, Christine Hanrahan, Tom Hanrahan, Phyllis Hansen, Ruth Hansen, Chuck Hansman, Feriel Hanyk, John Harcus, Doug Harding, John Hardy, Lynn Hardy, Douglas Hargrave, Helen Hargrave, Paul Harpley, Allan Harris, Allen Harris, Chris Harris, Jim Harris, Ross Harris, James Harrison, Mary Harrison, Hendrik Hart, Ron Hartford, Margaret Hartley, Robert Hartley, Alf and Laura Hartwick, G.T. Harvey, Jeff Harvey, Jim Harvey, John L. Harvey, J. Stanley Hastings, Sheila and David Hatch, David Hawke, Gwen Hawke, Juliana Hawke, Bob Hawker, George Hawkes, Rick Hawkins, Tom Hayman, Todd Haynes, Jean Heagy, Ian Heales, Bob Healey, Carl E. Hearn, Miles Hearn, Mildred Heatherton, K. Joan Hebden, Muriel Held, Fred Helleiner, Stephen Helleiner, Vin G. Helwig, Bill Henderson, Margaret Hendrick, Donald G. Hendriks, Kurt Hennige, Greg Henry, Iain Hensby, Gary Henson, Bill Henwood, Vonnie Heron, Al Herriot, Jim Heslop, W.E. Hewitt, Leo Heyens, D. Heyland, Steve Hiard, Verna Higgins, Gordon Higginson, Tom Hilditch, A.W. Hill, Jane Hill, Mrs. P.C. Hill, Ronald G. Hill, William Hill, Willie Him, Tom Hince, Fred Hindle, Frances Hindmarsh, Norm Hissa, Doug Hodgins, S.M. Hofstetter, Lambert Hogenbirk, P. Hogenbirk, Aileen Hogg, Mr. and Mrs. J. Hogg, George Holborn, James Holcroft, Chris Holding, George Holland, Bill Hollinshead, Mrs. Hollo, Margo Holt, Phyllis Holtz, William M. Honsberger, June Hooey, Kenneth Hook, Bob and Mary Hooper, Barb Hoover, Richard Hopf, C.A. Hopkins, Sharon Hopper, Nick Hordy, Bill Houghton, Gerry Howe, Laura Howe, Doug Howell, Heather Howkins, P. Hubbard, Larry Hubble, Bob Hubert, Dorothy Hughes, Jim Hughes, Tjerk Huisman, Ken Hulley, James Hunt, Muriel Hunt, Susan Hunt, Evelyn Hunter, Gilbert H. Hunter, Carmen Hunting, G. Huot, June Hurley, Joe Hurst, David Hussell, Helen Hutchinson, Bob Huxley, Ian Hyatt, Bruce Hyer, Colleen Hyslop

Sharon Illingworth, Wayne Ilton, Helen and Spencer Inch, Don Inkpen, Ian Ireland, Nancy Ironside, H.J. Irwin, Kathy Irwin

Peter Jaber, Warren Jacklin, Judy Jackson, Ross D. James, Pat Jamieson, Ernie Jantzen, Al Jaramillo, W.R. Jarmain, Grant Jarvis, Ron Jasiuk, Beth Jefferson, Dave Jeffrey, Elsa

Jennings, Herb Jennings, Jack Jennings, Mark Jennings, R. Jennings, Roy John, Ernie Johns, Cindy Johnson, Don Johnson, E. Johnson, John Johnson, Joseph W. Johnson, Mark Johnson, David H. Johnston, Marion Johnston, Pauline Johnston, Harry Johnstone, Helen Johnstone, Dave Jolly, Barry Jones, Ben Jones, James E. Jones, Ken and Sheelagh Jones, Michael Jones, Paul Jones, Scott Jones, Randy Jorgensen, Helen Juhola, N. Juhtund

Andreas Kaellgren, D. Kains, Eva Kaiser, J. Kaiser, Herb Kale, Miriam Kalliomaki, Freda Kamstra, James Kamstra, John Keast, Ross and Anne Keatley, J.R. Keddie, J.G. Keenlyside, P. Kehoe, Alice Kelley, John Kelley, Fred Kempf, Shelley Kempf, Dan and Jan Kennaley, David Kennedy, Judith Kennedy, Lynn Kennedy, P. Kennedy, Bruno Kern, Donald J. Kerr, Harry Kerr, Wayne Kerr, W.L. Kershaw, Mike Kilby, Rick Killeen, Dave Kinelly, Bill and Sharon King, Frank Kingdon, Ron Kingswood, Don Kirk, Mac Kirk, Mrs. Kirkpatrick, C. Kitney, C.A. Kitson, Erik Kiviat, J. Klatt, Vernon M. Kleen, Joe Kleiman, Shirley Klement, Rose Klinkenberg, R.A. Klips, Richard Knapton, Donna Knauber, George W. Knoeklein, R. Knowles, David Knowlton, David Knox, Ken Knox, Mairi Kobylanski, Rudolf Koes, G.B. Kolenosky, Don Komarechka, Indrek Kongats, Karl Konze, Marie-Luise Kopp, Ivar Kops, Joan Kotanen, Peter Kotanen, Philip Kotanen, Dan Kott, S. Kozak, James Krasovec, Staff Kratz, Bernd Krueger, Howard Krug, John Kruse, Mark Kubisz, Alis Kuhn, Albert Kuhnigk

M. L'Estrange, Robert M. Labre, Bernie Ladouceur, Diane Ladouceur, Mrs. Blair Laidlaw, A.D. and Betty C. Laing, R. Lake, Wendy Lalouette, Larry Lamb, Anne Lambert, Kathie Lambert, David Lamble, John Lamey, Bill Lamond, H. Lancaster, A. Landon, Charity Landon, Mia Lane, Marc Langlois, Jim Langstaff, Dave Larocque, Dale Larson, Jeff Larson, Brian Lasenby, Jocelyn Lauber, Eva Laubitz, Sarah B. Laughlin, Bob Laughton, Denise Lauzon, Pat Lavoie, Lewis Laweder, Betty Learmouth, Cheryl Learn, Peter Leaver, M. Lecocq, Danny Lee, Gerry Leering, A. Legge, P. Lehman, P. Lehmann, Grace E. Leigh, Mrs. D. Lein, C. Lemieux, O. Lemke, John Lemon, Margaret Lemon, T.N. Lemon, Steven Leonard, Norman Leppan, John A. Lester, R. Levoy, Dennis Lewington, Robert H. Lewis, Victoria Lewis, Ed LeBlanc, Joseph Libera, Ilme Liepins, M. and B. Lindeman, K.M. Lindsay, Bill Linklater, Alf Liston, Mrs. H. Litt, Peter Litt, Judy Little, Kathy Little, Allan Liversage, Helen J. Lloyd, Myron Loback, Tom Lobb, Ross Lock, Evelyn B. Loker, B. and S. Long, Charles Long, Frank Longstaff, Ernie Lord, Lynn Lougheed, D. Love, R. Lowe, Mrs. W. E. Luckner, Rick Ludkin, Jack Ludwig, Cherry Luening, H.G. Lumsden, Andrew Lynch, Karen and Merrill Lynch, Ron and Lynda Lyons

Ada MacDonald, Anne MacDonald, Georgina MacDonald, Irene MacDonald, John MacDonald, Ken J. MacDonald,

John Reynolds, Madeline A. Richard, Jim Richards, Mervyn J. Richards, N. Richards, Ian Richardson, Lynne Richardson, Peter Richardson, Percy W. Richter, Alf H. Rider, Eric Ridgen, Ron Ridout, J. Riley, Christopher Rimmer, K. Rinta, J.D. Rising, Chris Risley, Don Rivard, Tony Roach, Roxanne Roberti, J.O.L. Roberts, R. Roberts, R. Robertson, Vic Robins, Janice Robinson, Susan Robinson, Dorothy Robson, Alison Rodrigues, E. Rodriguez, Mrs. H. Roestenberg, Bruce Rogers, Ruby Rogers, Pan Rogerson, Randy Romano, Isabella Rombach, Gisela Roos, D. Ropke, Sam S. Rosa, Carl Rose, Michele Rose, Paul Rose, Steve Rose, J.D. Roseborough, S. and D. Rosenberger, Cary Ross, Donald M. Ross, Ken Ross, Enid Rotman, Mr. and Mrs. S. Rounding, F. Routledge, Betty Rowe, Kayo J. Roy, Bess and Walter Ruch, Dave Ruch, Ken Rumble, Mike Runtz, Dennis Rupert, Marg and Art Rusnell, Dorothy Russell, Hilya Rydholm, Ann Rynard

John Sabean, Tim Sabo, Doug Sadler, John Sadowski, Vicki Sahantien, Toni Salvadori, George and Mat Samuel, Chris J. Sanders, Lawrence Sanders, Al Sandilands, John Sankey, Don Sankney, Jack Satterly, Peter Satterly, Helen Saunders, Jack Saunders, Jane Saunders, Mark Sauvé, I. Saver, L. Scales, Don Scallen, K. and P. Schaeffer, D. and J. Schaus, John Schmelefske, Alice Schmidt, Dan Schneider, Ann Schoen, M. and D. Schoenefeld, Marion Scholz, R. Schryer, David Schuford, John Schwindt, Alex Scott, D.M. Scott, George Scott, John A. Scott, R. and D. Scott, Kenneth Scovell, Ron Scovell, A. Sculthorpe, D.K. Sealey, Art Sedore, Jack Seigel, Katharine Selander, Marilyn Semph, Gilles Seutin, David Sevigny, John D. Sewell, Frank Sexsmith, Fred Shantz, Norman Shantz, Mirek Sharp, Doreen Shaver, Art Shaw, Catherine Shaw, Nigel Shaw, Dave Shepherd, Arnet Sheppard, Tom Sheppard, Jerry Sherk, John Sherrin, Jim Shruder, Joey Shultes, Jacqui Shykoff, Bev Sidney, L. Sieciechowicz, Eunice Sikorski, D. Simkin, Wayne Simkin, Roger and Jonathon Simms, Arnie Simpson, Gerry Simpson, John Simpson, Allan Sinclair, Langis Sirois, Hilda Sivell, P.D. Skaar, Betty Skare, Jeff Skevington, Peter Skierszkan, Mrs. Aagot Skjaveland, Peter Skoggard, Doug Sleeman, Scott Slocombe, Medley Small, Phyllis Smeeth, Gord Smellie, Allan J. Smith, A.P. Smith, Blake Smith, David Smith, David A. Smith, Emlyn B. Smith, G. Smith, Helen Smith, James A. Smith, Jim Smith, Judith Smith, Marcia Smith, Marion E.E. Smith, Mary Smith, Merle Smith, Paul Smith, Paul D. Smith, Randy Smith, Roy Smith, Shelia Smith, Thelma Smith, Tom Smith, William Smith, Wright Smith, Rick Snider, Ross Snider, Warren Soloman, Ernest Somers, Peter Somerville, Bruce A. Sorrie, W.B. Spaulding, Brian Specht, J. Murray Speirs, Stan Splichal, Terry Sprague, Terry Spratt, Ronald M. Sprigings, James B. Spruce, Chris Spytz, D. St. Hilaire, Mark Stabb, Paul Stalker, Bob Stamp, Richard Stankiewicz, Earl Stark, Richard Steadman, R. Stephens, Bill, Luce and Charron Stephenson, Denis Stevens, John Stevens, D.G. Stevenson, Ian Stewart, Jeff Stewart, Robert B. Stewart,

W.E. Stewart, William Stewart, Timothy Stiles, Robert Stitt, Marten Stoffel, Bill Stone, Irene Stoneman, Edna Storoschulz, Tim Story, Greg Stott, Marian Stranak, Dan Strickland, M. Strickland, K.J. Stryland, Dan Stuckey, Cynthia Suhay, Don Sutherland, June Sutherland, Linda Sutterlin, Brian and Bronwen Sutton, Rob and Margo Swainson, David Swales, Maudie Swalm, Erik Swanback, Mrs. John Szabo, Nina Szpakowski

Richard D. Tafel, Pat Tafts, Lee Talbot, Craig Tallman, Lloyd Taman, Paul Tapley, James Tasker, R.R. Tasker, Ben Taylor, Berys M. Taylor, Deborah Taylor, Doug Taylor, Grace Taylor, Harold Taylor, Martin and Dixie Taylor, Neil Taylor, Paul Taylor, Peter Taylor, Phil Taylor, Roger Taylor, Steve Taylor, Betty Teachman, Ned Teachman, E. and M. Teager, Gaston Tessier, J.B. Theberge, R.G. Thobaben, Peter Thoem, Ailsa P. Thomas, Katie Thomas, V.G. Thomas, A.L. Thompson, Claudia Thompson, John L. Thompson, Lynne Thompson, Paul Thompson, David Thomson, D.C. Thomson, George Thomson, Harry and Sheila Thomson, John and Joan Thomson, H. Thoonen, Greg Thorn, Kirk Thorsteinson, Mick Throssell, Ron Tidrow, Nancy Tilt, Mike Tobin, D.W. Tomlinson, Jeannie Torrance, L. Torvi, Richard Toth, Renee Touw, Alan Tower, Brian Townes, Ron Tozer, Brian Track, Frank Tremayne, Phyllis Tremblay, Erland and Barbara Troup, Horace Troup, George Tuck, E. Tull, Gina Turff, Mike and Elizabeth Turner, Helmut Twardowski, Donald J. Tyerman, Rob Tymstra, Keith Tyndall

Phil Underwood, Dick Ussher

J. Vaillancourt, Alain Vallieres, George Vance, Jan Vanden Heuvel, Norma Vanderlon, Jan and Liza Vandermeer, S. VanderVeld, Hans Van Der Zweep, Peter Van Dijken, Clayton Van Horn, Garfield Van Horn, W. Van Kempen, Vince Van Vlymen, Bert and Ingrid Van Wout, C.G. Van Zyll de Jong, Clay Vardy, J.C. Varey, Peter Varty, Richard Veit, Gordon Verch, Richard L. Verch, Willis Verch, M. Vermeer, P. Vernon, Tim Vetter, Douglas C. Vincent, Ken Vogen, Adolf Vogg, Dennis R. Voigt, Bill Volkert, Peter Von Bistram, Jack and Barbara Von Haer

George Wagner, David G. Wake, Bill Walker, Bob and June Walker, C.A. Walker, Dorie Walker, Harry Walker, Peter Walker, Phil Walker, W.H. Walker, Jean Wallace, Mr. and Mrs. C. Wallwork, Dalton Walpole, Robert Walters, Kenneth Walton, J.B. Walty, Barbara Ward, Ed Ward, Hal Ward, J. Warnica, Gwen Warren, Mike Waschenko, Scott Watkin, Allan Watson, Hugh Watson, Marilyn Watson-Boissonneault, Patrick Weatherhead, Bruce and Betty Webster, Marg Webster, Chris Wedeles, Susan Weilandt, Bob Weir, John Weir, Ron D. Weir, Brian Weller, Phil Weller, Wayne Weller, Dan Welsh, Sharon Welsh, Steve Wendt, Allen Werden, Martin Wernaart, Chip Weseloh, John West, M.M. West, Richard West,

Christopher Westcott, R. Westerhoff, Fran Westman, R. Westmore, Joan Wheatley, Peter Whelan, Douglas Whelpdale, Karen Whistler, Alex White, Anne White, B.A. White, Carolyn White, David White, Emilie White, Harold A. White, Charlie Whitelaw, Bob and Dan Whittam, Steven Whitteker, Mark Wiercinski, Peter Wigham, Geoff Wigzell, Allan Wilcox, Don Wilkes, David and Mary Wilkins, Marty Wilkinson, Cleon L. Williams, H. Williams, Judith Williams, Lillian Williams, Anne Williamson, Colin Williamson, Harry Williamson, Jean Williamson, A.E. Wilson, Audrey Wilson, Janice Wilson, Phyllis Wilson, Reid and Margaret Wilson, William Wilson, Stevenson Winder, Joan Winearls, Keith Winterhalber, Harold Wirth, Joe Witalis, Margaret Withers, W. E. Witzel, Craig Witzke, Bob Wood, C. and S. Wood, Gordon Wood, K. and Gwen Warren Wood, Jim and Pat Woodford, P. Allen Woodliffe, Frankie Woodrow, Terrie Woodrow, Peter Woods, Alan Wormington, Rob Worona, Bob Wright, David Wright, Glenda M. Wright, Jo Wright, Laurie S. Wright, Sharri (Clark) Wright, Peter Wukasch, Bryan K. Wyatt, Karen Wylie-Bellhouse

Gus Yaki, Liz Yerex, Chris and John Yip, Roger Young, Bob Yukich

Cazmir Zielony, Craig Zimmerman, J.N. Zolkiewski

Introduction

The goal of the Ontario Breeding Bird Atlas project was to produce accurate and up-to-date maps of the distribution of the birds breeding in the province during the period 1981 through 1985. The methods used were based on those which had proved so successful in the breeding bird atlases of Britain and Ireland (Sharrock 1976), Holland (Teixeira 1979), and elsewhere. To achieve the project's goals, Ontario was divided into a grid and volunteer naturalists were encouraged to visit every grid unit and collect information on the birds breeding there. After 5 years and 180,000 hours of field work, the data had been collected and were ready for mapping. The results of that remarkable effort are the basis of this book.

Ontario is an enormous landmass of 1,068,587 square kilometres. Most of the population is concentrated in the south, although there are some cities and towns spread throughout road- and rail-accessible parts of the province, and small settlements scattered in the half of the province north of road or rail access. The small population of the north and the distances from southern population centres meant that it would be impossible to cover the whole province on the scale that could be managed in the south. Therefore, southern Ontario was divided into 10 x 10 km 'squares' and northern Ontario was divided into 100 x 100 km 'blocks.' (The boundary between 'southern' and 'northern' Ontario, as defined for the purposes of this Atlas, is shown on the 'southern Ontario' species maps throughout the book.) The intent of the project was to perform at least a specified minimum amount of field work in each square and block.

During the 5 year field period, data were collected throughout Ontario and on some of the islands under the jurisdiction of the Northwest Territories (NWT) in James and Hudson Bays; islands within blocks containing land in Ontario were covered, as was Akimiski Island, the eastern portion of which is in a block with no land in Ontario. The Twin Islands (NWT) were not covered. The data from the islands that belong to the Northwest Territories have been combined with data from the Ontario section of the block. They are presented in the species maps and statistical summaries. Some data were collected in the Quebec, New York and Michigan sections of squares that are partly in Ontario, but only the data from the Ontario sections of these squares

are used in this book.

The materials presented in the following sections provide general information helpful in interpreting the species maps. Those who are familiar with the geography of Ontario and atlas techniques should have no difficulty in skipping several of these introductory sections and going straight to the maps and species accounts. However, an initial perusal of the section on "How to use this book" will be useful as it explains the format of the maps and diagrams, content of the species accounts and abbreviations used in the text. For those less familiar with the province, the section on the "Biogeography of Ontario" provides background information which will aid in understanding many of the distributions depicted in the maps. Many physiographic and biogeographical terms used in the species accounts are defined in that section. The next section, "Conduct of the survey", describes how the data on the maps were collected and processed, what the breeding levels shown on the maps mean in terms of atlassers' observations, and how adequate coverage was obtained across the province. This is followed by a section which provides an assessment of the "Final coverage of the province" obtained after 5 years of field work and an "Overview of atlas results" which contains composite information that is not readily apparent from individual species accounts. The maps and species accounts form the main section of the book, divided into 233 species (and two hybrids) recorded in southern Ontario, 57 species recorded only in northern Ontario and, in a final section, 8 miscellaneous species which were either not recorded by the atlas or do not merit full treatment for other reasons.

Following the species maps and accounts is a section on "Applications of atlas data" in which examples show that the usefulness of the atlas data base extends beyond the production of maps. Appendices contain information on the administrative structure and operation of the project (Appendix A), on estimating abundance (Appendix B) and on data processing and error checking methods (Appendix C). These appendices will be of interest to those who wish to find out more about some of the technical and administrative aspects of the project. Appendix D lists species officially designated as rare, threatened, endangered, extirpated and extinct in Ontario and Canada.

Biogeography of Ontario

An understanding of the biogeography of Ontario is important in interpreting the distributions of breeding birds; most of the species accounts in this book discuss the relationship between breeding range and extent of suitable habitat. Because many of the features mentioned here are referred to in the following sections, we present this information first.

Climate and Topography

Ontario has a cool temperate climate exhibiting the extreme range of temperatures found in continental climes. However, temperatures are moderated near the Great Lakes. Figures 1 and 2 show mean daily temperature for the year and mean annual precipitation, and Figure 3 shows altitude and topography in both southern and northern Ontario. These features are important in determining habitat and hence bird distribution. Note, for example, that the Algonquin Highlands are at a high elevation and are cooler than surrounding areas, while the Grey Co area in southwestern Ontario is high, cool, and wet. Both areas have a larger coniferous element in their forests than adjacent areas, and Grey Co, although largely agricultural, has extensive wetlands, making it ideal for species such as the Common Snipe (see species map).

Physiographic Regions

There are 3 physiographic regions in Ontario: the Hudson Bay Lowland, the Canadian Shield, and the area to the south and east of the Shield in southern Ontario (Figure 4).

The Hudson Bay Lowland extends inland about 150 to 300 km from the marine shores of James and Hudson Bays to the edge of the Canadian Shield. It is underlain with sedimentary bedrock and is characterized by extremely flat topography and poor drainage. Small lakes and ponds dot the Lowland, and river mouths are often 'braided' with many small islands. It includes other small islands within a few kilometres of the coast, including East Pen and West Pen Islands near the Manitoba border. The Hudson Bay Lowland is divided into the Tundra and the Hudson Bay Lowland Forest zones described below.

The Canadian Shield is the most extensive physiographic region of Ontario. The 'Shield', as it is sometimes referred to in the text, extends from the southern edge of the Hudson Bay Lowland to southern Ontario. It consists largely of metamorphic and igneous bedrock underlying thin soils of poor fertility. Ontario has 250,000 lakes, most of which are on the Canadian Shield. Shield lakes generally have rocky shorelines and, being low in nutrients and productivity, are often clear. The rocky shorelines of Shield lakes are not conducive to the production of marshes. Near towns, these lakes often receive sewage discharge, which makes them more productive, thereby allowing some birds to nest in areas where they might not otherwise do so.

The Canadian Shield is not uniform throughout. One major feature is the former bed of Lake Ojibway-Barlow, which forms the relatively flat Clay Belt area. The Clay Belt is divided into the Little and Great Clay Belts, extending from Haileybury to Englehart, and from Timmins to north of Cochrane and west of Kapuskasing, respectively. The deeper soils of this area permit agriculture, although the Great Clay Belt is far less agriculturally developed than is the Little Clay Belt.

The Frontenac Axis is an extension of the Canadian Shield which crosses the St. Lawrence River northeast of Kingston. The Axis is a forested area interspersed with areas of marginal agriculture where there are patches of deeper soil. Consequently, the area has high habitat diversity, and atlas squares there tend to have high species totals. Because the southern Shield edge is usually not abrupt, there is often interspersion of habitats all along that transition zone, and the number of bird species is consequently relatively high.

Unless otherwise specified in the text, 'off the Shield', or 'south of the Shield', or 'south and east of the Shield' refer to that part of southern Ontario not underlain by the Canadian Shield. South and east of the Shield in southern Ontario are areas with deeper soils and flatter land more suitable for agriculture, underlain by sedimentary bedrock. Two other major topographic features are worthy of note (Figure 3). The Niagara Escarpment is a long, interrupted limestone escarpment extending from near Niagara Falls to the tip of the Bruce Peninsula and Manitoulin Island. The escarpment and adjacent areas are relatively heavily wooded, and the cliffs provide good nesting habitat for species such as the Turkey Vulture. The Oak Ridges Moraine extends westward 200 km from about 15 km north of Presqu'ile Provincial Park to the Niagara Escarpment near Orangeville, roughly paralleling the north shore of Lake Ontario. Because it has well-drained soils and deeply rolling hills, this area is less intensively farmed and so has more extensive tracts of woodland than adjacent areas. These topographic features are evident on the distribution maps of some species.

Forest Regions

The forest regions discussed here are essentially those described in *Forest Regions of Canada* by Rowe (1972), although some of his terminology has been modified for the purpose of this publication (Figure 4). The five main forest zones are, from north to south, the Tundra, Hudson Bay Lowland Forest, Boreal Forest, Great Lakes-St. Lawrence Forest, and Carolinian Forest zones. Much of the information below is extracted from Rowe (1972).

Tundra
A narrow strip of arctic tundra occurs along the Hudson Bay coast and the contiguous northwest coast of James Bay. It is underlain by continuous permafrost and sedimentary rock. It consists of dry uplands with lichens and heath plants, low-lying fens with grasses and sedges dominant, and uplifted beach ridges, often with numerous ponds and lakes. Dwarf willows and birches occur in sheltered areas. Stunted spruces grow singly or in patches along river banks and occur with increasing frequency inland towards the transition with the Hudson Bay Lowland Forest at the 'tree-line'. The most extensive areas of tundra habitat are on the eastern section of the Hudson Bay Coast near Cape Henrietta Maria and southward about 40 km along the James Bay coast, and on the western section of the Hudson Bay coast in the vicinity of the Manitoba border. Between these two areas is a strip of coastal tundra varying in width from 2 to 15 km. The Lapland Longspur is an example of a species associated with tundra.

Hudson Bay Lowland Forest
The Hudson Bay Lowland Forest region is an extensive and poorly known area. It extends southward from the tundra and the James Bay shore to an elevation of about 150 m at the edge of the Canadian Shield, and is characterized by extremely flat topography and poor drainage. It is dominated by open fens and bogs interspersed with treed fens and bogs of black spruce and tamarack. Ground cover is often lichen in dry areas and sphagnum moss in wet areas. River banks and other well-drained areas have denser forests of white spruce, balsam fir, trembling aspen, balsam poplar and white birch. The southern areas of the Lowland are generally more densely forested. In the more northerly and westerly areas of the Lowland, the percentage of peatland and of lakes is higher and the prevalent forest is of stunted open-grown black spruce and tamarack. The Greater Yellowlegs is typically associated with the Hudson Bay Lowland Forest zone in Ontario.

Boreal Forest
The Boreal Forest zone is generally underlain by the Precambrian bedrock of the Canadian Shield, and as such is rich in lakes. The forests are primarily coniferous (black spruce and white spruce, balsam fir, jack pine and tamarack) but white birch, trembling aspen and balsam poplar also occur. The proportion of spruce and tamarack declines to the south. The boundary between this and the Great Lakes-St. Lawrence Forest zone is not clearly defined; instead there is a broad transition area with trees characteristic of both regions, and there are outlying pockets of each forest type within the main boundary of the other region. Forest cutting is prevalent in road-accessible areas of the Boreal Forest. The Three-toed Woodpecker is primarily a bird of the Boreal Forest zone.

Great Lakes-St. Lawrence Forest
The forests in the Great Lakes-St. Lawrence Forest zone are a mixture of coniferous and deciduous trees, characterized by sugar maple and yellow birch, white and red pines, and hemlock. They also contain red maple, basswood, white elm, white cedar, beech, red oak, and white ash. Boreal tree species are inter-mixed with these species, especially along the transitional area between the two zones. The Canadian Shield underlies the northern part of this forest, with sedimentary rock under the southern section.

The Great Lakes-St. Lawrence Forest zone is divided into several forest sections, two of which are of particular interest (Figure 4). The Timagami Forest section is an area of transition with the Boreal Forest, with a higher proportion of boreal tree species. Because it extends southward into 'southern' Ontario (north of Sudbury and North Bay) its influence is apparent on the distribution maps of some species, such as the Boreal Chickadee. The Algonquin Highlands Forest section is also mentioned frequently in species accounts. It too shows a greater boreal component than do most other parts of the Great Lakes-St. Lawrence Forest zone. It is obvious on many species maps because of the presence of birds which are rare further south, or the absence of species which are common all around but not on the highlands themselves. Most of this area is contained within Algonquin Provincial Park. Although there is little agriculture in this section, there is considerable forest cutting, as is true of much of the rest of the Great Lakes-St. Lawrence Forest zone on the Shield.

The southern edge of the Canadian Shield bisects the Great Lakes-St. Lawrence Forest zone in southern Ontario. South and east of the Shield, the forest has been largely removed for agriculture and settlement. The remaining woodland occurs mostly in woodlots, with a few more extensively forested areas on poor agricultural land.

The Veery and Chestnut-sided Warbler occur throughout the Great Lakes-St. Lawrence Forest zone, though both also occur somewhat farther north.

FIGURE 1: *Mean daily temperature (°C) for the year*

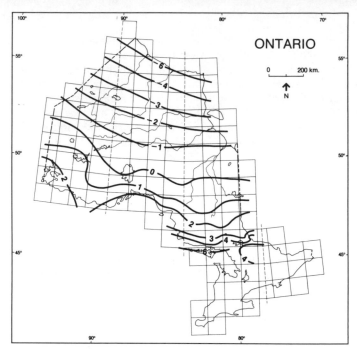

ONTARIO

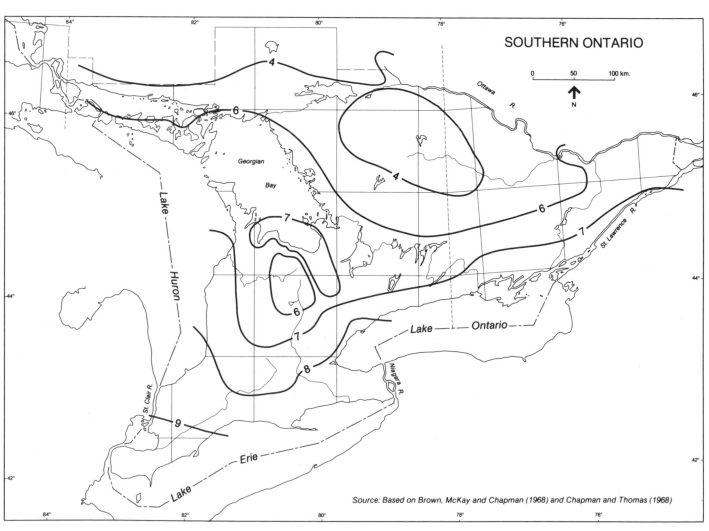

SOUTHERN ONTARIO

Source: Based on Brown, McKay and Chapman (1968) and Chapman and Thomas (1968)

FIGURE 2: *Mean annual precipitation (cm)*

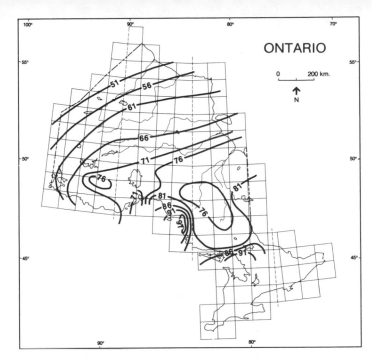

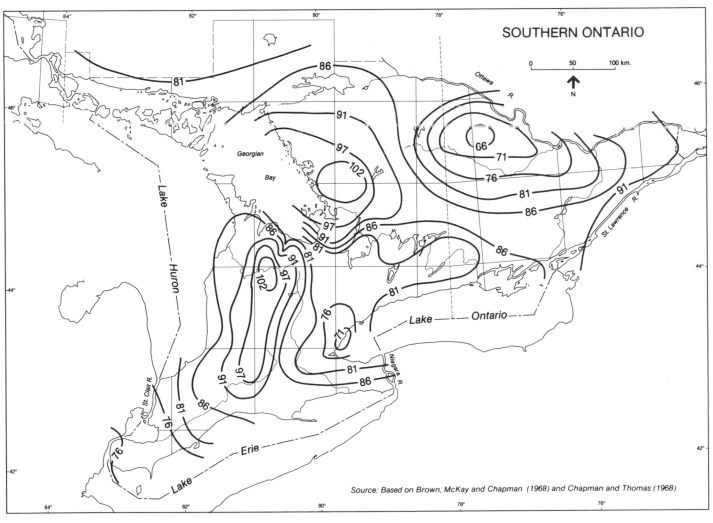

Source: Based on Brown, McKay and Chapman (1968) and Chapman and Thomas (1968)

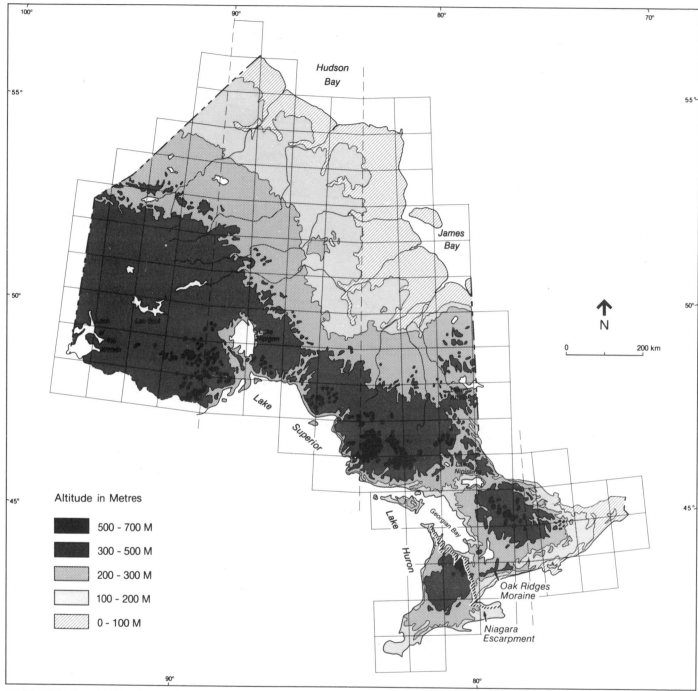

Altitude in Metres

- ██ 500 - 700 M
- ██ 300 - 500 M
- ▓▓ 200 - 300 M
- ░░ 100 - 200 M
- ▨▨ 0 - 100 M

FIGURE 3: *Altitude and some topographic features of Ontario*

Carolinian Forest

The Carolinian Forest zone, known as the Deciduous Forest in the eastern US, occurs in southernmost Ontario. Its forest communities are dominated by broad-leaved trees. Beech and sugar maple, along with basswood, red maple, red oak, white oak and bur oak, commonly occur. Also found are Canada's main concentrations of black walnut, sycamore, swamp white oak, and shagbark hickory, along with butternut, bitternut hickory, rock elm, and blue beech. Characteristic trees of this region include tulip tree, black cherry, Kentucky coffee-tree, and sassafras. Even more than the southern part of the Great Lakes-St. Lawrence Forest zone, the Carolinian Forest zone has been highly modified by human activity. Very few extensively wooded areas remain. Small woodlots are widespread, however, except in the extreme southwest, where little wooded land remains. The Red-bellied Woodpecker and the Carolina Wren are characteristic birds of the Carolinian Forest zone.

Ontario's Shorelines

Much of Ontario is bounded by water: the Great Lakes in the south, and Hudson Bay and James Bay in the north. For the most part, the shores of Lake Erie, Lake Ontario and southern Lake Huron are highly developed. The remaining extensive marshes and woodlands are associated primarily with promontories, many of which are protected as parks. The shores of Georgian Bay, northern Lake Huron and the North Channel (between Manitoulin Island and the mainland) are rocky, and although subject to cottage development, are still much less disturbed than the shores of the lower Great Lakes. The Lake Superior shoreline is also rocky; most of the shore is quite remote and receives little direct disturbance.

Lakes Erie and Ontario have only a few islands, most of which receive little disturbance and are used by nesting colonial birds such as cormorants, herons, gulls and terns. Georgian Bay, the waters off the Bruce Peninsula, the North Channel, and parts of Lake Superior, have many small rocky islands. Those with extensive open areas are often used by nesting colonial birds. Most of the islands are heavily treed and have an avifauna similar to that of the adjacent mainland.

The shores of James and Hudson Bays are extremely flat, forming an extensive coastal plain. Tundra habitat has been described above; on the western shore of James Bay south of the tundra, however, the Hudson Bay Lowland forest is separated from the coast by 1 to 2 kilometres of marshy habitat interspersed with beach ridges which parallel the coastline. These extensive marshy areas, which are dominated by sedges, are home to several species usually associated with the prairies: Marbled Godwit and Northern Shoveler are two examples.

Land-use

Prior to the arrival of European settlers in the early 18th century, the area that is now Ontario was mostly unbroken forest of the types described above. Some areas were cleared by native peoples, and there were some limited areas of long-grass prairie in the extreme southwest of the province, and also some prairie in the Rainy River area in the northwest. Modification of the landscape by people has become a very important factor in the distribution of Ontario's birds, at least in road- and rail-accessible areas.

All but a few remnants of the grassland areas have been converted to agricultural use, as has much of the forest in southern Ontario south of the Canadian Shield. The Great Lakes-St. Lawrence Forest region overlaps the southern edge of the Canadian Shield. Although forest types differ slightly on and off the Shield, bird distributions are often markedly affected by the Shield edge; not because of the forest type, but because of the differences in land-use on either side of the Shield edge (e.g., Eastern Meadowlark).

Figure 5 shows the distribution of agricultural land in Ontario. The southwestern counties have the smallest remaining proportion of woodland. The Shield is largely forested. Shallow soils mean that little agriculture occurs there: cleared land is located mostly near main highways. As a result, the distribution of the Eastern Meadowlark follows agricultural land along Highway 11 northward across the Shield from Orillia to North Bay. The more extensive areas of agriculture between the cities of Sudbury and North Bay are also evident on the map for that species. There is little agriculture in northern Ontario away from the Clay Belts and the Thunder Bay and Rainy River areas.

Forest harvesting is a major industry in Ontario, and is an important factor in the distribution and abundance of the province's breeding birds. Forestry is the primary activity in roughly 50% of Ontario (1/2 million square kilometres). Forest cutting is most concentrated in the highly productive area in the southern Boreal Forest and the northern Great Lakes-St. Lawrence Forest regions. The present annual cut is approximately 225,000 hectares, and since World War 2 about 50% of the productive Boreal Forest has been cut.

Although the rate of silvicultural treatment is increasing, much of the land now cut is left to regenerate naturally. Consequently, a major effect of cutting in upland areas of the Boreal Forest has been the reduction of 'upland' spruce and jack pine, and the proliferation of early successional shrub-dominated communities that develop into deciduous stands of birch and aspen, often with a significant component of balsam fir. On wetter sites, from which 'lowland' black spruce has been removed, trees have not regrown and have been replaced by alder swales. Thus, there is a major reduction in the area of mature coniferous forest on both upland and lowland sites.

Populations of birds such as the Alder Flycatcher, Chestnut-sided Warbler, Mourning Warbler and White-throated

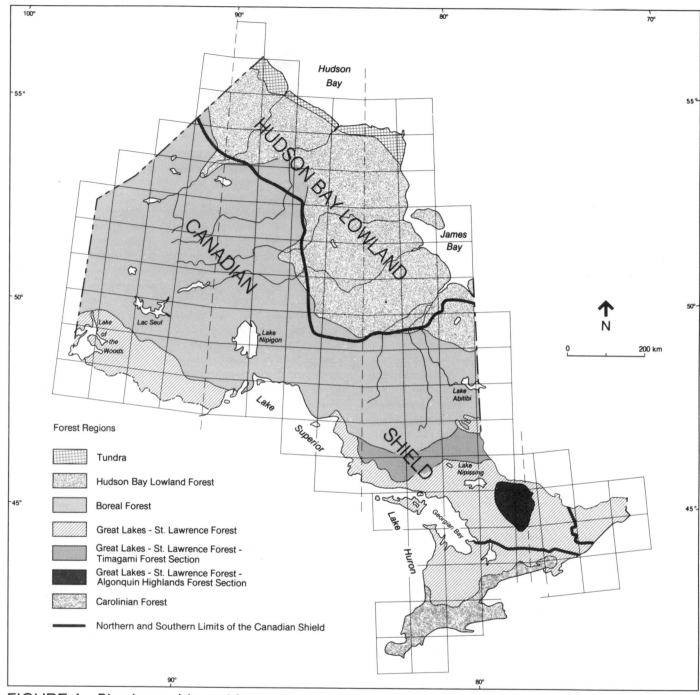

Forest Regions

Tundra
Hudson Bay Lowland Forest
Boreal Forest
Great Lakes - St. Lawrence Forest
Great Lakes - St. Lawrence Forest - Timagami Forest Section
Great Lakes - St. Lawrence Forest - Algonquin Highlands Forest Section
Carolinian Forest
Northern and Southern Limits of the Canadian Shield

FIGURE 4: *Physiographic and forest regions, and selected forest sections of Ontario*

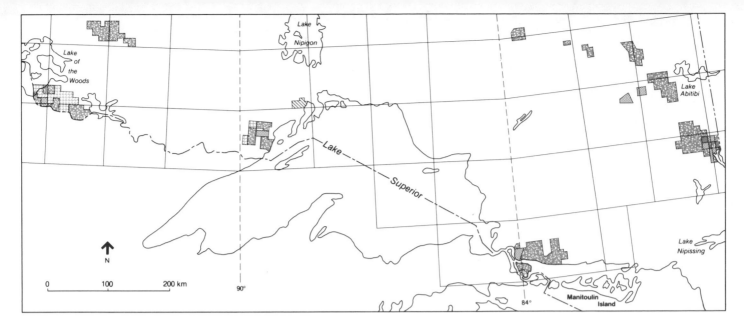

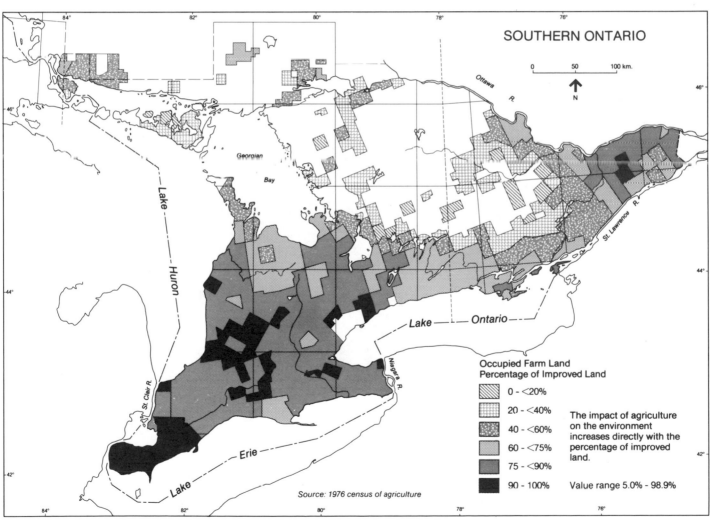

Occupied Farm Land
Percentage of Improved Land

▨	0 - <20%
▦	20 - <40%
▦	40 - <60%
░	60 - <75%
▓	75 - <90%
█	90 - 100%

The impact of agriculture on the environment increases directly with the percentage of improved land.

Value range 5.0% - 98.9%

Source: 1976 census of agriculture

FIGURE 5: *Improved land as a percentage of farm land, 1976*

9

Sparrow, which favour early successional habitats, and the Veery, Red-eyed Vireo, Black-and-white Warbler, Magnolia Warbler, American Redstart and Rose-breasted Grosbeak, which favour mid-successional habitats, have benefitted from these forest cutting and regeneration patterns. Birds such as the Downy and Hairy Woodpeckers, Yellow-bellied Fly-catcher, Boreal Chickadee, Golden-crowned Kinglet, Brown Creeper, Winter Wren, Swainson's Thrush, Blackburnian and Black-throated Green Warblers, which favour mature forest areas, are reduced by forest harvesting activities.

Wetlands are frequently drained for human settlement and agriculture. Of 2.3 million hectares of wetlands originally found in southern Ontario south of the Shield, less than 12% remain today (Rowntree 1979). The area in which the greatest proportion of wetland has been drained is the extreme southwest; the lack of wetland habitat in that area is evident on the species distribution maps of many wetland species, such as the Virginia Rail and the Black Tern.

Sewage ponds, and water impoundments created upstream from flood control dams, provide breeding habitats for some species. Wilson's Phalarope and Ruddy Duck have recently extended their ranges eastward into southern Ontario. Some water impoundments have created extensive marshes and other wetlands that add considerably to the diversity of species in a square. Perhaps the best example is Luther Lake, also known as Luther Marsh, (Figure 6a). The square containing this large wetland, which is important to a number of rare wetland species such as the Red-necked Grebe, has among the highest number of breeding species of any square in the province.

Human settlement in itself creates habitats that benefit some species. Figure 6 shows many of Ontario's larger cities,

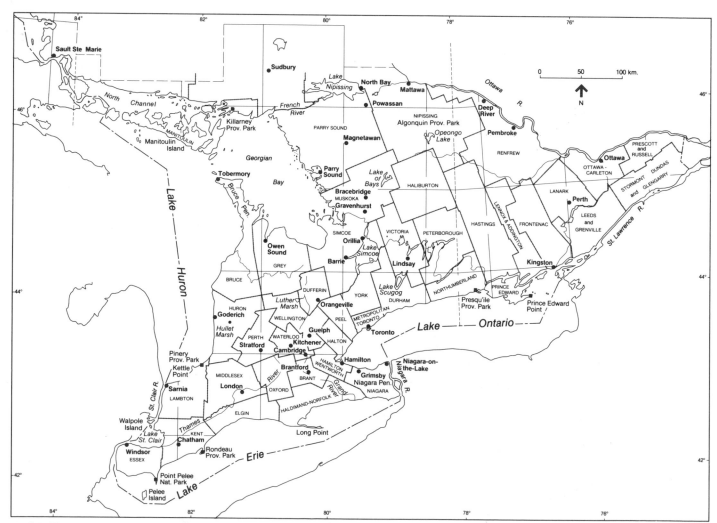

FIGURE 6A: *Counties, Regional Municipalities, Districts, urban centres and other localities in southern Ontario*

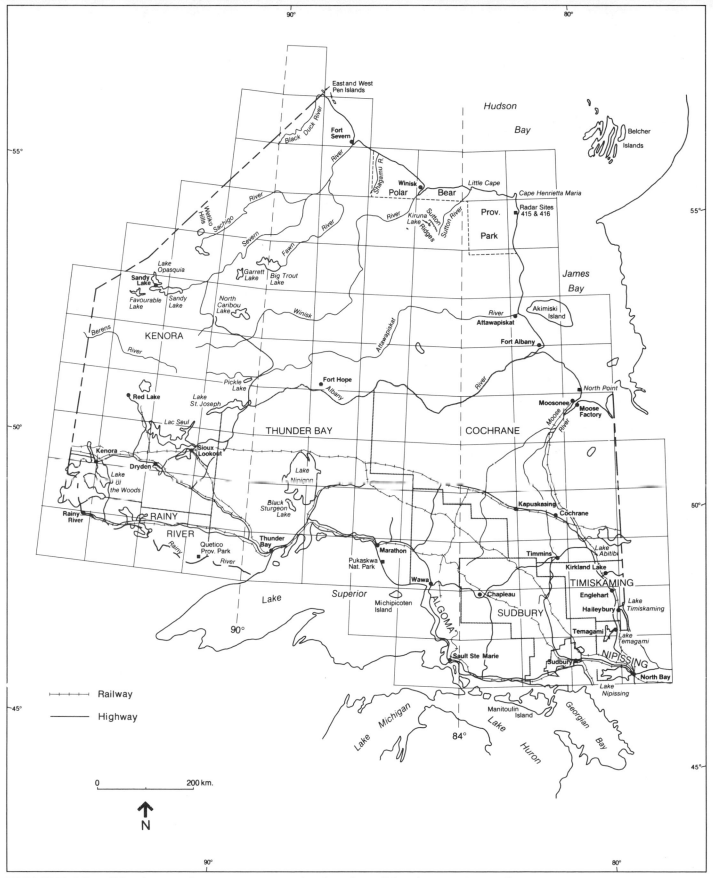

FIGURE 6B: *Districts, urban centres and other localities in northern Ontario*

11

towns and other localities mentioned in the text. The Barn Swallow map shows how that species was found in association with human settlements even in the most remote areas, and that in the north it was not often reported anywhere else.

There is extensive cottage development along lakeshores and riverbanks on the Shield, principally south of Algonquin Provincial Park and Parry Sound, and on the Bruce Peninsula. In the text, this area is sometimes referred to as 'cottage country'. Cottages, and the roads and power line corridors required to service them, have created a considerable amount of edge and open habitat affecting the composition of the bird community (Clark *et al.* 1983). As the map of its distribution in southern Ontario shows, the Ruby-throated Hummingbird appears to do particularly well in cottage country.

Another human influence that is evident on species distribution maps is the establishment of conifer plantations on cleared land south of the Shield. The maturing trees create pockets of coniferous habitat similar to those typically found farther north. Some of these sites were found to contain species such as Red-breasted Nuthatch, Golden-crowned Kinglet, and Magnolia and Yellow-rumped Warblers, whose main distribution is farther north.

Except around small coastal communities, the Hudson Bay Lowland has been little affected by direct human activity, although the effect of air-borne pollutants is unknown. The same is largely true on the Canadian Shield north of road or rail access.

In summary, Ontario's habitat has been profoundly affected by human activity in southern Ontario, moderately affected to the northern extent of the road and rail system, and scarcely at all north of that. Land-use practices have proved a boon to some species at the same time as being severely detrimental to others.

Conduct of the survey

The Grid System

The data collection units for the Ontario Breeding Bird Atlas project were based upon the Universal Transverse Mercator (UTM) grid system. Unlike quadrangles based on a latitude-longitude system, UTM squares are the same size and shape everywhere. Therefore, data can be readily compared between locations using the same system. UTM squares are clearly defined on the Canadian National Topographic Series maps at the scales of 1:25,000, 1:50,000 and 1:250,000. Ontario's 100 km UTM grid system is shown in Figure 7. In this Atlas, large units of 100 x 100 km are often referred to simply as 'blocks'; the smaller units of 10 x 10 km, 100 of which make up each of the blocks, as shown in Figure 7, are often referred to as 'squares'.

One drawback to the UTM system is that incomplete 'squares' are created along the 6^0 zone lines (i.e., those at 78^0, 84^0, and 90^0W longitude), because of the incompatibility of a square grid system with the surface of a spherical planet. To avoid having units very different in size from full-sized squares, slivers of less than 1/3 of a square along zone lines were joined to the adjacent full square, those between 1/3 and 2/3 of a square were joined across the zone line, and those larger than 2/3 of a square stood on their own. Hence, no square along a zone line was smaller than 2/3 of a full square or larger than 1 1/3 squares. Blocks were treated similarly, as shown on the species maps except that, for aesthetic reasons, the most northerly composite block on the 78^0W zone line in southeastern Ontario was retained although it was slightly smaller than 2/3 of a full block. All other squares and blocks stood on their own, even if they contained only a small area of land along a lakeshore, or on an island.

Because the great majority of Ontarians live in the southern part of the province, and because access to much of the north is very difficult, it was not feasible to collect data on the same grid scale throughout the province. Therefore, the province was divided into 2 sectors; southern Ontario (and the area around Thunder Bay) was surveyed and mapped by 10 km squares, and northern Ontario was atlassed by 100 km blocks. The boundary between these areas is shown on the southern Ontario species maps throughout the book. The location of that boundary changed substantially as work progressed, and was not finalized until after the 1985 season. The northern boundary of southern Ontario is the limit of the contiguous area in which essentially every 10 km square was covered adequately. Originally, it was hoped that all squares north to the Lake of the Woods area on the Manitoba border would be covered, and that the area north of there would be atlassed by 50 x 50 km squares. That plan was modified after the first year when it became obvious that there were far too few atlassers to cover northern areas that thoroughly; a system similar to that finally employed was then adopted. The final system stretched resources to the limit, but represented an attainable goal as long as the province's naturalists were enthusiastically involved throughout, and logistic support could be obtained.

Coverage Goals

The goal of the project was to attain at least a prescribed minimum degree of coverage in each square and block, so that coverage was spread as evenly as possible within the northern and southern sectors. Field tests in southern Ontario 1980 suggested that 16 hours of field work would be sufficient to find 75% of the species that breed in a square - a result very similar to that found in Britain. Therefore, the minimum standard of acceptable coverage was originally set at 16 hours per square. Over time, however, it became apparent that less experienced birders were not finding 75% of the species present in 16 hours, and that even highly experienced birders were not reaching that goal in squares with restricted road access or with a very high percentage of woodlands. Coverage goals were therefore modified. The Technical Advisory Committee provided Regional Coordinators with regional summaries of species richness based on results of the first two years of field work. Coordinators were then asked to assess the diversity of habitat in each square, and, based upon their experience, to estimate the total number of species breeding in each square. Adequate coverage was then established as the recording of breeding evidence for 75% of that number of species. However, a minimum of 16 hours of coverage was still required in all squares except those with

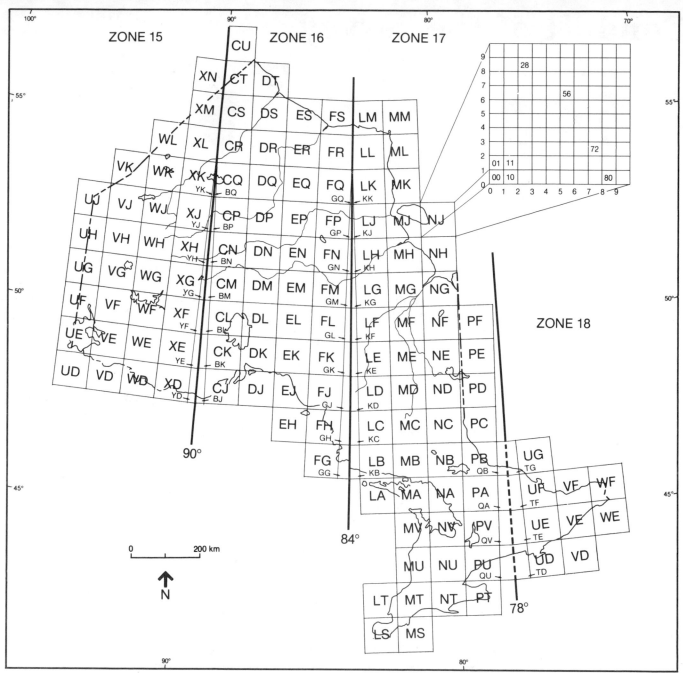

FIGURE 7: *UTM zone, block and square system in Ontario*

less than 10 square kilometres of land along Great Lakes shorelines or provincial boundaries.

Adequate coverage of northern blocks was more arbitrarily defined as a minimum of 50 hours of atlassing; this translated to roughly one week of atlassing by a crew of 2 to 4 observers in remote blocks. The goal in northern Ontario was to increase knowledge of the bird distribution in this little known area by covering at least major habitats in each block. Whenever possible, data were recorded on the basis of 10 km squares within the block, but attempts were not made to 'adequately cover' each 10 km square that was visited. Species totals were not established as goals because very little information was available upon which to base estimates of species richness, particularly in areas without road or rail access.

Atlas Regions

For administrative purposes, the province was divided into 46 regions, as shown in Figure 8, each of which had a Regional Coordinator. The names and tenure of Regional Coordinators are listed in Table 1. Regional Coordinators were responsible for working closely with atlassers to ensure that the goals and methods of the project were understood, and to ensure that all squares and blocks were adequately covered. They also enlisted atlassers, assigned atlassers to squares, collected and checked data, kept master cards for each square, checked data returned from the atlas office, screened Unusual Species Report Forms (USRFs), and reported to atlassers on progress in their regions. Regional Coordinators were responsible for the establishment of species number goals for each square in their region and for ensuring that adequate coverage was subsequently obtained. They were the vital link between project organizers and the atlassers, and they deserve a great deal of credit for the success of the project.

Data Collection Methods

To reduce variation among observers, a standard set of rules was established, based on those used in the *Atlas of Breeding Birds in Britain and Ireland* (Sharrock 1976). Instructions for Ontario participants were given in a Guide for Participants (Eagles and Cadman 1981, 1983). These were updated through a newsletter sent to all atlassers three times a year, through communications to Regional Coordinators, and at presentations and meetings across the province. Atlassers were provided with data cards to record their observations, and Unusual Species Report Forms on which to document observations of rare or out-of-range species.

Part of a completed data card is shown in Figure 9. Generally, observers filled out one card for each square for every year they collected data, although more than one observer might send in a card for the same square. On the front of each card, the atlasser filled in information on the zone-block-square, region, year, atlasser number (assigned by atlas

headquarters), dates and hours of observation. The bulk of the card was devoted to a list of the province's known breeding birds, next to which there were 5 columns in which breeding evidence and abundance were recorded in coded form.

The categories of breeding evidence, which were only slightly modified from those in the British and Irish Atlas, are listed in Table 2. There were 16 categories of breeding evidence, combined into 3 levels, and there was an additional category, in its own level, for birds simply observed in the breeding season with no evidence of breeding. These 4 levels, namely, species observed, and possible, probable and confirmed breeding, correspond to the 4 columns headed O, PO, P, and CO, respectively, on the data card. Using the codes given in Table 2, the atlasser reported the observed breeding evidence on the data card. The sequence of the categories in Table 2 indicates increasing confidence that breeding actually occurred. Only the highest level of breeding evidence and the highest category observed within that level was recorded for each species.

Species listed under 'species observed' were not considered to be breeding birds, were not entered into the computer data file and are not mapped in this book, although a few such reports are discussed in the text.

The category 'Nest-building' (coded 'N') was upgraded from probable to confirmed breeding for most species, at the end of the project. Exceptions were those species which are known to construct nests or cavities in which eggs may not be laid, or to use nest material in performing displays that are not necessarily directly involved in nest construction. Records of falcons, woodpeckers, and wrens were treated in this manner, as were those of the Common Raven and the American Crow. Records of nest-building in species not known to build nests, such as shorebirds (*Charadriidae* and *Scolopacidae*), owls, and Brown-headed Cowbirds were downgraded to SH (Table 2) unless there was evidence for a higher breeding category.

The fifth column on the data card, next to the species name, was reserved for an estimate of abundance in that square. A code was recorded in that column to indicate how many pairs of each species the atlasser estimated were breeding in the square. The following codes and categories were used.

Abundance Estimate Codes

Code	Estimated number of pairs of a species in a square
1	1 pair
2	2 - 10 pairs
3	11 - 100 pairs
4	101 - 1,000 pairs
5	1,001 - 10,000 pairs
6	more than 10,000 pairs

Estimation of abundance was encouraged, but not required,

TABLE 1: *The Regional Coordinators of the Ontario Breeding Bird Atlas*

Region	Name
1	Alan Wormington (1981 - 1982)
	Paul Pratt (1981 - 1986)
2	Allen Woodliffe (1981 - 1986)
3	Alfred Rider (1981 - 1986)
4	Bill Girling (1981 - 1983)
	Dave Martin (1984 - 1986)
5	David Hussell (1981 - 1982)
	Roy Smith (1983)
	Dave Shepherd (1984 - 1986)
6	Joan Kotanen (1981 - 1986)
7	Paul Eagles (1981 - 1986)
8	Martin Parker (1981 - 1986)
9	Peter Middleton (1981 - 1982)
	Tom Murray (1983 - 1986)
10	Doug Harding (1981)
	Peter VanDijken (1982 - 1986)
11	Bob Curry (1981 - 1986)
12*	Dave Broughton (1981 - 1982)
	Judith Kennedy (1983 - 1985)
13	Dave Hawke (1981 - 1986)
14	Connie Cochrane and Jack Siegel (1981 - 1986)
15	Roy John (1981)
	Ted McDonald (1982 - 1983)
	Doug Sadler (1984 - 1986)
16	Doug Sadler (1981 - 1986)
17	Bill Pratt (1981 - 1983)
	Geoff Carpentier (1984 - 1986)
18	Bob Bowles (1981 - 1986)
19	Dennis Barry (1982 - 1986)
20	Terry Sprague (1981 - 1986)
21	Ron Weir (1981 - 1986)
22	Don Sutherland (1981 - 1986)
23	Brian Morin (1981 - 1986)
24	Bruce DiLabio (1981 - 1983)
	Christine Hanrahan (1984 - 1986)
25	Michael Biro (1981 - 1986)
26	Jacques Bouvier (1981 - 1986)
27	Ron Tozer (1981 - 1986)
28	Alex Mills (1981 - 1986)
29	Harry Johnstone (1981)
	Jim Glenday (1982)
	Gary Clay (1983 - 1984)
	George Hawkes (1985 - 1986)
30	Chris Blomme (1981 - 1986)
31	Heather Baines (1981 - 1986)
32	Chris Bell (1981 - 1986)
33	Don Ferguson (1981 - 1986)
34	Bill McIlveen (1982 - 1986)
35	Chris Sanders (1981 - 1984)
	Tom Marwood (1985 - 1986)
36	Tom Baxter (1981 - 1986)
37	Michael Jones (1981 - 1986)
38	Margaret Hartley (1982)
	Jil Campling (1983 - 1984)
	Nick Escott (1985 - 1986)
39	Dave Elder (1983 - 1986)
40	Laura Howe (1981)
	Stephen McLeod (1981 - 1986)
41	Percy Richter (1982 - 1986)
42	Ted Armstrong (1982 - 1984)
	Rob Cunningham (1985 - 1986)
43	Paul Prevett (1982)
	Ken Abraham (1983 - 1986)
	Dave Shepherd (1983 - 1986)
44	Mike Cadman (1981 - 1986)
45	Chris Risley (1982 - 1984)
	Anne Lambert (1985 - 1986)
46	Margaret Bain (1982 - 1986)

*Note: Region 12 (Toronto) was split into 3 regions in 1982: regions 12, 45, and 46.

FIGURE 8: *Atlas regions in Ontario*

1. Essex
2. Kent
3. Lambton
4. London
5. Long Point
6. Huron/Perth
7. Waterloo/Wellington
8. Bruce
9. Grey/Dufferin
10. Halton/Peel
11. Niagara
12. Toronto
13. Barrie
14. Lindsay
15. Durham
16. Peterborough
17. Trenton
18. Muskoka
19. Haliburton
20. Prince Edward
21. Kingston
22. Thousand Islands
23. Cornwall
24. Ottawa
25. Perth
26. Pembroke
27. Algonquin
28. Parry Sound
29. Nipissing East
30. Nipissing West
31. Sudbury East
32. Sudbury West
33. Manitoulin
34. Spanish
35. Sault Ste Marie
36. Eastern Superior
37. Pukaskwa
38. Thunder Bay
39. Quetico
40. Lake of the Woods
41. Kirkland Lake
42. Cochrane
43. Moosonee
44. Big Trout Lake
45. Richmond Hill
46. Whitby

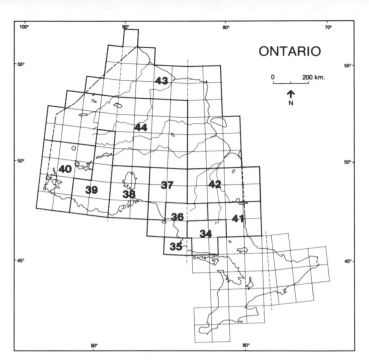

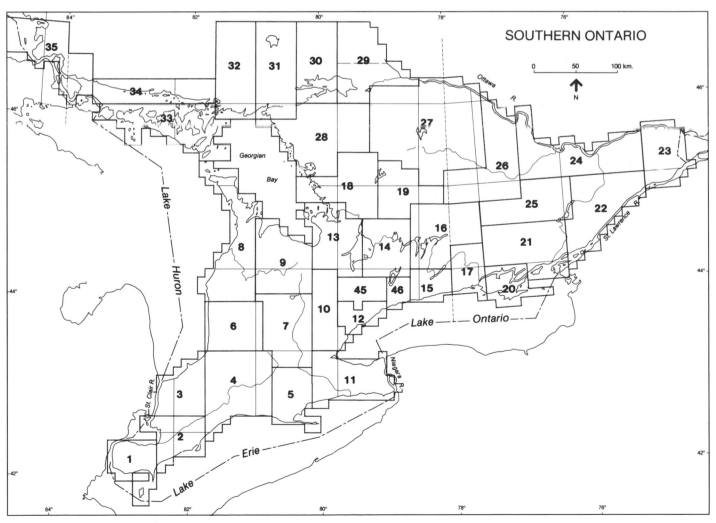

TABLE 2: *Breeding evidence levels, categories and codes*

Code	
	Species Observed
X	Species observed in its breeding season (no evidence of breeding).
	Possible Breeding
SH	Species observed in its breeding season in suitable nesting habitat.
SM	Single male(s) present, or breeding calls heard, in suitable nesting habitat in breeding season. N.B. The code was intended to be reserved for *singing* males, but an error was made in printing the Participants' Guide and not detected until the code was already in use. In the text of this book this category of breeding evidence is most often referred to as 'singing male'.
	Probable Breeding
P	Pair observed in suitable nesting habitat in nesting season.
T	Permanent territory presumed through registration of territorial behaviour (song, etc.) on at least two days, a week or more apart, at the same place.
D	Courtship or display, including interaction between a male and a female or two males, including courtship feeding or copulation.
V	Visiting probable nest site.
A	Agitated behaviour or anxiety calls of an adult.
B	Brood patch on adult female or cloacal protuberance on adult male (for birds examined in the hand).
N	Nest-building or excavation of nest hole.
	Confirmed Breeding
DD	Distraction display or injury feigning.
NU	Used nest or egg shells found (occupied or laid within the period of the survey). Use only for unique or unmistakable nests or shells.
FY	Recently fledged young (nidicolous species) or downy young (nidifugous species), including young incapable of sustained flight. Use with caution, especially at the edge of a square, as some young may move a considerable distance while remaining dependent upon the parents.
AE	Adults leaving or entering nest sites in circumstances indicating occupied nest (including high nests, nest holes, or nest boxes, the contents of which cannot be seen).
FS	Adult carrying faecal sac or food for young. Be careful with this, especially at the edge of a square, since some birds collect food a long distance from the nest or feed recently fledged young which have moved away from the locality of the nest. Also, note the difference between this and courtship feeding (D) or food storage. (In the text, this category of breeding evidence may be referred to as 'carrying food').
NE	Nest containing eggs. If the nest also contains a cowbird egg(s), record NE for both the host and the cowbird.
NY	Nest with young seen or heard. If a young cowbird is found in a nest, record NY for both the host and the cowbird.

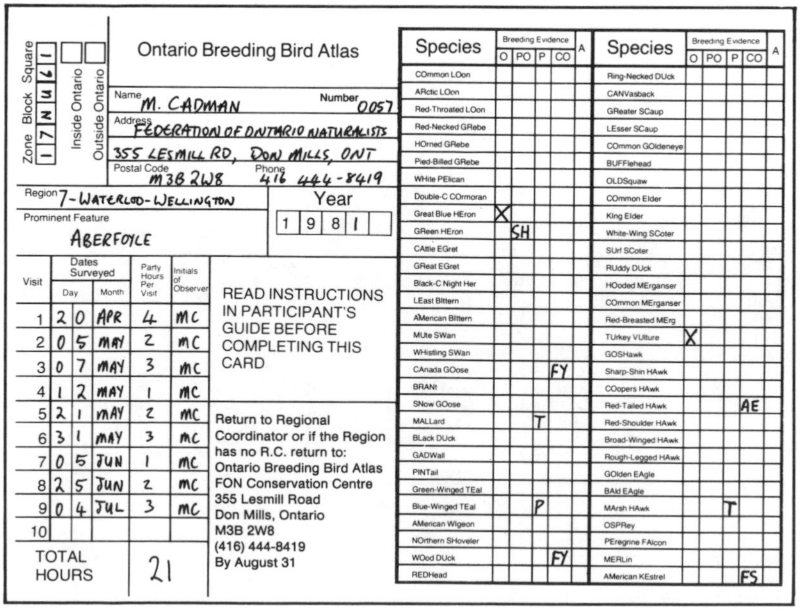
Ontario Breeding Bird Atlas

Zone 17 Block NU Square 61 | Inside Ontario | Outside Ontario

Name M. CADMAN Number 0057
Address FEDERATION OF ONTARIO NATURALISTS
355 LESMILL RD, DON MILLS, ONT
Postal Code M3B 2W8 Phone 416 444-8419
Region 7-WATERLOO-WELLINGTON
Prominent Feature ABERFOYLE
Year 1981

Visit	Dates Surveyed Day	Month	Party Hours Per Visit	Initials of Observer
1	20	APR	4	MC
2	05	MAY	2	MC
3	07	MAY	3	MC
4	12	MAY	1	MC
5	21	MAY	2	MC
6	31	MAY	3	MC
7	05	JUN	1	MC
8	25	JUN	2	MC
9	04	JUL	3	MC
10				
TOTAL HOURS			21	

READ INSTRUCTIONS IN PARTICIPANT'S GUIDE BEFORE COMPLETING THIS CARD

Return to Regional Coordinator or if the Region has no R.C. return to:
Ontario Breeding Bird Atlas
FON Conservation Centre
355 Lesmill Road
Don Mills, Ontario
M3B 2W8
(416) 444-8419
By August 31

Species	Breeding Evidence O	PO	P	CO	A
COmmon LOon					
ARctic LOon					
Red-Throated LOon					
Red-Necked GRebe					
HOrned GRebe					
Pied-Billed GRebe					
WHite PElican					
Double-C COrmoran					
Great Blue HEron	X				
GReen HEron		SH			
CAttle EGret					
GReat EGret					
Black-C Night Her					
LEast BIttern					
AMerican BIttern					
MUte SWan					
WHistling SWan					
CAnada GOose				FY	
BRANt					
SNow GOose					
MALLard			T		
BLack DUck					
GADWall					
PINTail					
Green-Winged TEal					
Blue-Winged TEal			P		
AMerican WIgeon					
NOrthern SHoveler					
WOod DUck				FY	
REDHead					

Species	Breeding Evidence O	PO	P	CO	A
Ring-Necked DUck					
CANVasback					
GReater SCaup					
LEsser SCaup					
COmmon GOldeneye					
BUFFlehead					
OLDSquaw					
COmmon Elder					
KIng Elder					
White-Wing SCoter					
SUrf SCoter					
RUddy DUck					
HOoded MErganser					
COmmon MErganser					
Red-Breasted MErg					
TUrkey VUlture	X				
GOSHawk					
Sharp-Shin HAwk					
COopers HAwk					
Red-Tailed HAwk				AE	
Red-Shoulder HAwk					
Broad-Winged HAwk					
Rough-Legged HAwk					
GOlden EAgle					
BAld EAgle					
MArsh HAwk			T		
OSPRey					
PEregrine FAlcon					
MERLin					
AMerican KEstrel				FS	

FIGURE 9: *Part of a completed atlas data card*

because many atlassers were not confident of their ability to make such estimates. In addition, atlassers making relatively brief visits to a square may have had insufficient information upon which to base abundance estimates. Detailed suggestions for making estimates were given in the atlas newsletter, and are reprinted in modified form in Appendix B.

Abundance estimates are based on varying degrees of coverage of squares, and there will have been variation among atlassers in arriving at estimates. Therefore, the data presented in the Atlas have been screened by the Data Review Committee. Estimates believed to be too high (*i.e.*, at a level of abundance not considered feasible in Ontario) have been removed. Unfortunately, it was not possible to screen out estimates believed to be too low. Therefore, estimates are generally considered to be under- rather than overestimates. They are best thought of as providing only a broad assessment of abundance, and are best used as a means of comparing different species. They are not always believed to be an accurate or even an approximate indication of absolute numbers.

Data Processing and Verification

Data were thoroughly checked to ensure that the stored records, maps and summaries in this book are accurate. A brief account of the checking procedures follows, but a more complete description of the process is provided in Appendix C.

At the end of each field season, atlassers sent their data cards to Regional Coordinators. Coordinators checked for missing or incorrect data before transcribing any new information to a master card for each of the squares in their region. They also requested Unusual Species Report Forms for rare species. Cards and USRFs, together with the Regional Coordinators' comments, were then sent to the atlas office.

The information on the cards was entered into the computer. Data on cards were 'keypunched,' error-checked by computer, corrected, and entered into raw data files on computer tape. Data from different years were combined to make computer-generated 'master cards' for each square, which were sent to Regional Coordinators for verification. Corrections were then made throughout the system, as necessary. A master computer file was created containing the highest level of breeding evidence and abundance reported for each species in each square. The data from the master file are presented in this book: on the species maps, and on the statistical summaries of breeding evidence and abundance on the map pages and in the species accounts. The highest abundance estimate was chosen because it was assumed that the person making the highest estimate had found more birds than the person submitting lower estimates, *i.e.*, the lower estimates could be based upon the fact that the person making the estimate had simply missed birds.

For the first 3 years of the project, the decision as to which species were rare enough to require documentation was left largely to each Regional Coordinator. In 1984, a Data Review Committee was formed to undertake a more systematic verification of rare species records. Ideally, this process should have begun in the first year of the project, but the computer capability necessary to facilitate the process was not then in place. Earlier implementation of the process would have made the job easier. About 5,000 records ultimately required USRFs - roughly 1% of all records submitted. The process was difficult and time-consuming, but resulted in a much more reliable data base. We are confident that it contains few significant errors.

Ensuring Good Coverage of a Square or Block

To cover a square or block properly, it was necessary for an atlasser to visit all habitat types. Atlassers purchased topographic maps (at a scale of 1:50,000) to identify the square's boundaries, and to help them locate various habitats. Aerial photos were sometimes used to identify different habitats, particularly in the far north where maps at a scale of 1:50,000 were not available.

Atlassers were encouraged to visit the square at all seasons, to ensure that early and late breeders were not missed, and to find used nests when the leaves were off the trees. Large stick nests could be marked and revisited during the breeding season to establish ownership, while some nests were readily identified to species (*e.g.*, Cliff Swallow, Northern Oriole), allowing confirmation of breeding even when the birds were not present. Field work at night, and in the evening and early morning, was also recommended so that nocturnal and crepuscular species would be as fully recorded as possible.

The breeding season for most species in Ontario is concentrated from late May through July, so most field work was undertaken during that period. Song is most pronounced from late May to mid-June. Most species are attending young in or out of the nest by early July. Thus, atlassers were advised to spend the early part of the breeding season concentrating on recording all species present in breeding habitat and pinpointing their location so that later visits could be made to upgrade the evidence to probable or confirmed levels.

In some regions, teams of people specialized in finding species which are early or late breeders, nocturnal or crepuscular, or particularly difficult to find. This activity assisted greatly in adding to our knowledge of poorly known birds such as the owls and rails.

Each atlasser was issued a card identifying him or her as an atlasser to assist in speaking to landowners for purposes of gaining access to private property. A sign for the dashboard of the atlasser's parked car was also used to help avoid problems with landowners and police.

Ensuring Coverage of all Squares and Blocks

Atlassers were encouraged from the beginning to cover squares in less populated areas as well as 'their' squares close to home; this was emphasized in the latter years of the project. A few neglected areas were selected for large-scale 'square-bashes' to which all atlassers were invited. Areas receiving coverage of this type included western Manitoulin Island, Killarney Provincial Park on northern Georgian Bay, Michipicoten Island in Lake Superior, northwestern Algonquin Provincial Park, the Barry's Bay area southeast of Algonquin Park, and the Marten River area north of North Bay. Even within fairly well-populated regions, local 'square-bashes' were held on evenings as well as weekends.

Northern Ontario presented special problems related to its distance from population centres, and the lack of road or rail access to many areas. Travel grants were made available through the James L. Baillie Memorial Fund and the Northern Scientific Training Program, and free air travel was arranged through the Ministry of Natural Resources, the Ontario Geological Survey, and Bell Canada which uses aircraft in servicing its facilities throughout northern Ontario. Austin Airways (a commercial airline serving the north's remote communities) provided free flights on a space-available basis for 2 years, and then charged atlassers a nominal 25% of its commercial rate during the last year of the project.

Our northern atlassing needs for skilled birders with outdoor survival skills was advertised widely throughout North America and Britain in birding and ornithological journals and newsletters. Over 300 letters were received in response to this appeal in the first year it was attempted (1983), providing most of the people needed to cover all the blocks in the north. Applicants were provided with a more complete application form and two information/instructional booklets - one on canoe-atlassing and a more general one on atlassing remote areas. Applications were screened by the northern committee, which attempted to ensure that every team, if not every individual, had the experience necessary to survive in the potentially dangerous, remote conditions of the north. Volunteers came from as far away as Britain, Florida and Louisiana to take advantage of this unique opportunity.

Most blocks along the coasts of James and Hudson Bays were covered by people who were taken by helicopter or small plane to coastal villages or other remote spots near the shore. Inland blocks were covered by people who were flown, usually in float planes, to remote locations and picked up a week or two later, or by teams of atlassers in canoes who travelled the large river systems. All but one canoe-atlassing team consisted of 4 people in 2 canoes. This represented the best compromise between the logistical problems of transporting larger groups and the potential hazards of using fewer than two canoes. One canoe-atlassing group, and most other groups that were dropped off in one spot and picked up there later, consisted of only 2 people.

The atlas office took advantage of government-funded student employment and other employment development programs to hire summer employees to collect data where volunteers were not available. In 1985, three naturalists' clubs (Sudbury Ornithological Society, Thunder Bay Field Naturalists, and the Pembroke Area Bird Club) also hired student field workers. From 1982 through 1985, a total of 39 summer positions were created (4 to 18 per year) and these people were paid to spend May through August collecting data. Without this assistance, it would not have been possible to cover adequately all the squares in southern Ontario.

Throughout the project, every effort was made to ensure that volunteers were kept up to date on coverage achieved and work still required. The major means of communication was through the atlas newsletter, but Regional Coordinators kept people abreast of local progress and needs by holding meetings and distributing their own regional summaries. These regional reports, summarizing goals, local coverage attained and planned activities, were an important factor in motivating volunteers and maximizing their efficiency.

A meeting of Regional Coordinators was held in the spring of 1984. Over the fall and winter of 1984-85, a series of 'super-region' meetings was held, at which Regional Coordinators and some active volunteers met and discussed how to ensure the coverage of their part of the province. These meetings were helpful in encouraging volunteers to travel to other nearby regions where there were still inadequately covered squares. Prior to the final field season, in the spring of 1985, a General Meeting of atlassers was held to make clear which squares and blocks still needed coverage, and to assign as much of this work as possible.

Final coverage

Atlasser Effort

A total of 1,351 people contributed data cards to the Ontario Breeding Bird Atlas. Because many of these atlassers were accompanied by others whose names were not provided on atlas cards, the number of volunteers may well have exceeded 2,000. Atlassers reported 123,879 hours of data collection. Since many of these hours were accumulated by groups of atlassers, the actual number of person-hours of field work probably far exceeded the reported total. Some cards, especially those from squares in which the atlasser lived, did not record an estimate of hours of field work. Besides the hours of intensive atlassing, many a record was noted while atlassers drove to work or mowed the lawn, making the actual hours of field work difficult to determine. Extrapolating from cards with hours reported, we estimate that approximately 25,000 more hours of field work can be added to the above total. Adding 15 minutes of travel time for every hour of field work makes the total estimated commitment to field work over 180,000 hours. In addition, atlassers attended training sessions and meetings, studied instruction manuals, and completed data cards and Unusual Species Report Forms. All of this added a considerable but unknown number of volunteer hours to the contribution of the atlassers.

The volunteer effort was not spread evenly over the 5 year period. Table 3 summarizes atlassing effort, and shows that all measures of effort increased as the survey progressed: the number of atlassers, hours of field work per atlasser and, as a result, the total hours of field work spent atlassing each year. The major increase in the number of participants came between the first and second year, while the number of hours per atlasser took a large jump between the third and fourth years. The average number of hours of field work per atlasser over the 5 year period was 92.

TABLE 3: *Atlassing effort*

The number of atlassers per year

	1981	1982	1983	1984	1985
Southern Ontario	379	643	706	629	722
All of Ontario	433	730	828	720	811

Total number of atlassers contributing data = 1351

The number of field hours per year

	1981	1982	1983	1984	1985
Southern Ontario	11,400	20,702	21,894	23,003	28,523
All of Ontario	13,436	23,446	26,543	27,827	32,627
Hours per atlasser	31.0	32.1	32.1	38.6	40.2

Total number of hours in the south = 105,522
Total number of hours in Ontario = 123,879
Average hours per atlasser = 92

Square and Block Coverage

Coverage, as measured by hours of field work in each square and block, was not spread evenly across the province (Figure 10). Most atlassers reside in and around large urban centres in southern Ontario, so coverage was most concentrated in those areas. As the project developed, coverage spread outward from the population centres to more remote areas, eventually filling in the gaps in southern Ontario and the rest of the province.

Figure 10a shows that the heaviest coverage was throughout southern Ontario, in the vicinity of the Clay Belts, in some blocks around Lake Superior, and at Moosonee, the

FIGURE 10: *Hours reported in (A) each block in all of Ontario and (B) each square in southern Ontario*

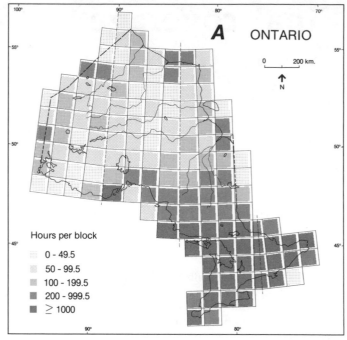

A ONTARIO

0   200 km.

N

Hours per block

0 - 49.5
50 - 99.5
100 - 199.5
200 - 999.5
≥ 1000

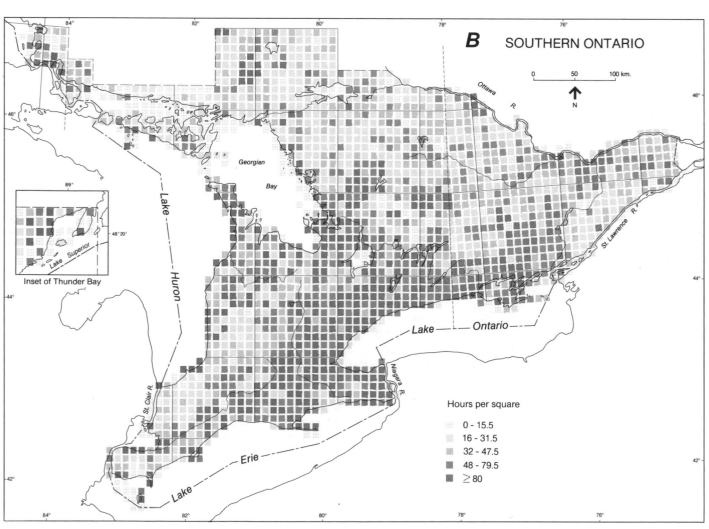

B SOUTHERN ONTARIO

0   50   100 km.

N

Ottawa R.

Georgian Bay

Lake Huron

Lake Superior

Inset of Thunder Bay

St. Lawrence R.

Lake — Ontario

Niagara R.

St. Clair R.

Lake — Erie

Hours per square

0 - 15.5
16 - 31.5
32 - 47.5
48 - 79.5
≥ 80

FIGURE 11: *Percentage of confirmed breeding records in (A) each block in all of Ontario, and (B) each square in southern Ontario*

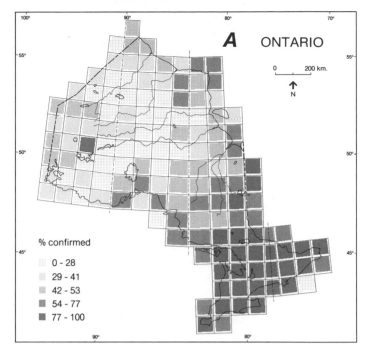

A ONTARIO

0 200 km.

N

% confirmed

0 - 28
29 - 41
42 - 53
54 - 77
77 - 100

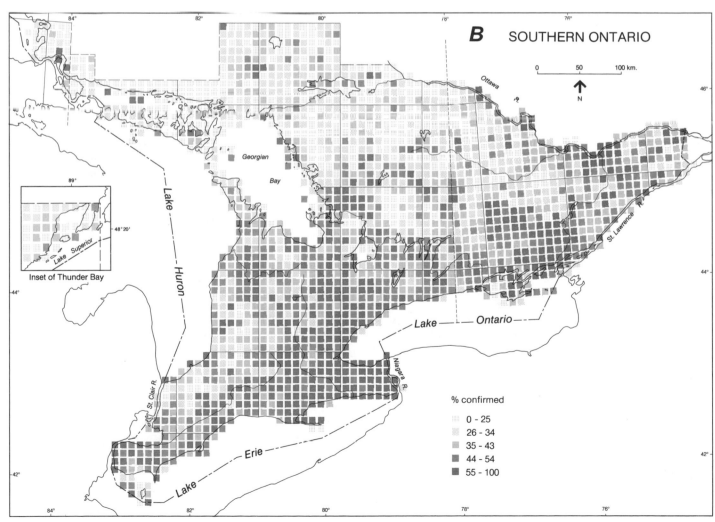

B SOUTHERN ONTARIO

0 50 100 km.

N

Georgian Bay

Inset of Thunder Bay

48°20'

Lake Superior

89°

Lake Huron

Lake Ontario

St. Clair R.

Niagara R.

Lake Erie

St. Lawrence R.

Ottawa

% confirmed

0 - 25
26 - 34
35 - 43
44 - 54
55 - 100

Sutton Ridges, Sachigo Lake, and the Winisk and Sutton River areas in the far north. Block 15WG, which is immediately east of Red Lake, was not visited on the ground during the atlas period, so no hours of coverage are shown. However, data for 4 species were collected by MNR employees flying over the block. Coverage was obtained in all the other blocks in northern Ontario, although no hours were recorded on data cards from 5 other blocks, as shown on the map.

In southern Ontario, coverage was generally most concentrated within 70 km of the north shore of Lake Ontario and the northeastern shore of Lake Erie, with heavy coverage also on the Bruce Peninsula and near London, Ottawa, the Orillia-Bracebridge area, Sudbury, Sault Ste Marie and Thunder Bay. Other particularly well-covered squares are scattered throughout southern Ontario, and often contain provincial or national parks. Park employees often contributed data, and free camping was provided for atlassers in provincial parks.

When the percentage of species confirmed in each square and block is mapped (Figure 11), the pattern is generally quite similar to that on the map of hours of coverage, as might be expected. On Figure 11a, the percentage of confirmed records is high in southern Ontario, intermediate in the more heavily populated areas of the north and low throughout much of the remote north. A few blocks on the coasts of James and Hudson Bays have fairly high percentages of confirmed records despite having relatively few hours of coverage. This is largely a result of their small land area, restricted habitat diversity, and open tundra habitat where confirmation of breeding is relatively easy to obtain. In southern Ontario, the percentage of species confirmed in each square and the number of hours of coverage have very similar patterns, indicating that more field work leads to a higher confirmation rate.

Figure 11 shows that most records in squares on the Canadian Shield in southern Ontario and in blocks throughout northern Ontario are of only possible or probable breeding. This is also apparent on species maps throughout the book, and partly reflects the fact that it is more difficult to confirm breeding for birds which nest in heavily wooded areas than it is to confirm those in open agricultural areas. It is also a result of the fact that coverage in some of those areas often barely exceeded the minimum requirement, with many squares and remote blocks being visited in only one year. Nevertheless, field work was undertaken primarily during the peak of the breeding season. Therefore, possible and probable records from these areas are very likely to represent birds which could have been confirmed had more time been available.

Figure 12 shows the number of species found in each block in Ontario and in each square in southern Ontario. The average number of species reported per block throughout the province was 107, and in southern Ontario the average per square was 88. Both of these averages incorporate data from partial squares and blocks along shorelines or provincial or international boundaries; the average values for blocks and squares representing complete land area would be slightly

higher.

Calculations reveal that after about 70 hours of field work the number of additional species found in a southern Ontario square was negligible; the average number of species reported in 70 or more hours was 97. Therefore, assuming that 97 species is very close to the average number of species actually present in a square, and knowing that the average number actually reported was 88, the atlas field work can be estimated to have reported an average of close to 92% of the species present. This total is well above the minimum acceptable coverage level of 75% set by atlas organizers, and results from the fact that the average number of hours of coverage per square (57) far exceeded the minimum requirement of 16 hours.

From the above calculation, it is clear that in southern Ontario about 8% of species were missed in an average square. Whereas a record on the map is a reasonably sure indication of the presence of a species, a blank square does not necessarily indicate that the species was absent. It is very difficult to prove that a species is absent from a square, but the absence of a record is likely to mean that it is at least uncommon or difficult to find. Likewise, the absence of atlas records from a county or other jurisdiction (or from the province) does not necessarily mean that the species did not breed there in the atlas years, but only that it was not found and reported by atlassers.

In the remote north (i.e., in blocks without road or rail access), coverage was often restricted to a few squares within each block (see Figure 13). These were usually the most readily accessible squares, such as those with large lakes or rivers upon which float planes could put down safely, at small communities with airstrips, or along the coastal tundra where balloon-tired planes and helicopters could land. Except on dry coastal tundra, travel on foot in the remote north is severely restricted, so canoes were essential for gaining access to new squares with different habitat types. Consequently, coverage of river-edge and lake-edge habitat was usually better than coverage of habitats which had to be reached on foot, and this is probably reflected in the species maps.

It should also be noted from Figure 13 that coverage of blocks containing coastal areas in far northern Ontario was sometimes restricted to, or concentrated in, the squares of the immediate coastal strip. In these cases, treed areas inland from the coast may not have received sufficient attention, with the result that the species frequenting those habitats are likely underrepresented. Where coverage shows a regular pattern of alternating squares in northern blocks, those squares contain data collected by Canadian Wildlife Service biologists performing systematic aerial waterfowl censuses. As a result of that work, waterfowl are relatively heavily represented in those blocks. Because coverage in the remote north was almost completely restricted to the months of June and July, species which are most easily found early or late in the season, such as the owls, are likely to be underrepresented. Conversely, because nearly all nights in the remote north were spent in tents, crepuscular and nocturnal species

FIGURE 12: *Number of species recorded in (A) each block in all of Ontario, and (B) each square in southern Ontario*

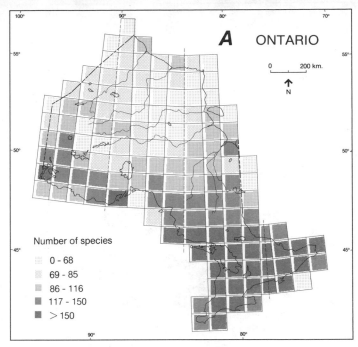

A ONTARIO

0 ___ 200 km.

N

Number of species

- 0 - 68
- 69 - 85
- 86 - 116
- 117 - 150
- > 150

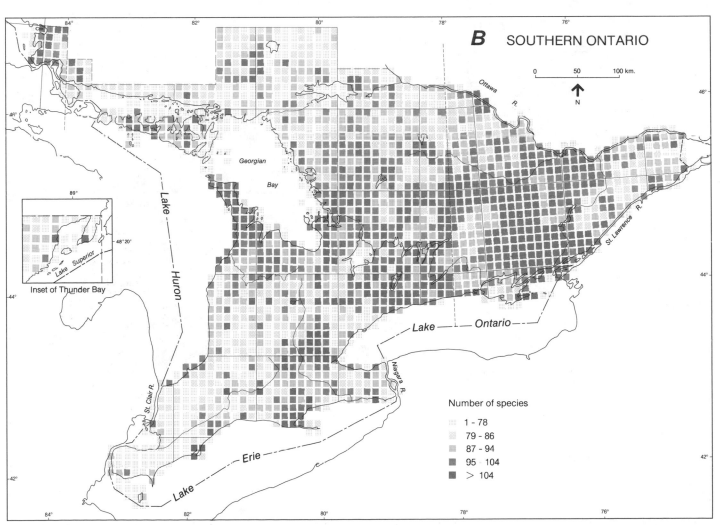

B SOUTHERN ONTARIO

0 ___ 50 ___ 100 km.

N

Ottawa R.

Lake Superior

Georgian Bay

Lake — Huron

Inset of Thunder Bay

St. Clair R.

Lake — Erie

Niagara R.

Lake — Ontario

St. Lawrence R.

Number of species

- 1 - 78
- 79 - 86
- 87 - 94
- 95 - 104
- > 104

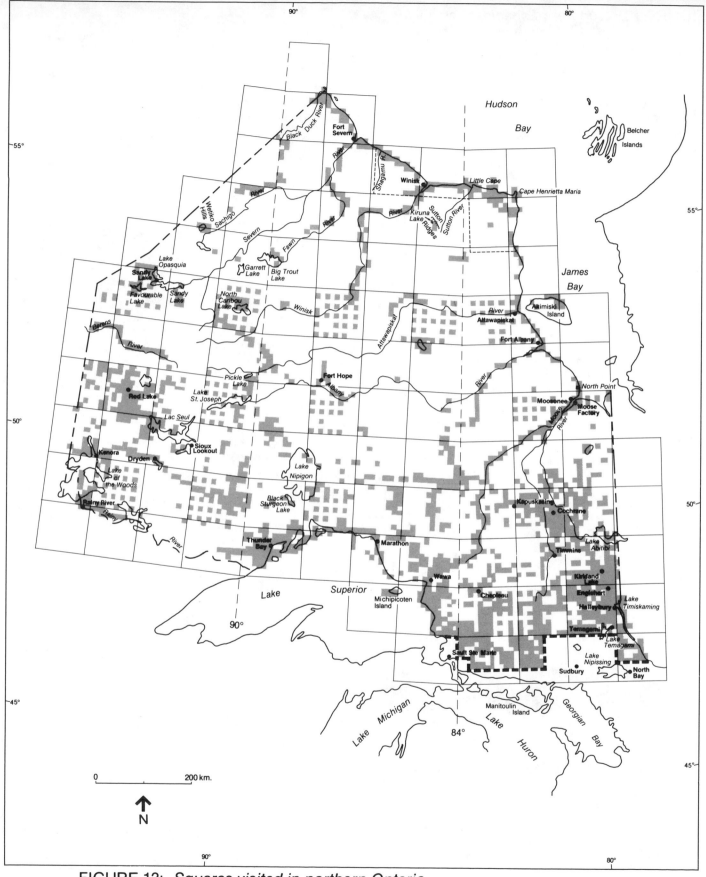

FIGURE 13: *Squares visited in northern Ontario*

FIGURE 14: *Distribution of abundance estimates.*

(A) *The number of squares in each block in Ontario with at least one abundance estimate*

(B) *The number of species for which abundance estimates were provided in each square in southern Ontario*

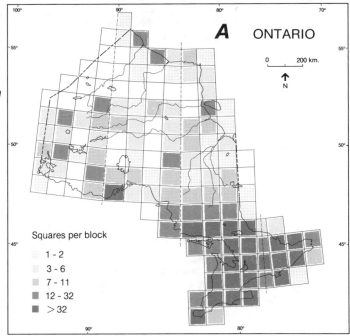

A ONTARIO

Squares per block

- 1 - 2
- 3 - 6
- 7 - 11
- 12 - 32
- > 32

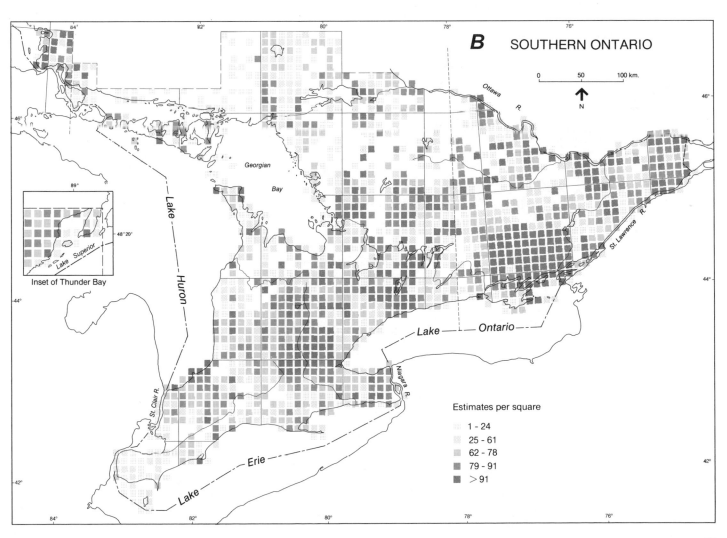

B SOUTHERN ONTARIO

Inset of Thunder Bay

Estimates per square

- 1 - 24
- 25 - 61
- 62 - 78
- 79 - 91
- > 91

active in June and July may well have been noted proportionately more often than was the case farther south.

In summary, coverage in the road- or rail-accessible northern blocks was generally more comprehensive than in remote blocks. Relatively easy access, and the fact that atlassers lived in some of those blocks, meant that most habitats were visited and that a high percentage of the species which bred there was likely reported. Farther north, where travel was more restricted, fewer different habitats and fewer examples of each could be covered, so the maps are probably a less complete reflection of the species actually breeding there.

Species Coverage

Not all species are equally well represented on the maps. The most fully represented species tend to be conspicuous, well known to atlassers, common, active (and usually vocal) in the daytime in June and early July, and found in readily accessible habitats. Generally, the more of these criteria a species meets, the greater the likelihood of its being found by atlassers.

Owls are probably not shown in all the squares in which they occur because they are nocturnal, active early in the season, and generally poorly known. Considerable 'owling' effort by atlassers has greatly improved our knowledge of the distribution of Ontario's owls, particularly in southern Ontario, but with the result that the distributions shown on the maps are greatly influenced by the degree of 'owling' effort. The same can be said for other nocturnally active species. Many of these appear to be particularly well represented north of Kingston, for example, as a result of an extremely diligent effort on the part of local atlassers.

Atlas results - an overview

Species Richness

Breeding evidence is presented for 292 species and 2 hybrids (Lawrence's and Brewster's Warblers). Of these, 57 species were reported only in northern Ontario, 39 (plus the 2 hybrids) were found only in southern Ontario, and 196 were found in both sectors. The pattern of species richness is closely tied to the biogeography of Ontario, and some of the more conspicuous patterns are discussed here.

The number of species generally declines from south to north (Figure 12), a pattern that was expected prior to the atlas project. A finer breakdown of the data reveals that in some cases the Hudson and James Bay coastal blocks contain more species than those on the Hudson Bay Lowland away from the coast. Coastal blocks with lower species totals, such as those along northern James Bay, contain only small areas of land, and are mostly of one habitat type: tundra. The high species totals on the coast are partly the result of the additional coverage these blocks received. However, they also reflect the presence of human settlements, which create habitat suitable for additional species. Moreover, coastal blocks often contain tundra habitats and Hudson Bay Lowland Forest as well as the tree-line ecotone in between.

The blocks of the interior of the Hudson Bay Lowland contain the fewest species of any area of the province, because of the area's uniform habitat. Habitat diversity increases on the north end of the Canadian Shield where dry soils, rocky outcroppings and deeper lakes are interspersed with low-lying bog or fen habitats characteristic of the Lowland. Species totals are still quite low in that region, partly reflecting low coverage.

Southward from approximately 50^0N all blocks are road- or rail-accessible. North of that line, road access is also available to the towns of Red Lake (block 15VG) and Pickle Lake (block 15XH), both of which are in blocks with high species totals (123 and 99, respectively). The higher species totals in these blocks are probably a reflection of easier access to all habitats, more hours of coverage, and the extra species which are suited to the modified habitats of these areas. In this regard, it is interesting to note that in block 15UG, which is immediately west of Red Lake but which has no road access or human settlements, there were only 113 species reported

from 371 hours of field work, whereas in the block containing Red Lake (15VG), 123 species were reported in only 185 hours. Moosonee, which is also accessible by rail, has the highest species total north of 50^0N (137 species). The high species total in the Moosonee block is the result of good coverage and ease of access, combined with the diversity of habitats, including coastal salt marshes, the large river, the southern Hudson Bay Lowland Forest, and the town itself. Higher species totals throughout the rest of northern Ontario are generally in blocks with large towns and resident atlassers, again reflecting the combined effect of more coverage, easier access, and the addition of species that benefit from human activity.

Block species totals are highest in southern Ontario, the area of the province in which every 10 km square was covered. Species totals for squares in southern Ontario should indicate the pattern of species richness fairly accurately, because of the high level of coverage achieved.

The area between Kingston and Ottawa has an extensive area of contiguous high species totals, and the area between Kingston and the southern end of Georgian Bay also has many species-rich squares. While those areas had good coverage, as shown in Figures 10 and 11, the number of species found is exceptionally high. The area includes the diverse habitats of the Frontenac Axis and the edge of the Canadian Shield, where agricultural land and wooded areas are interspersed; because the farm land in this area is of poor quality, much of it is abandoned, creating additional habitat diversity. The high number of species on the Bruce Peninsula also reflects habitat diversity.

High species totals occur in the Guelph-Waterloo-Hamilton area, which is at the boundary of the Carolinian and Great Lakes-St. Lawrence Forest regions. This area also includes part of the Niagara Escarpment and has considerable habitat diversity associated with the glacial landforms west of the escarpment. The three most southerly rows of blocks are dominated by agriculture and other human influences. The low percentage of forest cover in these blocks is reflected in the lower species totals. However, north of Long Point in Haldimand-Norfolk RM, there are some fairly extensive tracts of Carolinian Forest and some large and mature conifer plantations. As a result, some squares in this area support more typically northern bird species, and have higher species totals than most squares in the Carolinian Forest zone.

Low species totals occur in most squares in the extreme southwest, and in a band of squares extending northeast to near Orangeville. Although the reasons for this have not yet been studied, the low percentage of forested land remaining and the extensive wetland drainage in that area are likely to be contributing factors. High species totals were reported from squares in and around Pinery and Rondeau Provincial Parks, indicating that large numbers of species can still be found in the extreme southwest in the few squares with diverse habitats including extensive woodland.

Species totals in groups of squares northeast of Sudbury and south of Lake Nipissing are probably low as a result of insufficient coverage rather than actual low species richness. These squares have minimal road access, so coverage was especially difficult. Partial squares along shorelines are also low in species.

Because species richness depends on both biogeography and on completeness of coverage, we have tried to discuss the relative importance of these factors for each part of the province. We conclude that species richness in southern Ontario is a good reflection of habitat diversity. The overall picture of species richness in northern Ontario mainly reflects habitat diversity as well, but the lack of coverage in some areas may have underrepresented the true number of species present there.

Abundance Estimates

The number of abundance estimates provided varied greatly among squares and among regions. Figure 14 shows the number of squares in each block in which at least one abundance estimate was provided, and the number of species for which abundance estimates were provided in each square in southern Ontario.

Highlights of the Results

The maps in this book are the most important product of the atlas project. Nevertheless, certain results were new or surprising, and others are difficult to extract from the maps themselves. Some of these are noted below.

- New Ontario breeding species confirmed during the atlas period were: Cinnamon Teal, Canvasback, Bohemian Waxwing, Northern Shrike, Harris' Sparrow, and Snow Bunting. Additionally, an unmated female California Gull laid a clutch of eggs on Toronto's Leslie St. Spit in 1981, and a male Mountain Bluebird mated with a female Eastern Bluebird near Port Stanley in 1985.

- Several species known to breed previously in the province were not confirmed as breeding during the atlas period: Cattle Egret, King Rail, American Avocet, Marbled Godwit, Pectoral Sandpiper, Stilt Sandpiper, Chuck-will's-widow, Gray-cheeked Thrush, Kirtland's Warbler, Dickcissel, and Lark Sparrow.
- Species previously recorded as breeders but not reported at all during the atlas period were: Greater Prairie-Chicken, Black Guillemot, Passenger Pigeon and Bewick's Wren.
- The species recorded in only one square in the province were: Cattle Egret, Ross' Goose, King Eider, American Avocet, Pectoral Sandpiper, Purple Sandpiper, California Gull, Glaucous Gull, Snowy Owl, Western Kingbird, Northern Wheatear, Mountain Bluebird, Kirtland's Warbler, Lark Sparrow and Harris' Sparrow.
- The 10 species found in the most squares in southern Ontario were:

Species	Number of Squares (out of 1824)
Song Sparrow	1801
American Robin	1786
Red-winged Blackbird	1784
Tree Swallow	1783
Cedar Waxwing	1779
Northern Flicker	1777
Eastern Kingbird	1775
Chipping Sparrow	1764
Red-eyed Vireo	1762
Common Grackle	1762

- The 10 species found in the most squares in northern Ontario were:

Species	Number of Squares (out of 1834 visited)
White-throated Sparrow	925
Common Loon	877
Common Raven	757
Red-eyed Vireo	705
American Robin	695
Northern Flicker	691
Yellow-rumped Warbler	691
Cedar Waxwing	659
Swainson's Thrush	645
Spotted Sandpiper	598

- The 11 species found in the most blocks in the whole province were:

Species	Number of Blocks (out of 137)
Herring Gull	129
Mallard	128
Tree Swallow	128
Spotted Sandpiper	127
Swamp Sparrow	127
White-throated Sparrow	127
Yellow Warbler	126
Yellow-rumped Warbler	122
Northern Flicker	121
Red-eyed Vireo	121
Northern Waterthrush	121

- Some species were found more frequently than expected, such as the Sandhill Crane, Eastern Screech-Owl, Bohemian Waxwing, Louisiana Waterthrush, Hooded Warbler (though it is still very rare), and Orchard Oriole.
- Range expansions to the north were documented for some species, including Turkey Vulture, Northern Cardinal, and, of course, House Finch. In particular, several species which were not thought to extend beyond the Canadian Shield were found to breed on the Hudson Bay Lowland, *e.g.*, Winter Wren and Ovenbird. Locations to the south of the previously known range were found for a number of species, especially those which are taking advantage of maturing coniferous plantations in the south. These include Red-breasted Nuthatch, Golden-crowned Kinglet, and Magnolia Warbler. Expansion of range to the east is shown among several species more typical of the prairies, including Northern Shoveler, Gadwall, Wilson's Phalarope and Clay-colored Sparrow.

- The 14 squares in which the most species were reported were:

Square	Region	Number of Species	Number of Hours	Prominent Feature
17MS28	02	146	256	Rondeau Bay
18UE73	21	143	55	Canoe Lake
18TD87	17	142	239	Presqu'ile Prov. Park
18UE93	21	142	63	Chaffey's Locks
18UE72	21	137	48	Gould Lake
18VF22	24	136	723	South March
17PU37	45	136	865	Ballantrae
18QV38	26	135	91	Paudash Lake
17LA76	33	135	118	Gatacre Point
17MS27	02	134	135	Rondeau South Beach
17NU46	07	134	579	Luther Lake
17MV85	08	134	55	Wiarton
18UD46	20	134	69	Plover Fields
18UG00	27	134	293	Deep River

How to use this book

Species Treated in Maps and Texts

Two hundred and eighty-three species of birds have been accepted as breeding in Ontario by the Ontario Bird Records Committee (OBRC) (Wormington and James 1984). In addition, 3 species with atlas records (Mountain Bluebird, Bohemian Waxwing and Snow Bunting) were accepted as confirmed breeders by the Atlas Data Review Committee, based on documentation submitted by atlassers. The relevant records of those species have not yet been considered by OBRC, but they are treated here as accepted breeders pending their decisions. This gives a total of 286 breeding species.

In the following sections, maps are presented for all except 4 of the 286 breeding species. The 4 exceptions are species that were not reported during the atlas years (1981-85), namely Greater Prairie-Chicken, Black Guillemot, Passenger Pigeon (extinct) and Bewick's Wren, and they are described only with brief accounts of their history and status in Ontario. Information is also presented for 12 additional species that have not been accepted as confirmed breeders by OBRC or the Atlas Data Review Committee and each of these is identified by an asterisk following the English name in the species account. Ten of those 12 are species for which atlassers reported evidence of probable or possible breeding that was accepted by the Atlas Data Review Committee as meeting the atlas criteria. Records for those 10 species are mapped; they are: Yellow-crowned Night-Heron, Purple Gallinule, Purple Sandpiper, Glaucous Gull, Snowy Owl, Western Kingbird, Northern Wheatear, Kentucky Warbler, Summer Tanager and Hoary Redpoll. No breeding evidence was reported for the other 2 unaccepted species during the atlas years and their status is outlined only briefly. They are the Brant, which was previously accepted as a breeder, but was removed from the Ontario list during the atlas period, and the Trumpeter Swan, which may have formerly bred in Ontario and is currently the subject of an introduction program by MNR. Finally, in addition to the 298 species accounts, maps and texts are presented for 2 hybrids: Brewster's Warbler and Lawrence's Warbler.

Arrangement of Species Accounts and Maps

The species accounts in the Atlas are divided into 3 sections. The first and largest section contains the 233 species (plus the 2 hybrids) for which breeding evidence was reported during the atlas period in southern Ontario; many of these species also occurred in northern Ontario. The second section contains the 57 species for which breeding evidence was reported only in northern Ontario. The third section contains 6 species that were not reported during the atlas period, 4 of which have been previously accepted as breeders in Ontario and 2 of which may have nested formerly (see previous section), plus the California Gull and Mountain Bluebird. The latter 2 species are not given the comprehensive treatment afforded to the others, because their breeding records each involve only one individual which was not mated to a bird of the same species (see the species accounts for details). Within each section, species are listed in taxonomic order (AOU 1983).

The 3 sections (referred to here as 'southern Ontario species', 'northern Ontario species' and 'miscellaneous species') are presented differently. For each species recorded in southern Ontario, a full page of maps and statistics is presented, including a southern Ontario map and an Ontario map, southern Ontario and all Ontario breeding evidence statistics, and abundance estimates (see below). The maps are accompanied by a full page of text on the facing page. For species recorded exclusively in northern Ontario, the Ontario map, all Ontario breeding evidence statistics and the northern Ontario abundance estimates are given on the same page as the text. The miscellaneous species have no maps, and only brief written accounts.

Maps

Ontario Map

The Ontario map shows the outline of the province, the 100 km UTM blocks, the 78^0, 84^0 and 90^0W longitude zone lines, and major rivers and lakes. The highest level of breeding evidence reported in each block is the one shown on the map. The 3 levels of breeding evidence are depicted using 3 shades of orange, the darkest depicting confirmed breeding, the medium density showing probable breeding, and the lightest indicating possible breeding. The orange shaded area doesn't quite fill the whole block. Therefore, in blocks on the provincial boundary with only a small area of land in Ontario, the orange shaded area may not actually touch land. Note also that a similar-sized area is shaded whether or not part of the block is outside the province; other than from some islands just off the coast of James and Hudson Bays that are under the jurisdiction of the Government of the Northwest Territories, data from outside the province were not used in this Atlas.

The number of blocks in which the species was reported is given beneath the map, along with the percentage of the province's 137 blocks in which the species was reported. That total is then broken down into the number of blocks in which possible, probable and confirmed breeding were reported, and the percentages that each level contributes to the total. For example, if a species was reported in 50 (or 37%) of Ontario's 137 blocks, the breeding evidence summary accompanying the Ontario map might read as follows:

BREEDING EVIDENCE
Reported in 50 (37%) of 137 blocks
    Possible breeding 15 (30%)
    Probable breeding 10 (20%)
    Confirmed breeding 25 (50%)

Southern Ontario Map

The southern Ontario map shows the outline of that part of the province, the 100 km UTM blocks, and the 78^0 and 84^0W UTM zone lines and, again, a few major rivers and lakes are provided for reference. The northern boundary of the map is the northern limit of the contiguous area in which every 10 km square was adequately covered. Islands longer than about 3 km are shown on the map, but the few records from squares that contain only smaller islands may appear to be in the middle of lakes. An inset on the left side of the map shows the squares which were covered around the city of Thunder Bay in northern Ontario. The highest level of breeding evidence reported in each 10 km square is shown on the map. Breeding evidence and other data are presented as described above for the Ontario map, but by 10 km squares. The total number of squares in which the species

was recorded in southern Ontario is given, together with the same number expressed as a percentage of the 1,824 10 km southern Ontario squares. Thunder Bay squares are not included in the southern Ontario statistics.

Abundance Estimates

Abundance estimates are shown in histogram form, with separate histograms for estimates from northern and from southern Ontario. The horizontal axis of the histogram is divided into 6 sections corresponding to the 6 categories of abundance estimates. The height of each bar shows the percentage of all abundance estimates in that category. Because atlassers were not required to provide abundance estimates, they were not provided in all squares. Therefore, beside each histogram, the number of squares in which estimates were provided is given, along with the number of squares in which the species was recorded. As abundance was estimated only for squares, the number of *squares* is given, not the number of *blocks*.

Species Accounts

In the species accounts, English and scientific names follow the most recent (6th edition) American Ornithologists' Union Check-list of North American Birds (AOU 1983) and its 35th supplement (AOU 1985). Subspecific nomenclature follows Godfrey (1986) unless otherwise indicated. French names follow Ouellet and Gosselin (1983), as used in Godfrey (1986). An index of English, French and scientific names is provided at the back of the book. French names from La Société Zoologique de Québec (1983), as well as those used in Godfrey (1986), are listed in the index. References given in the text are listed in the 'Literature cited' section at the back of the book. The scientific names of plants mentioned in the text are also listed at the back of the book, in the 'List of plant names.' Plant nomenclature is generally according to Hosie (1969) for trees, and Gleason (1952) for other plants.

In the text, the term 'southern Ontario' or 'the south' refers to the area defined on the southern Ontario map. 'Northern Ontario' or 'the north' is everything north of 'southern Ontario', including Thunder Bay. Areas within southern and northern Ontario (*e.g.*, northwestern Ontario) are referred to using the terms shown in Figure 15. Each shaded square or block on the southern Ontario or Ontario map may be referred to as a 'record' in the species accounts. Also, as part of the vernacular of atlassing, accounts may say something like "the species is difficult to confirm," meaning that it is difficult to observe behaviour corresponding to one of the confirmed breeding categories defined in Table 2. The terms 'fledged young', and 'brood', when used in the species accounts, refer to young birds out of the nest, but incapable of sustained flight, as defined in Table 2. Furthermore, because abundance estimates were not provided for every square in which a species was reported, when cases of high abundance estimates are mentioned in the text, the reader should understand that these are only examples of areas of high abundance: there may be other squares with equally high or even higher actual abundance of the species for which abundance estimates were not provided.

FIGURE 15: *Atlas definitions for various parts of Ontario*

Abbreviations Used in the Text

AOU - American Ornithologists' Union

BBS - Breeding Bird Survey. A cooperative program of the USFWS and CWS. Roadside surveys carried out yearly. See Robbins *et al.* (1986).

Blue List - refers to the Blue List of species of concern; published in *American Birds*

CBC - Christmas Bird Count. Coordinated by the National Audubon Society and published annually in *American Birds*.

Co - County

COSEWIC - Committee on the Status of Endangered Wildlife in Canada

CWS - Canadian Wildlife Service

FON - Federation of Ontario Naturalists

ha - hectare (100 x 100 m); there are 100 ha per square kilometre

LPBO - Long Point Bird Observatory

MNR - Ontario Ministry of Natural Resources

Nat. Park - National Park

NWT - Northwest Territories

NYBBA - New York Breeding Bird Atlas

ONRS - Ontario Nest Records Scheme. Coordinated by the ROM. See Peck and James (1983, 1987).

Prov. Park - Provincial Park

ROM - Royal Ontario Museum

RM - Regional Municipality

US - United States of America

USRF - Unusual Species Report Form

USFWS - United States Fish and Wildlife Service

UTM - Universal Transverse Mercator Grid System

Species accounts

S.L. HOUSE

COMMON LOON
Huart à collier
Gavia immer

The Common Loon is a haunting and popular symbol of Ontario's wilderness lakes, easily recognized by its large size, conspicuous black and white plumage, and eerie calls. People who rarely notice other birds often know the Loon.

Except for small numbers in northwestern Europe, the Common Loon breeds exclusively in North America, from the northern US northward to beyond the tree-line. In Ontario, this species generally breeds on the lakes of the Canadian Shield and northward, with a few pairs occurring south of the Shield in southern Ontario. Most Loons winter along the Atlantic, Pacific, and Gulf coasts and the coasts of northwestern Europe (Palmer 1962). Although it will breed on lakes as small as 5 ha, larger lakes are usually used (Palmer 1962, Alvo *et al.* in press). Larger lakes may support several pairs, normally in visually separated bays, while smaller lakes have a single pair. Nests are often found on small islands or promontories, very close to the water's edge; sometimes they are located on old muskrat houses or on floating islands in marshes. Alvo (1981) has suggested that the proportion of Common Loons nesting in marshes might be on the increase due to lower disturbance there by boaters. Nevertheless, they often nest in areas frequented by boaters and fishermen (Heimberger *et al.* 1983, Alvo 1981).

Although the species is common throughout its Ontario range where lakes are of suitable size, Common Loon numbers have been reduced around population centres and in agricultural areas. The species is now largely absent from much of Ontario's Carolinian Forest zone, where at least some nesting used to occur (Snyder and Logier 1931, Peck and James 1983). At Long Point on Lake Erie, Loons were known to nest in the marshes early in the century (Snyder and Logier 1931), and occasional records still occur there (McCracken *et al.* 1981). Loons will accept artificial nesting islands, and can coexist with moderate development and recreational activities if people are well informed (Heimberger *et al.* 1983). A more serious threat than human recreation on nesting lakes may be the effects of air and water pollution (Fox *et al.* 1980, Frank *et al.* 1983, Barr 1986). There is some evidence that the incidence of successful breeding may be reduced on lakes affected by acid precipitation (Alvo *et al.* in press).

Low-lying nests are vulnerable to destruction by powerboat wash and man-made changes in water levels, and eggs are frequently damaged when frightened adults flush from the nest. Once eggs are hatched, chick survival appears less affected by human disturbance (Heimberger *et al.* 1983). Territorial pairs show conspicuous vocal and visual displays, and nests are fairly easily located by systematic search. Breeding is most easily confirmed through the sighting of unfledged chicks during their 10-11 weeks of growth in mid- to late summer: 71% of confirmations were of young birds. Even small chicks under 2 weeks old can be readily seen, often riding on a parent's back. Over 75% of probable breeding records were pairs, and many of these likely represented breeding attempts. In many cases, the presence of a pair indicates that breeding occurred in past years or will in the future, since pairs return to the same lake year after year. Some birds recorded as possible breeders in suitable habitat (about 30% of all records) may have resulted from the sighting of one bird while its mate was incubating - this is especially true in areas that were visited only briefly by atlassers. However, some of these were probably non-breeders. Unmated birds often flock together on larger lakes and can appear outside the normal breeding range (*e.g.*, along the Hudson Bay coast).

The atlas maps show that, although scattered nesting occurs in southwestern and far southeastern Ontario, the Common Loon is primarily a bird of the Canadian Shield portion of the Great Lakes-St. Lawrence Forest zone and the Boreal Forest zone. The paucity of records along the James Bay coast may be due to a lack of suitable lakes in that area, and the relative scarcity of confirmed breeding records in far northern Ontario probably reflects a lower population density of Common Loons rather than low atlas coverage. Peck and James (1983) showed only 4 breeding records on the Hudson Bay Lowland, but overall, their distribution of records is very similar to that shown by the atlas data.

Common Loons are still common and widely distributed in Ontario. The density of the breeding population is determined by territorial requirements and the distribution of suitable lakes and rivers. In 88% of the southern Ontario squares and 67% of the northern squares for which atlassers estimated abundance, fewer than 11 pairs were estimated to be present. The densely populated squares (11 to 100 pairs) were nearly all in the Great Lakes-St. Lawrence Forest zone, presumably reflecting the greater proportion of highly suitable habitat in that area.-- *E.H. Dunn*

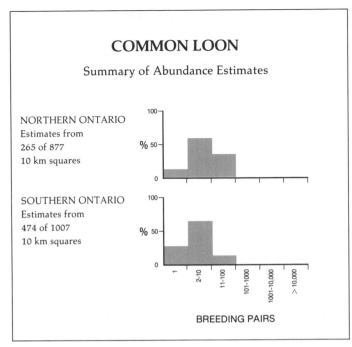

COMMON LOON

Summary of Abundance Estimates

NORTHERN ONTARIO
Estimates from
265 of 877
10 km squares

SOUTHERN ONTARIO
Estimates from
474 of 1007
10 km squares

BREEDING PAIRS

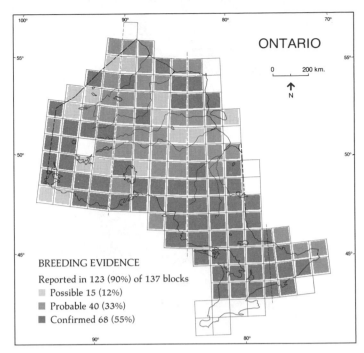

ONTARIO

0   200 km.

N

BREEDING EVIDENCE
Reported in 123 (90%) of 137 blocks
Possible 15 (12%)
Probable 40 (33%)
Confirmed 68 (55%)

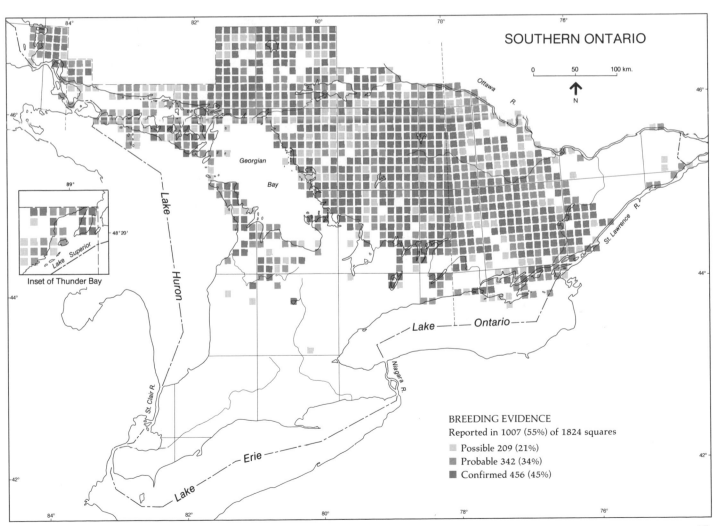

SOUTHERN ONTARIO

0   50   100 km.

N

Inset of Thunder Bay

BREEDING EVIDENCE
Reported in 1007 (55%) of 1824 squares
Possible 209 (21%)
Probable 342 (34%)
Confirmed 456 (45%)

PIED-BILLED GREBE
Grèbe à bec bigarré
Podilymbus podiceps

Ontario's most common grebe, the Pied-billed, is most familiar during migration, when it is found in open water, often arriving just after the ice melts. It chooses densely vegetated wetlands during the breeding season, where visibility is restricted, but has a powerful voice to aid in advertising its territory.

The breeding range of the Pied-billed Grebe extends from the extreme southern NWT south through Central America, the West Indies and South America to Chile and Argentina. It occurs from coast to coast in southern Canada. In Ontario, the Pied-billed Grebe has been found breeding north to Fort Albany and Sandy Lake (James *et al.* 1976), but it is far more frequently encountered in the south. Breeding habitat consists of ponds with considerable shore and emergent vegetation, marshes with areas of open water, and marshy inlets and bays (Palmer 1962, Faaborg 1976). It occasionally winters as far north as the Great Lakes (Godfrey 1986).

There is little to indicate that the Ontario breeding range of the Pied-billed Grebe has changed considerably as long as records have been kept. Baillie and Harrington (1936) described its breeding range as north to Cochrane District and Thunder Bay, but there had been little ornithological work north of these areas at that time. Farther south, however, some local distributional changes have been noted. Quilliam (1973) could not explain a decrease in birds in the Kingston area in the 1960's, and breeding evidence was not reported at Point Pelee during the atlas period, although Stirrett (1973) called it a regular summer resident there. Low water levels have been blamed for local decreases elsewhere, and the gradual filling in of marshes and ponds by vegetation will make them unsuitable for breeding. The draining of wetlands in southern Ontario has no doubt caused a loss of breeding habitat and in Pied-billed Grebe numbers, although

this may be somewhat compensated for by the recent increase in water impoundments.

The Pied-billed Grebe is most conspicuous early in the breeding season. Its characteristic and far-carrying call can be heard 1.6 km or more away under favourable conditions (Taverner 1953), and showy courtship displays, consisting of splashy dashes of the female followed by the male, occur in open water areas at this time (Glover 1953). After this brief period, however, especially when emerging cattails begin to restrict visibility, locating breeding pairs is difficult. The Pied-billed Grebe is territorial, defending an arc of about a 50 m radius around the nest (Glover 1953), though it will feed communally outside that area. Its nest, a floating platform of marsh vegetation, is anchored to emergent vegetation usually within 5 m of open water (Glover 1953), so that the birds can approach the nests from beneath the surface. Unless one is willing to enter the cattail marsh, nests are usually difficult to find: nests were reported in only 32 squares. Both parents incubate, and feed the young. Observation of the curiously striped young proved to be by far the easiest way to confirm breeding; three-quarters of southern Ontario confirmations and 8 of 9 of northern Ontario confirmations were obtained in this manner.

The maps show confirmed breeding north to Moosonee and west to Thunder Bay, and possible breeding north to Garrett and Bearskin Lakes near Big Trout Lake. These most northerly records were from June or July and so very likely represent breeding birds. It is apparent, however, that the Pied-billed Grebe is thinly spread north of a line from North Bay to Thunder Bay. In fact, the species was not found at all in much of its presumed range in northcentral Ontario (Godfrey 1986), an absence which is consistent with the lack of previous breeding records from that area (Peck and James 1983). Atlas data reveal somewhat more frequent occurrences between Cochrane and North Bay; the Regional Coordinator for the Kirkland Lake area describes it as "quite local, returning to favoured habitat." Southern Ontario data reveal more concentrated observations north of Kingston and south of Ottawa. This is in agreement with historical observations which considered the birds to be common in various parts of eastern Ontario (Macoun and Macoun 1909, Lloyd 1923, Toner *et al.* 1942). It was less frequently reported elsewhere in Ontario, with records tending to be somewhat grouped, presumably indicating areas with more suitable habitat. Regional Coordinators from agricultural areas generally agreed that the maps give an accurate and complete picture of the Pied-billed Grebe's breeding locations, but Coordinators from the Shield, where access is more difficult, felt that birds were probably missed in some squares.

Abundance estimates reflect the greater abundance of Pied-billed Grebes in southern Ontario. More than 10 pairs were thought to occur in several squares of the Lake St. Clair Marshes, at Long Point, Luther Lake, and Hullett Marsh, and in several squares north of Kingston. In the north, only one square was estimated to have more than 10 breeding pairs - that being near Timmins.-- *M.D. Cadman*

PIED-BILLED GREBE

Summary of Abundance Estimates

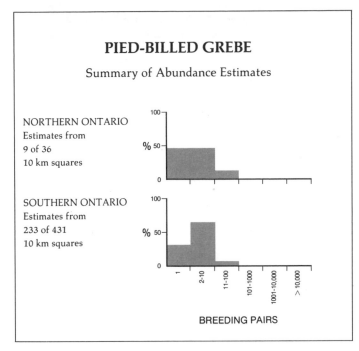

NORTHERN ONTARIO
Estimates from
9 of 36
10 km squares

SOUTHERN ONTARIO
Estimates from
233 of 431
10 km squares

BREEDING PAIRS

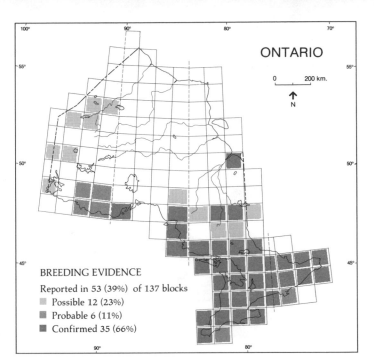

ONTARIO

0   200 km.

N

BREEDING EVIDENCE

Reported in 53 (39%) of 137 blocks

Possible 12 (23%)

Probable 6 (11%)

Confirmed 35 (66%)

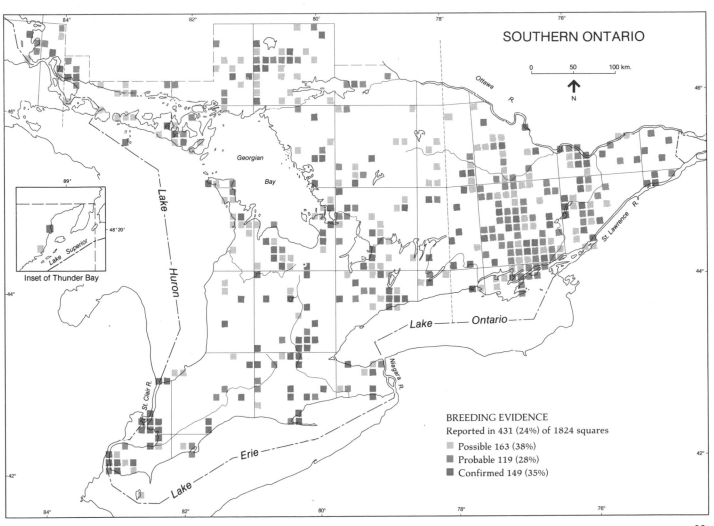

SOUTHERN ONTARIO

0   50   100 km.

N

Inset of Thunder Bay

BREEDING EVIDENCE

Reported in 431 (24%) of 1824 squares

Possible 163 (38%)

Probable 119 (28%)

Confirmed 149 (35%)

HORNED GREBE
Grèbe cornu
Podiceps auritus

The Horned Grebe is a locally common migrant but is a very rare breeder in Ontario. In Canada, it breeds from southern Quebec west to British Columbia and north to the arctic coast. The Horned Grebe reaches its greatest abundance in the prairie pothole region (Sugden 1977). Marshes with deep water and an interspersion of vegetation and open water are preferred. Usually only one pair breeds on a small pond (Sugden 1977), but larger wetlands can include more birds, including Pied-billed Grebes (Nudds 1982). On good sites, competition with Pied-billed Grebes may restrict the habitat used by Horned Grebes (Ferguson and Sealy 1983). It winters mainly along the Atlantic and Pacific coasts of North America.

McIlwraith (1886) reported that the Horned Grebe bred in the marshes of Lake St. Clair in southern Ontario. The known range expanded as breeding records were established across the south with reports from Bracebridge, Leeds Co, Point Pelee (Macoun and Macoun 1909), Ottawa, and Lake Nipissing (Baillie and Harrington 1936). Summer records from the far north suggested that the Horned Grebe might be breeding along the western Hudson Bay coast (Baillie and Harrington 1936, Manning 1952). Godfrey (1986) shows an Ontario range extending from Manitoba eastward in a narrowing band across to North Bay. Overall, however, there is little information to establish a definite historical range. Prior to the atlas period, the last known nesting occurred in 1938 (Peck and James 1983).

The Horned Grebe does not have the loud vocalizations that typify the Pied-billed Grebe. It is reasonably easy to observe in its open water habitat, but is very secretive near its nest. The elaborate courtship displays, the young being fed by adults and the young riding on parents' backs are behavioural aspects of the Horned Grebe that are often observed. Nests reported in Ontario were floating structures in shallow water, and were often in exposed situations (Peck and James 1983).

The atlas maps show a range for the Horned Grebe that is quite different from that shown in Godfrey (1986). Neither historical nor current records support a wide distribution across the Boreal Forest zone of central Ontario, which is so unlike typical prairie nesting habitat. The species was reported from at least 7 squares in the south during summer months, but most of these were considered late migrants or non-breeding summering birds and are not mapped. The majority of these records occurred in the extreme south, where most historical records indicate that the Horned Grebe bred previously (Palmer 1962, Peck and James 1983). The 2 records on the map are at the historic locations in the St. Clair and Rondeau Prov. Park marshes. The Regional Coordinator for the St. Clair marshes feels that nesting could easily have been "overlooked in the 10,000 hectare marsh about Walpole Island" on Lake St. Clair (A. Woodliffe pers. comm.). These records suggest that birds lingering after most migrants have gone may still, at least occasionally, breed in scattered spots across the south.

The only confirmed record is a report of a family group on a man-made pond at Fort Severn near the Hudson Bay coast in extreme northwestern Ontario. This record suggests a small population from there to the west, contiguous with western populations at Churchill, Manitoba, but it is unlikely that there will be any much further east along the north coast. Some records of summering birds suggest that this species may begin to breed in sewage lagoons in southern Ontario, as have so many other prairie pothole species.

Given the paucity of firm breeding evidence, it is likely that the Ontario breeding population is under 10 pairs per year.-- *P.F.J. Eagles*

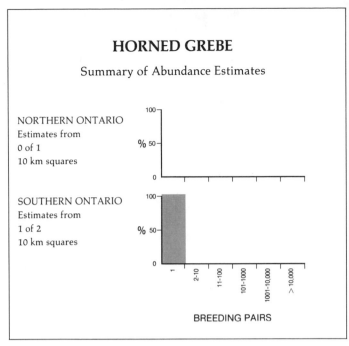

HORNED GREBE

Summary of Abundance Estimates

NORTHERN ONTARIO
Estimates from
0 of 1
10 km squares

SOUTHERN ONTARIO
Estimates from
1 of 2
10 km squares

%

%

100

50

0

100

50

0

1   2-10   11-100   101-1000   1001-10,000   >10,000

BREEDING PAIRS

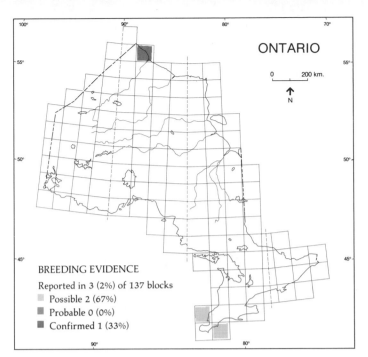

ONTARIO

0   200 km.

N

BREEDING EVIDENCE

Reported in 3 (2%) of 137 blocks

☐ Possible 2 (67%)
☐ Probable 0 (0%)
☐ Confirmed 1 (33%)

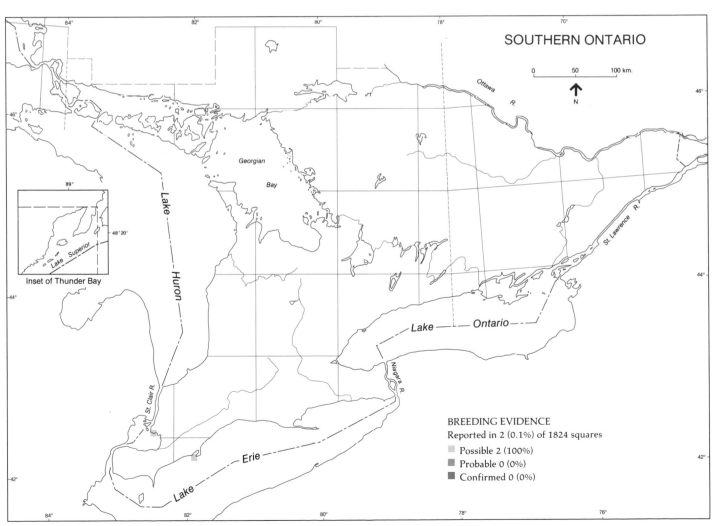

SOUTHERN ONTARIO

0   50   100 km.

N

Ottawa R.

Georgian Bay

89°

48°20'

Lake Superior

Inset of Thunder Bay

Lake Huron

St. Clair R.

Lake — Ontario

Niagara R.

St. Lawrence R.

Lake — Erie

BREEDING EVIDENCE

Reported in 2 (0.1%) of 1824 squares

☐ Possible 2 (100%)
☐ Probable 0 (0%)
☐ Confirmed 0 (0%)

RED-NECKED GREBE
Grèbe jougris
Podiceps grisegena

The Red-necked Grebe is distributed across the western two-thirds of Canada; it has a distinct western affinity but can be found as far east as northeastern Ontario and southwestern Quebec (Godfrey 1986). Elsewhere, the North American subspecies (*P. q. holboelli*) breeds in the northwestern US (including Alaska) and the eastern Palearctic, while the nominate race breeds as a disjunct population in the western Palearctic. Major habitat requirements include medium-sized but permanent freshwater lakes and marshes, protected marsh areas, or bays on larger lakes. Nests are usually located within extensive fringes of emergent vegetation (*e.g.*, cattails, bulrushes). The species is not widely distributed or common in Ontario, generally being restricted to specific traditional breeding areas. The Red-necked Grebe winters along the Atlantic and Pacific coasts as far south as the southern US.

Early in the 20th century, the species was considered by Baillie and Harrington (1936) to be a "rare summer resident of extreme western Ontario, west of Lake Superior." However, ONRS data suggest that it was more widespread even then; several former nesting sites for one or several pairs of grebes have been gradually lost over time. Specific examples from the ONRS include 2 nests in Simcoe Co in 1902, and 10 reported nests in Halton Co in the 1940's, including 2 on Lake Ontario in 1943. The range and population of the Red-necked Grebe appear to have declined over the past several decades, particularly in southern areas that have been increasingly developed since the turn of the century. A notable exception to this trend is Luther Marsh (Wellington and Dufferin Counties), where the Red-necked Grebe first bred in 1960, following the creation of the reservoir in the 1950's (Boyer and Devitt 1961), and where it continues to breed. These losses in southern Ontario appear to be directly related

to the loss of breeding habitat, specifically to shoreline development. The species bred colonially at Lillabelle Lake, near Cochrane, during the 1960's; Creighton (1965) reports finding 16 nests, at least 11 of which were found in 1963. Breeding has been confirmed as far north as Favourable Lake (Peck and James 1983).

The Red-necked Grebe is an uncommon waterbird with a distinctive profile and appearance. Its large size and long red neck allow relatively easy spotting and identification. These factors, along with its habits of calling during the breeding season, and of frequenting larger water bodies, help ensure that, if present, it will be found, especially in traditional breeding areas. Nevertheless, a large proportion of suitable lakes within the breeding range of the species were not visited by atlassers, so there may well be remote breeding locations as yet undiscovered. The nest itself is usually located on dead mats of vegetation near the edge of open water (Peck and James 1983); however, nests are frequently anchored to small, emergent clumps or submergents (De Smet 1982). While nests are not usually conspicuous, except perhaps before the growth of emergent vegetation, their predictable location should ensure frequent success in locating nests during specific searches. Nevertheless, only 2 nests were reported during the atlas period: one at Sioux Lookout and one on Manitoulin Island. Confirmation of breeding was aided by the fact that fledged young quickly take to open water, and hence are readily visible. Fledged young were reported from 3 squares. The observation of pairs in suitable habitat during the breeding season was the most commonly reported breeding category, constituting 36% of all records.

Atlas records support the northwestern and southcentral Ontario distribution of the species, as documented by Baillie and Harrington (1936), Cringan (1957) and Peck and James (1983). Although breeding was not confirmed at Lillabelle Lake during the atlas period, a probable breeding record was reported. The discovery during the atlas period of new breeding locations in 2 blocks northwest of Cochrane and probable breeding north of Lake Huron, were notable. Also notable was the pair in the far northwest at Garrett Lake near Big Trout Lake. These breeding locations, while covered generally by Godfrey's (1966) map, were previously unreported in the literature and the ONRS. Luther Marsh remains the only breeding area south of Manitoulin Island.

Abundance estimates were provided for only one square in each of southern and northern Ontario, and both indicate one breeding pair of birds. De Smet (1982) has estimated the Ontario population to be more than 200 breeding adults based on historical evidence and the assumption that many breeding areas remain to be discovered. The Red-necked Grebe is rare throughout its Ontario range.-- *E.R. Armstrong*

RED-NECKED GREBE

Summary of Abundance Estimates

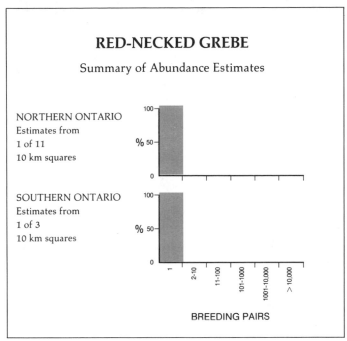

NORTHERN ONTARIO
Estimates from
1 of 11
10 km squares

SOUTHERN ONTARIO
Estimates from
1 of 3
10 km squares

%

BREEDING PAIRS

1   2-10   11-100   101-1000   1001-10,000   > 10,000

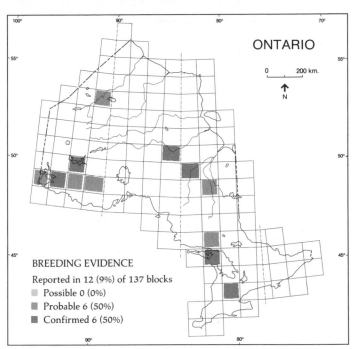

ONTARIO

0   200 km.

N

BREEDING EVIDENCE
Reported in 12 (9%) of 137 blocks
Possible 0 (0%)
Probable 6 (50%)
Confirmed 6 (50%)

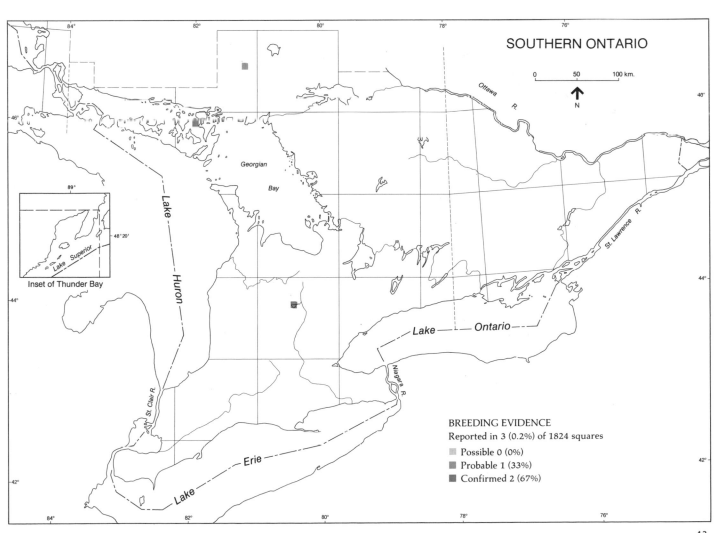

SOUTHERN ONTARIO

0   50   100 km.

N

Ottawa R.

Georgian Bay

Lake Huron

Lake Superior

89°

48°20′

Inset of Thunder Bay

St. Clair R.

Lake Erie

Niagara R.

Lake Ontario

St. Lawrence R.

BREEDING EVIDENCE
Reported in 3 (0.2%) of 1824 squares
Possible 0 (0%)
Probable 1 (33%)
Confirmed 2 (67%)

43

DOUBLE-CRESTED CORMORANT
Cormoran à aigrettes
Phalacrocorax auritus

In the last 10 to 14 years, Double-crested Cormorant numbers have been increasing rapidly in much of its range (Vermeer and Rankin 1984). From 1973 to 1981, the Great Lakes population increased from 125 to 2,273 pairs, or at an annual rate of well over 40% (Ludwig 1985). In many areas of the Canadian Great Lakes, cottagers report they are now seeing Cormorants almost daily, whereas 10 years ago they rarely, if ever, saw it.

The Double-crested Cormorant is a North American species that breeds across much of the continent except in the mountains and most areas north of 60°N latitude (AOU 1983). In Canada, its main breeding range extends from the Pacific coast, through the Canadian prairies and the Great Lakes area, along the St. Lawrence River to the Maritime Provinces and southern Newfoundland (Godfrey 1986). In Ontario, it is seen along the shores of large lakes and less commonly at smaller inland lakes. It usually nests on undisturbed islands and feeds almost exclusively on fish.

In Ontario, the Double-crested Cormorant has probably bred at Lake of the Woods for hundreds of years; it was recorded there about 1798 (James 1956). The first documented nesting from that area occurred in Minnesota in 1915 (Roberts 1936). The earliest recorded nestings in Ontario occurred in 1920 at Black Bay, Lake Superior (Dear 1940) and in 1924 at Lake Nipigon (Snyder 1928). The species probably moved eastward from Lake of the Woods into Lake Superior and Lake Nipigon between 1900 and 1920 (Peck and James 1983). The species was known to nest in Lake Huron's North Channel by 1931 (Snyder *et al.* 1942) and was breeding on Lakes Ontario and Erie by 1938 and 1939, respectively (Baillie 1947a, Core 1948). By the late 1940's and early 1950's, it had become so common that control measures were insti-

tuted to reduce suspected competition with commercial and sport fisheries (Omand 1947). This was the beginning of a 25 year decline in cormorant numbers on the Great Lakes. When control measures were removed, numbers continued to decline. The unsuspected side-effects of DDT and other pesticides had begun to take their toll. Cormorants were very sensitive to DDT-induced eggshell thinning, resulting in widespread egg breakage and reproductive failure (Anderson *et al.* 1969, Weseloh *et al.* 1983). With declining recruitment, breeding populations continued to decrease. In the mid- to late 1970's, after DDT and other long-lived toxic pesticides had been banned, cormorant numbers began to increase. By the early 1980's, populations were growing rapidly (Ludwig 1985, Price and Weseloh 1986), and new nesting colonies were being found annually (Clark *et al.* 1983, Blokpoel and Harfenist 1986).

The Double-crested Cormorant is a large, black, diurnal bird which nests in colonies, but often flies and feeds singly or in small flocks; it does not carry food externally back to its young or perform distraction displays, but it ranges widely to feed. Hence, it is a species which is easy to see but whose nesting is difficult to confirm if one does not have access to a boat. Fortunately, the species usually nests on the same islands in successive years, so historical data are important for the atlasser. Determining the breeding status is further complicated by non-breeding birds. If a new colony is to be established, cormorants will often take up residence at the site for a few years before they actually nest. The easiest and perhaps only sure way to confirm nesting is to locate the nesting colony.

The distribution of the confirmed breeding records clearly shows that nesting Double-crested Cormorants are confined to large bodies of water. In southern Ontario, there were no confirmed breeding records away from the Great Lakes or its connecting channels. In northern Ontario, in addition to those from the Great Lakes, there were confirmed atlas records from Lake of the Woods, Lake Nipigon, Lake Abitibi, and Little Sachigo Lake in the far northwest. At the latter location, there was a single nest with 3 eggs, which was about 500 km north of any previously known nestings in Ontario. The breeding site on the St. Lawrence River near Cornwall was also unknown prior to the atlas period.

The Double-crested Cormorant is one of the few species whose status changed dramatically during the atlas years. In 1981, the CWS found 907 nests at 18 sites on the Great Lakes in Ontario. In 1985, 2,221 nests were found at those same sites (145% increase) plus an additional 1,138 nests at 12 new or previously unknown sites in the same area, for a total 1981 to 1985 population increase of 270%, or a 39% average annual increase. A partial survey of the southern part of the Ontario side of Lake of the Woods in 1984 tallied 3,659 nests at 6 sites, indicating that the population there is probably even larger than on the Ontario Great Lakes.-- *D.V. Weseloh*

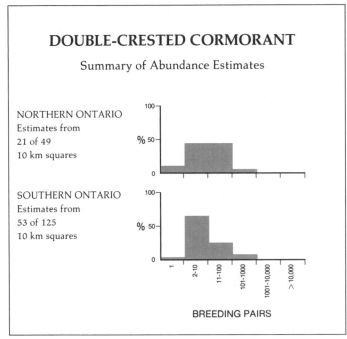

DOUBLE-CRESTED CORMORANT

Summary of Abundance Estimates

NORTHERN ONTARIO
Estimates from
21 of 49
10 km squares

SOUTHERN ONTARIO
Estimates from
53 of 125
10 km squares

BREEDING PAIRS

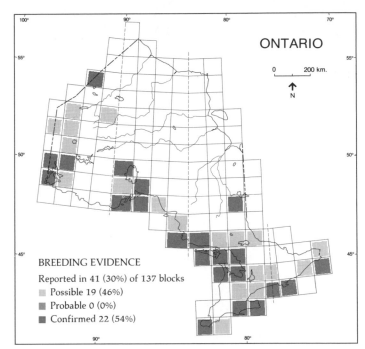

ONTARIO

0   200 km.

N

BREEDING EVIDENCE

Reported in 41 (30%) of 137 blocks

Possible 19 (46%)

Probable 0 (0%)

Confirmed 22 (54%)

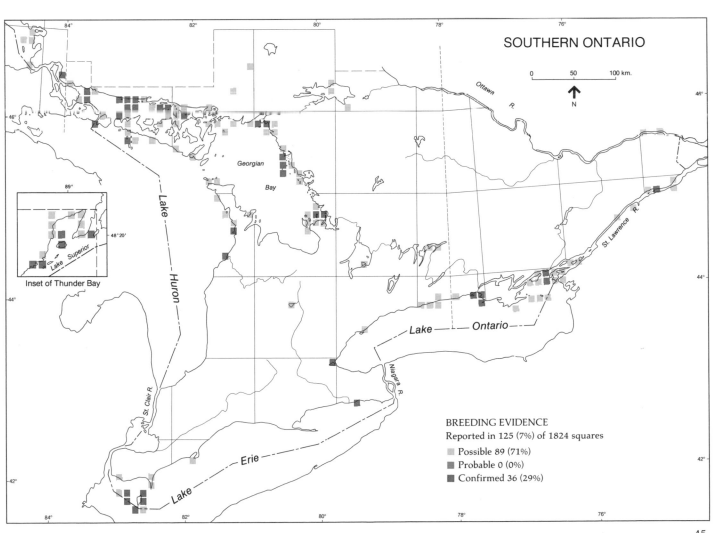

SOUTHERN ONTARIO

0   50   100 km.

N

BREEDING EVIDENCE

Reported in 125 (7%) of 1824 squares

Possible 89 (71%)

Probable 0 (0%)

Confirmed 36 (29%)

Inset of Thunder Bay

AMERICAN BITTERN
Butor d'Amérique
Botaurus lentiginosus

The American Bittern is distributed over nearly all of the North American continent and adjacent islands, breeding mainly from the southern Northwest Territories to the mid-southern US. It breeds throughout Ontario in suitable wetland habitat - mainly cattail and other marshes, swamps, thickets of alder or willow, and sometimes in nearby hay or pasture fields. Its winter range extends north to southwestern British Columbia and the middle US.

Baillie and Hope (1947) wrote that the American Bittern was "tolerably common" (at Sudbury in 1899), and Snyder *et al.* (1941) reported, "It is a common breeding bird of Prince Edward Co, from one to twelve noted daily by us during the summer of 1930;" the same could probably be said today. However, Saunders and Dale (1933) described the species as "a common summer resident in all suitable localities" (in Middlesex Co in the southwest), adding that "their well known pumping song is a characteristic sound from marsh and bog." Unfortunately, that is far from the case today. The drainage of wetlands for agriculture and settlement has left few suitable locations inland in the southwest, with most atlas records being from the few large Great Lakes shoreline marshes. The total southern Ontario population has undoubtedly declined because of habitat loss.

The American Bittern is a shy, retiring bird of the marshes and bogs. Thoreau called it the "Genius of the Bog". Although diurnal, the American Bittern is more active early in the morning and in the evening, but is far more often heard than seen. In spring, its loud, 3-syllable call (often written 'pump-er-lunk') is frequently heard in its marshland habitat, and often carries for as much as a kilometre (Terres 1980). A high proportion (over 40%) of all atlas records were of singing males or territorial birds, which is indicative of a species most readily located by its call. The low proportion

of confirmed records is further testament to the secretive habits of the species. Nests are often built over water, but may be built on old muskrat houses or some such mat of floating vegetation in a mass of cattails, bulrushes, sedges or even among bushes of cranberry, leatherleaf, buttonbush, or willow. Nests with eggs or young were reported in 54 squares, and fledged young were reported in 57. A family of American Bitterns may stay together for a prolonged period as several birds are sometimes seen together in late summer and early fall.

The American Bittern likely breeds in the majority of squares and blocks in which only possible or probable breeding was reported. The pattern of distribution shown on the map of Ontario reveals that, although the species occupied most blocks along the Hudson and James Bay coasts, it was only sporadically reported in the interior of the Hudson Bay Lowland and in the northern Boreal Forest region. In the southern part of the Boreal Forest region, it was reported in almost all blocks, suggesting that lack of coverage may have been a problem further north. The relative lack of suitable marshes would be another possible explanation for its absence in much of the far north. The rich coastal marshes evidently provide highly suitable breeding habitat; only 2 of 70 abundance estimates of more than 10 pairs per square were from northern Ontario - and both of those were from squares on the James Bay coast.

In southwestern Ontario, the American Bittern is primarily restricted to the large shoreline marshes, including Pt. Pelee, Rondeau, Long Point, and the St. Clair marshes. The frequency of occurrence increases to the northeast until the distribution of records is essentially contiguous in Prince Edward Co and north towards Ottawa. The edge of the Canadian Shield is not discernible on the map, indicating that suitable wetlands are widespread both on and off the Shield in this area, and west to Georgian Bay. The species is also widely distributed east of the Shield, suggesting that more wetlands remain in that area than in the southwest.

Abundance estimates indicate that the American Bittern is generally fairly common in squares where it was reported. The concentration of records in southeastern Ontario corresponds with the relatively high densities reported there on Breeding Bird Surveys (Speirs 1985); BBS data show another area of high density near Sudbury, which is also in agreement with the concentration of atlas records in that area.-- *D. Bucknell*

AMERICAN BITTERN

Summary of Abundance Estimates

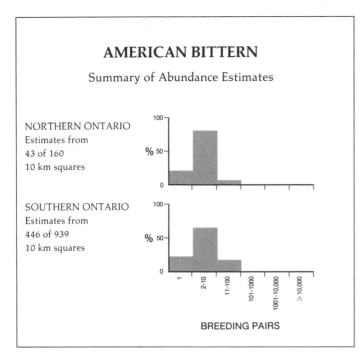

NORTHERN ONTARIO
Estimates from
43 of 160
10 km squares

SOUTHERN ONTARIO
Estimates from
446 of 939
10 km squares

%

100

50

0

%

100

50

0

1 2-10 11-100 101-1000 1001-10,000 > 10,000

BREEDING PAIRS

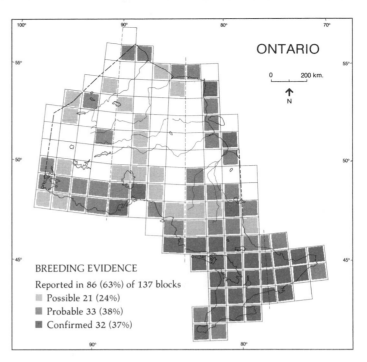

ONTARIO

0   200 km.

N

BREEDING EVIDENCE

Reported in 86 (63%) of 137 blocks

Possible 21 (24%)
Probable 33 (38%)
Confirmed 32 (37%)

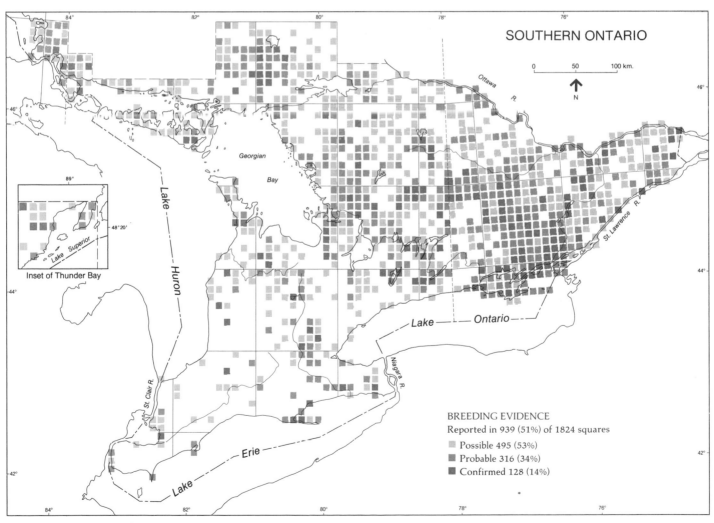

SOUTHERN ONTARIO

0   50   100 km.

N

Georgian Bay

Lake Huron

Lake Superior

Inset of Thunder Bay

89°

48°20′

St. Clair R.

Lake Erie

Lake Ontario

Niagara R.

Ottawa R.

St. Lawrence R.

BREEDING EVIDENCE

Reported in 939 (51%) of 1824 squares

Possible 495 (53%)
Probable 316 (34%)
Confirmed 128 (14%)

47

LEAST BITTERN
Petit Butor
Ixobrychus exilis

The Least Bittern has an extensive breeding range in North, Central, and South America. In North America, it extends southward from southeastern Canada, occurring throughout the eastern US, except for in the Appalachian region. There are scattered populations in the extreme western US. In Ontario, the Least Bittern can be found in the marshlands of the southern part of the province, north to a line extending from about Sault Ste Marie to Lake Nipissing. Atlas field work located birds in the Lake of the Woods area, where the species was not previously known to breed. Moderately dense cattail stands and dense reed beds in the larger marshes of the lower Great Lakes provide the most extensive habitat for the Least Bittern in Ontario, although it can be found in relatively small marshes, even within the limits of some of our larger population centres (Bent 1926).

In general, the Ontario range of the Least Bittern does not seem to have changed much in the past few decades. However, the species has certainly decreased in number within its range, as is unfortunately true for most birds dependent on wetlands.

It is a shy, retiring species. A quick glimpse is all a birder usually gets, as it flushes from the edge of the cattails, with legs dangling and neck extended, only to drop into a thicker area of cattails some distance away. However, it is fairly vocal, and if one is familiar with the soft call, this species can be easily detected on a calm late spring or early summer evening. Confirming breeding is relatively difficult, and often requires specific investigation for that purpose. Usually a canoe or chest waders, or both, are necessary, as the nests are most frequently just inside a relatively dense stand of vegetation, in water up to one metre deep. The small platform nest is made of cattails, and is usually supported by cattail stems. The adults often bend the tops of adjacent veg-

etation over the nest as a canopy, providing some concealment from above as well as a certain amount of shade from the intense sun. If the adult is off the nest, the bluish, unmarked eggs readily stand out. However, young at the nest may be quite difficult to see, as, after 4 or 5 days, they often climb out among the surrounding vegetation and, if threatened, will imitate a wind-blown cattail just as the adults do. Nests with eggs or young make up 57% of breeding confirmations.

The map of Least Bittern distribution strongly resembles that of the Black Tern, at least in southern Ontario. In the southwest, the squares with breeding evidence tend to follow the shoreline where suitable marsh habitat remains. Many of these marshes are quite large, and likely support the bulk of the provincial Least Bittern population. The inland records for southwestern Ontario likely originate from small river systems, water impoundments, and occasional sewage lagoons still containing cattail vegetation. Most squares with Least Bittern records are near the southern edge of the Canadian Shield. The large number of small lakes and rivers, and the deeper, more fertile soils off the Shield, apparently create wetland habitat suitable for both Least Bitterns and Black Terns. Prior to the atlas, there were no published records of Least Bittern breeding in northwestern Ontario. However, 3 blocks close to the Ontario-Manitoba border in the Lake of the Woods area had evidence of possible breeding of Least Bittern. This apparent extension is not surprising, as a breeding population has long been known in southeastern Manitoba (Bent 1926, Godfrey 1986) and nearby Minnesota. The new records probably reflect the greater intensity of bird study in that area rather than a true range extension.

The abundance estimates of the Least Bittern may be somewhat conservative. Squares containing the larger marshes of the province, such as those at Walpole Island, Lake St. Clair, Point Pelee, Rondeau, and Long Point, were reported to have 11 to 100 pairs in some squares. The only other squares that were reported to have more than 10 pairs were those containing Hullett Marsh (Huron Co), Lake Scugog, and Varty and Camden Lakes near Kingston. However, Kushlan (1973) noted that several researchers studying marshes on a similar latitude found an average density of more than one nest per hectare, and some colonial nesting situations gave densities much higher than that. This author has observed nests as close together as 4 m in the Rondeau marsh. There is certainly the possibility, then, that squares containing some of Ontario's largest marshes could accommodate more than 100 pairs, and even that some of the smaller marshes might support more than 10 pairs. Nevertheless, the Least Bittern is rare provincially. It is a species of concern in much of its range and has been on the "Blue List" from 1979 through 1986 (Tate 1986).-- *P.A. Woodliffe*

LEAST BITTERN

Summary of Abundance Estimates

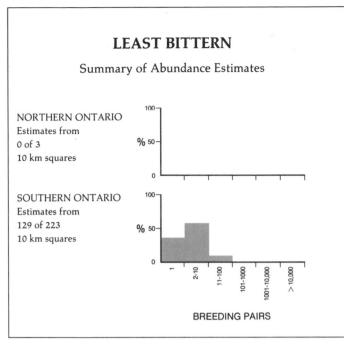

NORTHERN ONTARIO
Estimates from
0 of 3
10 km squares

SOUTHERN ONTARIO
Estimates from
129 of 223
10 km squares

%

BREEDING PAIRS

1 2-10 11-100 101-1000 1001-10,000 >10,000

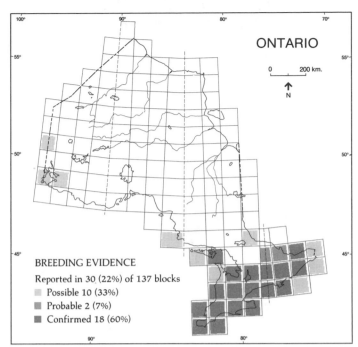

ONTARIO

0 200 km.

N

BREEDING EVIDENCE
Reported in 30 (22%) of 137 blocks
Possible 10 (33%)
Probable 2 (7%)
Confirmed 18 (60%)

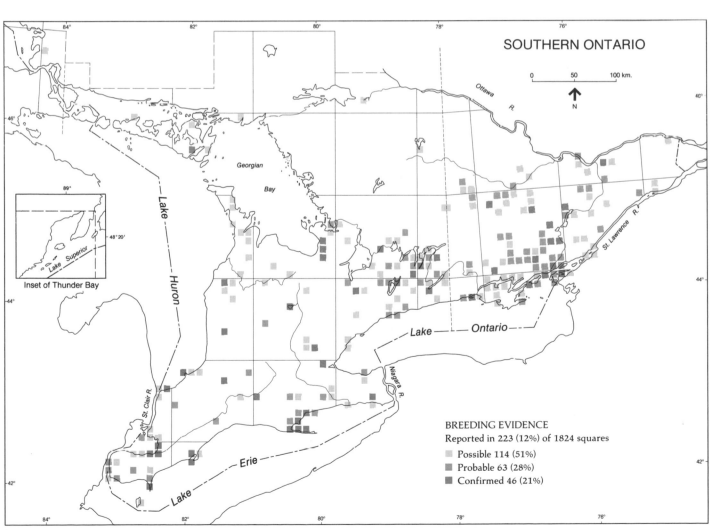

SOUTHERN ONTARIO

0 50 100 km.

N

Georgian Bay

Lake Huron

Lake Superior

48°20'

89°

Inset of Thunder Bay

St. Clair R.

Lake Erie

Niagara R.

Lake Ontario

St. Lawrence R.

Ottawa R.

BREEDING EVIDENCE
Reported in 223 (12%) of 1824 squares
Possible 114 (51%)
Probable 63 (28%)
Confirmed 46 (21%)

GREAT BLUE HERON
Grand Héron
Ardea herodias

This large, colonial species is well known to most people. The species is common throughout the northern US and in Canada north to about 54°N, although numbers are low in prairie and extensively agricultural regions. Though most Great Blue Herons winter south of the province, a few remain in areas of open water in southern Ontario. The Great Blue Heron feeds on fish, amphibians and invertebrates, hunting in shallow water of streams and lakes. It occurs wherever such wetland habitat is of good quality, including areas with large human populations, as long as there are relatively isolated woodlots suitable for nesting.

A province-wide inventory of Great Blue Heron colony sites, taken in the period 1978 through 1981, and a partial census in 1980 and 1981, estimated that a minimum of 13,000 pairs nested in at least 376 colonies (Dunn *et al.* 1985). Colony size averaged 35 pairs, with some, in southern Ontario, numbering up to several hundred pairs. Colony size was smaller to the north. The average density of 23 breeding pairs per 1,000 km² of range was comparable to those in Quebec and the Maritimes, where extensive censuses have also been undertaken. Where their Ontario ranges overlap, the Great Blue Heron has been known to share colonies with nesting Green-backed Herons, Black-crowned Night-Herons, Great Egrets, and Cattle Egrets.

Large stick nests are usually built high in trees (10 m to 15 m, Peck and James 1983). Deciduous trees are favoured in southern Ontario, while conifers are more often used farther north. Dead trees, particularly those killed by spring flooding or beaver ponds, are frequently chosen as colony sites, and heron excrement may eventually kill live trees. One to several nests may be built in a single tree, and occupied trees are usually adjacent to each other. These characteristics usually make colonies obvious and relatively easy to census;

some can be observed from low-flying planes. Colonies are sometimes abandoned, particularly small ones (Dunn *et al.* 1985), so the presence of nests in the non-breeding season cannot be taken as confirmation of breeding unless whitewash, egg-shells and/or dead chicks are also present. However, breeding is easily confirmed, as adults are constantly coming and going to and from colonies during the breeding season, and nestlings produce copious quantities of conspicuous whitewash. Nest(s) with young was by far the most frequently used category of confirmed breeding, making up half of all confirmations. Certain atlas breeding categories were modified for the Great Blue Heron. As Great Blue Herons may congregate regularly at feeding sites a long distance from the nesting area, all atlas codes for 'pair' and 'territory' for this species were downgraded to 'suitable habitat'.

The atlas maps show a distribution similar to that of the 1978-81 census (Dunn *et al.* 1985). Heronries are especially common along the southern edge of the Canadian Shield, where productive wetlands and isolated wooded areas are particularly abundant. No colonies were located in the Hudson Bay Lowland. The absence of relatively undisturbed woodlots in Essex Co and around Toronto is evident from the southern Ontario map. Great Blue Herons were confirmed as breeding in 448 squares in southern Ontario and 91 squares in the north. Estimates of over 100 pairs were provided in 12 southern Ontario squares, and these herons are probably even more abundant than estimated in the 1980-81 census.

The Great Blue Heron is about as common in Ontario as anywhere in Canada. It can coexist with humans as long as suitable undisturbed nesting woods and wetland feeding sites are available. A few large, relatively isolated colonies are located in southern Ontario, where extensive agriculture has undoubtedly reduced the species' original range through the removal of woodlots. Extensive water pollution could be a threat in the future.-- *E.H. Dunn*

GREAT BLUE HERON

Summary of Abundance Estimates

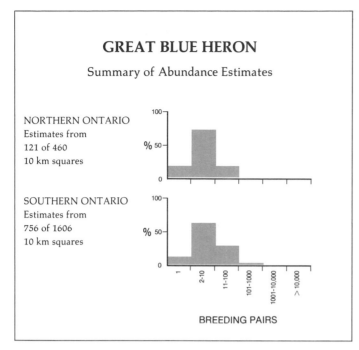

NORTHERN ONTARIO
Estimates from
121 of 460
10 km squares

SOUTHERN ONTARIO
Estimates from
756 of 1606
10 km squares

%

%

1 2-10 11-100 101-1000 1001-10,000 >10,000

BREEDING PAIRS

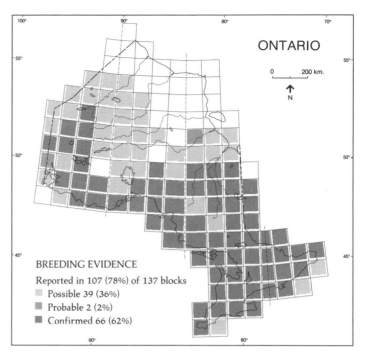

ONTARIO

0   200 km.

N

BREEDING EVIDENCE
Reported in 107 (78%) of 137 blocks

Possible 39 (36%)
Probable 2 (2%)
Confirmed 66 (62%)

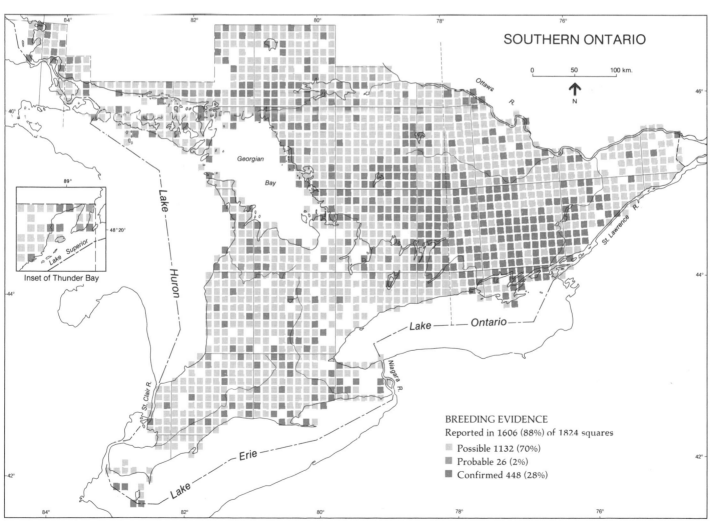

SOUTHERN ONTARIO

0   50   100 km.

N

Georgian
Bay

Lake
Huron

Ottawa R.

St. Lawrence R.

Lake — Ontario

Niagara R.

Lake — Erie

St. Clair R.

Inset of Thunder Bay
Lake Superior

BREEDING EVIDENCE
Reported in 1606 (88%) of 1824 squares

Possible 1132 (70%)
Probable 26 (2%)
Confirmed 448 (28%)

GREAT EGRET
Grande Aigrette
Casmerodius albus

The Great Egret, although widespread throughout the tropical, subtropical, and temperate zones of the world, is nowhere very numerous, and breeding areas are often widely separated. The species frequents marshes and open areas of fresh, brackish, and sometimes salt water, and breeds in treed swamps, thickets, and occasionally low vegetation on islands or mainland. It breeds in both the New and Old Worlds, but is largely absent in much of the Western Palearctic, and its status in Africa and the East Indies is poorly known. In North America, it breeds from extreme southern Saskatchewan, Ontario, and Quebec south through much of the US, except for the interior plains, and winters only as far north as the southern US. In Ontario, the Great Egret breeds only in the southern portion of the province.

The first indication of breeding of the Great Egret in Ontario was a sight record of a nest on East Sister Island, Lake Erie, in Essex Co, in 1953. Subsequently, other observations of nestings were made on this island in various years up to 1961 (Baillie 1963b). Breeding in Ontario was finally confirmed by photographic evidence at the East Sister Island location in June 1975 (Peck 1976). In 1959 and 1960, Great Egrets were reported breeding at a mainland heronry south of Amherstburg, in Essex Co. Between 1974 and 1980, it was noted breeding on Pelee Island, Lake Erie, and in 1977, another colony was discovered in the southwestern part of Walpole Island in Lake St. Clair, in Lambton Co (Peck and James 1983).

The marked tendency of this colonial breeding species to wander widely throughout the year, as well as during the post-breeding season, accounts for a large number of sightings in areas where it does not breed. However, this same wandering tendency is perhaps largely responsible for the gradual, northward extension of its breeding range in Ontario; only records of birds considered to be in suitable breeding habitat are shown on the atlas maps. Its various reported nestings are usually at established heronries of other herons and egrets, notably Great Blue Herons, Cattle Egrets, and Black-crowned Night-Herons. The presence of such heronries may also be a factor in inducing Great Egrets to expand their nesting in the province.

During the atlas survey period, breeding was again confirmed in Essex and Lambton Counties. In 1985, there were 10 to 20 nests on Middle Island, just south of Pelee Island, and just prior to the atlas period, in 1980, there were at least 108 occupied nests on East Sister Island (D.V. Weseloh pers. comm.). In 1981, at least 32 nests were occupied on East Sister Island. The colony in the southwestern part of Walpole Island has fluctuated between 20 and 35 active nests each year since 1977, and a new colony of about 5 nests was reported from further east in the Walpole Island area during the atlas period (P. A. Woodliffe pers. comm.). Both Walpole Island colonies are in buttonbush and cattails.

In addition, in 1984, an adult was seen at a nest in a Great Blue Heron colony at Luther Lake in Wellington Co, and, in 1985, a pair was found at a nest, also in a Great Blue Heron colony, on Nottawasaga Island, Georgian Bay, in Simcoe Co. In 1984 and 1985, even more northerly breedings were confirmed on Nickerson Island, Lake St. Francis, in Quebec. This latter location is very close to Glengarry Co in Ontario, and is the most northerly breeding recorded in eastern Canada, to date. The other records of occurrence in southern Ontario may have been wandering and non-breeding individuals. However, Great Egrets are now seen regularly in the breeding season on the Bruce Peninsula (M. Parker pers. comm.), and one was seen standing in a tree in a Great Blue Heron colony on the peninsula in June of 1986 (M. Cadman pers. comm.).

Concurrent with the Great Egret's northward breeding expansion in Ontario, is a continuing increase in observations and undoubtedly in its provincial population.-- *G.K. Peck*

GREAT EGRET

Summary of Abundance Estimates

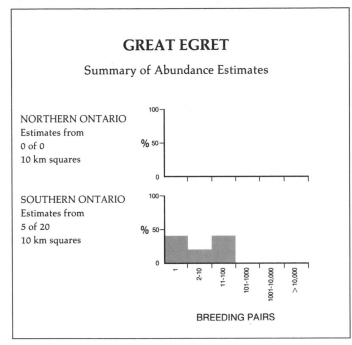

NORTHERN ONTARIO
Estimates from
0 of 0
10 km squares

SOUTHERN ONTARIO
Estimates from
5 of 20
10 km squares

BREEDING PAIRS

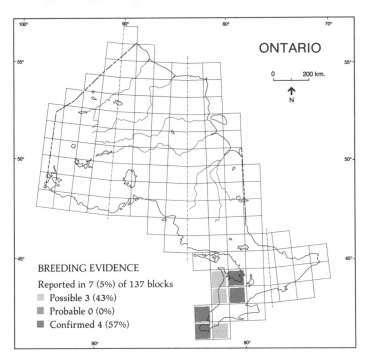

ONTARIO

0   200 km.

N

BREEDING EVIDENCE

Reported in 7 (5%) of 137 blocks

Possible 3 (43%)

Probable 0 (0%)

Confirmed 4 (57%)

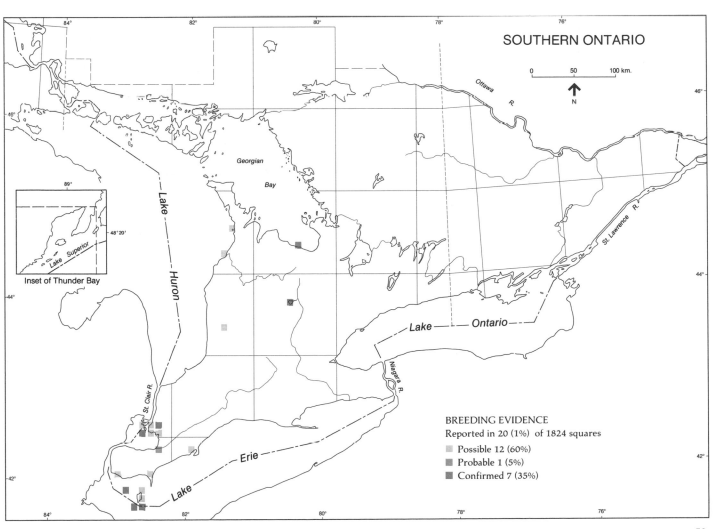

SOUTHERN ONTARIO

0   50   100 km.

N

Inset of Thunder Bay

BREEDING EVIDENCE

Reported in 20 (1%) of 1824 squares

Possible 12 (60%)

Probable 1 (5%)

Confirmed 7 (35%)

CATTLE EGRET
Héron garde-boeufs
Bubulcus ibis

The dramatic range expansion of the Cattle Egret is one of the most fascinating stories in ornithology. The species now occurs on every continent except Antarctica, and has even appeared on remote oceanic islands. Its African range began to expand about 1900, and it colonized South America naturally about 1930, although birds had been seen there as early as 1877. It appeared in North America in the early 1940's, and began breeding in Florida as early as 1953 (Rice 1956, Palmer 1962). Further expansion along the Gulf and Atlantic coasts followed, and the first bird in Canada was reported from Newfoundland in 1952 (Godfrey 1966). Typical feeding habitats are in pastures with livestock, or on agricultural land along watercourses and in wet areas, and it is in these localities that pioneering birds are most often seen. Nesting is usually in a heronry with other species. In winter, the bird retreats to the southern US and Central and South America.

The early Ontario occurrences were summarized by Baillie (1963a). Following the sighting of a bird in Norfolk Co in 1956 and numerous subsequent spring and fall reports, nests were found in Luther Marsh in 1962, closely followed by nesting at Presqu'ile Prov. Park. In both locations, the birds bred in existing heronries: a mixed colony at Luther, and with Black-crowned Night-Herons at Presqu'ile. Ontario naturalists braced themselves for the invasion; but it never came! Although Cattle Egrets have appeared, sometimes in numbers, in Ontario every year since 1963, the species has never become well established as a breeding bird. In 1968 there were 2 nests on Pigeon Island in Frontenac Co, and in 1970 1 nest on Amherst Island, Lennox and Addington Co (ONRS). In the mid-1970's there were up to 15 nests on Pelee Island and other nests were reported on East Sister Island, Essex Co (ONRS, Blokpoel and McKeating 1978). The pattern seems to have been one of nesting in small numbers in established heronries for a year or two, and then disappearing. Most sightings are in the spring or fall, with late summer reports that could coincide with post-breeding dispersal from US colonies.

Peck and James (1983) summarized the rather limited Cattle Egret nesting information (50 nests) available from Ontario. While the species usually nests in established heronries, most often with Black-crowned Night-Herons, it has bred singly, or in small homogeneous groups. Nests are typical of the herons: untidy structures of twigs, positioned at various heights in trees or shrubs.

It seems unlikely that a wider expansion of this species in the province would go undetected. It is a conspicuous bird when feeding in fields with cattle, and the birds of the Presqu'ile heronry, for example, could usually be found readily in pastures north of the colony. Finding actual nests can be more difficult as heronries are often relatively inaccessible; however, in recent years many of the more major colonies along the lower Great Lakes have been visited regularly by biologists engaged in pesticide monitoring programs, and in the course of the Ontario Heronry Inventory (Dunn *et al.* 1985). For the same reason, the 50 nest records cited above from Peck and James could well represent a relatively high proportion of total nests in the province since 1962.

As with other rarer heron species, breeding season sightings of Cattle Egrets can be difficult to interpret, for spring stragglers often remain into June only to disappear, and later in the summer, post-breeding dispersal can bring wandering birds north, creating ambiguity as to their origin. Eight breeding season records were submitted to the atlas, the most northerly being from the Lake Simcoe area. Only the single record shown on the map, on Walpole Island, was judged to be a potential breeding bird, based on the habitat in the area and the proximity to nearby heronries. Records were also submitted for 2 squares in Essex Co; it is feasible that these birds could have been nesting on the nearby Lake Erie islands, as herons from those colonies are known to fly to the mainland to feed (P. Pratt pers. comm.).-- *C.E. Goodwin*

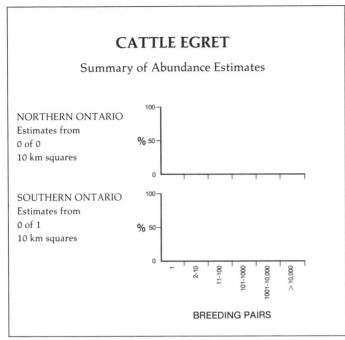

CATTLE EGRET

Summary of Abundance Estimates

NORTHERN ONTARIO
Estimates from
0 of 0
10 km squares

SOUTHERN ONTARIO
Estimates from
0 of 1
10 km squares

BREEDING PAIRS

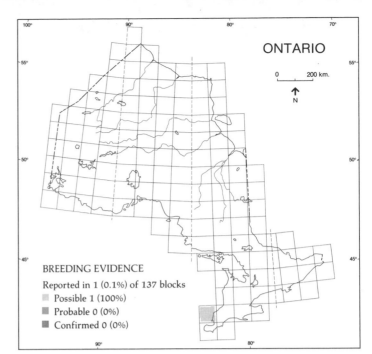

ONTARIO

0   200 km.

N

BREEDING EVIDENCE
Reported in 1 (0.1%) of 137 blocks
  Possible 1 (100%)
  Probable 0 (0%)
  Confirmed 0 (0%)

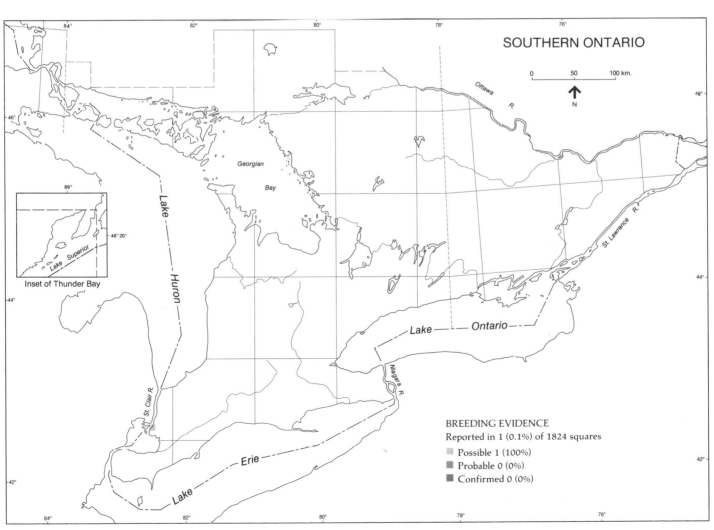

SOUTHERN ONTARIO

0   50   100 km.

N

Inset of Thunder Bay

BREEDING EVIDENCE
Reported in 1 (0.1%) of 1824 squares
  Possible 1 (100%)
  Probable 0 (0%)
  Confirmed 0 (0%)

GREEN-BACKED HERON
Héron vert
Butorides striatus

Formerly the Green Heron (*B. virescens*), the name change reflects the merger of the familiar North and Central American bird with the wider-ranging Striated Heron of South America and the Old World (AOU 1983). The North American breeding range of the Green-backed Heron includes most of the eastern US and extends north along the Pacific coast to southwestern British Columbia. It is absent from most of the Rockies and the western Great Plains, as well as northern New England, but has been expanding its range (Palmer 1962). In eastern Canada, breeding is limited mainly to southern Ontario, with small pockets in southwestern Quebec and southern New Brunswick (Godfrey 1986). Its winter range extends from the southern US to northern South America.

This is a heron of woodland pools and brushy drainage ditches, and can often be found in surprisingly unpromising locations provided there is a small area of wetland with heavy cover nearby. Nevertheless, Palmer's (1962) suggestion that there is "scarcely a stream, swamp, or shoreline where it may not be found" is certainly not true of most of Ontario, and atlassers failed to find the species even in some squares in agricultural southern Ontario where the bird was to be expected.

The Green-backed Heron is a rather solitary bird and is easily overlooked. The tangled pools it frequents are often difficult to survey properly, and the bird's dark plumage and its stealthy hunting behaviour make it even more inconspicuous. Once alarmed, the bird flies off with an abrupt squawk, but gives little hint of whether it has a nest nearby. In flight, however, it is often conspicuous and most distinctive, thus readily yielding possible breeding records. Its nests are often located singly; they can be situated away from wetlands, although many are placed over water. Birds can also be found nesting, usually at lower levels, in heronries with other species. A typical nest is a flimsy platform built in a deciduous shrub or a thicket of dense conifers; the degree of nest concealment varies widely (Palmer 1962).

The maps show that the range of the species is largely confined to the areas south and east of the Canadian Shield, but they also provide some support for the suggestion that the Green-backed Heron has expanded its range. At the turn of the century, ornithologists regarded it as uncommon, with its main centre of abundance in southwestern Ontario, but also considered it to be expanding its range (Macoun and Macoun 1909). Thirty years later, Baillie and Harrington (1936) reported it as a not very common breeder north to Middlesex, York, and Grenville Counties, but by 1983, Peck and James record it as nesting widely throughout the agricultural areas of southern Ontario. The atlas confirms this distribution, but also now shows scattered breeding season records on the Canadian Shield, including one at Lake of the Woods. Superficially, it would seem an ideal candidate to colonize the wooded Shield wetlands, but a closer study of even these marginal nestings suggests that most of them are still associated with pockets of agricultural land. The increase of the species has hardly been dramatic and indeed, in light of the enormous increase in field work in the last 50 years, a case could be made that the bird was possibly more common in the past than the published data showed.

Over 60% of atlas records relate only to single or paired birds in suitable breeding habitat. Almost half of the confirmed nestings are based on sightings of the fledged young, leggy creatures which can be relatively easy to find. It is highly likely that the vast majority of the possible and probable records in fact reflect actual nestings within the squares.

Abundance estimates fall between 2 and 10 pairs per square in 68% of cases, indicating that the species is thinly dispersed across the agricultural south. Higher estimates were reported in 99 squares, but these too are spread across the range of the species, showing no obvious pattern.-- *C.E. Goodwin*

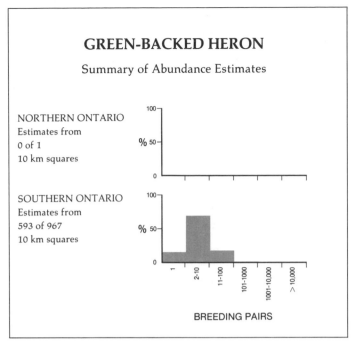

GREEN-BACKED HERON

Summary of Abundance Estimates

NORTHERN ONTARIO
Estimates from
0 of 1
10 km squares

SOUTHERN ONTARIO
Estimates from
593 of 967
10 km squares

%

100

50

0

100

50

0

1 2-10 11-100 101-1000 1001-10,000 >10,000

BREEDING PAIRS

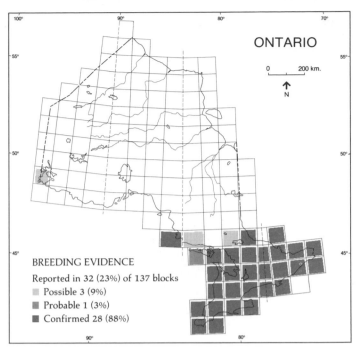

ONTARIO

0   200 km.

N

BREEDING EVIDENCE
Reported in 32 (23%) of 137 blocks
Possible 3 (9%)
Probable 1 (3%)
Confirmed 28 (88%)

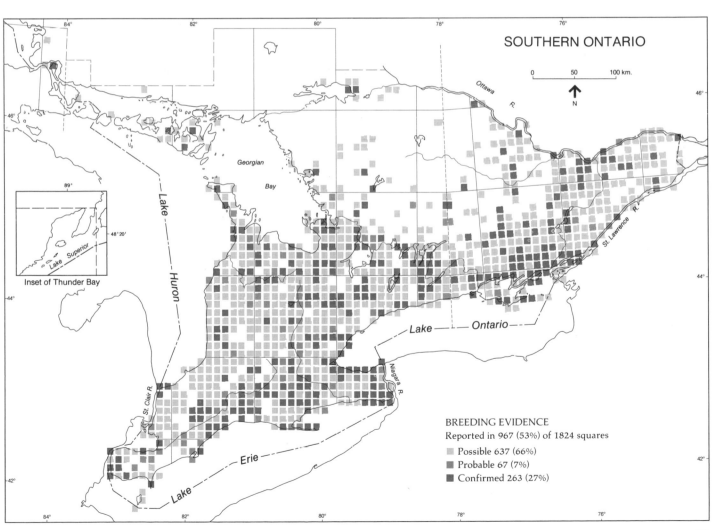

SOUTHERN ONTARIO

0   50   100 km.

N

Georgian
Bay

Lake Huron

Lake Superior

Inset of Thunder Bay

Ottawa R.

St. Lawrence R.

Lake — Ontario

Niagara R.

St. Clair R.

Lake — Erie

BREEDING EVIDENCE
Reported in 967 (53%) of 1824 squares
Possible 637 (66%)
Probable 67 (7%)
Confirmed 263 (27%)

BLACK-CROWNED NIGHT-HERON
Bihoreau à couronne noire
Nycticorax nycticorax

This elegant heron is widely distributed in the Old World as well as in the Americas. In the US, it seems to be equally at home in coastal salt marshes, mangroves, deciduous woodland swamps, and cattail marshes. Its Canadian range is quite limited, and in Ontario it is confined to the south, extending in small numbers north to Manitoulin Island and Temagami (Peck and James 1983). Even within this restricted range, the atlas map reveals a very localized distribution, with most of the heronries located near the shores of the lower Great Lakes or associated with larger inland bodies of water. It winters south from the central US.

Black-crowned Night-Herons are very gregarious, and most authorities imply that they always nest colonially; however, in Ontario, Peck and James (1983) report 4 isolated nestings (out of some 8340 nests): most Ontario nests were in deciduous shrubs or trees, although nests were reported in 2 species of conifer, and one colony was located on matted vegetation in a cattail marsh. In the latter, the nests were constructed of dead leaves and dead cattails. More usual sites were on islands, on wooded river banks, and in swamps (Peck and James 1983).

The population of this species has fluctuated considerably. In 1936, the bird was well established only in Lambton and Essex Counties, with a few pairs nesting at Hamilton and Toronto (Baillie and Harrington 1936). By the 1950's, there were over 100 nests in the Toronto heronry, and the birds were apparently widespread along the lower Great Lakes (Baillie 1951). Expansion continued through the 1950's, with a small colony appearing even in Nipissing District in 1961 (Woodford and Lunn 1961). The late 1960's and early 1970's brought a change in the fortunes of the species, with reported declines coinciding with the pesticide-induced

declines in the numbers of other fish-eating species and of raptors (*e.g.*, Arbib 1972, 1973, and 1974). For example, the Pigeon Island heronry near Kingston declined from 78 nests in 1968 to 15 in 1973 (Quilliam 1973), a pattern which seemed to be repeated in many Great Lakes heronries at that time. Recent years have seen a recovery in numbers of Black-crowned Night-Herons, and some evidence - supported by the atlas findings - that the species is expanding its range once again, with nesting confirmed on 2 small islands off Manitoulin Island. The creation of headlands on the Great Lakes shoreline, upon which poplars quickly grew, has created new nesting habitat in Toronto and Hamilton.

Although these herons are a large, relatively conspicuous species whose diurnal roosts are often well known to local observers, establishing proof of their nesting can present serious difficulty. The presence of birds in a heronry does not constitute evidence of breeding, as non-nesting herons will regularly frequent colonies of other species. On the other hand, the relative inaccessibility of many heronries and their susceptibility to disturbance can lead to nests being overlooked or their numbers underestimated.

The possible and probable breeding records shown on the map refer to birds seen in suitable breeding habitat, but could also apply to individuals or roosts of non-breeding and post-breeding birds. Such birds could well have dispersed from an established colony many kilometres distant. Wandering birds have occurred north even to northern Lake Superior (Gunn 1957), and during the atlas period a bird was reported in Hearst District. It would be inappropriate, however, to dismiss all such records as dispersal. The species does appear to be expanding, and it is possible, particularly towards the edge of its range in eastern Ontario, that isolated birds do indeed represent an increase in isolated nestings, or perhaps are the forerunners to nesting in future years.

Very large colonies have been reported, but the largest well-documented Ontario heronry, on Pelee Island, was estimated at some 900 pairs in 1971, although in most years estimates were considerably lower. Peck and James (1983) calculated the average size of 68 colony records in the province to be 123 nests. In only 4 of the 50 squares in which estimates were reported were the numbers present estimated to exceed 100 pairs. These estimates were from squares containing the large colony formerly on Pelee Island, which moved to Middle Island in 1980 (P. Pratt pers. comm.), the colony on East Sister Island (280 nests in 1981, P.A. Woodliffe pers. comm.), the colony on Nottawasaga Island in southern Georgian Bay, and the 2 colonies in the same square on Muggs Island and the Leslie Street Spit, Toronto. In the short term, this adaptable species seems likely to continue its recovery along the lower Great Lakes shorelines and through the wetlands and rivers of agricultural southern Ontario. Further expansion into the wetlands of the Canadian Shield and the Boreal Forest zone is more problematical, and, in the longer term, even its gains in the south may be jeopardized by progressive losses of habitat.-- *C.E. Goodwin*

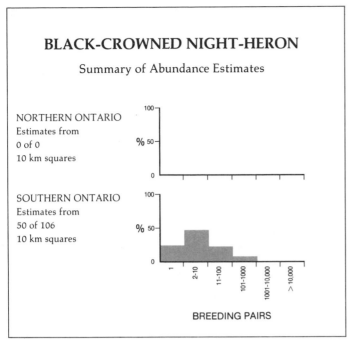

BLACK-CROWNED NIGHT-HERON

Summary of Abundance Estimates

NORTHERN ONTARIO
Estimates from
0 of 0
10 km squares

SOUTHERN ONTARIO
Estimates from
50 of 106
10 km squares

% 100 50 0

% 100 50 0

1   2-10   11-100   101-1000   1001-10,000   >10,000

BREEDING PAIRS

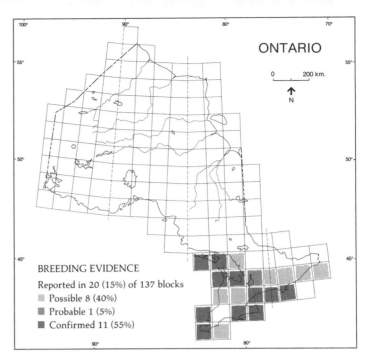

ONTARIO

0   200 km.

N

BREEDING EVIDENCE
Reported in 20 (15%) of 137 blocks

Possible 8 (40%)
Probable 1 (5%)
Confirmed 11 (55%)

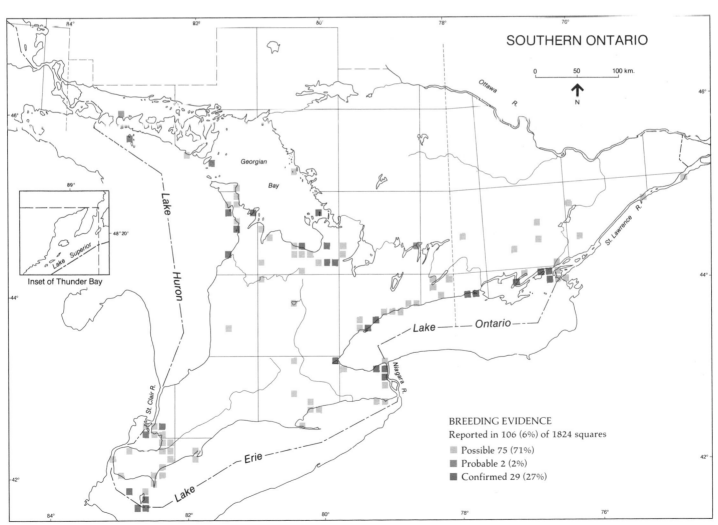

SOUTHERN ONTARIO

0   50   100 km.

N

Georgian Bay

Lake Huron

Lake Superior

89°   48°20'

Inset of Thunder Bay

St. Clair R.

Ottawa R.

St. Lawrence R.

Lake — Ontario

Niagara R.

Lake — Erie

BREEDING EVIDENCE
Reported in 106 (6%) of 1824 squares

Possible 75 (71%)
Probable 2 (2%)
Confirmed 29 (27%)

YELLOW-CROWNED * NIGHT-HERON
Bihoreau violacé
Nycticorax violaceus

Unlike the familiar Black-crowned Night-Heron, the Yellow-crowned is confined to the New World, and over much of its range it is mainly a coastal species. In North America, it occurs year-round in Florida and the Gulf States, and its breeding range extends north along the Atlantic coast to Massachusetts and inland through the Mississippi valley north to Ohio and southern Michigan. The birds that find their way into southern Ontario probably originate from the Mississippi valley, where the population has recently expanded northwards (Palmer 1962).

The Yellow-crowned Night-Heron has been reported occasionally in Ontario since the late 1800's, with some 10 sightings during each of the spring and fall periods. Assessment of these reports has always been difficult because of the similarity of the immature birds to young Black-crowned Night-Herons, together with the lack of good illustrations of plumages of the latter in its first 2 years of life in the standard bird identification guidebooks. Breeding season occurrences have been few, and mainly confined to the Lake Erie shoreline. On 23 May 1954, a single bird was seen near a nest in a colony of other herons on East Sister Island in Lake Erie, and for some time the species was listed as an Ontario breeding bird on the basis of this inconclusive evidence (Peck and James 1983).

It is likely that an Ontario nesting attempt by the Yellow-crowned Night-Heron would be in a heronry with other species, although this species is reported as being less gregarious than the Black-crowned Night-Heron, and isolated nests have occurred at the periphery of its range. It frequents a rather wide range of habitats, and is equally catholic in its choice of nest sites, although US nests are often in willows close to water (Palmer 1962). Hence, a nest could be difficult to locate, particularly as the species is shy, secretive, and often crepuscular.

There were 2 breeding season sightings of single adult birds in southern Ontario during the atlas period, but no evidence of nesting or of paired birds. However, the pattern of birds appearing and staying in the area during the breeding season might be a prelude to eventual nesting. The heavy dependence of Yellow-crowned Night-Herons upon a crustacean diet may be one factor slowing their northward expansion.

The map shows records in 2 squares in southwestern Ontario, both in locations where breeding would be most likely to occur. Paul Pratt found an adult bird at Ojibway Park in June 1982. The bird was carrying food as it flew towards a small oak woodlot which borders a series of ponds. No nest was found, but the species breeds in adjacent Wayne Co in Michigan (P. Pratt pers. comm.). On 7 June 1985, Gordon Cameron found an adult bird at Bolton. The bird was observed and photographed many times up to 3 July. It was often seen picking frogs at the edge of a small artificial lake. The data, then, are suggestive, but confirmation of Ontario nesting is still awaited.-- *C.E. Goodwin*

YELLOW-CROWNED NIGHT-HERON

Summary of Abundance Estimates

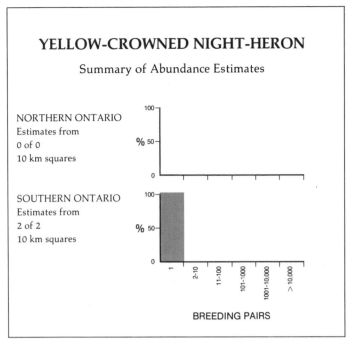

NORTHERN ONTARIO
Estimates from
0 of 0
10 km squares

SOUTHERN ONTARIO
Estimates from
2 of 2
10 km squares

% 100

% 50

% 0

100

50

0

1   2-10   11-100   101-1000   1001-10,000   >10,000

BREEDING PAIRS

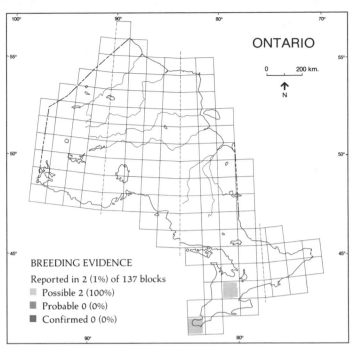

ONTARIO

0   200 km.

N

BREEDING EVIDENCE

Reported in 2 (1%) of 137 blocks

Possible 2 (100%)
Probable 0 (0%)
Confirmed 0 (0%)

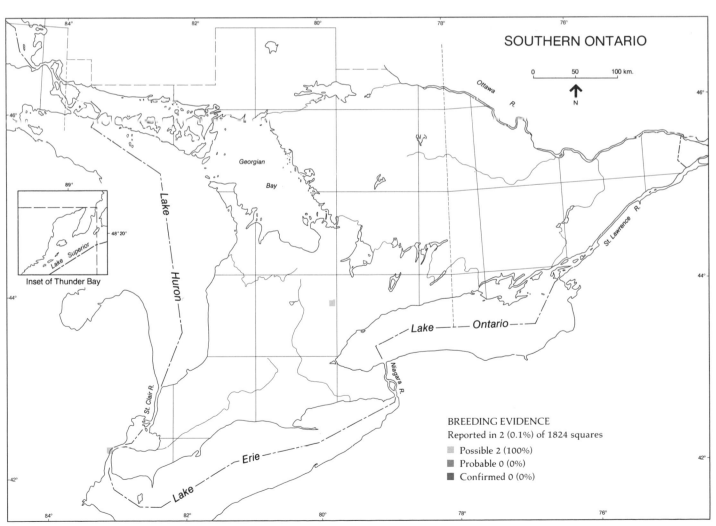

SOUTHERN ONTARIO

0   50   100 km.

N

Ottawa R.

Georgian Bay

Lake Huron

St. Clair R.

Lake Erie

Niagara R.

Lake Ontario

St. Lawrence R.

89°

48°20'

Lake Superior

Inset of Thunder Bay

BREEDING EVIDENCE

Reported in 2 (0.1%) of 1824 squares

Possible 2 (100%)
Probable 0 (0%)
Confirmed 0 (0%)

MUTE SWAN
Cygne tuberculé
Cygnus olor

Juliana Hawke

Male Mute Swans are the largest waterfowl in the world, being slightly heavier than Trumpeter Swans. Because of its spectacular size and white colour, this species is a great favourite with the public.

The Mute Swan is a native of Eurasia, with a discontinuous distribution across that continent (Cramp and Simmons 1977). It has been widely exported from Europe, and releases and escapes have established feral populations in Australia, New Zealand, South Africa, and North America. On this continent, it first bred in the wild about 1910 on the Hudson River, and by 1986 numbered about 5,300 in 11 Atlantic states (Allin 1987). In Michigan, feral breeding first occurred in 1919 or 1920, and there are now close to 1,500 birds in the state (J. Martz pers. comm.). There are about 100 in Wisconsin (W. Wheeler pers. comm.), and a scattering of birds in many other states. In Canada, escapes from captivity are frequent and birds may be encountered anywhere. In Ontario, the first published nesting occurred at Georgetown, Halton Co in 1958 (Peck 1966). There are now about 120 feral birds in the province. Twenty-five pairs nested in the Lake Ontario marshes between Bowmanville and Hamilton in 1985. Birds breeding between Bowmanville and the Toronto harbour area move southwest to winter between the Humber River mouth and Oakville. A few leave the area altogether and it is not known where they go.

In Ontario, the Mute Swan nests most frequently in cattail marshes, where its building activity results in a large structure up to 3 m in diameter surrounded by a moat. Some nests are reused and added to year after year. The earliest eggs are laid 1 to 11 April. They are deposited 48 hours apart and the clutch numbers 4 to 10 eggs (Lumsden unpubl.). Incubation lasts 35 to 38 days and cygnets fledge in 120 to 150 days (Scott 1972). A few Ontario birds have a period as long as 5.5 months during which an atlasser can confirm breeding. Their broods, unlike most other swan species, break up in late fall and rarely stay together during the winter.

Mute Swans are rather silent birds, uttering hisses and snorts which cannot be heard at any great distance. They are capable of trumpeting (Cramp and Simmons 1977), but very rarely do so. Mute Swans are very easy to atlas because they return to their territories early in spring, and the male patrols, with wings up, back and forth on the water near the nest. Most first nests are easy to find, but second nests may be hidden in extensive stands of cattail up to 2.5 m tall. Male Mute Swans can be very aggressive toward other water birds, and a few may physically attack people near the nest. Most incubating females sit very tight and usually do not leave the nest until the observer is within 25 m; some can even be caught on the nest for banding.

The Mute Swan is a grazer on submerged aquatic vegetation. It seldom leaves the water to graze on land. It chooses shoal water, preferably where there are extensive beds of pondweeds, or marshes with an abundance of algae and duckweed. It may feed as much by night as by day, especially during hot weather. Both sexes pull up vegetation from the bottom for the cygnets to feed upon.

In Ontario, Mute Swans are strongly conditioned to humans and do not hesitate to nest close to dwellings. The favour with which people regard Mute Swans is reflected in the fact that over 600 birds were held in captivity in Ontario under permit from the CWS in 1985. About 30% of the squares with Mute Swan records were removed from the atlas data bank because they were thought to refer to captive individuals.

The atlas map shows that this species is largely restricted to shoreline marshes in the urbanized area along the western end of Lake Ontario, and along the north shore of Lake Erie. The high proportion of confirmed squares indicates the ease with which Mute Swans can be atlassed. Although wild Mute Swans may colonize areas other than those shown on the map in the future, sightings of the species away from the shores of Lakes Erie and Ontario at present are most likely to be of captive birds. Abundance data indicate a maximum of no more than 10 pairs in any one 10 km square.-- *H.G. Lumsden*

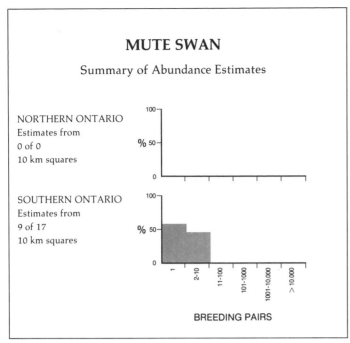

MUTE SWAN

Summary of Abundance Estimates

NORTHERN ONTARIO
Estimates from
0 of 0
10 km squares

SOUTHERN ONTARIO
Estimates from
9 of 17
10 km squares

%

BREEDING PAIRS

1 2-10 11-100 101-1000 1001-10,000 > 10,000

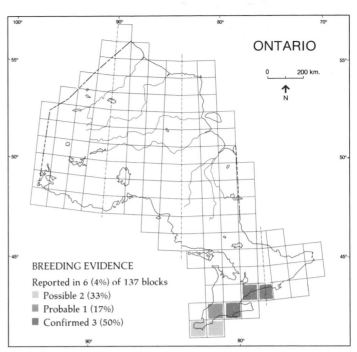

ONTARIO

0   200 km.

N

BREEDING EVIDENCE

Reported in 6 (4%) of 137 blocks

Possible 2 (33%)
Probable 1 (17%)
Confirmed 3 (50%)

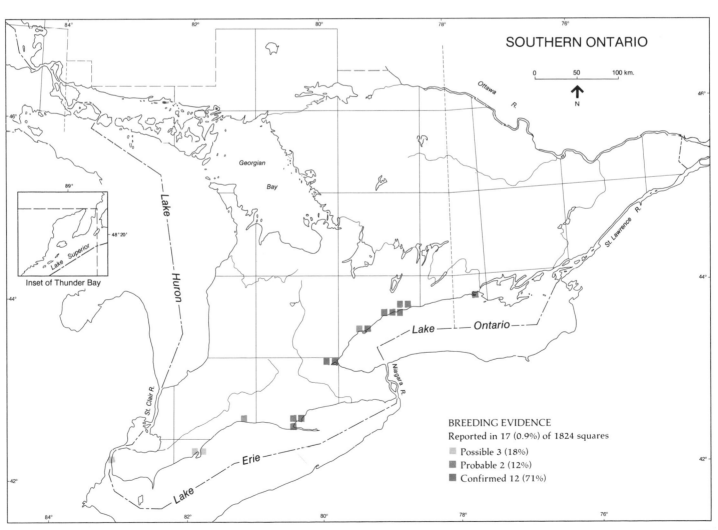

SOUTHERN ONTARIO

0   50   100 km.

N

Georgian Bay

Lake Huron

Lake Superior
89°
48°20'
Inset of Thunder Bay

St. Clair R.

Lake Erie

Niagara R.

Lake Ontario

Ottawa R.

St. Lawrence R.

BREEDING EVIDENCE

Reported in 17 (0.9%) of 1824 squares

Possible 3 (18%)
Probable 2 (12%)
Confirmed 12 (71%)

CANADA GOOSE
Bernache du Canada
Branta canadensis

The restoration of the Canada Goose to the Great Lakes region is one of the conspicuous conservation success stories in North America. The 1945-6 winter inventory in the Mississippi Flyway recorded only 49,000 birds, most of which bred in Ontario (Hanson 1950). Harvest control and stocking have since rebuilt these populations to 1,127,500 birds in December 1985 (Mississippi Flyway December Goose Survey).

The Canada Goose originally bred from the central US to the Atlantic and Pacific coasts, north to Baffin Island and Alaska. It occasionally breeds in western Greenland and formerly on the Commander and Kurile Islands. It has been introduced to western Europe and New Zealand (AOU 1983). One race, the Aleutian Canada Goose, is endangered. In southern Ontario, the Canada Goose breeds in marshes, wooded swamps, and shores of ponds, lakes, and rivers. It usually chooses peninsulas, islands, muskrat houses, or hummocks in marshes for nest sites. Occasionally, it nests among willows or boulders close to a shore or rarely in elevated sites 0.5 to 2.5 m off the ground in the fork of a large tree (Peck and James 1983). In muskeg country in the Hudson Bay Lowland, it nests in bog or fen, with 80% of the nests on islands or peninsulas in ponds 0.4 to 2 ha in size (Raveling and Lumsden 1977). In permafrost country near the Hudson Bay coast, it nests predominantly in low shrub fen (Abraham and Hendry 1985), on the elevated rims of ponds and lakes, or on ice-cored hummocks (Lumsden unpubl.).

During the early years of settlement in Ontario, breeding Canada Geese were extirpated throughout what is now the southern agricultural belt, except perhaps in the Lake St. Clair area. The Canada Goose was reintroduced by aviculturists in the Holstein area in Grey Co, in the vicinity of Holiday Beach Prov. Park south of Windsor, and in the Guelph area in the early 1930's. Releases were made on the Toronto Islands about 1961. In 1968, the MNR started a widespread restoration program in southern Ontario (Lumsden 1981), which has resulted in a population of about 60,000 birds in 1985. North of the Great Lakes, geese were introduced to the western Rainy River District in 1982 and first bred there in 1984. Goslings were stocked in the Thunder Bay area and near Sudbury in 1984 and 1985.

The Canada Goose is usually a diurnal feeder and is conspicuous and easy to find. Where breeding densities are substantial, there is frequent interaction among pairs and noisy honking often reveals the location of territories. Family groups remain together throughout the fall and winter. They break up only at the onset of a new breeding cycle. Pairs usually return in succeeding years to the same nest site in which they were previously successful. They may move to a new nest site in the vicinity after a nesting failure. This behaviour may help the atlasser to confirm breeding.

In southern Ontario, the incubating female usually sits very tight and does not leave the nest until an intruder is close. Males guarding a nest are conspicuous, and sometimes aggressive. Broods are conspicuous and are frequently seen grazing on farmland on shores or swimming on ponds, rivers, and lakes. In the muskeg, Canada Geese are much warier, nests are very hard to find, and broods are extremely secretive where cover is adequate. Broods make up 96% of confirmations in the north, the remainder being nests with eggs. The maps indicate quite clearly the ease with which breeding Canada Geese can be confirmed. Sixty-seven percent of southern Ontario records were of confirmed breeding, as were 82% of Ontario's block records. The species is still spreading and many of the 'possible' and 'probable' squares will have nesting geese in the years to come.

The atlas maps show clearly the occurrence of 2 populations of breeding geese, one in and around the fringes of the Hudson Bay Lowland, and one in southern Ontario. The southern population is of recent occurrence due to the reintroduction and spread of birds mentioned above. The scarcity or absence of breeding geese in the Parry Sound - Algonquin Prov. Park area reflects the infertile nature of the soils, and the resultant poor habitat. Much of the country immediately north of North Bay and Thunder Bay is similar, and is not likely to be widely colonized, although some spread can be expected in the richer sites near farmland. The Canada Goose is common within its range in both southern and northern Ontario, with generally higher densities occurring in the north.-- *H.G. Lumsden*

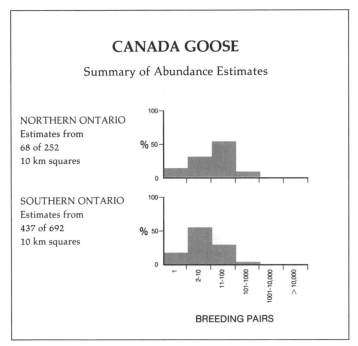

CANADA GOOSE

Summary of Abundance Estimates

NORTHERN ONTARIO
Estimates from
68 of 252
10 km squares

SOUTHERN ONTARIO
Estimates from
437 of 692
10 km squares

%

100

50

0

100

50

0

1   2-10   11-100   101-1000   1001-10,000   >10,000

BREEDING PAIRS

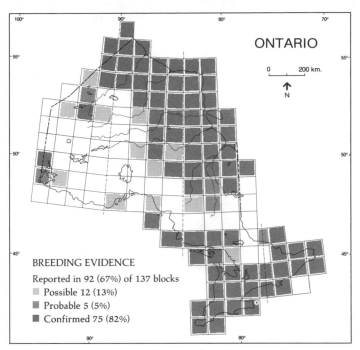

ONTARIO

0   200 km.

N

BREEDING EVIDENCE
Reported in 92 (67%) of 137 blocks

Possible 12 (13%)
Probable 5 (5%)
Confirmed 75 (82%)

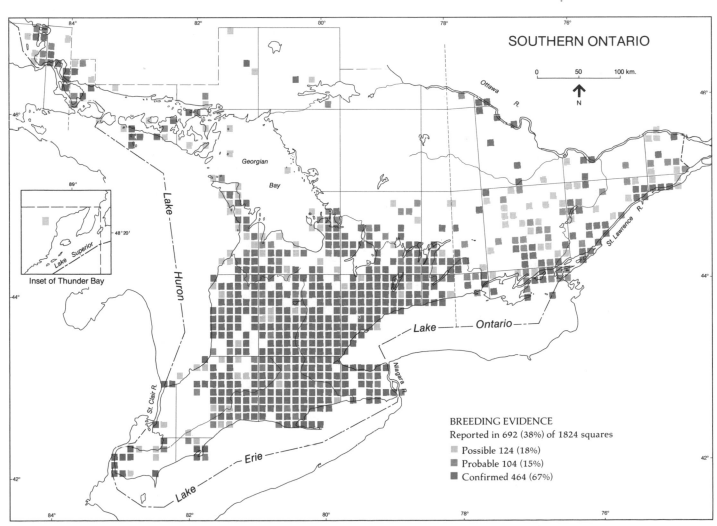

SOUTHERN ONTARIO

0   50   100 km.

N

Inset of Thunder Bay

BREEDING EVIDENCE
Reported in 692 (38%) of 1824 squares

Possible 124 (18%)
Probable 104 (15%)
Confirmed 464 (67%)

WOOD DUCK
Canard branchu
Aix sponsa

The Wood Duck is Canada's only perching duck. The drake is the most colourful of North American waterfowl, yet, because of the association of the species with secluded swamplands, the urban public remains relatively unfamiliar with it.

The Wood Duck breeds in eastern North America, including Cuba, to central Manitoba, northern Ontario, southern Quebec, and the Maritime Provinces. It also occurs along the Pacific coast. Small numbers winter in southern Ontario, but the winter range extends south to central Mexico. Beaver ponds, mature, wooded swamps, temporary woodland pools, and shallow wetlands with emergent vegetation and forested borders all provide suitable habitat. Its habit of nesting in swamps and woodlands allows the Wood Duck to exploit areas that may have limited nesting habitat for other duck species.

Baillie and Harrington (1936) reported Wood Duck sightings north of Lake Superior (49°N), but the northern limit on published range maps has consistently been restricted to southcentral Ontario (46°N) and the Kenora district (Bellrose 1980, Godfrey 1986). Peck and James (1983) noted isolated records at Lake Nipigon (50°N). Relative to past literature, then, the atlas suggests a northward range extension (to 53.5°N) for the species. It is perhaps more plausible to propose that the atlas data reflect the genuine original range of the Wood Duck, with the apparent expansion having resulted from increased field work in northern Ontario. It was a threatened species at the turn of the century, but the entire population benefited from the Migratory Bird Convention Act of 1917. Hunting seasons in Canada and the US were closed until 1941. Wood Duck bag limits were restricted in Ontario until 1970, but the Wood Duck is now the second most frequently shot duck in the province. Nest box projects

have met with limited success in this province; about 7% of nest boxes in Ontario are used by Wood Ducks, whereas occupancy in some southern states approaches 100% (H.G. Lumsden pers. comm.). The species has good potential to increase its numbers because it generally out-produces other duck species (Bellrose 1976). The beaver, which assists in creating ideal Wood Duck habitat, has also returned to its former strength in the last 75 years, and may have exceeded former numbers in the last 30 years. These have likely been factors in ensuring a full population recovery.

Because broods seek invertebrate food in shallow retreats with overhanging vegetation, locating them becomes difficult, but fledged young represent about one-third of all records in southern Ontario (76% of confirmations). In addition, hens and broods are among the wariest of waterfowl, fleeing at the first sight of a human. In just 11% of all records, nesting cavities were located. Locating occupied natural cavities in swamplands is a task that is complicated by the tendency of incubating hens to sit tight when approached.

The maps thoroughly reflect the breeding distribution of the Wood Duck in southern Ontario. In the north, however, they induce speculation, but do not confirm, that the Wood Duck breeds in varying abundance north of the Great Lakes-St. Lawrence Forest zone. There are atlas records near the Manitoba border at Island Lake in the northwest and near the Moose River in the northeast, although these isolated northern records could possibly be birds which have moved north to moult. The northernmost breeding confirmation occurs west of Lac Seul. Peck and James (1983) mapped 2 breeding records on the west shore of Lake Nipigon, although none were reported there during the atlas period. An examination of the ranges of the cavity-nesting Hooded Merganser, Common Goldeneye, and Bufflehead leaves the impression that Wood Ducks might be able to find suitable nest sites in northern habitats. Intriguingly, the range of the cavity-excavating Pileated Woodpecker most closely resembles that of the Wood Duck.

Abundance estimates indicate that the Wood Duck is fairly common in its range in both southern and northern Ontario. Most estimates (56%) are of 2 to 10 pairs per square.-- *M. Biro*

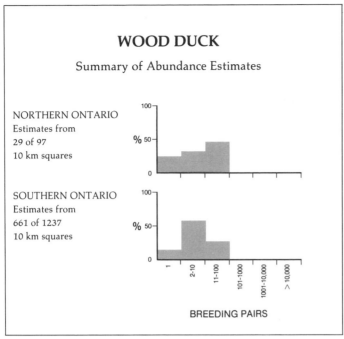

WOOD DUCK

Summary of Abundance Estimates

NORTHERN ONTARIO
Estimates from
29 of 97
10 km squares

SOUTHERN ONTARIO
Estimates from
661 of 1237
10 km squares

BREEDING PAIRS

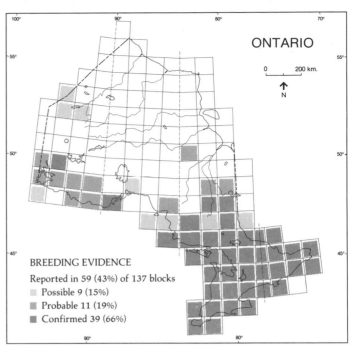

ONTARIO

0   200 km.

N

BREEDING EVIDENCE
Reported in 59 (43%) of 137 blocks

◻ Possible 9 (15%)
▨ Probable 11 (19%)
◼ Confirmed 39 (66%)

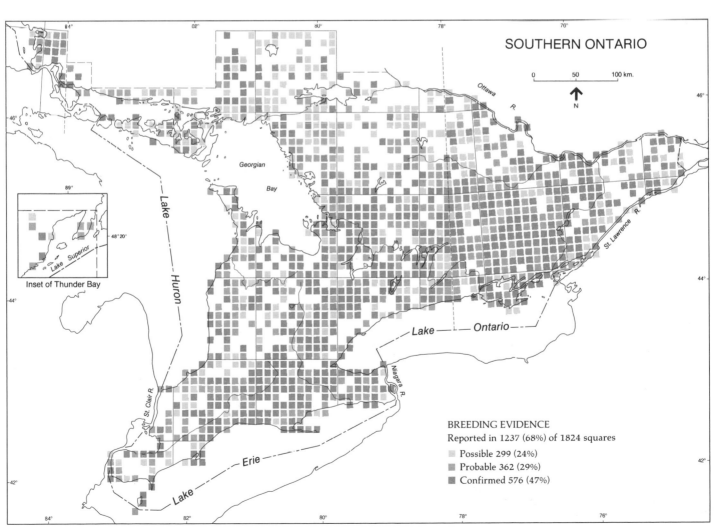

SOUTHERN ONTARIO

0   50   100 km.

N

Inset of Thunder Bay

BREEDING EVIDENCE
Reported in 1237 (68%) of 1824 squares

◻ Possible 299 (24%)
▨ Probable 362 (29%)
◼ Confirmed 576 (47%)

GREEN-WINGED TEAL
Sarcelle à ailes vertes
Anas crecca

This small duck breeds throughout Canada south of the tree-line and in the northwestern and central US. The Green-winged Teal and its Eurasian counterpart, the Common Teal, are now considered to be a single species (AOU 1983). The Common Teal, which is recognizable in the field, nests in northern and central Europe, Asia, and Alaska, and occurs occasionally in Ontario, although there is no evidence of nesting. In Ontario, the Green-winged Teal nests in the southern portion of the province, in the Hudson Bay Lowland, and occasionally in the Boreal Forest zone. It winters primarily in the US, but small numbers may linger in southern Ontario.

The Green-winged Teal usually nests in upland areas, often among dense stands of grasses or brush. Nests may be near water, but are usually 36 to 100 m away from the wetland edge (Bellrose 1980, Peck and James 1983). Marshes, bogs, and beaver ponds are appropriate areas to raise broods, provided that suitable nesting habitat is adjacent. Unlike the Blue-winged Teal, it is seldom found in close association with humans.

Early ornithologists were uncertain of the status of this species in Ontario. McIlwraith (1886) stated that it bred north of the US but was chiefly a migrant in Ontario, and Baillie and Harrington (1936) thought that it was a common breeder in the James Bay area but less common along Hudson Bay. They felt that it probably bred regularly in extreme northwestern Ontario, and considered a single nesting record from Ottawa to be an error. Godfrey (1966) mapped it as breeding throughout Ontario, except for agricultural southern Ontario, and considered it to be very local except in the James and Hudson Bay areas. Peck and James (1983) found it to breed sparingly throughout the province. The apparent extension of the nesting range into southern Ontario is partially due to an increase in the number of observers, but also probably represents a true range expansion. As in a number of other ducks, the creation of suitable habitat in sewage ponds and water impoundments is likely a contributing factor in their spread into southern Ontario. Despite the range expansion, CWS breeding pair surveys suggest a decline in Ontario's Green-winged Teal population since 1970, and there is much concern in recent years about the declining population in eastern North America (D. Dennis pers. comm.).

The Green-winged Teal is difficult to find during the nesting season. The male is easily identified, but courtship takes place in April and May (as well as in autumn and winter), before most atlassing was undertaken. Soon after the eggs are laid, perhaps as early as 12 June, the drake abandons the female, and may travel considerable distances away from the nest site (Bellrose 1980). Therefore, a drake in suitable habitat in late June or July is not necessarily evidence that nesting has occurred in that particular square; 8% of southern Ontario and 4% of northern Ontario records are of single males. Nests are extremely difficult to find, being only approximately 15 cm in diameter and hidden among dense vegetation. Hens do not flush from the nest until the last second, so one must either be very fortunate or work very hard to find a nest. During the incubation period, the hen completely covers the nest with down when she leaves it, making location of a nest almost impossible if the female is not attending it; nests with eggs or newly hatched young were reported in 12 squares.

Though more widespread, the pattern of distribution of the Green-winged Teal resembles that of the Northern Pintail: scattered in the southwest; somewhat of a concentration around the southern fringe of the Canadian Shield and in southeastern Ontario; a cluster of squares around Sudbury and in the blocks of the Clay Belt; sporadic occurrence through the remainder of the Boreal Forest and the northwest section of the Great Lakes-St. Lawrence Forest zone, with a concentration of records on the Hudson Bay Lowland. The Green-winged Teal is less restricted to coastal blocks of the Lowland than is the Pintail, presumably indicating a stronger preference for muskeg habitat. South of the Lowland, as with a number of other ducks, low wetland productivity is likely an important factor in the Green-winged Teal's scarcity on the Canadian Shield and throughout much of the Boreal Forest zone.

The Green-winged Teal population appears to be increasing south of the Shield in southern Ontario because of a general range expansion, and as a response to new habitat in the form of inland reservoirs and sewage lagoons. Abundance estimates suggest higher breeding densities in the north than in southern Ontario. CWS survey data indicate a southern Ontario population of 11,000 pairs (D. Dennis pers. comm), so the atlas estimates should be considered conservative.-- *A. Sandilands*

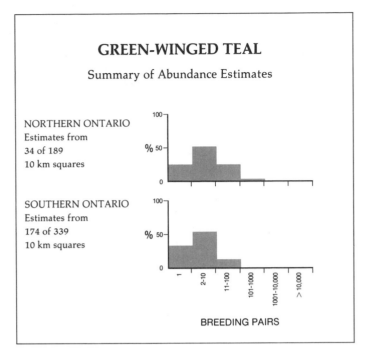

GREEN-WINGED TEAL

Summary of Abundance Estimates

NORTHERN ONTARIO
Estimates from
34 of 189
10 km squares

SOUTHERN ONTARIO
Estimates from
174 of 339
10 km squares

%

100

50

0

100

50

0

1   2-10   11-100   101-1000   1001-10,000   >10,000

BREEDING PAIRS

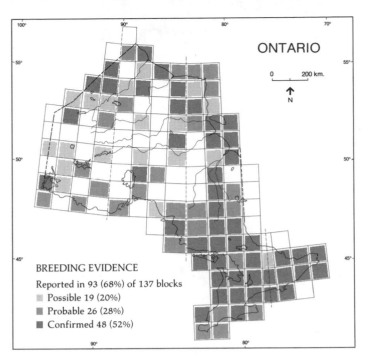

ONTARIO

0   200 km.

N

BREEDING EVIDENCE
Reported in 93 (68%) of 137 blocks

Possible 19 (20%)
Probable 26 (28%)
Confirmed 48 (52%)

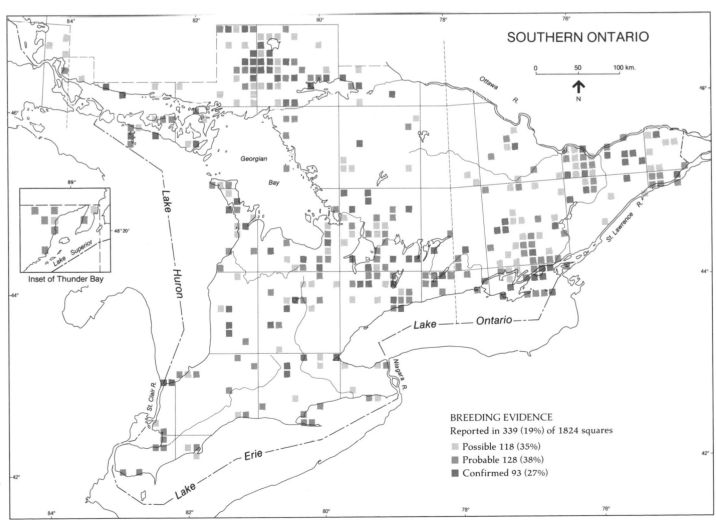

SOUTHERN ONTARIO

0   50   100 km.

N

Inset of Thunder Bay

Georgian Bay

Lake Huron

Lake Superior

Ottawa R.

St. Lawrence R.

Lake Ontario

Niagara R.

St. Clair R.

Lake Erie

BREEDING EVIDENCE
Reported in 339 (19%) of 1824 squares

Possible 118 (35%)
Probable 128 (38%)
Confirmed 93 (27%)

AMERICAN BLACK DUCK
Canard noir
Anas rubripes

The American Black Duck breeds in the northeastern US south to the Carolinas, and throughout the provinces of eastern Canada as far north as the tree-line. In Ontario, it achieves its greatest density in the mixed-wood forests of the Canadian Shield, while its distribution is currently sporadic farther south. A more uniform distribution of birds is present in the Boreal Forest zone. Nesting occurs in a wide variety of sites and may be in herbaceous growth, in the zone of sweet gale-leatherleaf-sedge associated with wetlands, or directly under conifers. The habitat in which nests are most commonly located in Ontario is perhaps on islands in lakes or around beaver ponds.

According to Palmer (1976a), a westward expansion of the American Black Duck has occurred within the past century; formerly, the breeding range did not extend west much beyond Lake Erie and central Ontario. The current distribution has not changed greatly in recent decades. However, dramatic reductions in breeding density have occurred, especially in southern Ontario. Canadian Wildlife Service surveys suggest that the American Black Duck population declined by 80% in southern Ontario between 1951 and 1981; at the same time, the Mallard population expanded eight-fold.

Once said to be the most common and widely distributed duck in Ontario (Baillie and Harrington 1936), it now ranks well behind the Mallard in abundance. Increased Mallard numbers in Ontario have contributed greatly to the decline of the American Black Duck, as the two species freely hybridize, and subsequent generations of hybrids tend to resemble Mallards more than this species. Retention of American Black Duck genes probably has assisted the Mallard to expand farther into American Black Duck range where factors such as the blood parasite *Leucocytozoan* previously prevented their invasion. The American Black Duck is relatively resistent to the parasite, while many Mallards readily succumb to it.

The American Black Duck is relatively easy to confirm as a breeding species. The presence of pairs during May is a fairly reliable indicator of nesting. By June, however, 2 American Black Ducks together are likely 2 males in a moulting or pre-moulting area - a fact that may not have been known to atlassers, so some misuse of the 'pair' breeding category may have occurred. The female usually flushes from the nest when an observer is several metres distant. Nests range from easy to impossible to locate. Those on islands are easily located, while those in floating sedge-leatherleaf bogs present a formidable task. The first broods hatch in mid-May in southern Ontario. However, renesting in the north may result in broods hatching as late as early July. Thus, July is the best month to confirm breeding, as essentially all broods are present then. By mid- to late July, the first broods are flying, but they usually remain associated with the natal pond and may readily be distinguished from adult females by their bright, unfaded plumage.

The maps give a good general indication of the distribution of the species in Ontario. The American Black Duck is widespread, but becomes sporadic in the far northwestern Boreal Forest zone, which is close to the northwestern extremity of its breeding range (Godfrey 1986). More specific searching would likely result in more squares with confirmed American Black Duck records in central Ontario. There is a possibility that records in some of the southern squares, especially in southwestern Ontario, really represent hybrids that were identified as this species.

CWS surveys revealed that breeding densities range from 7 pairs per 100 km^2 in northwestern Ontario to 28 pairs per 100 km^2 elsewhere on the Canadian Shield. Peak numbers of 31 pairs per 100 km^2 occurred in portions of the Clay Belt in northern Ontario (Dennis and North 1984). CWS surveys in 1981 suggested that 10 pairs were present per 100 km^2 in southern Ontario, with greatest densities in the eastern part of the province. Atlas abundance estimates are generally in agreement with these figures. The American Black Duck continues to decrease in numbers in Ontario, and the decrease is most severe in the southern portion of the province. The species at present is still fairly common on the Canadian Shield, common in the Boreal Forest sections of Ontario, and uncommon in the south.-- *D. Dennis*

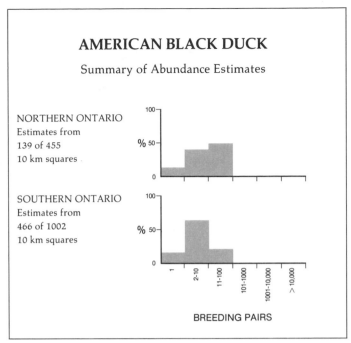

AMERICAN BLACK DUCK

Summary of Abundance Estimates

NORTHERN ONTARIO
Estimates from
139 of 455
10 km squares

SOUTHERN ONTARIO
Estimates from
466 of 1002
10 km squares

%

100

50

0

100

50

0

1 2-10 11-100 101-1000 1001-10,000 >10,000

BREEDING PAIRS

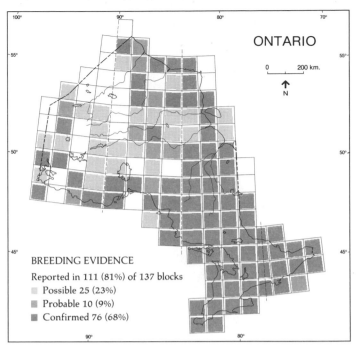

ONTARIO

0   200 km.

N

BREEDING EVIDENCE

Reported in 111 (81%) of 137 blocks

░ Possible 25 (23%)
▒ Probable 10 (9%)
▓ Confirmed 76 (68%)

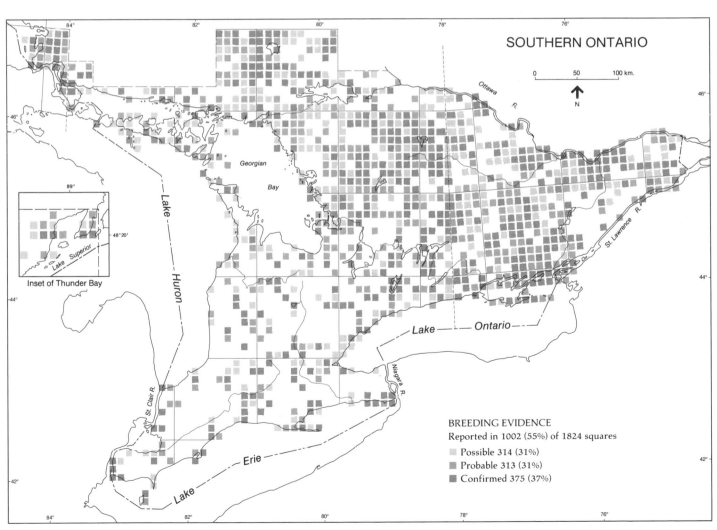

SOUTHERN ONTARIO

Ottawa R.

0   50   100 km.

N

Georgian Bay

Lake Huron

Lake Superior

48°20'

Inset of Thunder Bay

St. Clair R.

Lake Erie

Niagara R.

Lake — Ontario

St. Lawrence R.

BREEDING EVIDENCE

Reported in 1002 (55%) of 1824 squares

░ Possible 314 (31%)
▒ Probable 313 (31%)
▓ Confirmed 375 (37%)

MALLARD
Canard colvert
Anas platyrhynchos

The Mallard is the most abundant and widespread puddle duck in Ontario. Depending upon the environment it is inhabiting, it can be exceedingly wary or very tame. Because of its ability to live in close proximity to humans in urban situations, it is the only species of duck known to many who have only a passing familiarity with birds. It has often been domesticated, and is the ancestor of most domestic ducks.

Its world-wide range includes Europe and Asia in addition to North America: it was recently introduced to Australia, New Zealand, and Hawaii. During the breeding season, it occurs in western North America eastward to Ontario, Quebec, the New England states, and, recently, the Atlantic Provinces. Its Ontario range includes the entire province, but there are gaps in its distribution in the interior on the Canadian Shield, especially in the Algonquin Highlands.

The Mallard is one of the most adaptable ducks in its choice of nesting habitat. Typically, it prefers the edges of marshes, grassy meadows, islands, small ponds, and ditches. Nests may be a considerable distance from water, and they are occasionally found in unusual spots such as in tree roots, on top of stumps, in tree hollows, on lawns, in abandoned buildings, or on a variety of man-made structures (Peck and James 1983). Although ponds and lakes are common on the Shield, shorelines are generally steep and rocky, providing poor cover for nesting.

This species has exhibited a dramatic population increase in Ontario. Baillie and Harrington (1936) reported it to be a sparse nester in southwestern and northwestern Ontario. The history of this species in Wellington Co is probably typical of its status in the province: Klugh (1905) considered it to be a scarce migrant, Soper (1923) an occasional migrant, and Brewer (1977) a common breeding species. Up until the late 1940's, it was a rare to uncommon nester in northern Well-

ington Co, but became common to abundant in the 1950's and 1960's (Day 1955, Boyer and Devitt 1961). The increase in numbers is a result of a general expansion of its range eastward, plus successful competition for habitat with the American Black Duck. Widespread stocking of Mallards in eastern North America may have influenced this spread.

The Mallard is relatively conspicuous during the nesting season and is not a difficult species to confirm as a breeding bird. In urban environments or where it is abundant, nests are often easy to find, but confirmation may be more difficult where it is less common, as drakes are absent and the females unobtrusive. Broods may appear as early as mid-May and, as young are flightless for approximately 6 weeks, this is the optimum time to confirm nesting: 73% of confirmations were of broods. Depending upon the environment, broods may be conspicuous and unconcerned about humans, or they may be wary. Hens often give a splashing and vocal display to distract predators as the young disappear into the vegetation.

The atlas data reveal that the Mallard is widespread across Ontario. In fact, only the Herring Gull was reported in more blocks. In northern Ontario, gaps in Mallard distribution and blocks without confirmation are likely the result of insufficient field work, except in the Cape Henrietta Maria area, where it may truly be absent as a breeding bird. In southern Ontario, the Mallard was reported in almost every square off the Canadian Shield. Gaps in occurrence, especially those in the extreme southwest, may be the result of the near absence of wetland habitat. The Mallard was reported in most of the Shield country of southern Ontario, but its occurrence there is more sporadic, and it is absent from many squares, particularly in the Algonquin Highlands. Its absence is likely explained by the fact that most ponds in that area have steep banks, contain little aquatic vegetation, and are surrounded by forest.

Atlas data show that the Mallard is common throughout the province, though perhaps more so in the south, presumably as a result of higher wetland productivity. Most Ontario squares were thought to contain 2 to 10 (45%) or 11 to 100 (46%) breeding pairs of Mallards.-- *A. Sandilands*

MALLARD

Summary of Abundance Estimates

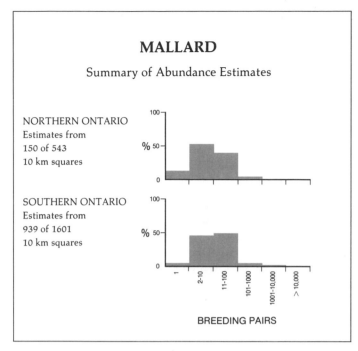

NORTHERN ONTARIO
Estimates from
150 of 543
10 km squares

%

SOUTHERN ONTARIO
Estimates from
939 of 1601
10 km squares

%

BREEDING PAIRS

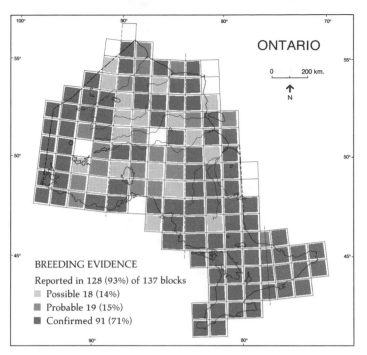

ONTARIO

0   200 km.

N

BREEDING EVIDENCE

Reported in 128 (93%) of 137 blocks

Possible 18 (14%)
Probable 19 (15%)
Confirmed 91 (71%)

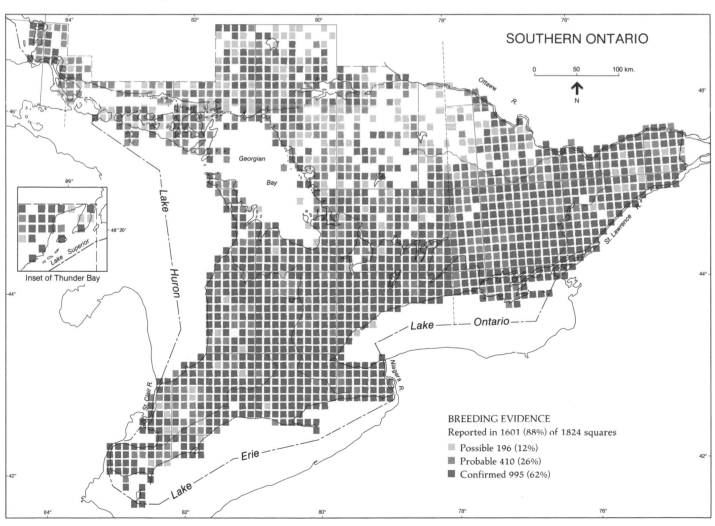

SOUTHERN ONTARIO

0   50   100 km.

N

Inset of Thunder Bay

Georgian
Bay

Lake Huron

Ottawa R.

St. Lawrence R.

Lake — Ontario

Niagara R.

St. Clair R.

Lake Erie

Lake Superior

BREEDING EVIDENCE

Reported in 1601 (88%) of 1824 squares

Possible 196 (12%)
Probable 410 (26%)
Confirmed 995 (62%)

NORTHERN PINTAIL
Canard pilet
Anas acuta

The drake Northern Pintail, with its bold pattern and long tail, is one of the more striking of our dabbling ducks. During spring migration, flocks may number in the thousands.

In North America, its breeding range includes the northwestern states, all of western Canada north to the arctic archipelago, and portions of Ontario, Quebec, and the Maritime Provinces. It also nests in Greenland, Iceland, and northern Eurasia. It winters as far north as southern Canada. Few Northern Pintail remain to nest in the accessible portion of Ontario. In this province, it is most common on the Hudson Bay Lowland, but it also occurs in agricultural southern Ontario and rarely in the Boreal Forest zone.

The Northern Pintail nests in shallow prairie marshes, fens, and tundra and also in meadows, in old orchards, and on islands. The nest is invariably on the ground and is placed close to water, with the maximum recorded distance from open water in Ontario being 90 m (Peck and James 1983). Bellrose (1980), however, reported that it nested farther from water than other ducks; normally nests were within 100 m of water, with the average being about 40 m, although distances of up to 1.6 km were measured. Areas of sparse or low vegetation are selected for the nest, and it may even be located on bare ground such as an agricultural field. Its preference for flat, open areas near water explains its rarity on the Canadian Shield and the more heavily forested areas of the Hudson Bay Lowland.

The breeding range of the Northern Pintail has not changed dramatically, although it appears to have increased in numbers in southern Ontario. McIlwraith (1886) stated that most pairs summered in the north, but that breeding had been recorded in the Lake St. Clair area. Baillie and Harrington (1936) considered it common along the west shore of James Bay and in the extreme western portions of Ontario.

They knew of only one positive nest record for southern Ontario, which was in Kent Co, but broods had been reported at Long Point in 1916. It is interesting to note that breeding was not confirmed in the Lake St. Clair or Long Point areas, or in the northwest during the atlas period. In the Kingston area, where some of the larger southern Ontario concentrations occur, it was first reported nesting in 1949 but later became a regular, uncommon breeder (Quilliam 1973). The overall population trend from 1957 to 1973 was an increase of 10% (Bellrose 1980). The southern Ontario increase may have been assisted by the creation of new nesting habitat in the form of sewage lagoons.

The Northern Pintail is one of the earlier nesting dabblers, and may initiate nesting in early April. Pursuit flights may reveal the presence of breeding birds in the early stages of pair formation. Almost as soon as incubation starts, drakes desert the hens and may fly a considerable distance away from the nest site for the flightless period. The nest is relatively easy to find for a ground-nesting species, and incubating hens do not sit particularly tight: nests were reported in 16 squares. Distraction displays were a significant factor, making up 7% and 26% of confirmed records in the south and north, respectively. Immediately after the eggs hatch, the hen may lead the young up to 700 m from the nest site to more favourable rearing areas (Sowls 1955) and broods may move as far as 1.25 km from the nest site prior to gaining the power of flight (Evans *et al.* 1952). Broods are relatively conspicuous, and the hen is easy to identify, so that June and July are the best months to confirm breeding for the Northern Pintail. As is true of most ducks, broods were by far the most frequently used category of confirmed breeding, making up 70% of total confirmations.

The data collected for the atlas appear to be representative of the accepted distribution of this species in Ontario. The breeding range shown on the maps is similar to that reported in other published works, except that the atlas has better defined the Northern Pintail's distribution in the southern portion of the province. Most inland records in southern Ontario are associated with sewage lagoons or man-made impoundments.

Atlas data verify that the Northern Pintail is rare on the Canadian Shield, except perhaps around Sudbury, and does not breed over much of the north. In southern Ontario, it is local and uncommon, with the greatest concentration of breeding records being near Kingston and in extreme southeastern Ontario. In southern Ontario squares in which estimates were provided, the abundance estimate was one pair in 50% of squares and 2 to 10 pairs in 45% of squares. Estimates are generally higher in the blocks adjacent to Hudson and James Bays, where the Northern Pintail is considered to be a common breeding bird, with 2 squares having estimates of over 100 breeding pairs.-- *A. Sandilands*

NORTHERN PINTAIL

Summary of Abundance Estimates

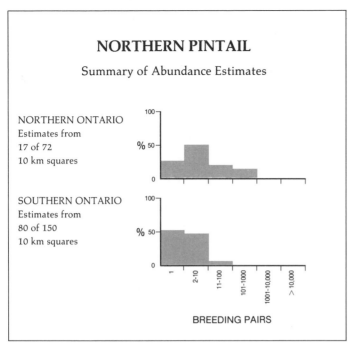

NORTHERN ONTARIO
Estimates from
17 of 72
10 km squares

SOUTHERN ONTARIO
Estimates from
80 of 150
10 km squares

%

100

50

0

%

100

50

0

1

2-10

11-100

101-1000

1001-10,000

> 10,000

BREEDING PAIRS

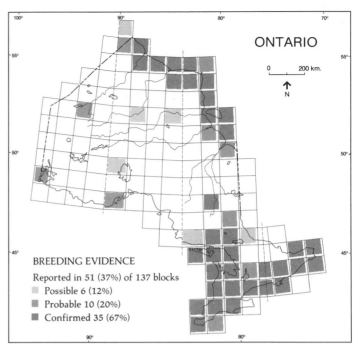

ONTARIO

0   200 km.

N

BREEDING EVIDENCE

Reported in 51 (37%) of 137 blocks

Possible 6 (12%)

Probable 10 (20%)

Confirmed 35 (67%)

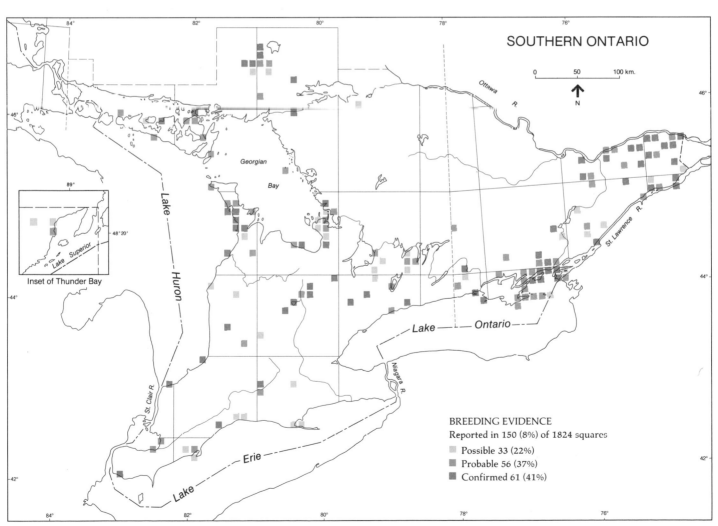

SOUTHERN ONTARIO

0   50   100 km.

N

Georgian

Bay

Ottawa R.

St. Lawrence R.

Lake — Ontario

Lake — Huron

St. Clair R.

Niagara R.

Lake — Erie

Inset of Thunder Bay

Lake Superior

89°

48°20'

BREEDING EVIDENCE

Reported in 150 (8%) of 1824 squares

Possible 33 (22%)

Probable 56 (37%)

Confirmed 61 (41%)

BLUE-WINGED TEAL
Sarcelle à ailes bleues
Anas discors

This small, fast-flying duck is highly migratory, being the only North American waterfowl species to winter predominantly in South America. The return of the beautifully-marked males is one of the most satisfying indicators of the progress of spring. Its breeding range, which is centred on the prairies, extends in a broad belt across the continent from the Atlantic seaboard to the northwestern Pacific coast, and from the tree-line south to the limits of the prairies and grasslands. In southern Ontario, the Blue-winged Teal is a common species usually found in open habitat, where it prefers marshes, sloughs, margins of winding rivers, and even streams and stock ponds in pasture land. The nest is usually constructed of grass or other vegetation and concealed in tall grass near the water. In the Boreal Forest zone, where numbers are much lower, the species is most often encountered in the larger cattail marshes associated with lakes and slow-moving rivers.

Baillie and Harrington (1936) considered the Blue-winged Teal to breed sparingly in southern and far-western Ontario and had no positive breeding record for the north. Since that time, this teal has increased in abundance in southern Ontario to the point that, among waterfowl, it is now second only to the Mallard in breeding density (Collins 1974, Ross *et al.* 1984). This rise, which has been essentially continuous at least from 1951 to 1976, has been attributed to increased clearing of land for agriculture (Cringan 1960); as well, this duck, like most waterfowl, has likely benefitted from the increased beaver populations which have created favourable new habitat during that period (Novak 1976). In northern Ontario, many new breeding records have been recorded during the atlas period. It is difficult to determine to what extent this represents an increase in abundance, as suggested by Schueler *et al.* (1974), or whether it reflects a better

understanding of the range of the species. One would, however, expect some influence from the rapidly rising southern population.

Given its preference for open habitat, the Blue-winged Teal is a relatively easy duck to atlas, particularly in spring, when the males are highly visible during nest initiation. Confirmation of nesting mostly arose through sightings of females with young broods (74% of confirmed records in southern Ontario, 82% in the north), which are most easily observed early in the morning. This is a reliable criterion given the propensity of the species for nesting beside small water bodies which limit the travel of young broods. Its habit of nesting near the brood-rearing water body facilitates finding nests with eggs (14% of southern Ontario confirmed records, but 0% in the north). Nests can be found either through zigzag searches along the shoreline or by dragging a rope between 2 observers over the habitat.

The atlas project has proven useful in determining the extent of this species' presence in northern Ontario, where it is thinly distributed. The map shows a strong association with the Great Lakes-St. Lawrence Forest zone and the northern Clay Belt of the Boreal Forest zone, which is probably indicative of land clearing for agriculture. Other records are more sporadic but imply some preference for the Hudson Bay Lowland. Also, a patch of records along the northwest border of Ontario may represent a small zone of influence from the much higher breeding densities on the prairies. In southern Ontario, the distribution is largely continuous throughout the agricultural areas off the Shield, except in the areas of most intensive agriculture (Kent, Lambton, Middlesex, Huron, and Perth Counties, the Niagara Peninsula, and east of Ottawa), presumably because of a paucity of suitable wetlands.

The breeding density of the Blue-winged Teal in southern Ontario has been measured at 26 pairs/100 km^2 in 1976 (Ross *et al.* 1984) although population trends at that time indicated a continuing rise. Interestingly, the modal value of atlas abundance estimates for southern Ontario was much lower (2 to 10 pairs/100 km^2), which would suggest an underestimate of this common species. In northern Ontario, population densities determined by CWS surveys (K. Ross unpubl.) are low (0 to 6 pairs/100 km^2), which corresponds reasonably well with the atlas abundance estimates. Considering the recent decline of this species on the prairies, due to drought and other inclement breeding conditions (USFWS and CWS 1986), it is difficult to predict when declining habitat quality, particularly in southern Ontario, will have an effect on the Blue-winged Teal in this province, although it has been shown to react quickly to local habitat change (Ross *et al.* 1984).-- *R.K. Ross*

BLUE-WINGED TEAL

Summary of Abundance Estimates

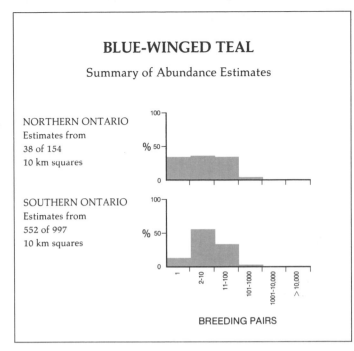

NORTHERN ONTARIO
Estimates from
38 of 154
10 km squares

SOUTHERN ONTARIO
Estimates from
552 of 997
10 km squares

BREEDING PAIRS

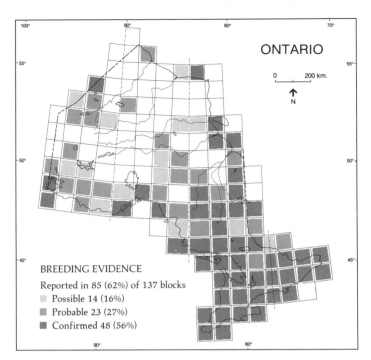

ONTARIO

0   200 km.

N

BREEDING EVIDENCE
Reported in 85 (62%) of 137 blocks

Possible 14 (16%)
Probable 23 (27%)
Confirmed 48 (56%)

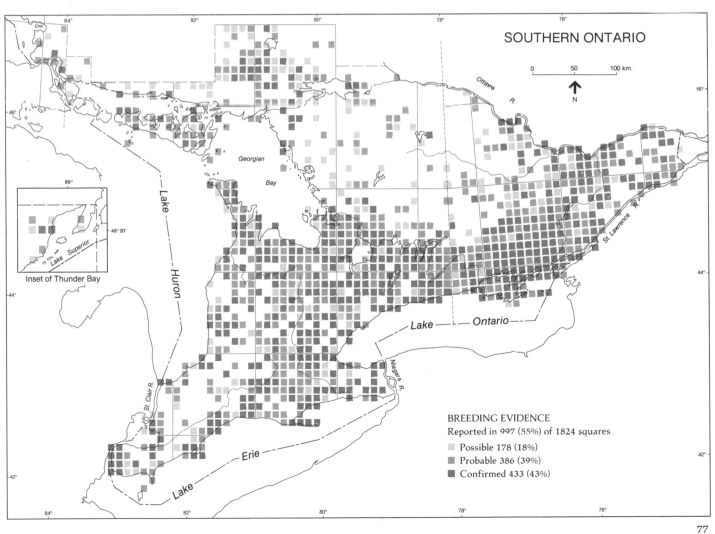

SOUTHERN ONTARIO

0   50   100 km.

N

Inset of Thunder Bay

BREEDING EVIDENCE
Reported in 997 (55%) of 1824 squares

Possible 178 (18%)
Probable 386 (39%)
Confirmed 433 (43%)

CINNAMON TEAL
Sarcelle cannelle
Anas cyanoptera

juliana Yawke

The Cinnamon Teal is one of the most recent additions to Ontario's list of breeding birds. Few people have been fortunate enough to see this species in Ontario, and nesting was not reported in the province until 1983.

As a breeding species, it is restricted to North and South America. In North America, it is found from the plains to the west coast, and it normally nests in Canada only in southern British Columbia and Alberta. Extra-limital summer records have been reported from Saskatchewan and Manitoba, and, more recently, Ontario.

In western North America, it is typically associated with prairie sloughs and other shallow water bodies. Dense vegetation is essential for nesting and for brood rearing. Nests are often built under mats of the previous year's dead vegetation, among cattails, bulrushes, or tall weeds. Usually nests are within 75 m of water, but they have been found as far away as 200 m.

Until very recently, the Cinnamon Teal was only a rare wanderer in Ontario during the spring and fall. James *et al.* (1976) reported 2 spring and 3 autumn records, all from southwestern Ontario. The first reported summer birds were a pair found at the Amherstburg sewage lagoons, 30 km south of Windsor, on 28 May 1983. The nest was discovered by Alan Wormington on 24 June. This constituted Ontario's first breeding record (James 1984b). The following year, a pair was present at the Townsend sewage lagoons east of Simcoe. The pair was observed from 23 June to 9 July 1984. When the female was no longer observed on the lagoon, a search for a nest turned up the carcass of the female. She had been killed by flying into a wire. No nest was found (R. Curry pers. comm.). Between 29 May and 11 June 1985, Alan Wormington observed an adult male Cinnamon Teal at Hillman Marsh in Essex Co. On 1 July 1985, Paul Lehman found

an adult male Cinnamon Teal on Sable Island in the Lake of the Woods.

Bellrose (1980) stated that numbers of Cinnamon Teal have fluctuated over a range of 50% during the period from 1960 to 1972, but there was no evidence of any lasting trend. The recent nestings in Ontario may be due to the fact that sewage lagoons provide habitat similar to prairie sloughs.

The drake Cinnamon Teal is unmistakable in breeding plumage, but during the eclipse plumage (June to August), it is similar to the female, possibly being slightly redder (Bellrose 1980). The female Cinnamon Teal is almost indistinguishable from the female Blue-winged Teal, but has a slightly larger bill and more pronounced facial markings. These field marks, however, are not entirely reliable. The voices of the 2 species of teal are also similar, so identification can be a problem.

Pair formation in the Cinnamon Teal often occurs prior to migration, but pairing and courtship may extend into June. The drake Cinnamon Teal is more attentive to the hen than are the drakes of many other duck species. It has been estimated that 25% of the drakes wait until the second week of incubation before deserting the hen, 63% have left by the end of the third week, and 12% remain until the ducklings hatch (Oring 1964). The more extended presence of the drake, which likes to loaf on conspicuous sites such as muskrat houses, helps in providing evidence that this species is breeding.

Nests are extremely difficult to find, as they are usually in tall, dense growth and often concealed under dead vegetation. The hen sits tight during incubation and may enter and leave the nest by access tunnels under matted vegetation. When she is absent, the nest is completely covered over with down and concealed from view. Broods tend to stay close to thick emergent cover and quickly disappear if alarmed. The hen, however, may perform a broken-wing display, which can alert an observer to the presence of a brood. This may be one of the more difficult dabblers to confirm because of problems in identifying hens and finding nests and broods.

Prior to the atlas, this species was not known to nest in Ontario. During 2 separate years of the atlas field work, at least one pair was found at sewage lagoons. The Cinnamon Teal should, at present, be considered a very rare nesting bird in southwestern Ontario. Only time will tell if these were isolated extra-limital nesting attempts or if the species will become firmly established as a breeding bird.-- *A. Sandilands*

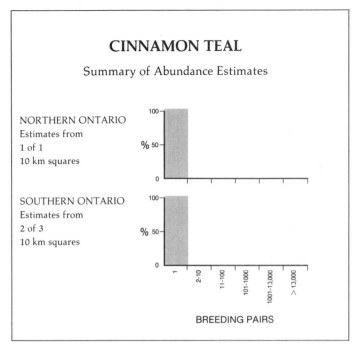

CINNAMON TEAL

Summary of Abundance Estimates

NORTHERN ONTARIO
Estimates from
1 of 1
10 km squares

SOUTHERN ONTARIO
Estimates from
2 of 3
10 km squares

BREEDING PAIRS

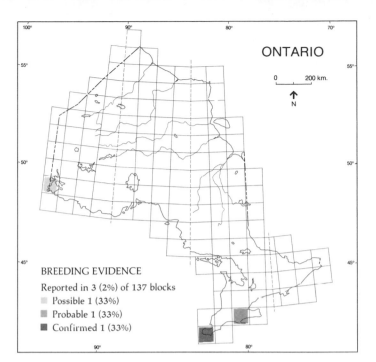

ONTARIO

0   200 km.

N

BREEDING EVIDENCE
Reported in 3 (2%) of 137 blocks
Possible 1 (33%)
Probable 1 (33%)
Confirmed 1 (33%)

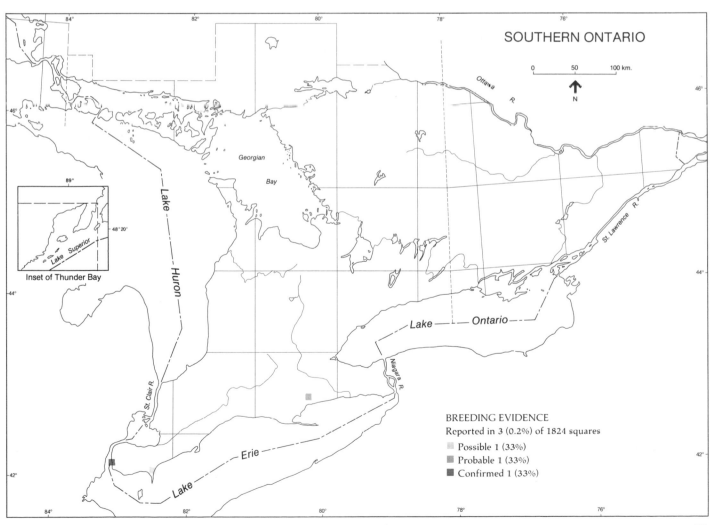

SOUTHERN ONTARIO

0   50   100 km.

N

Inset of Thunder Bay

BREEDING EVIDENCE
Reported in 3 (0.2%) of 1824 squares
Possible 1 (33%)
Probable 1 (33%)
Confirmed 1 (33%)

NORTHERN SHOVELER
Canard souchet
Anas clypeata

The Northern Shoveler breeds in Europe, Asia, and North America. On this continent, it is a widespread breeder from Alaska through the western provinces to the northwestern and midwestern states. The greatest nesting densities and abundance in North America are attained on the Canadian prairies. Ontario is on the eastern periphery of its main breeding range, although there are minor nesting areas in Quebec and the Maritime Provinces. In Ontario, most nesting records are from scattered locations in the southern agricultural areas, but it is also known to breed on the Hudson Bay Lowland and in the Lake of the Woods area of northwestern Ontario. There are a few unconfirmed records from the Boreal Forest zone. The North American birds winter from the southern US to northeastern South America.

The Northern Shoveler prefers grassy areas for nesting, and selects short grass areas in preference to tall grasses (Girard 1939). In the absence of grassy habitat, it will nest in meadows, hay fields, and, more rarely, bulrushes. It avoids weedy patches and clumps of woody vegetation, areas preferred by many other species of dabbling ducks. Few nest are very close to water, and most of those are on islands. Nests are typically 20 to 60 m from water, but may be as far away as 1.6 km. The Northern Shoveler's preference for areas with short grass explains its general absence from the Canadian Shield.

The Northern Shoveler appears to have increased as a breeding species in the province. McIlwraith (1886) did not consider it to be a nesting species, and Baillie and Harrington (1936) knew of only one Ontario report of its nesting, near Dunnville. James *et al.* (1976) considered it to be a rare local summer resident in the south and northwest. Peck and James (1983) also reported it to be rare and irregular in southern Ontario, but suspected breeding along the James and Hudson Bay coasts. The atlas confirms that it does nest in the northern coastal areas. The increase in known breeding numbers and range in Ontario may be due to 2 factors: the creation of habitat at sewage lagoons and impoundments (primarily in the south), and better observer coverage in the north. It is unknown if there has been an actual increase in numbers on the northern Lowland. Overall in North America, the Northern Shoveler has increased in the past 20 years, but there has been a slight decline in the Atlantic Flyway (Bellrose 1980).

Nesting is initiated in the first half of May, and both sexes are conspicuous during this period. Mate fidelity varies among individuals, with some drakes abandoning the hen on the first day of incubation and others remaining until the young are hatched (Oring 1964). The best times to locate nesting evidence for the Northern Shoveler are early in the nesting season, when the drake is still present, and after hatching, when the hen is attending to the brood. Pairs in suitable breeding habitat make up 89% and 85% of probable breeding records in southern and northern Ontario, respectively. The hen may lead the brood 1 to 2 km from the nest site to a suitable rearing pond. Young may take 5 to 9 weeks (probably 6 to 7 in Ontario) to acquire the power of flight, so they may be present from mid-June until early August. Nests with eggs or newly hatched young were reported from 5 squares. Sowls (1955) found that 42% of hens returned the following year to nest and, because of a high mortality rate, concluded that this probably represented most of the survivors. In Ontario, there are few, if any, places where the Northern Shoveler can confidently be predicted to be a nesting species. In areas where it does nest, confirmation of breeding is not overly difficult.

The maps reveal a pattern of distribution similar to that of the Northern Pintail, Green-winged Teal, and American Wigeon. There is generally an avoidance of much of the Boreal Forest zone south of the Hudson Bay Lowland and the Canadian Shield in southern Ontario, with a slight concentration of records south of the edge of the Shield and in southeastern Ontario, and scattered records in the southwest. Many of the records in the southwest are from squares with large, highly productive marshes, but sewage lagoons and water impoundments provide ideal nesting habitat, and are also of considerable importance throughout southern Ontario. As in these other species, there is a concentration of records around Sudbury and in the Clay Belt. The distribution of the Northern Shoveler on the Hudson Bay Lowland is primarily restricted to coastal blocks, where flat marshy areas with many ponds and short vegetation provide habitat apparently preferred over inland muskeg areas.

The highest abundance estimates (11 to 100 pairs) were reported from Hullett Marsh, in Huron Co, and from 2 squares on the James Bay coast. The low estimates and small number of records generally indicate that the Northern Shoveler is, at best, uncommon in Ontario.-- *A. Sandilands*

NORTHERN SHOVELER

Summary of Abundance Estimates

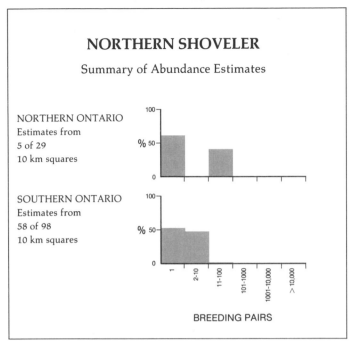

NORTHERN ONTARIO
Estimates from
5 of 29
10 km squares

SOUTHERN ONTARIO
Estimates from
58 of 98
10 km squares

BREEDING PAIRS

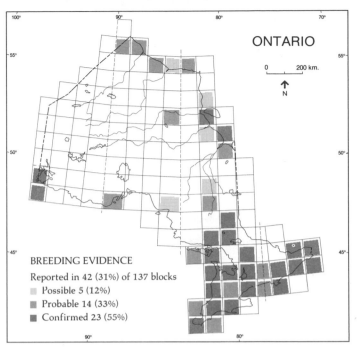

ONTARIO

0   200 km.

N

BREEDING EVIDENCE
Reported in 42 (31%) of 137 blocks
Possible 5 (12%)
Probable 14 (33%)
Confirmed 23 (55%)

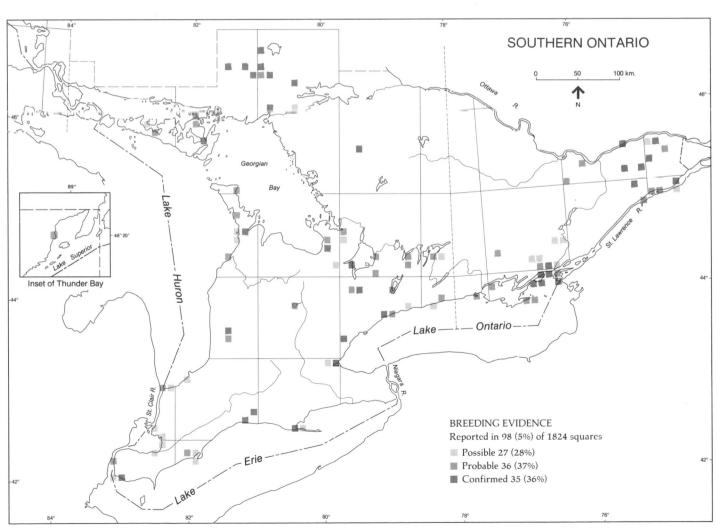

SOUTHERN ONTARIO

0   50   100 km.

N

Inset of Thunder Bay

Georgian Bay

Lake Superior

Lake Huron

Lake Erie

Lake Ontario

Ottawa R.

St. Lawrence R.

Niagara R.

St. Clair R.

BREEDING EVIDENCE
Reported in 98 (5%) of 1824 squares
Possible 27 (28%)
Probable 36 (37%)
Confirmed 35 (36%)

GADWALL
Canard chipeau
Anas strepera

Juliana Hawke

The Gadwall has a holarctic distribution, which includes Europe and Asia as well as North America. In North America, it attains its greatest abundance from the prairies westward. In Ontario, it is most frequent in the south, but also occurs sporadically on the Canadian Shield and the Hudson Bay Lowland. Small numbers winter as far north as the southern Great Lakes.

For nesting, the Gadwall prefers dry, open areas with good grass or herbaceous cover near water. Islands are preferred nesting sites, and often support higher than normal nesting densities. Such sites are occasionally shared with colonies of gulls or terns. Nests are depressions in the ground lined with various types of vegetation; they are often constructed at the base of small shrubs or trees, or among clumps of goldenrod or some other tall herb that affords concealment. Because of its preference for open, vegetated areas near water, it is found primarily in agricultural southern Ontario. In recent years, it has been found nesting at several sewage lagoons in this area.

The Gadwall is a relatively recent arrival as a breeding species in Ontario. The first Ontario nest was found in 1955 in the Lake St. Clair area (Baillie 1963b), and during the same year it was reported nesting at Luther Marsh (Day 1955). These were the only 2 Ontario nesting locations listed by Godfrey (1966), but it has since become much more widespread in Ontario. In some areas, such as Luther Marsh, and parts of the Toronto waterfront, it is locally abundant, outnumbering Mallards in nesting density. The population increase is a result of a general eastward expansion, aided in Ontario by impoundments and sewage lagoons, which create habitat like prairie potholes.

From an atlassing perspective, the Gadwall is not exceptionally difficult to confirm as a breeding species. During the courting season in late March and early April, drakes are conspicuous and often vocal. The hen flushes easily off the nest, occasionally at some distance. The eggs are not normally covered over when the hen leaves the nest, so it is not unusual to find unattended nests. Confirmation of nesting is easiest after the young have left the nest, which may occur from late June through July. The hen and her brood tend to hug the shoreline or edge of an island, usually where there is good emergent cover such as cattails or willows. During this period, the hen is very wary and will quickly disappear with her young into the vegetation if disturbed.

Once the Gadwall has nested in a certain locality, it is likely that it will return in subsequent years. As a pioneering species, it appears to occur in an area at extremely low densities initially, but it may continue to increase in numbers until the population spills over into adjacent areas. In a 30 year period, the population at Luther Marsh has grown from one to in excess of 100 nesting pairs.

Although it has been recognized for some time that the Gadwall is expanding in Ontario, atlas data have greatly assisted in documenting the present range and abundance of this species. It is still local in distribution, being common in a few small areas but virtually absent in adjacent suitable habitat. Southeastern Ontario, along the St. Lawrence River, is where it is most widely distributed. In the Kingston area, an estimated 70 pairs were present during the atlas period, although the species was unknown as a breeding bird there until 1972 (Quilliam 1973). It occurs sporadically along the shorelines of the Great Lakes to just north of Georgian Bay and at some southern Ontario inland impoundments and sewage lagoons. Atlas data show an extensive range expansion into northern Ontario, though breeding in the north remains local. Breeding was confirmed in the vicinity of Lake Nipigon, Thunder Bay, and about 100 km north of Red Lake, and unconfirmed records were obtained from the James and Hudson Bay coasts. Nesting was previously reported only as far north as the Wawa area (Peck and James 1983).

Only the Luther Marsh square had an estimate of more than 100 pairs. However, estimates of 11 to 100 pairs were provided for the square containing the Mountsberg Conservation area near Guelph, 2 squares along the Toronto waterfront, and 7 squares along the St. Lawrence River from Kingston eastward. Abundance estimates indicate that the Gadwall is fairly common in squares where it was reported in southern Ontario, while it is rare in northern Ontario. If the trend of the past 3 decades continues, it is likely to become more widespread in both the south and the north.-- *A. Sandilands*

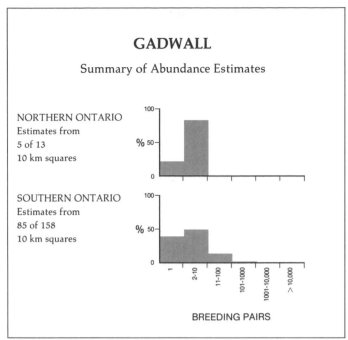

GADWALL

Summary of Abundance Estimates

NORTHERN ONTARIO
Estimates from
5 of 13
10 km squares

SOUTHERN ONTARIO
Estimates from
85 of 158
10 km squares

BREEDING PAIRS

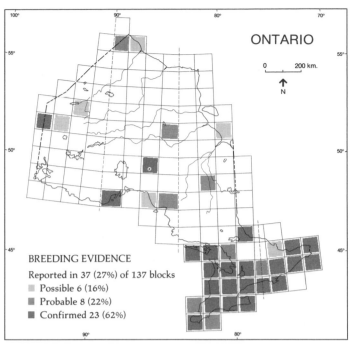

ONTARIO

0   200 km.

N

BREEDING EVIDENCE
Reported in 37 (27%) of 137 blocks

Possible 6 (16%)
Probable 8 (22%)
Confirmed 23 (62%)

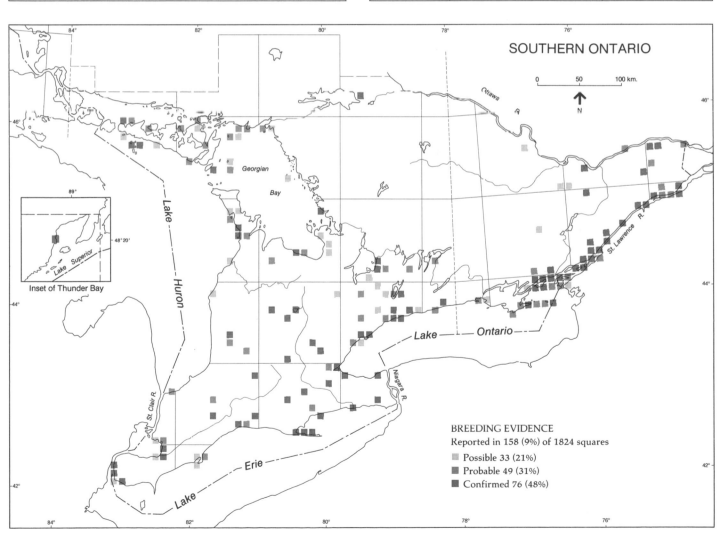

SOUTHERN ONTARIO

0   50   100 km.

N

Inset of Thunder Bay

BREEDING EVIDENCE
Reported in 158 (9%) of 1824 squares

Possible 33 (21%)
Probable 49 (31%)
Confirmed 76 (48%)

AMERICAN WIGEON
Canard siffleur d'Amérique
Anas americana

The American Wigeon nests throughout much of North America but predominantly west of Hudson Bay. It is one of the more northerly of the dabblers, ranging from the arctic to the southern prairies. In Ontario, it is most abundant along the James and Hudson Bay coasts, but also occurs sporadically in southern Ontario. Its winter range extends from southern Canada to northern South America.

The American Wigeon nests in a variety of habitats, but usually in upland sites associated with water. Islands are often selected for nest sites, but it also nests in bogs, marshes, meadows, and occasionally open coniferous woods. Normally, nests are 15 to 50 m from water, but they have been found almost in water and as far as 400 m away. Almost invariably, the nest is in a thick clump of vegetation or at the base of a shrub. Bellrose (1980) concluded that the American Wigeon was sparsely distributed within the Boreal Forest zone despite the presence of a large number of water bodies. He estimated that there were only 0.5 to 0.6 birds per km^2 in the Boreal Forest zone as far east as Ontario. In northwestern Ontario, his estimate was 0.04 American Wigeon per km^2. Its avoidance of the Boreal Forest zone is apparent in Ontario, where its distribution is primarily along the northern coasts and in the agricultural south.

The American Wigeon has increased and expanded its range in Ontario. No data are available to determine if numbers in the Hudson Bay area have changed, as this population went largely unnoticed until recently (Smith 1944, Manning 1952). In the remainder of the province, no nests were recorded until 1934 (Baillie 1960). It is now thinly distributed in agricultural areas, and is usually associated with the shorelines of the Great Lakes, inland impoundments, and sewage lagoons. Once a core population becomes established, the species often becomes locally common. The North American population of the American Wigeon has fluctuated greatly in the past 3 decades. Peak numbers occurred in 1970 and a low in 1963. All surveys indicate that the overall population has experienced a slight decline (Bellrose 1980).

The American Wigeon is not particularly difficult to confirm as a breeding species. Although pair bonding may begin on the wintering grounds, it often does not occur until the nesting area is reached, so birds are often conspicuous throughout May. Drakes are not particularly dutiful, with 60% leaving the hen during the first week of incubation and the remainder departing during the following week (Bellrose 1980). Broods stay within the vicinity of the nest site, and the hen accompanies her young until they are almost fully grown. Broods are likely to be seen from late June into August. Seventy-eight percent of southern Ontario breeding confirmations, and 100% of northern confirmations, were of broods. Bellrose (1980) stated that nests were extremely hard to find, and Peck and James (1983) reported that only 21 nests had been found in Ontario. On islands at Luther Marsh, the dragged rope technique was effective, with 7 nests being discovered in 4 days.

Atlas data have greatly improved our knowledge of American Wigeon distribution and abundance in the province. It was previously known to nest in the Hudson Bay area, Luther Marsh, Mud Lake, and in the vicinity of Toronto (Godfrey 1986). During the atlas period, breeding was confirmed in several blocks of the Hudson Bay Lowland, many of which are well inland from the coast. The pattern of southern Ontario breeding records, those in the Clay Belt, and the paucity of records in much of the Boreal Forest zone, show similarity with the atlas data for the Northern Pintail and the Green-winged Teal. South of the Hudson Bay Lowland, atlassers identified many previously unknown nesting sites and documented the range expansion of the American Wigeon into southeastern Ontario. The majority of recent breeding records in the province are associated with impoundments and sewage lagoons, and the southern Ontario population may continue to expand as a result of these new habitats.

Abundance estimates suggest generally more breeding pairs per square in northern Ontario than in the south, but sample sizes are small. Estimates, and the small number of records overall indicate that the American Wigeon is an uncommon breeding bird in Ontario.-- *A. Sandilands*

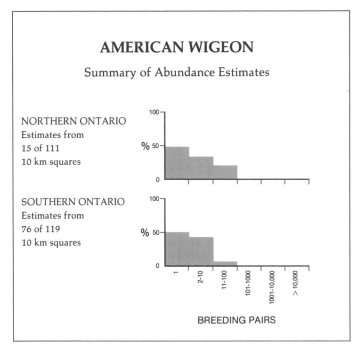

AMERICAN WIGEON

Summary of Abundance Estimates

NORTHERN ONTARIO
Estimates from
15 of 111
10 km squares

SOUTHERN ONTARIO
Estimates from
76 of 119
10 km squares

%

BREEDING PAIRS

1   2-10   11-100   101-1000   1001-10,000   > 10,000

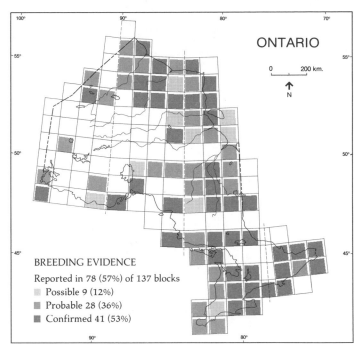

ONTARIO

0   200 km.

N

BREEDING EVIDENCE

Reported in 78 (57%) of 137 blocks

■ Possible 9 (12%)
■ Probable 28 (36%)
■ Confirmed 41 (53%)

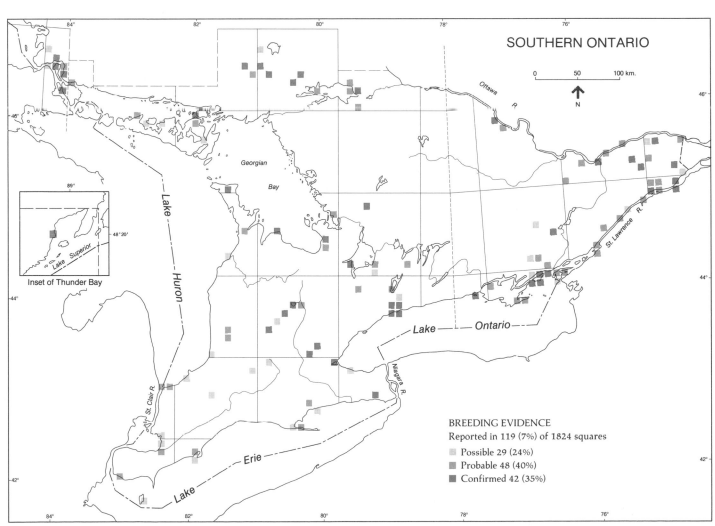

SOUTHERN ONTARIO

0   50   100 km.

N

Ottawa R.

St. Lawrence R.

Georgian Bay

Lake Huron

Lake Superior

Inset of Thunder Bay

89°

48°20'

Lake   Ontario

Niagara R.

St. Clair R.

Lake   Erie

BREEDING EVIDENCE

Reported in 119 (7%) of 1824 squares

■ Possible 29 (24%)
■ Probable 48 (40%)
■ Confirmed 42 (35%)

CANVASBACK
Morillon à dos blanc
Aythya valisineria

Juleana Hanke

The Canvasback nests only in North America, primarily in the western half of the continent, ranging from the Mackenzie River delta south through the prairies. In Ontario, it has been found nesting only in the southwest and northwest, at the extreme eastern extent of its range. It winters in the US and Mexico.

Canvasbacks typically choose small, shallow ponds for nesting, but prefer deep, permanent water bodies for feeding and courtship (Bellrose 1980). To be suitable for nesting, ponds must be lined with cattails or bulrushes. Almost all nests are placed in emergent vegetation above water, but they may also be on dry ground near the water's edge.

The historical breeding status of the Canvasback in Ontario is uncertain, as earlier authors did not consider it to be a nesting species. There are, however, reports of its nesting on Lake St. Clair as early as 1897 (Baillie 1962). Few species have had as much controversy over their breeding status as the Canvasback. Baillie (1962) added it to the list of Ontario breeding birds on the basis of records from Lake St. Clair in 1948, 1952, and 1953 in addition to the 19th century record. Peck (1976) and Peck and James (1983) considered it to be a hypothetical breeder, as there was no material evidence that it had ever nested in the province. Finally, a brood was photographed at Luther Marsh in 1983, to add it officially to the list of Ontario's breeding birds (James 1984b). Broods were also seen at Luther Marsh in 1965 (Goodwin 1965) and 1981, and adults were seen during the summers of 1957, 1973, 1974, 1976, and 1981 to 1983 (Sandilands 1984). Atlas data reveal that it also nested at Lake St. Clair in 1983 and at Berens Lake, 90 km north of Red Lake, in 1984. The Ontario population is so small that trends in numbers are meaningless. Overall, the Canvasback has not fared well, with a decline of 53% in North American population

between 1955 and 1974 (Bellrose 1980). Open hunting seasons are still held throughout North America, but with reduced bag limits. In years of extremely low populations, seasons were closed in the mid-1930's and early 1960's and 1970's.

Canvasbacks usually do not pair until they reach the nesting areas, so courtship behaviour may be observed in late April and early May. Nests are bulky and therefore may be relatively easy to find among emergent vegetation, although no nests have yet been found in Ontario. Drakes usually desert hens as soon as incubation starts, but they may remain in the general area for another 2 weeks. After fledging, the young may be led a considerable distance by the hen to another pond. Broods are very mobile, and, on the prairies, rarely spend more than a week on the same pond. Hens desert the young once they are about 3 weeks old. Confirmation is easiest when the hen is with the brood in late June to July.

During the atlas period, in addition to confirming the Canvasback in its historic nesting areas, atlassers found another breeding site in northwestern Ontario. It was not expected on the Canadian Shield because of its preference for relatively fertile marshes. Other studies, however, have reported it nesting in a variety of emergent vegetation such as bur-reeds and sedges in addition to cattails and bulrushes. The use of heath shrubs or sedges for nesting habitat may explain the one isolated nesting in northern Ontario. Although nests have never been found at Luther Marsh, the area in which the Canvasback has usually been seen suggests that it may nest in flooded leatherleaf.

The Canvasback is a very rare breeder in Ontario, with the total number of pairs probably numbering less than 10 a year. It is also very local, appearing mostly in larger marshes. Bellrose (1980) stated that this species has a strong instinct to return to its nesting and rearing areas. Therefore, successful breeding at Lake St. Clair, Luther Marsh, and elsewhere augur well for maintenance or even an increase of those populations.-- *A. Sandilands*

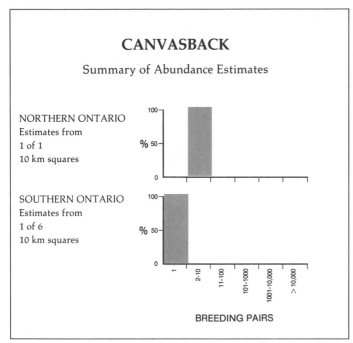

CANVASBACK

Summary of Abundance Estimates

NORTHERN ONTARIO
Estimates from
1 of 1
10 km squares

SOUTHERN ONTARIO
Estimates from
1 of 6
10 km squares

BREEDING PAIRS

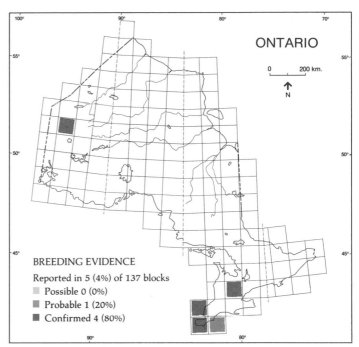

ONTARIO

0   200 km.

N

BREEDING EVIDENCE

Reported in 5 (4%) of 137 blocks

Possible 0 (0%)
Probable 1 (20%)
Confirmed 4 (80%)

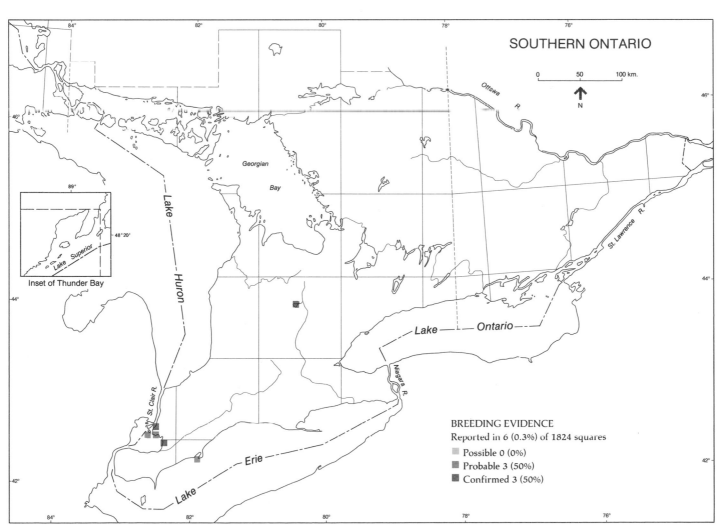

SOUTHERN ONTARIO

0   50   100 km.

N

Georgian
Bay

Lake Huron

Lake Ontario

Ottawa R

St. Lawrence R

Niagara R

Lake Erie

St. Clair R.

Inset of Thunder Bay
Lake Superior
48°20'
89°

BREEDING EVIDENCE

Reported in 6 (0.3%) of 1824 squares

Possible 0 (0%)
Probable 3 (50%)
Confirmed 3 (50%)

REDHEAD
Morillon à tête rouge
Aythya americana

The Redhead nests only in North America, and is most common on the prairies. In Ontario, it is primarily a migrant, but occurs in small numbers in the agricultural south during the breeding season. The winter range extends from the lower Great Lakes and the New England states south to Mexico, Guatemala, and parts of the West Indies.

Cattail or bulrush marshes and fens are preferred nesting sites, although the species will nest on islands and in meadows. Nests are usually close to water, and are often constructed in emergent vegetation above shallow water. The maximum distance a nest has been found from water is 266 m (Lokemoen and Duebbert 1973), but the majority are within 2 m (Bellrose 1980). The Redhead is more particular about the vegetation it nests in than are most species of ducks. On the prairies, hardstem bulrush is preferred, followed by cattails and sedges. In Ontario, leatherleaf is often a preferred species of vegetation for nesting. Because of its preference for bulrushes and cattails, it is found primarily in the south, where habitats containing these plants are more common.

The Redhead is semi-parasitic, and will occasionally lay eggs in other duck's nests. In Ontario, it has been reported parasitizing nests of Gadwall, Blue-winged Teal, Ring-necked Duck, and Lesser Scaup (Peck and James 1983). On the prairies, the Canvasback is the prime host, but Mallards, Cinnamon Teal, and Ruddy Ducks are also victimized (Bellrose 1980).

The Redhead appears to have undergone a slight population increase in Ontario. It bred in the Lake St. Clair area as early as 1877, and this is still by far the largest breeding population in the province. Very few Redheads nested on Lake St. Clair in 1949. A large increase took place when the Walpole Island marshes were dyked and pumped (H. Lumsden

pers. comm.). Since 1949, a number of new, small populations have arisen, most notably at Luther Marsh, Mud Lake near Port Colborne, and Lake St. Francis near Cornwall. New nesting locations were discovered during atlas field work at Mitchell, and at Cache Bay on Lake Nipissing. Additionally, a pair of birds were observed near Red Lake in northern Ontario, where it is unknown as a nesting species. In the Long Point area, it first appeared as a breeding bird in the 1960's, and its increase in population in this area may have been assisted by an introduction attempt in 1976 (McCracken *et al.* 1981). Overall, the Redhead population has never been strong in Ontario.

From an atlassing perspective, the Redhead is a moderately difficult species to confirm as a breeding bird. Caution must be exercised in assessing the significance of a sighting of an adult. The majority do not nest until their second year and only about half of the mature adults will nest in a given year. Most drakes desert the hen almost as soon as incubation starts, so the presence of a single drake may not be an indication that nesting has occurred in that area. Nests are not overly difficult to find, as they are usually among emergent vegetation, and some are quite bulky with elaborate ramps from the water to the nest. Nests were reported in 3 atlas squares. The species is easiest to confirm during the brood season, even though the hen usually deserts the brood when they are still at an early age; broods were reported in 13 squares. In nests that have been parasitized, young Redheads are obvious in the broods of other birds, being rather more yellow as downy ducklings than other species. Some caution is needed in identification because hens are easily confused with female Ring-necked Ducks.

Atlas data have improved our knowledge of the breeding distribution of the Redhead. Confirmed breeding was reported from the Lake St. Clair marshes, the Long Point marshes, Mud Lake, Presqu'ile, Mitchell, Luther Lake, Tiny Marsh, and the Cache Bay location mentioned above. Breeding evidence was reported elsewhere in new habitat created by sewage lagoons or impoundments. With the creation of these new nesting areas and the Redhead's ability to pioneer, it may become more widespread and common in Ontario.

With the exception of the Lake St. Clair area and Luther Lake, where abundance estimates were of 11 to 100 pairs per square, and the squares near Cornwall and Red Lake, each of which were estimated to contain 2 to 10 pairs, abundance estimates were of 1 breeding pair per 10 km square.-- *A. Sandilands*

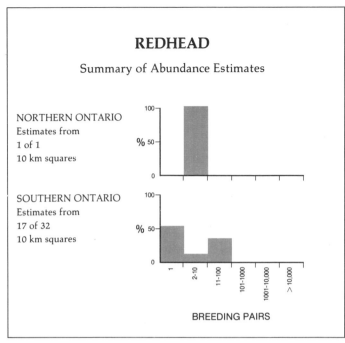

REDHEAD

Summary of Abundance Estimates

NORTHERN ONTARIO
Estimates from
1 of 1
10 km squares

SOUTHERN ONTARIO
Estimates from
17 of 32
10 km squares

BREEDING PAIRS

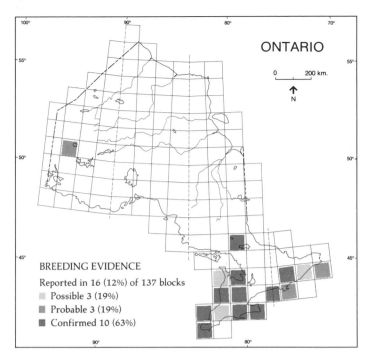

ONTARIO

0   200 km.

N

BREEDING EVIDENCE

Reported in 16 (12%) of 137 blocks

Possible 3 (19%)
Probable 3 (19%)
Confirmed 10 (63%)

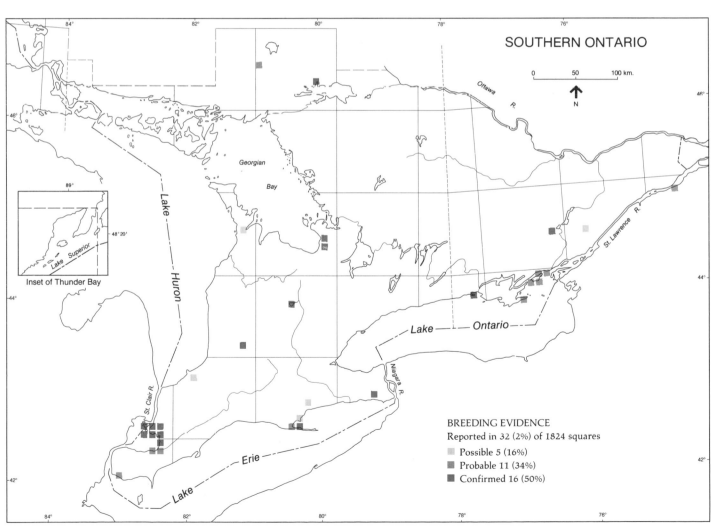

SOUTHERN ONTARIO

0   50   100 km.

N

Georgian

Bay

Lake

Huron

Lake Superior

48°20'

89°

Inset of Thunder Bay

Ottawa R.

St. Lawrence R.

Lake — Ontario

Niagara R.

St. Clair R.

Lake — Erie

BREEDING EVIDENCE

Reported in 32 (2%) of 1824 squares

Possible 5 (16%)
Probable 11 (34%)
Confirmed 16 (50%)

RING-NECKED DUCK
Morillon à collier
Aythya collaris

Widespread south of the tree-line in Canada, the Ring-necked Duck also breeds in states bordering the Great Lakes and in the northeastern US. In Ontario, the species commonly breeds in the Canadian Shield and Clay Belt regions and occurs regularly on the Hudson Bay Lowland. The species is among the most abundant breeding ducks in central and northern Ontario, but rarely breeds in the agricultural areas of southwestern and southeastern Ontario. Although present in a wide range of habitats, the Ring-necked Duck is essentially a species of freshwater sedge-meadow marshes or bogs, frequenting shallower water and denser vegetation than other diving ducks.

The Ring-necked Duck has undergone a rapid eastward expansion in its breeding range. Prior to the 1930's, the species nested in the west and mid-west, with northwestern Ontario and Michigan constituting the eastern limit of its range, except for localized records in northern Maine and New Brunswick (Mendall 1958). During the next 2 decades, nesting birds were recorded in the Thunder Bay District, the Abitibi Lake area, Muskoka, Algonquin Prov. Park, the Ottawa River valley, and the marshes north of Lake Ontario (Baillie and Harrington 1936). Beavers create good Ring-necked Duck habitat. An increase in beaver numbers, and loss of habitat in the southern portion of the range due to drought and agricultural drainage, may be factors in the rapid increase in breeding numbers in wooded habitat in Ontario. Although the species was common in eastern Ontario in the past, atlas surveys found only sporadic evidence of breeding. Stricter controls on beaver populations and extensive agricultural drainage may have contributed to their recent decline there.

In forested regions, Ring-necked Ducks favour small wetlands (less than 4 ha) at least partially surrounded by woody vegetation (D. McNicol unpubl.). Highly favoured are shallow sedge-marsh meadows, fens, or bays of lakes with abundant emergent and floating-leaved vegetation. Such habitats are rich in the aquatic invertebrates upon which ducklings feed, and afford protection for the incubating female and downy young. On the Hudson Bay Lowland, the species prefers scattered muskeg pools (R.K. Ross unpubl.). These small, often isolated habitats are not readily sampled by the atlasser. Early in the season, both sexes are conspicuous, with individuals or pairs often observed loafing in suitable nesting habitat. An unbalanced sex ratio (3 males to 2 females, Bellrose 1976) may inflate the frequency of possible and probable breeding records early in the season, when most atlassers are in the field. Additionally, since peak nest initiation occurs later than most ducks (late May, D. McNicol unpubl.), efforts to survey the species early in the breeding season rarely provide evidence of confirmed breeding. Females generally nest in natural depressions on islets of floating marsh plants or on clumps of marsh vegetation close to open water, and are secretive while incubating (Mendall 1958). The sites are difficult to locate, and, once hatched, the downy young are difficult to find amongst the dense vegetation.

The atlas maps confirm the broad distribution of the species in accordance with its previously reported breeding range. Discontinuities in breeding evidence, particularly in the Lac Seul and Lake Nipigon regions of northwestern Ontario, are inconsistent with historical records (Baillie 1950a) and probably reflect insufficient coverage of breeding habitat in hard-to-access areas. On the Hudson Bay Lowland, confirmed breeding would be difficult to obtain given the patchy distribution of the birds and censusing difficulties. Nevertheless, the absence of records on the northern Lowland suggests very low numbers or an absence of birds. In southern Ontario, most records are on the Canadian Shield, although, even there, records are sporadic away from the Algonquin Highlands and the Sudbury area north into the Timagami forest section. With more time, it seems certain that breeding could have been confirmed in all of the interior northern blocks and central Ontario squares where the species was found. Records of confirmed breeding at Luther Marsh and Tiny Marsh suggest suitable conditions for breeding in productive marshes of southern Ontario as well.

Random aerial surveys of breeding pairs conducted by the CWS, within the atlas period, confirm the uniform breeding distribution of the species, and indicate that atlas abundance estimates are generally too low. CWS data show relative densities ranging from 11 to 22 pairs/100 km^2 in northwestern Ontario, 13 to 21 pairs/100 km^2 in northcentral Ontario, 18 to 27 pairs/100 km^2 in northeastern Ontario, and 19 to 20 pairs/100 km^2 in southcentral Ontario; the species nests at relatively low densities in the southern Hudson Bay Lowland, ranging from 4 pairs/100 km^2 near Moosonee to 11 pairs/100 km^2 near Attawapiskat.-- *D. McNicol*

RING-NECKED DUCK

Summary of Abundance Estimates

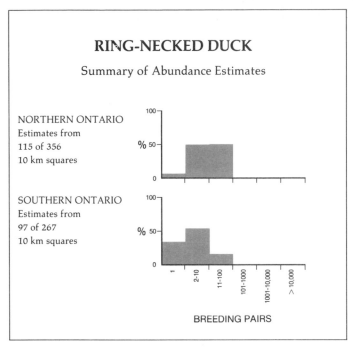

NORTHERN ONTARIO
Estimates from
115 of 356
10 km squares

SOUTHERN ONTARIO
Estimates from
97 of 267
10 km squares

BREEDING PAIRS

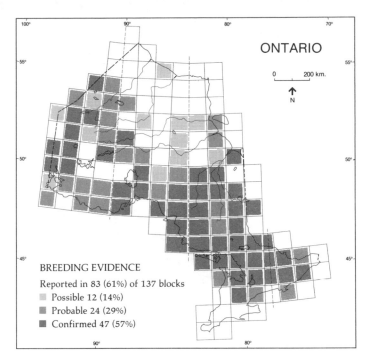

ONTARIO

0   200 km.

N

BREEDING EVIDENCE

Reported in 83 (61%) of 137 blocks

Possible 12 (14%)

Probable 24 (29%)

Confirmed 47 (57%)

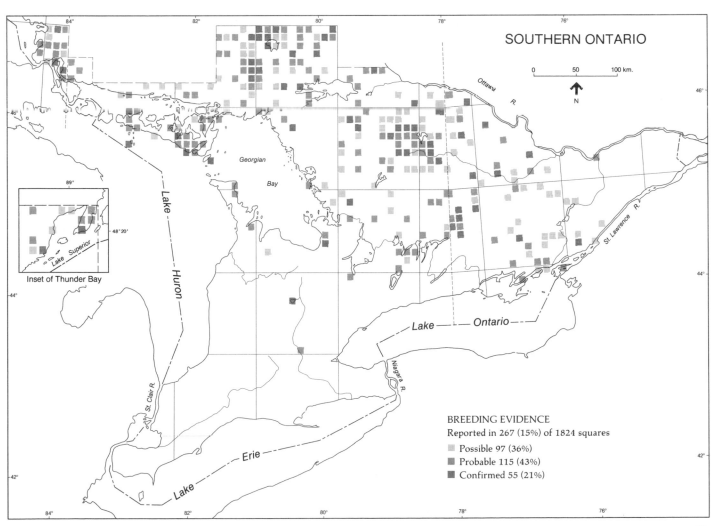

SOUTHERN ONTARIO

0   50   100 km.

N

Inset of Thunder Bay

BREEDING EVIDENCE

Reported in 267 (15%) of 1824 squares

Possible 97 (36%)

Probable 115 (43%)

Confirmed 55 (21%)

LESSER SCAUP
Petit Morillon
Aythya affinis

The Lesser Scaup breeds exclusively in North America, and is the smaller of the 2 species of scaup which are commonly called 'bluebill' or 'broadbill'. It has a wide distribution, occurring from western Labrador to Alaska. The greatest breeding density occurs in the Alaskan interior and the Northwest Territories. Densities gradually decline southeastward throughout the prairies until only scattered colonies of birds breed in the Great Lakes states and southern Ontario (Bellrose 1976). In Ontario, the highest densities are present on the Hudson Bay Lowland. Nesting habitat includes tundra ponds in the extreme north near the Hudson Bay coast, boreal wetlands farther inland, and the productive marshes inland from, and associated with, the southern Great Lakes. Nesting normally occurs on a dry site that is sheltered by grasses, sedges, or other plant cover, and is usually near water. Occasionally, nests are located on floating mats of vegetation.

The Lesser Scaup is said to be extending its breeding range eastward in Canada (Palmer 1976b). Baillie and Harrington (1936) suggested that the species nested regularly in the larger marshes along the southern border of Ontario east to Toronto. Because its habitat is relatively stable, continental Lesser Scaup populations have remained rather constant over the years, especially when compared with populations of other species such as the Redhead. It seems likely, therefore, that there has been little change in the status of the bird in Ontario in recent decades. Nevertheless, atlas records suggest that the Lesser Scaup did not breed in several southern Ontario locations where nesting had previously been confirmed (Peck and James 1983).

The Lesser Scaup is one of the species of ducks most likely to cause difficulties for atlas volunteers conducting field surveys. According to Palmer (1976b), a substantial number of yearling birds do not breed. In addition, Lesser Scaup have an unbalanced sex ratio heavily skewed to males; Bellrose (1976) suggests that the Lesser Scaup has the largest surplus of drakes of all the common game ducks. Many unpaired drakes also remain south of the normal breeding range as late as early summer. These non-breeding birds may be observed regularly in June along the Toronto waterfront as well as farther east along the Lake Ontario shore and the St. Lawrence River to the Quebec border. Observations of paired birds and lone drakes are seldom valid indicators of breeding in southern Ontario; only those considered to be in potential breeding habitat are shown on the maps.

Pairs and lone male Lesser Scaup are readily observed because they normally inhabit open water in wetlands with little emergent vegetation. Nests are started later than those of most ducks and the first broods will hatch in mid-June, while the last broods will hatch in early August. Thus, most sightings of broods can be made during the month of July. Because of the large number of pairs that are non-breeders, intentional nest searches in areas with a pair present are often fruitless. Nests with eggs or young were reported in only 2 squares; only 7 nests had ever been reported in the province prior to 1983 (Peck and James 1983). Broods, which generally use open water areas more often than areas with dense emergent vegetation, provide the best means to confirm breeding of Lesser Scaup, and were reported in 10 squares.

Atlas data indicate that the Lesser Scaup breeds widely but sparsely throughout the province. This was expected based on earlier sightings, but breeding had not previously been confirmed much north of the Albany River (Peck and James 1983). Records are scattered across northern Ontario, in no obvious pattern, suggesting a thinly spread population of a species that is difficult to find on its breeding grounds; it is likely that the Lesser Scaup breeds in a larger number of northern blocks than those in which it was reported. The very small number of confirmed records in the south is indicative of its rarity there. None of those confirmed records were in Great Lakes shoreline marshes.

Abundance estimates indicate that the Lesser Scaup is generally more common in squares in which it occurs in northern Ontario than it is in the south.-- *D.G. Dennis*

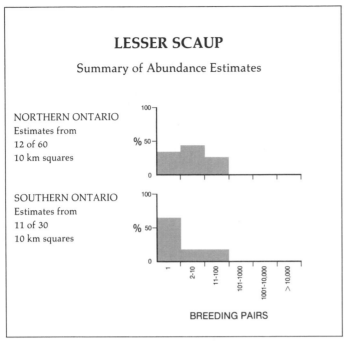

LESSER SCAUP

Summary of Abundance Estimates

NORTHERN ONTARIO
Estimates from
12 of 60
10 km squares

SOUTHERN ONTARIO
Estimates from
11 of 30
10 km squares

%

%

100

50

0

100

50

0

1

2-10

11-100

101-1000

1001-10,000

>10,000

BREEDING PAIRS

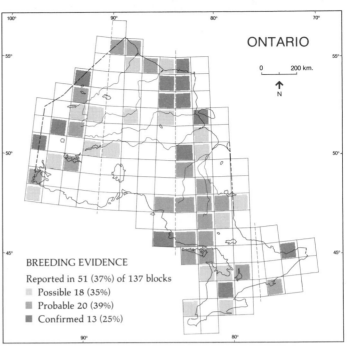

ONTARIO

0   200 km.

N

BREEDING EVIDENCE

Reported in 51 (37%) of 137 blocks

Possible 18 (35%)

Probable 20 (39%)

Confirmed 13 (25%)

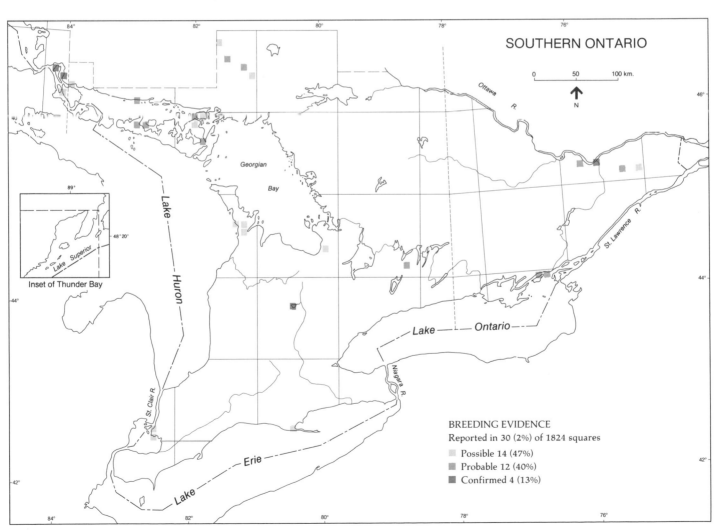

SOUTHERN ONTARIO

0   50   100 km.

N

Inset of Thunder Bay

Georgian Bay

Lake Huron

Lake Superior

Ottawa R.

St. Lawrence R.

Lake Ontario

Niagara R.

St. Clair R.

Lake Erie

BREEDING EVIDENCE

Reported in 30 (2%) of 1824 squares

Possible 14 (47%)

Probable 12 (40%)

Confirmed 4 (13%)

COMMON GOLDENEYE
Garrot à oeil d'or
Bucephala clangula

This conspicuous and hardy duck will winter in Ontario as far north as it can find open water. It is one of the few North American ducks which nests in tree cavities.

The Common Goldeneye, known to hunters as the 'whistler', is widely distributed across northern Eurasia and North America (AOU 1983). In Ontario, it breeds in the northern part of the Great Lakes-St. Lawrence Forest zone, in most of the Boreal Forest zone, and sporadically on southern areas of the Hudson Bay Lowland. The northern limit of breeding distribution is governed partly by the availability of trees over 30 cm in diameter with cavities of sufficient size to permit nesting. Females take broods to slow-flowing rivers and lakes with extensive shoal water up to 2 m deep, and with marshy points and margins. Highly favoured are stands of bulrush, growing in water 1 m deep. Such sites are rich in the larger, more mobile invertebrates on which ducklings feed.

Disturbance from cottagers has reduced breeding densities at the southern fringes of its range. Throughout its range in central Ontario, breeding densities are undoubtedly much lower than they were in pre-logging days. The Common Goldeneye no longer breeds on many lakes where riparian cutting has removed the older trees with cavities.

Common Goldeneye nest in cavities are often high (2 to 20 m), and difficult to access, as many dead aspens and birches are unsafe to climb. The incubating bird often sits tight and may not flush even when the tree is pounded. Sometimes down filaments stick to the entrance, indicating occupation. Nests with eggs or young were reported in 23 squares. There seems to be little preference for cavities in any particular tree species. In many areas, cavities are scarce and the Common Goldeneye may try to use, and may die in, summer cottage chimneys. It will readily use nest boxes. Nest

initiation dates are late in years of excessively high runoff and high water levels, and in years with late breakup. In the Elk Lake area near Englehart, first eggs were laid from 16 to 30 April, and the last nests were started between 17 and 27 May. Farther south, it may start nesting earlier. Renesting is very rare in Ontario after loss of the first clutch (H.G. Lumsden unpubl. data).

Common Goldeneyes do not normally nest until they are 2 years old. However, yearlings may prospect for nest sites in June, well after all breeding females have started to incubate. Seeing a female entering a cavity at this time does not necessarily indicate breeding. When they hatch, the ducklings remain in the nest cavity for at least 24 hours, during which time their coordination improves. They then jump, not climb, from the bottom of the cavity through the entrance and fall to the ground without injury (H.G. Lumsden unpubl. data). Broods may be highly mobile and travel as much as 20 km from the cavity in which they hatched. Isolated ducklings left behind by their mother en route may be found along lake shores and even in gravel pits. In suitable habitat, these deserted ducklings seem to survive quite well. Gang-brooding also occurs, with up to 26 ducklings following a single female. Broods comprised 87% of confirmed atlas records.

Virtually all breeding adults and subadults undertake moult migrations to the north, and pass the flightless period at the mouths of rivers and over shoal water on Hudson and James Bays. Common Goldeneyes seen north of 54^0N after mid-May may very well have been moult migrants. In southern Ontario, confirmed breeding records were reported from Lake Nipissing, in the Timagami Forest section north of Sudbury, along the shores of Manitoulin and adjacent islands, and north of Sault Ste Marie. Breeding was not confirmed farther south in the Algonquin, Parry Sound, Renfrew, and Glengarry Co areas, where the species has bred in the past (Peck and James 1983).

CWS surveys are basically in agreement with atlas data in suggesting that the Common Goldeneye is the second or third most abundant breeding duck in central and northern Ontario, though the absolute numbers in the atlas estimates may be too low. A summary of CWS surveys shows that it is most common in northwestern Ontario, where breeding densities of 45 pairs/100 km^2 were recorded in ground surveys. There were 25 pairs/100 km^2 on the Canadian Shield in northcentral Ontario, and 28 pairs/100 km^2 in the Clay Belt in northeastern Ontario (Dennis 1974b). South of the French and Mattawa Rivers, densities were much lower and were less than 2 pairs/100 km^2 (Dennis 1974a). Aerial surveys indicated 11 pairs/100 km^2 in the Sault Ste Marie area, and 2 pairs/100 km^2 on the Hudson Bay Lowland near Moosonee.-- *H.G. Lumsden*

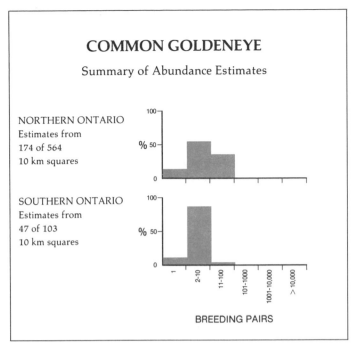

COMMON GOLDENEYE

Summary of Abundance Estimates

NORTHERN ONTARIO
Estimates from
174 of 564
10 km squares

SOUTHERN ONTARIO
Estimates from
47 of 103
10 km squares

% (vertical axis, 0 to 100)

BREEDING PAIRS

(horizontal axis: 1, 2-10, 11-100, 101-1000, 1001-10,000, > 10,000)

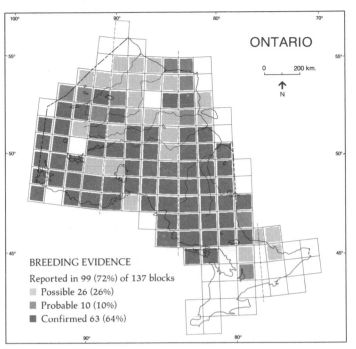

ONTARIO

0   200 km.

N

BREEDING EVIDENCE
Reported in 99 (72%) of 137 blocks

Possible 26 (26%)
Probable 10 (10%)
Confirmed 63 (64%)

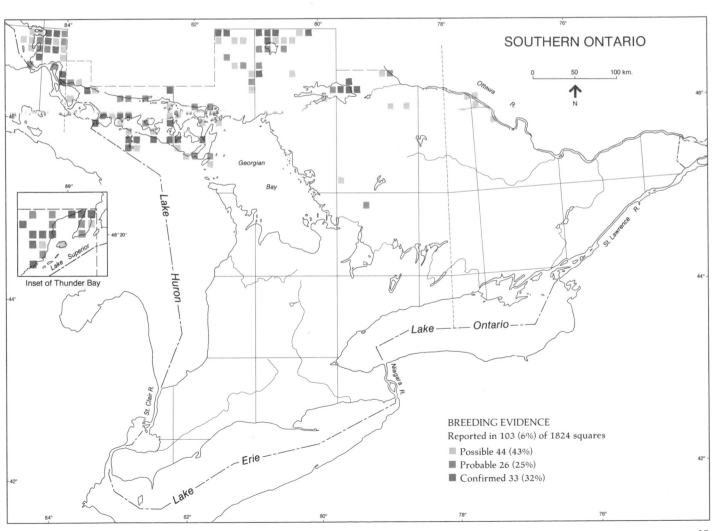

SOUTHERN ONTARIO

0   50   100 km.

N

Lake Huron

Lake Superior

Georgian Bay

Ottawa R.

St. Lawrence R.

Lake Ontario

Niagara R.

Lake Erie

St. Clair R.

Inset of Thunder Bay

BREEDING EVIDENCE
Reported in 103 (6%) of 1824 squares

Possible 44 (43%)
Probable 26 (25%)
Confirmed 33 (32%)

BUFFLEHEAD
Petit Garrot
Bucephala albeola

The Bufflehead is the smallest of the sea ducks, and one of the smallest of all North American waterfowl (Bellrose 1976). An indication of its diminutive size is that this little cavity-nester's favoured nest site is an old Northern Flicker hole (Erskine 1972).

The Bufflehead has a strictly North American distribution, breeding largely in western Canada and Alaska, north to the limit of trees. In Ontario, it has a broad range but its distribution is very scattered, with breeding being restricted to lake and pond habitats in the northern forested portions of the province. The Bufflehead winters from the Aleutian Islands, Great Lakes and Maritime Provinces south to Mexico and the coast of the Gulf of Mexico (Godfrey 1986).

Little evidence exists to suggest any changes in the Bufflehead population in Ontario. Erskine (1972) concluded that the continental population had declined through the 1800's and early 1900's, but winter populations on the Great Lakes increased from the 1920's to 1970's. A similar increase in winter populations was noted in the Toronto region between 1929 and 1976 (Goodwin *et al.* 1977).

The Bufflehead is a difficult species to confirm as breeding. It breeds at low densities in remote areas of the province, by lakes and ponds that seldom can be reached by canoe. Courting parties are conspicuous in late winter and early spring, and they also occur during spring migration. Although pairs and males on home ranges are conspicuous, they are present there for less than 3 weeks, allowing atlassers only a brief period to document nesting. In the south, most pairs later disappear without breeding.

Finding a Bufflehead nest is a rare occurrence in Ontario (Peck and James 1983), requiring dedicated pursuit and inspection of many cavities, or considerable good fortune. In 1983, a female was observed entering a presumed nest hole in a tree in a burned area along the Winisk River, near the Hudson Bay coast. The best of all periods to document breeding is the 8 week brood period, but secretive behaviour at this time and remoteness make this difficult too; only one brood was reported during the atlas period, that being of 5 young with a female in a small woodland pond near Lake Nipigon, in 1985.

Atlas data have only slightly improved our knowledge of the Bufflehead's breeding distribution. Two new confirmed breeding locations were found, and possible or probable breeding was reported from most areas where nesting had been shown previously (Peck and James 1983). The heart of the Ontario range, as reflected by the atlas maps, is in north-central and northwestern Ontario, extending northward onto the Hudson Bay Lowland in far northern Ontario. However, breeding birds of the Boreal Forest and Hudson Bay Lowland are undoubtedly underrepresented because of access problems, particularly because much of the coverage in the far north was from major rivers and coastal tundra habitats not conducive to breeding by this species.

Bellrose (1976) estimated 30,000 Bufflehead in the western half of Ontario, with a density of less than 0.1 bird/km^2. The few abundance estimates obtained during the atlas period tend to confirm this low-density, scattered distribution across northern Ontario.-- *K. Abraham*

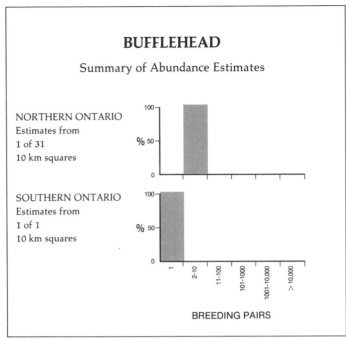

BUFFLEHEAD

Summary of Abundance Estimates

NORTHERN ONTARIO
Estimates from
1 of 31
10 km squares

SOUTHERN ONTARIO
Estimates from
1 of 1
10 km squares

% 100 50 0

% 100 50 0

1 2-10 11-100 101-1000 1001-10,000 >10,000

BREEDING PAIRS

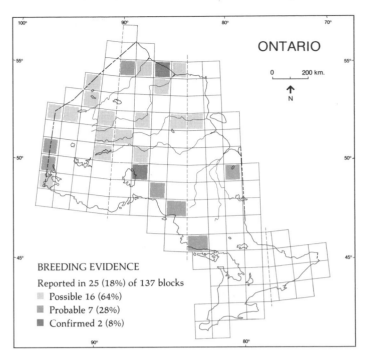

ONTARIO

0   200 km.

N

BREEDING EVIDENCE
Reported in 25 (18%) of 137 blocks

Possible 16 (64%)
Probable 7 (28%)
Confirmed 2 (8%)

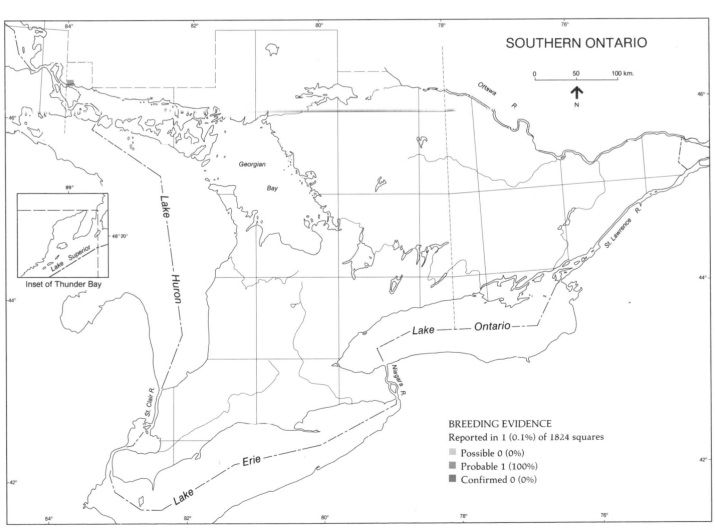

SOUTHERN ONTARIO

0   50   100 km.

N

Georgian Bay

Lake Huron

Lake Superior
48°20'
89°
Inset of Thunder Bay

Ottawa R.

St. Lawrence R.

Lake — Ontario

Niagara R.

St. Clair R.

Lake — Erie

BREEDING EVIDENCE
Reported in 1 (0.1%) of 1824 squares

Possible 0 (0%)
Probable 1 (100%)
Confirmed 0 (0%)

HOODED MERGANSER
Bec-scie couronné
Lophodytes cucullatus

The Hooded Merganser is the smallest of our 3 mergansers and the only one indigenous solely to North America. It nests in tree cavities in southeastern Alaska and across southern Canada, south to the northwestern and eastern parts of the US. In Ontario, the species is found primarily on secluded woodland ponds and rivers. James *et al.* (1976) considered the Hooded Merganser to be an uncommon local summer resident across the province north probably to James Bay. It winters largely to the south of Canada, as far as Mexico.

There appears to be little information indicating noticeable differences between the past and present distribution of the Hooded Merganser. Around the turn of the century, it had not been proved to breed in the province (nor in the Maritime Provinces or Quebec), but nesting was suspected in the southwest and in the Parry Sound and Muskoka areas (Macoun and Macoun 1909). Baillie and Harrington (1936) reported that the Hooded Merganser was not common, and that it was seldom recorded as a summer resident of Ontario; they surmised that it probably occupied the greater part of the province during the nesting season in suitable remote localities.

The Hooded Merganser is sparsely distributed, and is easily overlooked because it breeds on remote waterways. Furthermore, the timid and retiring nature of this duck makes it difficult to find, even in suitable habitat. It was found in less than one-third of the southern Ontario squares and in slightly over half of all blocks in the province. Single birds and pairs in breeding habitat represented 45% and 18% respectively of all records. The species is particularly hard to locate during the incubation period in May and June. The nest itself is located at a height of 6 to 10.5 m in a tree cavity or, if available, in a nest box, at the edge of, or over, water

(Peck and James 1983). The male deserts the nesting area during the long period of incubation, which averages approximately 31 days (Morse *et al.* 1969). During this period, the female spends most of her time on the nest, so the species is readily overlooked. In 253 visits to Hooded Merganser nests in southwestern Quebec, Bouvier (1974, unpubl.) found females on nests 191 times (75%). It is assumed that confirmed breeding data are underestimated because the female and her young are very secretive and remain well hidden in thick cover; the Hooded Merganser probably breeds in most squares where only possible or probable breeding was reported.

Although the maps may be incomplete, they give a reasonable picture of the breeding distribution of the Hooded Merganser. In southern Ontario, most records occurred on the Canadian Shield, with a definite concentration north of Kingston, presumably in an area of particularly suitable habitat. The density of records declines in the Lake Nipissing and French River areas, but increases again around Sudbury and north into the more coniferous and less disturbed Timagami Forest section. The species was found sporadically in the predominantly agricultural areas east and south of the Canadian Shield, south to Rondeau Prov. Park. The cluster of squares around the Guelph area is notable, and may be at least partly the result of an active Wood Duck nest box program there. In northern Ontario, the Hooded Merganser occurred mainly within the Great Lakes-St. Lawrence Forest zone, extending into the southern section of the Boreal Forest zone mainly in the Clay Belt, where breeding was confirmed in most blocks. There are only a few widely scattered records north of 50°N, the northernmost confirmed breeding occurring at about 52°N in the vicinity of the Berens River.

Abundance estimates suggest that the Hooded Merganser is more common within its range in northern Ontario than it is in the south. Only one square south of the Shield was estimated to contain more than 10 pairs, that being the square containing the Mountsberg Conservation Area near Guelph. Most estimates of more than 10 pairs were from squares in the area between Sudbury, Sault Ste Marie, Chapleau and Kirkland Lake.-- *J. Bouvier*

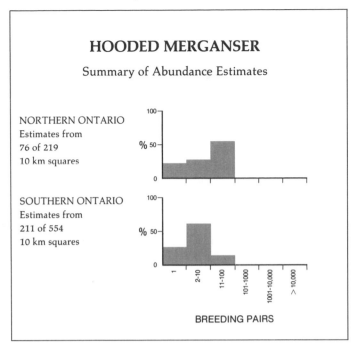

HOODED MERGANSER

Summary of Abundance Estimates

NORTHERN ONTARIO
Estimates from
76 of 219
10 km squares

SOUTHERN ONTARIO
Estimates from
211 of 554
10 km squares

BREEDING PAIRS

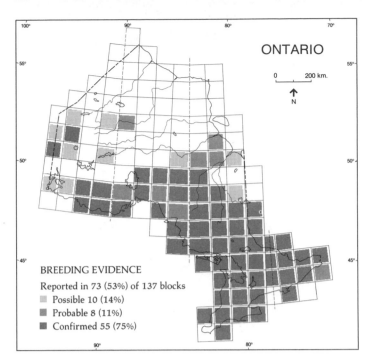

ONTARIO

0   200 km.

N

BREEDING EVIDENCE
Reported in 73 (53%) of 137 blocks

Possible 10 (14%)
Probable 8 (11%)
Confirmed 55 (75%)

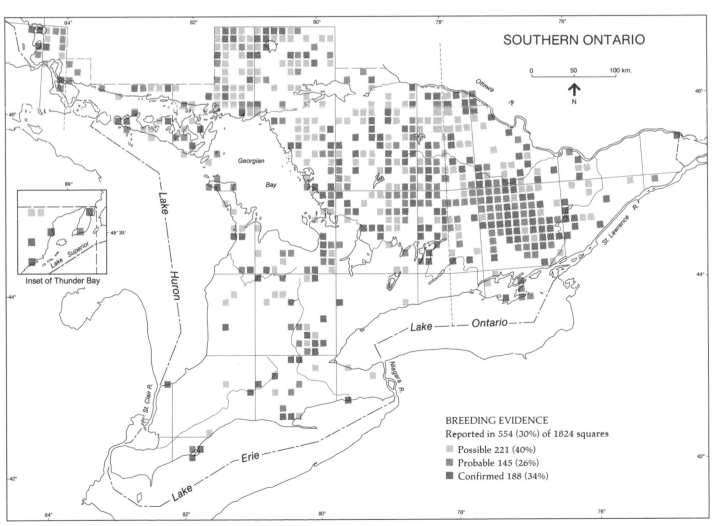

SOUTHERN ONTARIO

0   50   100 km.

N

Inset of Thunder Bay

BREEDING EVIDENCE
Reported in 554 (30%) of 1824 squares

Possible 221 (40%)
Probable 145 (26%)
Confirmed 188 (34%)

COMMON MERGANSER
Grand Bec-scie
Mergus merganser

The Common Merganser has a holarctic distribution which overlaps the Boreal Forest and montane regions of Eurasia and North America. It is among the most common breeding ducks in central and northern Ontario. The species nests regularly on the Hudson Bay Lowland but infrequently in southwestern and southeastern Ontario. In summer, this fish-eating species favours the cool, clear, freshwater lakes, rivers, and streams of the Canadian Shield. A large number winter on the Great Lakes.

Scattered records in southern Ontario counties, including Lambton (1936), Simcoe (1920 and 1935), Peterborough (1897), and Frontenac (1903 and 1906) (Baillie and Harrington 1936), suggest that isolated breeding outside its normal range is a regular occurrence. However, even on the Canadian Shield, human activity may be gradually impinging on the breeding range of this species in Ontario. Destruction of nest trees and shoreline cover by logging and cottage development, disturbance by power boats, and reduction of fish stocks by excessive fishing or as a result of environmental pollution may severely restrict the southern distribution of this species in the future.

The Common Merganser is an early spring migrant, often returning, already paired, at first signs of breakup (Bent 1923). The female selects the nesting location, favouring cavities in trees and snags but sometimes using crevices in cliffs and well-protected sites beneath boulders and brush (Bellrose 1976). The species readily accepts nest boxes and occasionally nests in chimneys. Nests are usually located adjacent to clear lakes or running water but may be located along small tributaries far from main waterways. Common Mergansers hatch in mid-June, with the young being capable of diving soon afterwards. The shallow, near-shore areas of large undisturbed lakes and rivers provide good visibility and abundant

fish for food, and overhanging vegetation provides necessary protection for the preflight young. Mixing of broods occurs regularly, with one or more females often attending large groups of 20 or more ducklings. Compared to many other ducks, the Common Merganser is readily observed by atlassers. If disturbed, both adults and young will 'run' on the surface of the water, seldom moving ashore for protection. Their preference for fishing in shoal areas of large bodies of water, often in large groups, combined with the tendency of broods to move progressively downstream, makes them conspicuous during the peak atlassing period in June and July. In hard-to-access regions of northern Ontario, confirmed breeding records were common along major lake and river canoe routes. Broods constituted 89% of all confirmed records.

The maps illustrate the broad distribution of the Common Merganser throughout the Boreal Forest and Great Lakes-St. Lawrence Forest regions, and confirm its status as a regular breeder on the Hudson Bay Lowland. While most earlier distribution maps (Godfrey 1966, Bellrose 1976, Palmer 1976b) exclude the species from the Hudson Bay Lowland, breeding evidence was reported in all but a few of the blocks in that area during the atlas period. It seems certain that, with more time, confirmed breeding would be obtained ubiquitously there. The species is common and widespread in areas adjacent to the Great Lakes north of 44^0N as well as on most large rivers and lakes in the interior, occasionally occupying narrow rivers and small lakes. The southern limit of its main breeding range extends along the southern edge of the Canadian Shield and the Bruce Peninsula through the Kawartha Lakes to the Ottawa River, east of Ottawa. Nesting is sporadic outside this range; breeding was confirmed along the St. Lawrence and Rideau River systems, and on several large rivers in southwestern Ontario. A female with 5 young on the south beach at Rondeau Prov Park on 30 June 1984, and a female with 7 young at the mouth of Clear Creek (Kent Co) on 13 June 1982, constitute the 2 most southerly confirmed breeding records.

Random aerial surveys of breeding pairs, conducted by the CWS within the atlas period, indicate that the species nests at relatively low densities on the Hudson Bay Lowland (2 to 3 pairs/100 km^2) (R.K. Ross unpubl.). Estimates from the Clay Belt at Cochrane and from the Kirkland Lake area are higher at 6 and 8.33 pairs/100 km^2, respectively. Elsewhere in the Canadian Shield of northeastern and southcentral Ontario, breeding densities are relatively uniform, ranging from 8 to 10 pairs/100 km^2. The highest densities in the province were recorded in northwestern Ontario at Kenora (28 pairs/100 km^2) and Red Lake (26 pairs/100 km^2), where large numbers of relatively undisturbed lakes teeming with fish still abound. Atlas abundance estimates are generally in agreement with the CWS survey values.-- *D. McNicol*

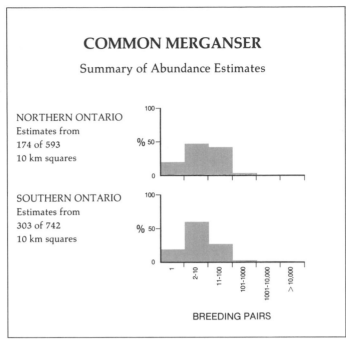

COMMON MERGANSER

Summary of Abundance Estimates

NORTHERN ONTARIO
Estimates from
174 of 593
10 km squares

SOUTHERN ONTARIO
Estimates from
303 of 742
10 km squares

%

BREEDING PAIRS

1 2-10 11-100 101-1000 1001-10,000 >10,000

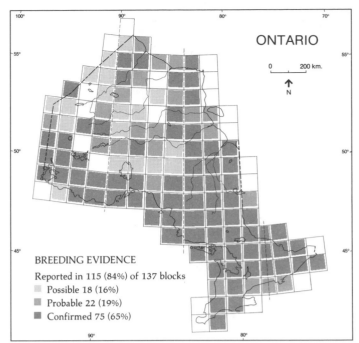

ONTARIO

0   200 km.

N

BREEDING EVIDENCE
Reported in 115 (84%) of 137 blocks
Possible 18 (16%)
Probable 22 (19%)
Confirmed 75 (65%)

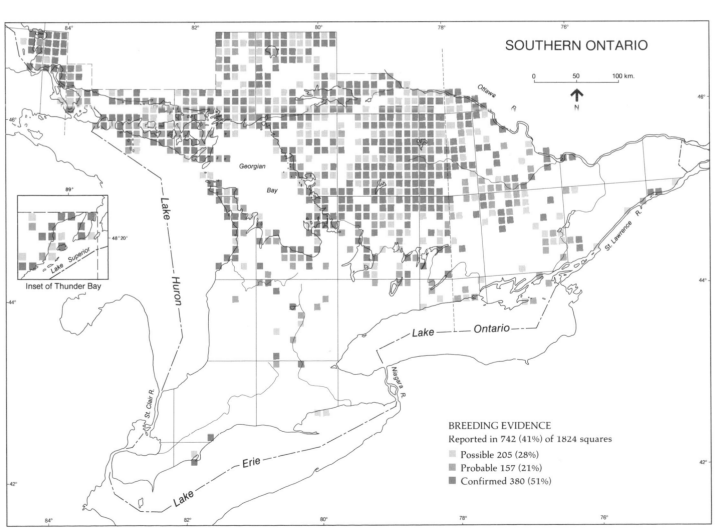

SOUTHERN ONTARIO

0   50   100 km.

N

Inset of Thunder Bay

Lake Superior
48°20'
89°

Lake Huron

Georgian Bay

Ottawa R.

St. Lawrence R.

Lake — Ontario

Niagara R.

St. Clair R.

Lake — Erie

BREEDING EVIDENCE
Reported in 742 (41%) of 1824 squares
Possible 205 (28%)
Probable 157 (21%)
Confirmed 380 (51%)

RED-BREASTED MERGANSER
Bec-scie à poitrine rousse
Mergus serrator

A holarctic species, the Red-breasted Merganser breeds from Eurasia, across Iceland, Greenland, and North America. On this continent, it breeds from the Atlantic Coast to the Pacific Coast, and from the Arctic Ocean south through the Boreal Forest to the northern parts of the Prairie Provinces, and to the Great Lakes and the St. Lawrence River. In Ontario, the Red-breasted Merganser has a widespread but sporadic distribution, nesting from eastern Lake Ontario north to the coasts of Hudson and James Bays. It winters primarily along the Atlantic coast. Although it has been known to nest in a wide variety of habitats (Bellrose 1976), in Ontario the Red-breasted Merganser exhibits a decided preference for rocky islands or shorelines of lakes and rivers (Peck and James 1983).

Historical changes in the status and distribution of this species in Ontario are difficult to determine. McIlwraith (1894) considered the Red-breasted Merganser to be "rather more numerous than [the Common Merganser]." This is certainly not the case today and suggests that several of the early ornithologists may have confused these similarly plumaged species. Although Baillie and Harrington (1936) acknowledged that little was known about the breeding of the Red-breasted Merganser in Ontario, they believed it to be "decidedly less common than the American [Common] Merganser and ... apparently of somewhat more northern distribution." This situation generally holds true today, and it appears unlikely that the species has undergone any significant population or range changes during the past century.

There are several aspects of the breeding ecology of this species which strongly influence its 'atlassability'. Pair bonds have usually been established prior to the birds' arrival on their breeding ground (Bellrose 1976). During the several weeks which elapse between the selection of a nest site and the onset of egg-laying, males and females are frequently seen together. The observation of pairs accounted for 31% of all atlas records. Males desert their mates soon after incubation begins, usually in late June, congregating in large offshore rafts for the duration of the summer. They undergo a moult of their flight feathers and attain eclipse plumage, thereby closely resembling females. One-year-old males may also be in a female-like plumage. During much of the breeding period, therefore, it is extremely difficult for observers to distinguish between the sexes of this species, and there is also the possibility of confusion between female Red-breasted and female Common Mergansers.

Attaining confirmed breeding status for this species was particularly difficult. Unlike its cavity-nesting relative, the Common Merganser, the Red-breasted Merganser builds its nest on the ground. It almost never occupies open sites, preferring instead to situate its nest under dense shrubbery, rocks, or driftwood (Strong 1912, Bengston 1970). These well concealed nests are extremely difficult to locate; only 4% of all atlas records involved the discovery of nests. The Red-breasted Merganser nests considerably later than the Common Merganser. Three-quarters of all Common Merganser nests typically hatch before the first Red-breasted Merganser nest hatches (Bellrose 1976), so merganser broods observed in August are probably Red-breasteds. Fledged young account for 87% of confirmed breeding records.

Atlas data show that the species is widely distributed across Ontario. It was reported in almost every block on the Hudson Bay Lowland, but only sporadically on the northern Canadian Shield. As coverage in both areas was concentrated along waterways, it is safe to assume that the Red-breasted Merganser currently breeds at higher densities on the Lowland than on the northern Shield. Farther south, records are concentrated along the shores of Lake Superior, and in southern Ontario there is an obvious preference for the shores of Lake Huron, including Georgian Bay, along which there are numerous islands suitable for nesting. The presence of breeding birds on Lake Nipissing and in the Thousand Islands area also suggests a preference for extensive bodies of water with rocky islands. Elsewhere, records are scattered, mostly along the Great Lakes shorelines. Some of these locations, such as Long Point (McCracken *et al.* 1981), have historical breeding records of Red-breasted Mergansers.

The highest abundance estimates (more than 100 pairs per square) were from the Thunder Bay area and northern Lake Superior, but high estimates (more than 10 pairs per square) were provided for a few squares in northern Georgian Bay and along the shore of Manitoulin Island.-- *D.M. Fraser*

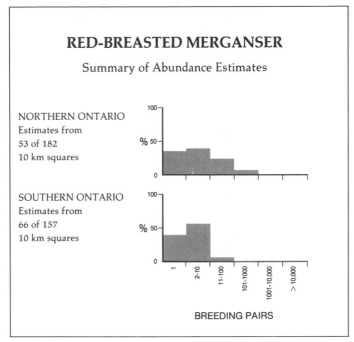

RED-BREASTED MERGANSER

Summary of Abundance Estimates

NORTHERN ONTARIO
Estimates from
53 of 182
10 km squares

SOUTHERN ONTARIO
Estimates from
66 of 157
10 km squares

BREEDING PAIRS

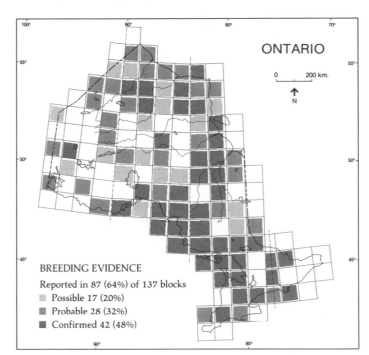

ONTARIO

0   200 km.

N

BREEDING EVIDENCE
Reported in 87 (64%) of 137 blocks

Possible 17 (20%)
Probable 28 (32%)
Confirmed 42 (48%)

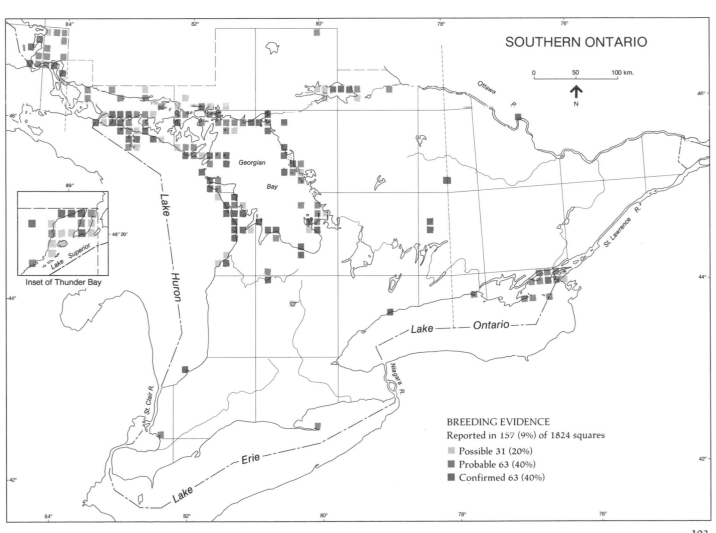

SOUTHERN ONTARIO

0   50   100 km.

N

Inset of Thunder Bay

BREEDING EVIDENCE
Reported in 157 (9%) of 1824 squares

Possible 31 (20%)
Probable 63 (40%)
Confirmed 63 (40%)

103

RUDDY DUCK
Canard roux
Oxyura jamaicensis

Indigenous to the Americas, this colourful species is slowly expanding its range in North America, and has been introduced in Britain. Primarily a 'prairie' species, the Ruddy Duck is normally found in, and was formerly confined to, the western portions of North America. It now breeds from central Mexico northward to the Great Plains, with scattered occurrences north to the subarctic (Bellrose 1976), and east to the Maritime Provinces (Godfrey 1986). While preferred habitat throughout its range is sloughs in prairie grasslands and parkland, it can adapt to a wide variety of other 'open' habitats; mainly those associated with fresh water (*i.e.*, ponds, lakes, and lagoons with emergent vegetation). Nests are woven structures, attached to cattails, bulrushes, bur-reeds, and sedges. They are usually slightly elevated above shallow water, situated in the marshy borders of suitable water bodies. The Ruddy Duck winters primarily from southern British Columbia and the middle US southward.

The Ruddy Duck was not originally a native of eastern Canada. The first reported nestings, on the St. Clair Flats, were published in the 1880's, but they were based on somewhat sketchy details and were lacking specimen or photographic evidence (Baillie 1962). In 1948, a brood was observed in the Lake St. Clair area, near Mitchell's Bay in Kent Co, and the first authenticated breeding record was obtained in 1949 with collection of downy young at nearby Walpole island, Lambton Co (Lumsden 1951). Reports from the Lake St. Clair area persist even today, although breeding was not confirmed there during the atlas period. Subsequent nestings have come from Welland Co, Thunder Bay District, Luther Marsh in Wellington and Dufferin Counties (Baillie 1962), where it has become a regular breeder since the late 1950's, and Cranberry Marsh in Durham RM (Richards 1977). Although the Ruddy Duck is still widely scattered and irreg-

ular south of the Canadian Shield, it is more abundant and widespread than previously, partly because of the proliferation of sewage ponds and lagoons throughout the province.

Somewhat colonial by habit, the Ruddy Duck is still difficult to detect because of its shy, retiring nature; only during a short courtship display period is it obvious. This species tends to return in successive years to the same places, as long as major alterations to habitat are not evident. Egg-laying occurs from late May to late August, with most nests being occupied from early to late June (Peck and James 1983); nests were reported in only 2 squares during the atlas period; all other confirmations were of broods. The Ruddy Duck is sometimes known to lay eggs in the nest of the American Bittern, and in nests of other waterfowl.

Breeding evidence was established in 45 squares (with only 13 confirmed) representing 19 Ontario blocks. Records are most concentrated in the extreme southwest. Although specific details were not requested in that area, 5 records were in squares containing sewage ponds (*e.g.*, Bright's Grove, Blenheim, Forest). Away from the extreme southwest, most of the records are from sewage lagoons. These include the lagoons south of the Shield at Mitchell, Townsend, Markdale, Port Perry, Lindsay, Casselman, and Alfred, near Sudbury at Garson and Chelmsford. In northern Ontario, both records were from sewage ponds: Englehart in the east and Rainy River in the west. Nesting was confirmed at water impoundments at Luther Lake and Hullett Marsh (Huron Co), and at a more natural site (Wiltsie Lake) northeast of Kingston. Breeding was not noted at Thunder Bay during the atlas period, though the Ruddy Duck has nested there in the past (Peck and James 1983).

Because the atlas data summarize observations over a 5 year period, the maps show more records than would normally occur in any one year; there are only a few reliable breeding locations for the Ruddy Duck in Ontario. Of the 9 abundance estimates of 2 or more pairs per square, only one, at Luther Lake (where 11 to 100 pairs were estimated), is outside the extreme southwest.-- *J.M. Richards*

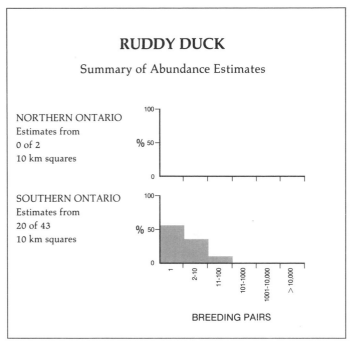

RUDDY DUCK

Summary of Abundance Estimates

NORTHERN ONTARIO
Estimates from
0 of 2
10 km squares

SOUTHERN ONTARIO
Estimates from
20 of 43
10 km squares

BREEDING PAIRS

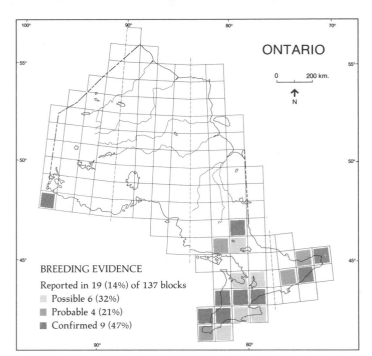

ONTARIO

0   200 km.

N

BREEDING EVIDENCE
Reported in 19 (14%) of 137 blocks
Possible 6 (32%)
Probable 4 (21%)
Confirmed 9 (47%)

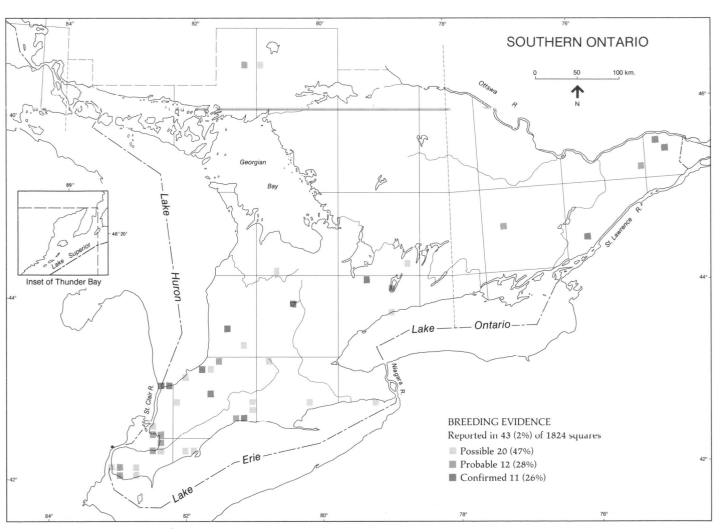

SOUTHERN ONTARIO

0   50   100 km.

N

Georgian Bay

Lake Huron

Lake Superior
48°20'
Inset of Thunder Bay

Ottawa R

St. Lawrence R.

Lake — Ontario

Niagara R.

St. Clair R.

Lake — Erie

BREEDING EVIDENCE
Reported in 43 (2%) of 1824 squares
Possible 20 (47%)
Probable 12 (28%)
Confirmed 11 (26%)

TURKEY VULTURE
Urubu à tête rouge
Cathartes aura

The Turkey Vulture is a wide-ranging New World species breeding from southern Canada to southern South America. Its winter range has recently expanded north to New Jersey and beyond (Brown and Amadon 1968), with a few birds being reported recently on Christmas Birds Counts in southern Ontario. In Canada, it nests primarily south of the Boreal Forest, with Ontario's population being concentrated in southern Ontario and between Thunder Bay and the Manitoba border. Bottomland hardwood forests and thickets and rocky cliffs (Jackson 1983) are the preferred nesting habitat, though various types of terrain, except heavy unbroken forest, are used (Godfrey 1986).

The Turkey Vulture has been undergoing a major range expansion into northeastern North America since the 1920's. Possible explanations include a long-term warming trend, increased food resulting from white-tailed deer expansion (and increases in deer road-kills), and deterioration of habitat in the southern part of its range (Wilbur 1983). In Ontario, it was common near Lake St. Clair as early as 1823 (Macoun and Macoun 1909). With the exception of the Lake of the Woods area, it was restricted to the southwestern counties until the late 19th century (McIlwraith 1886). It was first reported near Guelph in 1901 (Klugh 1905), in Prince Edward Co in 1944, and at Kingston in 1950 (Quilliam 1973). Snyder (1951) described it as casual north to Georgian Bay and the Orillia area. Turkey Vultures were reported more frequently north of Lake Simcoe during the 1950's and 1960's (Devitt 1967). The first Parry Sound record occurred in 1954 and the first breeding season record for the Magnetawan area was in 1981 (Mills 1981). Several atlas records of confirmed breeding near Sudbury represent a northern extension of the breeding range described by James *et al.* (1976). There was evidence of population increase and range expansion in the northwest in

the 1930's and 1940's (Snyder 1953). Atlas data show the Turkey Vulture to be more widespread to the north and west of Thunder Bay than indicated in previous reports (Godfrey 1986), but the low level of breeding evidence and the lateness of the season of a few of those records make it questionable whether all of the more northerly records in that area represent breeding birds.

There are few Ontario birds so readily identified at great distances, or which are more likely to be recorded if present. Conversely, nests are notoriously difficult to find; only 2% of all atlas records are of nests with eggs or young. The Turkey Vulture has been known to nest in caves, rock-piles, hollow logs and stumps, on the ground next to a log, or in thickets (Jackson 1983), and nests are usually in undisturbed sites. The birds visit the nesting site infrequently after the eggs are laid, changing incubating duties only every 24 hours, and feeding young only 2 or 3 times per day (Davis 1983).

The Turkey Vulture may not be as widespread a nesting bird as the southern Ontario atlas map suggests. Individual birds cover considerable distances when foraging, and are therefore likely to be recorded in squares other than their home square. Furthermore, non-breeding birds, including immatures, occur in Ontario during the breeding season, and these birds could be recorded as possible breeders. Although the northern range expansion is continuing, the distribution of the Turkey Vulture remains patchy, with conspicuous gaps in the extreme southwest, Prince Edward Co, east of the Shield in southeastern Ontario, and in the Algonquin Highlands. Curiously, except where the Niagara Escarpment provides extensive woods and, perhaps more importantly, cliffs, there are few records between Toronto and Stratford. The paucity of records in the southeast is difficult to explain given the concentration of records on the adjacent Frontenac Axis and Canadian Shield proper and the widespread use of agricultural land elsewhere. Southeastern Ontario is at the northeastern limit of the Turkey Vulture's breeding range in North America, so it may be that the species has simply not yet expanded into that area in numbers. In the Algonquin area, as in most of northern Ontario, the lack of agricultural land, which provides open country for foraging, may be responsible for the scarcity or absence of the species. If this is the case, the Turkey Vulture might be expected to continue to expand northward into the agricultural Clay Belt area. A Turkey Vulture reported in Winisk in 1983 was deemed to be a wandering non-breeder, as were birds reported previously from Fort Severn and Moose Factory (Baillie and Harrington 1936).

When considered in conjunction with previous data, the atlas results show that the Turkey Vulture's range is continuing to expand in Ontario, and suggest that the species has never been more common here.-- *M.D. Cadman*

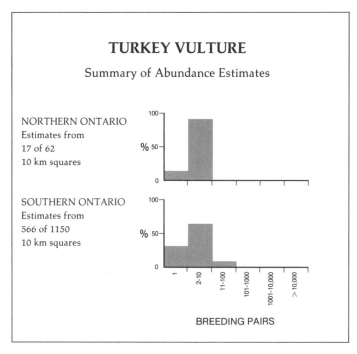

TURKEY VULTURE

Summary of Abundance Estimates

NORTHERN ONTARIO
Estimates from
17 of 62
10 km squares

SOUTHERN ONTARIO
Estimates from
566 of 1150
10 km squares

%

%

100

50

0

100

50

0

1 2-10 11-100 101-1000 1001-10,000 > 10,000

BREEDING PAIRS

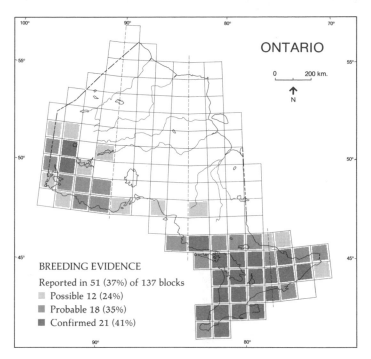

ONTARIO

0   200 km.

N

BREEDING EVIDENCE
Reported in 51 (37%) of 137 blocks
Possible 12 (24%)
Probable 18 (35%)
Confirmed 21 (41%)

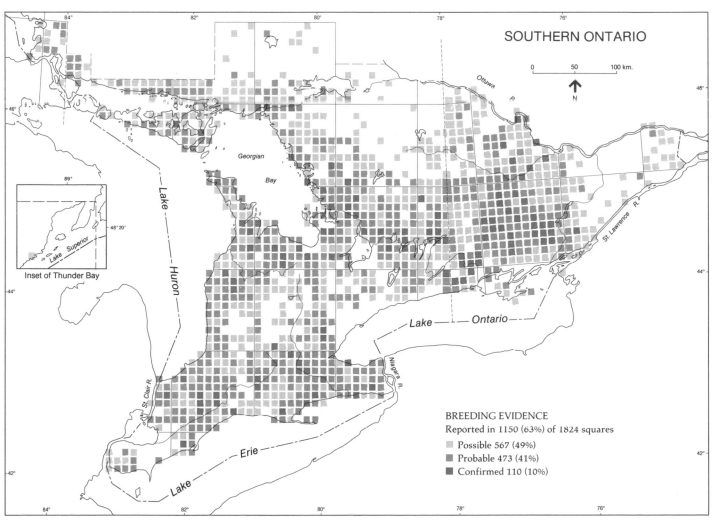

SOUTHERN ONTARIO

Georgian
Bay

Ottawa
R.

Lake
Huron

Lake
Superior

Inset of Thunder Bay

St. Clair R.

Lake   Erie

Lake

Niagara
R.

Lake — Ontario

St. Lawrence R.

0   50   100 km.

N

BREEDING EVIDENCE
Reported in 1150 (63%) of 1824 squares
Possible 567 (49%)
Probable 473 (41%)
Confirmed 110 (10%)

OSPREY
Balbuzard
Pandion haliaetus

The Osprey is the only species of the genus *Pandion*, and is named after the legendary King of Athens. To watch this magnificent bird plunge directly from great heights into water feet first and then rise to flight carrying a live fish is an exciting sight.

The species is cosmopolitan, breeding in the temperate or tropical regions of all continents except South America. It breeds generally throughout Canada, but is absent from the arctic and parts of the arid prairies. It nests throughout Ontario, as far north as the Hudson and James Bay coasts. Except in migration, the Osprey is invariably associated with lakes, rivers, or sea-coasts, since its principal food is fish. Its need for water determines its breeding distribution; in Ontario, it occurs most commonly in association with the lakes and rivers of the Boreal and the Great Lakes-St. Lawrence Forest zones south to the edge of the Canadian Shield. Ospreys winter in the extreme southern US and southward to Argentina, Paraguay, and Peru.

During the last century, although it was rarer than in the Maritimes, the Osprey was distributed throughout Ontario; it bred beside lakes and rivers in the less settled parts and was seldom found breeding west of Toronto in the south (McIlwraith 1894, Macoun and Macoun 1909). By the 1930's, additional field work showed it to be a frequent breeder from the east end of Lake Ontario north to James Bay and west to Manitoba (Baillie and Harrington 1936). By the 1960's, a serious decline in nest productivity occurred in the US, as a result of toxic chemical sprays which were washed into water courses, and ingested by fish that were then eaten by the Osprey. These DDT residues affected the estrogen hormone which controls calcium and egg shell thickness, resulting in thinner shells and broken eggs (Peakall 1970). Following tight restrictions on the use of DDT, pesticide resi-

dues declined in the breeding areas, and the number of Osprey young per occupied nest increased consistently between 1968 and 1981 (Tate and Tate 1982). Having been placed on the Blue List in 1972, the Osprey was removed from the list in 1981 (Tate 1981). The dramatic decline in Osprey reproductive success in the US was not noted in Canada, where overall numbers appear to have remained stable (Fyfe 1976); Fyfe considered its abundance in Ontario to be from low to medium in comparison to populations in other provinces.

Pairs are easily found when they return in April and May as they hunt over water, dive and swoop in courtship flights, or perch in a lookout tree. The nest is a massive structure usually placed in a tree near the water's edge. Where suitable trees are absent, the Osprey will use artificial nesting platforms as well as other fabricated structures such as hydro poles or towers. Occasionally, the nest tree may be a few hundred metres from the water. The same nest is occupied in successive years and additional sticks added. The male provides all the food for the female during the approximately 32 to 33 days of incubation, as well as for the entire family for an additional month while the young are in the nest. The young fledge about 52 or 53 days after hatching, return to the nest for feeding and roosting for another week, and can be found nearby for some time after that. Nests with young make up 40% of southern Ontario's confirmed records.

Given its conspicuousness, the Osprey was probably reported from the great majority of squares and blocks in which it occurred. Nevertheless, the absence of records from some of the far northern blocks is likely due to insufficient coverage; there is apparently suitable habitat throughout the north. It was reported in 112 of Ontario's 137 blocks, which is the most for any raptor, and was confirmed in 72 blocks, which is among the highest of any species. In southern Ontario, most records were from the Canadian Shield with its abundance of lakes; it was found in 607 squares, very few of which were southwest of a line joining Toronto and the southern end of Georgian Bay. Nest platforms on Guelph Lake, Luther Lake, and in the Kawartha Lakes area provided some of the most southerly confirmed breeding records. Its near absence on the Bruce Peninsula, which has many apparently suitable lakes, is difficult to explain. Confirmed breeding was reported in 43% of southern Ontario squares in which the species was recorded, which, among raptors, is exceeded only for the American Kestrel, the Red-tailed Hawk, and the Common Barn-Owl.

In squares for which abundance estimates were provided, the Osprey was considered to be uncommon in both northern and southern Ontario, although a higher percentage of southern Ontario squares were thought to contain only one pair. Four of 5 squares with estimates of over 10 breeding pairs are in the Haliburton-Bancroft area; the other is near Chapleau in the north.-- *R.D. Weir*

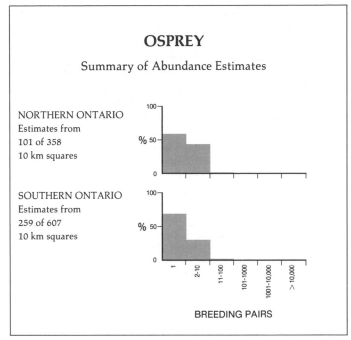

OSPREY

Summary of Abundance Estimates

NORTHERN ONTARIO
Estimates from
101 of 358
10 km squares

SOUTHERN ONTARIO
Estimates from
259 of 607
10 km squares

%

BREEDING PAIRS

1 2-10 11-100 101-1000 1001-10,000 >10,000

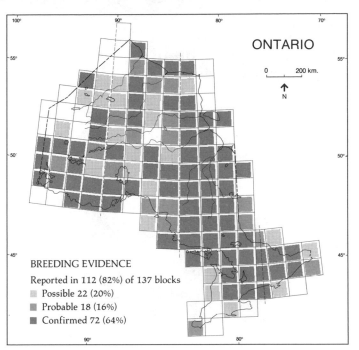

ONTARIO

0   200 km.

N

BREEDING EVIDENCE
Reported in 112 (82%) of 137 blocks

Possible 22 (20%)
Probable 18 (16%)
Confirmed 72 (64%)

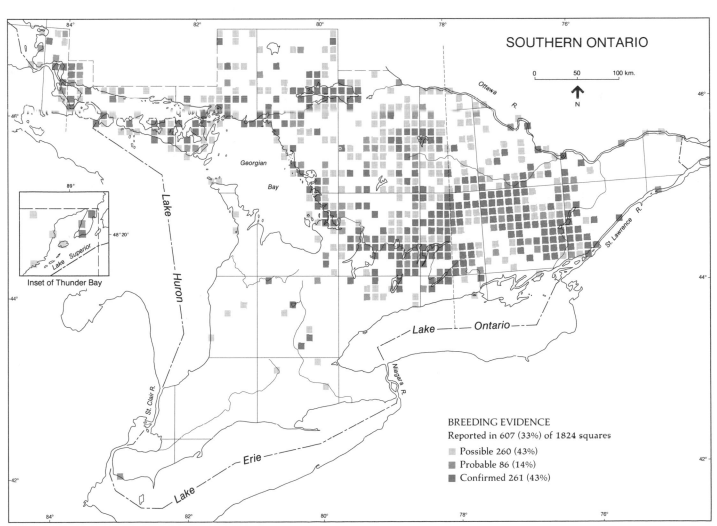

SOUTHERN ONTARIO

0   50   100 km.

N

Ottawa R.

Georgian Bay

Lake Huron

Lake Superior

Inset of Thunder Bay

89°

48°20'

St. Clair R.

Lake Erie

Niagara R.

Lake Ontario

St. Lawrence R.

BREEDING EVIDENCE
Reported in 607 (33%) of 1824 squares

Possible 260 (43%)
Probable 86 (14%)
Confirmed 261 (43%)

BALD EAGLE
Pygargue à tête blanche
Haliaeetus leucocephalus

The Bald Eagle nests throughout North America in a diversity of habitats in close proximity to water. It is primarily a fish-eater and hence is found in greatest abundance in areas of high fish productivity. Current populations, particularly in Ontario, are a fraction of former numbers over much of the range.

The Bald Eagle can now be found in near-pristine numbers only in Alaska, coastal British Columbia, boreal Canada, and parts of the Maritime Provinces. In southern Ontario, it was a common breeder in its preferred habitat up to the early 1950's (Weekes 1974). In the Rideau Lake country, C. Broley studied as many as 15 nests within a 32 km radius of his cottage (Broley 1952). In the next decade, southern and central Ontario's Bald Eagle population was reduced to fewer than 10 pairs. Shooting, habitat destruction, and reproductive failure caused by the pesticide DDT, were largely responsible for the decline. In an attempt to restore this population, 16 fledglings were released by CWS at Long Point on Lake Erie between 1983 and 1985. In northwestern Ontario, nesting density has remained high through present times. Although the northern birds also suffered pesticide-induced reproductive failure, the population has recovered rapidly since DDT was banned in 1972 (Grier 1982).

The Bald Eagle's large size and conspicuous adult plumage make it relatively easy to find. Adults are resident on their territories from early spring before ice-melt until the late fall. Large stick nests built in trees are used year after year. Where human disturbance is low, nest-trees are usually a few metres from shore and are easily visible from a boat; inland nests are considerably more difficult to find. Adults typically perch conspicuously at the very tops of trees near the nest. Otherwise, there are no predictable behaviours of undisturbed birds that help in locating nests. Generally, incubating birds lie flat and are not easily flushed, while parents with young vocalize and fly above human intruders.

Several factors may complicate atlassing for Bald Eagles. A single pair may alternately occupy 2 or more nests in successive years, possibly several kilometres away from each other. Also, breeding birds may use some lakes for foraging while nesting a few kilometres away. Bald Eagles occasionally occupy a territory, with or without a nest, but do not breed there (Gerrard *et al.* 1983), and other non-breeding eagles may congregate in areas of favourable habitat. Observation of young-of-the year does not imply local production, for immatures wander; birds hatched as far away as Florida may be seen in Ontario in the summer. These situations may explain why so much of the breeding evidence in southern Ontario is of possible or probable status.

The distribution of Bald Eagles in northwestern Ontario, studied extensively since 1966 (Grier 1985), is probably more continuous than illustrated. Eagle numbers in northeastern Ontario have not been studied. The distribution of birds in southern Ontario is accurate as shown for confirmed breeders, but possible and probable breeding evidence presented here may well not represent breeding at all. Reproduction of pairs in southern Ontario has been extremely low, with a total annual production of zero to 5 fledglings for the period 1969 to 1983 (McKeating 1985). Reproductive success seems to have improved since 1981. Current distribution across the province depends on the availability of high quality habitat, which must provide both protection from human disturbance and abundant uncontaminated prey.

Approximately 500 to 1,000 pairs of Bald Eagles nest in northwestern Ontario (Grier 1985). The density of breeding birds is greatest in the Lake of the Woods area, and declines dramatically northeast of Red Lake. In southern Ontario, the number of pairs is no doubt fewer than 2 dozen, and likely only half as many. Eagles are long-lived, so single pairs sighted year after year may be responsible for clusters of reports. In 1973, the Bald Eagle was classified by MNR as endangered in Ontario.-- *G.R. Bortolotti*

At the request of the Regional Coordinators, 5 records (2 confirmed, 2 probable, and 1 possible) in blocks 18UE and 18UF in southeastern Ontario are not shown on the southern Ontario map.

BALD EAGLE

Summary of Abundance Estimates

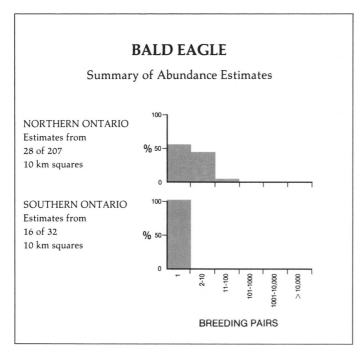

NORTHERN ONTARIO
Estimates from
28 of 207
10 km squares

SOUTHERN ONTARIO
Estimates from
16 of 32
10 km squares

%

BREEDING PAIRS

1 2-10 11-100 101-1000 1001-10,000 >10,000

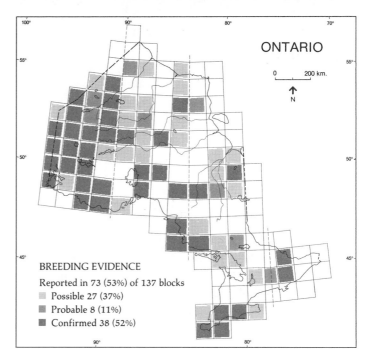

ONTARIO

0   200 km.

N

BREEDING EVIDENCE

Reported in 73 (53%) of 137 blocks

Possible 27 (37%)
Probable 8 (11%)
Confirmed 38 (52%)

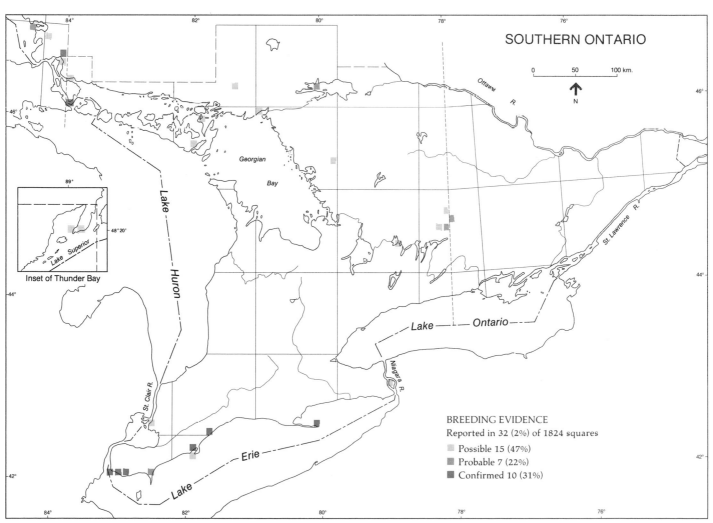

SOUTHERN ONTARIO

0   50   100 km.

N

Georgian Bay

Lake Huron

Lake Superior

Inset of Thunder Bay

St. Clair R.

Ottawa R.

St. Lawrence R.

Niagara R.

Lake — Ontario

Lake — Erie

BREEDING EVIDENCE

Reported in 32 (2%) of 1824 squares

Possible 15 (47%)
Probable 7 (22%)
Confirmed 10 (31%)

NORTHERN HARRIER
Busard Saint-Martin
Circus cyaneus

The Northern Harrier is a holarctic species, breeding in Eurasia from the British Isles and Mediterranean areas, east to Siberia. In the Americas, where, until recently, the species was known as the Marsh Hawk, it nests from western Alaska to Newfoundland, primarily south of the tundra, south to the south-central US. Its range includes all of Ontario. In Ontario, it frequents open country, hunting over marshes, bogs, meadows, and agricultural fields. In some parts of the province, it is a common breeding bird, and even nests within urban centres where suitable habitat remains. Although a few birds stay in southern Ontario during mild winters, the species is migratory and most move south in autumn to the US or as far as northern South America, returning in March and April.

The Northern Harrier probably benefitted greatly from the creation of new habitat with the clearing of forests. It appears always to have been common within Ontario during historical times (McIlwraith 1894, Macoun and Macoun 1909). Baillie and Harrington (1936) considered it one of the commonest and most widely distributed hawks in the province. However, in response to falling numbers, it has been on the Blue List since 1972; BBS data show a significant decline in eastern North America between 1965 and 1979 (Robbins *et al*. 1986). The species has been affected adversely by intensive agricultural development and land reclamation as marshes have been filled.

Pairs of Northern Harriers are easily spotted during the nesting season as they hunt over open terrain. The males return to nesting areas and begin their aerial acrobatics often before the females arrive. A male's display flight consists of up to 70 graceful dives from a height of 25 metres to within a metre of the ground before it recovers to swoop upwards again. This display takes place in open areas, and is readily seen from a considerable distance because of the contrasting blue-grey back and white underside of the male. The nest is placed on the ground among grasses or weeds, or in cattail and reed marshes. In dry areas, the structure may be small, but in moist locations it is more substantial. Usually 4 to 6 eggs are laid, and are incubated solely by the female; the male makes frequent visits to the nest to feed her. Nests are fairly easy to find by watching the adults once the eggs have hatched. However, nests with eggs or young were reported in only 74 squares; breeding was most often confirmed by observation of adults carrying food. The male seldom visits the nest when young are present, but rather flies by and calls to his mate, who joins him in flight. The male drops the food, which is caught in the air by the female. Whenever a human intruder approaches the nest, the normally quiet pair becomes noisy and aggressive, and performs a distraction display.

The distribution of the Northern Harrier shown on the atlas maps corresponds fairly closely to the description provided in Peck and James (1983). Its breeding range extends north to the shores of Hudson and James Bays, where the flat, open terrain is ideal for nesting. Its absence from many blocks inland from the coast on the Hudson Bay Lowland and in the Boreal Forest zone may be partly due to a lack of observer coverage, although its numbers are likely low there because of a paucity of suitable habitat. The relatively low productivity of wetlands and the lack of clearing for agriculture mean that there is little suitable habitat in these northern areas, or on the Canadian Shield in southern Ontario. In southern Ontario, it was recorded in 64% of the squares, placing it third after the American Kestrel and Red-tailed Hawk, the most frequently reported raptors in the south. Its absence from many squares in the Carolinian Forest zone and around Toronto is the result of the relative sparsity of suitable wetlands, at least partly because of the steady conversion of wetland to agriculture and housing.

Abundance estimates indicate that the Northern Harrier occurs in about the same numbers in squares in which it was reported in southern and northern Ontario. It was generally uncommon in the squares in which estimates were provided, but 11 or more pairs were estimated to occur in 37 squares. Thirty-six of these squares are in southern Ontario, north of the Carolinian Forest zone and south of the Shield (including one at Sault Ste Marie); the other is from Cochrane.-- *R.D. Weir*

NORTHERN HARRIER

Summary of Abundance Estimates

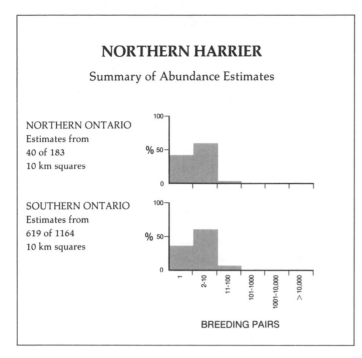

NORTHERN ONTARIO
Estimates from
40 of 183
10 km squares

SOUTHERN ONTARIO
Estimates from
619 of 1164
10 km squares

BREEDING PAIRS

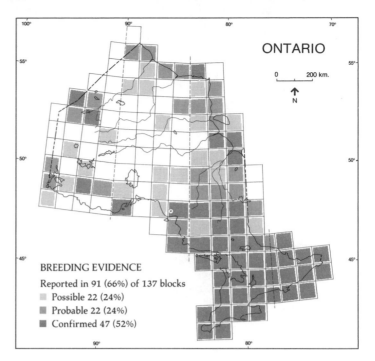

ONTARIO

0   200 km.

N

BREEDING EVIDENCE
Reported in 91 (66%) of 137 blocks

Possible 22 (24%)
Probable 22 (24%)
Confirmed 47 (52%)

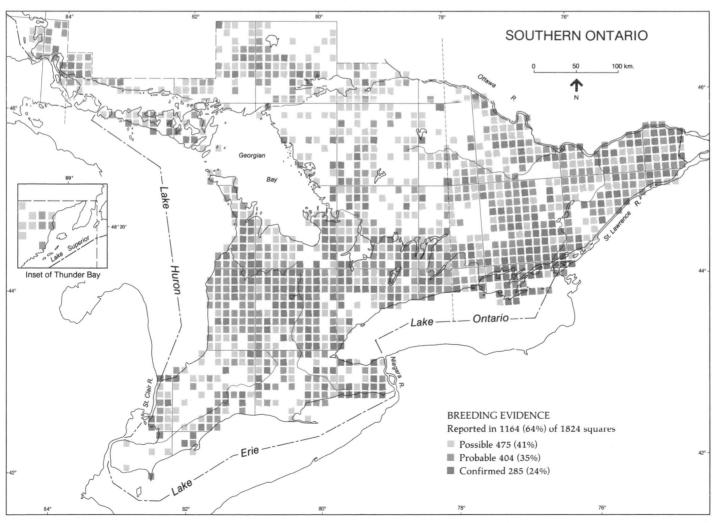

SOUTHERN ONTARIO

0   50   100 km.

N

Inset of Thunder Bay

BREEDING EVIDENCE
Reported in 1164 (64%) of 1824 squares

Possible 475 (41%)
Probable 404 (35%)
Confirmed 285 (24%)

113

SHARP-SHINNED HAWK
Epervier brun
Accipiter striatus

The Sharp-shinned Hawk is the smallest and most common of the 3 Accipiter hawks occurring in Ontario. This bold and dashing species moves at great speed in its chase for the small birds that make up most of its food. It is most often seen on migration, flapping and sailing in the open.

The Sharp-shinned Hawk occurs only in the Americas, breeding in almost all of the forested parts of Canada south through most of the US and into parts of Mexico (Brown and Amadon 1968). It breeds generally throughout Ontario, except perhaps some parts of the far north on the Hudson Bay Lowland. Its favoured nesting habitat consists of dense, wet, coniferous woods or bogs. If in mixed woods, a conifer is usually selected as a nest tree. Small numbers of birds winter as far north as southern Canada, but most go further south, some as far as Costa Rica.

McIlwraith (1894) described the species as a common summer resident in southern Ontario during the 19th century, an assessment supported by Macoun and Macoun (1909). Baillie and Harrington (1936) modified its status to "fairly common", and well distributed in summer north to Thunder Bay and Abitibi. James *et al.* (1976) assessed it as rare and local in summer north to Moosonee and Sandy Lake, although Peck and James (1983) considered its northerly breeding limit to be the Albany River. The decline in breeding numbers appears to have occurred mainly in heavily settled parts of Ontario, where habitat alteration has occurred. The overall impact of this drop in breeding numbers is difficult to assess, but it is likely that the population of the species is stable, but lower than that of the last century. Elsewhere in North America, its breeding numbers are also lower and the species has been on the Blue List since 1972 (Tate and Tate 1982). Humans are the main enemy of the Sharp-shinned Hawk through destruction of nesting

habitat, use of pesticides, and formerly, shooting.

Nesting Sharp-shinned Hawks are usually difficult to locate, mainly because of their silent, unobtrusive behaviour around the nest. An old crow's nest is occasionally used, but more typically, a new nest is constructed annually in a tree within a thick grove, often along the edge of a forest. When approached closely, the female may slip away unnoticed. Thus, nests are easily overlooked. Sometimes, however, both adults become aggressive and noisy. Their shrill mating calls are given during April and May in the area where the nest is to be built, and adults may actually be seen carrying sticks. Incubation lasts 34 to 35 days. Down caught on nearby twigs, or moulted adult feathers on the ground, may serve as clues to nest location; the female's moult begins in May, the male's soon after. Hunting does not occur in the vicinity of the nest site. The resulting peaceful atmosphere is likely their strategy to avoid attracting attention. Following fledging, the family group remains together in the general vicinity of the nest for a few weeks.

The atlas maps extend the known range of the Sharp-shinned Hawk north onto the Hudson Bay Lowland and show it to be the most widespread of Ontario's Accipiters. Breeding was confirmed in 5 Lowland blocks, with possible breeding evidence reported in several other blocks, especially in the far northwest. The low density prey populations of the open fen and bog habitat of the Lowland are probably not suitable for breeding Sharp-shinned Hawks, but along thickly wooded river banks and other well-drained locations, where bird densities are higher, the Sharp-shinned Hawk can apparently breed. The species was reported in virtually every block in the Great Lakes-St. Lawrence Forest zone, but was reported more sporadically through the Boreal Forest south of the Hudson Bay Lowland. It is difficult to assess whether the absence of records is due to insufficient coverage, a relative scarcity of birds, or some combination of the two. The lack of past nesting evidence in much of the north (Peck and James 1983) would suggest that the species is scarce there, but the concentration of atlas records around the relatively well-covered Moosonee area suggests that increased coverage would reveal birds in more blocks, at least those in the southern Boreal Forest. Deforestation and the relative lack of coniferous habitat are likely explanations for the absence of the Sharp-shinned Hawk in extreme southwestern Ontario and much of the Niagara Peninsula.

Abundance estimates indicate that the Sharp-shinned Hawk is uncommon in squares in which it was reported in both southern and northern Ontario.-- *R.D. Weir*

SHARP-SHINNED HAWK

Summary of Abundance Estimates

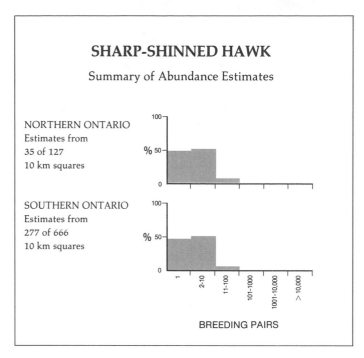

NORTHERN ONTARIO
Estimates from
35 of 127
10 km squares

SOUTHERN ONTARIO
Estimates from
277 of 666
10 km squares

%

100

50

0

%

100

50

0

1 2-10 11-100 101-1000 1001-10,000 >10,000

BREEDING PAIRS

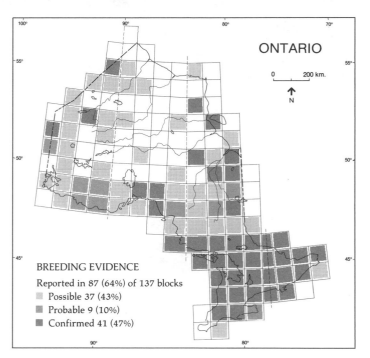

ONTARIO

0   200 km.

N

BREEDING EVIDENCE

Reported in 87 (64%) of 137 blocks

Possible 37 (43%)

Probable 9 (10%)

Confirmed 41 (47%)

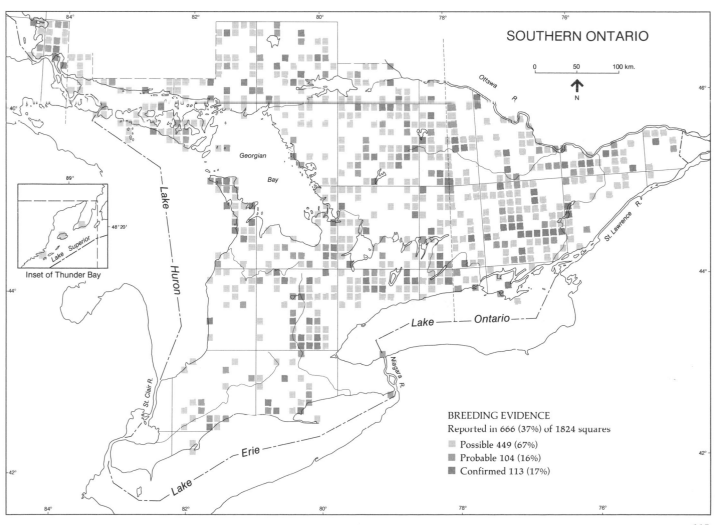

SOUTHERN ONTARIO

0   50   100 km.

N

Georgian
Bay

Ottawa
R.

St. Lawrence R.

Lake — Ontario

Niagara
R.

Lake   Superior

48°20'

89°

Inset of Thunder Bay

Lake

Huron

St. Clair R.

Erie

Lake

BREEDING EVIDENCE

Reported in 666 (37%) of 1824 squares

Possible 449 (67%)

Probable 104 (16%)

Confirmed 113 (17%)

COOPER'S HAWK
Epervier de Cooper
Accipiter cooperii

The Cooper's Hawk is like a larger version of the Sharp-shinned Hawk and has the most southerly breeding distribution of our 3 Accipiter species. Like the others, it has a low, swift flight, and relies on the element of surprise to capture its prey, principally medium-sized birds. Its short wings and long tail enable it to control movement precisely as it darts among forest branches with impunity.

The species breeds all across southern Canada, except in the extreme south of Saskatchewan, and generally throughout the US south to northwestern Mexico. In Ontario, its breeding range extends throughout the mixed or deciduous forests of the Great Lakes-St. Lawrence and Carolinian Forest zones. Although most individuals leave the province in fall, small numbers winter in southern Ontario.

One hundred years ago, the Cooper's Hawk apparently bred only sparingly in southern Ontario, from the St. Lawrence valley westwards (Macoun and Macoun 1909). Baillie and Harrington (1936) classed it as less common and a more southerly breeder than the Sharp-shinned Hawk. James *et al.* (1976) described it as a rare summer resident in the south and west, north to Kenora, Lake Nipigon and North Bay, while Peck and James (1983) defined the northerly edge of its range to be the latitude of northern Lake Superior (49°N). Its population was judged by Fyfe (1976) to be stable or in slight decline in Ontario, a status similar to his assessment of the Sharp-shinned Hawk population. Blue-listed for North America in 1972 (Arbib 1972), the Cooper's Hawk has remained so, although Tate and Tate (1982) wrote that its numbers were increasing in several regions. The main reason for its decline appears to have been destruction of forest habitat needed for nesting, but pesticide contamination of its prey and persecution by shooting may also have contributed (Penak 1983). Shooting was of most significance before 1945,

but pesticides such as DDT, which are still in use south of the US, may still be a problem.

Nests of Cooper's Hawks are not easily located. Generally, pairs are well dispersed. Adults return to their woods in March and April, when they may be seen in courting flights. Dense woods are chosen, often near pools of water or streams, and the substantial nest structure of twigs is placed 6 to 20 m up in a deciduous or pine tree. Occupied nests were reported in 32 squares and used nests in 4. Most of the incubation is done by the female, during which time the male makes frequent visits to feed her. It is during this stage of the breeding cycle in May or June that the male 'sings' as part of the songbird chorus at dawn. A modulated soft version of its cackling call is given for a short period after dawn from a perch near the nest site. Other telltale signs of the nest include the extension of the female's tail feathers over the edge of the nest and the presence of a nearby plucking perch that may be an abandoned nest. Unfortunately for atlassers, the young make virtually no noise while in the nest, but they do give hunger calls once they fledge. If disturbed on eggs, the female will usually slip silently away, but once the young have hatched, she becomes noisy and aggressive, and may attack intruders. While young are in the nest, the male provides most of the food, and, for a short period after fledging, the young return to the nest to get food brought by the parents. The family party remains together for up to 2 months after fledging. There is some evidence that Cooper's Hawks will nest near human activity where nesting habitat and food are available (Penak 1983). Egg dates in Ontario are 27 April to 8 July (Peck and James 1983).

The atlas maps show a breeding distribution confined mainly to southern Ontario, as would be anticipated from previous reports (Godfrey 1986). The most northerly record was of an adult carrying food for young across a lake north of Sandy Lake, about 600 km north of the previous most northerly confirmed record (Peck and James 1983). The majority of the records are on the southern Canadian Shield, although few of these are confirmed records, likely because of the difficulty of finding nesting areas in extensive woodlands. The species is absent from much of the Algonquin Highlands. South of the Shield, the records are more thinly distributed, but a higher proportion of these are of confirmed breeding because of the relative ease of finding nesting areas in smaller woodlots. The most southerly confirmed record is of a nest in Rondeau Prov. Park, but the lack of forest for nesting in much of the extreme southwest and on the Niagara Peninsula is evident from the absence of the species in these areas.

Abundance estimates indicate that the Cooper's Hawk is uncommon, where reported, in both southern and northern Ontario.-- *R.D. Weir*

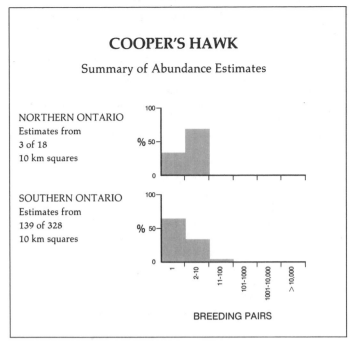

COOPER'S HAWK

Summary of Abundance Estimates

NORTHERN ONTARIO
Estimates from
3 of 18
10 km squares

SOUTHERN ONTARIO
Estimates from
139 of 328
10 km squares

%

BREEDING PAIRS

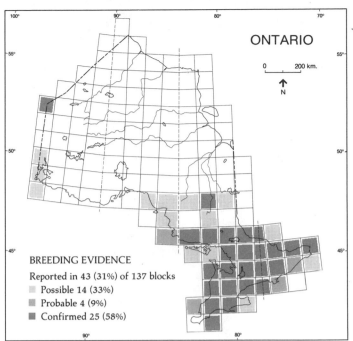

ONTARIO

0   200 km.

N

BREEDING EVIDENCE
Reported in 43 (31%) of 137 blocks
▫ Possible 14 (33%)
▨ Probable 4 (9%)
■ Confirmed 25 (58%)

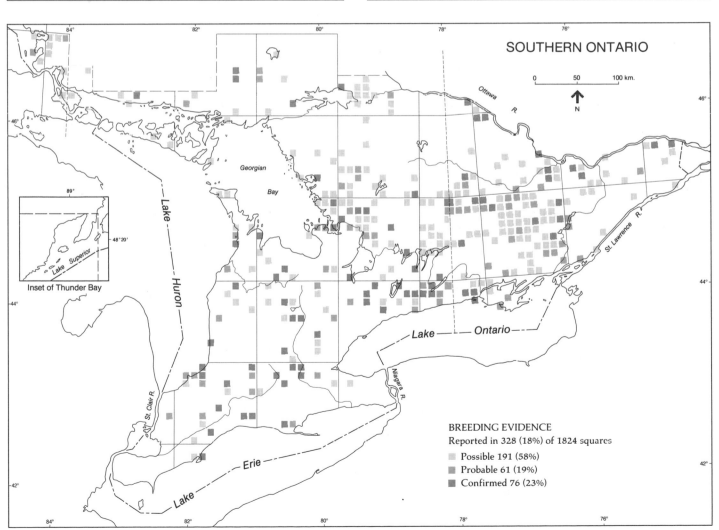

SOUTHERN ONTARIO

0   50   100 km.

N

Georgian
Bay

Lake Huron

Lake Superior

Inset of Thunder Bay

Ottawa R.

St. Lawrence R.

Lake — Ontario

Niagara R.

St. Clair R.

Lake — Erie

BREEDING EVIDENCE
Reported in 328 (18%) of 1824 squares
▫ Possible 191 (58%)
▨ Probable 61 (19%)
■ Confirmed 76 (23%)

NORTHERN GOSHAWK
Autour des palombes
Accipiter gentilis

The Northern Goshawk is the largest of our Accipiters. It is a powerful raptor rivalling the Great Horned Owl in ferocity, yet it mates for life, is a devoted parent, and can be remarkably gentle with its mate and young.

The species is distributed through the northern hemisphere of both the Old and New Worlds, north to near the tree-line. In North America, its breeding range extends south along the Rocky Mountains into the western US, and south in the east to New England. In Eurasia, it breeds south to the Mediterranean, Tibet, and Japan. Within Ontario, it breeds throughout the forested areas in sparse numbers, frequenting either deciduous, mixed, or coniferous woodland. In eastern Canada, the Northern Goshawk is mainly sedentary, but is subject to cyclical irruptions south from its breeding range when food supplies fail. Birds breeding in Ontario's far north may be regular migrants.

McIlwraith (1894) wrote of irregular visits by Northern Goshawks during winter into southern Ontario, while Macoun and Macoun (1909) refer to many birds being shot during such irruptions. Baillie and Harrington (1936) classified it as a rare breeder of central and northern Ontario, extending south to Muskoka District and east to Frontenac Co. Forty years later, James *et al.* (1976) also considered it a rare breeder. There is no firm evidence that the Northern Goshawk ever nested in the extreme south of the province (R.D. James pers. comm.), although it may have done so. The forests through most of the remainder of its Ontario range have been relatively undisturbed, especially on the Canadian Shield, so suitable habitat is widespread. In addition to forest fragmentation, other human threats to the species include indiscriminate killing and the raiding of nests by falconers.

The Northern Goshawk is normally secretive and could have been easily overlooked by atlassers, especially where field work did not begin until after early April, when the noisy courtship period has passed. The female returns to the large nesting territory by early March and begins her screaming calls to attract her former mate. Once he appears, their display flights and noisy calling take place over the nesting territory. Nest building, undertaken by the male, is confined to the early morning. At this time, the pair will perform its screaming duet on most days just before sunrise. This very noisy period is the best time to locate the nest, which is a bulky, untidy platform of sticks placed some 8 to 25 m up in a crotch of a deciduous or coniferous tree within a dense stand. Incubation lasts about 35 to 36 days. After the onset of incubation, the pair becomes silent, and the male visits the nest frequently to feed his mate, flying quietly within the canopy. Egg dates for Ontario nests are from 16 April to 2 June (Peck and James 1983). The male continues to bring all the food during the 41 days needed by the young to fledge. Unlike the other Accipiters in Ontario, the Northern Goshawk hunts near the nest site (as well as far from it). The adults may scream at, and will physically attack, people climbing to their nest, and can therefore be dangerous. This agitated behaviour probably helped atlassers find nests which might otherwise have been missed; 40% of confirmed breeding records are of nests with young.

The distribution of the Northern Goshawk shown on the maps generally conforms with the expected range, except perhaps that more records than anticipated were reported in southern Ontario, and fewer were reported from the north. The species was found in 47% of Ontario's blocks, with most of the deficiencies occurring north of settled and road-accessible areas. The wide gaps in northern Ontario are at least partly due to low observer coverage for this difficult and sparsely distributed species, but the paucity of records suggests that densities are lower in the Boreal Forest. Nevertheless, the scattered records on the northern Canadian Shield and on the Hudson Bay Lowland suggest that the species breeds throughout the north. In southern Ontario, only one record was reported south of 43°N, in the primarily deciduous forests of the Carolinian Forest zone. This is consistent with the distribution of known nest records (Peck and James 1983) and suggests that the species may not breed in that area. It is known to nest in reforestation plots, and, in fact, the most southerly record is from such a location, as are several other records south of the Canadian Shield. The 298 squares where the Northern Goshawk was found in southern Ontario represent only 16% of all squares in that sector, thereby indicating that the species breeds sparingly. Although the species is likely underrepresented even in the south, the map probably gives a fairly accurate representation of its distribution.

Abundance estimates indicate that the Northern Goshawk is uncommon to rare in squares in which it was reported in both southern and northern Ontario.-- *R.D. Weir*

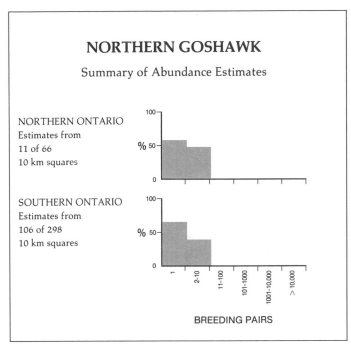

NORTHERN GOSHAWK

Summary of Abundance Estimates

NORTHERN ONTARIO
Estimates from
11 of 66
10 km squares

SOUTHERN ONTARIO
Estimates from
106 of 298
10 km squares

%

100

50

0

100

50

0

1   2-10   11-100   101-1000   1001-10,000   >10,000

BREEDING PAIRS

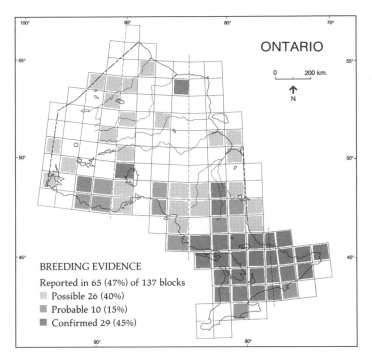

ONTARIO

0   200 km.

N

BREEDING EVIDENCE
Reported in 65 (47%) of 137 blocks

Possible 26 (40%)
Probable 10 (15%)
Confirmed 29 (45%)

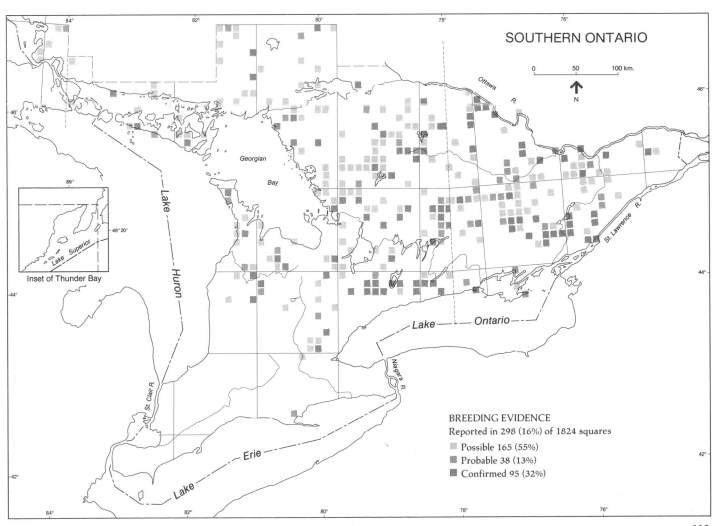

SOUTHERN ONTARIO

0   50   100 km.

N

Ottawa R.

St. Lawrence R.

Georgian Bay

Lake Huron

Inset of Thunder Bay

Lake Superior

Lake   Ontario

Niagara R.

St. Clair R.

Erie

Lake

BREEDING EVIDENCE
Reported in 298 (16%) of 1824 squares

Possible 165 (55%)
Probable 38 (13%)
Confirmed 95 (32%)

RED-SHOULDERED HAWK
Buse à épaulettes
Buteo lineatus

The Red-shouldered Hawk is a shy woodland Buteo whose numbers have declined alarmingly since the 1950's over much of eastern North America. The earliest migrants offer a striking sight as their reddish breast feathers contrast with the white winter snows still lying on the ground.

The species occurs only in North America, reaching the northerly edge of its breeding range in southern Ontario, extreme southwestern Quebec, and southern New Brunswick. Migrants begin to arrive from the southern US by mid-March and continue to enter Ontario until late April mainly via the land bridges through Windsor, Niagara Falls, and the Thousand Islands area. It frequents mixed and deciduous forests or swamps adjacent to lakes or streams, habitats it often shares with the Barred Owl. Mature maple-beech-hemlock woodlands are most frequently used for nesting in the Waterloo Region, and elsewhere (Sharp and Campbell 1982). Replacement of Red-shouldered by Red-tailed Hawks in the Waterloo Region was associated with reduction of tree density and crown diameter, implicating selective cutting in woodlands as the cause of the decline of Red-shouldered Hawks in that part of Ontario (Bryant 1986). Because of the loss of mature woodlands in southwestern Ontario, the species has almost vanished as a breeder in that area.

During the 19th century, the Red-shouldered Hawk was the most common hawk in southern Ontario (McIlwraith 1894). By 1900, it was still described as quite a common breeder, outnumbering the Red-tailed Hawk in the south by a ratio of 4 to 1 (Macoun and Macoun 1909). However, by 1936, the conversion of forest to farmland in the southwest had led to a declining population (Baillie and Harrington 1936). As large tracts of woods disappeared from southern Ontario, Red-shouldered Hawk numbers steadily declined. By the 1950's, it was less common than the Red-tailed Hawk,

which prospered with the habitat alteration. The situation today is very different from that of 50 years ago: population declines throughout North America resulted in the species being placed on the Blue List in 1972 (Arbib 1972), where it has remained ever since.

Red-shouldered Hawks return to their nesting woods in early spring while snow is still on the ground and leaves are unopened. It is at this time that they are most easily detected as pairs become very noisy and engage in courtship flights. A pair will soar in great circles and make spectacular dives above their woodlot, at which time their loud cries of 'kee-aah' are repeated. Two to 4 eggs are laid. Once incubation is well underway, and leaves are fully opened, these hawks become more quiet, secretive, and difficult to detect. Most of the hunting at this time is done within the canopy, and the birds rarely fly out into the open. In squares where atlassing began after mid-May, the species was probably underrecorded. The nest is also difficult to find once leaves have opened as it is placed between 5 and 25 m high in dense areas of the woods. The nest is bulky and composed of sticks and twigs placed in the main crotch or fork near the trunk. Beech, maple, and ash are preferred nesting trees.

The atlas maps show the Red-shouldered Hawk to be confined mainly to southern Ontario, the only more northerly record being a pair of birds near Englehart, 40 km south of Kirkland Lake, in 1981. Peck and James (1983) show an even more northerly nest record from near Timmins. The majority of Red-shouldered Hawk atlas records are clustered on the southern Canadian Shield, with records being most concentrated between Kingston and the Ottawa River, and in the Muskoka-Parry Sound area. The forests and wetlands of the southern Shield no doubt provide more extensive suitable habitat than is available elsewhere in its Ontario range. The relative sparsity of records on the rest of the southern Shield is difficult to explain and may be because of a relative paucity or absence of birds, perhaps due to some subtle differences in habitat, or it may be because of underrecording. Off the Shield, records are scattered in eastern Ontario, but concentrations of records are evident in the Oak Ridges Moraine area (north and east of Toronto), in Bruce and Grey Counties, and in the Milton-Cambridge-Waterloo area. These are relatively well-forested areas with adequate wetlands, as is southern Haldimand-Norfolk RM near Long Point, where one of the more southerly clusters of records occurs. The most southwesterly records are from the extensively wooded Skunk's Misery area.

Abundance estimates indicate that the Red-shouldered Hawk is generally uncommon to rare in squares where it was reported. Estimates of more than 10 pairs were provided for 10 squares north and northwest of Kingston, suggesting that breeding densities are highest in that area. COSEWIC and the MNR have designated the species as "rare" in Canada and Ontario, respectively.-- *R.D. Weir*

RED-SHOULDERED HAWK

Summary of Abundance Estimates

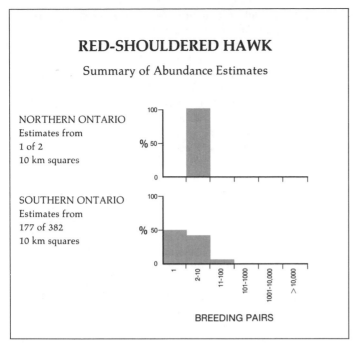

NORTHERN ONTARIO
Estimates from
1 of 2
10 km squares

SOUTHERN ONTARIO
Estimates from
177 of 382
10 km squares

BREEDING PAIRS

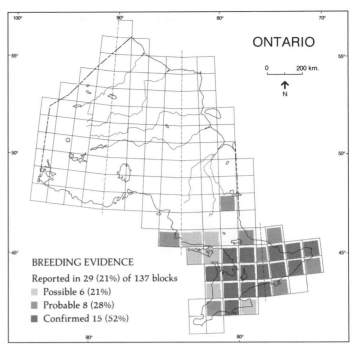

ONTARIO

0   200 km.

N

BREEDING EVIDENCE

Reported in 29 (21%) of 137 blocks

Possible 6 (21%)
Probable 8 (28%)
Confirmed 15 (52%)

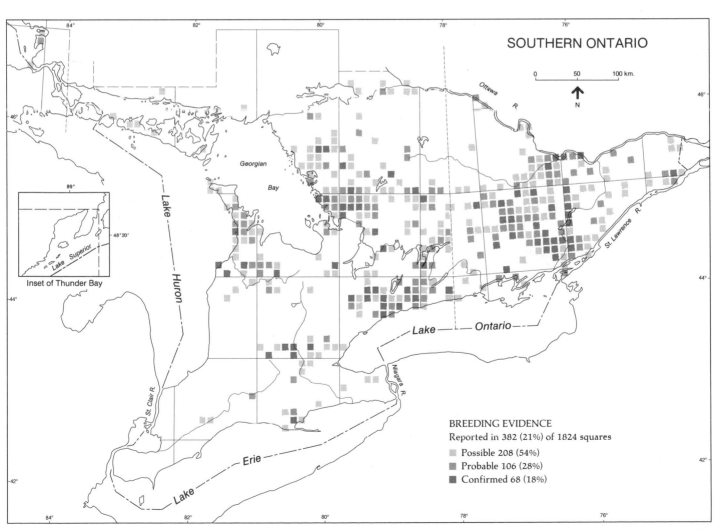

SOUTHERN ONTARIO

0   50   100 km.

N

Inset of Thunder Bay

BREEDING EVIDENCE

Reported in 382 (21%) of 1824 squares

Possible 208 (54%)
Probable 106 (28%)
Confirmed 68 (18%)

BROAD-WINGED HAWK
Petite Buse
Buteo platypterus

This quiet and retiring raptor of the forests is seldom seen in open country during the nesting season, yet spectacular concentrations are observed during spring and autumn migration. It migrates in large flocks and on some days every September, thousands are seen near the shores of Lake Erie, at Hawk Cliff, south of St. Thomas, and at Holiday Beach in Essex Co, as they make their way south to their wintering grounds in Central and South America.

The Broad-winged Hawk does not occur outside the Americas. Its breeding haunts are dense deciduous or mixed woods from Alberta east across mid-Canada to the Maritimes and south to the US Gulf Coast. In Ontario, it nests where suitable habitat exists north to the central Boreal Forest zone. A few individuals are known to wander north in summer to Moosonee on James Bay.

The species was widely distributed in Ontario during the 19th century and was considered to be the commonest breeding hawk from Renfrew Co to Parry Sound District (Macoun and Macoun 1909). MacClement (1915) agreed with this assessment and noted it to be especially plentiful in Muskoka District. Baillie and Harrington (1936) judged it to be the most common hawk in central and northern Ontario, but found it to nest less commonly south of Middlesex Co. Fyfe (1976) assessed the population of the Broad-winged Hawk in Canada as stable and its abundance as high. In April, tens of thousands of Broad-winged Hawks enter Ontario from Minnesota, Michigan, and New York. Particularly large numbers are seen at Grimsby, and at Ivy Lea, at the east end of Lake Ontario. Most continue further north; the depletion of forest in much of southwestern Ontario has led to the virtual disappearance of the species as a nester there.

Broad-winged Hawks are more difficult to detect during the nesting season, at which time they are shy, and well dispersed. Almost all of their hunting is done within the forest canopy, or adjacent to forested areas. Adults return to their nesting sites about the time leaves open, hindering the detection of the nest by observers. Immediately upon arrival, courting flights can be seen above the treetops and mating calls can be heard from within the forest. Once incubation of the 2 to 4 eggs begins, the adults become quiet and are difficult to find. Most often, the nest is built within a densely forested area and placed 7 to 12 m up in the main crotch of a deciduous tree, although conifers may be used. The nest may be either small and inconspicuous, or may consist of a highly visible bulky platform of sticks decorated with leaves. When an adult is discovered, it will often sit silently, allowing a close approach before taking flight. Nest sites are most often betrayed when adults, worried by visitors, complain dolefully, but the birds remain unaggressive. Their plaintive high-pitched 'peeo-wee-ee' alarm is known to fool novice observers who mistake it for the similar-sounding Eastern Wood-Pewee. Repeated observation of an adult in an area (such as perching sites in dead trees or on electric transmission lines) often identifies an area where nest searches may be successful. Breeding was most frequently confirmed by observing adults carrying food for young (37%), or by locating newly fledged young (20%); family groups remain together for several weeks after fledging.

The distribution of the Broad-winged Hawk shown on the atlas maps includes nearly all of the Canadian Shield, except for a few blocks immediately south of the Hudson Bay Lowland. In northern Ontario, it is likely that its range is more or less correct as shown. The open muskeg habitat of the Lowland is not suitable for a woodland species. Off the Shield in southern Ontario, the species occurs in most squares on Manitoulin Island and on the Bruce Peninsula. It is more sporadic around the southern and eastern edge of the Shield, and is a rare breeder in the southwest, where the proportion of forested land is lower. Areas with more extensive forest off the Shield, such as southern Haldimand-Norfolk RM, parts of the Niagara escarpment, the Oak Ridges Moraine, and Pinery Prov. Park, are frequented by Broad-winged Hawks, as is evident on the map. Prior to the atlas period, there was little evidence of breeding in much of the southwest (Peck and James 1983).

Abundance estimates indicate that numbers of breeding pairs are similar in squares in which estimates were provided in northern and southern Ontario. The species is fairly common within its range in both sectors.-- *R.D. Weir*

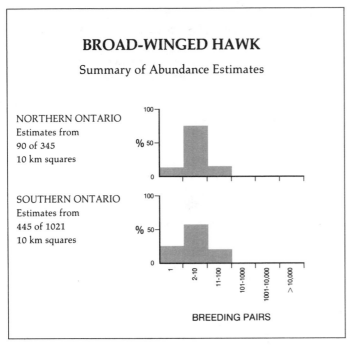

BROAD-WINGED HAWK

Summary of Abundance Estimates

NORTHERN ONTARIO
Estimates from
90 of 345
10 km squares

SOUTHERN ONTARIO
Estimates from
445 of 1021
10 km squares

% (axis labels: 0, 50, 100)

x-axis: 1, 2-10, 11-100, 101-1000, 1001-10,000, >10,000

BREEDING PAIRS

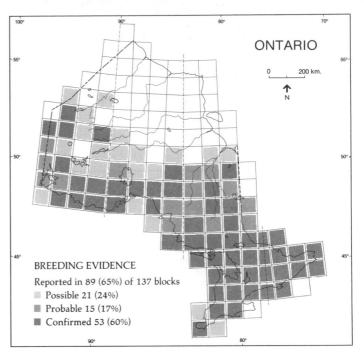

ONTARIO

0   200 km.

N

BREEDING EVIDENCE
Reported in 89 (65%) of 137 blocks

Possible 21 (24%)
Probable 15 (17%)
Confirmed 53 (60%)

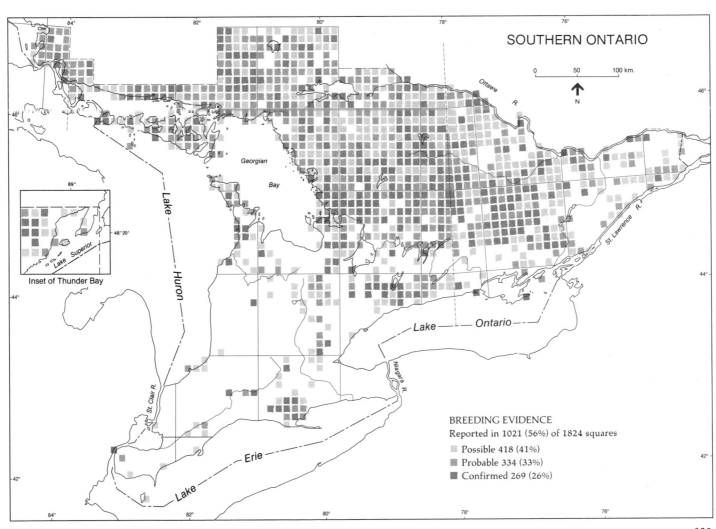

SOUTHERN ONTARIO

0   50   100 km.

N

Georgian
Bay

Lake Huron

Lake Superior

89°

48°20'

Inset of Thunder Bay

Ottawa R.

St. Lawrence R.

Lake — Ontario

Niagara R.

St. Clair R.

Lake — Erie

BREEDING EVIDENCE
Reported in 1021 (56%) of 1824 squares

Possible 418 (41%)
Probable 334 (33%)
Confirmed 269 (26%)

RED-TAILED HAWK
Buse à queue rousse
Buteo jamaicensis

The Red-tailed Hawk is one of the most widely distributed and best known raptors in North America. It is unique to the New World, breeding across most of Canada north to the tree-line, and south through the US, West Indies, and Central America to Panama. The favoured breeding habitat for the Red-tail is dry woodland near open country, where the birds hunt for food. Frequently, such habitat occurs in areas adjacent to agricultural land, along railroad lines and highways. The Red-tail is also found in wet woodlands and hedgerows, and nests can be found in isolated trees in fields. Red-tailed Hawks and Great Horned Owls often share the same habitat. The former hunts by day and the latter by night. The Red-tailed Hawk winters as far north as southern Ontario, southern British Columbia, and the Maritime Provinces (Godfrey 1986).

It is not surprising that, as Ontario's forests have been reduced to woodlots and hedgerows by human settlement, the Red-tailed Hawk has increased its numbers to become one of the province's most numerous hawks during all seasons of the year. Although found throughout Ontario during the 19th century, it was most often seen during spring and autumn migration (McIlwraith 1894). By 1900, it was rare along the St. Lawrence River and pairs bred only in scattered locations in the south, being outnumbered by Red-shouldered Hawks by a ratio of 4 to 1 (Macoun and Macoun 1909). The Red-tailed Hawk had increased by the 1930's to become a fairly common breeding bird north to Parry Sound and Thunder Bay Districts (Baillie and Harrington 1936). In the Waterloo region, the increase of Red-tailed Hawks at the expense of Red-shouldered Hawks was associated with a reduction of tree density and crown diameter in woodlots, probably as a consequence of selective cutting (Bryant 1986). In the 1970's, it was a common summer resident across the province, north to the Hudson Bay Lowland (James *et al.* 1976). Peck and James (1983) suggested it may breed as far north as forest trees are found, which may well have always been the situation in the north, where there is no obvious reason to suspect a change in the population or range of this species.

The Red-tailed Hawk is easily located during the breeding season as it hunts over open fields and meadows, soars over its nest woods or waits patiently on an open commanding perch from which it searches for food. It usually feeds on rodents and other small mammals, but is also an opportunistic feeder. A patient observer may see one of the birds hopping awkwardly along the ground snatching grasshoppers from a freshly mown hayfield or carrying a snake to its nest. The nest is most often built in a deciduous tree. However, when conifers are selected, pine is the overwhelming choice (Peck and James 1983). Tree nests are usually placed in the main crotch or near the tree top, although some have been found as low as 5 m. In those parts of its range where tree sites are scarce, hydro poles or cliff edges are used. Like the birds, nests are easily located, and make up the majority of southern Ontario (63%) and northern Ontario (58%) breeding confirmations. Before leaves open, the well-built bulky platform of sticks stands out in the tree tops. Once incubation of the 2 to 4 eggs is under way, the male makes frequent visits to the nest to feed the incubating female. When disturbed, the pair will circle overhead uttering a harsh drawn-out squeal: 'kee-ee-ee'. As the nesting season advances, the young in the nest may be heard peeping or imitating their parents. Egg dates vary between 3 March and 15 July (Peck and James 1983).

The maps show the Red-tailed Hawk to be widely distributed across Ontario, though only sporadically on the Hudson Bay Lowland and in some northerly areas of the Boreal Forest. The most northerly confirmed breeding record for the atlas, and for the province to date, was of a nest with young from an island in the Winisk River. The species was reported in almost every block in the Great Lakes-St. Lawrence and Carolinian Forest zones. However, the extensive woodlands of the Boreal Forest zone present relatively few of the open dry areas preferred by the Red-tailed Hawk, so the distribution is less contiguous there, especially north of cleared land. As shown on the southern Ontario map, the frequency of occurrence is highest in agricultural areas: records are more scattered and confirmation of breeding is more difficult to obtain away from the open landscapes created by agriculture. The Red-tailed Hawk occurs in virtually every square south and east of the Canadian Shield, but is more sporadic on the heavily forested Shield itself.

As might be expected from the higher proportion of agricultural land in the south, abundance estimates suggest that the Red-tailed Hawk is more common in the south than it is in the north. However, in only 11% of all squares with abundance estimates were there thought to be more than 10 pairs.-- *R.D. Weir*

RED-TAILED HAWK

Summary of Abundance Estimates

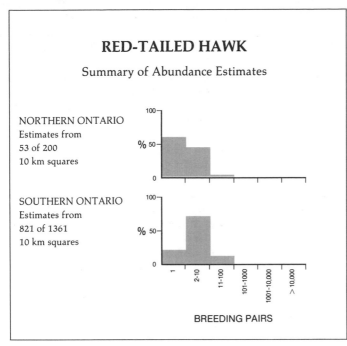

NORTHERN ONTARIO
Estimates from
53 of 200
10 km squares

SOUTHERN ONTARIO
Estimates from
821 of 1361
10 km squares

BREEDING PAIRS

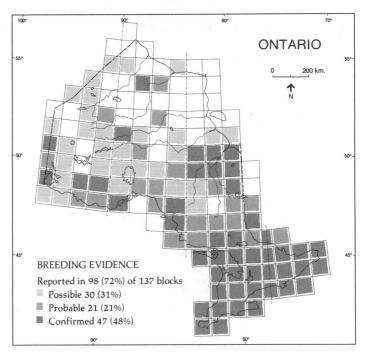

ONTARIO

0   200 km.

N

BREEDING EVIDENCE
Reported in 98 (72%) of 137 blocks

▫ Possible 30 (31%)
▪ Probable 21 (21%)
◾ Confirmed 47 (48%)

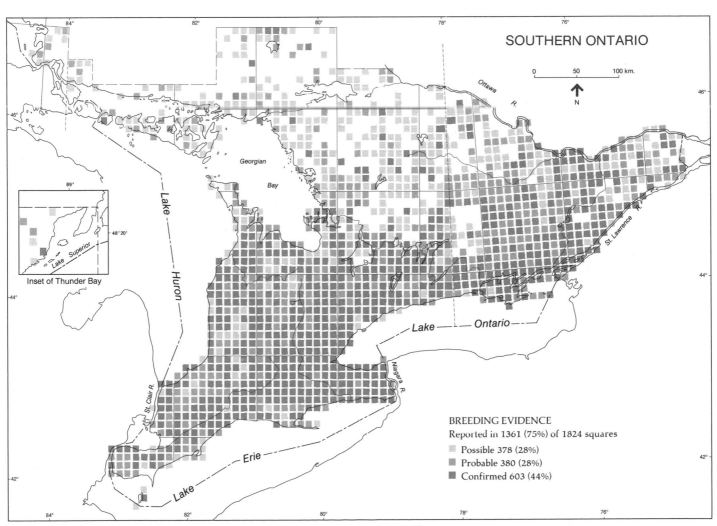

SOUTHERN ONTARIO

0   50   100 km.

N

Georgian Bay

Lake Huron

Lake Superior

48°20'

Inset of Thunder Bay

St. Clair R.

Lake Erie

Niagara R.

Lake — Ontario

Ottawa R.

St. Lawrence R.

BREEDING EVIDENCE
Reported in 1361 (75%) of 1824 squares

▫ Possible 378 (28%)
▪ Probable 380 (28%)
◾ Confirmed 603 (44%)

AMERICAN KESTREL
Crécerelle d'Amérique
Falco sparverius

The American Kestrel vies with the Red-tailed Hawk for top honours as the most numerous, and at the same time, the most widely distributed hawk in Ontario. This colourful species, the smallest of our hawks, is often seen hovering on the wind as it surveys the ground below.

The American Kestrel is confined to the Americas, breeding throughout North America from the tree-line south through Central and South America to Tierra del Fuego. It breeds throughout Ontario, north to Moosonee and west to Sandy Lake (Peck and James 1983), and, as discovered during the atlas period, occasionally on the Hudson Bay Lowland north to the Hudson Bay coast. It favours open country, especially grassland and woodland edges, but shuns dense forest. It can frequently be seen hunting over the grassy medians and verges of four-lane highways. Unlike most hawks, the American Kestrel chooses natural cavities or woodpecker holes for the nest, and, because of this, has a very high fledging rate - up to 87% (Cade 1982). That it tolerates human activities is shown by its willingness to nest in building cavities and nest boxes, and tree cavities even within city parks. Nest box programs have aided the species tremendously, especially where natural nest sites have been destroyed.

The American Kestrel has been a common bird in Ontario for as long as records have been kept, but it was no doubt less common before the creation of open spaces when woods were cleared for settlement and agriculture. Macoun and Macoun (1909) described it as the commonest of the small hawks breeding in the province. Baillie and Harrington (1936) judged it to be fairly common and well distributed north to James Bay. James *et al.* (1976) classified it as a common summer resident across the province. Its population in Canada was assessed by Fyfe (1976) as almost stable every-

where. Because of declines in breeding numbers in several US states, the American Kestrel was on the Blue List from 1972 to 1981 (Tate 1981), but it was removed from the list in 1982 (Tate and Tate 1982).

The species is sexually dimorphic. The slightly larger, rufous-winged female can easily be distinguished from the smaller, grey-winged male; 12% of all records are of pairs. The American Kestrel is easily located during the breeding season as it perches in conspicuous spots atop trees, fenceposts, or wires, or as it hovers over open areas. The male provides virtually all the food for the female during the 29 to 30 day incubation period, and for the family in the subsequent 30 days while the young are in the nest. Nest holes are readily found by watching the male carry food. The family party remains together in the vicinity of the nest for several weeks after fledging, and most breeding confirmations were obtained at that time; 39% of confirmations were of young birds. Both parents are noisy, especially when disturbed, and their excited high-pitched 'killy-killy-killy' is a likely sign that a nest or young are nearby. The percentage of confirmed records in southern Ontario (49%) is the highest for any of Ontario's hawks.

During the atlas period, the known Ontario breeding range of the American Kestrel was extended to the Hudson Bay coast when a nest was discovered in an airport hangar at Winisk. A possible breeding record was reported from the town of Fort Severn, also on the north coast, but the only other record from the northwestern section of the Hudson Bay Lowland was one pair seen along the edge of an aspen forest on the shore of the Fawn River. The lack of trees sufficiently large to provide nest sites may be a limiting factor on the Lowland. Records are sporadic in the northern Boreal Forest zone south of the Lowland, most likely as a result of very low numbers of American Kestrels because of the lack of suitable open habitat in that area. The same may be said of the Algonquin Highlands, the shore and islands of Georgian Bay, and much of Parry Sound District in southern Ontario. The American Kestrel was reported in nearly all squares south and east of the Shield and in agricultural areas on the Shield itself.

Though most abundance estimates are in the 2 to 10 pairs category in both the north and south, 24% of southern Ontario's estimates are of more than 10 pairs, whereas only 15% of northern records are in that category. This suggests that the American Kestrel is more common in southern Ontario, presumably because there are more extensive areas of prime breeding habitat in this area.-- *R.D. Weir*

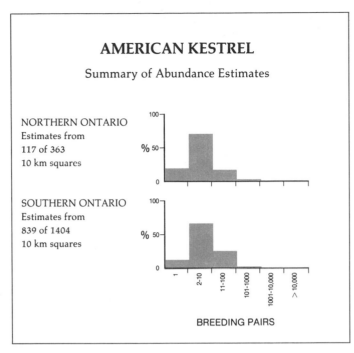

AMERICAN KESTREL

Summary of Abundance Estimates

NORTHERN ONTARIO
Estimates from
117 of 363
10 km squares

SOUTHERN ONTARIO
Estimates from
839 of 1404
10 km squares

%50 / 100

BREEDING PAIRS

1   2-10   11-100   101-1000   1001-10,000   >10,000

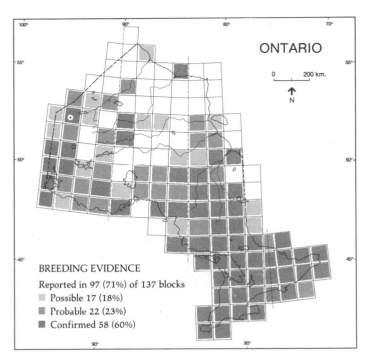

ONTARIO

0   200 km.

N

BREEDING EVIDENCE
Reported in 97 (71%) of 137 blocks

Possible 17 (18%)
Probable 22 (23%)
Confirmed 58 (60%)

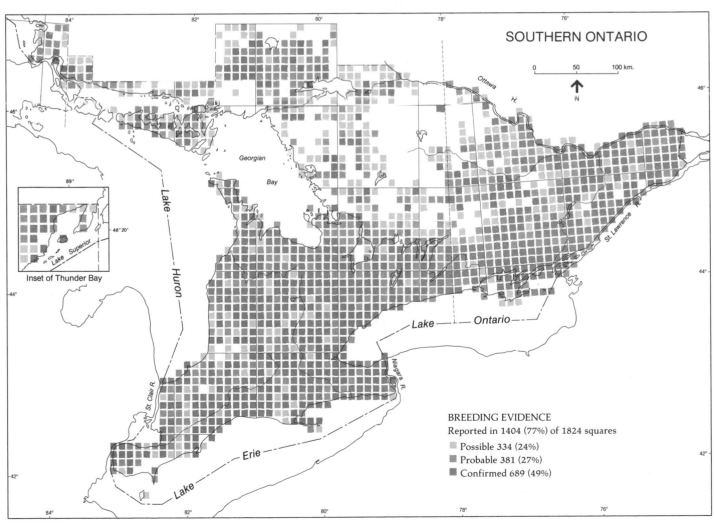

SOUTHERN ONTARIO

0   50   100 km.

N

Ottawa R.

St. Lawrence R.

Lake Huron

Lake Superior

Georgian Bay

89°
48°20'

Inset of Thunder Bay

St. Clair R.

Niagara R.

Lake Ontario

Lake Erie

BREEDING EVIDENCE
Reported in 1404 (77%) of 1824 squares

Possible 334 (24%)
Probable 381 (27%)
Confirmed 689 (49%)

127

MERLIN
Faucon émerillon
Falco columbarius

The Merlin is swift and dashing, as is typical of its genus. It is not a well known falcon in North America. It usually nests away from human settlement, although some birds are now breeding in Thunder Bay (Escott 1986) and in some prairie cities. It is also vastly outnumbered by its slightly smaller cousin, the American Kestrel, and is overshadowed by the publicity afforded the endangered Peregrine Falcon.

The species breeds from northern Europe eastwards to Siberia and across virtually all of Canada north to the tree-line. In Ontario, it is almost absent as a breeder south of about 45°N. During the nesting season, its haunts are variable, including open forests, second-growth conifers, mixed woodlands, and plantations. One of the essential requirements, however, is open country nearby for hunting. Its winter range in the New World extends from southern Canada to northern South America.

Relatively little is known of the abundance of the Merlin in Ontario during the 19th century. It was noted only as a migrant by McIlwraith (1894), who wrote of the difficulty in defining the breeding range of a bird that is "rare everywhere." However, Macoun and Macoun (1909) considered it abundant along the James Bay coast from Cape Henrietta Maria to Moose Factory. Baillie and Harrington (1936) defined its breeding range as all of northern and central Ontario, an assessment confirmed by Peck and James (1983). The remoteness of much prime breeding habitat makes it difficult to determine numbers of breeding Merlins accurately. During the 1950's and 1960's, Merlin numbers across much of North America declined as the species suffered reproductive failure because of organochlorine residues ingested either on the breeding grounds or in the wintering areas farther south (De Smet 1985). Fyfe (1976) placed its numbers in Ontario in the rare to medium category, but thought the species was

either declining or stable. The Merlin was on the American Birds Blue List for 10 years from 1972 to 1981 (Tate 1981), but was then removed as its numbers began to rise again in response to the restricted use of hazardous pesticides. However, it remains a species of "special concern" (Tate 1986), and its population levels remain below those of the pre-1950's, suggesting that the species has not yet fully recovered (De Smet 1985).

There is considerable evidence of Merlin nesting territories being used for many years in succession, suggesting nest site fidelity. Males generally arrive during April or May, a few days to a month before the females. Merlins usually, if not always, use old nests of other birds, especially crows, or those of squirrels; 30 of 31 nests reported to the ONRS were in coniferous trees; the other was on a cliff-ledge (Peck and James 1983). Incubation lasts 28 to 32 days, during which time the male calls the female off the nest to give her food, which is transferred in the air. The male continues to supply all the food for her and for the young in the early stages of brooding; adults carrying food for young formed the most frequently used confirmed breeding category. Adults are noisy around the nest and fly about advertising their presence with their shrill chatter, which is louder and harsher than that of the Sharp-shinned Hawk; agitated behaviour was the most frequently reported probable breeding category. Nevertheless, the nest itself may still be difficult to find. Young are capable of making their first flight at 25 to 30 days of age, but they remain near the nest and are dependent upon their parents for a few days to a month.

The maps indicate that all but a very few confirmed breeding records occur north of a line from Algonquin Prov. Park to Manitoulin Island. The most southerly cluster of records is in northern Grey Co, which has extensive coniferous forests. Other records are concentrated along the shores of the Great Lakes, Lake Nipissing, Lake Temagami, and the Ottawa River, showing the preference of the species for coniferous habitats adjacent to open areas. The lack of breeding records south of 45°N represents no change since the 19th century. The pattern of records suggests that the Merlin is widespread throughout the northern reaches of the Great Lakes-St. Lawrence Forest region and the Hudson Bay Lowland, but is relatively scarce or perhaps even absent throughout the rest of Ontario's Boreal Forest region. Although somewhat unexpected, the lack of records in the Boreal Forest conforms with the distribution plotted by Peck and James (1983). Although it may be partly caused by inadequate coverage, it is probably mainly indicative of the true situation, because most atlas field work in the Boreal Forest was along rivers and lakeshores, where Merlins are easily detected, if present.

Abundance estimates suggest that the Merlin is more common in its northern Ontario range than it is in the south. However, it is uncommon to rare in both areas.-- *R.D. Weir*

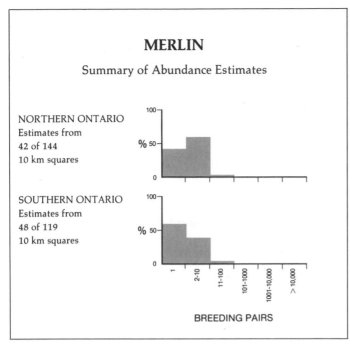

MERLIN

Summary of Abundance Estimates

NORTHERN ONTARIO
Estimates from
42 of 144
10 km squares

SOUTHERN ONTARIO
Estimates from
48 of 119
10 km squares

BREEDING PAIRS

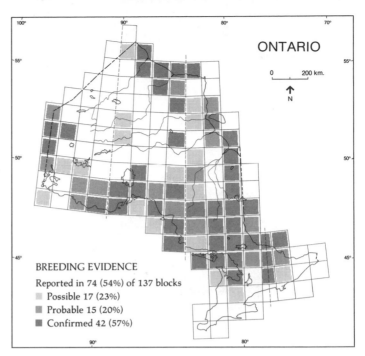

ONTARIO

0   200 km.

N

BREEDING EVIDENCE
Reported in 74 (54%) of 137 blocks
■ Possible 17 (23%)
■ Probable 15 (20%)
■ Confirmed 42 (57%)

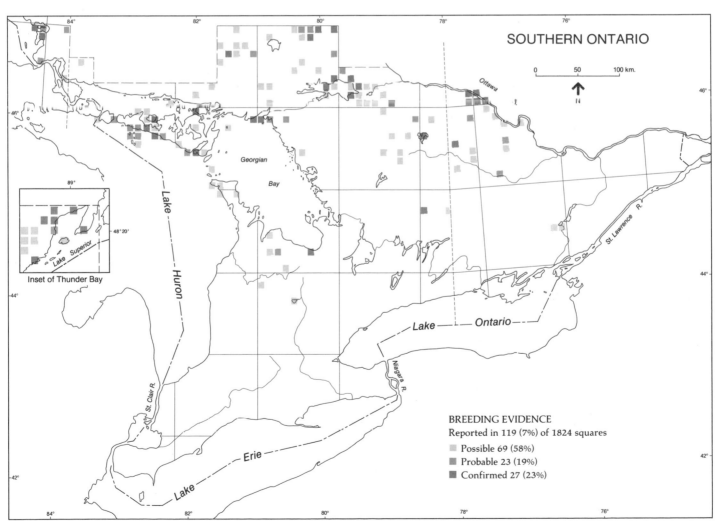

SOUTHERN ONTARIO

0   50   100 km.

N

Georgian
Bay

Lake
Huron

Lake — Ontario

Lake — Erie

St. Clair R.

Niagara R.

St. Lawrence R.

Ottawa

Inset of Thunder Bay

Lake Superior

89°
48°20'

BREEDING EVIDENCE
Reported in 119 (7%) of 1824 squares
■ Possible 69 (58%)
■ Probable 23 (19%)
■ Confirmed 27 (23%)

PEREGRINE FALCON
Faucon pèlerin
Falco peregrinus

The Peregrine Falcon, one of the world's fastest birds, has been clocked at speeds of up to 300 km/h as it stoops at its avian prey. It is a magnificent flier that has been coveted by falconers for centuries. Today, however, it is more widely known in North America and Europe as a species endangered by persistent organochlorine compounds, especially DDT, which accumulate in adults as a result of their diet of small birds. Infertile eggs, the inability to lay, and eggshell thinning leading to broken eggs in the nest, have been the consequences (Hickey 1969).

The Peregrine is nearly cosmopolitan, breeding on 5 continents. Within North America, its range and numbers have been reduced greatly since the 1950's, to the extent that is has been almost extirpated from southeastern Canada. Three subspecies are known in Canada. *F. p. anatum* breeds from the tree-line in Canada south to Mexico. It was seriously depleted throughout its range by pesticide contamination.

Macoun and Macoun (1909) considered the Peregrine Falcon to breed rarely in central Ontario, including Leeds and Hastings Counties, although MacClement (1915) noted that it bred only in the northern part of the province and was common around Hudson Bay. Baillie and Harrington's (1936) assessment, based on more field work, was that it was a rare breeder in Ontario wherever suitable rock cliffs were available. It appears as though the species was never common in this province during this century. Breeding without human help has not been noted in Ontario since 1963. The program of reintroduction began in Algonquin Prov. Park in 1977 as a project of the MNR, using young Peregrines raised in captivity at a CWS breeding facility. The MNR's Toronto release project began in 1981, and subsequent releases from various tall buildings took place annually through 1985. Other urban releases have taken place in Ontario at Brockville, Arnprior

and Brock University in St. Catharines. The greatest threat to the continued existence of this species is chemical contamination as a result of human activity. In some parts of the Peregrine's breeding range, taking of eggs or young for use in falconry poses an additional threat.

Nesting habitat of the Peregrine most commonly includes cliffs and crags, especially those situated near water. Tall buildings are accepted by birds nesting in urban settings. Some eyries, situated on cliff faces or skyscraper ledges, have a long history of continued use by Peregrines. Courtship display is elaborate, lasting up to 10 weeks in southern areas. Incubation requires about 32 days and is assisted by the male. Female Peregines are very aggressive in defence of the nest. Young fledge some 35 to 40 days following hatching, after which they remain dependent for some time upon their parents.

The only confirmed breeding record during the atlas period, despite a systematic check of historical breeding sites, was of 2 released birds which nested on a church tower in Arnprior in 1983. The story of this ill-fated nesting is well known; the female was found shot nearby on 20 August and the 2 young were not seen after that date (Rigden and Lang-Runtz 1984). A nest was found in northern Ontario in 1986, with indications that the birds had been present in 1985. This location and those of 2 other records of birds in suitable habitat are not shown on the atlas maps for fear of potential disturbance of the sites.

F. p. anatum is classified as endangered in Canada by COSEWIC. The Peregrine Falcon is also listed as endangered in Ontario by the MNR, and is protected by the Endangered Species Act, and the Game and Fish Act.-- *R.D. Weir*

PEREGRINE FALCON

Summary of Abundance Estimates

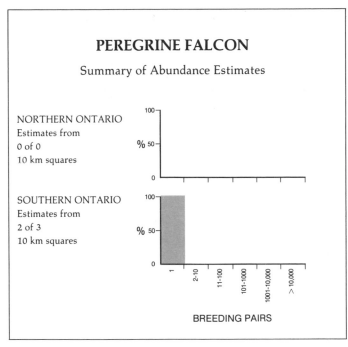

NORTHERN ONTARIO
Estimates from
0 of 0
10 km squares

SOUTHERN ONTARIO
Estimates from
2 of 3
10 km squares

BREEDING PAIRS

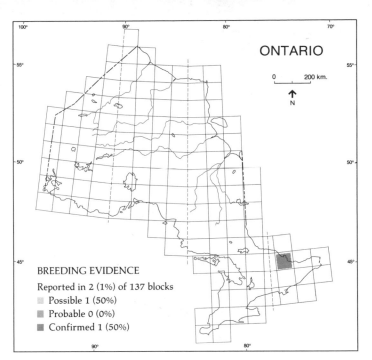

ONTARIO

0   200 km.

N

BREEDING EVIDENCE
Reported in 2 (1%) of 137 blocks
■ Possible 1 (50%)
■ Probable 0 (0%)
■ Confirmed 1 (50%)

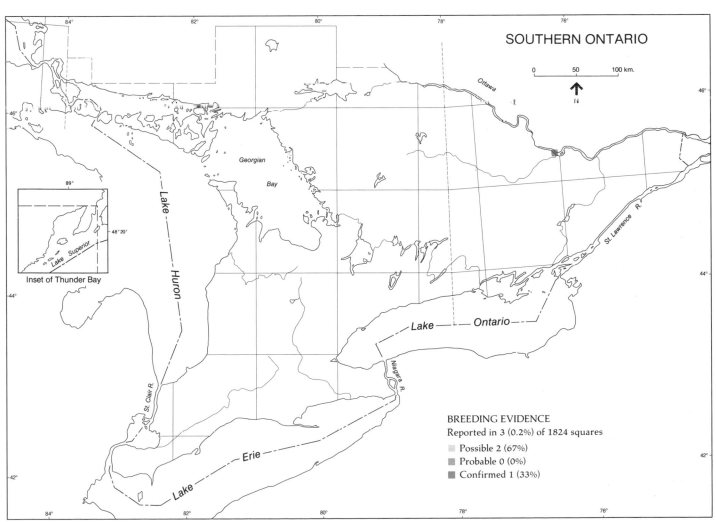

SOUTHERN ONTARIO

0   50   100 km.

N

Georgian Bay

Lake Huron

Lake Superior

Inset of Thunder Bay

St. Clair R.

Niagara R.

Lake — Ontario —

Lake — Erie —

Lake

Ottawa

St. Lawrence R.

BREEDING EVIDENCE
Reported in 3 (0.2%) of 1824 squares
■ Possible 2 (67%)
■ Probable 0 (0%)
■ Confirmed 1 (33%)

131

GRAY PARTRIDGE
Perdrix grise
Perdix perdix

The Gray Partridge is an introduced species, which thrives on agricultural land in those areas where land-use and relatively light rainfall provide suitable habitat. It is a sociable species, often living in coveys of up to 20 birds during the winter.

The Gray Partridge ranges widely in Europe and Asia as far east as Chinese Turkestan (Cramp and Simmons 1980). Attempts to introduce it to North America started in the late 1700's, but success did not come until the late 1800's. It is now established as a year-round resident in 19 states and 9 provinces. Gray Partridges do best in areas with less than 10 cm of rain in June and July (Twomey 1936). In much of southern Ontario, rainfall is within this limit; therefore land-use and frequency of ice storms appear to determine survival. The Gray Partridge thrives on rich soils and, in Ontario, usually occupies areas where the soil is clay, clay-loam, and sandy loam. Although the highest densities are found in southeastern Ontario on the flat clay plain of the Nation River valley, it also does well on gently rolling topography.

When Gray Partridges were abundant in the 1950's and 1960's, the best range in Prescott, Stormont, and Dundas Counties supported one bird per 4 ha, and age ratios suggested that this was one of the most productive populations in the world (Dawson 1963). At that time, mixed farming was the dominant land-use. Fields were small, snake-rail fences were common, and drainage ditches were overgrown with weeds and grasses. Gray Partridges usually nest in hay fields, among long grass in fence rows, and in weedy corners of fields. Most important of all for this species, haying operations started about 28 June, after the peak of the hatch (Dawson 1963). By the time that field work for the atlas was in progress, much had changed. Agriculture has intensified

and fall plowing is widespread. Small fields and snake-rail fences are largely gone, and alfalfa is now grown for livestock, and is cut during the period from 10 to 15 June, when most Gray Partridge nests are still being incubated. Densities have declined sharply through eastern Ontario as a result. Changes in land-use have improved Gray Partridge densities in the sandy parts of the Haldimand-Norfolk RM. In the 1950's, these soils were devoted to tobacco farming, and there were no birds. From 1962 to 1984, the acreage of shell corn and soybeans increased by a factor of 6 and the birds have moved in and are now increasing.

The first introductions were made in Ontario in 1909. Between 1927 and 1938, 3,832 Gray Partridges were shipped from the provincial game farms to nearly every county and territorial district in the province. Most of these plantings failed. However, by 1962, birds survived in 25 counties (Dawson 1963), in Timiskaming District, and in the Thunder Bay area. Atlas data suggest that now only 17 of these counties are occupied. Birds still survive near the grain elevators at Thunder Bay (they were reported again in 1986, J. McNicol pers. comm.) but were missed by atlassers; the Timiskaming birds have gone.

In spring, males utter a harsh metallic 'scirl' in advertisement and threat (Cramp and Simmons 1980), which serves to space out pairs. They do not defend conventional territories like most other birds (Jenkins 1961). Gray Partridge nests are hard to find because the female covers the nest when she is absent. She will readily renest if the first clutch is destroyed, and 20% of the chicks hatch later than 19 July. Nests are well concealed, and the female sits very tight, which may explain why only 5% of the atlas records are of nests. Incubation lasts for 23 to 25 days, and then the broods disappear into growing crops and pasture. They are very secretive and silent at this time of year, but an atlasser may occasionally see a brood crossing a road. Sometimes a dusting place is found containing feathers, tracks, or droppings, which indicate that chicks are present.

The Gray Partridge is almost certainly underrepresented on the maps because of its secretive behaviour during the breeding season. Breeding was confirmed in 47% of the squares in which it was found. Because the birds breed as yearlings, one can be confident that nearly all of the squares with possible and probable records contain breeding pairs. It is possible that additional areas have a few birds missed by the atlas. Gray Partridges are relatively easy to find in winter, when they enter farm yards to feed on manure piles. The birds are very sedentary, and any square in which they are found will almost certainly contain a breeding pair the next summer.

Of the 15 abundance estimates of more than 10 pairs, all but one is in extreme southeastern Ontario, indicating that the population there is still larger than the one in the southwest.-- *H.G. Lumsden*

GRAY PARTRIDGE

Summary of Abundance Estimates

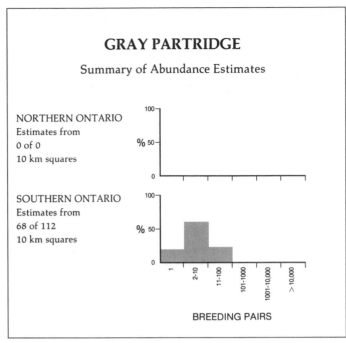

NORTHERN ONTARIO
Estimates from
0 of 0
10 km squares

SOUTHERN ONTARIO
Estimates from
68 of 112
10 km squares

100

% 50

0

100

% 50

0

1 2-10 11-100 101-1000 1001-10,000 > 10,000

BREEDING PAIRS

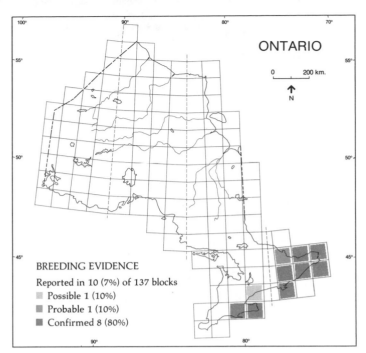

ONTARIO

0   200 km.

↑
N

BREEDING EVIDENCE

Reported in 10 (7%) of 137 blocks

　Possible 1 (10%)
　Probable 1 (10%)
　Confirmed 8 (80%)

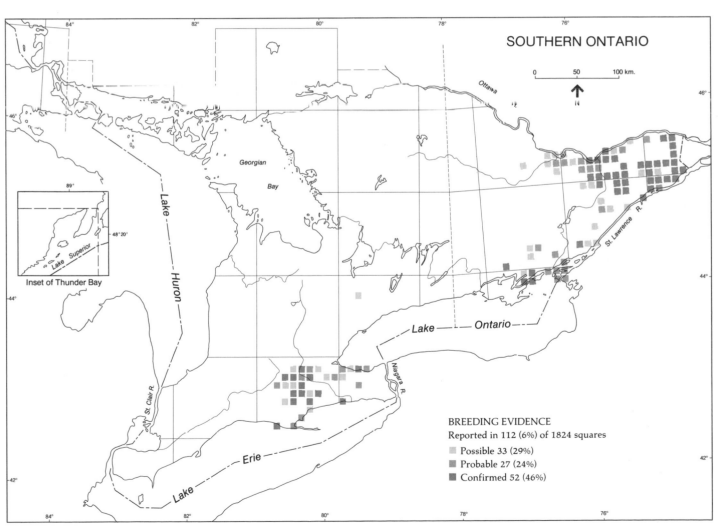

SOUTHERN ONTARIO

0   50   100 km.
↑

Georgian
Bay

Lake Huron

Lake Superior
48°20'
89°
Inset of Thunder Bay

St. Clair R.

Lake —— Ontario ——

Ottawa

St. Lawrence R.

Niagara R.

Lake —— Erie ——

Lake

BREEDING EVIDENCE

Reported in 112 (6%) of 1824 squares

　Possible 33 (29%)
　Probable 27 (24%)
　Confirmed 52 (46%)

RING-NECKED PHEASANT
Faisan de chasse
Phasianus colchicus

The cock Ring-necked Pheasant is one of the most colourful of Ontario birds. Modern land-use has not been kind to the bird, and it has been reduced to a fraction of its former numbers.

In Eurasia, the Ring-necked Pheasant ranges from the delta of the Volga River to Korea and Taiwan. It has been successfully introduced in Europe, Hawaii, Japan, New Zealand, North America (Vaurie 1965), and Australia (Pizzey 1980). In Canada, it has been introduced to all provinces except Newfoundland. There is virtually no winter survival of Ring-necked Pheasants in those parts of Ontario having more than 178 cm of snow. Up to 127 cm can be considered possible Ring-necked Pheasant range. Frequency of ice storms also affects survival, as does the distribution of cropland, woodland, marsh, and scrub. When these factors are put together, one can delineate the area (south of the line on the map) where the species can be expected to survive on its own (Clarke and Braffette 1947). A large number of breeders throughout southern Ontario stock pheasants, which accounts for the numerous and widely scattered squares with records of this species north of the winter survival line. A few of these plantings may persist because of artificial winter feeding and unusual habitat combinations.

In winter, Ring-necked Pheasants shelter in woodlots, scrub, or dense cattail marshes. Croplands growing shell corn and soybeans are needed for winter food. For nesting, the species needs grassland cover; in many studies, fence rows proved to be most attractive, followed by hayfields, but dense stands of weeds were preferred on Pelee Island (Stokes 1954). Broods remain in cover in growing crops, swales, or the edges of marshes. They use stubble when the small grain crops are cut, and move to corn fields in the fall. When cover, winter food, and nesting habitat are spaced far apart, survival of the Ring-necked Pheasant is spotty and densities are low.

Pheasants were first released in Ontario in the late 1800's, and the first success recorded was on Point Pelee, where they were abundant by 1901. About the same time, the bird became established on the Niagara Peninsula following releases in 1897. The first hunting season was in 1910. In 1919, the province imported eggs, and in 1922 established a hatchery and distributed eggs to farmers. In 1927, 6 week old poults were shipped all over the province, and Pelee Island probably received its first birds. Subsequently, Ring-necked Pheasants became established at the southern end of Lake Simcoe and in the Lake Ontario counties as far east as Prince Edward Co and Kingston (Clarke and Braffette 1947, Snyder 1951).

Following World War II, the introduction of herbicides and elimination of fence rows and hedges degraded the habitat, and Ring-necked Pheasant populations declined. Pelee Island, which had fall populations of 33,500 to 38,200 in 1949 and 1950 (Stokes 1954), now has a natural recruitment of only a few hundred birds (D. McTavish pers. comm.), and the famous pheasant hunt is maintained by stocking.

The Ring-necked Pheasant is difficult to atlas, although the species is not as secretive as the Northern Bobwhite or Gray Partridge. From March to May, cock pheasants are conspicuous when they crow and whirr their wings on their territories. Crowing is at its highest intensity 15 to 45 minutes before sunrise; it declines on rainy days, and its frequency is highest from 0^0 to 10^0C. Cock pheasants may crow in response to explosions, thunder, and tape recordings. The Ring-necked Pheasant is a persistent renester, and active nests with eggs may be found from late April to August. Most nesting occurs in May (Peck and James 1983). Broods were by far the most frequently reported category of confirmed breeding, constituting 82% of such records. Broods do not break up until the young are about 14 weeks old.

The maps give a good picture of the distribution of the Ring-necked Pheasant in Ontario. The large blank areas within the region of winter survival are real. Intensive agriculture has created large areas where the species can no longer survive.

Abundance estimates indicate that the Ring-necked Pheasant is fairly common within its range. However, wild pheasants harvested by hunters in the 1980 to 1981 small game hunter report survey (MNR files) constituted about 47,000 birds. This suggests that atlas abundance estimates are generally too low.-- *H.G. Lumsden*

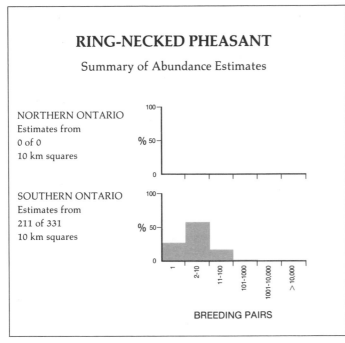

RING-NECKED PHEASANT

Summary of Abundance Estimates

NORTHERN ONTARIO
Estimates from
0 of 0
10 km squares

SOUTHERN ONTARIO
Estimates from
211 of 331
10 km squares

BREEDING PAIRS

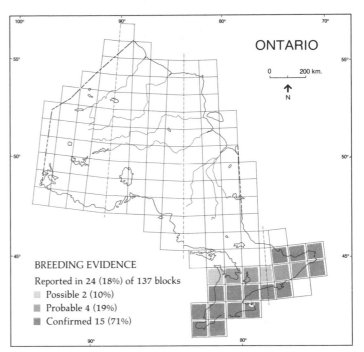

ONTARIO

0   200 km.

N

BREEDING EVIDENCE
Reported in 24 (18%) of 137 blocks
Possible 2 (10%)
Probable 4 (19%)
Confirmed 15 (71%)

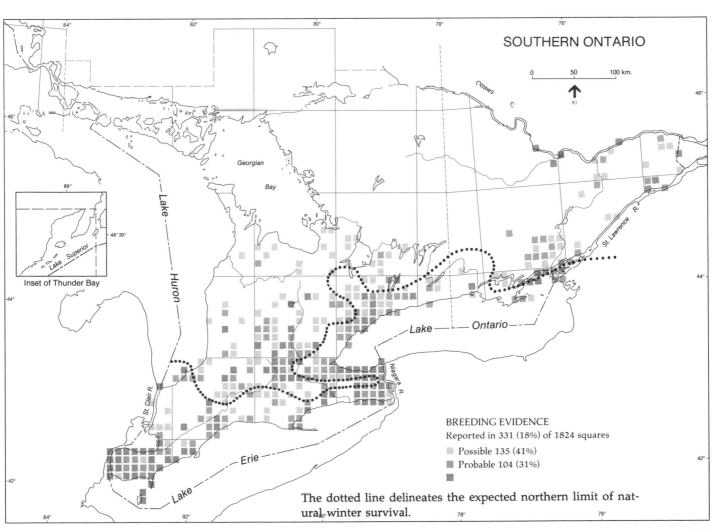

SOUTHERN ONTARIO

0   50   100 km.

N

Inset of Thunder Bay

BREEDING EVIDENCE
Reported in 331 (18%) of 1824 squares
Possible 135 (41%)
Probable 104 (31%)

The dotted line delineates the expected northern limit of natural winter survival.

SPRUCE GROUSE
Tétras du Canada
Dendragapus canadensis

The Spruce Grouse is remarkably approachable, and has exciting courtship and aggressive displays. It occurs only in North America, where its distribution roughly coincides with the Boreal Forest zone. Since its successful introduction to Newfoundland, it is a permanent resident of all Canadian territories and provinces except Prince Edward Island. The Spruce Grouse may attain its greatest abundance in Ontario, where it occupies a variety of jack pine and black spruce forests (Szuba and Bendell 1983). Formerly thought to prefer mature forests and spruce bogs, it is now known to be most abundant in upland black spruce and, particularly, pure young jack pine forests with a ground cover of ericaceous plants such as blueberry and trailing arbutus.

The past and present range of the Spruce Grouse in Ontario probably coincide, except in the extreme south. Historical records show that the southern limit of its range in the province has receded northward and westward. Bull (1936) claimed it was "plentiful" in Peel Co before 1900 but disappeared along with the forests. Several authors list specimens and sightings obtained from 1866 to 1910 near Lake Simcoe and as far south as Guelph (Rand 1948, Devitt 1967, Mills 1981, Speirs 1985). Baillie and Harrington (1936) said it once nested near Ottawa. Its disappearance from its former range is undoubtedly the result of modern land-use practices having altered suitable habitats, and possibly because the Spruce Grouse is a popular game bird.

Many northern blocks have no sightings, suggesting that the Spruce Grouse may be easily missed by atlassers. It does not flush readily (hence its other name, 'fool hen'), and is cryptically coloured when not displaying. It is usually quiet, but territorial males perform loud 'flutter flights,' and females give loud aggressive calls in late April and early May. At that time, both sexes respond to tape-recorded calls of female

Spruce Grouse. However, the short display period ends before most migratory birds and atlas volunteers have returned to the breeding range. Nests were found in only 11 squares, which is not surprising because the inconspicuous ground nests are extremely difficult to find, the cryptic hens not flushing until almost touched. Breeding is easiest to confirm after hatch in June; chicks were observed in 84% of squares with confirmed records. Because of nest depredation and chick mortality, 50% of the hens in a population may be broodless in some years. Males and broodless hens moult in June and July, becoming so secretive that they are difficult to locate even with excellent pointing dogs trained specifically to find Spruce Grouse. Unsuccessful breeding attempts are likely to be missed, particularly where populations are sparse.

Prior to the atlas period, the Spruce Grouse was generally thought to be absent south of Algonquin Prov. Park. However, atlas field work located single birds in suitable habitat in the vicinity of Bon Echo Prov. Park and Weslemkoon, 40 km and 30 km southeast of Bancroft, respectively. Nevertheless, as expected, the majority of southern Ontario records are from Algonquin Prov. Park, mostly from squares along the Highway 60 corridor at the south end of the Park. A few records are scattered between there and the more coniferous Timagami Forest Section north of Sudbury. The greatest concentration of confirmed Spruce Grouse records extends from that latitude northward to the Hudson Bay Lowland to the south of James Bay in northeastern Ontario. The degree of confirmation there probably reflects abundance; it is in this area that populations are thought to exceed 50 birds/km^2 in prime habitat, 2 to 4 times denser than has been reported elsewhere in North America (Szuba and Bendell 1983). The species is widespread throughout the rest of northern Ontario, but sporadic in northcentral and northwestern Ontario and on the Hudson Bay Lowland. The pattern of occurrences suggests that the Spruce Grouse probably occurs in most northern blocks, but was overlooked because of insufficient coverage. The open bog and fen habitat of the Lowland is probably not well suited to Spruce Grouse, though the more heavily wooded sections on well drained land appear to be. The most northerly confirmed atlas record is of a female with chicks in spruce forest on the banks of the Fawn River, although breeding was previously confirmed north to Fort Severn (Peck and James 1983).

Abundance estimates indicate, as might have been predicted, that the Spruce Grouse is considerably more common in northern Ontario than in the south.-- *K.J. Szuba and B.J. Naylor*

SPRUCE GROUSE

Summary of Abundance Estimates

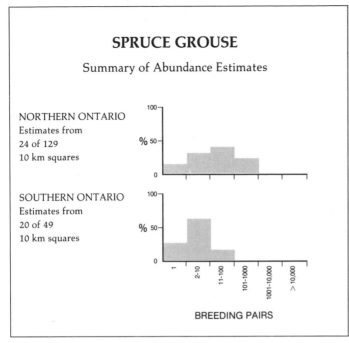

NORTHERN ONTARIO
Estimates from
24 of 129
10 km squares

SOUTHERN ONTARIO
Estimates from
20 of 49
10 km squares

BREEDING PAIRS

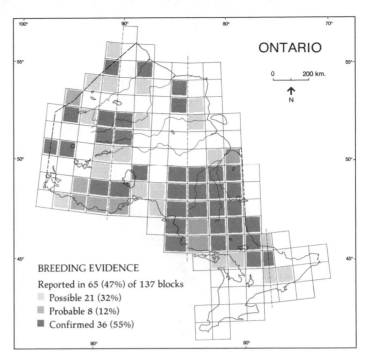

ONTARIO

0   200 km.

N

BREEDING EVIDENCE
Reported in 65 (47%) of 137 blocks
Possible 21 (32%)
Probable 8 (12%)
Confirmed 36 (55%)

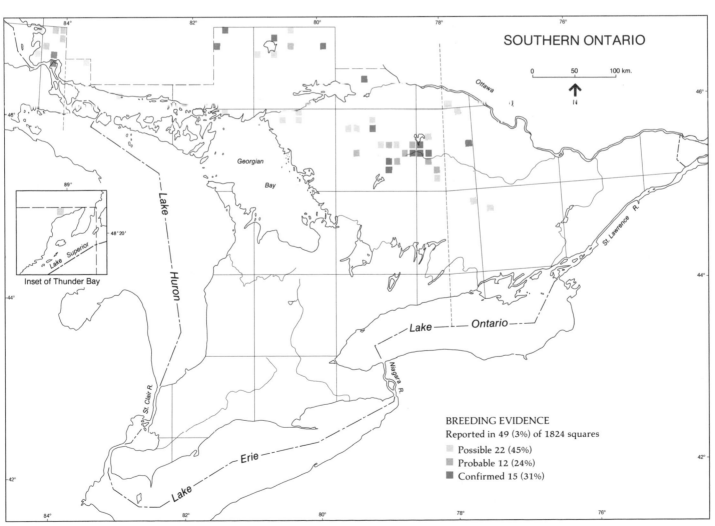

SOUTHERN ONTARIO

0   50   100 km.

N

Inset of Thunder Bay

BREEDING EVIDENCE
Reported in 49 (3%) of 1824 squares
Possible 22 (45%)
Probable 12 (24%)
Confirmed 15 (31%)

RUFFED GROUSE
Gélinotte huppée
Bonasa umbellus

The Ruffed Grouse is commonly and erroneously referred to by a number of vernacular names, including 'partridge' and 'pheasant' (Johnsgard 1973). Its name is derived from the presence of elongated dark feathers on the neck, which, when elevated during display, resemble a mediaeval ruff. This species is well known to many Ontario residents because of its status as a game bird.

The Ruffed Grouse is widely distributed throughout North America. Its range extends from the east to the west coast and from Alaska as far south as South Carolina. This range comprises a surprisingly wide variety of forest community types, from temperate coniferous rain forest to relatively dry deciduous forest (Johnsgard 1973). Important to its occurrence is the presence of poplar and birch trees. In Ontario, it is a widespread, permanent resident across most of the province, with the exception of the extreme southwestern regions (where suitable habitat is now probably greatly reduced) and the Hudson Bay Lowland (where unsuitable habitat predominates) (James *et al.* 1976, Speirs 1985). Two distinct colour phases can be observed among Ruffed Grouse, and 4 of the 11 recognized subspecies (Johnsgard 1973) have been identified within the province (James *et al.* 1976).

The Ruffed Grouse is known for dramatic changes in its population density; historically, numbers have fluctuated on an approximately 10 year cycle (Johnsgard 1973). Such fluctuations, though still apparent in well forested northern parts of its range, are non-existent in the agricultural south, where the species is now largely restricted to disjunct woodlots. Where the cycle does take place in Ontario, the peak occurs around the end of the decade, so numbers should have been high in the early years of the atlas project. The decline is quite rapid, so populations would be expected to be low from at least the middle of the project onwards (H.G. Lums-den pers. comm.).

Because of its secretive nature, the Ruffed Grouse is difficult to observe under most conditions. However, signs of breeding activity become apparent in late March with the start of 'drumming' by the males. This unusual sound is not produced vocally, but by the beating of the wings and serves to attract the females to the drumming log near which mating occurs. Drumming is concentrated at dawn and dusk, but can be heard during the day, and all through the night. The female builds a well concealed nest, usually at the base of a tree or under a fallen log, where the large clutch of about 11 buff-coloured eggs is laid. After an incubation period of about 23 days, the downy young hatch and leave the nest within a few hours. The young are able to fly when about 10 days old. However, the brood remains intact with the female until the chicks are about 12 weeks old. Apart from display and copulation, the male grouse plays no part in the nesting cycle. The females, difficult to find at most times, are particularly evasive during the breeding season; one must nearly step on the hen before she will leave the nest. However, females are obtrusive and vocal when their brood is approached, often drawing attention to their location when young would not otherwise have been suspected. Family groups are quite readily located when the young are close to fledging, which explains why 78% of confirmed breeding records were based on observations of broods.

The impact of agriculture and urbanization is reflected on the southern Ontario map. The species is largely absent from the extreme southwest of the province, along the Niagara River, and in the heavily urbanized Toronto region. By contrast, its absence from much of the Hudson Bay Lowland reflects a historic distribution based on an absence of appropriate woodland. Because it is incapable of prolonged, continuous flight, it has been unable to colonize offshore islands in Georgian Bay, Lake Erie, and elsewhere. The Ruffed Grouse likely breeds in most of the squares where possible and probable breeding were reported; it is sedentary and resident, so its occurrence in any square or block during the breeding season indicates that breeding is likely.

In addition to being a widely distributed species, the Ruffed Grouse is also common. This is supported by the atlas data, which show that the estimated density varied between 2 and 100 pairs in about 80% of all squares for which estimates were provided. The estimated abundance is very similar in northern and southern Ontario, suggesting that the species is fairly evenly distributed across its Ontario range.-- *A.L.A. Middleton*

RUFFED GROUSE

Summary of Abundance Estimates

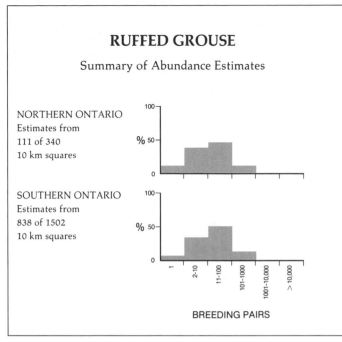

NORTHERN ONTARIO
Estimates from
111 of 340
10 km squares

SOUTHERN ONTARIO
Estimates from
838 of 1502
10 km squares

%

100

50

0

%

100

50

0

1   2-10   11-100   101-1000   1001-10,000   >10,000

BREEDING PAIRS

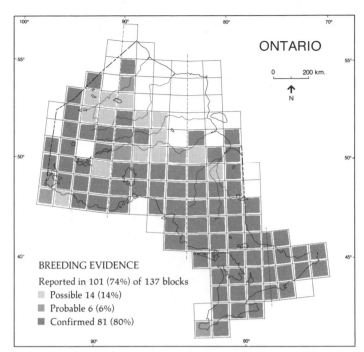

ONTARIO

0   200 km.

N

BREEDING EVIDENCE

Reported in 101 (74%) of 137 blocks

Possible 14 (14%)
Probable 6 (6%)
Confirmed 81 (80%)

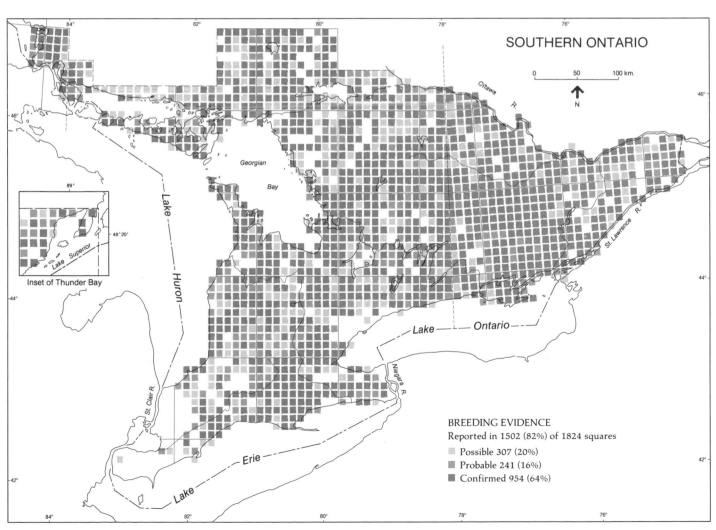

SOUTHERN ONTARIO

Ottawa R.

0   50   100 km.

N

Georgian Bay

Lake Huron

Lake Superior

48°20'

89°

Inset of Thunder Bay

St. Clair R.

Lake Erie

Niagara R.

Lake Ontario

St. Lawrence R.

BREEDING EVIDENCE

Reported in 1502 (82%) of 1824 squares

Possible 307 (20%)
Probable 241 (16%)
Confirmed 954 (64%)

SHARP-TAILED GROUSE
Gélinotte à queue fine
Tympanuchus phasianellus

The Sharp-tailed Grouse ranges through the central and western US into the mountains and north to Alaska. In Canada, it is found from west-central Quebec to the central Yukon and locally from Ontario to British Columbia. In Ontario, there are 2 races that are indistinguishable in the field. The Northern Sharp-tailed Grouse (*T. p. phasianellus*) ranges throughout the Hudson Bay Lowland and as far south as Upsala and the north shore of Lake Superior. The Prairie Sharp-tailed Grouse (*T. p. campestris*) has a disjunct distribution in release areas (see below) on Manitoulin Island, and in the Sault Ste Marie area, the western Rainy River District, and the Dryden area. The species is absent as a breeding bird from a broad strip of forested country east of Lake Superior north of the settled shores of Lake Huron to roughly Matheson, Foleyet, and White River. Two habitat types are occupied in Ontario. The Northern race breeds on muskeg in relatively open fen and bog with shrubs and scattered trees. The Prairie form inhabits grasslands and shrubby areas on limestone plains, the fringes of agriculture, abandoned pastures, and logged or burned-over areas.

The Prairie Sharp-tailed Grouse may not have been native to Ontario prior to settlement. The Sault Ste Marie area and Manitoulin Island were invaded from Michigan in the late 1940's (Lumsden unpubl.). There seems to be no record of the species in the western Rainy River District before settlement. It is likely that Sharp-tailed Grouse spread there from Minnesota and perhaps moved from Manitoba to the Dryden area as settlement progressed.

The northern populations fluctuate widely in numbers and are migratory in parts of their range. In the autumns of 1865, 1896, and 1932, exceptional irruptions took place, and the birds moved south in very large numbers. In 1932, the birds spread over 100,000 km^2 south of their normal breeding grounds, reaching Manitoulin Island, Bracebridge, and Bancroft (Snyder 1935, 1951). Northern Sharp-tailed Grouse are not adapted to grassland habitats and they soon disappeared from those areas to which they had irrupted. In the southern fringes of its breeding range, the Northern race is probably more abundant now than it was before settlement. Logging and fires have opened up the forests and provided more extensive open, brushy habitat. However, more intensive farming in the Rainy River-Dryden area has reduced the abundance of the Prairie race from that reported by early settlers.

Seven attempts have been made to introduce the Sharp-tailed Grouse to southern and eastern Ontario. The first release was made at Galt in 1894 (Snyder 1935). In 1925-26, more were stocked near Flesherton (Baillie 1947b). Following the irruption of 1932-33, birds were trapped and held for later release at a number of sites in southern Ontario. None survived from these 3 plantings (MNR files). Four recent releases were made of Prairie Sharp-tailed Grouse trapped in winter in the Rainy River District. Balsam Lake Prov. Park (Victoria Co) received birds in 1963 and 1964, Spencerville (Grenville Co) in 1970 and 1971, Camden Lake (Lennox and Addington Co) in 1972 and 1973, and Point Petre (Prince Edward Co) in 1975. The birds spread and still survive southwest of Balsam Lake near Beaverton (A. O'Donnell pers. comm.), in the Camden Lake area (T. Humberstone pers. comm.), and in the Spencerville area (W. McClure pers. comm.), but were missed by atlassers. Near Point Petre, atlassers recorded possible breeding in one square.

The Sharp-tailed Grouse is a 'lek' species. In Ontario, 2 to 20 males may use the same dancing ground annually. The location is usually on a knoll in the open or in a lightly treed area in the bush. Males display morning and evening in spring and fall. The most intensive display occurs from late March to May, 45 minutes before dawn, and ceases at sunup. The cooing and cackling calls can be heard 1.6 km away. The peak of mating occurs in the week of 25 April in the southern part of its range and later in the north. Nests are well hidden in cover, and the clutch, which averages 12 eggs, hatches in 21 days. The female sits tight during the second half of incubation and usually performs a vigorous distraction display when disturbed with chicks. Outside of the lek season, the Sharp-tailed Grouse is secretive and easily overlooked. Atlassers found no nests, but did locate broods in 26 squares.

The map does not outline the distribution of the Northern Sharp-tailed Grouse well. Its habitat is widely distributed, and it likely breeds in most if not all blocks north of 48°N, except in the tundra area of Cape Henrietta Maria and near the Pen Islands. The Prairie Sharp-tailed Grouse is fairly well represented on the maps. The 3 abundance estimates of more than 10 pairs per square are from the Sault Ste Marie area.-- *H.G. Lumsden*

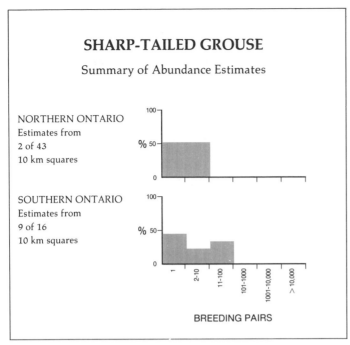

SHARP-TAILED GROUSE

Summary of Abundance Estimates

NORTHERN ONTARIO
Estimates from
2 of 43
10 km squares

SOUTHERN ONTARIO
Estimates from
9 of 16
10 km squares

%

BREEDING PAIRS

1 2-10 11-100 101-1000 1001-10,000 >10,000

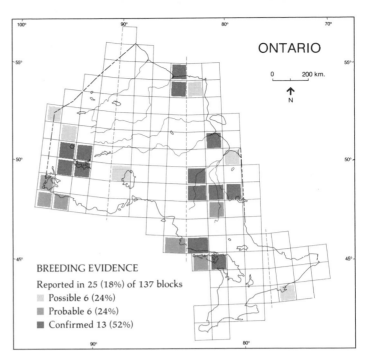

ONTARIO

0   200 km.

N

BREEDING EVIDENCE
Reported in 25 (18%) of 137 blocks
Possible 6 (24%)
Probable 6 (24%)
Confirmed 13 (52%)

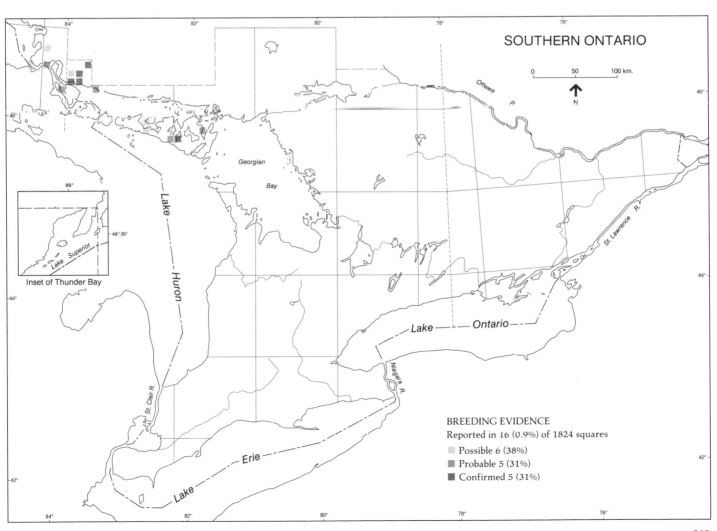

SOUTHERN ONTARIO

0   50   100 km.

N

Inset of Thunder Bay

Lake Huron

Georgian Bay

Ottawa R.

St. Lawrence R.

Lake Ontario

Niagara R.

St. Clair R.

Lake Erie

BREEDING EVIDENCE
Reported in 16 (0.9%) of 1824 squares
Possible 6 (38%)
Probable 5 (31%)
Confirmed 5 (31%)

WILD TURKEY
Dindon sauvage
Meleagris gallopavo

The Wild Turkey returned to southern Ontario in 1984 through introduction of wild-caught birds from the US. It had disappeared from much of its historical range in eastern North America by the late 19th century (Williams 1981). The decline of the eastern Wild Turkey is generally attributed to loss of hardwood habitat (Clarke 1948, Williams 1981). However, re-established birds in the US have proved very adaptable, occupying a wide range of forest and adjacent agricultural lands. Agricultural crops, particularly waste corn or other grains, are important to the birds. A source of year-round water and seeps, and associated green foliage are other characteristics of good Wild Turkey habitat. Although the bird is extremely wary, the wide range of habitat the species now occupies in the US suggests that unregulated hunting also contributed to its decline (Williams 1981).

In Ontario, the Wild Turkey was extirpated in the early 1900's, with the last birds possibly persisting until 1909. The original range of the species was in southernmost Ontario, extending to just east of Toronto and to just north of Barrie (Alison 1976). Numerous attempts to reintroduce the Wild Turkey by government, sportsmen's groups and interested individuals, through the release of pen-reared stock, failed to establish lasting or expanding populations. Those flocks that persisted are reportedly maintained through the winter by artificial feeding (Alison 1976). However, since the early 1940's, the capture and immediate relocation of wild birds proved extremely successful in restoring the Wild Turkey to abundance in many areas of the US.

The Wild Turkey is a difficult species to atlas, because of its wariness and relatively secretive habits. The birds spend much of their time in wooded cover, usually out of sight of observers. However, male Wild Turkeys (gobblers) call or 'gobble' in the spring to attract females (hens). The distribu-

tion of the birds can therefore be assessed from gobbling counts. Nests are on the ground in cover, and are difficult to find. A nest with eggs was reported in one square, and used nests were reported in two. Hens with broods can sometimes be observed feeding in open areas, particularly in fields, when the broods become older; 70% of breeding confirmations were of broods. Wild Turkeys are usually most visible in the late fall or winter, when they form flocks that can sometimes be observed feeding in fields.

All mapped atlas data originates from birds that were known or believed to have survived in the wild for at least a year. Nevertheless, interpretation of atlas results requires that a distinction be made between breeding evidence for reintroduced wild-caught birds and that for birds which originated from the release of pen-reared stock. Only breeding data for the area of Haldimand-Norfolk RM and Elgin Co in southwestern Ontario (5 contiguous squares), and for the area of Hastings and Northumberland Counties (3 contiguous confirmed records) in southeastern Ontario, should be considered as breeding evidence for truly wild birds. All other breeding evidence reported for this species derives from the release of pen-reared stock and does not represent truly wild populations. Release of pen-reared stock to the wild is illegal and is discouraged because of potential genetic and disease effects on wild populations.

It is expected that wild-caught birds will spread or be introduced to many parts of southern Ontario within a few years. From 1984 to 1986, a total of 253 wild-caught birds from various US states have been stocked in southern Ontario in 5 different locations. Stockings of 47 and 49 Michigan birds, in 1984 and 1985, respectively, in the easternmost sites near Trenton resulted in an estimated population of 700 birds by late 1985. The area south of Delhi in Haldimand-Norfolk RM was stocked with 27 Missouri birds in 1984. An additional 14 and 15 Iowa birds (originally from Missouri stock) were released in the same area in 1985 and 1986, respectively. By the fall of 1986, a conservative estimate of 175 birds was made for this population. However, numbers are probably higher, because of the wary or secretive nature of the Wild Turkey.

In early 1986, a total of 70 New York and 31 Vermont Wild Turkeys (of the same stock) were released in 3 different areas of southcentral Ontario. A total of 34 birds were released in Short Hills Prov. Park near Fonthill, 35 birds in Tosorontio Township, Simcoe Co, and 32 birds in South Dumfries Township, Brant Co. All 3 of these stockings appear to have been successful, with broods reported at each of the 2 former sites and sightings of birds reported at the latter site.-- *J.L. Harcus*

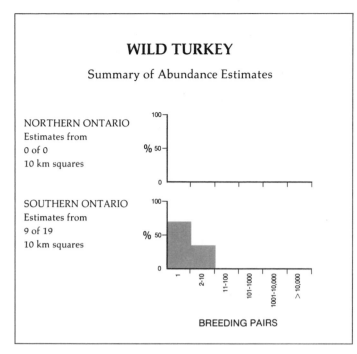

WILD TURKEY

Summary of Abundance Estimates

NORTHERN ONTARIO
Estimates from
0 of 0
10 km squares

SOUTHERN ONTARIO
Estimates from
9 of 19
10 km squares

BREEDING PAIRS

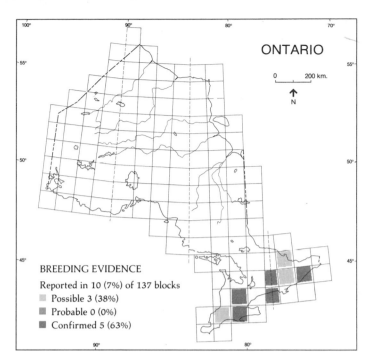

ONTARIO

0   200 km.

↑
N

BREEDING EVIDENCE

Reported in 10 (7%) of 137 blocks

▢ Possible 3 (38%)
▨ Probable 0 (0%)
▪ Confirmed 5 (63%)

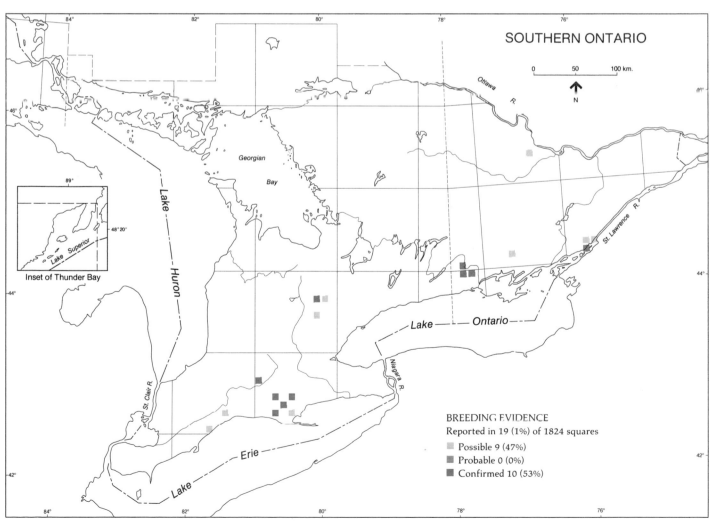

SOUTHERN ONTARIO

0   50   100 km.

↑
N

Inset of Thunder Bay

BREEDING EVIDENCE

Reported in 19 (1%) of 1824 squares

▢ Possible 9 (47%)
▨ Probable 0 (0%)
▪ Confirmed 10 (53%)

NORTHERN BOBWHITE
Colin de Virginie
Colinus virginianus

The Northern Bobwhite ranges in North America from the Atlantic coast to the Great Plains, and south to Mexico. It reaches the northern limit of its distribution in southern Ontario. It has been successfully introduced to 4 western states, British Columbia, the West Indies, and New Zealand (AOU 1983). The Northern Bobwhite needs grassland such as prairie, hayfields, drainage ditches, and road-side and pond verges for nesting cover. It also needs woody cover in the form of thickets, tangles of vines, and shrubs on fence rows, or the borders of woodlots. These provide refuge from predators, shelter from inclement weather, and food in the form of weed seeds and fruits. Croplands growing corn, soybeans, and small grains provide important winter foods for the Northern Bobwhite. The bird must also have access to green material such as clover and grass in winter to satisfy its vitamin A requirements (Nestler 1946). Since the species is remarkably sedentary, these 3 habitat elements, grassland, brushy cover, and cropland, must be in close proximity to one another.

Before European settlement, the Northern Bobwhite was probably present in Kent and Essex Counties, where there were thousands of hectares of long-grass prairie. Bobwhites could have thrived on the brushy and forested edges of those prairies. The earliest report is that of the Commandant of the French settlement at Detroit, M. de Lamothe Cadillac, who mentioned an abundance of game including quail in a letter dated 1701 (Lajeunesse 1960). Settlement changed the distribution of the species, as the unbroken forests elsewhere were cleared, and the Northern Bobwhite colonized to the east and north. By 1810, it had spread to Norfolk Co, and by 1816 to Hamilton. By 1850, it had reached Owen Sound, southern Muskoka, and as far east as Kingston. Then hard winters and more intensive agriculture caused retreat, and by 1904 there were few left north of Hamilton and London (Clarke 1954). Herbicides and modern farming methods have caused further decline and shrinkage of the range of the species.

The Northern Bobwhite was imported and released as early as 1884 (Clarke 1954), but later was bred on game farms and was stocked in large numbers. Today, commercial breeders all over the province are licensed to raise and release Northern Bobwhites, and the birds may appear anywhere.

The Northern Bobwhite is easily found with hunting dogs, but, except in spring, it is not a conspicuous bird to the atlasser. From late March to June, the familiar 'bob-bob-white' whistling call is mostly uttered by unmated males, and does not necessarily indicate breeding. However, unmated males often associate with pairs, and paired males occasionally call (Klimstra 1950). The highest intensity of whistling occurs in the early morning and evening, and males will readily respond to tape recordings. Whistling birds were the most frequent indication of the presence of Northern Bobwhites (35% of records). The Northern Bobwhite is a persistent renester, and eggs may be found from May to September (Peck and James 1983). Nests are usually well concealed in cover near open or cultivated land, and were reported in only one atlas square; all other breeding confirmations were of broods. Because the Northern Bobwhite moves very little, a covey seen in fall or winter was almost certainly raised nearby.

The maps show clearly the concentration of Northern Bobwhites in Middlesex, Lambton, and Kent Counties. It is interesting to note that, within the Carolinian Forest zone, the distribution of the Northern Bobwhite is essentially complementary to that of the Ring-necked Pheasant. Competition may be a factor; the habitat carrying capacity seems to apply to the total number of competing gallinaceous birds, rather than to each species separately, but it is not clear how this competition works (Edminster 1954). Habitat differences may also be important. The scarcity of Northern Bobwhites in Essex Co is due to the absence of woody cover and grassland, but the presence of cattail in dredge cuts and marshes allows the Ring-necked Pheasant to persist. North of a line running from Hamilton through London, virtually all of the Northern Bobwhite records are likely to have resulted from stocking, and there can be little expectation of persistence and winter survival there.

Holdsworth (1973) estimated in 1972-73 that there were 1,055 covies (or 116 Northern Bobwhites per 10 km square) in winter in the 5 counties of Middlesex, Elgin, Lambton, Kent, and Essex. Converting atlas figures to winter estimates (75% pairs successful, 3.38 immatures per adult, Edminster 1954), there could have been 20 to 135 birds per 10 km square during the atlas years. The fact that Holdsworth's estimate is closer to the upper range of the atlas data suggests that there may have been a decline since 1972-73, which agrees with subjective assessments (P.A. Woodliffe pers. comm.).-- *H.G. Lumsden*

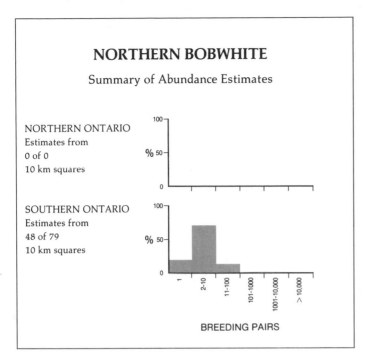

NORTHERN BOBWHITE

Summary of Abundance Estimates

NORTHERN ONTARIO
Estimates from
0 of 0
10 km squares

SOUTHERN ONTARIO
Estimates from
48 of 79
10 km squares

100

% 50

0

100

% 50

0

1   2-10   11-100   101-1000   1001-10,000   >10,000

BREEDING PAIRS

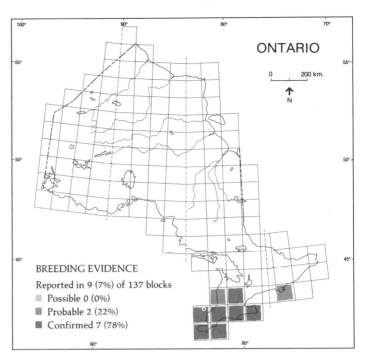

ONTARIO

0   200 km.

N

BREEDING EVIDENCE
Reported in 9 (7%) of 137 blocks
Possible 0 (0%)
Probable 2 (22%)
Confirmed 7 (78%)

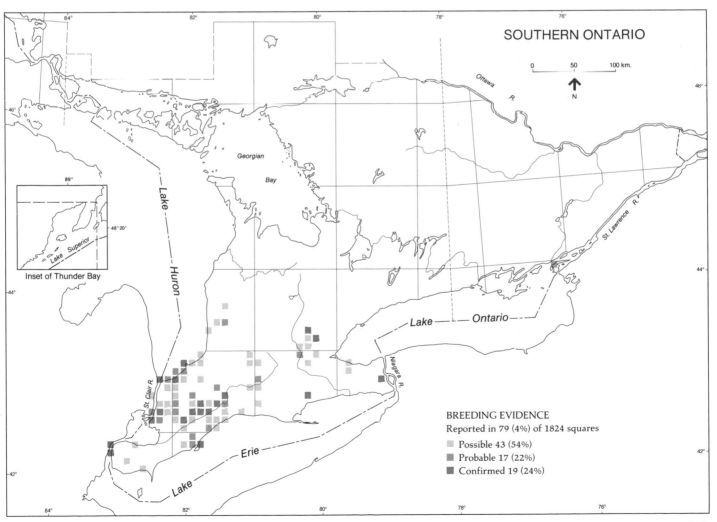

SOUTHERN ONTARIO

0   50   100 km.

N

Inset of Thunder Bay

BREEDING EVIDENCE
Reported in 79 (4%) of 1824 squares
Possible 43 (54%)
Probable 17 (22%)
Confirmed 19 (24%)

YELLOW RAIL
Râle jaune
Coturnicops noveboracensis

R. TUCKERMAN 1986

The shy and elusive Yellow Rail breeds across North America east of the Rocky Mountains, in a band extending from the northern US to the latitude of Hudson Bay. An isolated population also occurs in Mexico (AOU 1983). Its preferred breeding habitat consists of large, dense, grass and sedge marshes, where dead stems of the previous growing season form a mat-like canopy above shallow standing water (Bart *et al.* 1984). Only wetlands large enough to harbour a small group of territorial birds are likely to have Yellow Rails on a long-term basis (Anderson 1977). It is probably the least known member of its family in Ontario, and is among the most infrequently encountered of all the birds known to breed in the province. Taverner (1934) stated that "according to actual records, it is one of the rarest birds in Canada." The Yellow Rail winters primarily along the coasts of the US and Mexico.

Although the species was first recorded in Ontario (from the Severn River) as early as 1777 (Macoun and Macoun 1909), historical accounts of the distribution and abundance of the Yellow Rail in Ontario are sketchy, and it is difficult to assess whether its range has changed in recent years. McIlwraith (1894) considered it to be a rare breeder in the province, but reported records from Toronto and the eastern shore of James Bay. Baillie and Harrington (1936) knew of records ranging from the Lake St. Clair marshes in the south to Moose Factory in the north. The first nest of the Yellow Rail in Canada was discovered in the Holland Marsh in 1938 (Devitt 1939). It is therefore likely that the Yellow Rail historically had a widespread breeding distribution in the province. Today, following extensive draining of wetlands in southern Ontario, this portion of the population has been almost extirpated. The species has apparently always been most common in the extensive wetlands along the James and Hudson Bay coasts, but has not previously been known to breed in the Boreal Forest region.

The retiring habits of the Yellow Rail make it a particularly difficult species to census accurately. It is notoriously difficult to observe in its habitat, and, like other rails, is an expert at escaping quietly through the vegetation when disturbed. Therefore, it is rarely seen during the breeding season, and when it is, it might easily be confused with the more common Sora. It is best detected by means of its repetitive 'ticking' call, which is easily distinguished from the Sora's descending 'whinny' call. The call can be heard throughout the breeding season (most frequently prior to incubation), but is most often given towards dusk, and is perhaps best heard at night, when most other marsh birds are silent. The species is also known to respond well to imitations of its call (Devitt 1939). However, few atlassers would be familiar with its vocalizations, and, therefore, the distribution of the Yellow Rail may have been underestimated by the atlas data. It was detected in only 6 squares in southern Ontario, and from 17 blocks in the province as a whole. Breeding confirmation was particularly difficult to obtain, and was achieved in only 2 squares in the province. The low confirmation rate is a reflection of the difficulty of locating the nests, which are well concealed (often covered) in emergent marsh vegetation or under cut grasses (Bent 1926, Harrison 1975). Attempts to find nests are confounded by the tendency of the female to depart quietly when approached, and by the absence of agitated behaviour around the nest. Only one nest was found during the atlas period, that being at the Richmond Fen, 25 km south of Ottawa in 1982 (Jones 1982). The other confirmed breeding record was a sighting of fledged young on the west coast of James Bay, 30 km northeast of Moosonee.

Despite a concentrated effort during the atlas project, the range and abundance of the Yellow Rail remain somewhat obscure. The small number of records (31 squares) suggests that it is a rare breeder, but records from as far south as Cranberry Marsh near Whitby, as well as from far northern Ontario suggest that the species has a widespread but very scattered distribution across the province. Some southern Ontario records could have been non-breeding birds. The lack of records from the Boreal Forest zone (with the exception of a record from Sandy Lake in the northwest) likely reflects the general absence of suitable breeding habitat in this area, although some individuals may have gone undetected. The majority of records are from the blocks along the coasts of Hudson and James Bays from Moosonee to the Manitoba border. This coastal strip is clearly the stronghold of the species in Ontario, and abundance estimates suggest that it is locally common in this area.-- *D.R.C. Prescott*

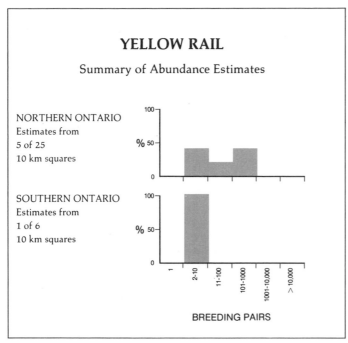

YELLOW RAIL

Summary of Abundance Estimates

NORTHERN ONTARIO
Estimates from
5 of 25
10 km squares

SOUTHERN ONTARIO
Estimates from
1 of 6
10 km squares

BREEDING PAIRS

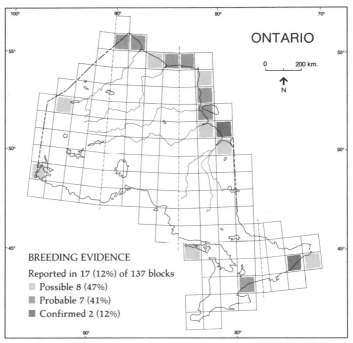

ONTARIO

0   200 km.

N

BREEDING EVIDENCE
Reported in 17 (12%) of 137 blocks
 Possible 8 (47%)
 Probable 7 (41%)
 Confirmed 2 (12%)

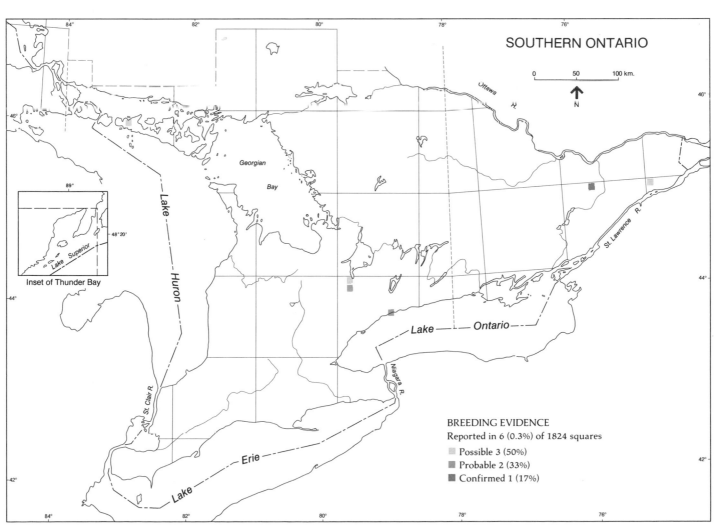

SOUTHERN ONTARIO

0   50   100 km.

N

Georgian
Bay

Lake Huron

Lake Superior
48°20'
Inset of Thunder Bay
89°

Ottawa R.

St. Lawrence R.

St. Clair R.

Niagara R.

Lake — Ontario

Lake — Erie

BREEDING EVIDENCE
Reported in 6 (0.3%) of 1824 squares
 Possible 3 (50%)
 Probable 2 (33%)
 Confirmed 1 (17%)

KING RAIL
Râle élégant
Rallus elegans

Considered by some authorities (see Ripley 1977) to be a subspecies or form of the Clapper Rail (*Rallus longirostris*), the King Rail is North America's largest rallid.

Its breeding range includes much of eastern North America, from the Atlantic coast west to the Great Plains, and from the Gulf of Mexico north to the Great Lakes region. It is most common in freshwater and brackish tidal marshes on the Atlantic and Gulf coastal plains (Terres 1980). In Ontario, it nests exclusively in freshwater marshes in the extreme south. Although rare throughout its Ontario range, it is somewhat more common and regular in the marshes associated with Lake St. Clair and Lake Erie. Large expanses of shallow-water marshes that merge with scrubby swales seem to be preferred in Ontario. The King Rail appears to be an ecotonal or 'edge' species in many respects. Hence, habitat tolerances are rather broad, ranging from mesic to wet conditions in cattail and grass/sedge marshes, shrubby swamps, and marshy borders of ponds and rivers (Peck and James 1983). In parts of the US at least, its distribution is reported to coincide with that of the muskrat (Meanley 1969), presumably because of the openings that the muskrat provides.

The atlas maps depict a distribution similar to that reported by Baillie and Harrington (1936) and Peck and James (1983). W. Carrell (in Snyder 1941) maintained that the Prince Edward Co marshes harboured the largest populations of King Rails in the province, but this has not been borne out by other information since that time or by the atlas data. Although atlas records strengthen prior evidence for breeding in a zone from the Bruce Peninsula east to Kingston, the province's northernmost confirmed breeding station remains as Oshawa (Speirs 1977). The atlas maps do not provide evidence for either a range extension or a contraction in Ontario. However, population declines have been noted in many parts of North America (Tate 1986). Drainage of wetlands has probably contributed to local declines in Ontario populations. Also, King Rail populations are apparently sensitive to local hunting pressures in the US, and they may also be adversely affected by pesticides (Ripley 1977).

This species is apt to breed in the same marshes in successive years. However, in true rallid fashion, the King Rail is typically secretive and difficult to detect. General inaccessibility to its marsh habitat adds to the difficulty of finding this species. It is seldom flushed and is heard more often than seen, particularly early in the breeding season. It can often be heard at night. Even so, it has a variety of vocalizations (Meanley 1957, 1969), many of which are unfamiliar to most atlassers. Like the Clapper Rail, it is probably quite responsive to taped broadcasts of its calls (Tomlinson and Todd 1973), but most atlas squares did not receive such specialized coverage. The atlas results were dominated by reports of birds seen in suitable habitat and territorial birds. As is true of all of Ontario's rallids, breeding is difficult to confirm. Although nests with eggs are reported to be aggressively defended against human intruders (Terres 1980), they are usually difficult to find. Peck and James (1983) were aware of only 12 nest records for Ontario. Breeding was not confirmed during the atlas period, but the literature record for Ontario indicates that fledged young are more readily found than are nests.

The map depicts with some accuracy the scattered distribution of the King Rail in southern Ontario, which is in large measure a reflection of the distribution of its habitat. Because it is on the northern fringe of its range in Canada, such a distribution was to be expected. However, for reasons given above, this species was probably underrecorded by atlassers. It is probably somewhat more common and slightly more widespread in southern Ontario than the map indicates.

The MNR has classified the King Rail as rare in Ontario, and this is borne out by the atlas results. It was reported in only 16 squares. Although abundance estimates were provided for only 4 squares, there is little doubt that the King Rail is numerically rare as well. One pair per square was the density most commonly estimated, the single estimate of 2 to 10 pairs being from a square containing part of the Walpole Island marsh. Breeding was confirmed at Walpole Island in 1986. Based on atlas results, the known Ontario population consisted of only about 25 pairs. However, considering detection difficulties, its population may be somewhat greater.-- *J.D. McCracken and D.A. Sutherland*

KING RAIL

Summary of Abundance Estimates

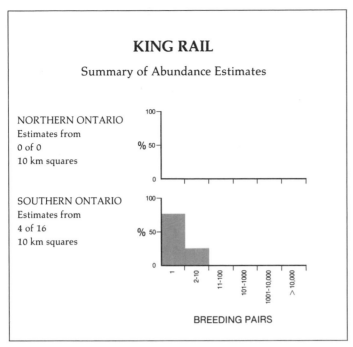

NORTHERN ONTARIO
Estimates from
0 of 0
10 km squares

SOUTHERN ONTARIO
Estimates from
4 of 16
10 km squares

BREEDING PAIRS

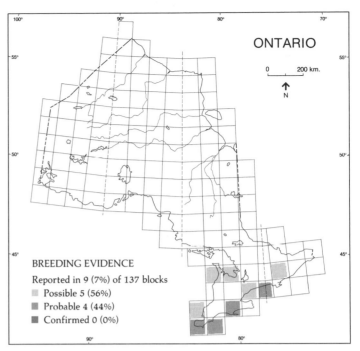

ONTARIO

0   200 km.

N

BREEDING EVIDENCE
Reported in 9 (7%) of 137 blocks
Possible 5 (56%)
Probable 4 (44%)
Confirmed 0 (0%)

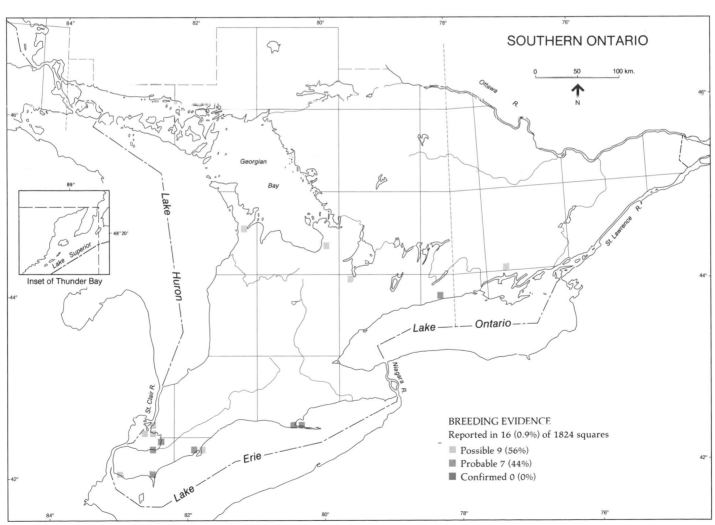

SOUTHERN ONTARIO

0   50   100 km.

N

Inset of Thunder Bay

Lake Superior

Georgian Bay

Lake Huron

Lake Ontario

Lake Erie

St. Clair R.

Niagara R.

Ottawa R.

St. Lawrence R.

BREEDING EVIDENCE
Reported in 16 (0.9%) of 1824 squares
Possible 9 (56%)
Probable 7 (44%)
Confirmed 0 (0%)

VIRGINIA RAIL
Râle de Virginie
Rallus limicola

R TUCKERMAN 1988

The Virginia Rail breeds from southern Canada, southward locally to southern South America, and winters mainly from the southern US southward. In Ontario, it is found primarily in marshy areas of the Great Lakes-St. Lawrence Forest and Carolinian Forest regions. It is widespread and locally common through southern Ontario, becoming much more sparsely distributed towards the northern edge of its range. Freshwater marshes of just about every conceivable size and description, ranging from roadside ditches to large cattail marshes, provide suitable habitat (Walkinshaw 1937, Peck and James 1983). However, it seems to favour shallow water marshes, pond edges, and sloughs that are dominated by cattails, grasses, and sedges. Walkinshaw (1937) reported that, in Michigan, it favoured water depths between 10 and 25 cm. Some interspersion of open water and vegetation cover seems to be an important habitat component. Nest location has been associated with ecotonal edges of dense vegetation cover (Glahn 1974). Although the 2 species often occur in the same marsh, the Virginia Rail usually occupies slightly drier habitat than does the Sora (Bent 1926).

The Virginia Rail's breeding range has probably not altered appreciably from that presented in historical accounts, but local depletions have probably occurred in southern Ontario because of habitat loss associated with the draining of wetlands.

The Virginia Rail is highly elusive and requires a special effort in order to assess its status in an atlas square. It is a species which is rarely flushed and it is heard far more often than seen. Even then, its variety of vocalizations (see Kaufmann 1983) may make it difficult for some atlassers to identify. Indeed, for many years there was some debate whether it was the source of the so-called 'ornithological mystery song' (Reynard 1974). Although secretive, it is usually highly responsive to tape-recorded broadcasts of its calls, particularly the pumping series of descending quacking notes (Glahn 1974, Griese *et al.* 1980). Use of tape-recordings invariably reveals the Virginia Rail to be much more common and widespread than would otherwise be guessed. Nests are very difficult to find, and most confirmations of breeding (70%) were based on reports of fledged young, presumably when an atlasser's attention was drawn to them by the adults' alarm calls. Bent (1926) noted that the adults are much bolder when the young are hatched, and Glahn (1974) similarly reported that nests were easiest to find when young were leaving them. Even so, the Virginia Rail was confirmed as breeding in only 26% of the southern Ontario squares from which it was reported, and breeding was not confirmed in the north. Problems in seeing the bird and the general inaccessibility of its habitat are no doubt responsible for those low figures.

The atlas maps show the same broad distribution as has been reported for this species in other accounts (Baillie and Harrington 1936, Godfrey 1986). Because of its secretive nature, the Virginia Rail is probably not quite as sparsely distributed as the maps indicate. In southern Ontario, it probably breeds wherever suitable habitat occurs. Such habitat is relatively sparse on the Canadian Shield and in intensively cultivated areas, (*e.g.*, much of the southwest, the Niagara Peninsula, and extreme southeastern Ontario). Like the Sora, the Virginia Rail's distribution seems to be concentrated in 2 broad belts (from Kingston north to Ottawa and from the Bruce Peninsula to eastern Lake Erie) and generally in association with the lower Great Lakes. Although habitat availability contributed to these patterns, regional differences in atlassing coverage should also be considered in their interpretation.

The Virginia Rail is difficult to survey, and its numbers were underestimated by atlassers; densities are greater than those indicated on the abundance histogram. It is locally common through southern Ontario, where it may be very common in some of the larger marshes. It is much less common and more thinly scattered in the northern part of its range.-- *J.D. McCracken*

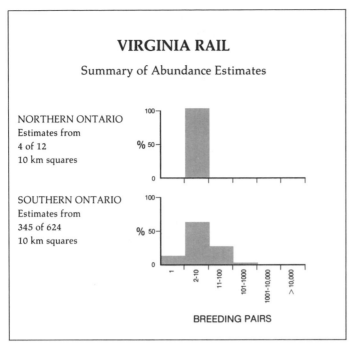

VIRGINIA RAIL

Summary of Abundance Estimates

NORTHERN ONTARIO
Estimates from
4 of 12
10 km squares

SOUTHERN ONTARIO
Estimates from
345 of 624
10 km squares

%

BREEDING PAIRS

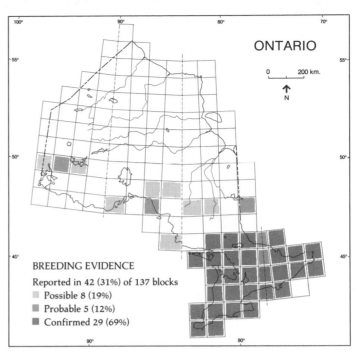

ONTARIO

0   200 km.

N

BREEDING EVIDENCE
Reported in 42 (31%) of 137 blocks

- Possible 8 (19%)
- Probable 5 (12%)
- Confirmed 29 (69%)

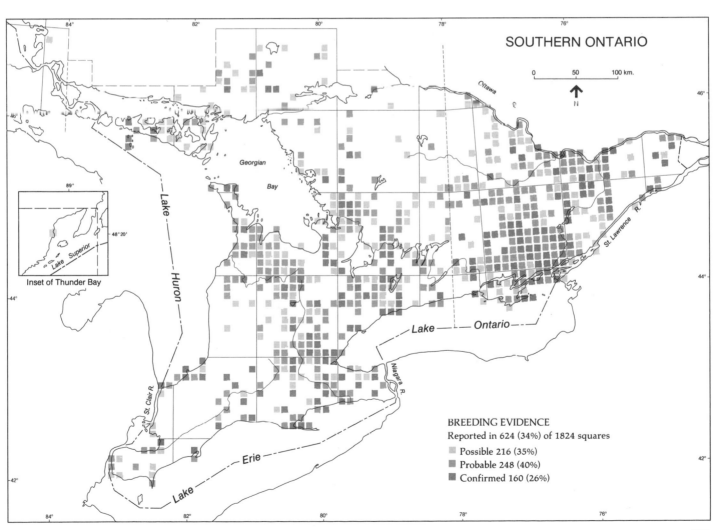

SOUTHERN ONTARIO

0   50   100 km.

N

Georgian
Bay

Lake Huron

Inset of Thunder Bay

Lake Superior

St. Clair R.

Lake Erie

Ottawa

St. Lawrence R.

Lake — Ontario

Niagara R.

BREEDING EVIDENCE
Reported in 624 (34%) of 1824 squares

- Possible 216 (35%)
- Probable 248 (40%)
- Confirmed 160 (26%)

SORA
Râle de Caroline
Porzana carolina

R.TUCKERMAN 1986

The Sora breeds in most of the southern half of Canada, south through much of the US. Its winter range extends from the southern US to northern South America, although it occasionally winters farther north. It is the most widespread species of rail in Ontario and can be found wherever suitable habitat occurs. Although it exhibits a preference for cattail in some areas, the Sora is not clearly associated with any particular plant species, and is quite opportunistic in its habitat selection (Lowther 1977, Johnson and Dinsmore 1985). Marshy habitats of all sizes and description are frequented, including cattail marshes, grassy marshes, bulrush marshes, bogs, fens, and wet meadows. Glahn (1974) found that nests were often situated at the edges of densely vegetated marsh communities. The Sora often occurs together with the Virginia Rail, but the Sora tends to prefer wetter situations (Bent 1926, Mousley 1937, Peck and James 1983). Water depth requirements are usually reported to lie between 15 and 30 cm.

The Sora's broad habitat tolerances are no doubt reflected in its wide distribution in the province. Although its breeding range has not altered from that shown in previous accounts (Baillie and Harrington 1936, Godfrey 1966), there have probably been local population declines in some areas of southern Ontario due to habitat loss.

As with other species of rails, the Sora's habitat and breeding behaviour make it difficult to find, and even harder to confirm as a breeding bird, without special effort. Hence, it was probably underrecorded by atlassers. The Sora is particularly elusive and difficult to flush, and it is heard far more often than seen. Although its 'whinny' is loud and distinctive and may be heard at any time of the day or night, it is by no means an incessant singer. Further, it does not always respond as well to taped broadcasts of its vocaliza-

tions as does the Virginia Rail (Glahn 1974, Griese *et al.* 1980). The nest is well concealed and is generally inaccessible because of the nature of the habitat. Even when found, it is not easily distinguished from the nest and eggs of the Virginia Rail. During the atlas field work, breeding was, by far, most frequently confirmed through the observation of fledged young. This is consistent with Glahn (1974), who found that Sora nests were easiest to locate when young were leaving them. The Sora was even harder to confirm (18% of squares in which it was reported in southern Ontario) than was the Virginia Rail (26%). Despite this low rate, the Sora likely breeds in all squares in which it was reported.

According to the atlas maps, the Sora is most evenly distributed off the Canadian Shield in southern Ontario, becoming quite scattered in distribution through the rest of the province. In northern Ontario, it was reported in every block along the James Bay coast, where suitable habitat is abundant. Possible breeding was reported as far north as the coast of Hudson Bay; this record is strengthened by Schueler *et al.* (1974), who noted that "it seemed to be common" in the vicinity of Winisk. Prior to the atlas, breeding had been confirmed as far north as Fort Severn (Peck and James 1983). Although breeding densities are lower and habitat is more scattered outside of its southern Ontario range, less thorough coverage of the northern portions of the province, combined with difficulties in detecting the bird, also help account for the sporadic northern distribution presented on the map. Nevertheless, the pattern of records is similar to that presented in Peck and James (1983), and suggests that the species is at best sparsely distributed in the Boreal Forest zone.

In southern Ontario, its range seems to be concentrated in the southeast (Kingston to Ottawa areas), in a belt from the Bruce Peninsula to eastern Lake Erie, and generally along the lower Great Lakes. A broadly similar pattern occurs in the distribution presented for the Virginia Rail. As with the Virginia Rail, the Sora was not reported in many squares in areas of intensive agriculture (*e.g.*, the southwest, the Niagara Peninsula, and far southeastern Ontario) nor in the highly disturbed area in and around Toronto. Habitat availability undoubtedly contributes to these patterns, but regional differences in intensity of altassing must also be considered in their interpretation.

The Sora is generally regarded as a common species throughout Ontario (James *et al.* 1976), but is most common in the larger marshes found in the southern portion of the province. Although it has a wider distribution, its breeding densities often do not seem to be as great as those of the Virginia Rail, though atlas abundance estimates show little difference.-- *J.D. McCracken*

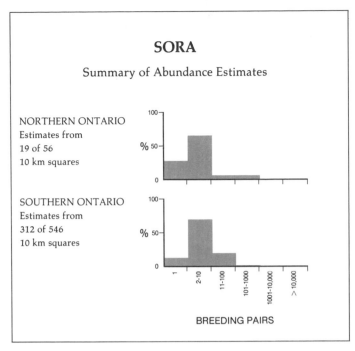

SORA

Summary of Abundance Estimates

NORTHERN ONTARIO
Estimates from
19 of 56
10 km squares

SOUTHERN ONTARIO
Estimates from
312 of 546
10 km squares

%

BREEDING PAIRS

1, 2-10, 11-100, 101-1000, 1001-10,000, >10,000

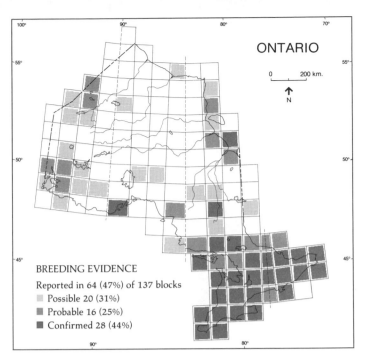

ONTARIO

0   200 km.

N

BREEDING EVIDENCE
Reported in 64 (47%) of 137 blocks

Possible 20 (31%)
Probable 16 (25%)
Confirmed 28 (44%)

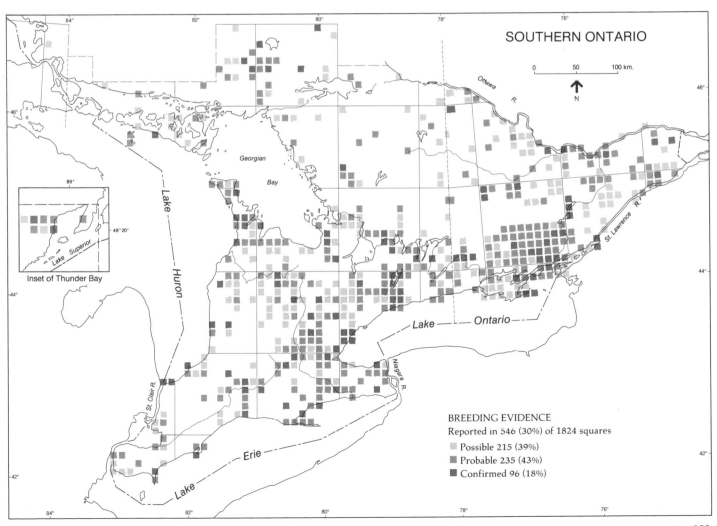

SOUTHERN ONTARIO

0   50   100 km.

N

Georgian Bay

Ottawa R.

Lake Huron

Lake Superior

Inset of Thunder Bay

48° 20'

89°

St. Clair R.

St. Lawrence R.

Lake — Ontario

Niagara R.

Lake — Erie

BREEDING EVIDENCE
Reported in 546 (30%) of 1824 squares

Possible 215 (39%)
Probable 235 (43%)
Confirmed 96 (18%)

PURPLE GALLINULE *
Gallinule violacée
Porphyrula martinica

The Purple Gallinule is a very rare visitor to the marshes of southern Canada. The species normally breeds from the US Gulf States and South Carolina south to Argentina. However, it often nests in locations that are isolated and far removed from the usual continuous range. Most of the North American breeders migrate to Central and South America for the winter, but some remain in the Gulf States.

As early as 1891, a bird was reported at Point Pelee (Speirs 1985), and in 1892, one was recorded in April at Pickering (Macoun and Macoun 1909). Since that time, individuals have been recorded in various places from southern Ontario to Newfoundland (Godfrey 1986). Most sightings have been in the spring or fall. The closest confirmed breeding record to Ontario appears to be one from Franklin Co, Ohio in June 1962. A pair of birds raised 5 young in that year (Thomson 1983). Franklin Co is approximately 250 km due south of Windsor.

The Purple Gallinule is somewhat similar to the Common Moorhen, but the adults of the two species are easily separable in the field. Despite its colourful plumage, the bird can be hard to see when hidden in thick vegetation such as that in a dense cattail marsh. The flight appears to be weak, rail-like, and slow. The bird barely clears the top of the marsh vegetation, and quickly drops down, out of sight.

The Purple Gallinule nests in herbaceous wetlands, usually marshes with reeds, cattails, wild rice, and pickerelweed. The nest is placed on a small, floating island of vegetation. It is made of dried and decaying leaves of plants, and is firmly interwoven with growing plants that provide a solid support and good concealment. The bird seems to prefer nest sites that are surrounded by water up to 3 m in depth. The species lays from 5 to 10 eggs, with 6 to 8 being usual. Such large clutches should help make broods readily visible. It is speculated that incubation may be aided by the heat given off by the decomposition of the vegetable matter in the nest (Bent 1926).

There were only 2 breeding season records of this species in Ontario during the atlas period. On 8 July 1981, an individual was observed at Point Pelee Nat. Park. In 1984, one was found on 3 May and stayed until at least 27 June, well into the breeding season, in the Stoney Point marsh along the southern shore of Lake St. Clair in Essex Co. The bird was seen by hundreds of observers, and it became something of a celebrity as birders rushed to get the species on their lists. There is no evidence that either individual had a mate. In the spring of 1986, single Purple Gallinules were observed at Stoney Point Marsh and Hillman Marsh, both in Essex Co (R. Hawker pers. comm.).

Conversations with the coordinators of the breeding bird atlas projects under way in Michigan, Ohio, Indiana, Illinois, Pennsylvania, and New York reveal that the closest confirmed breeding to Ontario in the last few years has been in extreme southern Illinois, where 1 or 2 pairs nest each year along the Ohio River (V. Klean pers. comm.). This location is approximately 800 km southwest of Windsor. No records of confirmed breeding have occurred in any of the other states mentioned, but a few spring birds have been observed.-- *P.F.J. Eagles*

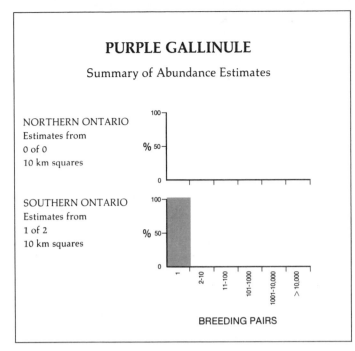

PURPLE GALLINULE

Summary of Abundance Estimates

NORTHERN ONTARIO
Estimates from
0 of 0
10 km squares

SOUTHERN ONTARIO
Estimates from
1 of 2
10 km squares

BREEDING PAIRS

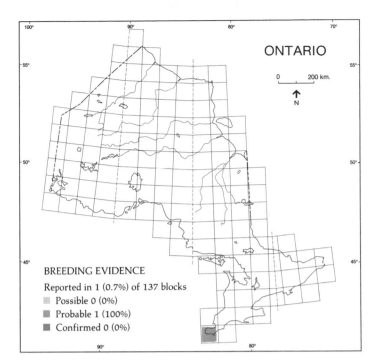

ONTARIO

0   200 km.

↑
N

BREEDING EVIDENCE

Reported in 1 (0.7%) of 137 blocks

Possible 0 (0%)
Probable 1 (100%)
Confirmed 0 (0%)

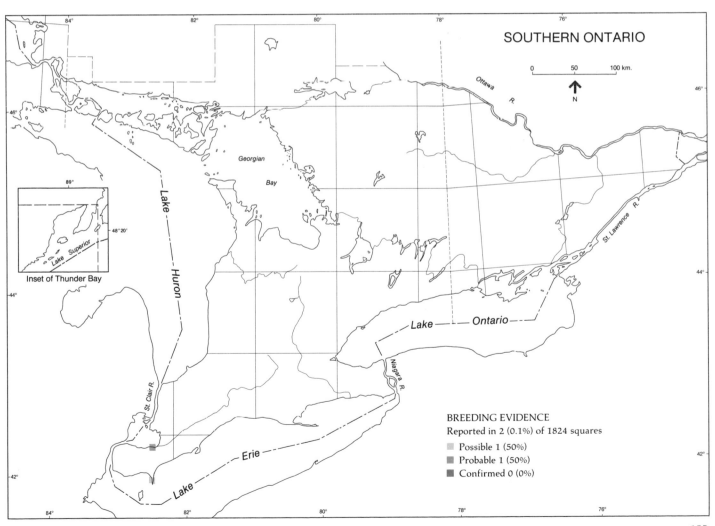

SOUTHERN ONTARIO

0   50   100 km.

↑
N

Inset of Thunder Bay

BREEDING EVIDENCE

Reported in 2 (0.1%) of 1824 squares

Possible 1 (50%)
Probable 1 (50%)
Confirmed 0 (0%)

COMMON MOORHEN
Poule-d'eau
Gallinula chloropus

The Common Moorhen (formerly Common Gallinule) has virtually a worldwide distribution, with the exception of Australia and Antarctica. In North America, it breeds from Central America northeastward through much of the US. Its Canadian breeding distribution is primarily limited to southern Ontario, where it is most widespread and common along Lakes Erie and Ontario and into the southeastern portion of the province. This pattern is presumably linked to habitat availability, but its absence from northern and western sections of Ontario may also be due in some measure to unfavourable climate. Breeding habitat consists of deep-water marshes and ponds of various descriptions, but cattail marshes with interspersed sheltered pools and channels are favoured. The species winters mainly from the southern US southwards.

Although the atlas data do not indicate that any change in its breeding range has taken place recently, the Common Moorhen has experienced local population declines due to habitat loss. These declines have been off-set slightly by the creation of suitable habitat at sewage lagoons and impoundments.

The Common Moorhen is not quite as elusive and retiring as are other members of the rail family, yet its breeding biology is still poorly known (Frederickson 1971). It is quite vocal and, for a rail, it is reasonably visible. It responds fairly well to taped broadcasts of its calls, as well as to those of other rails. Care must be taken, however, because many of its vocalizations closely resemble those of the American Coot. Nests and eggs may present similar problems in identification. At any rate, nests are generally inaccessible to atlassers and they may be difficult to locate. The majority of confirmed records (81%) are of fledged young. Despite the inaccessibility of its habitat, breeding was confirmed in 46%

of the squares in which the species was reported. This is a considerably higher rate of confirmation than was found for the other rail species and is indicative of the greater detectability of young Common Moorhens, which have a strong tendency to forage in the open.

The atlas maps confirm the distribution of this species as outlined by Baillie and Harrington (1936) and Godfrey (1966), and are very similar to Peck and James' (1983) map. James *et al.* (1976) state that there are summer records from as far north and west as the Thunder Bay area. Baillie and Harrington's (1936) most northerly record involves a bird collected at Lake Nipissing in 1921. Without comment, Schueler *et al.* (1974) include a remarkable summer record for Winisk on the Hudson Bay coast in 1965. The Common Moorhen's tendency to be more ubiquitous in southeastern Ontario is consistent with BBS data shown on the range map in Speirs (1985), which indicate that this species is most common in that area. Interestingly, a similar pattern is not displayed by the American Coot (a species whose habitat requirements are similar to those of the Common Moorhen), presumably because the American Coot has a more western affinity. Also, the preference of the latter species for deeper water may mean it is more restricted to larger marshes. As with several other marsh species (see for example, Black Tern and Least Bittern), the distribution of the Common Moorhen in the southwest is mostly limited to large and productive marshes on Lake St. Clair and the north shore of Lake Erie. However, breeding of this species was also confirmed in several 'inland' squares with smaller marshes and ponds.

Where the amount of habitat is small or marginal, the Common Moorhen often nests singly. However, it is loosely colonial where habitat is optimal or extensive. Abundance estimates indicate that it is locally common in southern Ontario. Of the 4 estimates of over 100 pairs in a square, 2 were from squares in the Long Point marshes, and 2 were from squares containing Camden and Varty Lakes, just northwest of Kingston.-- *J.D. McCracken*

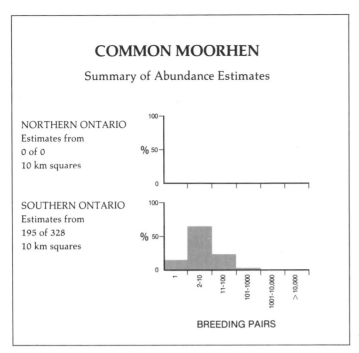

COMMON MOORHEN

Summary of Abundance Estimates

NORTHERN ONTARIO
Estimates from
0 of 0
10 km squares

SOUTHERN ONTARIO
Estimates from
195 of 328
10 km squares

%

BREEDING PAIRS

1 2-10 11-100 101-1000 1001-10,000 >10,000

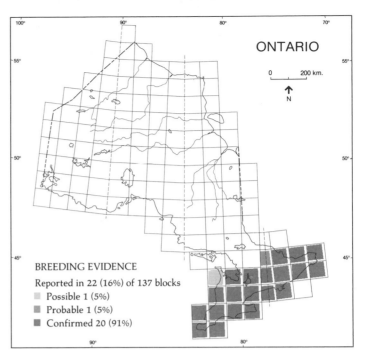

ONTARIO

0   200 km.

N

BREEDING EVIDENCE
Reported in 22 (16%) of 137 blocks
Possible 1 (5%)
Probable 1 (5%)
Confirmed 20 (91%)

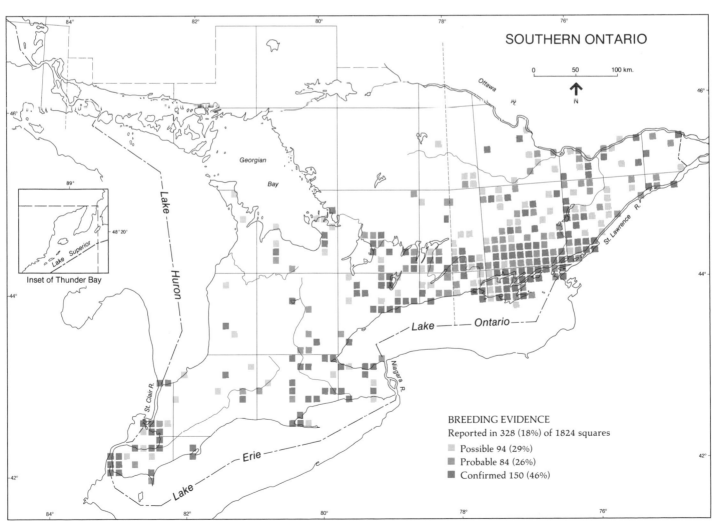

SOUTHERN ONTARIO

0   50   100 km.

N

Georgian Bay

Ottawa R.

St. Lawrence R.

Lake — Ontario

Niagara R.

Lake Erie

Lake Huron

St. Clair R.

89°
48°20'
Lake Superior
Inset of Thunder Bay

BREEDING EVIDENCE
Reported in 328 (18%) of 1824 squares
Possible 94 (29%)
Probable 84 (26%)
Confirmed 150 (46%)

AMERICAN COOT
Foulque d'Amérique
Fulica americana

R. TUCKERMAN 1986

The American Coot is a locally common breeding species across the southern half of Canada and generally southwestward through the US, Mexico, Central America, and the northern part of South America. In Ontario, it is generally confined to the southern and central portions of the province. Isolated populations occur in northern Ontario (James *et al*. 1976, Peck and James 1983). The American Coot's habitat consists of freshwater marshes, ponds, and sloughs of various sizes and vegetational descriptions. Peck and James (1983) concluded that it was most common in cattail marshes, and less common in willow swales and bogs. Expanses of deep open water are a critical habitat requirement (Friley *et al*. 1938, Gullion 1954, Sugden 1979). It is unaccountably absent from many apparently suitable marshes within its Ontario range (Baillie and Harrington 1936), resulting in a more scattered distribution than that displayed by the Common Moorhen. It winters as far north as the New England states on the east coast, southern British Columbia on the west coast, and southern Ontario in the middle of the continent.

The range of the American Coot, as described by atlas data, does not differ markedly from previous accounts (Baillie and Harrington 1936, Godfrey 1966). However, Baillie and Harrington noted that it was formerly more common in some areas. Populations have probably declined locally through southern Ontario because of habitat loss associated with the drainage of marshes. These losses have probably been off-set to some extent through the creation of suitable habitat at sewage lagoons and impoundments.

The vociferousness and semi-colonial nature of the American Coot aid in its detection, but other characteristics hamper atlassers' efforts to confirm breeding. Breeding was confirmed in 34% of squares with records of American Coot in southern Ontario, and in 14% of such squares in the north. This is a higher rate than was found for some of the other members of the rail family, but it is still low overall. The American Coot is quite retiring, and, during the breeding season, it is far more often heard than seen. Even then, its diverse vocal repertoire (see Gullion 1952) makes its calls difficult to distinguish from those of the Common Moorhen. Its habitat is not easily accessible to atlassers, and its nest and eggs are usually fairly well hidden and may also be hard to distinguish from those of the Common Moorhen. Nevertheless, it builds a large number of structures associated with nesting (Gullion 1954). As with other rail species, breeding was most often confirmed through the observation of fledged young.

The American Coot has a fairly spotty breeding distribution in Ontario. It is sensitive to habitat changes, particularly water levels (Weller and Spatcher 1965). Hence, local populations may fluctuate considerably from year to year, and the species may not necessarily be present every year in the same location. Although the American Coot may have been slightly underrecorded by atlassers because it is difficult to detect, the atlas map is particularly similar to the map produced by Peck and James (1983), in that the great majority of records come from south and east of the Shield in southern Ontario.

Although it ranges farther north and west, the American Coot is more sparsely distributed than the Common Moorhen. Although both species can often be found nesting in the same marsh, it tends to prefer sections which are more open and deeper. Such habitat is most likely to occur in larger marshes, so optimal habitat for this species is probably not widely available in Ontario, accounting for the scattered distribution. Also, the American Coot has a continental range which historically lies more towards the western portion of the continent. It is a hardy bird (Bent 1926), probably hardier than the Common Moorhen, and it is perhaps better able to cope with harsher climates found in the western and northern portions of its range in Ontario. Peck and James (1983) seem to concur, pointing out that lack of suitable habitat limits the distribution of the American Coot in Ontario more than does northern latitude. In this regard, it is interesting to note that the 2 Sudbury area records and the Rainy River area record were of birds in sewage lagoons, and the most northerly confirmed record was from a small urban lake which, until recently, received mine waste. These nutrient-laden water bodies apparently provide suitable habitat where most natural lakes in the north do not.

Abundance estimates indicate that the American Coot ranges from being locally rare to locally common in Ontario. It is rarest in the northern portion of its range, and most common in the larger marshes associated with the lower Great Lakes.-- *J.D. McCracken*

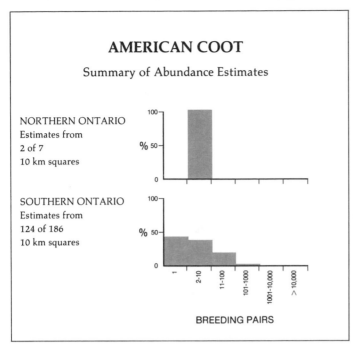

AMERICAN COOT

Summary of Abundance Estimates

NORTHERN ONTARIO
Estimates from
2 of 7
10 km squares

SOUTHERN ONTARIO
Estimates from
124 of 186
10 km squares

BREEDING PAIRS

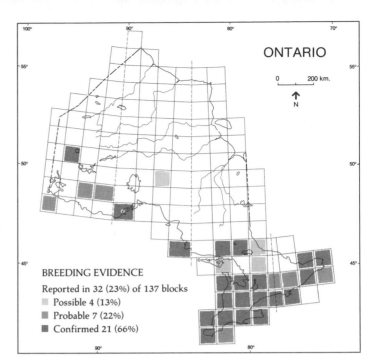

ONTARIO

0   200 km.

N

BREEDING EVIDENCE
Reported in 32 (23%) of 137 blocks

Possible 4 (13%)
Probable 7 (22%)
Confirmed 21 (66%)

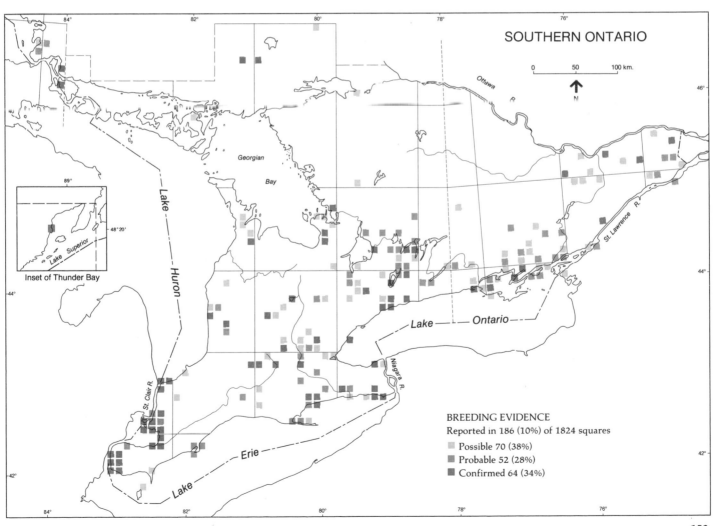

SOUTHERN ONTARIO

0   50   100 km.

N

Georgian
Bay

Lake
Huron

Lake   Superior
Inset of Thunder Bay

Ottawa   R.

St. Lawrence   R.

Lake   Ontario

Niagara   R.

St. Clair R.

Erie

Lake

BREEDING EVIDENCE
Reported in 186 (10%) of 1824 squares

Possible 70 (38%)
Probable 52 (28%)
Confirmed 64 (34%)

SANDHILL CRANE
Grue du Canada
Grus canadensis

R. TUCKERMAN 1986

Six subspecies of Sandhill Crane, of which one is endangered, range from the Caribbean to northern Canada and into eastern Siberia. In Canada, the species breeds locally in the northern prairies, widely in the muskeg, and across the tundra from Bylot and Baffin Islands to the Alaskan border (AOU 1983).

Ontario Sandhill Cranes generally breed in low shrub bog or fen in the Algoma District (Tebbel and Ankney 1982) and the Hudson Bay Lowland (Riley 1982). Peaty wetlands with an abundance of sphagnum, cattail, leatherleaf, sweet gale, and sedges provide typical nesting habitat. In these areas, the nest is typically no more than a few twigs and stems which fail to cover the substrate. More bulky nests are built in the cattail marsh habitats which are also used for breeding. The nest is surrounded by vegetation high enough to screen it from view and is usually within 2 m of water. In areas of human settlement, wetlands occupied by Sandhill Cranes are secluded by forest or distance and are generally located close to upland meadows, which are used for feeding by the broods (Tebbel and Ankney 1982).

In Ontario, Sandhill Cranes bred on Lake St. Clair until the early years of the 20th century. The last breeders in the south were reported to be present on the Hickory Ridge on Walpole Island about 1920 (Lumsden unpubl.). The Sandhill Crane survived throughout the settlement period in the upper peninsula of Michigan. It is not known if it survived in the adjacent Algoma District of Ontario. The first recent report in that area was of a pair seen dancing near Searchmount, north of Sault Ste Marie, in the early 1950's. There is no doubt that it has increased and spread, whether by colonization from Michigan, or increase from a few surviving pairs. In 1979, Tebble and Ankney (1982) estimated that about 225 cranes were present in the southern Algoma Dis-

trict. Atlassers found that cranes have recently colonized Manitoulin Island and had reached the Bruce Peninsula by 1985. The large muskegs in the western Rainy River District support a breeding population of cranes, which is probably part of the Sandhill Crane stock breeding in the adjacent area of Minnesota.

In the Hudson Bay Lowland, there have been substantial changes over the years in the status of Sandhill Cranes. In the 17th century, large numbers of cranes, presumably from the Lowland, stopped in Simcoe Co, just south of Georgian Bay, en route to and from their wintering grounds in New England. That population was extirpated in the 19th Century (Lumsden 1971). It seems likely that birds from farther west in the Lowland recolonized this vacant breeding range, but do not use the same migration route as did the area's previous occupants. This may account for the scarcity of migrating Sandhill Cranes in southcentral and southwestern Ontario despite the current abundance in the Hudson Bay Lowland. On the James Bay coast, cranes were rarely seen on survey flights from 1950 to 1970. A rapid increase was noted in the 1970's. During the atlas years, breeding was recorded, and over 200 cranes were seen along the coast between Ekwan Point and Hook Point in one day (K. Abraham pers. comm.). The spread continues. They reached the Cochrane area about 1967 and were first found breeding in 1979 (Hall-Armstrong and Armstrong 1982). Nesting northeast of Timmins and near Geraldton was confirmed in 1984.

The Sandhill Crane returns to its nesting grounds in early April. Laying starts in late April in Algoma, and in early May in the southern Hudson Bay Lowland. It is remarkably secretive near its nest, but calling at dusk and dawn may reveal its presence. It responds well to tape recordings when on territory (D. Ankney pers. comm.). The incubating bird sits very tight and may show some agitated behaviour after flushing. Nest searching or watching for broods is very time consuming, so it is not surprising that only 17% of records confirm breeding. Nests were found in only 7 squares, and broods were noted in 28.

Sandhill Cranes are widely distributed in the area north of 51^0N, where atlassing took place mostly in summer, when the breeding cycle would be well advanced. Breeders there were unlikely to be detected when they were silent and secretive and with their flightless young. It is likely that breeding takes place in all the 'possible' and 'probable' blocks and in most of the blocks north of 51^0N for which there are no atlas records. The presence of birds in the Rondeau Prov. Park marsh in 1983 and 1985 indicates the possibility of future nesting. Perhaps pioneers from the southern Michigan stock will nest there soon.

Abundance estimates suggest that breeding densities are higher in northern Ontario than in the south. All 6 estimates of more than 10 pairs per square are from north of 51^0N.-- *H.G. Lumsden*

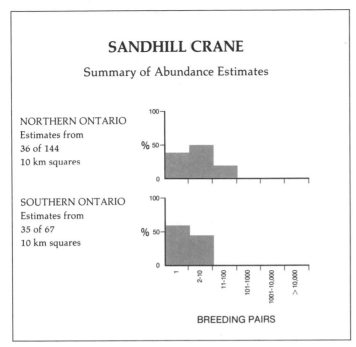

SANDHILL CRANE

Summary of Abundance Estimates

NORTHERN ONTARIO
Estimates from
36 of 144
10 km squares

SOUTHERN ONTARIO
Estimates from
35 of 67
10 km squares

% axis labels: 100, 50, 0

BREEDING PAIRS
(x-axis labels: 1, 2-10, 11-100, 101-1000, 1001-10,000, >10,000)

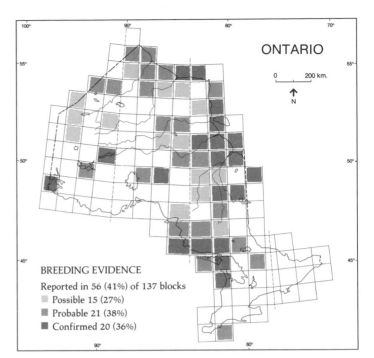

ONTARIO

0   200 km.

N

BREEDING EVIDENCE
Reported in 56 (41%) of 137 blocks
Possible 15 (27%)
Probable 21 (38%)
Confirmed 20 (36%)

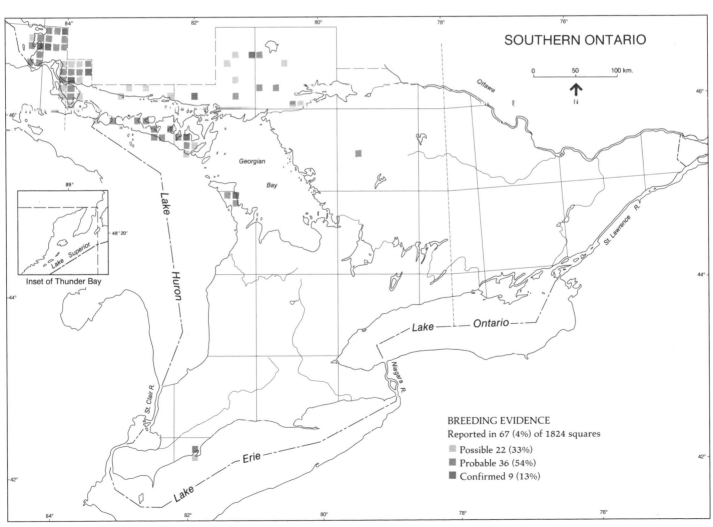

SOUTHERN ONTARIO

0   50   100 km.

N

Inset of Thunder Bay

Lake Superior

48°20'
89°

Lake Huron

Georgian Bay

St. Clair R.

Lake Erie

Niagara R.

Lake Ontario

St. Lawrence R.

Ottawa

BREEDING EVIDENCE
Reported in 67 (4%) of 1824 squares
Possible 22 (33%)
Probable 36 (54%)
Confirmed 9 (13%)

PIPING PLOVER
Pluvier siffleur
Charadrius melodus

S.L. HOUSE

Once, the soft, musical piping and "wild, plaintive cries" of this small beach-nesting plover could be heard in Ontario at widespread locations on the Great Lakes. Now the species maintains a mere toe-hold in Ontario, and it continues to decline seriously in other parts of its range.

Although a few winter south to the Caribbean, the Piping Plover breeds only in North America: along the Atlantic coast, on the Great Lakes, and in the Prairies and northern Great Plains. It nests on dry, sandy, unvegetated or sparsely vegetated beaches on the shores of large lakes and on the Atlantic coast. In the plains, it nests mainly on alkaline or mildly saline lakes and sloughs, where the substrate may be pebble beach, gravel, or mud (Bell 1978). Sandbars in rivers are also used in the US. These are dynamic habitats, being destroyed and recreated by storms, and affected by changing water levels.

Historic data concerning the population size are lacking, but it is known that hunting caused a marked reduction around the turn of the century. Once protected, the species appeared to recover, but in recent decades a substantial decline has been noted. The present world population is estimated at only 3,535 to 4,147 individuals, representing some 1,649 to 1,939 pairs, of which 716 to 1,001 occur in Canada (Haig and Oring 1985). There is little doubt that the recent decline is related primarily to the increasing recreational use of beaches. Destruction of nests and young by pedestrian and vehicular traffic and disturbance of nesting birds are the main problems. It is also possible that natural predation by gulls, crows, raccoons, or foxes has increased with clearing and urbanization (Cairns and McLaren 1980). Controlled water levels may be a further problem, reducing the quality or amount of nesting habitat in some areas.

The greatest decline has occurred on the Great Lakes,

where the species has been virtually extirpated from Illinois, Indiana, Ohio, Pennsylvania, New York, and Ontario, but remains as a breeder in Michigan, Wisconsin, and Minnesota. In Ontario, there are former breeding records from at least 18 locations on Lakes Erie, Huron, and Ontario, and the St. Lawrence River, and a historic population in the order of 152 to 162 pairs has been estimated (Russell 1983). The last known breeding in southern Ontario occurred in 1977 on the relatively undisturbed beaches of Long Point, which once supported one of the largest breeding concentrations on the Great Lakes (Snyder and Logier 1931). The thousands of summering gulls which have become a feature of Long Point's beaches in recent decades have been implicated as a factor in the decline of the Piping Plover there. High rates of predation by mammals (particularly raccoons) may also have been a factor.

Southern Lake of the Woods remains as the only regular breeding location in Ontario. There is a small population of a few pairs, which may even be absent in some years. A maximum count of 10 adults was made during the atlas period. In 1981, a nest with eggs was found, and, in 1983, recently hatched chicks were observed. Use of nesting areas by non-breeding gulls may be detrimental in some years. However, the presence of a larger population of about 40 to 50 individuals on the Minnesota side of Lake of the Woods provides some security to the Ontario population, especially in view of generally high reproductive success there (Wiens and Cuthbert 1984).

The presence of a pair of Piping Plovers at Wasaga Beach in 1981 was highly unusual. The last known nesting in this area took place in 1938 (Devitt 1967). The present level of human use makes it unlikely that the birds bred successfully, although the habitat was otherwise suitable. Similar isolated nesting attempts should be watched for, since in 1984 an unexpected nesting occurred at a former breeding site on the New York side of Lake Ontario (Kibbe and Boise 1984).

It is possible that Piping Plovers were missed on remote islands in Lake Huron or southeastern Lake Superior, although no such breeding locations are known. For any sites visited, a walk of the entire beach area should have revealed the presence of any territorial birds, since, although they are well camouflaged against the sand, their striking call notes given in response to human intruders make them fairly easy to detect.

The Piping Plover was legally declared "endangered" in Ontario in 1977. In 1978, COSEWIC designated it as "threatened" in Canada, but in 1985 the status was upgraded to "endangered", in light of documented declines in Ontario and Manitoba and on the Atlantic coast (Haig 1985). In the US, it is federally listed as endangered in the Great Lakes watershed and threatened in the remainder of its range, including the winter range (U.S. Federal Register 1985).-- *A.B. Lambert*

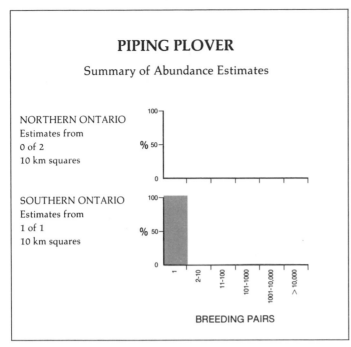

PIPING PLOVER

Summary of Abundance Estimates

NORTHERN ONTARIO
Estimates from
0 of 2
10 km squares

SOUTHERN ONTARIO
Estimates from
1 of 1
10 km squares

BREEDING PAIRS

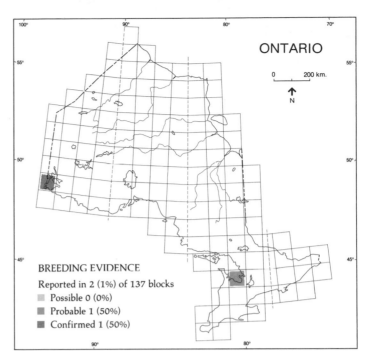

ONTARIO

0   200 km.

N

BREEDING EVIDENCE
Reported in 2 (1%) of 137 blocks
░ Possible 0 (0%)
▓ Probable 1 (50%)
█ Confirmed 1 (50%)

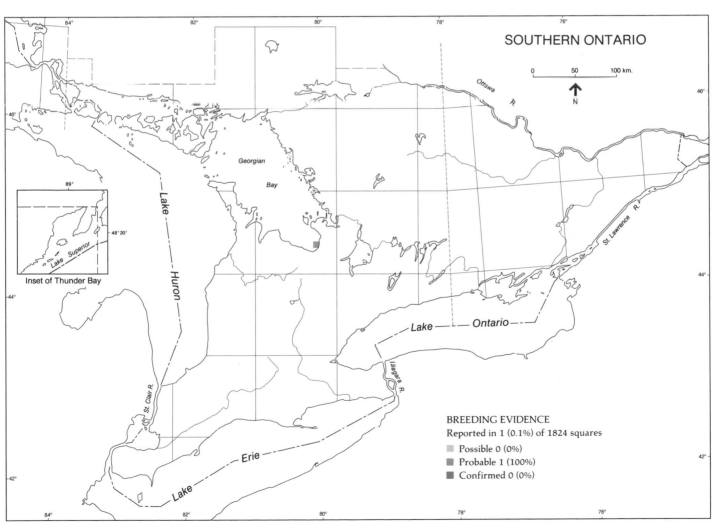

SOUTHERN ONTARIO

0   50   100 km.

N

Inset of Thunder Bay

BREEDING EVIDENCE
Reported in 1 (0.1%) of 1824 squares
░ Possible 0 (0%)
▓ Probable 1 (100%)
█ Confirmed 0 (0%)

KILLDEER
Pluvier kildir
Charadrius vociferus

S.L. HOUSE

The Killdeer is the most familiar of the North American shorebirds because of its close association with agriculture and human settlements. It breeds widely throughout North America in non-mountainous regions south of the tree-line to Mexico and the West Indies, as well as in western Peru and northwestern Chile. In Ontario, the Killdeer is found in disturbed sites throughout the province. The breeding habitats are open areas, not necessarily associated with water. These include meadows, pastures, cultivated fields, lake shores, river banks, orchards, and woodland clearings. The Killdeer commonly nests on airports, golf courses, lawns near houses, wastelands resulting from construction works, railroad rights-of-way, and occasionally gravel roofs (Cramp and Simmons 1983, Peck and James 1983). Because of the wide range of habitats used, the species has been found in nearly all of southern Ontario and much of the north. Killdeers breeding in Ontario winter in the central and southern parts of the US.

A century ago, the Killdeer was considered abundant on the prairies and common along the Great Lakes, but rare on both sea coasts (Chamberlain 1887, Macoun 1900, Taverner 1922a). By 1925, it was no longer considered rare in coastal British Columbia (Brooks and Swarth 1925), and, during the mid-1960's, it underwent a range expansion in Nova Scotia (Tufts 1961, 1973). Early accounts of the birds of Ontario mentioned nesting, primarily on ploughed and grazed fields throughout the province (McIlwraith 1894, Nash 1913, Taverner 1922a). The Killdeer has remained common over the last 100 years. It was reported nesting at Moose Factory in 1930, and was suspected to breed in suitable locations along the James Bay and southern Hudson Bay coasts in the 1940's (Manning 1952). It has also been reported in the breeding season on North Twin Island, NWT, in James Bay (Manning

1981).

The Killdeer is conspicuous, with its loud, raucous, and persistent call, and is difficult to confuse with other species. Its association with habitats used by humans, and its elaborate injury-feigning display, mean that this species is rarely missed by atlassers. The breeding season is long (April to July), and, because Killdeers can raise 2 broods in a season, atlassers had ample opportunity to confirm breeding. Breeding was confirmed most often by observation of downy young, or nests with eggs, but distraction displays account for 21% of confirmed records in southern Ontario and 35% of confirmations in the north. Distraction displays given near the end of the breeding season can sometimes be misleading, since Killdeers can continue this display after they have moved considerable distances from the breeding areas (pers. obs.). However, this is unlikely to have had a significant effect on the atlas data.

Atlas data show that the Killdeer is widespread in Ontario, breeding in virtually every block with road or rail access, and even in heavily forested areas of northern Ontario as far north as Hudson Bay. North of road or rail access, breeding locations are mostly associated with human settlements, but a few are not, notably along the Hudson Bay coast and elsewhere on the Hudson Bay Lowland. The lack of records in some northern blocks containing settlements may represent a lack of coverage, rather than an absence of Killdeers. Although the Killdeer is usually noisy and visible, some individuals, particularly during the laying period, will leave the nest site quietly and be inconspicuous to observers. Thus, unless the atlasser actually walked over potential nest areas such as airstrips and rights-of-way, the species could be missed. Squares and blocks in heavily forested regions (such as Algonquin Prov. Park and much of the Boreal Forest zone) may lack any breeding habitat for the Killdeer, and this is reflected in its absence from these areas on the maps.

Over half of all squares with Killdeers in southern Ontario contain abundance estimates, and most of these (53%) suggest breeding densities of 11 to 100 pairs per 10 km square. These estimates are generally low compared to densities based on counts of breeding Killdeers from rural areas in southern Ontario, e.g., 12.9 to 30 pairs per 100 ha (Nol and Lambert 1984). Densities in rural and suburban areas like southwestern and southeastern Ontario would be relatively high, while densities in far northern Ontario are likely very low, with perhaps fewer than 10 pairs per block. Atlas abundance estimates indicate that the Killdeer is a common breeding species in southern Ontario but is generally less common in the north.-- *E. Nol*

KILLDEER

Summary of Abundance Estimates

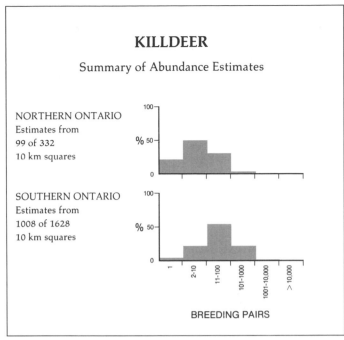

NORTHERN ONTARIO
Estimates from
99 of 332
10 km squares

SOUTHERN ONTARIO
Estimates from
1008 of 1628
10 km squares

%

BREEDING PAIRS

1   2-10   11-100   101-1000   1001-10,000   >10,000

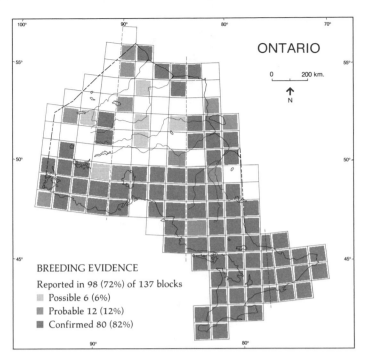

ONTARIO

0   200 km.

N

BREEDING EVIDENCE
Reported in 98 (72%) of 137 blocks
- Possible 6 (6%)
- Probable 12 (12%)
- Confirmed 80 (82%)

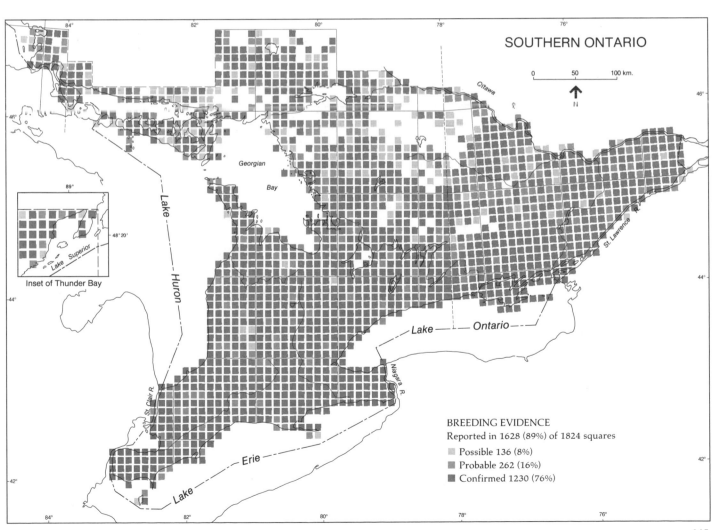

SOUTHERN ONTARIO

0   50   100 km.

N

Georgian Bay

Lake Huron

Lake Superior

Inset of Thunder Bay

St. Clair R.

Lake Erie

Niagara R.

Lake Ontario

Ottawa

St. Lawrence R.

BREEDING EVIDENCE
Reported in 1628 (89%) of 1824 squares
- Possible 136 (8%)
- Probable 262 (16%)
- Confirmed 1230 (76%)

SOLITARY SANDPIPER
Chevalier solitaire
Tringa solitaria

S.L. HOUSE

The Solitary Sandpiper has the rare habit among shorebirds of being an arboreal nester and using the old nests of other birds, such as the American Robin, Gray Jay, and Rusty Blackbird. It shares its arboreal nesting traits with only the Green Sandpiper *(Tringa ochrophus)* and (infrequently) the Wood Sandpiper *(Tringa glareola)* of Eurasia. The Solitary Sandpiper nests across the subarctic and Boreal Forest zones of North America. Historically, there are very few proven breeding records, but the atlas data have clarified the picture considerably, showing that it probably nests throughout northern Ontario south of the tree-line. It prefers open, wet woodlands in northern coniferous forests, nesting near wetlands, ponds, and lakes. The Solitary Sandpiper winters from the coastal southern US south to Chile, Brazil, and the Galapagos Islands.

Previous to the atlas, there was only one nest record of this species for Ontario - found at Sutton Lake, south of the Hudson Bay coast on 28 June 1964, in an old American Robin's nest (Schueler *et al.* 1974). There are additional records of downy young, but the delineation of the presumed nesting distribution of the Solitary Sandpiper in Ontario rests largely on observations of territorial behaviour during the breeding season.

The Solitary Sandpiper often shares nesting habitat with the Greater Yellowlegs, but it is not as loud or aggressive as that species. Given its more retiring nature and low population densities, it is not as easy to locate as either yellowlegs species. Atlassers who spent time in suitable wetland habitat, however, should have detected the Solitary Sandpiper, though never in large numbers. It is most conspicuous when performing display flights and 'song' shortly after arrival on the nesting grounds, or when agitated near the nest, or when with young. One bird, pretending to be crippled as part of its distraction display, almost fell into an atlasser's canoe (M. Peterson pers. comm.). About 27% of all records were of agitated birds or distraction displays; another 65% were of single birds or pairs in suitable habitat. This suggests that atlassers were rarely present early in the nesting season when birds were singing, displaying, and chasing each other.

Several other factors make atlassing somewhat difficult and could account for missed records. Breeding habitat is often difficult to access and atlas. The Solitary Sandpiper is also, as its name implies, non-gregarious and defends large, all-purpose territories about a pond (Oring 1973). Consequently, pairs are well scattered over suitable habitat. Nests are difficult to find, requiring either luck or intensive, time-consuming observation devoted to that task only. Flightless young are less difficult, but still not easy to locate. Three-quarters of all breeding confirmations were achieved by finding flightless young, presumably after atlassers were attracted by agitated or distraction behaviour.

The atlas data outline the Ontario nesting limits of this species, extending the previously mapped range south to the southern edge of the largely coniferous Timagami Forest section, just north of Sudbury. This is probably a result of better coverage rather than a southward range expansion. The few records south of Sudbury are of single birds in breeding habitat, and were carefully reviewed prior to publication because of the possibility of their being migrants; non-breeding Solitary Sandpipers often linger into early June in southern Ontario and many appear again in early July. There are a several blocks in the middle of northern Ontario where Solitary Sandpipers would have been assumed to nest but where their presence was not detected by atlassers. However, nesting evidence would not be expected, and was not found, in blocks containing only coastal tundra, such as some of those on the Hudson Bay and northern James Bay coasts. Also, the absence of records from the majority of blocks in the southern portion of northwestern Ontario suggests that the species is rare or absent as a breeding bird in much of that area; relatively little suitable breeding habitat is available in the mixed forests of the Quetico Forest section of the Great Lakes-St. Lawrence Forest zone.

As would be expected, abundance estimates suggest that the species is fairly common in the north, but is less common in southern Ontario, at the southern edge of its range.-- *R. Harris*

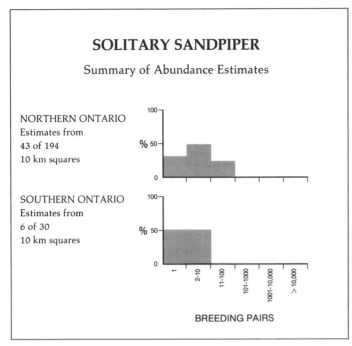

SOLITARY SANDPIPER

Summary of Abundance Estimates

NORTHERN ONTARIO
Estimates from
43 of 194
10 km squares

SOUTHERN ONTARIO
Estimates from
6 of 30
10 km squares

BREEDING PAIRS

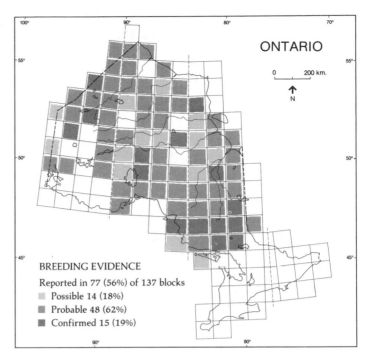

ONTARIO

0   200 km.

N

BREEDING EVIDENCE
Reported in 77 (56%) of 137 blocks
■ Possible 14 (18%)
■ Probable 48 (62%)
■ Confirmed 15 (19%)

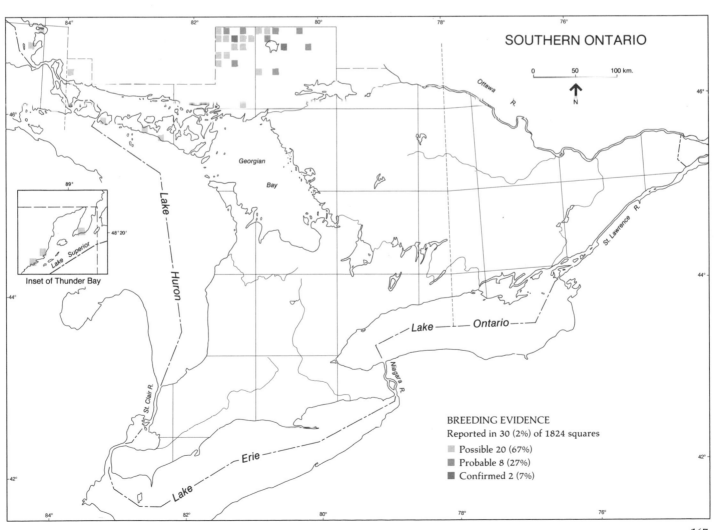

SOUTHERN ONTARIO

0   50   100 km.

N

Georgian Bay

Lake Huron

Lake Superior

Inset of Thunder Bay

Ottawa R.

St. Lawrence R.

Lake — Ontario

Niagara R.

St. Clair R.

Lake Erie

BREEDING EVIDENCE
Reported in 30 (2%) of 1824 squares
■ Possible 20 (67%)
■ Probable 8 (27%)
■ Confirmed 2 (7%)

SPOTTED SANDPIPER
Chevalier branlequeue
Actitis macularia

S.L. HOUSE

The Spotted Sandpiper is probably the best known and most widely distributed of all the New World sandpipers. It is a common summer resident throughout temperate North America, inhabiting a wide variety of habitats which range from saltwater mud flats to high alpine timberline regions (Johnsgard 1981). It occurs in all parts of Ontario and here, as elsewhere, it is usually found near ponds, lakes, rivers, streams, marshes, and other water bodies, where it hunts for aquatic and terrestrial insects. Like its Eurasian counterpart, the Common Sandpiper (*Actitis hypoleucos*), the Spotted Sandpiper is a very successful and adaptable species, and it is not unusual to see this bird using cultivated fields, sewage lagoons, and other man-modified environments.

There is no evidence to suggest that the status of the Spotted Sandpiper has changed significantly within historical times. Early ornithologists such as Elliot (1895) called it a very common species throughout North America, while Samuels (1880) believed that no other bird was distributed "so generally and abundantly" in the northeastern United States. More recent accounts, such as that by Taverner (1953), described it as the commonest summer sandpiper in Canada. Atlas data support the view that the Spotted Sandpiper is both abundant and widespread. The species was recorded in 1,638 of 1,824 squares in southern Ontario, and in 127 of 137 blocks throughout the province. In areas for which estimates of abundance are available, the Spotted Sandpiper was estimated to be only slightly less numerous than the ubiquitous and highly vocal Killdeer.

Most Ontario naturalists are well acquainted with the rapid anxious flight, the curiously vibrating wingbeat, and the emphatically uttered 'peet!' of the Spotted Sandpiper. Likewise, there are few who have not encountered this shorebird as it teeter-totters nervously from one pebble to

another along the shoreline, bobbing its tail constantly as it pecks and probes for food. These characteristic traits have earned the Spotted Sandpiper such local names as the 'teeter', 'tip-up', or 'peet-weet', and they make the Spotted Sandpiper both obvious and unmistakable. It is unlikely that individuals of this species are often missed or misidentified by observers. However, while the Spotted Sandpiper is relatively easy to find in any given square, certain facets of its breeding biology suggest that the actual number of nests or mated pairs is probably underestimated in the atlas data.

The Spotted Sandpiper is an unusual species in that the females are often polyandrous, *i.e.*, they will mate and breed with more than one male during the breeding season (Hays 1972, Oring and Knudson 1972, Oring *et al.* 1983). On Gull Island in New York State, Hays (1972) found that only 2 of 6 females remained monogamous throughout the breeding season, and one female produced successive clutches for 4 different males. Males are territiorial and they provide most or all of the care of eggs and brood, but females will move between nesting territories. Nests are often difficult to find, since they may be hidden in grass, among rocks, or even under decaying logs (Johnsgard 1981). For these reasons, it is not surprising that the Spotted Sandpiper was confirmed as breeding in only 39% of the southern Ontario squares in which it was reported; the actual incidence of breeding is likely much higher. However, when data are summarized on the block basis, only the Gray Jay was confirmed in more blocks than the Spotted Sandpiper; it frequently nests on river banks and lake shores so it was especially well-represented in atlas data for the north because these habitats were well covered.

A glance at the distribution map for the Spotted Sandpiper confirms that this shorebird is extremely widespread and successful in Ontario. There are no major gaps in its distribution, nor are there apparent areas of breeding concentration; the bird appears equally suited to the rugged topography of the Canadian Shield as to the urban areas and quiet streams and fields of southwestern Ontario. Indeed, the species seems to be distributed almost uniformly throughout the province, and it is probable that the observed variation in breeding status is attributable more to the difficulty of confirming breeding than to an actual absence of breeding birds.

Abundance estimates suggest that the Spotted Sandpiper is equally common in northern and southern Ontario.-- *A. Bryant*

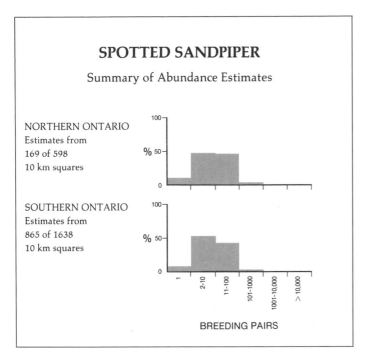

SPOTTED SANDPIPER

Summary of Abundance Estimates

NORTHERN ONTARIO
Estimates from
169 of 598
10 km squares

SOUTHERN ONTARIO
Estimates from
865 of 1638
10 km squares

BREEDING PAIRS

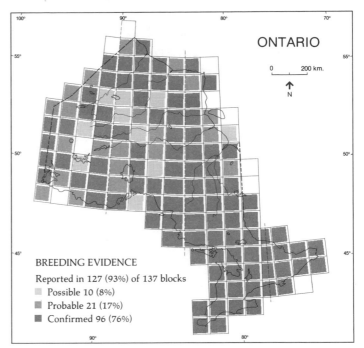

ONTARIO

0   200 km.

N

BREEDING EVIDENCE
Reported in 127 (93%) of 137 blocks

Possible 10 (8%)
Probable 21 (17%)
Confirmed 96 (76%)

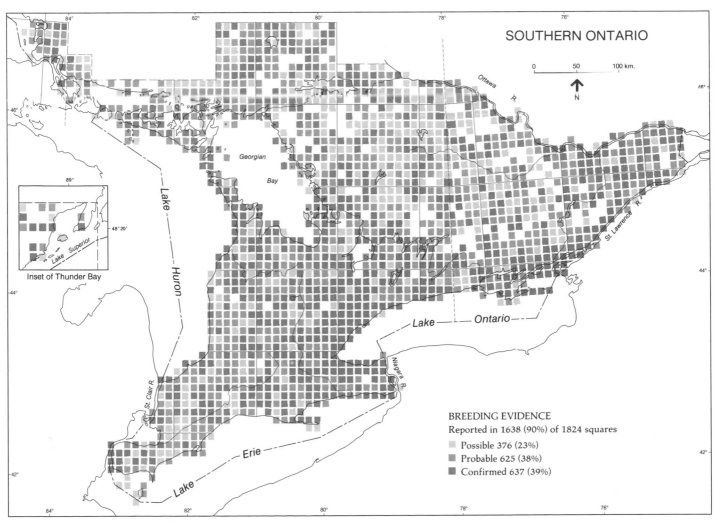

SOUTHERN ONTARIO

0   50   100 km.

N

Inset of Thunder Bay

BREEDING EVIDENCE
Reported in 1638 (90%) of 1824 squares

Possible 376 (23%)
Probable 625 (38%)
Confirmed 637 (39%)

UPLAND SANDPIPER
Maubèche des champs
Bartramia longicauda

S.L. HOUSE

The Upland Sandpiper breeds only in North America: from southern Canada south to Maryland in the east, and to Wyoming and southern Alaska in the west (Godfrey 1986). In Ontario, it breeds generally in agricultural land in the south, but there are isolated populations near Thunder Bay and Rainy River in the north. As a bird of dry grassland, it occurs at its greatest breeding density in prime habitat in North Dakota (Johnsgard 1981). Where almost no natural grassland remains, such as in Ontario, it is a bird of extensive, dry, grassy fields, wet meadows, hayfields, and old fields with little or no shrub or tree growth. Pastures are quite frequently used, but cultivated land is not. It winters in southern South America (Godfrey 1986).

As a grassland species, the Upland Sandpiper must have benefitted greatly from the creation of fields as the forests were cleared on the deep soils of Ontario in the early 1800's. The resultant population increase was soon limited by the market hunting of the late 1800's. In 1886, McIlwraith reported from his observations around Hamilton that this species "is now very seldom seen in Ontario, though older sportsmen tell us that in former times it was often observed in the pasture fields in spring and fall." This suggests that the bird was moving through the Hamilton area during migration and that the population was falling from an earlier level. The prevalent attitude towards the bird is best described by McIlwraith (1886) when he states, "It is abundant on the prairies ... where it will afford good sport and a table delicacy, to many a future settler." By 1919, Roberts (in Bent 1929) declared that, in Minnesota, "it is nearing extinction. Here and there an occasional breeding pair may yet be found, but they are lonely occupants of the places where their ancestors dwelt in vast numbers." Reports published by Macoun and Macoun (1909) indicate that its status in Ontario

had become equally precarious at this time. Since the early decades of this century, however, the species has been increasing in Ontario; BBS data show a significant increase in the Ontario population between 1965 and 1979 (Robbins *et al.* 1986).

The Upland Sandpiper is fairly large and easy to identify, and has a unique 'wolf-whistle' call that can be heard for up to a kilometre. It tends to stand visibly on fence posts, and its display flights are conspicuous. However, it is unobtrusive when not displaying, and so is easily missed, especially where populations are small and single pairs or small groupings of pairs are widely scattered. The same territories are occupied year after year, providing ample opportunity to confirm breeding once birds have been located. Nests are well-hidden, often in 15 to 25 cm long grass (Peck and James 1983), and are difficult to find; nests with eggs or young were reported in only 23 squares (8% of breeding confirmations). The young leave the nest shortly after hatching and stay with the parents for an extended period of juvenile growth. As a result, broods are more readily observed, making up 69% of breeding confirmations.

The atlas maps show that the Upland Sandpiper occurs primarily in agricultural land south and east of the Canadian Shield in southern Ontario. Within that landscape it is largely absent in several areas. In the extreme southwest, Essex and Kent Counties, it is rare and appears to be declining (P.A. Woodliffe pers. comm.), presumably because the land is so heavily utilized for cash crops. The species is also absent in the Toronto area, because of heavy urbanization, and north of Trenton, for reasons unknown to the author. In the more heavily agricultural areas of southwestern Ontario, the records are sparse, reflecting scarcity of pasture land habitat. The distribution in the remainder of the south is sporadic, although a higher proportion of pasture land is characteristic of areas where many records occur. The densest concentration of breeding Upland Sandpipers is found in southeastern Ontario, from Prince Edward Co eastward to Prescott and Glengarry Counties. This is an area of abundant amounts of old field habitat created by mixed farming and frequent farm abandonment. There are a few scattered records in areas of suitable field habitat on the Shield, with a small concentration in the agricultural land near Sudbury. In northern Ontario, a few records came from Thunder Bay and the Rainy River areas.

Atlas data show that the Upland Sandpiper is fairly common in southern Ontario, but rare and very local in the north. Abundance estimates of more than 10 pairs per square occur sporadically south of the Shield, but are concentrated in the atlas regions of Lindsay, Thousand Islands, Cornwall, and Ottawa. BBS data show a partially similar trend, with by far the highest breeding densities being reported east of Ottawa (Speirs 1985).-- *P.F.J. Eagles*

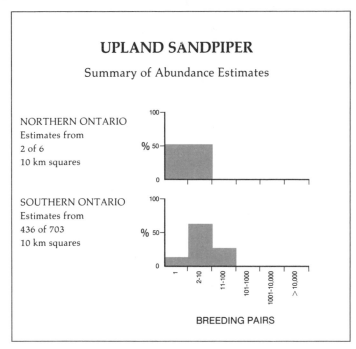

UPLAND SANDPIPER

Summary of Abundance Estimates

NORTHERN ONTARIO
Estimates from
2 of 6
10 km squares

SOUTHERN ONTARIO
Estimates from
436 of 703
10 km squares

%

BREEDING PAIRS

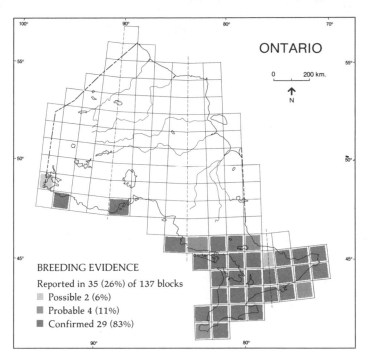

ONTARIO

0   200 km.

N

BREEDING EVIDENCE
Reported in 35 (26%) of 137 blocks
  Possible 2 (6%)
  Probable 4 (11%)
  Confirmed 29 (83%)

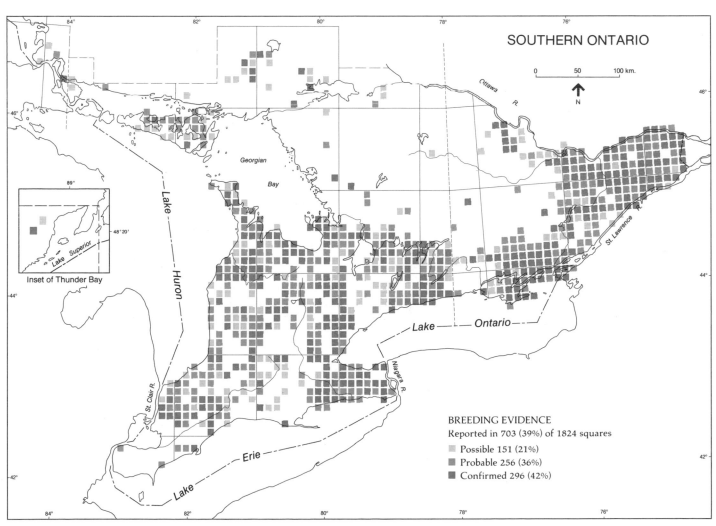

SOUTHERN ONTARIO

0   50   100 km.

N

Inset of Thunder Bay

Lake Superior

Georgian Bay

Lake Huron

Lake Erie

Lake Ontario

Ottawa R.

St. Lawrence

Niagara R.

St. Clair R.

BREEDING EVIDENCE
Reported in 703 (39%) of 1824 squares
  Possible 151 (21%)
  Probable 256 (36%)
  Confirmed 296 (42%)

COMMON SNIPE
Bécassine des marais
Gallinago gallinago

S.L. HOUSE

The Common Snipe is a holarctic species that is widespread in North America, breeding from the tree-line south to the central US. Though the winter range extends as far south as southern Brazil, a few regularly winter as far north as southern Ontario. It nests throughout Ontario, although only very locally in the extreme southwest. The Common Snipe is found in wetlands with soft organic soils and scattered low vegetation that provides cover but good visibility. Preferably, there should be lookout posts 1 m or more high (Cramp and Simmons 1983). Favoured habitats include various types of peatlands, freshwater marshes, bogs, fens, moist meadows, swamps, and areas of low alders and willows along water bodies (Tuck 1972, Johnsgard 1981). Sweet gale is a common nest cover.

Little information is available regarding historical changes in the Ontario range or abundance of this species. Baillie and Harrington (1936) merely stated that it "undoubtedly nests in suitable marshy habitat throughout the whole of Ontario." As with many other wetland species, however, it seems probable that the Common Snipe is now less abundant and widespread in southern Ontario, particularly the extreme southwest, where wetlands have been lost to agriculture and urban development.

Depending on the stage of the nesting season, the Common Snipe can be either very easy or extremely difficult to locate. When it is 'winnowing' and otherwise actively displaying, calling and flying about early in the season, the species is hard to miss. The display period can be both very active, because there is a high degree of promiscuity in this species, and prolonged, as different males may breed sequentially on one territory. Males will commonly use a conspicuous elevated perch, such as a fencepost or dead treetop, as a lookout. 'Winnowing' is most frequent in the evening. Once

this display period has passed, however, the Common Snipe becomes very inconspicuous. It is a secretive bird, nesting in habitat that is often hard to access. Nests are usually found by accidentally flushing the incubating bird at very close range. The Common Snipe sits tightly with its brood and, unlike many other shorebirds, does not display the agitated mobbing behaviour towards intruders that would signal its presence from a distance. Instead, it flushes at close range and performs an injury-feigning distraction display. Breeding was confirmed in only 16% of southern Ontario squares, and 10% of northern Ontario squares in which the species was reported; in both sectors, the majority of confirmed records were of young birds, nests, and distraction displays. Nests with eggs or young were reported from 56 squares in the whole province.

The atlas maps corroborate earlier accounts that the Common Snipe nests throughout Ontario. They are also in agreement with BBS data regarding the frequency of occurrence in different areas (Speirs 1985). In southern Ontario, the Common Snipe is most widespread on the Niagara Peninsula, in the cool, wet areas of Bruce, Grey, and Dufferin Counties, along the edge of the Canadian Shield, and throughout the southeast. The sparse distribution of records in southwestern Ontario and the Toronto-Hamilton-Lake Simcoe area undoubtedly reflects the genuine scarcity of the species there rather than atlassing difficulties. Loss of wetland habitat to farmland and urban development in these areas has been substantial. The Common Snipe likely nests in more squares in southern Ontario to the east and north of Georgian Bay than atlas data show. Suitable wetland nesting habitat is often more difficult to access there. Nevertheless, suitable habitat for the species is less abundant on the shallower soils of that part of the Shield and it probably occurs less frequently there.

The Common Snipe also undoubtedly nests throughout more of northern Ontario than atlas records indicate. It could be easily missed because of habitat inaccessibility or because most atlassing in remote northern areas was conducted after the conspicuous display period. Nevertheless, it is clear that the Common Snipe was reported in a higher proportion of blocks, and at higher levels of breeding evidence, on the Hudson Bay Lowland than it was on the northern Boreal Forest south of the Lowland. Since both the above atlassing difficulties applied equally throughout the north, these differences in abundance and distribution appear not to be artifacts of atlassing effort. The Hudson Bay Lowland, where wetland habitat abounds, is probably the region of greatest abundance of the Common Snipe in Ontario. Such habitat is less frequent on the Shield country to the south.

Abundance estimates indicate that the Common Snipe is a fairly common bird in squares where it was reported in both northern and southern Ontario, occurring, on average, at about the same breeding densities in both areas.-- *R. Harris*

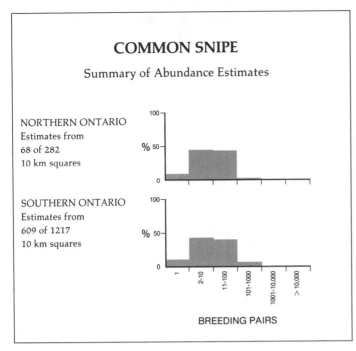

COMMON SNIPE

Summary of Abundance Estimates

NORTHERN ONTARIO
Estimates from
68 of 282
10 km squares

SOUTHERN ONTARIO
Estimates from
609 of 1217
10 km squares

%

100

50

100

50

1   2-10   11-100   101-1000   1001-10,000   >10,000

BREEDING PAIRS

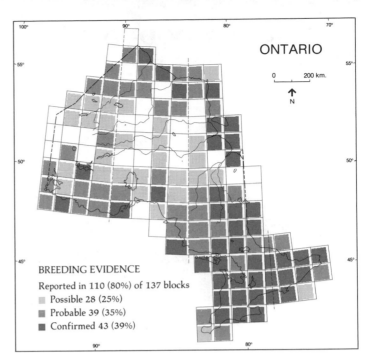

ONTARIO

0   200 km.

N

BREEDING EVIDENCE
Reported in 110 (80%) of 137 blocks
Possible 28 (25%)
Probable 39 (35%)
Confirmed 43 (39%)

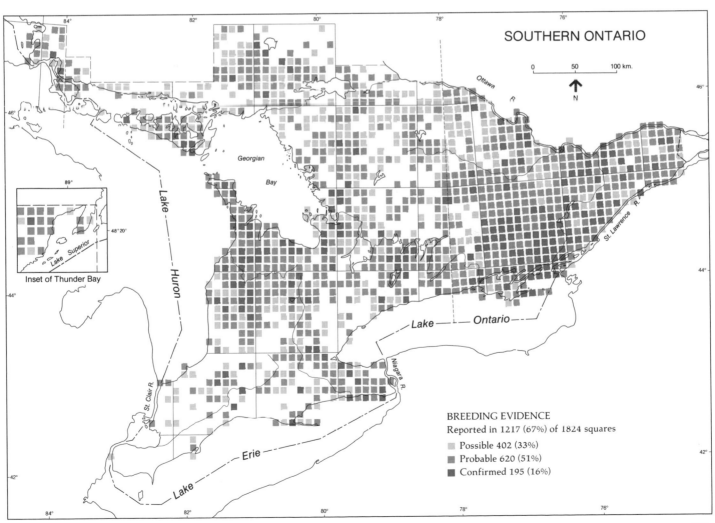

SOUTHERN ONTARIO

0   50   100 km.

N

Georgian Bay

Lake Huron

Lake Superior

Inset of Thunder Bay

St. Clair R.

Lake Erie

Niagara R.

Lake Ontario

Ottawa R.

St. Lawrence R.

BREEDING EVIDENCE
Reported in 1217 (67%) of 1824 squares
Possible 402 (33%)
Probable 620 (51%)
Confirmed 195 (16%)

173

AMERICAN WOODCOCK
Bécasse d'Amérique
Scolopax minor

S.L.HOUSE

This secretive bird is the only North American species of wader to have adapted entirely to life in the woodland. It is a nocturnal migrant, very secretive in the daytime, and is conspicuous and easy to find only in spring, when males perform their territorial display.

The American Woodcock breeds east of the Great Plains, almost throughout the forested area of eastern North America, and winters in the southeastern US. The American Woodcock probes for invertebrates, particularly earthworms, living in the surface layers of soil and litter. In its choice of habitat, soil type is important, and the highest densities of the species are found on sandy or sandy loam soils, alluvial soils, and gleisols. The one requirement for all soils favoured by this species is that they be moist. Waterlogging in spring and fall for short periods does not adversely affect their use by this species. Good habitat, however, neither dries out nor remains flooded throughout the summer. Ideal cover consists of an overstory composed of aspen, alder, and birch, in pure or mixed stands, sometimes with some cedar, hawthorn, willow, tamarack, and other conifers. This overstory should not constitute more than about 70% of the total canopy and should be less than 30 years old. The understory is generally composed of *Spiraea*, dogwoods, and shrub willows, with herbaceous cover such as jewelweed, goldenrod, and grasses that are not too dense. An important component of habitat is open, grassy clearings or abandoned fields, which can be as small as 20 m in diameter. Male American Woodcocks display in such clearings in spring. Because of the absence of earthworms and scarcity of invertebrates, the species avoids peat lands, even though the forest age and cover type may be favourable.

In Ontario, the American Woodcock increased in number during the 19th century, when logging by settlers removed stands of old trees and opened canopies. Earthworms were extirpated in Ontario by the advance of the glaciers (Reynolds 1977). Several European species of earthworms were introduced with the spread of agriculture, and now constitute an important source of food for the American Woodcock. With the intensification of agriculture, clearing and draining of wet woodland, and development of housing in formerly suitable covers, the habitat is being degraded and is disappearing. The bird is, therefore, no longer as abundant as it used to be. Data on recent trends in numbers of American Woodcock in Ontario, derived from singing-ground counts in spring, suggest an increase from 1971 to a peak in 1979, a decline to 1983, and then recovery to close to average levels in 1985 (Bateman 1985).

The American Woodcock has been described as being crepuscular in activity. However, it feeds during most of the daytime and roosts at night in grassy clearings. It is certainly most active and can be most readily found at dawn and dusk. Males display with a series of 'beezp' calls given from the ground in a clearing. This is followed by a steep twittering ascent, circling, increasingly excited twittering and erratic flight, then a plunge to the starting point. The display period lasts 30 to 40 minutes and starts about 21 minutes after sunset on a clear night and about 7 minutes earlier on cloudy evenings. In the morning, song flights start an average of 32 minutes before sunrise. Males display from arrival until early June.

The American Woodcock arrives in Ontario in March, and begins to nest in early April. However, most nests are started in May and some birds are still incubating what are probably replacement clutches in July (Peck and James 1983). The incubating female sits tight and sometimes can be touched on the nest. She may flush, with laboured flight, hanging feet, and fluffed under and upper tail coverts in a distraction display. The eggs take 20 to 21 days to hatch. Nests with eggs or young were found in a total of 139 squares, 133 of which were in southern Ontario, but locating broods was the most frequent means of confirming breeding, such evidence having been reported in 232 squares. The young, at least during their first few days, are fed by the mother. They can fly as early as 14 days of age (Pettingill 1936).

It is likely that many of the squares on the map are blank because atlassers were not able to cover suitable habitat during the short period at dusk and dawn when the males were displaying. The relatively low percentage of squares in which breeding was confirmed indicates the difficulty of finding nests or downy young; it is likely that nearly every square and block with woodland, as far north as 49°N, has breeding American Woodcocks. North of this latitude, suitable habitat is sparse and breeding is local. The most northerly record was of a calling male at Moosonee on 16 May 1983.

Abundance estimates indicate that the American Woodcock is fairly common in southern Ontario, but more sparsely distributed near the edge of its range in northern Ontario.-- *H.G. Lumsden*

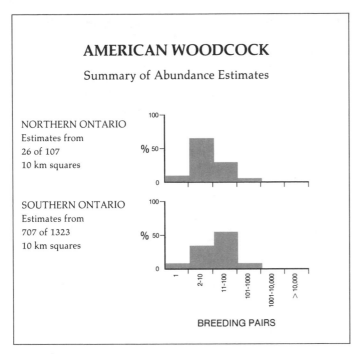

AMERICAN WOODCOCK

Summary of Abundance Estimates

NORTHERN ONTARIO
Estimates from
26 of 107
10 km squares

SOUTHERN ONTARIO
Estimates from
707 of 1323
10 km squares

BREEDING PAIRS

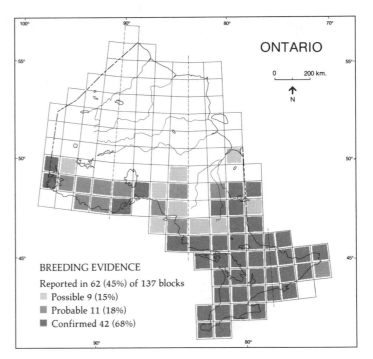

ONTARIO

0   200 km.

N

BREEDING EVIDENCE

Reported in 62 (45%) of 137 blocks

Possible 9 (15%)

Probable 11 (18%)

Confirmed 42 (68%)

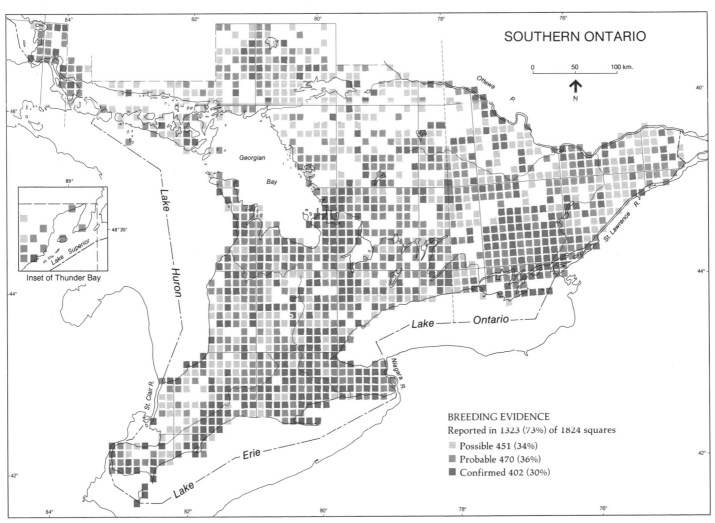

SOUTHERN ONTARIO

0   50   100 km.

N

Georgian Bay

Lake Huron

Lake Superior

Inset of Thunder Bay

48°20'

89°

Ottawa R.

St. Lawrence R.

Lake Ontario

Niagara R.

St. Clair R.

Lake Erie

BREEDING EVIDENCE

Reported in 1323 (73%) of 1824 squares

Possible 451 (34%)

Probable 470 (36%)

Confirmed 402 (30%)

175

WILSON'S PHALAROPE
Phalarope de Wilson
Phalaropus tricolor

S.L.HOUSE

Female phalaropes are more brightly plumaged than their mates. They compete among themselves for males, courting them actively, then leaving them to incubate the eggs and raise the young alone.

Wilson's Phalarope breeds in North America, principally in the western provinces and US states south to Kansas and California. Three small separate populations occur farther east: in southern Wisconsin, in contiguous areas of southern Ontario and southern Michigan (Johnsgard 1981), and at the south end of James Bay in northern Ontario. It winters in western and southern South America. Over most of its breeding range, Wilson's Phalarope is a bird of the prairies, where its breeding habitat consists of ponds and lakes that are close to wet meadow vegetation, as well as sloughs along intermittent streams (Johnsgard 1981). In southern Ontario, it is most often found in association with sewage lagoons with lush grassy or sedgy margins, or in marshes. The James Bay birds breed in freshwater coastal marshes (Morrison and Manning 1976).

Historical records suggest that Wilson's Phalarope has only recently become established in southern Ontario; McIllwraith (1886) cited only 2 records. Early reports were primarily from the southwest, the first nest being found at Dunnville in 1879 (Baillie and Harrington 1936). Scattered nestings and summer sightings began to occur with more regularity farther to the east and somewhat farther north during the following decades. By 1974, a nest had been found as far east as Montreal (Morrison and Manning 1976), and breeding was confirmed in New Brunswick in 1986. There are few early records from the Lake of the Woods area or farther north along the Manitoba border, although this may be because of a lack of coverage. The latter area is at the eastern edge of the main prairie range of the species and Goodwin (1982)

stated that Wilson's Phalaropes had been reported in the Lake of the Woods area in larger numbers than anywhere else in the province. The record of a male showing extremely agitated behaviour about 50 km northwest of Sandy Lake in 1983, and the recent breeding record from Churchill, Manitoba (Godfrey 1986) suggest that the species is continuing to expand its range to the north. The nesting population at the south end of James Bay was discovered in 1975 (Morrison and Manning 1976).

Wilson's Phalarope is a strikingly plumaged bird that is quite easy to spot, especially when feeding in the characteristic phalarope manner of spinning on the surface of the water to stir up invertebrate prey. Vocal communication is not as conspicuous as it is in more territorial shorebirds (Howe 1975a). Aerial courting chases are initiated by females (Howe 1975b) and render the birds fairly conspicuous early in the season. The female may stay near the nest, acting as a lookout, while the male incubates, and will join the male in hovering around intruders (Johns 1969). This species is often found in small 'colonies' (Hohn 1967); 61% of abundance estimates are of 2 or more pairs, suggesting small colonies. Where this occurs, females will form flocks which feed in the open while the males incubate. However, females most often leave the breeding area before the young hatch (Hohn 1967), and sometimes leave immediately after laying (Murray 1983). The males keep broods in dense cover until they are close to fledging. Nevertheless, broods were more frequently reported than nests - representing 63% and 26% of confirmed records, respectively.

Because birders and Wilson's Phalaropes seem to share an affinity for sewage ponds, it is probable that records will be biased towards such localities. 'Natural' locations were more likely to be overlooked in southern Ontario, but the map probably represents the majority of breeding locations. A few previously reported breeding locations, such as Sundridge (Parry Sound District) and Toronto's Leslie St. Spit (Speirs 1985) were not found to have Wilson's Phalaropes during the atlas period. However, 2 new locations near Luther Lake suggest that the Luther colony, which has been active with 8 to 10 pairs since 1958 (Brewer 1977), may be expanding. The most marked change in population has occurred in southeastern Ontario, where about 30 pairs per year breed around Kingston, primarily on Amherst Island (R.D. Weir pers. comm.), and to the southeast of Ottawa, where that region's first confirmed record occurred during the atlas period. All records east of Ottawa are from sewage lagoons (C. Hanrahan pers. comm.). Breeding was reported at Kelly Lake near Sudbury, with possible breeding recorded by atlassers at other sewage lagoons nearby.

Abundance estimates indicate that Wilson's Phalarope occurred locally in small numbers across southern Ontario. The 2 highest abundance estimates (11 to 100 pairs) were from the James Bay coast, indicating that the population there was at least as large as the 20 to 25 pairs estimated in 1977 by Sinclair (1978).-- *M.D. Cadman*

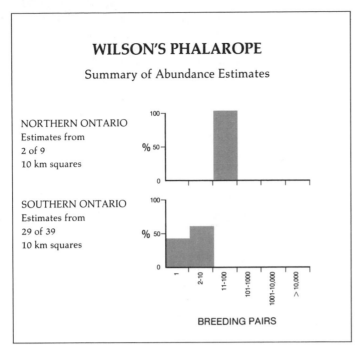

WILSON'S PHALAROPE

Summary of Abundance Estimates

NORTHERN ONTARIO
Estimates from
2 of 9
10 km squares

SOUTHERN ONTARIO
Estimates from
29 of 39
10 km squares

BREEDING PAIRS

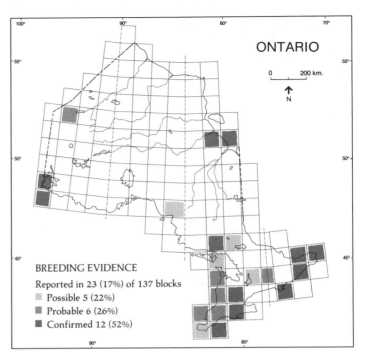

ONTARIO

0   200 km.

↑
N

BREEDING EVIDENCE
Reported in 23 (17%) of 137 blocks
Possible 5 (22%)
Probable 6 (26%)
Confirmed 12 (52%)

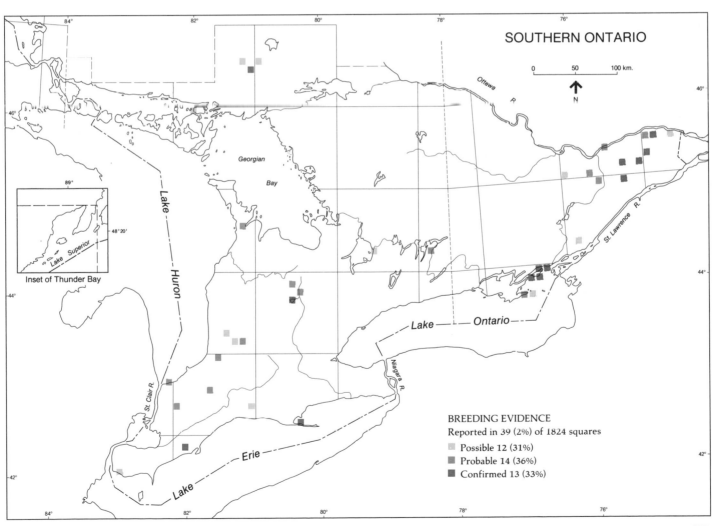

SOUTHERN ONTARIO

0   50   100 km.

↑
N

Georgian
Bay

Lake Huron

Lake — Ontario

Ottawa R.

St. Lawrence R.

Niagara R.

St. Clair R.

Lake — Erie

Inset of Thunder Bay

Lake Superior

BREEDING EVIDENCE
Reported in 39 (2%) of 1824 squares
Possible 12 (31%)
Probable 14 (36%)
Confirmed 13 (33%)

LITTLE GULL
Mouette pygmée
Larus minutus

S. L. HOUSE

A commonly disputed topic among Little Gull enthusiasts concerns the breeding location of Little Gulls seen in North America. It has been noted that there were too many sightings in North America to be accounted for by the few pairs known to nest here annually. Baillie (1963a) felt that most Little Gulls seen in North America were vagrants from eastern Siberia that had made contact with Bonaparte's Gulls in Alaska and migrated southeast with them. More recently, Bruun (1968), Johnson and Adams (1977), and Hutchinson and Neath (1978) suggested that most Little Gulls seen in North America originated in Europe and moved here as vagrants via a westward movement. However, new data from the atlas suggest that a larger proportion of the Little Gulls seen here may actually be nesting in North America. Hence, the role of vagrants in the North American population may have to be re-evaluated.

The Little Gull is chiefly a breeding bird of northern and central temperate Asia with scattered nestings in many countries of northern and central Europe (Cramp and Simmons 1983). In the New World, nesting has been reported only along the Great Lakes and the St. Lawrence River and on the Hudson Bay Lowland (McRae 1984). The Little Gull nests predominantly in marshes and builds its nest on a wet grassy knoll or on an islet of tangled, floating vegetation. It sometimes nests with other colonial waterbirds, especially Common, Arctic, Forster's, and Black Terns (Peck and James 1983, McRae 1984).

The first known nesting of the species outside of the Old World occurred in 1962 at Oshawa's Second Marsh (Scott 1963). Nine years later, in 1971, the first known young in the New World were produced and fledged at Cranberry Marsh, 13 km to the west of Second Marsh, on the north shore of Lake Ontario (Tozer and Richards 1974). Since that time, the number of reported nestings of the Little Gull in North America has increased slowly. It has nested in Wisconsin, Michigan, Quebec, Manitoba, and at least 3 additional sites in Ontario (Peck and James 1983, McRae 1984).

As with many other colonially nesting birds, if a Little Gull nesting territory can be located, finding the nest or confirming breeding should not be too difficult. Any adult Little Gulls observed over or near a marsh, especially along the Great Lakes shore, or any seen in association with terns in late May or June, should be considered to be potential breeders. Even from a distance, an adult Little Gull flying among dozens, or hundreds, of white terns is fairly readily identified because its diagnostic black underwing is easily seen. The behaviour of the adults near the nest is often characteristic. They will fly low overhead, perhaps swooping on the intruder (as Ring-billed and Herring Gulls would do), and give a low, soft, almost plaintive 'kew, kew, kew'. By watching from a distance, it may then be possible to see the adult(s) alight at or near the nest, if there is one. Nests with young were reported in 2 atlas squares. If Little Gulls are suspected of nesting as a single pair on a marsh, or in association with Black Terns, their white plumage should make finding the territory or nest relatively easy.

There were records of Little Gulls from 6 squares in southern Ontario and 7 squares in the north during the atlas period. Only 3 of these were of confirmed breedings: a nest on 25 June 1983 on North Limestone Island, Georgian Bay, a nest on 15 July 1985 in a marsh on the west shore of James Bay, and adults with newly fledged young on 24 July 1984 at the Limestone Rapids on the Winisk River. The locations presented on the distribution map attest to the preference of this species for nest sites near large expanses of open water. Also, all 6 southern Ontario records are from squares in which Common or Black Terns were reported as nesting during the atlas period. None of the northern Ontario sites were previously known. Judging from atlas records and reports from outside Ontario, the coastal lowland of James Bay and Hudson Bay may be the prime nesting area of the Little Gull in North America (McRae 1984, D. McRae and J. Richards pers. comm.). Nestings from this area are probably underrepresented on the atlas maps; there are extensive areas of coastal marsh which were not visited, and a large number of rivers where breeding could occur that were not covered during the atlas period.

The Little Gull is a rare breeding bird in the New World. The maximum number of nesting pairs at any one site in Ontario (and North America) has been 5 (Tozer and Richards 1974). During the atlas period, there were estimates of 2 to 10 pairs from 4 squares. These were from the Long Point marshes, the mouth of the Sutton River on Hudson Bay, a square on the James Bay coast 40 km north of the Albany River, and a square 40 km southeast of Moosonee.-- *D.V. Weseloh*

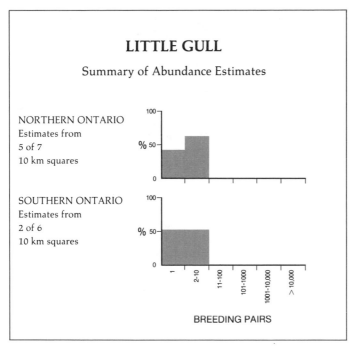

LITTLE GULL

Summary of Abundance Estimates

NORTHERN ONTARIO
Estimates from
5 of 7
10 km squares

100
% 50
0

SOUTHERN ONTARIO
Estimates from
2 of 6
10 km squares

100
% 50
0

1 2-10 11-100 101-1000 1001-10,000 >10,000

BREEDING PAIRS

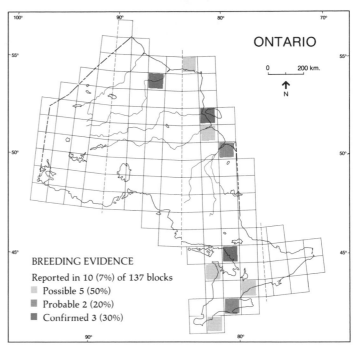

ONTARIO

0   200 km.

↑
N

BREEDING EVIDENCE
Reported in 10 (7%) of 137 blocks
Possible 5 (50%)
Probable 2 (20%)
Confirmed 3 (30%)

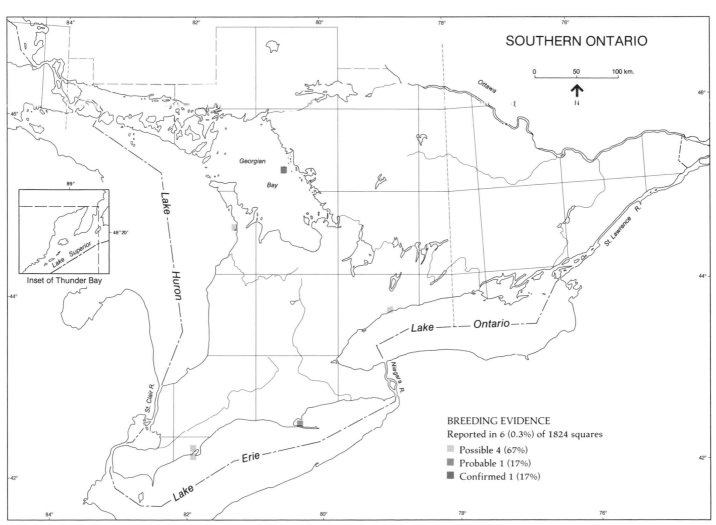

SOUTHERN ONTARIO

0   50   100 km.

↑
N

Inset of Thunder Bay

Lake Superior

48°20'
89°

Georgian Bay

Lake Huron

St. Clair R.

Lake Erie

Niagara R.

Lake — Ontario

Ottawa

St. Lawrence R.

BREEDING EVIDENCE
Reported in 6 (0.3%) of 1824 squares
Possible 4 (67%)
Probable 1 (17%)
Confirmed 1 (17%)

RING-BILLED GULL
Goéland à bec cerclé
Larus delawarensis

S.L. HOUSE

During the last 35 years, the Ring-billed Gull has increased its numbers, extended its breeding range, and established new colonies at urban, industrial, and rural sites in Ontario. This adaptable and opportunistic species thrives in man-made environments and has become a problem in many parts of the province (Blokpoel and Tessier 1986).

The Ring-billed Gull breeds in a belt across North America, covering the southern parts of the Canadian provinces and the northern parts of many northern American states. The Ontario gulls winter mainly along the Atlantic and Gulf coasts of the US (Southern 1974). The species nests throughout Ontario north to southern James Bay (Godfrey 1986). The Ring-billed Gull has adapted to nesting on both natural sites (usually small, partly vegetated islands) and artificial sites (dykes, breakwaters, dredge-spoil areas, slag dumps, industrial yards, etc.). It has also nested at sewage lagoons, on a garbage dump near Ottawa (Weir 1983), and on a roof near Owen Sound (Blokpoel and Smith, unpubl. data).

The Ring-billed Gull was probably fairly common in Ontario during the first half of the 19th century: John James Audubon (1840) referred to it as the "Common American Gull." It could not cope with the depredation by the people who colonized southern and central Ontario, and by 1920 only a few colonies were left on remote islands in Lake Huron. Little is known about its status between 1920 and 1940, but Baillie and Harrington (1936) supported the suggestion of a continuing localized population. Between 1940 and 1960, the Ring-billed Gull population of the entire Great Lakes was fairly stable, with an estimated 27,000 pairs in 1960 (Ludwig 1974). Since then, the Great Lakes population has quickly increased to an estimated 648,000 pairs in 1984. Of these, 427,000 pairs nested in the Ontario portion of the Great Lakes (Blokpoel and Tessier 1986). Ring-billed Gulls

are also increasing in numbers elsewhere in their breeding range (*e.g.*, in Quebec - Mousseau 1984). The main reasons for this phenomenal increase probably are a change in people's behaviour towards gulls, a change in the gulls' behaviour towards people, and an increase in food sources and artificial colony sites. Human persecution of the species has virtually stopped in Ontario, and, in turn, the gulls have lost most of their fear of people while exploiting many new opportunities in the human environment. The large and rapid population increase has resulted in many problems, especially in southern Ontario, where many people and gulls live close together. During 1985 and 1986, Ring-billed Gull colonies at 7 urban and industrial sites were reduced or eliminated by man to alleviate local problems. The burgeoning Ring-billed Gull numbers also have a negative impact on Common Terns and Caspian Terns because the gulls arrive earlier on shared breeding grounds and take over the terns' nesting habitat (Blokpoel and Tessier 1986).

Because Ring-billed Gulls frequently nest in large colonies, and build nests in open areas near large bodies of water, it is usually easy to locate breeding areas and confirm nesting. Scattered nestings involving only a few pairs are, of course, easier to miss. Since fledged chicks often disperse quickly from the colonies, the data for the distribution maps have been checked to ensure that all reports of 'fledged young' pertain to chicks that were too young to fly, or to fledglings in a colony. Thus, the map presents a reliable picture of the gulls' nesting distribution, except for the possibility that a few colonies in northern Ontario remained undetected because of difficult access.

As the maps show, most birds breed on the islands and shores of Lakes Huron, Erie, and Ontario, as well as the St. Lawrence River. Colonies are also found on rocky islands in Lake Superior, Lake of the Woods, southern James Bay, Lake Nipissing, and the Ottawa River. A colony at the mouth of the Attawapiskat River was reported prior to 1981 and again in 1986, but was not visited during the atlas period. Atlassers reported a few new colonies at inland locations in southern Ontario (*e.g.*, Lakes Simcoe and Muskoka and the Lindsay sewage lagoons). In northwestern Ontario, colonies were discovered at Bearskin Lake in 1984 and North Caribou Lake in 1985. This area was previously considered to be unoccupied by Ring-billed Gulls (Peck and James 1983).

Some gulleries on Lakes Erie and Ontario contained tens of thousands of breeding pairs in 1985, *e.g.*, Port Colborne mainland (48,000 nests), Hamilton Harbour spit (14,000), Tommy Thompson Park, also known as the Leslie Street Spit, (47,000), and High Bluff Island, Presqu'ile Prov. Park (33,000). Because the Ring-billed Gull is an adaptable, prolific, and long-lived species, it seems reasonable to predict that its numbers will continue to increase both in Ontario and elsewhere in its breeding range.-- *H. Blokpoel*

RING-BILLED GULL

Summary of Abundance Estimates

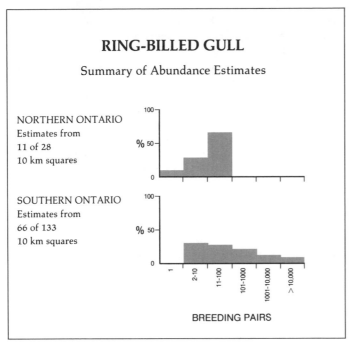

NORTHERN ONTARIO
Estimates from
11 of 28
10 km squares

SOUTHERN ONTARIO
Estimates from
66 of 133
10 km squares

BREEDING PAIRS

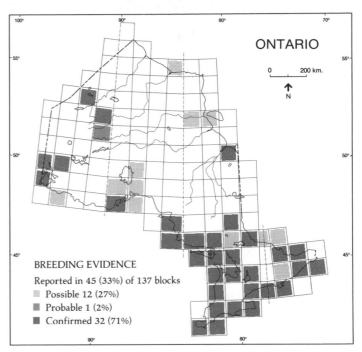

ONTARIO

0   200 km.

N

BREEDING EVIDENCE

Reported in 45 (33%) of 137 blocks

Possible 12 (27%)

Probable 1 (2%)

Confirmed 32 (71%)

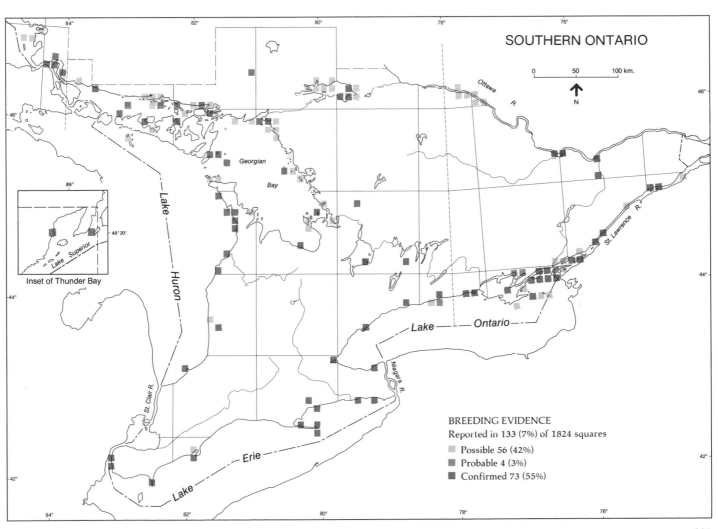

SOUTHERN ONTARIO

0   50   100 km.

N

Ottawa R.

Georgian
Bay

Lake Huron

St. Clair R.

Lake Erie

Niagara R.

Lake — Ontario

St. Lawrence R.

Lake Superior

48°20'

89°

Inset of Thunder Bay

BREEDING EVIDENCE

Reported in 133 (7%) of 1824 squares

Possible 56 (42%)

Probable 4 (3%)

Confirmed 73 (55%)

HERRING GULL
Goéland argenté
Larus argentatus

S.L.HOUSE

The Herring Gull is distributed through most of the arctic to mid-northern temperate latitudes of the world (Cramp and Simmons 1983). In North America, it breeds across the northern third of the continent, including all of Ontario (AOU 1983). The Herring Gull breeds colonially or in isolated pairs on undisturbed islands, peninsulas, or cliffs along sea coasts, lakes, and rivers. Hence, it is well suited for nesting in both northern and southern Ontario. The Herring Gull was the most widely recorded species in the province, having been reported in 129 blocks.

Egg collecting and possibly other forms of disturbance, such as the millinery trade of the late 1800's, had profound effects upon this species. In the early 1900's, Herring Gull populations on Lake Ontario may have been at an all-time low. The species was reported to be extirpated from former breeding areas around Pigeon Island, the Lower Duck Islands (Lake Ontario), and the northeast shores of Lake Ontario (Young in Macoun and Macoun 1909, Eaton 1910). The Migratory Bird Treaty of 1916 protected the species, allowing it to expand its range and breeding numbers. Nevertheless, in (western) Lake Erie, the Herring Gull was not known to nest until 1926 (Shipman 1927), perhaps representing a recolonization after earlier persecution. There has been an increase in the Great Lakes population since the 1940's, according to Christmas Bird Counts (Kadlec and Drury 1968). However, overall numbers wintering at Toronto and Hamilton show no obvious trend in the period 1931 to 1977 (Freedman and Riley 1980). The species was singled out for attention in the 1960's and 1970's because of problems with reproduction attributed to pollutants in the Great Lakes (Gilbertson 1974), but there is little evidence that the overall breeding range in Ontario has been changed recently.

The Herring Gull is large and conspicuous, and its fairly specific nesting habitat is quite well known. Offshore islands or inland lakes with relatively undisturbed islands, particularly if barren, are preferred by Herring Gulls for nesting. Because gulls use the same colony site year after year, historical data can be useful in locating nesting areas. Gulls do not perform distraction displays, but they give distinctive alarm calls and swoop at or even strike intruders near their nests. It is easy to see such large white birds incubating eggs on ground nests near the perimeters of islands or on raised platforms near water; hence, 61% of confirmed breeding records consist of occupied nests. By late May, the young will hide among rocks and vegetation when alarmed, and breeding is more difficult to confirm, at least where islands are inaccessible for closer search. Because young Herring Gulls fly widely soon after fledging, only flightless young are appropriately designated as 'fledged young'; 30% of confirmed breeding records are in that category. While most such records are, no doubt, legitimate records of downy or unfledged young, some of them may refer to flying birds.

The distribution of confirmed Herring Gull nesting records includes the islands and shorelines of the Great Lakes and their connecting rivers, the Canadian Shield, including Muskoka, Haliburton, the Kawarthas, Parry Sound District, and the Algonquin Highlands, and much of northern Ontario. On the Shield, birds nest widely, but in small numbers and often singly. Even where lakes are subjected to intense recreational use, they are often large enough to have secluded, remote, or protected islands, suitable for nesting. In any case, nesting usually begins well before the height of the recreational season. Inland breeding records in southwestern Ontario are from exceptional situations such as islands in Luther Lake and in a water-filled gravel pit northwest of London. Major nesting areas on the lower Great Lakes are known from Kingston, Brighton, Toronto, Hamilton, Port Colborne, off Pelee Island, the Detroit River, and the Bruce Peninsula, as well as at Niagara Falls, a colony which somehow went unrecorded by atlassers (Blokpoel 1977, Blokpoel and McKeating 1978, Weseloh *et al.* in press). Records from Long Point, Rondeau Prov. Park, and the southwestern shores of Lake Huron probably represent irregular and minor nestings. In these areas of the Great Lakes, nesting is generally solitary, sporadic, and unsuccessful.

The most frequently used abundance estimate (45%) was of 2 to 10 pairs per square: 13% were of single pairs. These estimates are low when compared with CWS surveys from a slightly earlier time, 1976 to 1980, but the CWS surveys excluded the large number of interior sites, most of which support only one or a few pairs. They showed that the largest Canadian Great Lakes colony (Chantry Island on Lake Huron) had 3,200 nests and 4 other colonies had approximately 1,000 nests or more: 3 off the Bruce Peninsula and one in western Lake Erie. Most colonies in that survey had from 11 to 100 nests, and approximately 9% of nestings were solitary (Weseloh *et al.* 1986).-- *D.V. Weseloh*

HERRING GULL

Summary of Abundance Estimates

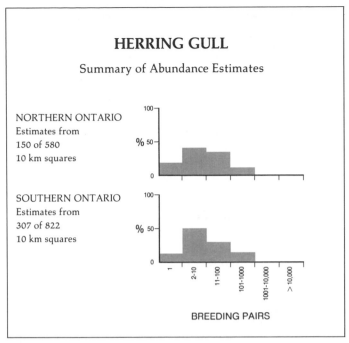

NORTHERN ONTARIO
Estimates from
150 of 580
10 km squares

SOUTHERN ONTARIO
Estimates from
307 of 822
10 km squares

%

BREEDING PAIRS

1   2-10   11-100   101-1000   1001-10,000   > 10,000

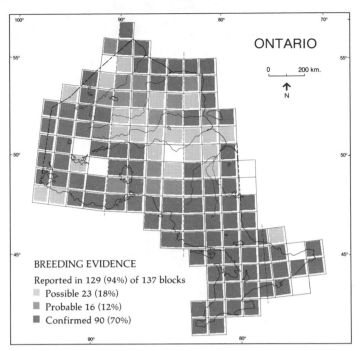

ONTARIO

0   200 km.

N

BREEDING EVIDENCE

Reported in 129 (94%) of 137 blocks

Possible 23 (18%)

Probable 16 (12%)

Confirmed 90 (70%)

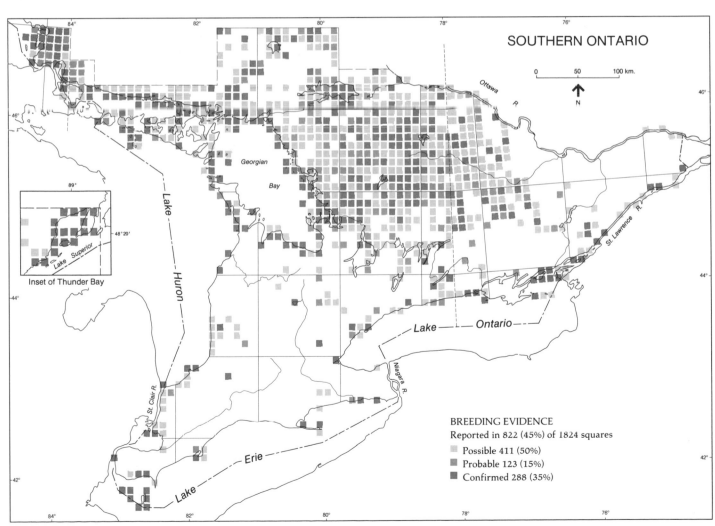

SOUTHERN ONTARIO

0   50   100 km.

N

Inset of Thunder Bay

Lake Superior

48°20'

89°

Georgian Bay

Ottawa R.

St. Lawrence R.

Lake Huron

Lake Ontario

Niagara R.

Lake Erie

St. Clair R.

BREEDING EVIDENCE

Reported in 822 (45%) of 1824 squares

Possible 411 (50%)

Probable 123 (15%)

Confirmed 288 (35%)

GREAT BLACK-BACKED GULL
Goéland à manteau noir
Larus marinus

S.L. HOUSE

The Great Black-backed Gull, the largest of all North American gulls, is, as the scientific name indicates, essentially a marine species that is normally found at or near the sea.

The species has a North Atlantic breeding range, nesting in northwestern Europe, Iceland, southern Greenland, and northeastern North America (Voous 1960). In Canada, the Great Black-backed Gull nests along the coast of the Atlantic Provinces, along the shores of the Gulf of St. Lawrence in Quebec, and at isolated locations on the Great Lakes (Godfrey 1986). The species usually nests on flat, rocky islands close to the sea coast, but occasionally on large inland lakes or on moorland near the coast (Voous 1960, Cramp and Simmons 1983). The North American population winters on the Great Lakes in limited numbers, and on the Atlantic coast from Newfoundland to North Carolina.

Since 1954, there has been occasional breeding by single pairs of Great Black-backed Gulls on Lakes Huron and Ontario, and it was suggested that the species seemed poised to invade the Great Lakes (Ludwig 1974). However, a review of the available information from 1946 to 1977 did not indicate that such an invasion was taking place (Angehrn *et al.* 1979). Nevertheless, its regular summer presence in small numbers in Ontario contrasts strikingly with the situation that existed before the 1950's, when it was strictly a winter bird here.

The Great Black-backed Gull remains a rare nesting bird in Ontario: confirmed nesting records refer to only 7 sites. Nesting was always by single pairs, and usually during only one breeding season (see Table). Of the 7 locations, 6 are on Lake Ontario, suggesting that the Great Black-backed Gull entered the Great Lakes from the Atlantic coast via the St. Lawrence River. Compared to the breeding of Herring and Ring-billed Gulls, some of the nesting of the Great Black-

backed Gull occurred late in the season. Reproductive success was unknown for 2 of the localities because follow-up visits were not made. At one locality, the eggs disappeared, but chicks were observed at the other 4 sites.

It is possible, but not very likely, that single pairs of Great Black-backed Gulls have nested occasionally between 1981 and 1985 at locations other than those listed in the Table. As shown on the maps, possible or probable breeding records occurred on Georgian Bay (Club Island), eastern Lake Erie (Port Colborne), and even on the Ottawa River (at Mattawa), in addition to 5 other Lake Ontario sites. The lower Great Lakes were frequently surveyed during the atlas years, but some of the innumerable, and often remote, islands in Lake Huron and Lake Superior were probably checked only once, if at all. Nevertheless, the maps are likely both accurate and complete.

The number of birds nesting in Ontario's portion of the Great Lakes may be slowly increasing, but as yet, there certainly has not been an invasion of Great Black-backed Gulls in that area. On the other hand, a small 'colony' of 3 to 5 pairs has become established on Little Galloo Island, south of Kingston in the US portion of Lake Ontario (Weseloh 1984). At this colony, young birds are believed to have fledged during 1981, 1982, and 1983.

Further monitoring of this and other colonies in the Great Lakes will indicate whether the Great Black-backed Gull will establish itself as a regular nester in Ontario. If the species were to become a common breeder, it would put pressure on any terns that share its nesting islands, because the species is a known predator of tern chicks.-- *H. Blokpoel*

Great Black-backed Gull confirmed records

Lake Ontario

Toronto Outer Harbour	1981	One pair. The nest had 2 eggs on 2 June, but was empty and deserted on 23 June.
Toronto Outer Harbour	1982	One pair. The nest had 3 eggs. First egg laid 23 May. Eggs disappeared by 11 June.
Snake Island south of Kingston	1983	One pair with one chick on 15 June.
Pigeon Island south of Kingston	1985	One pair with two chicks on 24 June.
Peter Rock between Port Hope and Cobourg	1985	One adult near a nest on 5 June. Nest had 3 eggs.
Gull Island Presqu'ile Prov. Park	1985	One pair with nest in July. Two flying chicks in August.
High Bluff Island Presqu'ile Prov. Park Prov. Park	1985	One pair near an empty nest on 18 May.

Lake Huron

Basswood Island near Oliphant	1983	One adult with one small young.

GREAT BLACK-BACKED GULL

Summary of Abundance Estimates

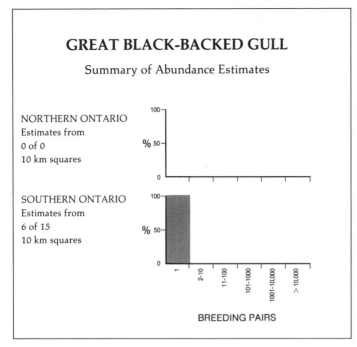

NORTHERN ONTARIO
Estimates from
0 of 0
10 km squares

SOUTHERN ONTARIO
Estimates from
6 of 15
10 km squares

% 50

1   2-10   11-100   101-1000   1001-10,000   > 10,000

BREEDING PAIRS

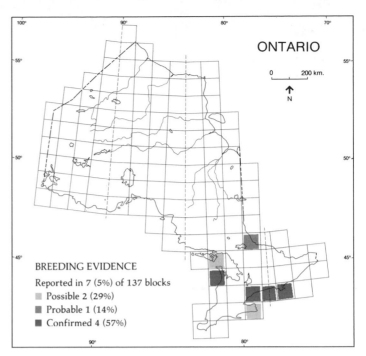

ONTARIO

0   200 km.

N

BREEDING EVIDENCE

Reported in 7 (5%) of 137 blocks

Possible 2 (29%)
Probable 1 (14%)
Confirmed 4 (57%)

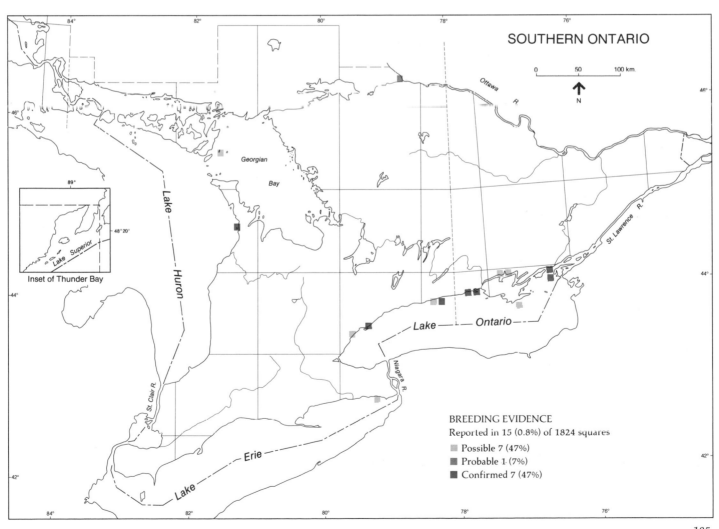

SOUTHERN ONTARIO

0   50   100 km.

N

Ottawa R.

Georgian Bay

Lake Huron

Lake Superior

Inset of Thunder Bay

St. Clair R.

St. Lawrence R.

Lake — Ontario

Niagara R.

Lake — Erie

BREEDING EVIDENCE

Reported in 15 (0.8%) of 1824 squares

Possible 7 (47%)
Probable 1 (7%)
Confirmed 7 (47%)

CASPIAN TERN
Sterne caspienne
Sterna caspia

S.L. HOUSE

Equipped with a heavy, scarlet bill and a hoarse, raucous call, the Caspian Tern makes a formidable and unmistakable impression. The largest of all tern species, it is slightly bigger than a Ring-billed Gull but smaller than a Herring Gull.

Worldwide, the Caspian Tern has an almost cosmopolitan, but highly disjunct, breeding range. It breeds on all continents except South America and Antarctica. Its nesting distribution extends from just south of the Arctic Circle in northern Canada and Europe, south through the temperate regions of North America and Eurasia, the tropics of Africa, and the temperate regions of Africa, Australia, and New Zealand as far south as 55°S (Voous 1960). In Canada, the breeding range of the Caspian Tern is mainly restricted to large lakes and coastal areas: Great Slave Lake, Lake Athabasca, Lakes Winnipeg and Winnipegosis, the Great Lakes, west-central James Bay, the Gulf of St. Lawrence, and the coasts of Newfoundland (Godfrey 1986). The Caspian Tern always nests on the ground, usually in colonies, but sometimes as single pairs. It prefers open habitat (either natural or artificial) near lakes, rivers, beaches, and shorelines. As do most other tern species, the Caspian Tern normally nests on islands to reduce predation by mammals. The nest substrate is often sand, small pebbles, or lightly vegetated soil.

Fifty years ago, the Caspian Tern was known to breed at only 4 localities in Ontario (2 in Georgian Bay, one in Lake Ontario, one in Lake Simcoe), but it was suspected that more nesting places existed in Georgian Bay, Lake Huron, and possibly elsewhere (Baillie and Harrington 1936). These authors also mentioned a specimen taken at Moose Factory on 31 May 1928. The breeding distribution of the Caspian Tern probably has not changed much during the last 5 decades but it has certainly become better documented.

In Ontario, the Caspian Tern has a wide breeding distribution: confirmed nesting in several squares in southern Ontario, probable nesting on Hudson Bay, and possible breeding in the Lake of the Woods area and on James Bay. The possible breeding records in James Bay on Akimiski Island and at the mouth of the Attawapiskat River agree with Peck and James (1983), who reported nesting in the Akimiski Strait. The probable breeding record at the mouth of the Winisk River suggests that breeding may also occur on the Hudson Bay shore.

In southern Ontario, almost all reports of breeding are on large bodies of water: Lake Ontario and Lake Huron, especially Georgian Bay and the North Channel. Two cases of confirmed nesting at inland locations were at Lake Simcoe and Kasshabog Lake. There are 2 main categories of breeding: (1) large, well established colonies located on remote islands in Georgian Bay, the North Channel, and Lake Ontario (Blokpoel 1977, Weseloh et al. 1986), and (2) scattered nesting by one or a few pairs for one or 2 years. Some of the well-established colonies have hundreds of nests and go back several decades (Ludwig 1965). The colony on South Limestone Island in Georgian Bay was first reported 70 years ago (Baillie and Harrington 1936). Most of the scattered, incidental nestings do not result in the establishment of permanent colonies, but the colony on the Leslie Street Spit in Toronto is an exception as it grew from 7 nests in 1976 to 197 nests in 1985. Another exception was a new colony of 70 nests in eastern Georgian Bay that was first reported in 1985. Confirmed breeding was reported in 1986 at a possible incipient colony in Hamilton Harbour. All of the established colonies and many of the scattered nestings were located on sites which were also used by nesting Ring-billed Gulls.

As the maps indicate, it is unlikely that the Caspian Tern nests in interior northwestern Ontario, preferring to nest on sea coasts and very large lakes. The map for southern Ontario properly presents the distribution of large colonies, but a few cases of breeding by single and/or late nesting pairs may have gone unnoted.

COSEWIC considers the Caspian Tern to be a rare species in Canada (WWF 1986). In Ontario, the Caspian Tern is considered rare by the MNR (I. Bowman pers. comm.). The breeding population on the Canadian portion of the Great Lakes increased during the 1970's and early 1980's (Blokpoel 1983), but there is increasing evidence that Ring-billed Gulls are encroaching on traditional nesting habitat of Caspian Terns. At some colonies, there is also competition with burgeoning numbers of Double-crested Cormorants and an increased risk of nest inundation during storms due to the recent high water levels of the Great Lakes. Further monitoring is necessary to determine whether the Caspian Tern will be able to maintain a healthy breeding population in the Great Lakes area.-- *H. Blokpoel*

CASPIAN TERN

Summary of Abundance Estimates

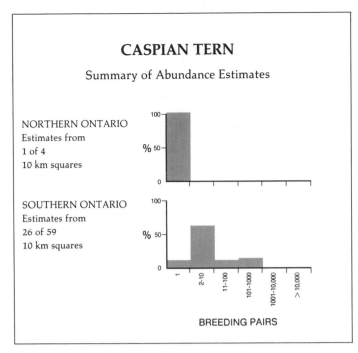

NORTHERN ONTARIO
Estimates from
1 of 4
10 km squares

SOUTHERN ONTARIO
Estimates from
26 of 59
10 km squares

BREEDING PAIRS

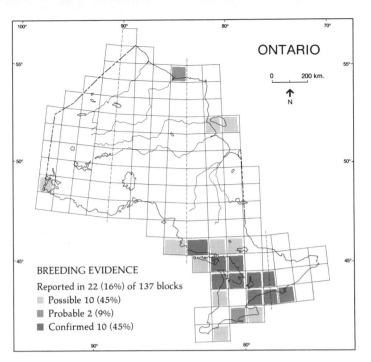

ONTARIO

0   200 km.

N

BREEDING EVIDENCE
Reported in 22 (16%) of 137 blocks
- Possible 10 (45%)
- Probable 2 (9%)
- Confirmed 10 (45%)

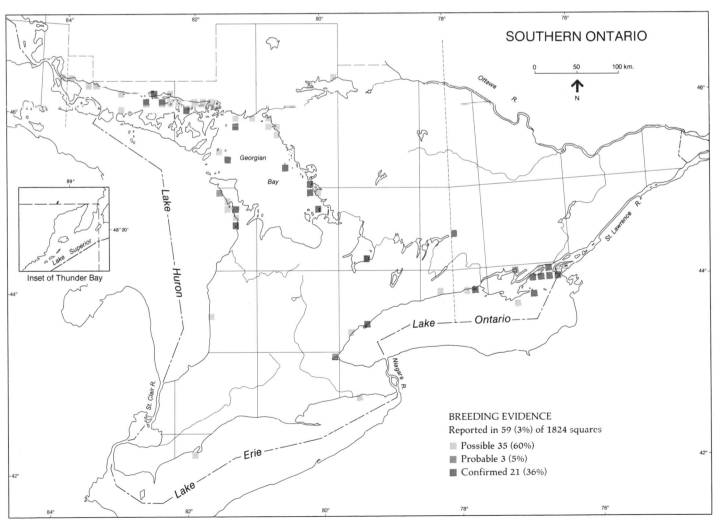

SOUTHERN ONTARIO

0   50   100 km.

N

Georgian
Bay

Inset of Thunder Bay
Lake Superior

Lake Huron

St. Clair R.

Lake Erie

Niagara R.

Lake Ontario

Ottawa R.

St. Lawrence R.

BREEDING EVIDENCE
Reported in 59 (3%) of 1824 squares
- Possible 35 (60%)
- Probable 3 (5%)
- Confirmed 21 (36%)

COMMON TERN
Sterne pierregarin
Sterna hirundo

S.L. HOUSE

The Common Tern has a mainly holarctic breeding range which forms a belt across Eurasia from the Bering Sea east to the Atlantic Ocean, and in North America from the Atlantic coast west to the Rocky Mountains (AOU 1983). It nests throughout Ontario except in the coastal areas of the Hudson Bay Lowland (Godfrey 1986). The North American population winters from the southern US southward through South America. Natural nesting sites include small, sparsely vegetated islands and open edges of sandy and gravelly beaches. The species also uses artificial nesting sites, including breakwalls, dredge spoil islands, and various off-shore structures such as flat-topped supports for navigational aids, and timber cribs.

Early accounts (McIlwraith 1894, Macoun and Macoun 1909) describe Common Tern breeding locations along the St. Lawrence River, in the Bay of Quinte, and in the vicinity of Lakes Erie, St. Clair, and Huron. The overall breeding range in Ontario does not appear to have changed much during the last 50 years. Baillie and Harrington (1936) mentioned that Common Terns nested throughout Ontario, although the northern limits of the range were still undetermined.

Common Tern nesting sites are easily detected because the birds usually nest in colonies, and off-duty birds normally mob intruders. Occasionally, terns nest as single pairs, in which case a nest could be missed if the incubating bird is sitting tight and the off-duty bird is foraging elsewhere. Actual nests comprise 23% of the records of this species. Separating Common and Arctic Terns is difficult in the field, but during the breeding season the 2 species can normally be told apart by knowledgeable observers. There is almost complete geographic separation of these 2 species in Ontario, and records from areas of potential range overlap were carefully scrutinized.

The maps provide evidence of breeding Common Terns from most of the province except for the north shore of Lake Superior (for which Peck and James (1983) also show no records), several blocks in northwestern and northcentral Ontario, the entire coast of Hudson Bay, and virtually all of the coast of James Bay, where breeding was restricted to the Moosonee area. Godfrey (1986) also shows the species nesting at the southern tip of James Bay. Atlas data show that the species does nest, albeit in small numbers, on the interior of the Hudson Bay Lowland. Some of the blank blocks on the northern Canadian Shield may reflect incomplete coverage rather than the absence of the species, but, given the evidence from the Shield in southern Ontario, the species is probably restricted to large lakes and rivers, and may not occur in every block. In southern Ontario, most breeding occurs along the shores of the Great Lakes, on the St. Lawrence and Ottawa Rivers, and in a conspicuous belt following the Trent-Severn Waterway between Trenton and the southeastern end of Georgian Bay. This belt contains Lake Simcoe, Rice Lake, and several other fairly large lakes between those two. The apparent absence of breeding Common Terns in much of the southern Ontario Canadian Shield away from the Great Lakes is surprising. However, there are records of breeding birds in several squares on Lake Nipissing, suggesting that the Common Tern prefers large bodies of water for nesting and/or that the small lakes on the Shield are an insufficient source of food. The lack of breeding records for southwestern Ontario is probably due to the general scarcity of suitable bodies of water.

In 1936, the Common Tern was considered a common breeder in the province (Baillie and Harrington 1936), and it probably still has that status. On a local level, many Common Tern colonies in southern Ontario suffer from human disturbance, encroachment by vegetation, washouts due to high water levels, and, increasingly, takeover by Ring-billed Gulls (Courtney and Blokpoel 1983). In parts of southern Ontario where there are relatively few suitable colony sites, the number of Common Terns is declining, but in areas where there are still many unused islands available for nesting, such as the North Channel of Lake Huron, populations are stable (Blokpoel and Harfenist 1986). The 2 abundance estimates of over 1,000 pairs per square were from the Leslie Street Spit and Port Colborne colonies.-- *H. Blokpoel*

COMMON TERN

Summary of Abundance Estimates

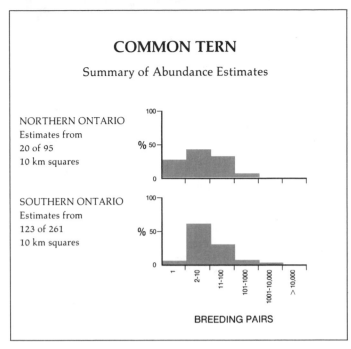

NORTHERN ONTARIO
Estimates from
20 of 95
10 km squares

SOUTHERN ONTARIO
Estimates from
123 of 261
10 km squares

%

BREEDING PAIRS

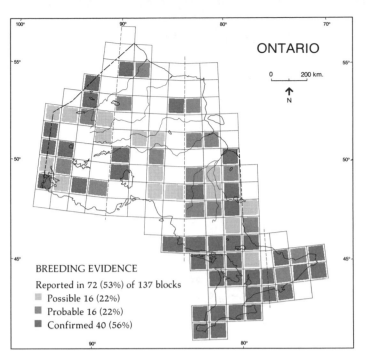

ONTARIO

0   200 km.

N

BREEDING EVIDENCE
Reported in 72 (53%) of 137 blocks

▢ Possible 16 (22%)
▨ Probable 16 (22%)
■ Confirmed 40 (56%)

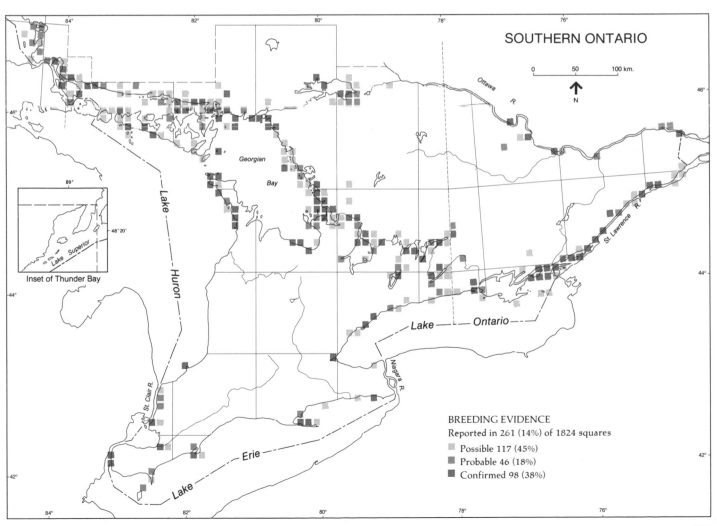

SOUTHERN ONTARIO

0   50   100 km.

N

Lake Huron

Georgian Bay

Ottawa R.

St. Lawrence R.

Lake — Ontario

Niagara R.

Lake — Erie

St. Clair R.

89°
48°20'
Lake Superior
Inset of Thunder Bay

BREEDING EVIDENCE
Reported in 261 (14%) of 1824 squares

▢ Possible 117 (45%)
▨ Probable 46 (18%)
■ Confirmed 98 (38%)

FORSTER'S TERN
Sterne de Forster
Sterna forsteri

S. L. HOUSE

Unlike some other closely related terns, the Forster's Tern appears to be increasing in numbers, or at least expanding its breeding range in the Great Lakes region. Within the memory of most Ontario birders, it has been absent or unknown as a breeding species in the province.

The Forster's Tern nests only in North America, occupying both fresh- and salt-water marshes through much of the interior, and along the Atlantic, Gulf, and Pacific coasts (AOU 1983). In Canada, it nests primarily in the Prairie Provinces, with smaller populations in interior British Columbia and southern Ontario (Godfrey 1986). Previously known Ontario nesting has been confined to Lakes Erie and St. Clair. The winter range is in the southern US and Mexico.

The historical breeding status of the Forster's Tern in Ontario is not well known, perhaps in part because few people have had the incentive to seek out its extremely inaccessible nest sites in deep-water marshes. There were nests in the Lake St. Clair marshes in the latter half of the 19th century (Morden and Saunders 1882). More recent documentation of breeding in Ontario, with the exception of Bent's (1921) mention of a Port Maitland record (which lacks details), has all been since 1970, and relates to observations at the St. Clair Marshes at Walpole Island and Mitchell's Bay, as well as Rondeau, Point Pelee and Long Point, where the first acceptably documented nesting in the province was reported in 1976 (Goodwin 1975, 1976, 1977, 1978, 1979b, 1981, Goodwin and Rosche 1971, Peck 1976). It is not certain whether, in the intervening years, the species was absent as a breeding bird or simply remained undetected. The evidence suggests an actual range expansion and population increase in recent years, which have been attributed to high water levels. Such water levels have a greater adverse impact on, and thus indirectly reduce the competition from, nesting

Common Terns (Scharf and Shugart 1984). Such an explanation accords well with Lake Erie water level data, which show almost continuously above-average levels from at least as far back as 1860 until 1890, and again during the past decade or more.

Marsh nesting necessarily restricts the breeding distribution of the Forster's Tern, and its association with only deep portions of large cattail marshes (Bergman *et al.* 1970) restricts it further. Adults vigorously and noisily defend nests with eggs, thereby drawing attention to the location of a nesting colony. Conspicuous aerial courtship behaviour, continuing well after nesting begins and showing a resurgence with hatching (McNicholl 1969), also makes it easy to find colonies, although sometimes the terns return to the previous year's site, engage in courtship behaviour, and then shift to a new area for nesting (McNicholl 1975, McCracken *et al.* 1981). Although young use the nest for several days after hatching, they will leave to hide in the water if they are disturbed (McNicholl 1971). Breeding can then be confirmed only by a careful search or by watching from a distance for the parents to feed the young.

Confirmed nesting records of the Forster's Tern during the atlas field work period at sites along Lakes Erie and St. Clair correspond well with published literature (Peck and James 1983, Speirs 1985). The Kettle Point nesting site on Lake Huron is a range extension. Although an observation in suitable habitat on Lake Simcoe was not sufficient to confirm nesting there, increases elsewhere suggest that this area, along with other large, deep-water cattail marshes, bears close watching. Evidence of possible breeding in western Ontario (Larus Lake in Woodland Caribou Prov. Park, Red Lake, and Lake of the Woods) requires confirmation, as there is no previous indication of nesting there, although Speirs (1985) reports sightings by A. Wormington on Lake of the Woods near an area with suitable habitat. The closest known colonies in Manitoba are at the southern end of Lake Winnipeg (Hatch 1972, Koonz and Rakowski 1985), east of which the species is rare to uncommon even as a transient (Taylor 1985).

Forty percent of the atlas records of this species are accompanied by abundance estimates. The 2 highest estimates, of 101 to 1,000 pairs, are for squares covering parts of Walpole Island and Long Point. In 4 additional squares adjacent to the one at Walpole Island, abundance estimates are in the 11 to 100 pairs range. A similar number is thought to be breeding in the outer part of Rondeau harbour. These estimates, together with the 3 remaining estimates of 2 to 10 pairs per square, closely match the known range of colony size (1 to 1,000 pairs; usually 10 to 150 or 200 pairs), as reviewed by McNicholl (1971) and Clapp *et al.* (1983). In Ontario, the Forster's Tern is locally common, but rare overall.-- *M.K. McNicholl*

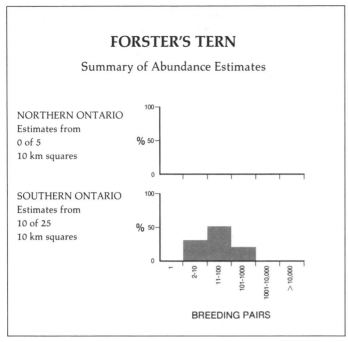

FORSTER'S TERN

Summary of Abundance Estimates

NORTHERN ONTARIO
Estimates from
0 of 5
10 km squares

SOUTHERN ONTARIO
Estimates from
10 of 25
10 km squares

BREEDING PAIRS

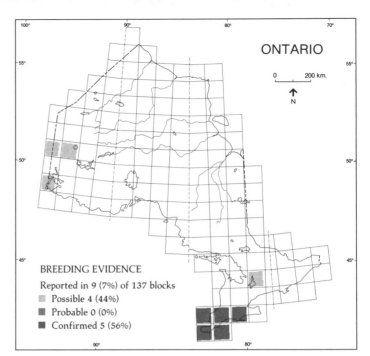

ONTARIO

BREEDING EVIDENCE
Reported in 9 (7%) of 137 blocks
Possible 4 (44%)
Probable 0 (0%)
Confirmed 5 (56%)

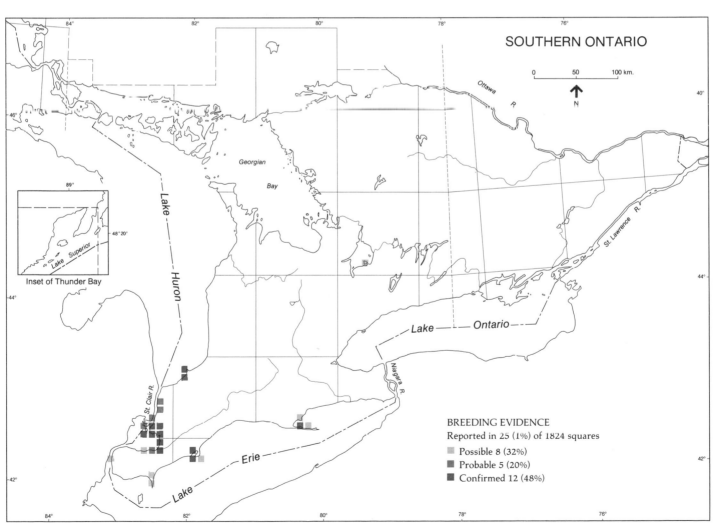

SOUTHERN ONTARIO

Inset of Thunder Bay

BREEDING EVIDENCE
Reported in 25 (1%) of 1824 squares
Possible 8 (32%)
Probable 5 (20%)
Confirmed 12 (48%)

BLACK TERN
Guifette noire
Chlidonias niger

S.L. HOUSE

This locally common denizen of the marsh is Ontario's smallest tern, showing off its striking black underparts only during the breeding season. Its breeding range covers marshes in the middle latitudes of North America and Europe. The North American population winters in the northern parts of coastal South America.

The Black Tern is easy to locate, as it spends much time in flight, hovering over open water and delicately picking up invertebrates and occasionally small fish. The species normally nests in loose aggregations of a few to 30 or more pairs, amidst moderately dense emergent marsh vegetation. Cattails are particularly favoured. Nests are made of dead vegetation, found on floating mats of plant material or on small patches of mud, and sometimes on muskrat houses. There is usually open water adjacent to the nest; water at nest sites is typically 0.5 to 1.5 m deep (Dunn 1979). Nesting requirements are quite specific, and much of the unoccupied but apparently suitable habitat throughout southern Ontario may in fact be unattractive to Black Terns. Birds return to nest in an area year after year as long as the habitat remains suitable, but once emergent vegetation becomes too dense or too sparse, or water levels change markedly, the birds move abruptly to new areas - a trait which is also common among other species of terns.

Flocks of Black Terns may feed several kilometres from a nesting area, but the patchy distribution of occupied nesting habitat suggests that the regular presence of terns in the breeding season is a good indication of nesting, either in the local area or at least in an adjacent square. Nesting areas in extensive marshes can be difficult to pinpoint from a distance, but persistent shrill alarm cries and dive-bombing greet the rare intruder who wades or boats into a nesting area, revealing the nearby presence of a nest or young. For people who do go to the trouble of penetrating a marsh, nests with eggs or young are fairly easy to find. They represent almost 20% of all the records across the province. Although chicks can sometimes be seen being fed by adults in open areas, most are fed near the nest for several weeks. At the first sign of disturbance, they leave the nest and hide, and can be very difficult to see.

Historical records indicate that the present distribution of the Black Tern in Ontario remains much as it was in the past (Baillie and Harrington 1936, Peck and James 1983). The map shows obvious concentrations around the southern edge of the Canadian Shield and the margins of the Great Lakes, reflecting the abundance of productive wetlands with extensive shallow areas and emergent vegetation. Such habitat is less common on the Shield, where rocky shores predominate and wetlands are less productive. In southwestern Ontario, the Black Tern is largely restricted in its breeding to a few, generally large and highly productive marshes, such as those at Point Pelee, Lake St. Clair, Rondeau, Long Point, and Luther Lake, and to a few suitable sewage lagoons. A number of the shallow, weedy, limestone-based lakes on the Bruce Peninsula and Manitoulin Island maintain breeding populations. In southeastern Ontario, marshes along the St. Lawrence and Ottawa Rivers are important nesting areas. Elsewhere in the province, suitable marshlands are scarce, and Black Tern nesting sites are correspondingly few.

Although Black Terns are locally common in Ontario, most squares (95%) in which abundance estimates were provided were thought to have fewer than 100 pairs, and 63% were thought to have no more than 10 pairs. It is difficult to assess changes in Black Tern abundance, because the species commonly abandons breeding sites when vegetation and water levels change. It is likely that there have been decreases over the past several decades with the extensive alteration and destruction of marsh habitat that have taken place, just as has occurred in western Europe (Cramp 1985). Disturbance from human recreational activity is minimal, since the species nests in habitat which is unsuitable for swimming, boating, or fishing.-- *E.H. Dunn*

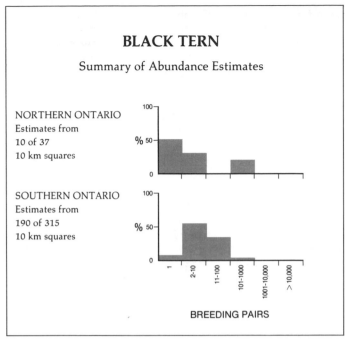

BLACK TERN

Summary of Abundance Estimates

NORTHERN ONTARIO
Estimates from
10 of 37
10 km squares

SOUTHERN ONTARIO
Estimates from
190 of 315
10 km squares

BREEDING PAIRS

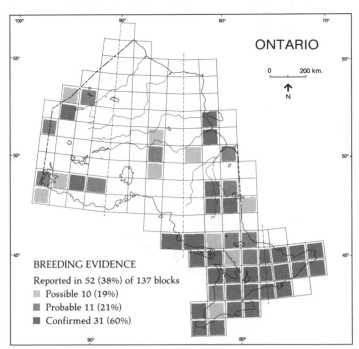

ONTARIO

0   200 km.

↑
N

BREEDING EVIDENCE
Reported in 52 (38%) of 137 blocks
Possible 10 (19%)
Probable 11 (21%)
Confirmed 31 (60%)

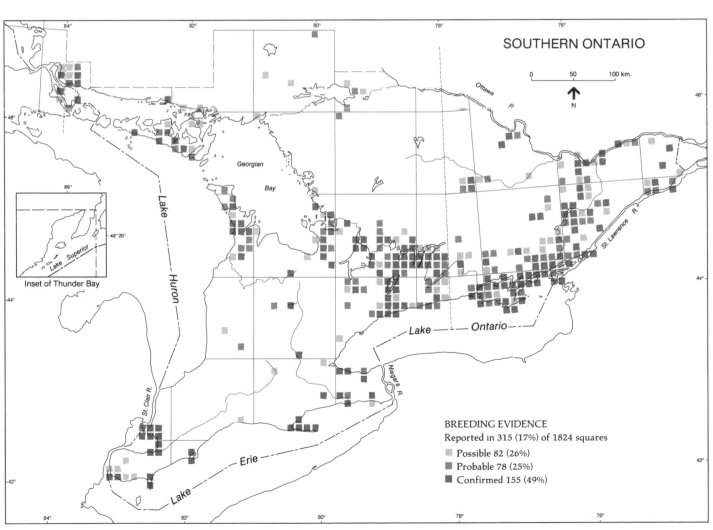

SOUTHERN ONTARIO

0   50   100 km.

↑
N

Georgian
Bay

Lake Huron

Lake Ontario

Lake Erie

Ottawa R.

St. Lawrence R.

Niagara R.

St. Clair R.

Inset of Thunder Bay

Lake Superior

89°

48°20'

BREEDING EVIDENCE
Reported in 315 (17%) of 1824 squares
Possible 82 (26%)
Probable 78 (25%)
Confirmed 155 (49%)

ROCK DOVE
Pigeon biset
Columba livia

The Rock Dove occurs across southern Canada, primarily in areas of human habitation. In Ontario, it is most abundant in the southern part of the province, occurring in cities, towns and rural areas. Primarily a seed eater, the Rock Dove feeds on waste grain, spilled grains around railway yards, and human handouts. The species is usually found in close association with human development.

This species is native to Eurasia and North Africa but has become widely established, through introduction, on all continents except Antarctica. The Rock Dove population in Canada is descended from feral domestic stock imported from Europe by early settlers. Populations are now naturalized and non-migratory. There has been little apparent change in populations in the recent past, although BBS data for southern Ontario and Quebec suggest a significant positive increase in population size from 1967 to 1983. It is difficult to establish longer-term trends, as for many years Rock Doves were not considered part of the natural avifauna and data were not routinely recorded. Habitat quality or quantity do not appear to have been altered significantly in recent years, although there is a long-term trend towards increasing grain corn acreage which could benefit the Rock Dove (Armstrong and Noakes 1983).

Being a resident of densely populated urban centres as well as agricultural areas, the Rock Dove is easy to locate, and to confirm as breeding. It perches, flies and forages in open areas, and has a readily identifiable call. The species has an extended nesting season, with occupied nests being reported from 27 January to 24 December, and nesting sites are usually quite visible; all Ontario records of nests are on artificial structures (Peck and James 1983). The major obstacle to successful confirmation of breeding relates, ironically, to the close relationship of the species with man. Many people raise free-flying 'pigeons', and this presents a problem in knowing whether observed birds or their nests are wild, especially in areas outside of the core range, such as in northeastern Ontario. As well, many fanciers train Rock Doves for racing and homing, often releasing them several hundred kilometres from home. Disoriented birds or those simply passing through an area may be noted and recorded as residents. Sightings in remote locations (Moosonee and Attawapiskat) have not been included in the atlas because of doubts about the origins of the birds. Observations in the possible breeding category, and particularly in the suitable habitat classification (18% of all records) are suspect for this reason, though domestic birds are greatly outnumbered by wild ones. Fortunately, a high degree of confirmed breeding was obtained, relating primarily to adults entering nests (42% of confirmed breeding records), nests with young or eggs (33%), and fledged young (20%). It is unlikely that many nests of tame pigeons were recorded as confirmed breeding.

The extent of the confirmed breeding range of the Rock Dove in Ontario, based on atlas data, substantially enlarges the breeding range based on previously reported nesting attempts (Peck and James 1983). The Rock Dove is present in virtually all squares south of the Canadian Shield, and in squares along all the major highways in the northern part of southcentral Ontario. In otherwise forested areas, the linear alignment of sightings is evident on the map where Highway 35 passes through Haliburton Co, along Highway 11 from Gravenhurst to North Bay, and along Highway 17 from North Bay to Sault Ste Marie and beyond. Breeding farther north is fairly widespread in road- or rail-accessible areas, but clumped, with confirmed breeding in the Kenora and Thunder Bay regions, and the Clay Belt region of northeastern Ontario; in all of these cases the occurrence of the Rock Dove likely reflects the presence of agricultural activities and other human development. The distribution of the Rock Dove essentially parallels that indicated by atlas data for the Mourning Dove in southern Ontario, and the distribution is quite similar even in northern Ontario.

Where it occurs, the Rock Dove is generally estimated to be a common to very common species in southern Ontario: estimates of over 1,000 pairs were offered for a considerable number of squares in the corn-growing regions of Ontario's agricultural lands.-- *E.R. Armstrong*

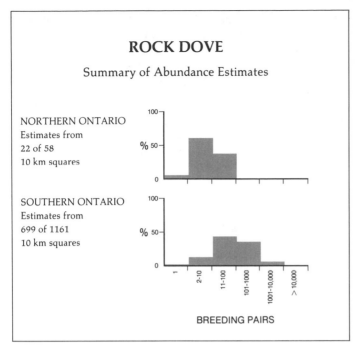

ROCK DOVE

Summary of Abundance Estimates

NORTHERN ONTARIO
Estimates from
22 of 58
10 km squares

SOUTHERN ONTARIO
Estimates from
699 of 1161
10 km squares

% 100

50

% 100

50

1 2-10 11-100 101-1000 1001-10,000 >10,000

BREEDING PAIRS

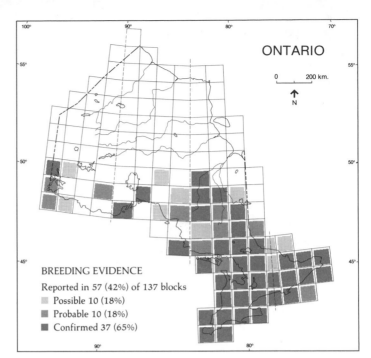

ONTARIO

0   200 km.

N

BREEDING EVIDENCE
Reported in 57 (42%) of 137 blocks
Possible 10 (18%)
Probable 10 (18%)
Confirmed 37 (65%)

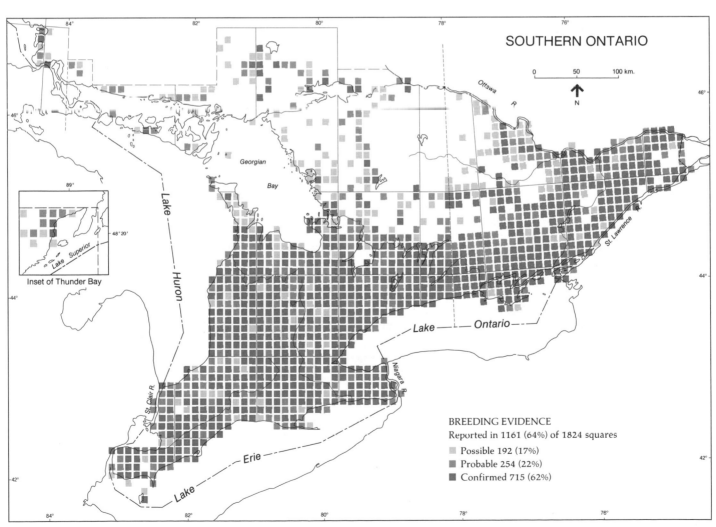

SOUTHERN ONTARIO

0   50   100 km.

N

Georgian
Bay

Lake
Huron

Lake

St. Clair R.

Lake   Erie

Niagara R.

Lake — Ontario

St. Lawrence R.

Ottawa
R.

89°

48°20'

Lake   Superior

Inset of Thunder Bay

BREEDING EVIDENCE
Reported in 1161 (64%) of 1824 squares
Possible 192 (17%)
Probable 254 (22%)
Confirmed 715 (62%)

195

MOURNING DOVE
Tourterelle triste
Zenaida macroura

The Mourning Dove has a widespread distribution throughout North America. In Ontario, it is most abundant in the rural areas of the southwest, but observations have been reported throughout the southern two-thirds of the province. The typical breeding habitat of this species consists of interspersed agricultural fields, woodlots or shelterbelts, and open areas. It is also common in some southern Ontario towns and cities, where mature coniferous trees provide suitable nesting habitat. The Mourning Dove reaches its highest densities in rural areas, reflecting its reliance on seeds in the diet, primarily from waste grain and weeds (Armstrong and Noakes 1981).

The Mourning Dove has been a common breeding species of southwestern Ontario since at least the early 20th century (Baillie and Harrington 1936). Earlier populations visited Ontario only during the breeding season (Saunders and Dale 1933). However, the size and range of wintering populations have increased rapidly and in linear fashion since the early 1950's (Armstrong and Noakes 1983). Although significant downward trends in breeding Mourning Dove populations have been recorded in recent years in the eastern US (Dolton 1976), southern Ontario populations have apparently increased significantly between 1967 and 1983 (Breeding Bird Survey data). Although the main breeding range is in the southern part of the province, occasional and widely dispersed observations to the north (*e.g.*, at Moosonee) and to the west have been reported as far back as 1929 (Baillie and Harrington 1936), and continue to the present, with one bird having been seen at Fort Severn in 1983.

The Mourning Dove is a relatively conspicuous bird, readily discovered in an area because of its habits of seeking gravel along roadsides and of perching on hydroelectric transmission lines. The most common category of breeding observations in the data base consists of the sighting of pairs. The distinctive cooing song of the Mourning Dove is readily identified by even relatively novice birders, can be heard over long distances, and is given through a long breeding period from March to September.

Loose colonies of nesting Mourning Doves occur in some areas of southern Ontario. Nests are not difficult to locate in agricultural areas, being relatively low in groves or shelterbelts of suitable trees, usually evergreens. Nests with eggs or young constitute 34% of all atlas records. Many confirmations of breeding (15% of the records) are the result of observation of fledged young. Confirmation of breeding was much rarer on the Canadian Shield than in the south, probably because of much lower population densities and less available habitat in that area. Moreover, it is an almost overwhelming task to search all of the extensive suitable nesting habitat for the nest itself.

The map confirms the breeding of the Mourning Dove in nearly every square in southern Ontario, south of Georgian Bay and the Canadian Shield. The core range and, to a large extent, the peripheral sightings reflect the distribution of agricultural land. Breeding is also probable or confirmed along the edge of the Canadian Shield itself (*i.e.*, from Georgian Bay to Ottawa) where suitable habitat was present, but records are thinly spread on most of the Shield. Areas of human settlement and development, as reflected in Mourning Dove distribution, can be detected in certain areas of the map. The corridors from Sudbury east toward North Bay and south toward Parry Sound, for example, are readily apparent, as the species was not recorded in surrounding undeveloped squares. The atlas has significantly extended the previously documented breeding range of the Mourning Dove (Peck and James 1983) to include a small but widespread population in northern Ontario. Breeding was confirmed or probable in a substantial part of Ontario north of Lake Huron, including the agricultural areas between Thunder Bay and the Manitoba border, and the Clay Belt of northeastern Ontario.

Abundance estimates were made for 62% of the squares for which there are breeding data in southern Ontario. Estimates of such a conspicuous species as this are likely to be more reliable than for many other species. Where the Mourning Dove occurred, it was often assessed as being common to abundant. The highest estimates are from the grain-growing districts of Lambton, Middlesex, Oxford, Perth, and Huron Counties. All 17 estimates in northern squares are lower (10 or fewer pairs per square).-- *E.R. Armstrong*

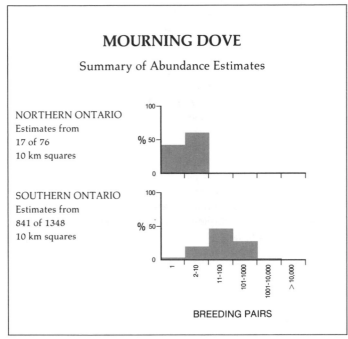

MOURNING DOVE

Summary of Abundance Estimates

NORTHERN ONTARIO
Estimates from
17 of 76
10 km squares

SOUTHERN ONTARIO
Estimates from
841 of 1348
10 km squares

%

BREEDING PAIRS

1   2-10   11-100   101-1000   1001-10,000   >10,000

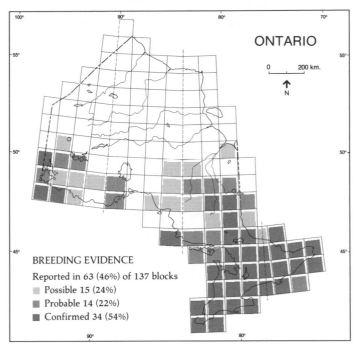

ONTARIO

0   200 km.

N

BREEDING EVIDENCE
Reported in 63 (46%) of 137 blocks
Possible 15 (24%)
Probable 14 (22%)
Confirmed 34 (54%)

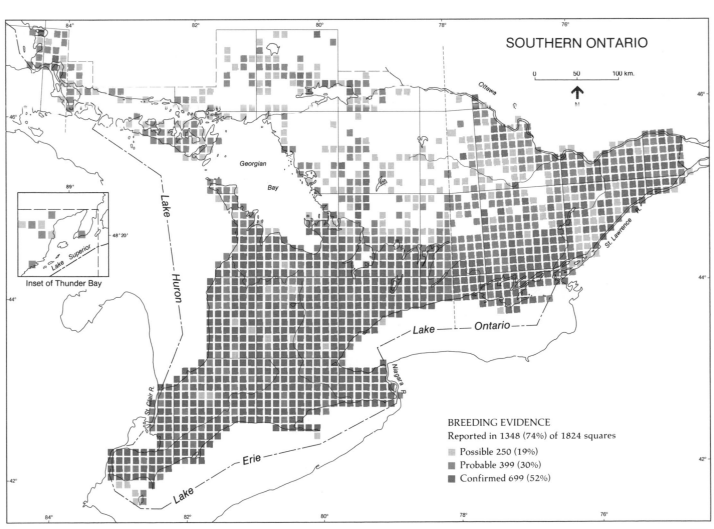

SOUTHERN ONTARIO

0   50   100 km.

N

Georgian
Bay

Ottawa

St. Lawrence R.

Lake — Ontario

Niagara P.

Lake Huron

Lake Erie

St. Clair R.

89°

48°20'

Lake Superior

Inset of Thunder Bay

BREEDING EVIDENCE
Reported in 1348 (74%) of 1824 squares
Possible 250 (19%)
Probable 399 (30%)
Confirmed 699 (52%)

BLACK-BILLED CUCKOO
Coulicou à bec noir
Coccyzus erythropthalmus

Unlike its southern counterpart, the Yellow-billed Cuckoo, this species occurs throughout a broad area of Ontario. The Black-billed Cuckoo breeds in eastern North America from southern Canada to the central and northern US and west to the Great Plains. Its Ontario range is largely confined to the southern half of the province. It winters in northwestern South America. Black-billed Cuckoo habitat in Ontario consists of dense, usually deciduous vegetation of low to medium height, particularly where the vegetation is interspersed with open areas. Such conditions occur widely wherever farming has been attempted and abandoned, where old fields are overgrown with hawthorns, and along hedgerows in actively farmed areas. Nevertheless, this species is nowhere so abundant that one can be confident of finding it even by spending a full day searching for it.

There is no evidence to define the Ontario range of the Black-billed Cuckoo when the province was still almost completely covered by forests, but the species undoubtedly benefitted from forest clearance. Its abundance and range, as described by McIlwraith (1894) and Macoun and Macoun (1909), differ very little from the present, though in the latter work it is stated that the bird was "formerly much more common than at present." Such a claim may be simply a reflection of the fact that, in some years, coinciding with massive regional outbreaks of forest tent caterpillars (*Malacosoma disstria*), Black-billed Cuckoos, which feed on the caterpillars, appear in abnormally large numbers in parts of Ontario where they are generally much more scarce or absent (Baillie and Hope 1947, Baillie 1952). Such irruptions appear to take place sporadically, occurring in different areas each year. Alternatively, Macoun and Macoun's statement may reflect a decline in suitable habitat through the southern part of the Black-billed Cuckoo's range, associated with an

intensification of agriculture. Writing in 1963, Todd stated, "It is apparently extending its range still farther to the northward - coincident probably with the opening up of the country." Snyder (1953) reported it to be rare in the western Rainy River District in 1929, though more numerous soon after, perhaps because of a caterpillar outbreak. It is now well established between Thunder Bay and Kenora, and BBS data indicate that the province's highest concentration occurs near Rainy River (Speirs 1985).

The Black-billed Cuckoo is not a particularly vocal nor an aggressive bird, and is prone to hide in thickets. The song, albeit a fairly loud one, is usually uttered only at long and irregular intervals, sometimes as a single rattling call disturbing the stillness of the night. The songs of the two species of cuckoos can be very difficult to differentiate with certainty. The time that it spends on its breeding grounds coincides with the period when deciduous foliage is at its fullest. Hence, being a retiring bird of deciduous shrubbery, it is not often seen even where fairly common. It seldom sits in exposed places and moves very little from perch to perch. When seen, it is most often flying quickly from one thicket to another, low over the ground. For these and other behavioural reasons, this species is remarkably inconspicuous for a bird of its size: 54% of records are of possible breeding, an extremely high proportion for such a widely distributed bird. It likely breeds in a number of squares where atlassers did not find it at all, and in most squares where only possible or probable breeding was recorded. This species has been known to parasitize the nests of Yellow-billed Cuckoos and other birds by laying its eggs in those nests (Clarke 1890, Peck and James 1983). Nests of Black-billed Cuckoos are seldom built before June and are sometimes occupied as late as early September. They are flimsy structures usually about 1.0 to 1.5 m off the ground in trees, shrubs, bushes, or vines; active nests were found in 55 squares.

The maps show that the Black-billed Cuckoo is primarily a bird of the Carolinian and Great Lakes-St. Lawrence Forest regions, but scattered records occurred within the southern Boreal Forest region, and there was even one report from North Point on the James Bay coast. With the exception of the area north of Kingston, where coverage was exceptionally thorough, and where gypsy moth (*Porthetria dispar*) infestations have occurred, there is no single area of southern Ontario where the Black-billed Cuckoo is concentrated. Records are somewhat more sparse on the Canadian Shield than they are farther south, presumably as a result of the lower proportion of agricultural land.

Abundance estimates provided by atlassers suggest that the Black-billed Cuckoo is uncommon in most areas where it is found. In southeastern Ontario, however, there are a few squares with estimates of over 100 pairs.-- *F.M. Helleiner*

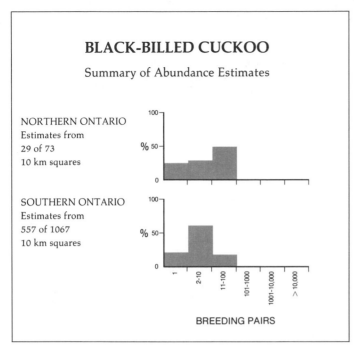

BLACK-BILLED CUCKOO

Summary of Abundance Estimates

NORTHERN ONTARIO
Estimates from
29 of 73
10 km squares

SOUTHERN ONTARIO
Estimates from
557 of 1067
10 km squares

%

BREEDING PAIRS

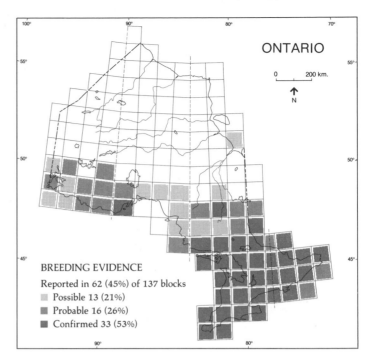

ONTARIO

0   200 km.

N

BREEDING EVIDENCE
Reported in 62 (45%) of 137 blocks
  Possible 13 (21%)
  Probable 16 (26%)
  Confirmed 33 (53%)

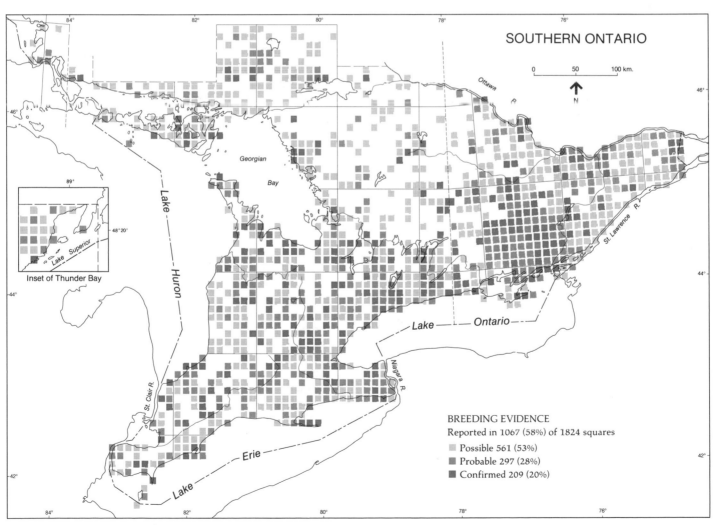

SOUTHERN ONTARIO

0   50   100 km.

N

Inset of Thunder Bay

Lake Superior

Lake Huron

Georgian Bay

St. Clair R.

Lake Erie

Niagara R.

Lake Ontario

Ottawa R.

St. Lawrence R.

BREEDING EVIDENCE
Reported in 1067 (58%) of 1824 squares
  Possible 561 (53%)
  Probable 297 (28%)
  Confirmed 209 (20%)

YELLOW-BILLED CUCKOO
Coulicou à bec jaune
Coccyzus americanus

The Yellow-billed Cuckoo is much the less known and less common of the 2 species of cuckoos found in Ontario. Because of this, and because of their similarity, any Ontario report of a Yellow-billed Cuckoo north of southern Ontario needs to be fully documented to be credible.

The Yellow-billed Cuckoo breeds throughout North America as far north as extreme southern Canada, and winters in South America (AOU 1983). In Ontario, it is a bird which occurs in the Carolinian Forest zone, and only the more southerly parts of the Great Lakes-St. Lawrence Forest zone. It breeds in deciduous habitats: shrubs and small trees usually in semi-open areas such as the fringes of agricultural land, orchards, or urban parkland, but occasionally in more continuous woodlots.

There is some evidence that the Yellow-billed Cuckoo was once more common than it is today, and that minor shifts in its range have also occurred. Prior to European settlement, the amount of semi-open land was limited, and the species may well have been rarer and more restricted in its distribution. McIlwraith (1894) stated that it was "rather scarce and not generally distributed" along our southern border, but "very common about Niagara Falls." Writing in 1909, Macoun and Macoun noted that it was a "rather common summer resident at Toronto," and added, "Twenty years ago this species was rather rare, but now it is more common than the black-billed." Breeding Bird Surveys show that this has not been the case in recent years. During the decade 1968-1977, there were on average 6 times as many Black-billed as Yellow-billed Cuckoos in southern Ontario (Speirs 1985). The intensification of agriculture, leaving fewer hedgerows or orchards and little roadside shrubbery, has occurred in much of the rural area within the range of this species in the past few decades, with the resultant loss of

much good Yellow-billed Cuckoo habitat. At the same time, at the fringes of the Canadian Shield, which have traditionally been largely beyond the range of the Yellow-billed Cuckoo (Baillie 1950c), though well within that of the Black-billed, much former agricultural land has been abandoned and has grown over with suitable Yellow-billed Cuckoo habitat.

Descriptions of the habits of the Yellow-billed Cuckoo frequently contain adjectives like "silent," "secluded," or "noiseless." Both its habits and its colouration make it a relatively inconspicuous bird, one which is likely to escape notice unless it happens to call. The call is quite loud, but is given infrequently and may be difficult to distinguish from that of the Black-billed Cuckoo. In view of this, it is perhaps surprising that the Yellow-billed Cuckoo was found in 351 squares in southern Ontario but not surprising that it was confirmed as a breeding species in only 42. The nest is a frail platform saddled on a horizontal limb or a fork of a shrub, vine, or sapling, at a height of 1 to 2 m (Peck and James 1983). There is some evidence that the Yellow-billed Cuckoo occasionally lays its eggs in the nest of another species (Bent 1940), but, unlike the Cuckoo (*Cuculus canorus*) of Eurasia, neither of the Ontario cuckoos is an obligate nest parasite.

The map shows the scarcity of this species in all but the most southerly parts of Ontario and the Kingston area. The main concentration of Yellow-billed Cuckoo observations is along the shores of Lake Erie. Almost two-thirds of the breeding confirmations are in the area south of a line through Hamilton and Sarnia. A secondary concentration in the area northwest of Kingston is unexpected in the light of most previously published information (*e.g.*, Baillie and Harrington 1936, Quilliam 1965). The first hint of an incipient range expansion there occurred in the late spring of both 1962 and 1963, when larger than usual numbers were observed (Quilliam 1965). The BBS from 1968 to 1977 provides the first firm evidence that the species really does breed commonly in the Kingston area (Speirs 1985). The area has had infestations of the gypsy moth (*Porthetria dispar*), which is a favoured food of this species (Bent 1940). The preponderance of suitable habitat due to succession on the abandoned farmland of that part of the Canadian Shield may also be a contributing factor. The northern limit of the Yellow-billed Cuckoo's range in Ontario is clearly defined by the atlas map. Although there are several published summer records of the species from northern Ontario (including northwestern Ontario), there are no breeding records for that area (Peck and James 1983, Godfrey 1986). Thus, the record of a Yellow-billed Cuckoo carrying nesting material on the outskirts of Sudbury in May of 1984 represents a significant northward range extension.

Abundance estimates provided by atlassers indicate that the Yellow-billed Cuckoo is uncommon or rare in most areas where it is found. However, in the traditional core of its Ontario range, south and west of London (mostly in Essex, Kent, and southern Lambton Counties), estimates of over 10 pairs were provided in 15 squares.-- *F.M. Helleiner*

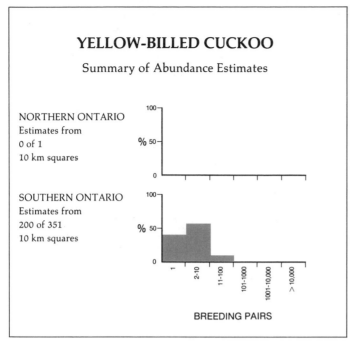

YELLOW-BILLED CUCKOO

Summary of Abundance Estimates

NORTHERN ONTARIO
Estimates from
0 of 1
10 km squares

SOUTHERN ONTARIO
Estimates from
200 of 351
10 km squares

BREEDING PAIRS

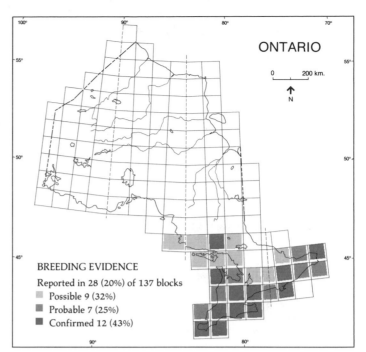

ONTARIO

0   200 km.

N

BREEDING EVIDENCE
Reported in 28 (20%) of 137 blocks
Possible 9 (32%)
Probable 7 (25%)
Confirmed 12 (43%)

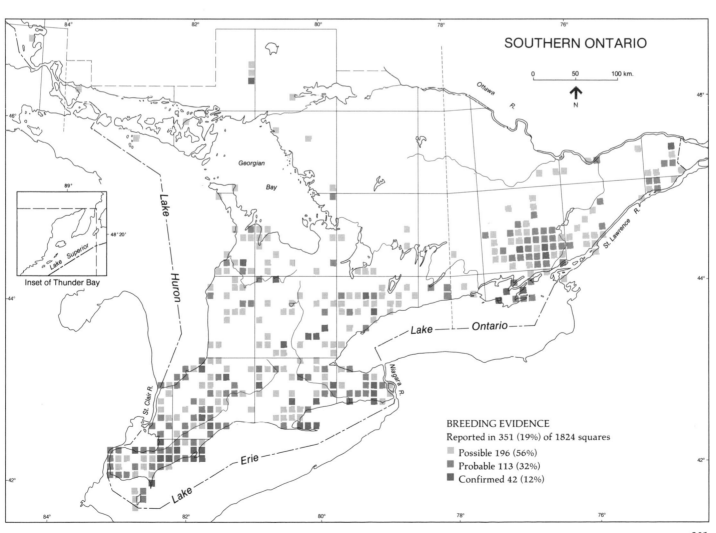

SOUTHERN ONTARIO

0   50   100 km.

N

Inset of Thunder Bay

BREEDING EVIDENCE
Reported in 351 (19%) of 1824 squares
Possible 196 (56%)
Probable 113 (32%)
Confirmed 42 (12%)

COMMON BARN-OWL
Effraie de clochers
Tyto alba

J.S.

The ghost-like habits and nocturnal activities of the Common Barn-Owl make it a difficult bird to find. The species is nearly cosmopolitan, occurring in both the tropical and temperate regions of the world. In the Americas, it breeds from the southern tip of South America northwards through most of the US, where it is most common in the southern states. In Canada, its breeding range penetrates only into the extreme southerly parts of British Columbia and Ontario. This species is currently the rarest of Ontario's breeding owls. The Common Barn-Owl frequents open country, favouring agricultural areas, where it hunts over open fields and orchards and around farms. Its very long wings are well adapted to flight in open areas and it shuns woodlands. By day, it roosts quietly in dark areas inside barns or other buildings (even occupied ones), towers, or hollow trees, often within cities, although it remains unknown to most people.

The Common Barn-Owl was rare during the 19th century in Ontario. McIlwraith (1894) considered it an accidental visitor from the south. It is possible that the species moved into Ontario only when the development of agriculture made suitable buildings and habitat available. Macoun and Macoun (1909) and MacClement (1915) list few records. Baillie and Harrington (1936) classified it as rare, breeding only in Kent, Essex, and Middlesex Counties; Speirs (1985) reports nesting at Queenston and at St. Catharines in the 1950's, while Peck and James (1983) noted nesting at Kingston and in Dundas Co in the St. Lawrence valley. Concern for falling numbers in the US led to its inclusion on the Blue List from 1972 to 1981 (Tate 1981), but it was removed from the list in 1982 (Tate and Tate 1982). Even though some withdrawal occurs in winter from the northerly edge of its range, probably the main factors preventing a more widespread colonization of Ontario by the Common Barn-Owl are snow cover in winter, which limits its ability to capture mice, and its intolerance of severe cold. There is evidence that even the low numbers in this province are slowly declining because of loss of farmland to urban sprawl (Campbell and Campbell 1984). Captive breeding projects have been used successfully to augment wild populations in Great Britain and the US. In Ontario, between 1974 and 1986, 182 young were raised at the Owl Rehabilitation Research Foundation at Vineland Station and released in southern Ontario. This species has been known to raise 2 broods per year (though not in Ontario), and it adapts well to nest boxes; thus, there is much opportunity for human assistance, although there is scant evidence that any of the released birds have survived.

Although no nest is actually constructed, the nest site may be either in a natural cavity of a tree or cliff or more commonly in a dark sheltered part of an abandoned building, water tower, barn loft, or church steeple. Nest boxes placed in these locations may be used. However, a prerequisite for any nest site is that there be open country nearby for foraging. Breeding Common Barn-Owls are best located by watching at dusk for adults carrying food, searching for shiny black pellets which indicate the presence of daytime roosts, or listening at night for territorial screams by the adults, and for noisy young that snore and hiss in the nest. Usually 2 to 6 eggs are laid in Ontario nests (Peck and James 1983). Incubation begins with the first egg and lasts for 30 to 34 days. Both adults continue to feed the young after fledging. The Common Barn-Owl shows a strong attachment to nest sites used in previous years, perhaps because of the shortage of suitable sites.

For a rare species that is difficult to locate even where it is common, finding the Common Barn-Owl in 6 squares and confirming it in 4 represents a creditable effort. All occurrences were in the south, with a concentration in and around the Niagara Peninsula, an area where no reports of nesting existed prior to 1950. This is also the area where young birds have been released. In the Essex-Kent-Lambton area, the historic core area of the range in Ontario, an extensive search, even including newspaper advertisements, revealed no recent nests. Only 29 nests had been reported to the ONRS up to 1980 (Peck and James 1983).

To estimate the breeding abundance of such a rare species is difficult. It is certain that there were at least 4 pairs in Ontario during the atlas period, and possibly 6. It is impossible to know how many additional pairs remained undetected, but there were probably very few. An estimated upper limit of 25 to 30 pairs has been suggested (Campbell and Campbell 1984).-- *R.D. Weir*

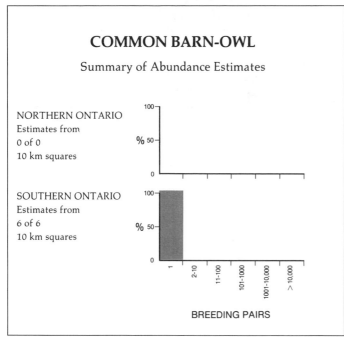

COMMON BARN-OWL

Summary of Abundance Estimates

NORTHERN ONTARIO
Estimates from
0 of 0
10 km squares

100
% 50
0

SOUTHERN ONTARIO
Estimates from
6 of 6
10 km squares

100
% 50
0

1   2-10   11-100   101-1000   1001-10,000   > 10,000

BREEDING PAIRS

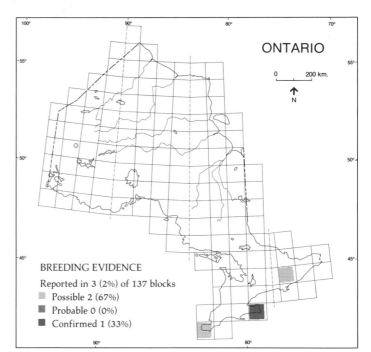

ONTARIO

0   200 km.

N

BREEDING EVIDENCE

Reported in 3 (2%) of 137 blocks

Possible 2 (67%)

Probable 0 (0%)

Confirmed 1 (33%)

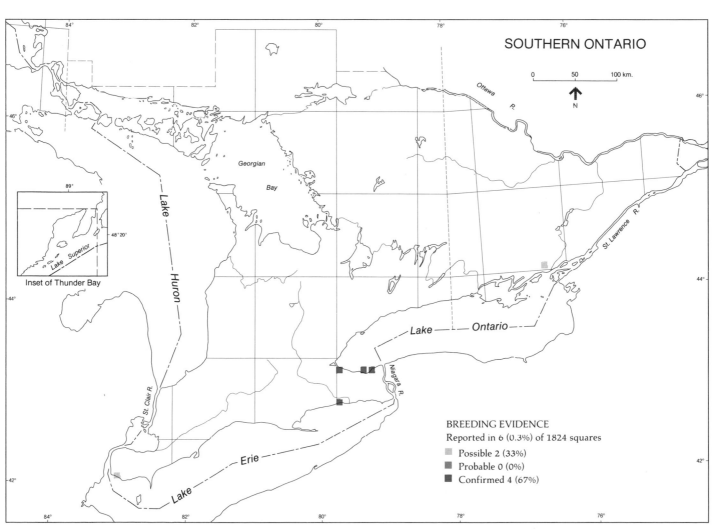

SOUTHERN ONTARIO

0   50   100 km.

N

Georgian
Bay

Lake

Huron

Ottawa   R.

St. Lawrence R.

Lake — Ontario

Niagara R.

St. Clair R.

Lake   Erie

89°

48°20′

Lake   Superior

Inset of Thunder Bay

BREEDING EVIDENCE

Reported in 6 (0.3%) of 1824 squares

Possible 2 (33%)

Probable 0 (0%)

Confirmed 4 (67%)

EASTERN SCREECH-OWL
Petit-duc maculé
Otus asio

J.S.

The Eastern Screech-Owl is one of the 3 small owls that occur in Ontario, and the most common breeding owl in southern urban centres. It is the only one that is polychromatic, coming in red, grey, and intermediate colour phases, which are independent of age, sex, and season.

This North American species breeds in Mexico, through most of the eastern US, and into southern Canada. It is a permanent resident of southern Ontario north only to about 46°N. The Eastern Screech-Owl frequents open woodland, orchards, and shade trees within cities and along streams. Mature deciduous growth is preferred. The species is exclusively nocturnal, waiting until after sunset to become active. During the day, it roosts in a tree cavity, making it almost impossible to find. Because the species is easily overlooked, its abundance within the province has been and continues to be difficult to assess.

During the 19th century, the Eastern Screech-Owl was restricted to southwestern Ontario and was common in Toronto, London, and Hamilton (McIlwraith 1894), where it was then the commonest owl. By the turn of the century, it was becoming more numerous at Kingston and was also noted at Ottawa (Macoun and Macoun 1909). By the mid-1930's, it was considered to be the commonest of all the nesting owls in extreme southern Ontario, breeding north to Muskoka District and Carleton Co (Baillie and Harrington 1936). James *et al.* (1976) described it as an uncommon permanent resident north to Ottawa and North Bay, but then, as now, records on the Canadian Shield were very scarce. Although the range of the species appears to have expanded over recent time, its numbers may actually have declined as the mature deciduous forests of southern Ontario were cleared for settlement. Certainly, there is evidence of unexplained short-term fluctuations in the past (McIlwraith 1894).

In the US, its habitat has also been reduced. In 1981, the Eastern Screech-Owl was placed on the Blue List because numbers were thought to have fallen in Ontario and 4 states in the US (Tate 1981). In the following year, it was delisted as populations appeared stable.

Nesting Eastern Screech-Owls are easily located during March, April, and May by imitating or playing tape recordings of their calls in areas of suitable habitat. If the sites where this species has been located are revisited over several weeks, its continued presence will define territory and probable breeding: 29% of records are in this category. This small owl is misnamed, for its usual call is not a screech but a melancholy low whistle that descends rapidly and ends with a soft tremulous whistle on a single, lower-pitched note. The call is given often during the mating season and only by night. The nest is placed in a tree cavity, an old woodpecker hole, or a nest box placed for Wood Ducks. To find the nest hole is difficult, and luck is a major factor. As a result, there are more records of probable than of confirmed breeding. Fussing Blue Jays and other small birds may betray the nest location as they pester the dozing owls.

The map shows that the Eastern Screech-Owl is primarily a bird of southern Ontario, south of the Canadian Shield. Its known range has only recently extended to the Sault Ste Marie area (Godfrey 1966, 1986), where several records of possible or probable breeding are shown. Since it is a non-migratory species, these records most likely represent breeding birds. In much of the range south of the Shield, the map has significant gaps, especially in Lambton and western Middlesex Counties, and along the north shore of Lake Ontario and the St. Lawrence River. In these areas, it is likely that nocturnal atlassing, using taped calls of owls, was more limited than in other areas, and that the species actually does occur there. Elsewhere, particularly in Essex, Huron, and southern Frontenac and Lennox and Addington Counties, atlassers who spent many hours at night searching woodlots for this species were able to show a much more widespread distribution than had previously been suspected. Surprisingly, large numbers of records were obtained in the Ottawa area, at the northern edge of the range of the species.

Abundance estimates indicate that 2 to 10 pairs of Eastern Screech-Owls were estimated to occur in most squares (32%) in which estimates were provided. Abundances of greater than 10 pairs per square were estimated in 22% of the squares, suggesting that the Eastern Screech-Owl is not nearly as rare as many naturalists thought it to be.-- *R.D. Weir*

EASTERN SCREECH-OWL

Summary of Abundance Estimates

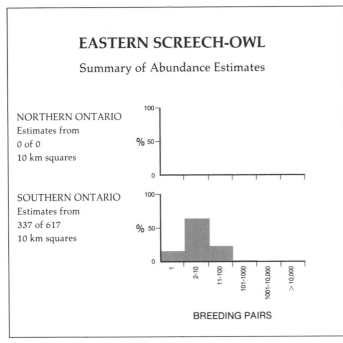

NORTHERN ONTARIO
Estimates from
0 of 0
10 km squares

SOUTHERN ONTARIO
Estimates from
337 of 617
10 km squares

100

% 50

0

100

% 50

0

1 2-10 11-100 101-1000 1001-10,000 >10,000

BREEDING PAIRS

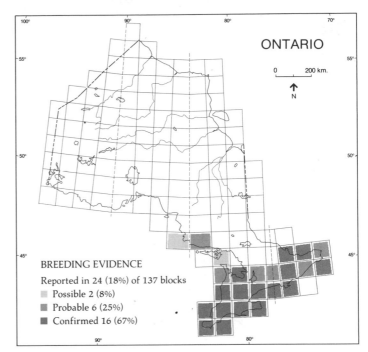

ONTARIO

0   200 km.

N

BREEDING EVIDENCE

Reported in 24 (18%) of 137 blocks

Possible 2 (8%)
Probable 6 (25%)
Confirmed 16 (67%)

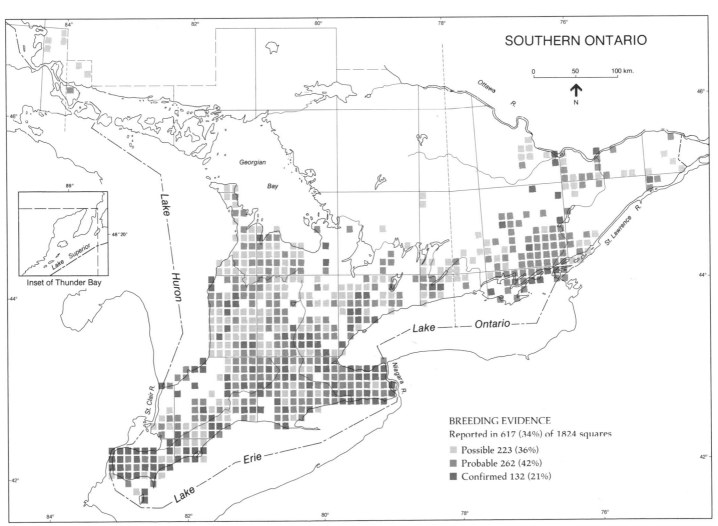

SOUTHERN ONTARIO

0   50   100 km.

N

Inset of Thunder Bay

Georgian Bay

Lake Huron

Lake Superior

St. Clair R.

Lake Erie

Niagara R.

Lake Ontario

Ottawa R.

St. Lawrence R.

BREEDING EVIDENCE

Reported in 617 (34%) of 1824 squares

Possible 223 (36%)
Probable 262 (42%)
Confirmed 132 (21%)

GREAT HORNED OWL
Grand-duc d'Amérique
Bubo virginianus

The Great Horned Owl is the most familiar of the 10 owl species known to breed in Ontario. It feeds upon insects, birds, and small to medium-sized mammals. The diet usually reflects what is most available, with small mammals constituting the greatest portion. This raptor is often persecuted by farmers concerned about attacks on poultry.

The Great Horned Owl nests only in the Western Hemisphere, from the tree-line in Canada's subarctic south throughout North, Central, and South America. In Ontario, the Great Horned Owl is absent only from the coastal Hudson Bay area. Its habitat consists of deciduous, mixed, or coniferous forests and wooded groves within agricultural lands that provide trees in which to nest and roost inconspicuously by day. It is usually present near dump sites, where it picks off rats with efficiency. This species frequently shares its habitat with the Red-tailed Hawk.

During the last century, the Great Horned Owl was distributed throughout Ontario and was common in Muskoka and Parry Sound Districts, but rare in Toronto (Macoun and Macoun 1909). Although Baillie and Harrington (1936) considered it to be a fairly common breeder north only to Cochrane and Thunder Bay, it has probably always been a common permanent resident throughout, extending north to the coastal Lowland (James *et al.* 1976). Its numbers have no doubt been reduced since the last century in those parts of its range where forests have been eliminated, but it continues to nest even within cities where sufficient groves of trees remain.

Like most owls, this species is nocturnal, but is easily detected by its deep, soft hoots given most often at dawn and dusk. Because of its large size, it is observed roosting during the day more often than any other Ontario breeding owl. It responds readily to imitations of its call and to that of the Eastern Screech-Owl as well. While the snow is deep and temperatures are well below freezing in January, prolonged hooting can be heard as territories are proclaimed and pairs court. At this time, they are most likely to be seen perched in the open. The Great Horned Owl occupies the old nest of a crow, hawk, or occasionally a squirrel or heron. A particular nest may be reused in consecutive seasons by the Great Horned Owl, but there is some preference for using nests on an alternative or intermittent basis with other birds (Peck and James 1983). Nests are usually in deciduous trees and can be found as early as February or March by looking for the incubating bird, whose silhouette in a leafless tree can sometimes be seen from a great distance. Young birds remain in the nest for up to 50 days after hatching, during which time the parents sometimes attack intruders. Breeding confirmation can also be achieved after the young leave the nest, usually by early June. For a few weeks they are nearly flightless, but noisy, often screaming loudly. The family party remains together on the nesting territory at least until the end of the summer. Six to 7 months per year are taken up by nesting duties, a longer time than almost all other single-brooded Canadian birds, thereby increasing the opportunities for atlassers to confirm breeding.

The Great Horned Owl occurs in all forest regions of the province south of the tundra. In southern Ontario, it was reported in 62% of the squares, and nesting was confirmed in 38% of these. These percentages are very much higher south of the edge of the Canadian Shield than they are to the north. The Barred Owl has a distribution which appears to be just the reverse of that. While the Barred Owl seems to require extensive woodlands, the Great Horned Owl is more widely distributed south of the Shield where most remaining woodland is in relatively small woodlots. It may well be that agricultural land-use creates habitat suitable for the Great Horned Owl but not for the Barred Owl. The scarcity of records of the Great Horned Owl in northern Ontario may be partly a result of inadequate coverage in those remote squares, particularly in the early months of the year, when breeding activity is most evident. However, the species is also probably much less numerous where continuous forests prevail; the sparsity of records in the Boreal Forest and Hudson Bay Lowland regions likely indicates relatively low density populations.

As might have been predicted from the frequency of records on the maps, abundance estimates show that the Great Horned Owl is more common in southern Ontario than in the north. Of 103 estimates of more than 10 pairs per square, none were in northern Ontario; only 21 were on the southern Ontario section of the Canadian Shield, including 13 on the Frontenac Axis northeast of Kingston, where agricultural land and fairly extensive wooded areas are interspersed. The rest of the high estimates were from agricultural southern Ontario.-- *R.D. Weir*

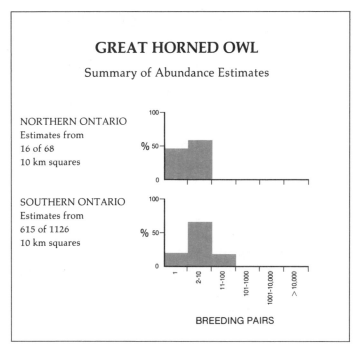

GREAT HORNED OWL

Summary of Abundance Estimates

NORTHERN ONTARIO
Estimates from
16 of 68
10 km squares

SOUTHERN ONTARIO
Estimates from
615 of 1126
10 km squares

BREEDING PAIRS

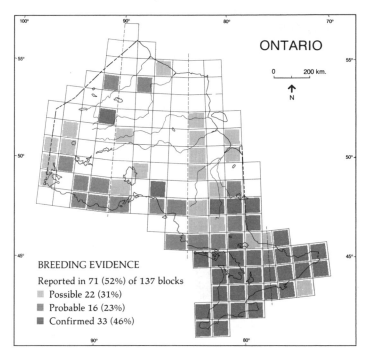

ONTARIO

0   200 km.

N

BREEDING EVIDENCE
Reported in 71 (52%) of 137 blocks

Possible 22 (31%)
Probable 16 (23%)
Confirmed 33 (46%)

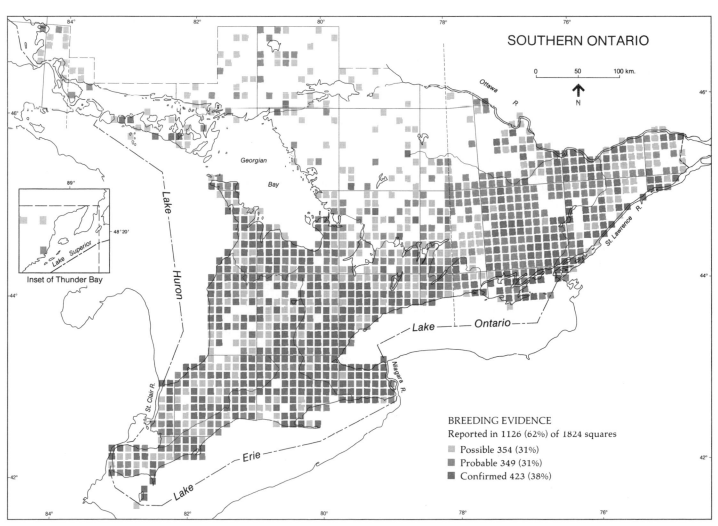

SOUTHERN ONTARIO

0   50   100 km.

N

Georgian
Bay

Lake
Huron

Lake Superior

Inset of Thunder Bay

Ottawa
R.

St. Lawrence R.

Lake — Ontario

Niagara R.

St. Clair R.

Lake — Erie

BREEDING EVIDENCE
Reported in 1126 (62%) of 1824 squares

Possible 354 (31%)
Probable 349 (31%)
Confirmed 423 (38%)

BARRED OWL
Chouette rayée
Strix varia

The characteristic and far-carrying nocturnal call of the Barred Owl ('Who cooks for you, who cooks for you-all') is one of the most familiar and treasured sounds of campers and residents on Ontario's Canadian Shield.

The Barred Owl is a New World species, resident from Central America northward through the eastern US. Its range reaches southern Canada and extends west into the Canadian Cordillera. In Ontario, it is a permanent resident of coniferous or mixed woodland primarily of the Great Lakes-St. Lawrence Forest region, and is partial to dense, moist forest, especially near lakes, streams, and swamps. The average home range of 9 radio-tagged Barred Owls in Minnesota was determined to be 229 ha (Nicholls and Warner 1972). There are few suitable woodlands of this size south of the Shield in Ontario, and the Barred Owl is consequently found only sparingly in this area. Along the southern sections of the Canadian Shield, and occasionally farther south, it shares its habitat with the Red-shouldered Hawk.

As is the case for most Ontario raptors, the range of the Barred Owl has changed since the last century as a result of human settlement. McIlwraith (1894) considered it to be not rare along Ontario's southern boundary. Macoun and Macoun (1909) noted it to be not common along the St. Lawrence River, rare in the London area, but common in the Parry Sound and Muskoka Districts. MacClement (1915) wrote of its becoming less common as thick swampy growths were destroyed near settled areas. By the 1930's, this formerly common breeder of southern Ontario had become rare, but Baillie and Harrington (1936) reported its occurrence throughout much of the rest of the province north to James Bay and west to Lake of the Woods - areas suffering a minimum of habitat change at the hands of people. More recently, it was still considered rare by James *et al.* (1976).

By night, the Barred Owl is one of the noisiest owls; loud courtship calling is especially obvious during March and April. By day, it is quiet and easily overlooked as it roosts in secluded shady retreats. Pairs are attached to favourite nesting sites and return in subsequent years. The Barred Owl does not build its own nest structure. Nests are usually in tree hollows, but occasionally an old nest of a crow or hawk is used. Eggs are laid during April or May, and, after hatching, the young are brooded closely by both parents for 3 weeks. When 4 or 5 weeks old, usually during June, the young climb out of the nest into the tree branches. During this time, adults make frequent trips carrying food, most often at night but sometimes by day. The one period of the year when the adults become quiet occurs while young are in the nest. The presence of an old nest used as a feeding platform, and located near the active nest, is also helpful in finding the nesting area. Around the occupied nest, adult Barred Owls will usually flush when disturbed, and are much less aggressive than the Great Horned Owl. Young in the nest call to their parents by night, which is a useful clue to their presence. Because cavity nests are difficult to find, breeding is more easily confirmed by finding the young birds. Fledged young account for 69% of the confirmed breeding records, while nests account for only 18%.

The maps show the Barred Owl to be almost extirpated as a breeder from areas of Ontario south of 44°N. A nest near Waterloo in 1983 and young birds in Backus Woods near Long Point in 1985 are the only confirmed breeding records in southwestern Ontario. The main concentrations of Barred Owls are in mixed forests on the Canadian Shield, where lakes, swamps, and forests abound. Although the map is probably not a complete picture, especially in areas of the Shield that are difficult to access, it appears that its range in southern Ontario has become even more restricted since the 1930's. However, there are counties in eastern Ontario, including Hastings to Frontenac, where numbers appear to be increasing, perhaps as the species spreads from the northwest. Breeding was confirmed in 62 squares, about one-quarter of them in that small area of the province. The northern range limit of the Barred Owl coincides roughly with the northern edge of the Great Lakes-St. Lawrence Forest region. Although there are a few records farther north, it is doubtful that such a vocal species could be overlooked if it were present in significant numbers.

Abundance estimates suggest that the Barred Owl is an uncommon, but not rare, bird in most of the areas where it occurs. In addition to the eastern Ontario areas already noted, there are estimates of more than one pair per square from a significant number of squares northeast of Peterborough (many with estimates of over 10 pairs) and in the Sault Ste Marie area.-- *R.D. Weir*

BARRED OWL

Summary of Abundance Estimates

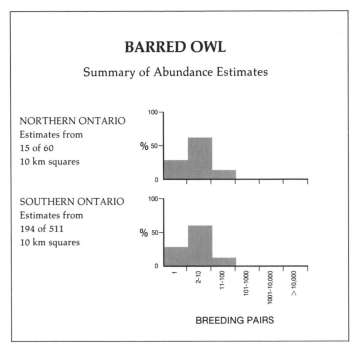

NORTHERN ONTARIO
Estimates from
15 of 60
10 km squares

SOUTHERN ONTARIO
Estimates from
194 of 511
10 km squares

BREEDING PAIRS

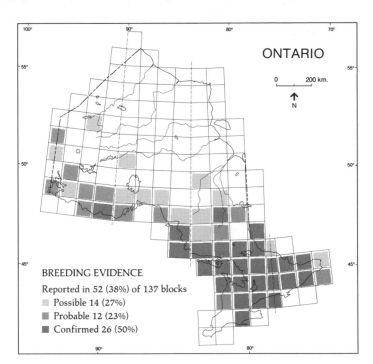

ONTARIO

0   200 km.

N

BREEDING EVIDENCE
Reported in 52 (38%) of 137 blocks
☐ Possible 14 (27%)
▨ Probable 12 (23%)
■ Confirmed 26 (50%)

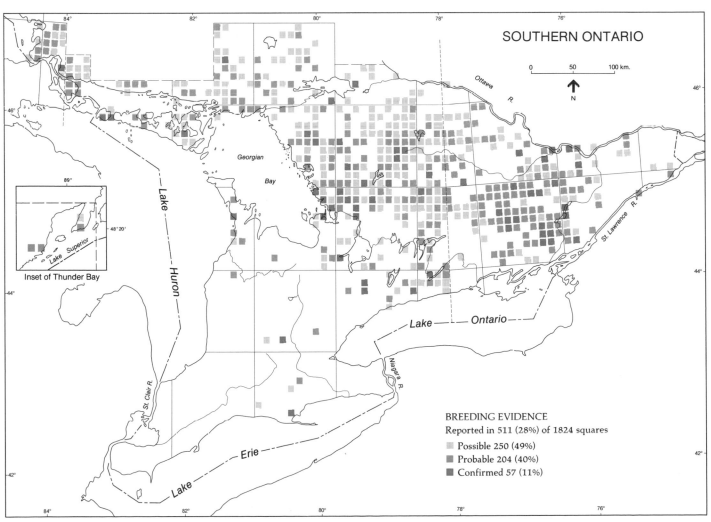

SOUTHERN ONTARIO

0   50   100 km.

N

Inset of Thunder Bay

BREEDING EVIDENCE
Reported in 511 (28%) of 1824 squares
☐ Possible 250 (49%)
▨ Probable 204 (40%)
■ Confirmed 57 (11%)

GREAT GRAY OWL
Chouette lapone
Strix nebulosa

This species is the mysterious "phantom of the northern forest" (Nero 1980). Its imposing appearance, relative rarity, and strong evocation of the remote northern wilderness make this species one of the 10 'most wanted' of North American birds (Tucker 1972). Vole population crashes in northern areas brought large numbers of this species to southeastern Ontario in the winters of 1978-1979 and 1983-1984, giving many naturalists their first viewing opportunity.

The Great Gray Owl is found across much of the Boreal Forest zone and its associated muskeg in Eurasia and North America. In North America, it occurs from the west coast eastward to Ontario and presumably into Quebec, although there are no breeding records from that province to date. From near the northern extent of spruce forest, its range extends southward along the mountains into the US as far as California. In Ontario, it is almost exclusively a bird of the north, at least in summer, with records from only as far south as the Sudbury area. Nests are frequently located in stands of poplar or spruce adjacent to, or comprising 'islands' within, extensive tracts of muskeg. Like most other owls, the Great Gray Owl does not construct its own nest, but uses abandoned crow, raven, or hawk nests, or even platforms constructed by humans. Open fens, bogs, and meadows are important hunting habitats.

Breeding of the Great Gray Owl was confirmed in Ontario in 1911, when young already able to fly were collected in Nipissing District (Baillie and Harrington 1936). No actual nests were found until 1964, and a total of only 3 nests were reported between then and the commencement of atlas field work (Peck and James 1983). The increased number of sightings and reported nestings in recent years presumably reflects expanded observer coverage rather than an increase in population or a change in range.

Unlike most owls, the Great Gray Owl is active during daylight, frequently feeds in open habitat, and is usually quite tame. Yet, this owl can be far from conspicuous. Although at times it will actively defend nests against human intruders, and is vocal in the breeding season, it can be surprisingly elusive in the vicinity of the nest (Nero 1980). Confirmation of nesting can also be difficult because the Great Gray Owl nests early in the season and does not make frequent visits to the nest. Time restrictions for atlassers visiting remote areas probably mean that the species is underrepresented on the maps: it likely bred in all squares with records. Although family groups were reported in 3 squares, no nests were found during the atlas period. Most records (31) were simply observations of birds in apparently suitable habitat in northern squares.

It seems likely that the Great Gray Owl breeds sparsely throughout northern Ontario (Godfrey 1986), except in the tundra zone along the Hudson Bay coast. Habitat conditions throughout the Hudson Bay Lowland and Boreal Forest zones certainly appear to be suitable. Undoubtedly, remoteness, problems with access, and the general physical difficulty of surveying much of the terrain over large areas of its range during the atlas project have resulted in underrepresentation on the map. However, the large gap on the map in the central part of far northern Ontario does raise the question of whether the species is really only relatively sparse in that area, or whether it may actually be absent.

Although there is considerable uncertainty and contradictory opinion about numbers, the Great Gray Owl is generally regarded as rare, and was so designated by COSEWIC (Nero 1979). Nonetheless, Nero (1980) suspects that the species is more common than generally believed. It was most evident in 2 blocks near Marathon (7 squares in one block, 3 in the other) and near Rainy River (4 almost contiguous squares in one block). Otherwise, it is limited to only 1 or 2 squares per block. No abundance estimates are available, which may reflect our lack of knowledge of this elusive bird.-- *J.P. Prevett*

GREAT GRAY OWL

Summary of Abundance Estimates

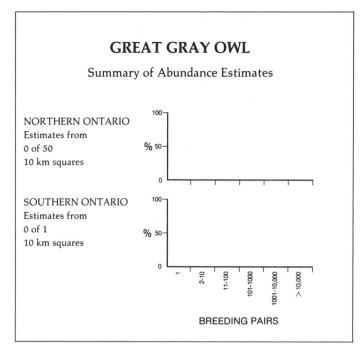

NORTHERN ONTARIO
Estimates from
0 of 50
10 km squares

SOUTHERN ONTARIO
Estimates from
0 of 1
10 km squares

%

100

50

0

%

100

50

0

1   2-10   11-100   101-1000   1001-10,000   >10,000

BREEDING PAIRS

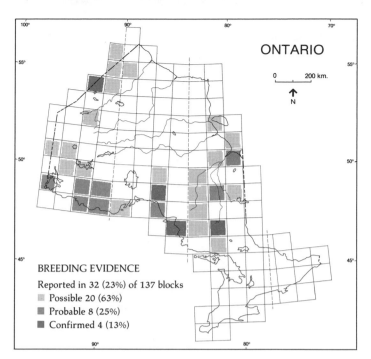

ONTARIO

0   200 km.

N

BREEDING EVIDENCE

Reported in 32 (23%) of 137 blocks

Possible 20 (63%)

Probable 8 (25%)

Confirmed 4 (13%)

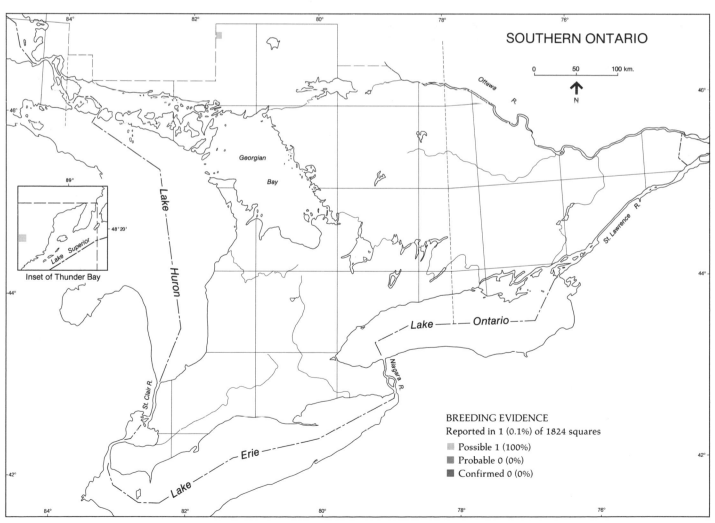

SOUTHERN ONTARIO

0   50   100 km.

N

Georgian Bay

Lake Huron

Lake Superior

Inset of Thunder Bay

89°

48°20'

Ottawa R.

St. Lawrence R.

Lake — Ontario

Niagara R.

St. Clair R.

Lake — Erie

BREEDING EVIDENCE

Reported in 1 (0.1%) of 1824 squares

Possible 1 (100%)

Probable 0 (0%)

Confirmed 0 (0%)

LONG-EARED OWL
Hibou moyen-duc
Asio otus

J.S.

Because it remains inactive and well hidden in dense cover by day, hunts exclusively at night, calls infrequently, and possesses a wide repertoire, much of which is unfamiliar to even the most experienced bird-watcher, the Long-eared Owl is the most difficult of Ontario's owls to locate during the nesting season.

The Long-eared Owl breeds in Eurasia southward to North Africa; in North America, its breeding range includes the southern half of Canada and most of the US. Its Canadian breeding range covers large parts of all the provinces from Nova Scotia to British Columbia. In Ontario, the species occurs in all of the south and probably all forested regions of the north, although the northern edge of its nesting range is poorly defined. The Long-eared Owl prefers either coniferous or mixed forest, and it chooses dense stands that may consist of reforestation planting, copses, or isolated groves in farmland. More often than not, the areas chosen are moist. The species migrates in autumn from most of the northerly part of its Ontario range and is more easily found in southern Ontario during winter, when it is gregarious, sometimes numbering 50 to 75 in a single woodlot. These observations, together with the considerable numbers detected as they migrate past banding stations along the lower Great Lakes, suggest that the species is more common than atlas records indicate.

It is not surprising, then, that the ornithologists of the last century found relatively few Long-eared Owls during the nesting season. McIlwraith (1894) and Macoun and Macoun (1909) wrote of its being not common anywhere in the province, and rare in eastern Ontario and Toronto. MacClement (1915) claimed that it frequented the shores of Hudson Bay in summer. Baillie and Harrington (1936) called it "a not uncommon summer resident of southern Ontario," and James

et al. (1976) considered it to be an uncommon summer resident as far north as Severn Lake and Moosonee. Because the Long-eared Owl is so poorly known, it is almost impossible to make a definitive statement comparing its numbers or distribution in the 19th century with those of the present day.

The Long-eared Owl seems temperamental when it comes to responding to calls of its own species. When it does answer, various shrieks, whistles, dove-like coos, hisses, barks, or hoots (up to 20 in succession) may be given. Like most owls, it does not construct its own nest, but rather uses an old structure built by hawks, crows, or squirrels. Old nests of the Red-tailed Hawk that are high in pine trees are often chosen. The 2 to 6 eggs are usually laid during early spring in the south, occasionally when snow is still on the ground. Incubation lasts about 28 days, during which time the female sits closely, with the male roosting nearby. The bird is not easily flushed and will usually freeze when intruders approach. Even when a nest with eggs is known to be present, the adult birds are not easily seen. They sit very still and their vertical stretch pose makes them appear like broken branches. During the 10 weeks after the young have hatched, the chances of locating a nest improve greatly. The young remain in the nest for up to 5 weeks, during which time their hunger calls, which sound like an unoiled gate hinge or squeaky wheel, can be heard by night. They climb out of the nest before they can fly, and are fed by the parents for an additional 5 weeks. If discovered with young even by day, the adults engage in a spectacular distraction display, which may consist of various threatening postures, fluttering along the ground, hissing, and sometimes an attack (Bent 1937, pers. obs.).

The maps show few occurrences of the Long-eared Owl. Within southern Ontario, the species was found in only 169 of the 1824 squares. It was confirmed as nesting in only 28 squares in all of Ontario: fledged young in 12 and a nest in 16 squares. The species may appear throughout the south wherever suitable habitat exists. The concentration of occurrences in squares near Kingston and Ottawa reflects a tremendous effort in locating the species by all-night searches. The picture in northern Ontario is even sketchier, where inaccessibility to the blocks compounds the problem of finding Long-eared Owls. Individuals were located in widely separated blocks, suggesting that the species likely occurs anywhere in-between as long as suitable habitat is available. The most northerly confirmed record is of a young bird found near a nest in the Sutton Ridges, near Hudson Bay. This record extends northward the confirmed breeding range of the species in Ontario as mapped by Peck and James (1983). There is no doubt that the maps give an incomplete picture of the true distribution of the species, which could be revealed only by a special survey.

Nobody knows how common the Long-eared Owl is in Ontario. Any extrapolation from the abundance estimates offered would yield a density which would likely be a gross underestimation.-- *R.D. Weir*

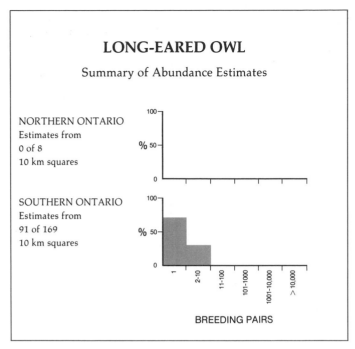

LONG-EARED OWL

Summary of Abundance Estimates

NORTHERN ONTARIO
Estimates from
0 of 8
10 km squares

SOUTHERN ONTARIO
Estimates from
91 of 169
10 km squares

%

100

50

0

%

100

50

0

1   2-10   11-100   101-1000   1001-10,000   >10,000

BREEDING PAIRS

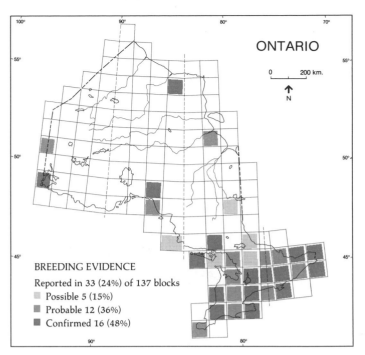

ONTARIO

0   200 km.

N

BREEDING EVIDENCE

Reported in 33 (24%) of 137 blocks

Possible 5 (15%)
Probable 12 (36%)
Confirmed 16 (48%)

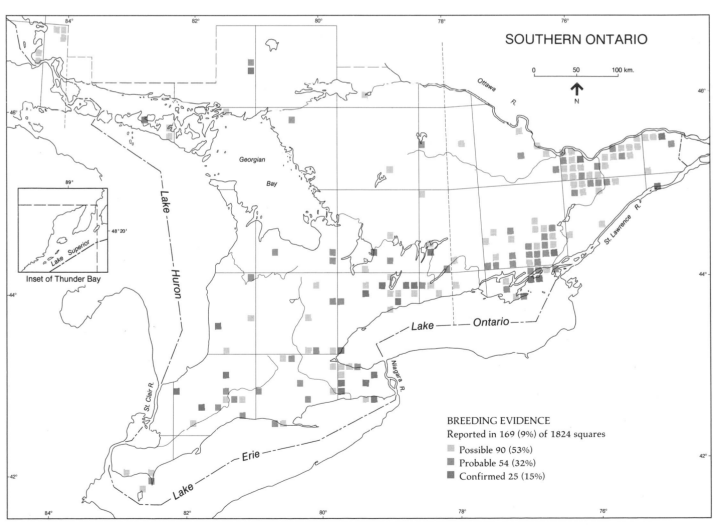

SOUTHERN ONTARIO

0   50   100 km.

N

Inset of Thunder Bay

Georgian
Bay

Lake Superior

Lake Huron

St. Clair R.

Ottawa R.

St. Lawrence R.

Lake — Ontario

Niagara R.

Lake Erie

BREEDING EVIDENCE

Reported in 169 (9%) of 1824 squares

Possible 90 (53%)
Probable 54 (32%)
Confirmed 25 (15%)

SHORT-EARED OWL
Hibou des marais
Asio flammeus

The Short-eared Owl is nearly cosmopolitan, being found on all continents except Australia and Antarctica. Its breeding range extends throughout a large portion of the US and all of Canada excluding the high arctic. Although it occurs at the northern and southern extremes of Ontario, its breeding distribution is very uneven and highly correlated with its specialized habitat requirements. The Short-eared Owl frequents open areas, especially meadows, bogs, marshes, and even sand dunes. It hunts by flying low over open spaces in much the same manner as the Northern Harrier. Forests are normally shunned, so the species is likely absent as a breeder from extensive areas of the Boreal Forest zone in this province. However, its habitat needs are met along the coastal area of Hudson and James Bays, as well as in some sections of agricultural southern Ontario..

McIlwraith (1894) noted that a few Short-eared Owls nested in southern Ontario but that most went farther north. Both he and Macoun and Macoun (1909) agreed that the species was more common in Ontario than the Long-eared Owl. Baillie and Harrington (1936) wrote that the Short-eared Owl bred in widely separated localities north to Hudson Bay, where it was common, but that it was a rare breeder elsewhere. The assessment by Peck and James (1983) was similar. Destruction of most wetlands in extreme southern Ontario since the 19th century has removed large areas of its breeding habitat. The 3 most southerly counties and RMs in which breeding was documented by Peck and James (1983) are Kent, Elgin, and Haldimand-Norfolk RM. The atlas survey revealed confirmed breeding in none of these areas, and of these, only in Haldimand-Norfolk RM was there any breeding evidence at all. A number of confirmed breeding records, based on sightings of young birds, appear on the map in areas of southern Ontario not reported by Peck and

James (1983): near Clinton (Huron Co), Beaverton (Durham RM), Prince Edward Co, near Ottawa, and near Alexandria (Glengarry Co).

The Short-eared Owl is both diurnal and nocturnal, and is most active around dusk, when adults hunt over fields and marshes. Courtship during April and May is conspicuous, consisting of the pair flying and soaring together. The male executes short dives and claps his wings below the body on the down stroke. Both sexes call frequently with a dog-like bark, but the species is unlikely to be found by its voice. The nest is a raised platform of dead grass placed among vegetation on wet ground in open areas, but to confirm nesting is not as simple as might be expected for a species nesting in the open. Because the female sits tightly, the nest is difficult to locate during incubation. The male carries food to her, which is a helpful clue. The young remain in the nest for 5 weeks after hatching, during which period the behaviour of the adults changes. Intruders near the nest are greeted by flying adults that feign injury by dragging themselves along the ground. After fledging, the young hide in the grass nearby for up to 6 weeks, during which time they are difficult to locate. Short-eared Owls migrate south from the northerly part of the breeding range, and may overwinter in large numbers where prey is abundant. At these sites, some birds invariably remain to nest as long as food lasts. Some sites, such as on Wolfe and Amherst Islands near Kingston, are used nearly every year, except during seasons when the vole population reaches a low in its 4 or 5 year cycle.

The range map confirms some of what was previously known about the distribution of the Short-eared Owl. Its presence along the Hudson and James Bay coasts, and its almost complete absence from the Boreal Forest zone (except for one sighting near Pickle Lake) are not surprising. However, its spotty and limited occurrence in southern Ontario is surprising and cause for concern. Although breeding confirmation is difficult to obtain (only 14 squares in the south), the species can be found fairly easily. Yet, it was located in only 87 squares in the province, making it one of the most limited in occurrence of Ontario's owls. The destruction of wetlands by drainage for agriculture and housing is likely the single most important factor responsible for its steady decline, and, like the Northern Harrier, the Short-eared Owl will inevitably suffer a further decline as long as wetlands continue to disappear.

In response to falling numbers in Ontario, New York, Ohio, Indiana, and parts of the west, the Short-eared Owl was placed on the Blue List in 1976, where it remains (Tate and Tate 1982). There is no doubt that breeding numbers of the Short-eared Owl are now much lower than in the 19th century, and the species can no longer be considered to be a more common breeder than the Long-eared Owl in southern Ontario. It is impossible to estimate accurately its numbers in the north, but they very likely exceed those in the south.-- *R.D. Weir*

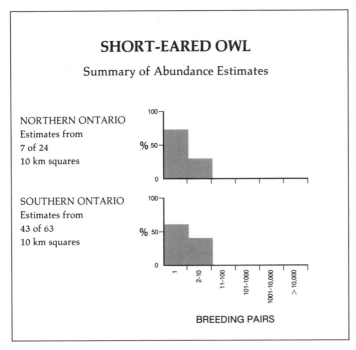

SHORT-EARED OWL

Summary of Abundance Estimates

NORTHERN ONTARIO
Estimates from
7 of 24
10 km squares

SOUTHERN ONTARIO
Estimates from
43 of 63
10 km squares

% 100 50 0

% 100 50 0

1   2-10   11-100   101-1000   1001-10,000   >10,000

BREEDING PAIRS

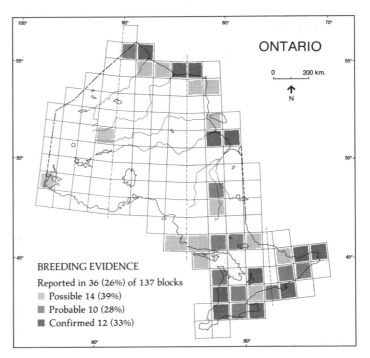

ONTARIO

0   200 km.

N

BREEDING EVIDENCE
Reported in 36 (26%) of 137 blocks

Possible 14 (39%)
Probable 10 (28%)
Confirmed 12 (33%)

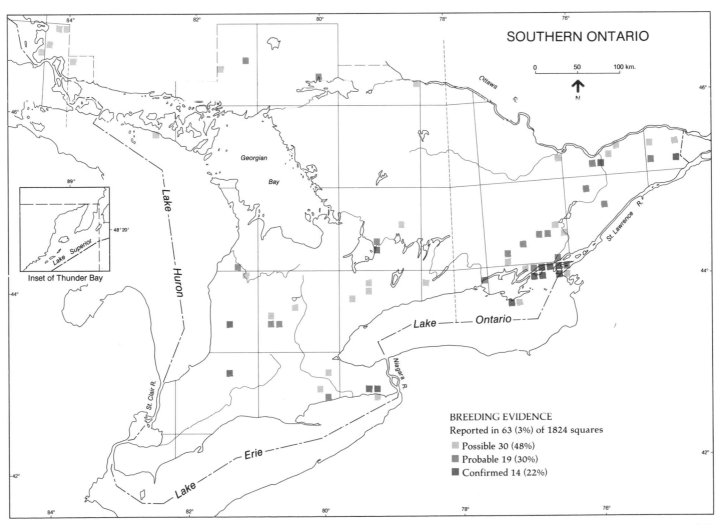

SOUTHERN ONTARIO

0   50   100 km.

N

Georgian Bay

Lake Huron

Lake Superior

Inset of Thunder Bay

89°   48°20'

Ottawa

St. Lawrence R.

Lake — Ontario

Niagara R.

St. Clair R.

Lake — Erie

Lake

BREEDING EVIDENCE
Reported in 63 (3%) of 1824 squares

Possible 30 (48%)
Probable 19 (30%)
Confirmed 14 (22%)

BOREAL OWL
Nyctale boréale
Aegolius funereus

J.S.

This is one of Ontario's most infrequently found owls, being small, northern, and nocturnal. It is a permanent resident in boreal forests across Eurasia and North America. In Ontario, its summer range has been assumed to extend from the tree-line in the north to about the latitude of Lake Nipigon and Kapuskasing (Godfrey 1986). This area is dominated by black and white spruce, balsam fir, trembling aspen, balsam poplar, and white birch, species associated with Boreal Owl habitat (Bondrup-Nielsen 1976). Occasionally, the Boreal Owl is irruptive, and southern Ontario then experiences invasions during the winter (Catling 1972).

The provincial breeding distribution of the Boreal Owl has never been well known. Reports prior to the atlas project suggested that it might have bred as far south as Quetico Prov. Park (Goodwin 1979a), Thunder Bay (Baillie 1950b), and Kapuskasing (Bondrup-Nielsen 1976), and as far north as the tree-line. The distribution of these records was consistent with what little was known of the breeding range elsewhere in North America. As of 1975, the year when Ontario's only known nest prior to the atlas period was found near Kapuskasing, there were but 20 known Canadian nest records (Bondrup-Nielsen 1976).

The Boreal Owl typically roosts in thick foliage by day, and is consequently not often encountered by accident. Access to many of the remote northern blocks was limited, and atlas work in any one block was often confined not only to one year, but also to a brief period within that year. Like the Northern Saw-whet Owl, the Boreal Owl restricts most of its vocal activity to a period in the spring, primarily March to May (Bondrup-Nielsen 1976), a time before most atlassers in the north began field work. The species uses tree cavities for nest sites, and, because its trips to and from the nest occur mainly at night, birds are likely to be found by day only if field observers check all potentially suitable nesting holes.

Although the provincial range of this owl is still far from being defined completely, atlas work has provided new information, including the discovery of a second nest, containing young, 13 km west of Atikokan, on 10 June 1984. The only other confirmed record was of fledged young near Timmins. Near the latitude of Lake Superior, the Boreal Owl is found in a number of blocks, at least as far south as the southern limit established prior to the commencement of atlas field work. The most interesting records, constituting possible range extensions, are those north of Lake Huron. For instance, one bird was heard calling continually for 2 minutes on 21 March 1982, in Fairbanks Township, about 30 km west of Sudbury, and several others were reported north of there in the same year. The present distribution of this species may therefore extend farther south than previously thought, where suitable habitat is available.

Atlas field work yielded very few abundance estimates, because of the difficulties reviewed above. Bondrup-Nielsen (1976) concluded, in his 1974-1975 study of the species in Ontario, that it was not so rare as thought and that, in preferred habitat with suitable nesting holes and ample food, it would prove to be fairly common.-- *A. Mills*

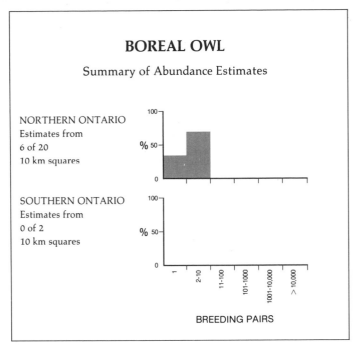

BOREAL OWL

Summary of Abundance Estimates

NORTHERN ONTARIO
Estimates from
6 of 20
10 km squares

SOUTHERN ONTARIO
Estimates from
0 of 2
10 km squares

BREEDING PAIRS

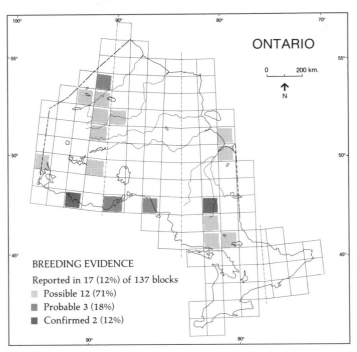

ONTARIO

0   200 km.

↑
N

BREEDING EVIDENCE
Reported in 17 (12%) of 137 blocks

■ Possible 12 (71%)
■ Probable 3 (18%)
■ Confirmed 2 (12%)

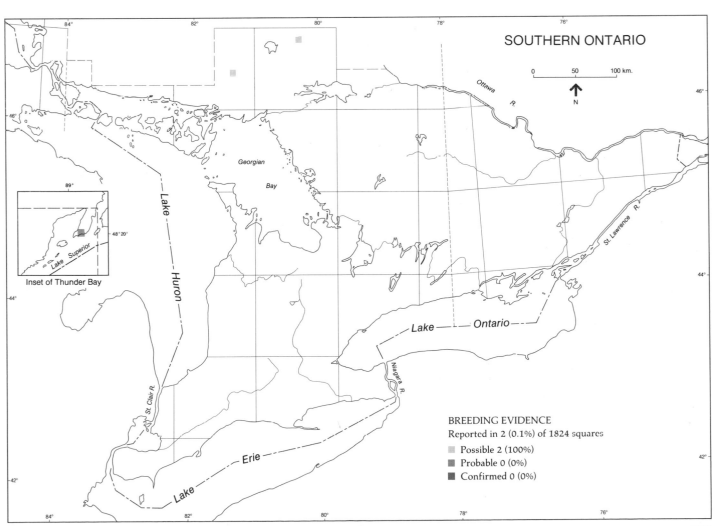

SOUTHERN ONTARIO

0   50   100 km.

↑
N

Georgian
Bay

Ottawa R.

St. Lawrence R.

Lake — Ontario

Lake
Huron

Lake

St. Clair R.

Niagara R.

Lake — Erie

89°

48°20'

Lake Superior

Inset of Thunder Bay

BREEDING EVIDENCE
Reported in 2 (0.1%) of 1824 squares

▦ Possible 2 (100%)
■ Probable 0 (0%)
■ Confirmed 0 (0%)

NORTHERN SAW-WHET OWL
Petite Nyctale
Aegolius acadicus

J.S.

The breeding distribution of the Northern Saw-whet Owl extends from southeastern Alaska, south-central Canada, and Nova Scotia south to Mexico in the west and Maryland in the east (Godfrey 1986). Breeding habitat varies geographically, but in Ontario the species frequents a variety of forest types, favouring coniferous or northern mixed woods.

Details of the breeding distribution of this species in Ontario have never been well known. McIlwraith (1894) believed that most bred south of Canada, commenting that the Saw-whet Owl "does not penetrate far into British America." We now know that it has been found as a breeder almost throughout the province, with the exception of far northern Ontario. Summer records suggestive of breeding have previously been known from Pelee Island, Fort Erie, and the St. Lawrence Islands in the south to Moosonee, Lake Nipigon, and Red Lake in the north (Baillie and Harrington 1936, James *et al.* 1976). In many cases, however, summer records in the far south or far north were suspected of being rather exceptional.

As the atlas project progressed, field workers met with increasing success in locating this species. Because it usually roosts by day in thick cover or tree cavities, and because of its small size, the Northern Saw-whet Owl is not often seen during the nesting season. Its most salient characteristic is a whistled call note repeated at night about 100 times per minute. It became apparent to atlassers that although the bird often called incessantly early in the season, it soon became quiet. Nesting birds can become more or less silent before the end of May. In subsequent years, especially in 1985, atlassers made a concerted effort in certain regions to conduct night work (often with a tape) to look for this and other nocturnal species.

Because the Northern Saw-whet Owl is migratory, in many cases judgment is required to designate the status of particular records. In Algonquin Prov. Park, where wintering birds are probably scarce, it is not uncommon to hear the species in numbers by the beginning of March. Other birds are still northbound in southern Ontario in mid-April (Catling 1971). The cessation of song, heretofore presumed to indicate the disappearance of the bird, may simply mean a pair is nesting and has no need for further vocal activity. In many cases, supported by the historically widespread breeding distribution, many records of birds heard calling were given the benefit of the doubt and assumed to be possibly breeding.

As with most other nocturnal species, this owl is not easy to confirm as a breeder. Activity at the nest is restricted to the night, and their inconspicuous tree cavity nests are not likely to be found by day unless observers check specific holes. Furthermore, eggs are often laid early in April, meaning that nesting can be over before the time when most atlas work was done; 16 of the 25 confirmed records were of young birds, while the other 9 were of nests.

The maps probably give an accurate indication of the breeding range of the Northern Saw-whet Owl in southern Ontario. It appears to be a sparse breeder in the southwest, which is consistent with the amount of suitable habitat. In the southern portion of the Canadian Shield, it appears to be particularly common, but this is probably a result of combining good habitat with good coverage in 1985. The patchy distribution probably reflects variation in coverage rather than in owl distribution. Considerable census efforts were made in Kingston, Ottawa, Haliburton, Peterborough, and Parry Sound regions, for instance, and these concentrations show up on the maps. Farther north, atlas data suggest that the Northern Saw-whet Owl is scarce, but this is probably an artefact of sampling. Most field work in the north was done after the peak of vocal activity for this species. The northern limit of its range is not well defined. Although most of the records are from within the Great Lakes-St. Lawrence Forest zone, there are a few from well within the Boreal Forest region, as there have been in the past. Obviously, the low percentage of confirmed breeding records is grossly unrealistic; most records probably are of nesting individuals, even though the evidence is lacking.

Abundance estimates are particularly difficult to establish for this species. The problem is compounded by apparent variations in numbers from year to year. The large number of records from 1985, which are partly related to the intense activity in searching for owls that year, led to a dramatic filling in of the map. By the end of 1984, 138 squares had records of the species, while, by the end of 1985, this total had risen to 459. The Kingston Regional Coordinator estimated that about 190 pairs per year nested in that region (60 squares). The Parry Sound Coordinator found 74 individuals in 38 squares in 1985, and the Ottawa Coordinators found it not uncommon to hear 3 birds calling at once, again in 1985.-- *A. Mills*

NORTHERN SAW-WHET OWL

Summary of Abundance Estimates

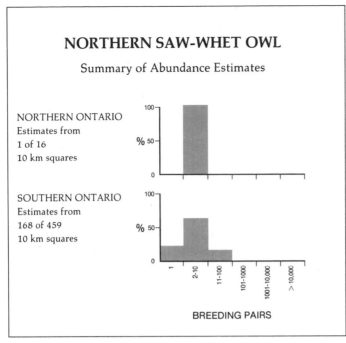

NORTHERN ONTARIO
Estimates from
1 of 16
10 km squares

SOUTHERN ONTARIO
Estimates from
168 of 459
10 km squares

BREEDING PAIRS

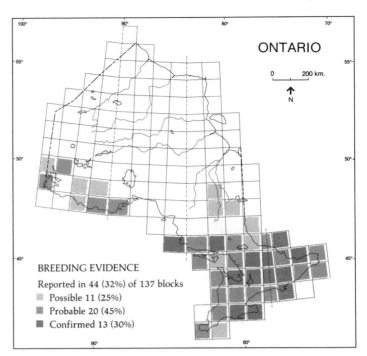

ONTARIO

0   200 km.

N

BREEDING EVIDENCE
Reported in 44 (32%) of 137 blocks
Possible 11 (25%)
Probable 20 (45%)
Confirmed 13 (30%)

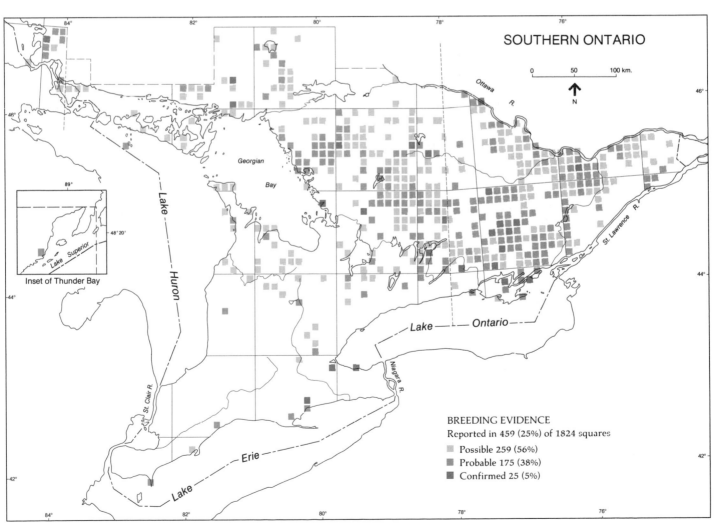

SOUTHERN ONTARIO

0   50   100 km.

N

Inset of Thunder Bay

BREEDING EVIDENCE
Reported in 459 (25%) of 1824 squares
Possible 259 (56%)
Probable 175 (38%)
Confirmed 25 (5%)

COMMON NIGHTHAWK
Engoulevent d'Amérique
Chordeiles minor

This is the familiar member of the 'goatsucker' family found over most Ontario towns and cities and through much of the countryside. Its dramatic aerial courtship or territorial flight is unique among our birds. As a breeding bird, the Common Nighthawk ranges across North America from near the tree-line in Alaska and Canada to Mexico, and locally south to Panama. It winters in South America (AOU 1983). The species breeds throughout Ontario, though only locally in many regions. Being an aerial insectivore like the swallows and feeding at significant heights, it can find suitable feeding almost anywhere, even in the most highly congested urban areas, but its breeding habitat is much more restricted.

The contemporary range of the Common Nighthawk is probably little different from that of 50 or 100 years ago. Macoun and Macoun (1909) reported the use of flat gravel roofs as nesting sites by southern Ontario birds, and indicated that the species was present, even "abundant", from Missinabi to Moose Factory in 1904. However, in recent years this historically common bird has exhibited a numerical decline through much of its continental range. The first 10 years of the *American Birds* Blue List featured the Common Nighthawk 6 times, based on the impressions of observers from the Pacific coast to the Atlantic seaboard (Tate 1981). Goodwin and Rosche (1970, 1974) have reported declines in both southern and northern Ontario breeding populations in recent years. Nevertheless, the species can by no means be considered rare at present, since it is still widespread and because large numbers of late summer migrants have been noted during the 1980's (*e.g.*, Weir 1984).

The Common Nighthawk is an easy species to identify. The erratic bouncing flight and the nasal calls that descend from the twilight skies are quite distinctive, and its loud 'booming' dives are especially noticeable. Its short summer visit at these latitudes may have resulted in confusion between breeding birds and late northbound migrants in early June, or southbound birds later in the summer, but it is doubtful that this has significantly affected the maps. Although it ranges more widely than the more strictly territorial songbirds (Armstrong 1965), it is likely that most observations were recorded in the square in which it actually breeds. Though easy to locate and identify, the Common Nighthawk is not easy to document as a nesting bird. It nests on bare ground in open areas, often merely clearings in relatively dense forest, though, by the time the young are hatched, there may be some cover of low vegetation. The nest itself is no more than a scrape, containing one or usually two eggs, on the ground or on the flat roof of an urban building (which often provides a suitable artificial nest-site). This nesting behaviour, its predominantly crepuscular and nocturnal habits, and its method of feeding the young by regurgitation reduce the possibilities for confirmation of breeding. Generally, to confirm breeding, the 'nest' must be found, though distraction displays also take place near the nest (Sutherland 1963, Gramza 1967). In urban habitats, finding a nest requires a special effort to check suitable rooftops. In natural habitats, the extensiveness of suitable open areas, the peregrinations of the adults, the effective camouflage of the eggs, the tendency of the incubating bird to sit tight on the eggs or young, and the fact that the young move within a few days of hatching all conspire to make nest-finding difficult.

The Ontario breeding distribution of the Common Nighthawk, as shown on the maps, conforms to that published elsewhere (Peck and James 1983, Speirs 1985, Godfrey 1986). Doubtless, most squares reporting the species host breeding pairs, even though only 11% of the records are of confirmed breeding. Unlike strictly nocturnal species, this bird can be active at any hour; consequently, its distribution map does not suffer as much from the problem of inadequate night coverage as do the maps of other truly nocturnal species. Gaps in the Ontario range generally occur where there are no suitable rooftops or where extensive agriculture or extensive heavy forest or ground cover preclude nesting. High densities in non-urban habitats occur where there are open sand plains, or, as is typical on the Canadian Shield, areas of extensive rock outcrops.

Abundance estimates indicate that 2 to 10 pairs of Common Nighthawks per square is the typical frequency, though frequencies of one pair and 11 to 100 pairs are common. Along the southern edge of the Canadian Shield and in the rocky habitats adjacent to Georgian Bay and Lake Huron, the species is particularly numerous. Because the species probably breeds throughout the north, a very significant proportion of Ontario's nesting population occurs there, even though densities may be low.-- *A. Mills*

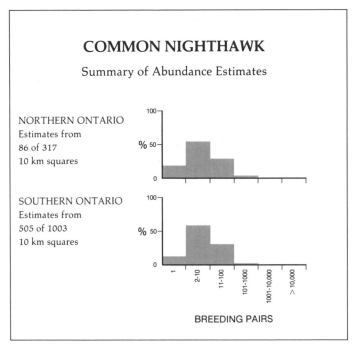

COMMON NIGHTHAWK

Summary of Abundance Estimates

NORTHERN ONTARIO
Estimates from
86 of 317
10 km squares

SOUTHERN ONTARIO
Estimates from
505 of 1003
10 km squares

%

100

50

%

100

50

1   2-10   11-100   101-1000   1001-10,000   >10,000

BREEDING PAIRS

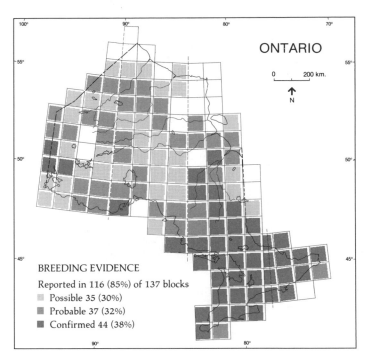

ONTARIO

0   200 km.

N

BREEDING EVIDENCE
Reported in 116 (85%) of 137 blocks
  Possible 35 (30%)
  Probable 37 (32%)
  Confirmed 44 (38%)

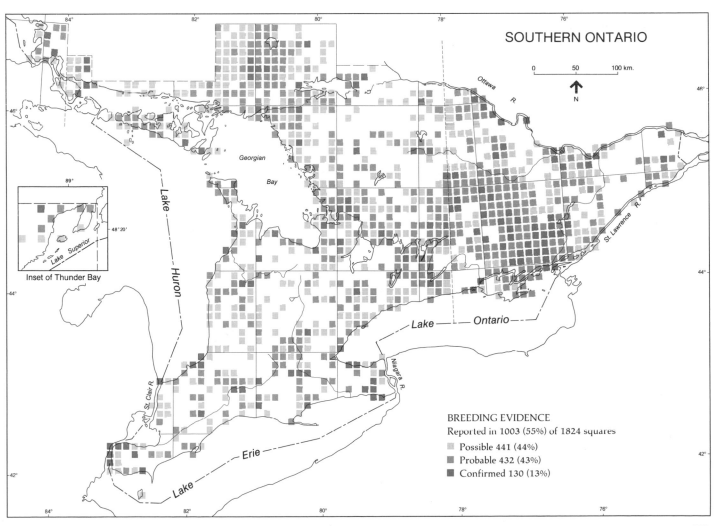

SOUTHERN ONTARIO

0   50   100 km.

N

Georgian
Bay

Lake
Huron

St. Clair R.

Lake   Erie

Niagara R.

Lake   Ontario

Ottawa R.

St. Lawrence R.

89°
48°20'
Lake   Superior
Inset of Thunder Bay

BREEDING EVIDENCE
Reported in 1003 (55%) of 1824 squares
  Possible 441 (44%)
  Probable 432 (43%)
  Confirmed 130 (13%)

CHUCK-WILL'S-WIDOW
Engoulevent de Caroline
Caprimulgus carolinensis

R. TUCKERMAN 1986

Historically associated with the American southeast, this large member of the 'goatsucker' family is a recent addition to Ontario's list of breeding birds. Its traditional range extended from the Gulf coast north to southern Ohio and from the Atlantic seaboard west to Kansas (Godfrey 1966). Through the 1970's, it extended its range northward to include, among others, a few localities in extreme southern Ontario. Summering birds, presumably nesting, have been found in Minnesota, Wisconsin, Michigan, and New England (reported variously in *American Birds)*, and the species has also consolidated its range and increased in numbers elsewhere. It is a nocturnal bird of either deciduous or pine woodlands of the Carolinian Forest zone of eastern North America. In its limited Ontario range, it occupies these same habitats. The provincial population is probably only a few pairs at present.

Except for a May 1906 specimen collected at Point Pelee and now in the Royal Ontario Museum (Fleming 1906), the Chuck-will's-widow was not reported in Ontario until May 1964, again at Point Pelee Nat. Park. Other isolated May reports come from Rondeau Prov. Park in 1970 and Point Pelee in 1971. In the spring and early summer of 1975, 2 birds were found regularly at Point Pelee, and breeding was suspected; another was at Pelee Island (Goodwin 1975). In the same year, the species first nested in New York State (Kane and Buckley 1975). Birds, sometimes more than one, were heard annually at Point Pelee through the remainder of the 1970's, and Rondeau Prov. Park has had at least one singing bird every year since 1976; 7 were found there in the summer of 1982. In the Long Point area, breeding has been suspected since 1978, especially at the mature pine plantation at St. Williams. Despite these many recent records from the north shore of Lake Erie, only one instance of confirmed

breeding has been documented, when a nest with 2 eggs was found at Point Pelee in June 1977 (Peck and James 1983).

Ontario records of the Chuck-will's-widow have not been confined to Pelee, Rondeau, and Long Point. One found in late May 1982 at Kitchener (not June as in Weir 1982) was presumed to be a migrant that had overshot its destination, as were one at Wheatley Prov. Park in May of 1979 or 1980 and a roosting female that was found in Kingston in mid-May 1984. Prince Edward Point on Lake Ontario hosted single calling birds in late May of 1976 and 1977, and 2 in late May 1978. They were not looked for later in the summer, though the 1976 bird was present at least into the first week of June (Sprague and Weir 1984). None was found in this area during the atlas period. A single male found at Perth Road (25 km north of Kingston) in June 1982 was at the only site away from Lake Erie shown on the atlas map.

The map is probably an accurate representation of the breeding range of the Chuck-will's-widow in the province. The song is loud and very distinctive and, further drawing attention to it, is uttered only at night. These features make overlooking it unlikely, provided that field workers covered potential areas under suitable conditions of twilight or moonlit night, when the species is most vocal. The probability of finding one in a square investigated only by day is low, however, since the cryptic plumage and diurnal inactivity render it very inconspicuous. A few birds may have been missed, therefore, but more extensive night coverage probably would not have resulted in a dramatically different documentation of the extent of the Chuck-will's-widow breeding range in Ontario.

No breeding was confirmed during the atlas period, though it almost certainly occurred at Rondeau and St. Williams, and perhaps elsewhere as well. Incubating birds are well camouflaged and are reluctant to leave the nest; one must almost step on the bird in order to flush it. In addition, territories are large, and activity at the nest is confined to the dark hours.

The Chuck-will's-widow is a recent arrival and has only a tenuous foothold as a breeder in the province. As with some other range extensions in the past, the northern limit may retreat again. Alternatively, the colonization of the past decade may be the incipient stages of a more dramatic extension to be realized in the future. The pure stands of pine being planted in southwestern Ontario and the continued existence of residual Carolinian forests may be a deciding factor. In the meantime, it remains a very rare bird, with probably fewer than 10 pairs breeding in the province.-- *A. Mills*

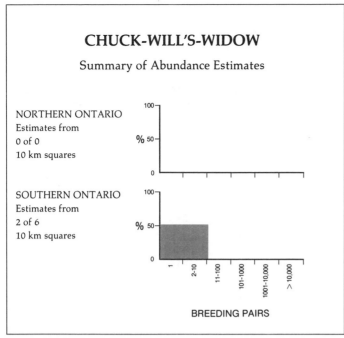

CHUCK-WILL'S-WIDOW

Summary of Abundance Estimates

NORTHERN ONTARIO
Estimates from
0 of 0
10 km squares

SOUTHERN ONTARIO
Estimates from
2 of 6
10 km squares

BREEDING PAIRS

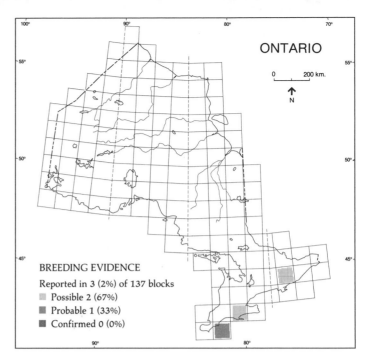

ONTARIO

0   200 km.

N

BREEDING EVIDENCE
Reported in 3 (2%) of 137 blocks
Possible 2 (67%)
Probable 1 (33%)
Confirmed 0 (0%)

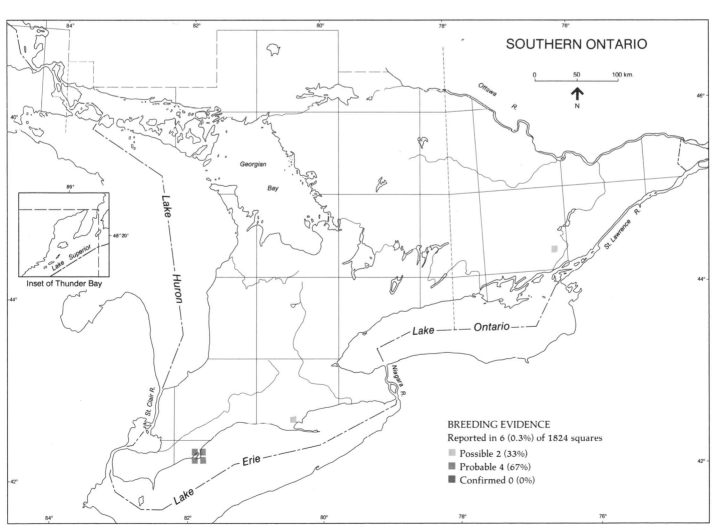

SOUTHERN ONTARIO

0   50   100 km.

N

Inset of Thunder Bay

BREEDING EVIDENCE
Reported in 6 (0.3%) of 1824 squares
Possible 2 (33%)
Probable 4 (67%)
Confirmed 0 (0%)

WHIP-POOR-WILL
Engoulevent bois-pourri
Caprimulgus vociferus

R. TUCKERMAN 1986

The loud song of the nocturnal Whip-poor-will, an incessant repetition of its name, is a familiar sound in some parts of the province. Most people associate it with vacation spots, with summer nights, and with the past. The species breeds from Nova Scotia and Saskatchewan to Honduras; northern birds retreat to the southern US and beyond for the winter (Godfrey 1986). The Whip-poor-will reaches its northern limit in the province. Summer individuals have been found as far north as Kenora and Kapuskasing in the past (Peck and James 1983), though the bulk of the provincial population has always been found south of Sudbury and North Bay. Where its favourite habitat of open woodlands with frequent clearings predominates, it can be quite common. For nesting, it prefers sites with shaded leaf litter (or pine needles) at wood edges or in forest clearings without any herbaceous growth.

The limits of the present range of the Whip-poor-will do not deviate from the limits of its historical range. Macoun and Macoun (1909) reported that it was common from Georgian Bay to Lake Erie around the turn of the century; a glance at the map reveals that few inhabit most of this area today. Over the course of this century, the Whip-poor-will has disappeared from many of its previous breeding localities including for example, 'cottage country' (Mills 1981) and Wellington Co (Brewer 1977). Habitat changes due to extensive agriculture in some areas and due to the maturation of the forest and better fire control in others are usually implicated in these disappearances and declines. The Whip-poor-will was placed on the continent's Blue List of declining, threatened, or vulnerable species for 2 of the List's first 10 years (Tate 1981), based on declines reported in many places in eastern North America.

This 'goatsucker' does not present serious identification difficulties. Being completely inactive by day and having remarkable camouflage, the Whip-poor-will is usually seen only if an observer stumbles upon one. The Chuck-will's-widow is much larger and rarer, and the superficially similar Common Nighthawk can be distinguished by plumage, shape, and habits. The Whip-poor-will's most salient feature is its song, which is distinctive, penetrating, repeatedly delivered, and restricted to heavy twilight and night. Because it is essentially only a voice and because the quiet females are virtually undetectable, many of the breeding evidence criteria used for other species are not useful in determining the breeding status of this species. The most frequently reported category of breeding evidence is that of territorial behaviour (42%); to confirm breeding, in most cases the 'nest' has to be found. This is challenging because breeding territories are large, incubating birds are difficult to spot and are very reluctant to flush, no nest is built, and the adults feed the young only infrequently, and by regurgitation.

Despite possible omissions from some squares due to insufficient field work at night, the general distribution shown on the map is an accurate representation of the provincial breeding range. Most blocks north to Sault Ste Marie and Kirkland Lake have Whip-poor-wills, though the southwest and the northern perimeter are sparsely populated. Outlying records from northcentral and northwestern Ontario suggest that there is a low-density breeding population in the southern part of the Boreal Forest zone. Dense populations occur from the Thousand Islands along the southern margin of the Canadian Shield and up the Georgian Bay shore to the Sudbury area. These rocky areas produce broken forests (often of pine and oak with a lot of juniper) that are ideal habitat. South of the Canadian Shield, some sandy areas such as Pinery Prov. Park produce similar habitat. Other pockets of breeding Whip-poor-wills in the south are associated with extensive plantations of pine. The area from northern Muskoka to North Bay has relatively few Whip-poor-wills, probably because of the extensive hardwood forests in that area, which do not have suitable breeding habitat.

Eighty percent of the abundance estimates for the Whip-poor-will range from 2 to 100 pairs per square, while about 14% are estimates of just one pair per square. Some of the highest estimates (100 or more pairs per square) are from the Frontenac Axis of the Canadian Shield, the rocky area to the north of Belleville and Kingston. Clearly, in some regions of the province the species has healthy populations; in other regions, numbers appear to be relatively low or even at precarious levels.-- *A. Mills*

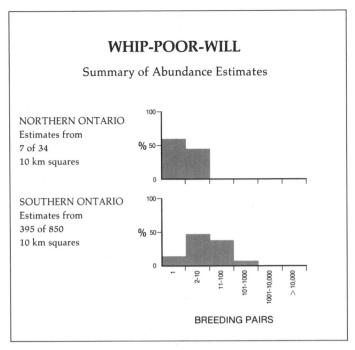

WHIP-POOR-WILL

Summary of Abundance Estimates

NORTHERN ONTARIO
Estimates from
7 of 34
10 km squares

SOUTHERN ONTARIO
Estimates from
395 of 850
10 km squares

BREEDING PAIRS

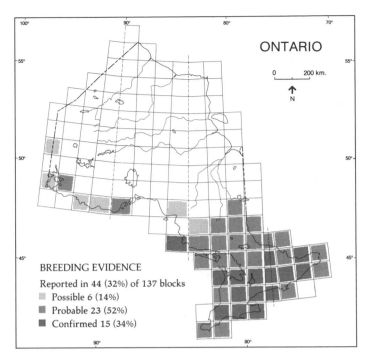

ONTARIO

0   200 km.

N

BREEDING EVIDENCE
Reported in 44 (32%) of 137 blocks

☐ Possible 6 (14%)
☐ Probable 23 (52%)
■ Confirmed 15 (34%)

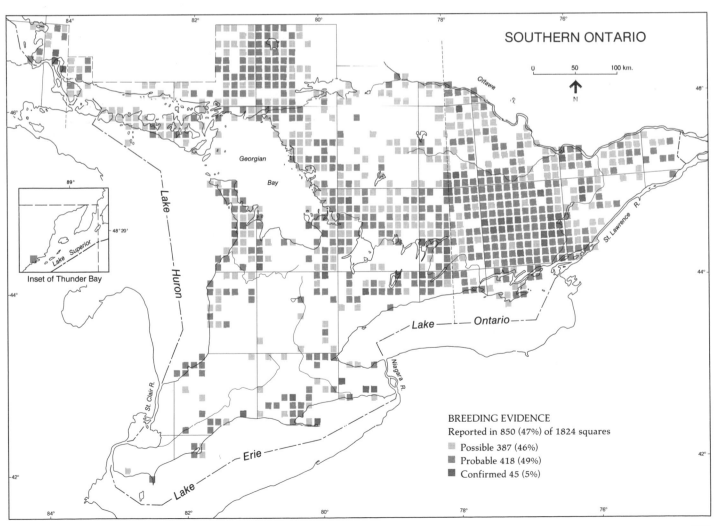

SOUTHERN ONTARIO

0   50   100 km.

N

Inset of Thunder Bay

Georgian
Bay

Lake
Huron

Lake   Superior

St. Clair R.

Niagara R.

Lake — Ontario

St. Lawrence R.

Ottawa

Lake — Erie

Lake

BREEDING EVIDENCE
Reported in 850 (47%) of 1824 squares

☐ Possible 387 (46%)
☐ Probable 418 (49%)
■ Confirmed 45 (5%)

CHIMNEY SWIFT
Martinet ramoneur
Chaetura pelagica

The only swift known to occur in Ontario is the Chimney Swift, so named because it has become adapted to human habitations, where it now uses the interior of chimneys almost exclusively as its nesting site. It is often seen, in small flocks, circling around urban skies at dusk on summer evenings.

The Chimney Swift is a species of the Americas, which is widespread in the summer across most of southeastern Canada and the eastern US. It winters in the upper part of the Amazon River basin and western Peru south to northern Chile (AOU 1983). Within the summer range, it is common in many urban centres, but less so in rural areas. This species originally nested in hollow trees, and undoubtedly still does so in the more remote, forested parts of Ontario, where no other suitable structures exist. Elsewhere, it makes use of the interior of derelict buildings and chimneys, both as nesting sites and as places to roost. During the daytime, Chimney Swifts remain almost continuously in flight, often at considerable heights, which is where they are most commonly seen feeding and displaying.

Historical records of the Chimney Swift in Ontario, while scarce, suggest that it formerly occupied much the same range as it does today, at least in southern Ontario. Macoun and Macoun (1909) noted that it was common at one location west of Lake Superior, and Speirs (1985) documents its occurrence at several northern Ontario locations from Cochrane and the Clay Belt in the east to Sioux Lookout and Redditt in the west between the 1930's and the 1960's. No change in its abundance has been documented.

The Chimney Swift is not a difficult species to find. When flying, it is fairly conspicuous against the sky, and often draws attention to its aerial acrobatics by giving its distinctive chattering call. Atlassers probably found a very high proportion of the breeding pairs in urban areas, especially those in squares in which atlassers lived. In the deep forests of central and northern Ontario, on the other hand, the species may have escaped detection more often. While the Chimney Swift is not a difficult bird to find, it is quite another matter to confirm that it is breeding in an area. The nest sites are usually inaccessible unless one has a means of crawling into chimneys, or at least peering into them from the top. One can usually hope to do no better than record the species as probably breeding, either by recording the frequent aerial displays of 2 or, more commonly, 3 birds, by observing pairs of swifts continuously flying around together, or by seeing the birds enter a chimney. Indeed, a high proportion (46%) of the southern Ontario squares are in the 'probable' category. It is safe to assume that the species actually does breed in most of those squares. In the north, where hollow trees are used as nesting sites, it is especially difficult to confirm that the species is breeding.

Although the map shows that there is no part of southern Ontario where the Chimney Swift does not breed, there is a surprisingly high number of squares where it was not reported when the atlas field work was being conducted. A close inspection of the map reveals that many of these squares are in well-covered rural areas with few sizable towns: Bruce, Grey, Dufferin, Renfrew, and Prince Edward Counties, and Manitoulin District, for example. In those areas, as well as on the map as a whole, one can almost pinpoint the urban centres by identifying the squares where the Chimney Swift was confirmed as breeding. Windsor, Sarnia, Goderich, London, Stratford, Orangeville, Bradford, Lindsay, Picton, Tobermory, Wiarton, and Bracebridge are all examples of urban centres that stand out from their rural surroundings by reason of a higher level of breeding evidence for the Chimney Swift. At the northern fringe of its range, Sudbury, Kirkland Lake, Thunder Bay, and Rainy River are all urban centres where the species was reported as at least probably breeding. The most northerly record is of birds in the vicinity of the town of Pickle Lake. For reasons that are not clear, the most solid concentration of squares where the species was recorded is in the forested area between Algonquin Prov. Park and Georgian Bay, an area with very little urban development.

Except in the forest, it is probably easier to estimate the abundance of breeding Chimney Swifts than to do so for most other species. Potential nesting sites are limited in number and quite conspicuous. Only 6% of the available estimates are of more than 100 pairs per square, indicating that the Chimney Swift is generally uncommon where it occurs.-- *F.M. Helleiner*

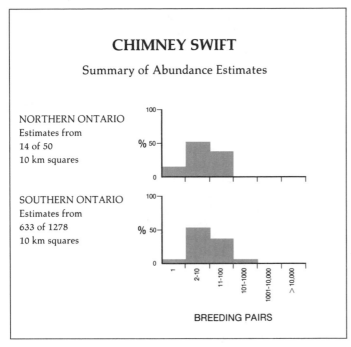

CHIMNEY SWIFT

Summary of Abundance Estimates

NORTHERN ONTARIO
Estimates from
14 of 50
10 km squares

SOUTHERN ONTARIO
Estimates from
633 of 1278
10 km squares

%

BREEDING PAIRS

1 2-10 11-100 101-1000 1001-10,000 >10,000

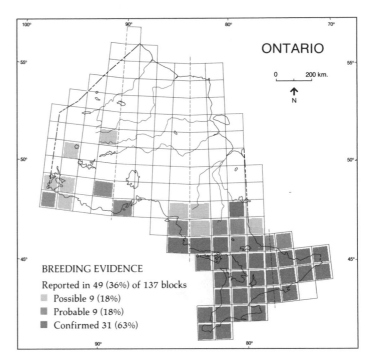

ONTARIO

0   200 km.

N

BREEDING EVIDENCE

Reported in 49 (36%) of 137 blocks

Possible 9 (18%)
Probable 9 (18%)
Confirmed 31 (63%)

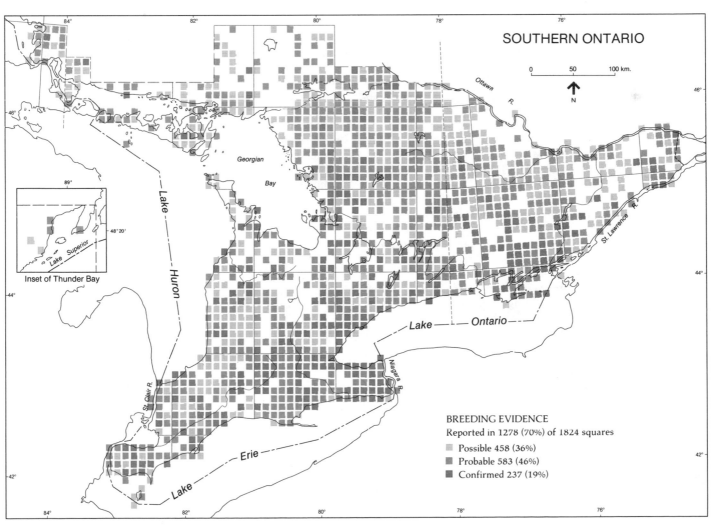

SOUTHERN ONTARIO

0   50   100 km.

N

Georgian
Bay

Lake
Huron

Lake Superior

48°20'

89°

Inset of Thunder Bay

St. Clair R.

Lake   Erie

Lake   Ontario

Niagara R.

Ottawa R.

St. Lawrence R.

BREEDING EVIDENCE

Reported in 1278 (70%) of 1824 squares

Possible 458 (36%)
Probable 583 (46%)
Confirmed 237 (19%)

RUBY-THROATED HUMMINGBIRD
Colibri à gorge rubis
Archilochus colubris

The smallest of northeastern birds, the Ruby-throated Hummingbird graces woodlands and suburban and rural gardens where flowers or feeders provide it with nectar. It breeds in eastern North America from the Gulf States to southern Canada, north in Ontario to Red Lake in the northwest and to Lake Abitibi in the northeast. It winters from the extreme southern US to Central America.

The Ruby-throated Hummingbird exploits a variety of habitats. It favours mixed woodland margins for nesting, though breeding territories may include dense forests of any type, wetlands, cultivated lands, or manicured grounds. Its northern distribution may be affected by its commensal relationship with the Yellow-bellied Sapsucker (Miller and Nero 1983), whose drilled holes in tree trunks provide the Hummingbird with a source of sap and some associated insect food.

A Ruby-throated Hummingbird can minimize energy-consuming foraging ventures by flying directly to dependable sucrose sources (Southwick and Southwick 1980). Careful observation of the female might therefore lead an observer to a sapsucker drill or to a nest. Nests are small, well camouflaged, and difficult to find; only 9% of all records involve nests. The male is relatively easy to find because of its conspicuous courtship display and vigorous defence of its territory. The male might also be located as it sits on an open perch near a favourite feeding site, but it takes no part in nesting activities. Hummingbird breeding was confirmed in only 18% of the squares in which the species was found, an indication of the difficulty in finding the highest categories of breeding evidence; fledged young, which often accompany their mothers to feeders, provided the most common means by which breeding was confirmed.

The atlas map of Ontario fully reflects the previously published breeding range. In northwestern Ontario, it indicates a range that extends slightly beyond Lac Seul, the northern limit stated by Peck and James (1983). The map shows that the species is found infrequently north of the Great Lakes-St. Lawrence Forest zone. It occurs more frequently in some settled areas of the southern Boreal Forest, such as at Cochrane and Red Lake. Previous reports from as far north as Sandy Lake and Attawapiskat (James *et al.* 1976), coupled with the more extensive distribution of its associate, the Yellow-bellied Sapsucker, lend support to speculation that the Ruby-throated Hummingbird breeds in low numbers beyond the mapped range. One bird was observed at Winisk, on the Hudson Bay coast, on 26 June 1983, but it was judged to be a vagrant, rather than a potentially breeding bird.

The Ruby-throated Hummingbird is rare in the extreme southwest of the province, probably because of a scarcity of wooded areas. Throughout the rest of southern Ontario, even in the hinterland of Algonquin Prov. Park and other uninhabited areas, the species is fairly evenly distributed. The concentration of confirmed breeding records along the southern margin of the Canadian Shield likely reflects a relatively high population density there. It is difficult to say whether this is a natural phenomenon or a result of the proliferation of hummingbird feeders at cottages and homes in this area. It seems likely that the species has benefitted from the opening up of the continuous forest for roads, power-line corridors, cottages, and towns. Certainly, nests are most easily found in habitat such as that created around cottages, where much underbrush has been removed such that nests in the lower tree branches are easily visible.

The Ruby-throated Hummingbird is uncommon in northern Ontario and fairly common through its southern Ontario range. In highly agricultural localities, it may be rare as a breeding bird. BBS data (Speirs 1985) show that the lowest population levels occur in the southeastern agricultural areas, in the extreme southwest, and between Brantford and Georgian Bay. Atlas records are also most thinly distributed in these areas.-- *M. Biro*

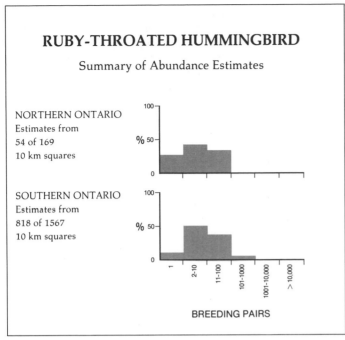

RUBY-THROATED HUMMINGBIRD

Summary of Abundance Estimates

NORTHERN ONTARIO
Estimates from
54 of 169
10 km squares

SOUTHERN ONTARIO
Estimates from
818 of 1567
10 km squares

BREEDING PAIRS

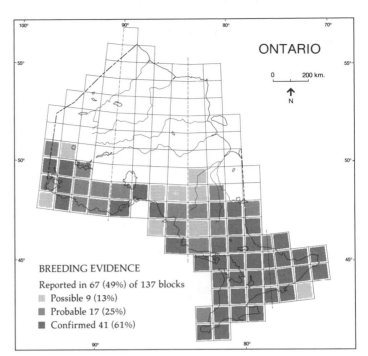

ONTARIO

0   200 km.

N

BREEDING EVIDENCE
Reported in 67 (49%) of 137 blocks
Possible 9 (13%)
Probable 17 (25%)
Confirmed 41 (61%)

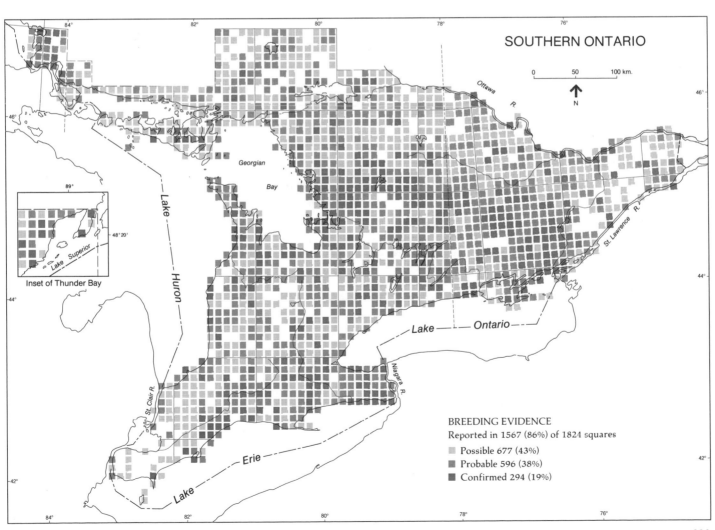

SOUTHERN ONTARIO

0   50   100 km.

N

Inset of Thunder Bay

Georgian Bay

Lake Huron

Lake Superior

Ottawa R.

St. Lawrence R.

Lake Ontario

Niagara R.

St. Clair R.

Lake Erie

BREEDING EVIDENCE
Reported in 1567 (86%) of 1824 squares
Possible 677 (43%)
Probable 596 (38%)
Confirmed 294 (19%)

BELTED KINGFISHER
Martin-pêcheur d'Amérique
Ceryle alcyon

The Belted Kingfisher, the only member of its family in Ontario, inhabits much of North America in the breeding season from central Alaska and northern Quebec south to the southern US. Small numbers winter along Ontario streams that have open water, but many move south, and then the species may be found as far south as the northern part of South America. It is widespread in the province, wherever it can excavate its burrows in sand, clay, or gravelly banks near a body of water with suitable fishing (Bent 1940). Most typically in Ontario, good habitat consists of eroded stream or river banks and lakeshore bluffs, but artificial sites such as gravel pits and road cuts close to a food source are also extensively used for nesting.

McIlwraith (1886), Baillie and Harrington (1936), and Godfrey (1966), all suggest that the Belted Kingfisher has been common from southern Ontario to Moosonee for at least 100 years. More recently, Peck and James (1983), Godfrey (1986), and the data presented here indicate that it occurs north to the Hudson Bay coast. There is little reason to believe that there has been a range expansion in the north; the lack of previous field work in remote northern areas is the more likely explanation of this northward extension of the known range. It may be even more widespread now in the rail- and road-accessible north since the creation of more gravel pits and road cuts. Other factors may now be limiting the bird's range. Since turbid and polluted water has been shown to interfere with proper fishing, leading to the abandonment of nest sites (Laycock 1984), erosion and chemical pollution in the open farmland of southern Ontario may have made some previously suitable areas unusable. This could be the reason for the absence of the Belted Kingfisher in a number of squares in the extreme southwest of the province.

The Belted Kingfisher is a conspicuous bird. Its character-istic profile, flight pattern, and rattling call are all readily identified at a distance. However, finding the actual nest can require a bit of persistence. Territories may be large, and nest sites are sometimes distant from feeding areas, particularly when the birds nest in gravel pits (Davis 1982). Fortunately, the large oval burrow entrance itself is distinctive, usually in an exposed slope and having twin ruts made by the birds' feet (Bent 1940). Both parents are active at the nest site for several months, and therefore provide many opportunities for viewing (Davis 1982). Not surprisingly, 54% of confirmed breeding evidence involved observations at the nest hole. The most frequently reported level of breeding evidence, in 23% of the records, consisted of pairs observed in suitable nesting habitat during the breeding season. For this species, that indicates a strong likelihood that the birds are breeding, because the male appears on its territory and chooses a nest site about one month before the female arrives. The pair excludes all other kingfishers from their territory (Davis 1982).

Since the Belted Kingfisher is usually easy to find, especially during the nesting season, the map of its breeding distribution is probably accurate. The absence of records and the low levels of breeding evidence reported in much of the far north suggests a lower density of breeding birds in that area than elsewhere. Birds should have been readily located if they were present because the coverage of much of that area was by canoe, most often along rivers. A lack of eroded banks for nesting may be a factor in undisturbed areas on the northern part of the Canadian Shield, but is not a factor on most of the Hudson Bay Lowland (see Bank Swallow map), where large banks line most of the rivers.

Throughout Ontario, most atlassers who reported on abundance estimated 2 to 10 pairs in a square. Davis (1982) found that Belted Kingfishers were limited in abundance partly by the size of their nesting territory but also by the availability of suitable nest sites. About 28% of the possible nest sites in his study area were not open to other kingfishers because they were defended by territorial birds which nested at another site. The Belted Kingfisher remains a familiar bird of Ontario waterways, but changes in abundance and distribution may occur with changes in habitat.-- *P. Read*

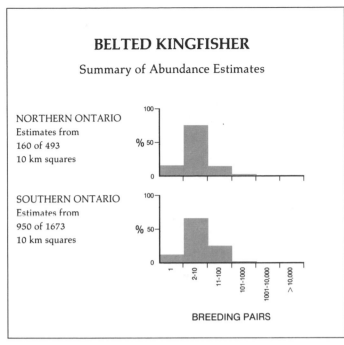

BELTED KINGFISHER

Summary of Abundance Estimates

NORTHERN ONTARIO
Estimates from
160 of 493
10 km squares

SOUTHERN ONTARIO
Estimates from
950 of 1673
10 km squares

BREEDING PAIRS

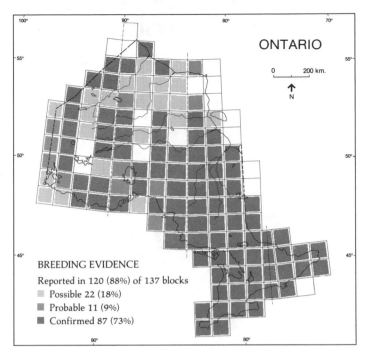

ONTARIO

0   200 km.

↑
N

BREEDING EVIDENCE

Reported in 120 (88%) of 137 blocks

Possible 22 (18%)

Probable 11 (9%)

Confirmed 87 (73%)

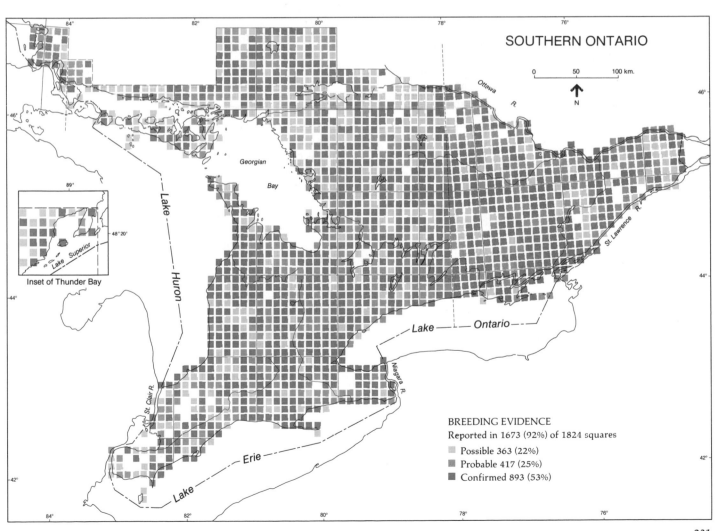

SOUTHERN ONTARIO

0   50   100 km.

↑
N

Georgian Bay

Lake Huron

Lake Superior

48°20'

89°

Inset of Thunder Bay

St. Clair R.

Lake Erie

Niagara R.

Lake — Ontario

Ottawa R.

St. Lawrence R.

BREEDING EVIDENCE

Reported in 1673 (92%) of 1824 squares

Possible 363 (22%)

Probable 417 (25%)

Confirmed 893 (53%)

RED-HEADED WOODPECKER
Pic à tête rouge
Melanerpes erythrocephalus

The Red-headed Woodpecker is a handsome bird, its striking plumage easily identified and well known by birders and many non-birders alike.

The Red-headed Woodpecker is limited to North America. It is widespread but uncommon in the central and eastern US. In Canada, this species is found in southern Saskatchewan, southern Manitoba, the Lake of the Woods area, southern Ontario, extreme southwestern Quebec, and southernmost New Brunswick. It prefers open deciduous woods and fields, pastures, city parks, river edges, and roadsides where scattered large trees occur (Bent 1939, Terres 1980, Peck and James 1983, Godfrey 1986). In Ontario, these areas are found throughout the Carolinian Forest zone and in agricultural areas in the Great Lakes-St. Lawrence Forest zone.

The Red-headed Woodpecker population has fluctuated over the years. Its numbers and range no doubt increased as the early settlers cleared the dense woodland once prevalent in southern Ontario (Macoun and Macoun 1909). The fencerows and woodlots in these new agricultural areas were ideal habitat. Since then, however, dead and dying trees and woodlots have been falling to the axe. This has reduced the available habitat, as well as the numbers of Red-headed Woodpeckers (Beardslee and Mitchell 1965, Peck and James 1983, Speirs 1985, Robbins *et al.* 1986). It has also been postulated that the "almost complete disappearance" of this species from Manitoulin Island during the 1950's was "intimately entwined with the use of pesticides" (Nicholson 1981). The introduction of the European Starling, whcih usurps its nest holes, and the increase in automobile traffic, which has caused many casualties along roadways, have been suggested as factors responsible for its general decline (Speirs 1985). The recent demise of the white elm resulting from the Dutch elm disease brought a minor resurgence of the Red-headed Woodpecker population by providing additional nesting sites and a supply of insects. The species apparently maintained its breeding range throughout the decades when its numbers were declining.

The Red-headed Woodpecker is a noisy, vocal species, and can be heard 'kwrring' from the upper levels of dead trees, throughout its breeding cycle. Once it arrives at its nesting ground in early to mid-May, it is highly visible, with its bold black and white back and wing pattern flashing as it goes from tree to tree. Because of this visibility and its preference for open areas and woodland edges, it is a relatively easy species to document for the atlas. Its recorded distribution is almost continuous on the map, and there was a high percentage of confirmed and probable breeding records. Several Regional Coordinators expressed the opinion that this species was underconfirmed in their regions, suggesting that it probably bred in most areas of the south where it was found. Others in the more northerly areas felt that some of the birds recorded there, although in suitable habitat, may have been extralimital wanderers.

South of the Canadian Shield, gaps in the recorded distribution of the Red-headed Woodpecker are evident in the areas of heavy urban development around Toronto and Hamilton, and in the most intensively agricultural area in Essex Co, presumably because of habitat destruction. As the amount of agricultural land decreases towards the edge of the Shield, the frequency of occurrence of this species becomes more scattered. Conversely, the Niagara Peninsula is evidently a preferred area, as shown by the concentration of confirmed breeding records there. In some respects, the atlas data agree very closely with the distribution of the Red-headed Woodpecker given in Peck and James (1983): away from the Ottawa area, the bird is not often encountered in far eastern Ontario; and it is not known to breed in the area north of North Bay or from Rainy Lake to Thunder Bay (Godfrey 1966, 1986).

The vast majority (91%) of estimates of Red-headed Woodpecker abundance are of no more than 10 pairs per square, implying that it is generally an uncommon bird in southern Ontario, albeit a widely distributed one.-- *P.A. Woodliffe*

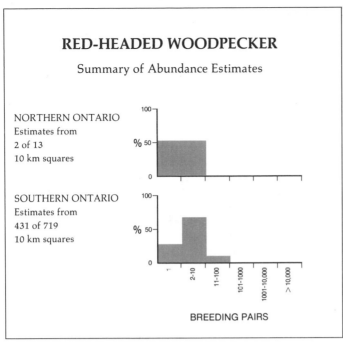

RED-HEADED WOODPECKER

Summary of Abundance Estimates

NORTHERN ONTARIO
Estimates from
2 of 13
10 km squares

SOUTHERN ONTARIO
Estimates from
431 of 719
10 km squares

BREEDING PAIRS

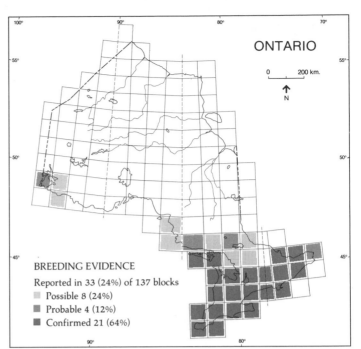

ONTARIO

0   200 km.

N

BREEDING EVIDENCE

Reported in 33 (24%) of 137 blocks

Possible 8 (24%)
Probable 4 (12%)
Confirmed 21 (64%)

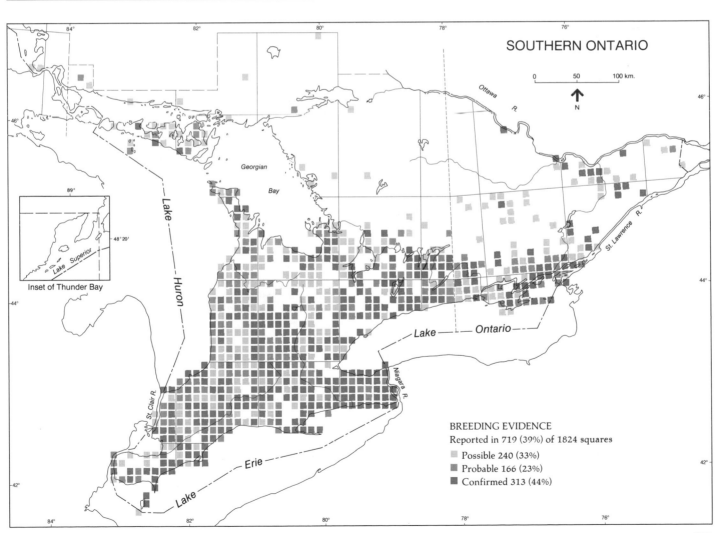

SOUTHERN ONTARIO

0   50   100 km.

N

Inset of Thunder Bay

BREEDING EVIDENCE

Reported in 719 (39%) of 1824 squares

Possible 240 (33%)
Probable 166 (23%)
Confirmed 313 (44%)

RED-BELLIED WOODPECKER
Pic à ventre roux
Melanerpes carolinus

J.S.

The Red-bellied Woodpecker is an attractive species, rarer and more restricted in range than any other Ontario woodpecker. The population has fluctuated greatly in the past 125 years and is currently experiencing an increase.

This exclusively North American species is a common resident in most of the southeastern US. Its breeding range in Ontario is restricted almost entirely to the Carolinian Forest zone, though small numbers were recorded during the atlas period as far east as eastern Lake Ontario and as far north as Manitoulin Island. Concentrations occur mainly in Kent, Lambton, Middlesex, and the southern part of Huron Counties, with more scattered records eastward to the Niagara River. In Ontario, this species is found mainly in mature deciduous woodland and occasionally in wooded residential areas (Peck and James 1983, Godfrey 1986). Its range appears to be expanding, but the bird is still considered rare in most areas.

The Red-bellied Woodpecker's numbers have fluctuated widely during its known history. It was "very abundant in the counties west of London" until about 1885, when its numbers declined (Macoun and Macoun 1909). Baillie and Harrington (1937) list birds north to Wellington Co and east to Halton Co. The bird commonly stays throughout the year in the same vicinity. Winter records of this non-migratory species have shown the numbers to be increasing. They were extremely rare until the early 1970's. Ten birds on a total of 5 Christmas Bird Counts, all but one in the extreme southwest, were recorded in 1973. Ten years later, a total of 64 birds were recorded on 10 counts (Rupert 1974, 1984). Only one bird was reported in Prince Edward Co prior to 1969, but there have been a dozen records there since that time, paralleling an increase of sightings in the Kingston area and in the adjacent area of New York State (Sprague and Weir 1984). A

pair of birds visited a nest hole at Prince Edward Point from 13 to 21 May 1983, but follow-up visits were not made to see if nesting was successful.

The Red-bellied Woodpecker is quite vocal while setting up territory. However, by mid-May, before many atlassers were in the field, it is considerably quieter, and, because it normally inhabits the denser, more mature, mosquito-infested wooded areas, it is not likely to be recorded as frequently as it actually occurs. Almost one-half of all records are of possible breeding: perhaps in the farthest reaches of its range only single birds may be present, as they pioneer new areas. It could be that, in the more typical southern parts of its range, observations were made at a point in the breeding cycle when one bird was incubating and only the off-nest mate was sighted. Thirty-two percent of the records are in the confirmed breeding category, and, of these, more than one-third were made by observing adults entering or exiting from a nest cavity.

As the maps indicate, the Red-bellied Woodpecker is primarily restricted to the Carolinian Forest region. Other records along the north shore of Lake Ontario and east of Lake Huron suggest that nesting may also be occurring outside the Carolinian zone. Within Lambton, Middlesex, and northern Kent Counties, where the records are most concentrated, gaps that occur between sightings are likely due to the quiet nature of the species during the time when most atlassing was done. There is a smaller concentration of records in the Grand River valley, extending almost as far north as Cambridge, and another along the northern edge of the Niagara Peninsula. The paucity of records elsewhere probably reflects smaller populations or an absence of birds. However, most of the Regional Coordinators who commented on the species felt that it was underrecorded and underconfirmed, for there is apparently suitable habitat throughout much of the Carolinian zone, with the exception of the intensively farmed areas of Essex and southern Kent Counties. There, woodlots of any sort are scarce, and those that do exist are in poor condition.

Of the squares with Red-bellied Woodpecker data, almost 60% have abundance estimates. Of these estimates, 49% are of one pair per square, 43% are of 2 to 10 pairs, and 9% are of 11 to 100 pairs. In the latter category, only the lowest end of the range would be likely. Except in a few favoured localities in southwestern Ontario, the Red-bellied Woodpecker remains a rare and irregular breeding bird.-- *P.A. Woodliffe*

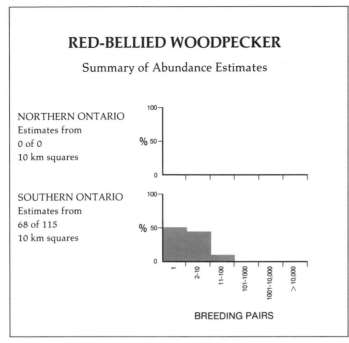

RED-BELLIED WOODPECKER

Summary of Abundance Estimates

NORTHERN ONTARIO
Estimates from
0 of 0
10 km squares

SOUTHERN ONTARIO
Estimates from
68 of 115
10 km squares

BREEDING PAIRS

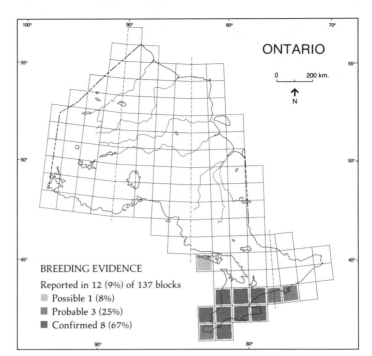

ONTARIO

0   200 km.

↑
N

BREEDING EVIDENCE
Reported in 12 (9%) of 137 blocks
▪ Possible 1 (8%)
▪ Probable 3 (25%)
▪ Confirmed 8 (67%)

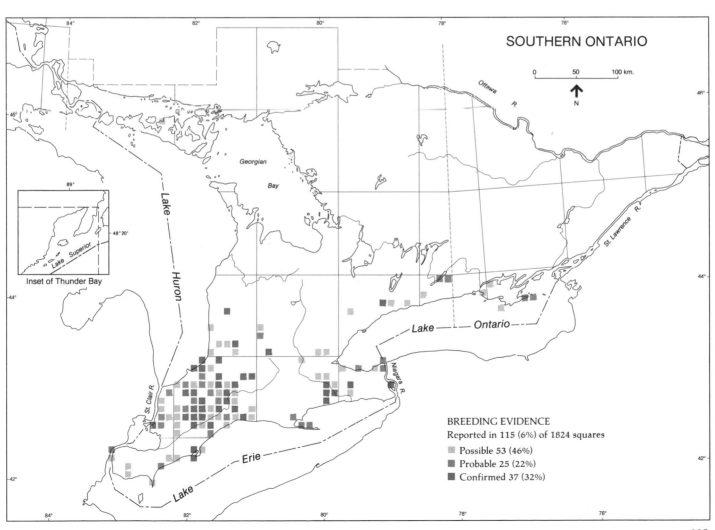

SOUTHERN ONTARIO

0   50   100 km.

↑
N

Georgian
Bay

Lake Huron

Lake Ontario

Ottawa R.

St. Lawrence R.

Niagara R.

St. Clair R.

Lake Erie

89°
48°20'
Lake Superior
Inset of Thunder Bay

BREEDING EVIDENCE
Reported in 115 (6%) of 1824 squares
▪ Possible 53 (46%)
▪ Probable 25 (22%)
▪ Confirmed 37 (32%)

YELLOW-BELLIED SAPSUCKER
Pic maculé
Sphyrapicus varius

There are four species of sapsucker in Canada which are known to hybridize and which have been the subject of considerable taxonomic study and debate (AOU 1985). The Yellow-bellied Sapsucker is the only one that occurs in eastern Canada.

This species breeds only in North America, from the Alleghenies in the southeastern states to southern Labrador and Newfoundland, and in almost all the deciduous and mixed forests of Canada east of the Rockies. For as long as records have been kept, it has been well established in Ontario, as far north as, but not including, the Hudson Bay Lowland. Its winter range extends from the central US to the West Indies and Central America (Godfrey 1986). The Yellow-bellied Sapsucker is most characteristic of the dry second-growth forests of southcentral Ontario, where plentiful nesting sites can be found. The preferred nest location is the dead heart of a living poplar tree (Philipp and Bowdish 1917), usually trembling aspen (Kilham 1971). The fungus *Fomes igniarius* causes heartwood to decay while the sapwood remains sound (Shigo and Kilham 1968). These conditions offer the Yellow-bellied Sapsucker an excellent nest-building situation. Adjacent white birches readily furnish sap, a staple in the bird's diet (Kilham 1964). Its breeding territory may contain several other species of trees, commonly yellow birch, eastern hemlock, red maple, and apple trees.

Historical information on the status of the Yellow-bellied Sapsucker in Ontario reads very much like the contemporary description. Macoun and Macoun (1909), for example, indicate that it "breeds plentifully in the counties of Leeds, Lanark and Renfrew," is "abundant in Algonquin Park," "rare" at Toronto and "not common" in the London area. According to Mills (1981), it is perhaps the most common

woodpecker during the summer in the Muskoka and Parry Sound districts.

The Yellow-bellied Sapsucker drills for sap in vertical or horizontal rows of small holes (about 0.5 cm in diameter) around the trunk or large limbs of a tree. These unmistakable holes, if still secreting sap, are a sign of its recent activity. Though it may not be immediately visible, the bird announces itself with piercing call notes or with its distinctive territorial or courtship drumming, on natural or artificial reverberating surfaces. Young Yellow-bellied Sapsuckers are also noisy, both in the nest and after fledging, and thus simplify the task of confirming breeding of this species. Nest trees are often used repeatedly over a period of years (Mousley 1916). Not surprisingly, therefore, nests were found in one-quarter of the squares for which there are atlas records. Eggs were seen in only 4 out of those 367 nests, many of the rest being too high to look into but not too high to hear or see the young.

The map clearly shows that the Yellow-bellied Sapsucker is almost totally absent as a breeding bird from extreme southwestern Ontario, the Niagara Peninsula, from the squares near Lake Ontario, and from parts of extreme southeastern Ontario. Although it does nest in other deciduous trees where aspen is either absent or not suffering from heart rot (Kilham 1971), these are sub-optimal conditions, occurring throughout Ontario's Carolinian Forest and in areas of intensive agriculture. In the Great Lakes-St. Lawrence Forest and in much of the Boreal Forest, the preferred host tree is ubiquitous, and so is the Yellow-bellied Sapsucker. That its range appears to be discontinuous in much of northern Ontario is probably an underrepresentation of its true status; at the time of year when most of the field work took place in the north, most of the bird's characteristic drumming activity and vocalization would have ceased. At the northern fringe of its range, much of the Hudson Bay Lowland may lack sufficiently large trees of the right species for nesting, although the map suggests that such trees are available in the southeastern part of the Lowland.

No other woodpecker is estimated at over 1,000 pairs per square as frequently as this one. Six of those 11 squares occur in a band across the centre of Haliburton Co. The Northern Flicker is the only woodpecker that exceeds this species in number of squares estimated at over 100 pairs. The highest concentration of Yellow-belllied Sapsuckers in the province extends across Muskoka and Haliburton. Estimates of over 100 pairs were made in about one-third of the squares in this core.-- *M. Biro*

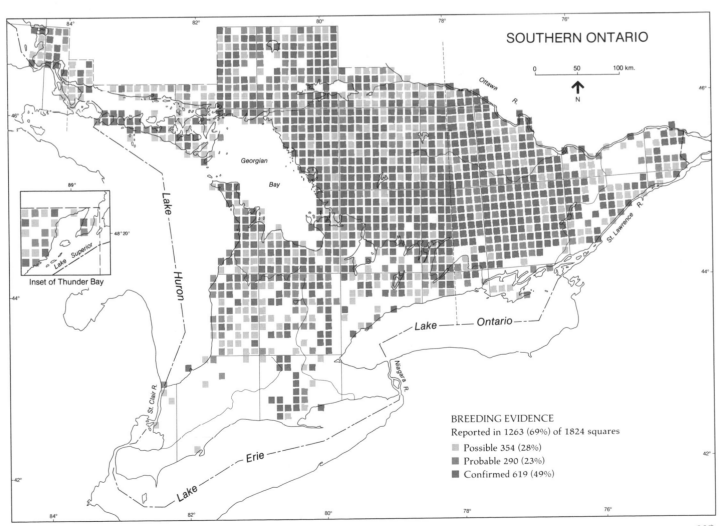

YELLOW-BELLIED SAPSUCKER

Summary of Abundance Estimates

NORTHERN ONTARIO
Estimates from
108 of 309
10 km squares

SOUTHERN ONTARIO
Estimates from
630 of 1263
10 km squares

%

BREEDING PAIRS

1 2-10 11-100 101-1000 1001-10,000 >10,000

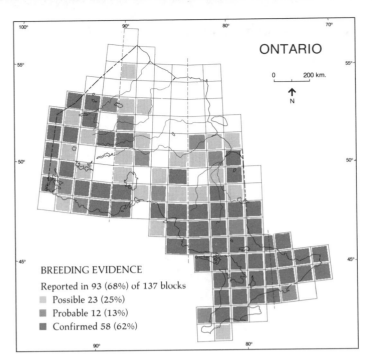

ONTARIO

0   200 km.

N

BREEDING EVIDENCE
Reported in 93 (68%) of 137 blocks

Possible 23 (25%)
Probable 12 (13%)
Confirmed 58 (62%)

SOUTHERN ONTARIO

0   50   100 km.

N

Lake Huron

Georgian Bay

Lake Superior

Inset of Thunder Bay

Ottawa R.

St. Lawrence R.

Lake Ontario

Niagara R.

St. Clair R.

Lake Erie

BREEDING EVIDENCE
Reported in 1263 (69%) of 1824 squares

Possible 354 (28%)
Probable 290 (23%)
Confirmed 619 (49%)

237

DOWNY WOODPECKER
Pic mineur
Picoides pubescens

J.S.

The Downy Woodpecker is the smallest member of the woodpecker family in the province. It is distinguished from the similar Hairy Woodpecker by its smaller size, stubby bill, and black spots along the white outer tail feathers. It is a regular visitor to bird feeders adjacent to wooded areas, especially in winter, when it is one of our most familiar birds.

The Downy Woodpecker occurs throughout the year in most of the woodlands of North America south of the treeline. In Canada, the northern limit of its range extends from the southern Yukon to southern James Bay and east to Newfoundland (Godfrey 1986). Within Ontario, it is generally distributed in the south, but the northern edge of its range is poorly defined. North of the Great Lakes-St. Lawrence Forest region, records are infrequent. It occurs in a wide variety of wooded habitats, from wilderness to urban ravines and parks, but prefers deciduous trees and tall shrubbery (Godfrey 1986); 78 of 81 woodland nests reported to the ONRS were in deciduous or mixed woods (Peck and James 1983).

Evidence suggests that the Downy Woodpecker has been a common resident in Ontario throughout historic time. Hadfield (1864) wrote, "It is so abundant that one can scarcely enter a wood without finding it." Baillie and Harrington (1937) described the Downy Woodpecker as "breeding commonly throughout its range," and regional bird summaries suggest that this remains true.

Though well known to Ontario naturalists because of its occurrence in urban and other disturbed areas and at feeders, the Downy Woodpecker is a secretive bird in the breeding season and can be surprisingly difficult to find at that time. Its nest is in a cavity excavated in a dead tree or a dead part of a living one. Its call note, its whinnying territorial call, and its drumming are heard only infrequently. On the other hand, the young are easy to detect by their vocalizations after they leave the nest, and 16% of the records were established by observing fledged young.

Atlas data confirm the breeding of the Downy Woodpecker in nearly every block to the northern limit of the Great Lakes-St. Lawrence Forest region, except between Thunder Bay and the Manitoba border, where most blocks show only probable breeding. Beyond that zone, where forest cover becomes increasingly coniferous, confirmation of breeding is found only sporadically, the majority of records coming from blocks with road or rail access. Confirmed breeding records occur as far north as Moosonee, Fort Albany, and Deer Lake (almost 200 km north of Red Lake). A few possible breeding records were obtained in the central portion of northern Ontario, as far north as Big Trout Lake. The distribution pattern of the records in the north indicates that, although the range of the Downy Woodpecker may extend nearly as far north as that shown by Godfrey (1986), it is certainly not as widespread or common in the Boreal Forest zone as it is farther south, probably because of its preference for deciduous woodlands. South of the Canadian Shield, the species is found in almost all squares, the few exceptions probably being deficient in woodlands. On the Shield, the absence of records in certain squares is probably due to its secretive nature during the breeding season, and, in some cases, difficulty of access.

Abundance estimates for the Downy Woodpecker indicate that it is usually a common bird in both southern Ontario and those parts of northern Ontario where it occurs. It is commoner than the Hairy Woodpecker in both parts of the province, but especially in the south, where 60% of estimates are of more than 10 pairs per square, compared with 46% in the case of the Hairy Woodpecker. Breeding bird plot studies in Durham Region have documented densities of 8 to 16 birds per 40 ha of woodland, and 2 birds per 40 ha in urban habitats (Tozer and Richards 1974). If extrapolated to a whole 10 km square, the woodland densities would produce values substantially higher than 99% of the estimates provided here. However, they are based on limited areas of habitat that are highly suited for the species, and there are few, if any, squares consisting entirely of such good habitat.-- *M. Parker*

DOWNY WOODPECKER

Summary of Abundance Estimates

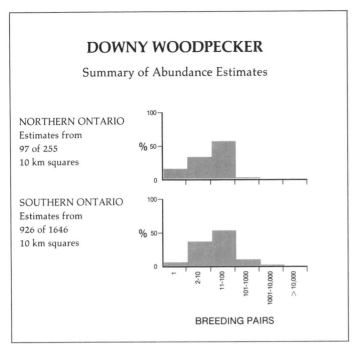

NORTHERN ONTARIO
Estimates from
97 of 255
10 km squares

SOUTHERN ONTARIO
Estimates from
926 of 1646
10 km squares

BREEDING PAIRS

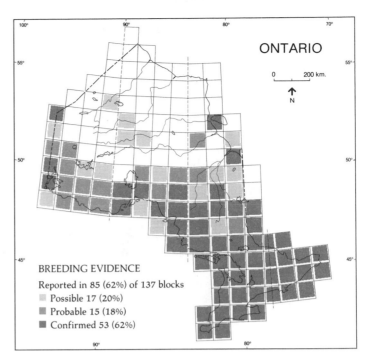

ONTARIO

0   200 km.

N

BREEDING EVIDENCE
Reported in 85 (62%) of 137 blocks
☐ Possible 17 (20%)
☐ Probable 15 (18%)
■ Confirmed 53 (62%)

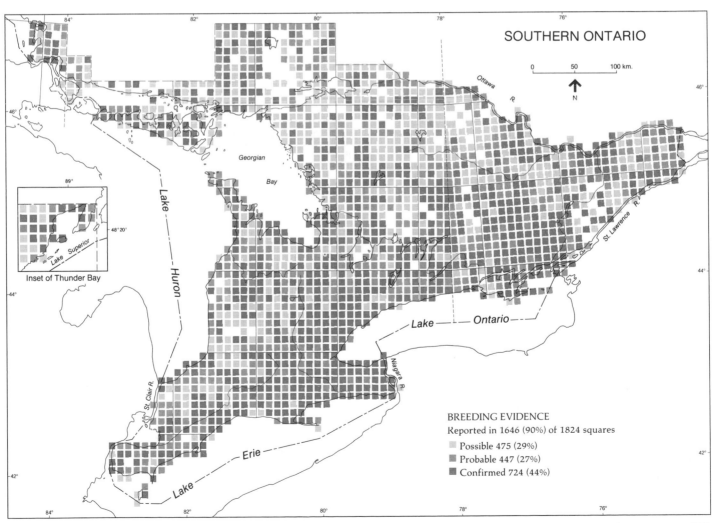

SOUTHERN ONTARIO

0   50   100 km.

N

Georgian Bay

Lake Huron

Lake Superior

Inset of Thunder Bay

48°20'
89°

St. Clair R.

Ottawa R.

St. Lawrence R.

Lake Ontario

Niagara R.

Lake Erie

BREEDING EVIDENCE
Reported in 1646 (90%) of 1824 squares
☐ Possible 475 (29%)
☐ Probable 447 (27%)
■ Confirmed 724 (44%)

HAIRY WOODPECKER
Pic chevelu
Picoides villosus

The Hairy Woodpecker is a familiar visitor to winter feeders in or near wooded areas, though it is not as common as the similar, but smaller, Downy Woodpecker.

The Hairy Woodpecker is a permanent resident throughout the woodlands of North America as far south as the extreme southern US and the Bahamas, and in the mountains to Panama. The northern edge of its range extends from the central parts of Alaska and the Yukon to central Quebec and Newfoundland (Godfrey 1986). It occurs across most of southern Ontario and north almost to Hudson Bay, which is somewhat farther north than the northern range limit of the Downy Woodpecker. The Hairy Woodpecker usually nests in deciduous, or preferably mixed, woodlands, but has also been reported nesting in campgrounds, in orchards, and along roadsides (Peck and James 1983).

There is little to indicate any change in the Ontario range of the Hairy Woodpecker as long as records have been kept. However, the extensive clearing of forests for settlement and agriculture has almost certainly reduced numbers in some areas, such as the extreme southwest (McIlwraith 1894), where few atlas records were obtained. There is some evidence that, in earlier times, this species may have outnumbered the Downy Woodpecker in southern Ontario, at least during the winter (Macoun and Macoun 1909), but this is certainly not the case in most areas today. If their relative abundance has indeed changed, it may be simply the result of an increase in the Downy Woodpecker, rather than a decrease in the Hairy.

The Hairy Woodpecker is usually easy to find and to identify. Both its call note and its territorial call are loud, piercing, and distinctive, and the tapping sound that it makes while feeding can be heard for a considerable distance. During the nesting period, however, the bird tends to be secretive. The nest is a cavity excavated into the trunk of a tree, usually a living deciduous tree, and usually at a height of 4.5 to 10.5 m (Peck and James 1983). The Hairy Woodpecker nests earlier in the season than the other common woodpeckers, the eggs usually having hatched before the end of May. Because much atlassing activity took place after that date, it is not surprising that only 5 records are of nests containing eggs, none of them in northern Ontario, where most atlassing took place from June onward. The number of nests with eggs is also low because of the difficulty of actually seeing the contents of the cavity nest. After the eggs hatch, the young birds become exceedingly noisy, uttering an uninterrupted squawking both in the nest and after fledging. Inaccessible nests and nests with young birds account for 17% of the records, and an additional 11% are of fledged young.

The maps show few records of the Hairy Woodpecker in the far northern section of the province, particularly near the Hudson Bay coast. Trees on the northern Lowland are mostly thinner than 12.5 cm, which is the minimum bole diameter recorded for a nest location in the ONRS data (Peck and James 1983). South of the Hudson Bay Lowland, the species occurred in almost every block and was confirmed as breeding in more than half of those in which it was reported. In southern Ontario, the Hairy Woodpecker is widely distributed, except for its absence from many of the squares in Essex Co and part of Metropolitan Toronto, both of which lack extensive wooded areas. Many Regional Coordinators suggested that the species is even more widespread than shown on the map, and that it likely breeds in most, if not all, of the squares where only possible or probable breeding were reported. This seems reasonable for a non-migratory species such as this one.

Abundance estimates indicate that the Hairy Woodpecker is fairly common in both southern and northern Ontario.-- *M. Parker*

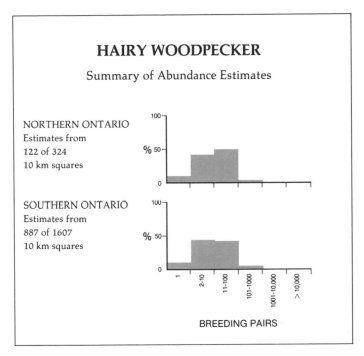

HAIRY WOODPECKER

Summary of Abundance Estimates

NORTHERN ONTARIO
Estimates from
122 of 324
10 km squares

SOUTHERN ONTARIO
Estimates from
887 of 1607
10 km squares

%

%

100

50

0

100

50

0

1 2-10 11-100 101-1000 1001-10,000 >10,000

BREEDING PAIRS

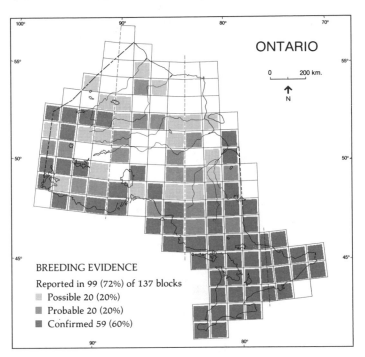

ONTARIO

0   200 km.

N

BREEDING EVIDENCE
Reported in 99 (72%) of 137 blocks
Possible 20 (20%)
Probable 20 (20%)
Confirmed 59 (60%)

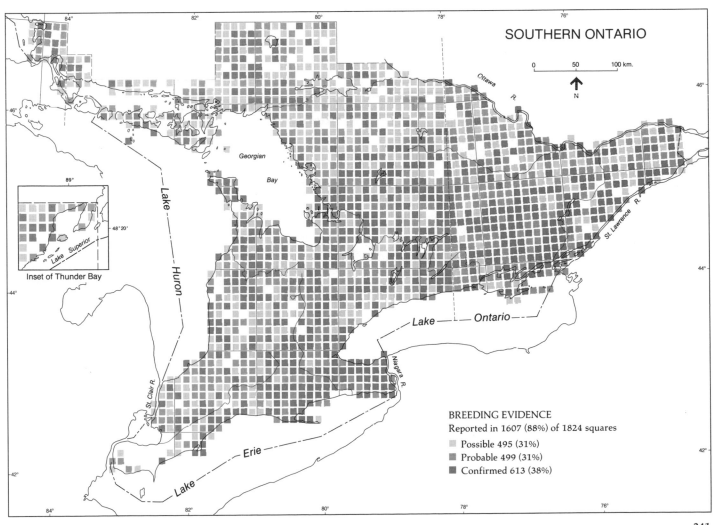

SOUTHERN ONTARIO

0   50   100 km.

N

Georgian Bay

Lake Superior

Inset of Thunder Bay

Lake Huron

St. Clair R.

Lake Erie

Niagara R.

Lake Ontario

Ottawa R.

St. Lawrence R.

BREEDING EVIDENCE
Reported in 1607 (88%) of 1824 squares
Possible 495 (31%)
Probable 499 (31%)
Confirmed 613 (38%)

THREE-TOED WOODPECKER
Pic tridactyle
Picoides tridactylus

J.S.

In many respects, the Three-toed Woodpecker is similar to the closely related Black-backed Woodpecker. Although their geographic ranges are similar and overlap extensively across the Boreal Forest of North America, that of the Three-toed Woodpecker is slightly more northern. Its southern range limit is not as far south in eastern Canada but farther south in the mountains of the western US. Unlike the Black-backed Woodpecker, it has a circumpolar distribution, including the Boreal Forest of Europe and Asia. In Ontario, the species is an uncommon resident in the forested northerly portions of the province, and is less common than the Black-backed Woodpecker. The preferred habitat includes primarily coniferous woodlands tending towards the wet conditions found in bogs (Peck and James 1983). The species occasionally moves southward and has appeared as a notable bird in several Christmas Bird Counts in southern Ontario.

One of the earliest records of the species in Ontario is that of a bird seen at Pic, Lake Superior in 1848 (Baillie and Hope 1943). Other sight records are scattered through the literature, but there is little to indicate significant changes in distribution or population. Widespread southerly movements of apparently non-breeding Three-toed Woodpeckers occurred in 1956 to 1957 (West and Speirs 1959) and 1963 to 1966 (Speirs 1985). It has been suggested that these irruptions were related to the development of the Dutch elm disease, which provided an increased resource of dead trees and bark beetles (Devitt 1967, Sadler 1983).

The Three-toed Woodpecker has the same general dark, inconspicuous colouring as the Black-backed species. It differs by having a white back, heavily barred with black, that blends well with the bark of its host trees. The bird is not particularly vocal and is more often heard drumming than uttering its own call. It is quite tame, showing little fear of humans. Most (61%) of the records presented here involve sightings of single birds in suitable breeding habitat. Nests are not often found. Prior to 1983, only 10 nests had ever been reported in Ontario (Peck and James 1983); nests were found in 8 squares during atlas field work, including 6 records of nests with young. The boggy habitat in which this species nests is less accessible than the edges of waterways or burns frequently chosen by the Black-backed Woodpecker, and the Three-toed Woodpecker is undoubtedly reported relatively less frequently because of this.

Atlas data indicate a patchy distribution across much of the Boreal Forest region, and a few scattered records north and south of there. It is almost exclusively a northern Ontario bird. Within the Boreal Forest, 2 groupings of records are evident on the map; one group lies roughly north of Pickle Lake and the other in the Sudbury, Algoma, and Cochrane Districts. These clusters are most likely artefacts of coverage; this woodpecker likely breeds in all blocks in the Boreal Forest south of the Hudson Bay Lowland, but probably less frequently and uniformly in the Lowland itself, where there is a scarcity of trees large enough for nesting in some areas. There is a confirmed record of a nest with young at Kiruna Lake in the Sutton Hills region of the northeastern Lowland, and two other nests with young were reported at Aquatuk Lake in the same region in 1964 and 1980 (Peck and James 1983, ONRS). These records and those for the Black-backed Woodpecker from the same region may indicate that there are relatively dense populations of both species in and around the Sutton Hills. The most interesting records are those from the Sudbury District and the area north of Kingston, which extend the breeding range for the Three-toed Woodpecker considerably south of the range indicated by Godfrey (1986). The discovery of the Sudbury District birds, which may well have been there for some time, is a direct result of the systematic coverage undertaken for this atlas. The 1981 extension of the breeding range to Frontenac Co, where the birds raised young at Opinicon Lake in June and July 1981, and the 1984 observation of a summering female near Bon Echo Prov. Park in Lennox and Addington Co, while quite unprecedented, should not be altogether surprising. Although populations in this region may not be permanent, wandering birds would frequently invade the area in winter, there is suitable habitat available, and the species does breed farther south in the Adirondack Mountains of New York State.

Abundance estimates confirm that the Three-toed Woodpecker is generally uncommon, but suggest that, in favourable areas, it reaches densities comparable with those of the Black-backed Woodpecker. One observer estimated that, in the summer of 1983, there were more than 10 breeding pairs in each of 4 nearly contiguous squares 75 km northwest of Hearst. These squares are located along the Kenogami River and its tributaries near the former hamlet of Mammamattawa.-- *W.D. McIlveen*

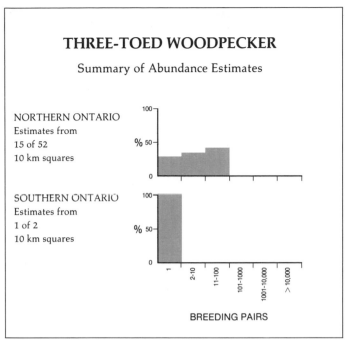

THREE-TOED WOODPECKER

Summary of Abundance Estimates

NORTHERN ONTARIO
Estimates from
15 of 52
10 km squares

SOUTHERN ONTARIO
Estimates from
1 of 2
10 km squares

BREEDING PAIRS

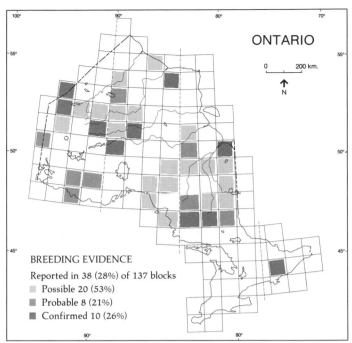

ONTARIO

0   200 km.

N

BREEDING EVIDENCE

Reported in 38 (28%) of 137 blocks

Possible 20 (53%)
Probable 8 (21%)
Confirmed 10 (26%)

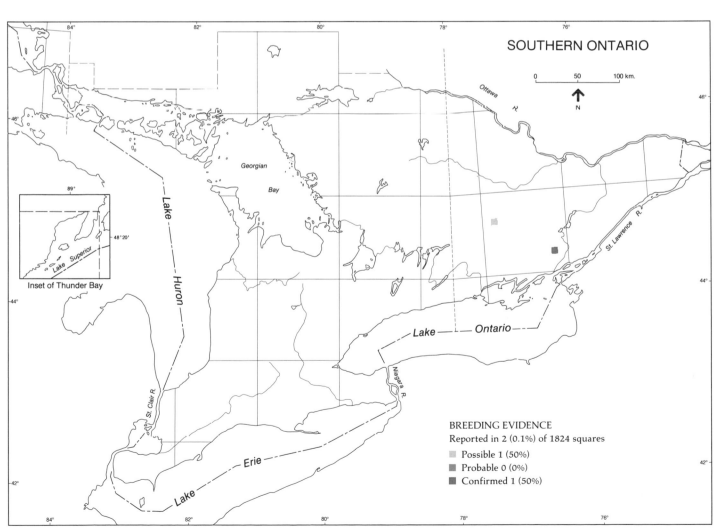

SOUTHERN ONTARIO

0   50   100 km.

N

Inset of Thunder Bay

BREEDING EVIDENCE

Reported in 2 (0.1%) of 1824 squares

Possible 1 (50%)
Probable 0 (0%)
Confirmed 1 (50%)

BLACK-BACKED WOODPECKER
Pic à dos noir
Picoides arcticus

The Black-backed Woodpecker is distributed across the Boreal Forest region of North America (Godfrey 1986). In Ontario, it is known to breed from the southern edge of the Canadian Shield to Hudson Bay, occurring throughout the Boreal Forest and south into the Great Lakes-St. Lawrence Forest region where there is suitable coniferous habitat. Small numbers, and larger numbers during occasional 'invasions', can be found south of the breeding range in winter. The species shows a preference for burned-over coniferous sites where trees are left standing. Such trees offer an excellent opportunity for colonization of the woody material by burrowing insects, which in turn attract the woodpeckers. The species is not particularly abundant, it frequents less populated areas, and its habitat is less than attractive for human activities. Hence, few people encounter the species unless they are looking for it.

There is nothing to suggest that there has been much permanent change in the breeding range or population of the Black-backed Woodpecker during historical times, although larger than normal numbers were found in southern Ontario during the 1950's and 1960's, albeit not generally in the breeding season (West and Speirs 1959, Quilliam 1965, Devitt 1967, Sadler 1983). The die-off of elms caused by Dutch elm disease has been implicated in that series of irruptions. Human impact on the status of this species has probably been minimal overall, although forestry practices likely have removed some stands of potential habitat. The impact of additional forest fires caused by human activities has been offset by the recent practice of fighting such fires, thereby reducing the size, though not the number, of burned-over areas. A shortage of suitable habitat does not appear to be a problem for the Black-backed Woodpecker.

The sombre colouration of this species, except for a small yellow crown patch on the male, does little to attract the attention of observers and offers excellent camouflage against the dark surfaces of burned trees. This woodpecker is not particularly vocal, although it does utter a sharp call note. Movement of the bird from tree to tree is often the only means by which it is discovered: 60% of all records are of possible breeding, a high percentage which is usually indicative of an uncommon species, or one whose breeding is difficult to confirm. The presence of this woodpecker is frequently indicated by its feeding activity, which produces patches of freshly exposed inner bark on the trunks of dead or unhealthy trees.

The Black-backed Woodpecker is very tame, allowing humans to approach quite closely as it continues with its activities. Fleming (1901) reported a colony of 6 or 7 nests in the Parry Sound District, but most nests appear to be solitary (Bent 1939). Conifers near open areas or water such as bogs, beaver ponds, or lake shores are preferred nest sites, but deciduous trees and utility poles are sometimes used (Peck and James 1983). Both parents tend the nestlings, which, like most woodpecker young, make a continuous chatter, which aids in locating the nest. Though nests with young were found in 21 squares, a nest with eggs was reported from only one square.

The maps indicate that the Black-backed Woodpecker can be found over much of the province except southeastern and southwestern Ontario, where suitable breeding habitat is practically non-existent. They also show isolated records south of the Canadian Shield (on Manitoulin Island, in Presqu'ile Prov. Park, and near Hatherton, in Grey Co). Although these records suggest that breeding does occur south of the Shield, confirmed breeding was not established in any of these instances. Elsewhere in southern Ontario, most records are from the Algonquin Highlands. It is evident that the species breeds in most of the blocks from the southern edge of the Shield into northeastern Ontario, as well as in several blocks in northwestern Ontario. The species likely breeds in all blocks in northern Ontario south of the Hudson Bay Lowland. The small, scattered spruces and tamaracks of the muskeg on the Lowland may be too thin to support nest cavities, as trees with a trunk thickness of 19 to 30 cm at breast height are needed for nesting (Peck and James 1983). However, larger trees are often present on well-drained river banks, and these might be expected to maintain small populations in most Lowland blocks. Breeding has been previously reported on the northeastern Lowland at Hawley, Aquatuk, and Sutton Lakes (Schueler *et al.* 1974).

Abundance estimates indicate that the Black-backed Woodpecker is uncommon throughout much of its range in the province. It is probably more abundant in the north, where most of the higher estimates were made, and it is fairly common in certain areas of prime habitat, such as around Lake Temagami (M. Cadman pers. comm.).-- *W.D. McIlveen*

BLACK-BACKED WOODPECKER

Summary of Abundance Estimates

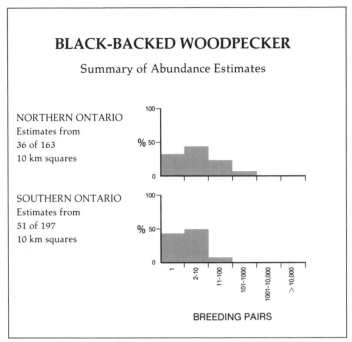

NORTHERN ONTARIO
Estimates from
36 of 163
10 km squares

SOUTHERN ONTARIO
Estimates from
51 of 197
10 km squares

% 50
% 50

BREEDING PAIRS

1 2-10 11-100 101-1000 1001-10,000 >10,000

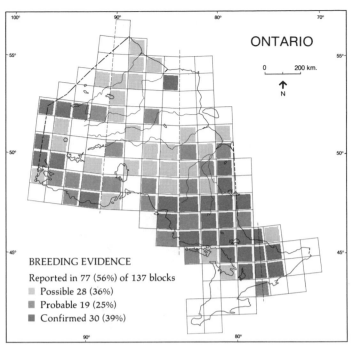

ONTARIO

0   200 km.

N

BREEDING EVIDENCE
Reported in 77 (56%) of 137 blocks
Possible 28 (36%)
Probable 19 (25%)
Confirmed 30 (39%)

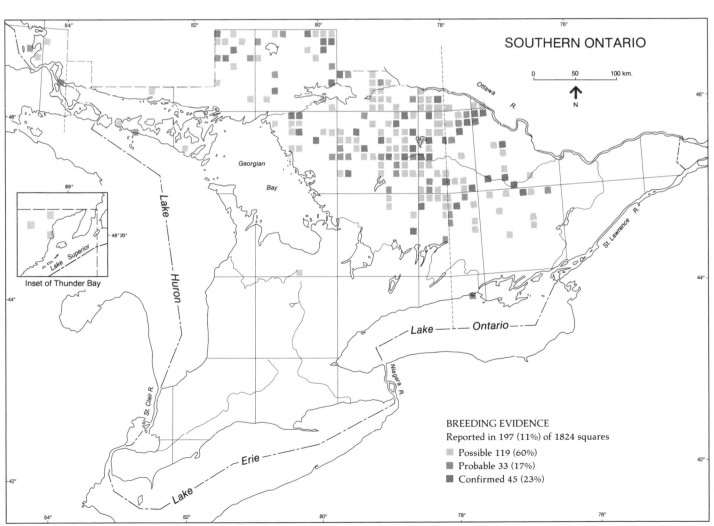

SOUTHERN ONTARIO

0   50   100 km.

N

Inset of Thunder Bay

BREEDING EVIDENCE
Reported in 197 (11%) of 1824 squares
Possible 119 (60%)
Probable 33 (17%)
Confirmed 45 (23%)

NORTHERN FLICKER
Pic flamboyant
Colaptes auratus

The Northern Flicker is a rather atypical woodpecker, both in its plumage and in its behaviour. A conspicuous white rump patch and its habit of ground-foraging, primarily for ants (Godfrey 1966), aid in making this species easy to identify in the field. With perhaps the exception of the Downy and Hairy Woodpeckers, whose familiarity is enhanced through regular visits to winter feeding stations, the Northern Flicker is the woodpecker best known to the general public.

The Northern Flicker breeds across almost all of North America, including Ontario, except in the most northern area beyond the tree-line. A few hardy individuals winter in the milder parts of the province, regularly appearing in Christmas Bird Counts (Speirs 1985). There are 3 closely-related races; however, only the 'yellow-shafted' form is prominent in this province. It breeds in a wide variety of habitats, including open deciduous, mixed, or coniferous forest edges, agricultural lands, beaver ponds, and other wetlands, provided that trees of sufficient size are available to accommodate the nest cavities.

There is no evidence that the distribution and abundance of the Northern Flicker have changed within Ontario in recent times. On one hand, the original clearing of land for agricultural settlement reduced the available woodland, but the creation of open woodland and forest edges may actually have increased the preferred habitat of the species. More recently, the expansion of residential areas into the rural environment may have caused local population and distribution changes, and competition from European Starlings probably has caused a significant reduction in the population in some areas.

Many aspects of the Northern Flicker's behaviour make it a conspicuous species. Territorial and mating displays which precede the nest-building period are sufficiently obvious and

vocal to attract the attention of observers (Stokes 1979): loud chases of stiffly posturing birds are a familiar springtime sight. Its nest holes are relatively easy to find as woodchips at the base of the tree betray the presence of the nest above. Furthermore, as the nests are often in open situations, they can be viewed from a considerable distance and the adults are consequently not inhibited in making visits to the nest: 21% of all records and 42% of confirmed breeding records are of adults entering nest holes that are too high to investigate their contents. Once on the nest or while feeding the young, an activity carried out by both parents (Bent 1939), the adults tend to be less audible, but the nestlings often reveal the nest by uttering a nearly constant crying sound. The noisy family groups that move about after the young leave the nest are perhaps the most conspicuous evidence of breeding.

The widespread distribution of the Northern Flicker in Ontario reflects its capacity to breed in a variety of habitats. There are records in virtually all squares in southern Ontario, except those partial squares on islands and along shorelines with more water than land. Only the American Robin was found in more squares across the whole province. There are confirmed breeding records for over 60% of all southern Ontario squares, a total far higher than for any other woodpecker. In the heavily forested area of southcentral Ontario, including Algonquin Prov. Park and Renfrew Co, the breeding evidence is not as strong as in more southerly areas, but this may merely reflect less thorough coverage due to difficulties of access. The species was recorded as far north as the Fort Severn area, but there are large areas of northern Ontario, especially in the Hudson Bay Lowland, where the species either was not encountered or showed only possible or probable breeding evidence. This situation no doubt is a partial reflection of the limited coverage in some blocks. However, the paucity of large trees in the Lowland, especially deciduous ones, is a likely factor in the apparently lower population levels in that area. The distribution pattern shown here confirms the previously known breeding range (Peck and James 1983) and fills in major gaps in northern Ontario.

In southern Ontario, over 10 pairs of Northern Flickers were estimated to be present in 78% of the squares for which observers provided abundance estimates. Thus, it is a common bird in most of the south. The existence of gaps in the distribution of the species in the northern parts of the province suggests that, although the species is generally present there, it does not always reach the population densities attained in the southern portions of the province. However, in squares with abundance estimates in the north, 66% had over 10 pairs, which is only slightly less than the corresponding figure in the south.-- *W.D. McIlveen*

NORTHERN FLICKER

Summary of Abundance Estimates

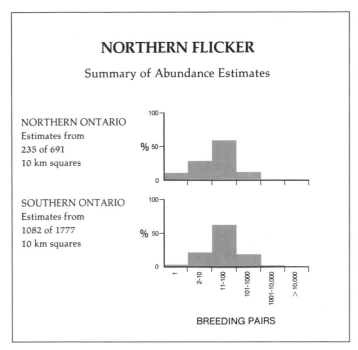

NORTHERN ONTARIO
Estimates from
235 of 691
10 km squares

100
% 50
0

SOUTHERN ONTARIO
Estimates from
1082 of 1777
10 km squares

100
% 50
0

1 2-10 11-100 101-1000 1001-10,000 >10,000

BREEDING PAIRS

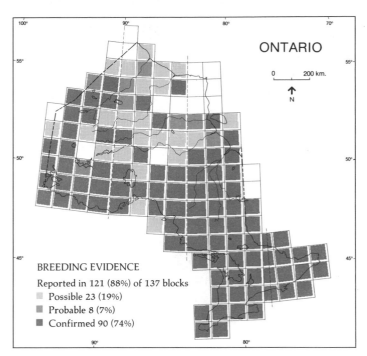

ONTARIO

0   200 km.

N

BREEDING EVIDENCE
Reported in 121 (88%) of 137 blocks
Possible 23 (19%)
Probable 8 (7%)
Confirmed 90 (74%)

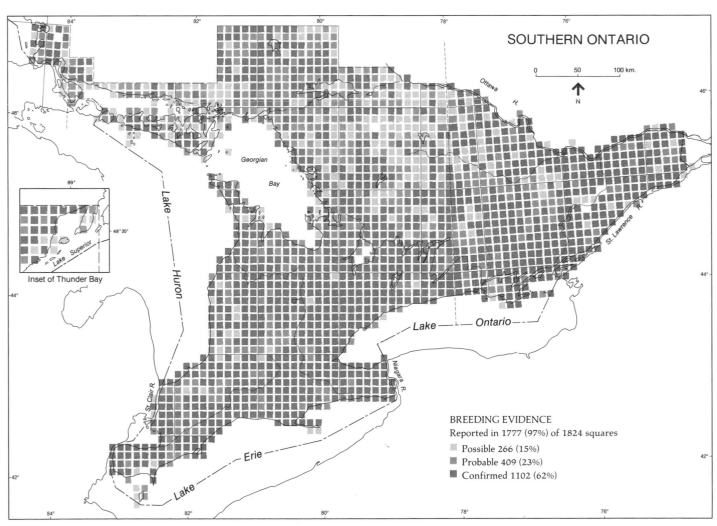

SOUTHERN ONTARIO

0   50   100 km.

N

Georgian
Bay

Ottawa R.

St. Lawrence R.

Lake Huron

Lake Superior

Inset of Thunder Bay

48° 20'

89°

Lake Ontario

Niagara R.

St. Clair R.

Lake Erie

BREEDING EVIDENCE
Reported in 1777 (97%) of 1824 squares
Possible 266 (15%)
Probable 409 (23%)
Confirmed 1102 (62%)

PILEATED WOODPECKER
Grand Pic
Dryocopus pileatus

For a bird so large, robust, and strikingly plumaged, the Pileated Woodpecker can be surprisingly elusive. It is most often seen flying over roadways or other open areas, and is always a welcome sight. As Hoyt (1957) states, "a glimpse of it remains a thrill for most who see it for the first time or the thousandth."

The Pileated Woodpecker occurs across North America, from the southwestern Mackenzie District and Great Slave Lake in the north, south to Texas and southern Florida. In Canada, it is present in all provinces and territories except Newfoundland, where it has never occurred, and Prince Edward Island, where it has been extirpated (Godfrey 1986). The Pileated Woodpecker occurs throughout forested parts of Ontario (Speirs 1985), generally south of the Hudson Bay Lowland. It is generally recognized to prefer extensive tracts of mature forest containing water and large diameter trees for cavity construction, but nests have also been reported in more open agricultural areas and parks containing tall trees. Peck and James (1983) indicate that nest trees in Ontario are 35.5 cm or larger (diameter at breast height). In winter, the Pileated Woodpecker may forage in a greater variety of habitats and has been observed at bird feeders. It is a resident species, and the pair defends a territory throughout the year (Kilham 1983).

The range of the Pileated Woodpecker contracted with the advance of settlement in eastern North America, and declines were obvious by 1900. The species was reduced in numbers throughout southern Ontario north to Muskoka and Haliburton (Devitt 1967), but probably remained near pre-settlement levels in northern Ontario. Factors contributing to the decline included extensive forest clearing, loss of forests to fire, shooting by sportsmen, and commercial sale (Devitt 1967, Freedman and Riley 1980). With fire protection,

improved game laws, and maturing of second growth forests, the Pileated Woodpecker population began to recover by the 1940's, and the species has now re-occupied much of its former range in Ontario.

The Pileated Woodpecker has a loud call and its slow, heavy drumming is distinctive. Combined with the large foraging holes made by this species, these characteristics indicate the presence of one or more birds. The low percentage of records for which breeding has been confirmed (17% in southern Ontario, 13% in the north) is indicative of the large territories of this species, its elusiveness, and its relatively low breeding density. Because the species nests in cavities located 3.5 m or more above the ground and uses separate cavities for roosting (Kilham 1983), confirmation of breeding through observation at cavities requires careful attention until feeding of the young commences (around mid-June). Like other young woodpeckers, the young of this species are highly vocal in the nest and often give away its location. The family group stays together into September (Hoyt 1957), enhancing opportunities for late-season atlassers to establish confirmed breeding.

The maps highlight gaps in the distribution of the Pileated Woodpecker in areas of intensive agricultural and urban land-use. The absence of the species in Essex Co and much of Kent Co and the Niagara Peninsula is particularly evident. It is also sparsely distributed in the remainder of southwestern Ontario as well as in the 5 counties of extreme southeastern Ontario. Although some authors have suggested that the Pileated Woodpecker has adapted to using smaller woodlots, it is apparent that the species is more frequent in areas with a higher percentage of woodland remaining. Speirs (1985) indicates that woodlots of 40 ha or more are required for this species. The northern limit of its distribution appears to be in the vicinity of 53°N in the northwest, and approximately 52°N along James Bay, and generally follows the edge of the Hudson Bay Lowland. The stunted trees of the muskeg are too small for Pileated Woodpecker nests.

The map probably represents the breeding distribution of the Pileated Woodpecker fairly accurately. Roughly half of the Regional Coordinators who offered opinions on its breeding status thought that the species was underconfirmed. For a non-migratory species, one can interpret that view as a suggestion that most of the records of possible and probable breeding actually do represent breeding locations.

Abundance estimates were provided for 45% of the Ontario squares in which the Pileated Woodpecker was recorded. Nearly 90% of estimates indicate a density of 1 to 10 pairs per square. Eleven or more pairs in a square were estimated for only 74 squares.-- *K. Dance*

PILEATED WOODPECKER

Summary of Abundance Estimates

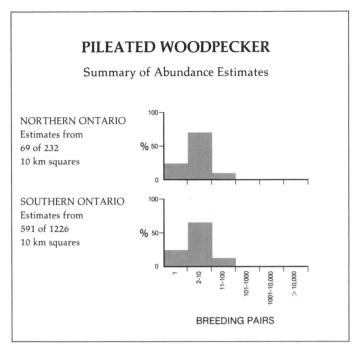

NORTHERN ONTARIO
Estimates from
69 of 232
10 km squares

SOUTHERN ONTARIO
Estimates from
591 of 1226
10 km squares

%

1 2-10 11-100 101-1000 1001-10,000 >10,000

BREEDING PAIRS

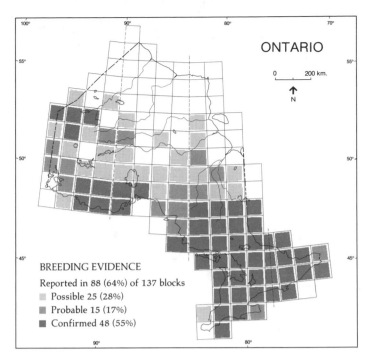

ONTARIO

0   200 km.

N

BREEDING EVIDENCE

Reported in 88 (64%) of 137 blocks

Possible 25 (28%)
Probable 15 (17%)
Confirmed 48 (55%)

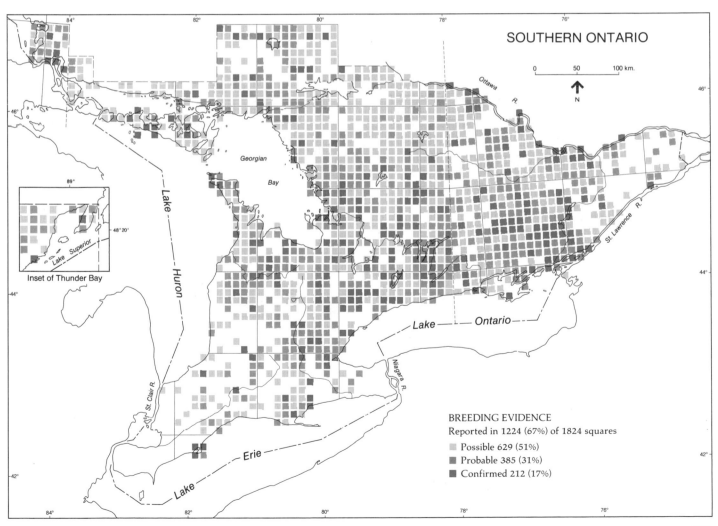

SOUTHERN ONTARIO

0   50   100 km.

N

Lake Huron

Georgian Bay

Ottawa R.

St. Lawrence R.

Lake Ontario

Niagara R.

St. Clair R.

Lake Erie

Inset of Thunder Bay

89°

48°20'

Lake Superior

BREEDING EVIDENCE

Reported in 1224 (67%) of 1824 squares

Possible 629 (51%)
Probable 385 (31%)
Confirmed 212 (17%)

OLIVE-SIDED FLYCATCHER
Moucherolle à côtés olive
Contopus borealis

Loud and distinctive, the 'quick-three-beers' song of the Olive-sided Flycatcher ranks among the most familiar summer sounds of northern Ontario and southern Ontario's 'cottage country'.

The species breeds only in North America. It is found in the coniferous forest and mixed forest regions across Canada, south into the US where suitable habitat exists, including higher elevations along mountain ranges. It winters in South America. In Ontario, the Olive-sided Flycatcher is widespread throughout the Boreal Forest zone north to the Hudson Bay Lowland, and south in the Great Lakes-St. Lawrence Forest to the vicinity of Lake Simcoe. Semi-open coniferous or mixed forest near water is the breeding habitat most often chosen by this species, often in wetlands with standing dead conifers. Spruce or tamarack bogs, forested edges of beaver ponds or rivers, and burns are favoured nest sites. Extensive bog and fen habitat along the southern edge of the Canadian Shield and the western shore of the Bruce Peninsula enable this species to breed south of the heart of its Ontario range.

Little evidence exists to suggest that the distribution or abundance of the Olive-sided Flycatcher has changed significantly within the past century. McIlwraith's (1886) classification of it as "rare in Ontario" was likely more a reflection of its status in the developed parts of southern Ontario, as relatively little was known about the largely unexplored Boreal Forest region at that time. By 1937, knowledge of the bird life of northern Ontario had improved, and Baillie and Harrington (1937) more accurately described it as a "not uncommon summer resident ... north from about Lat. 44^{0}N to James bay, and west ... probably ... to the western extremity of Ontario." However, at the turn of the century it was considered a "frequent summer resident at Mount Forest" by Klugh (1905). This indicates a diminution of its breeding

range in this part of Ontario, mirroring a trend for other northern species in the same area (D. Brewer pers. comm.).

The Olive-sided Flycatcher is easy to detect, but breeding is difficult to confirm. The song is well known and loud: 72% of records are of singing or territorial males. It chooses prominent tree-top locations for hunting and singing perches, and can often be seen from a considerable distance in its fairly open habitat. Since it is difficult to gain access to the bird's preferred wetland nesting habitat, and perhaps because it breeds late in the season, only 8% of the records are of confirmed breeding. Nests with eggs or young were found in only 11 squares in the province; nests whose contents could not be checked were discovered in an additional 10 squares. Even though the nest is a bulky structure, it is usually high enough (4 to 15 m) to be well concealed on a horizontal branch of a spruce or other conifer (Peck and James 1987). The majority (66%) of the few confirmed breeding records were of fledged young or adults carrying food or faecal sacs. The Olive-sided Flycatcher is a very late migrant, seldom arriving on territory before late May. Egg dates in Ontario range from mid-June to late July (James *et al*. 1976). An incubation period of about 2 weeks, combined with about 3 weeks for fledging (Bent 1942), means that eggs are often not hatched before mid-July and young are rarely fledged before the month's end. As atlas field work was largely concentrated in June and early July, coverage of many squares may have been finished before young were out of the nest.

The map generally confirms previous knowledge and suspicions about the distribution of the Olive-sided Flycatcher. In northern Ontario, its wide distribution extends north to near latitude 56^{0}N along the Severn River, although it occurs only sporadically on the Hudson Bay Lowland, mainly in the northeastern corner of the province. The gaps which appear on the map of northern Ontario can probably be attributed to the difficulty of travel in some areas. In southern Ontario, the Olive-sided Flycatcher is fairly widely distributed on the Canadian Shield, and on the Bruce Peninsula and Manitoulin Island. A territorial bird was recorded east of Newmarket (south of Lake Simcoe) and another north of Woodstock, both in 1985, marking the most southerly known territories of this species in Ontario. Its presence in 8 squares on the Bruce Peninsula also represents an extension of the range published by Godfrey (1986). Although summer records for the Olive-sided Flycatcher are occasionally reported south of the area described, most might be considered as early fall or late spring migrants, or non-breeding adults, based upon the habitat used by most Ontario birds. However, because the species also frequents farmland and orchards in Nova Scotia (Godfrey 1986), more southerly records should not be arbitrarily dismissed.

Abundance estimates suggest that this species is thinly distributed in almost all of its range in Ontario, but can reach densities of 100 pairs per square in some areas, especially between Muskoka and the Temagami area.-- *T. Cheskey*

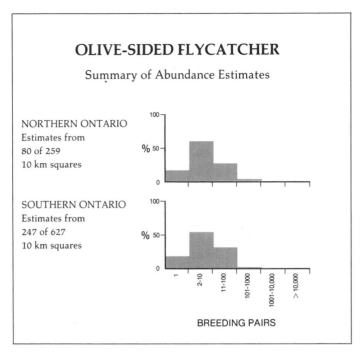

OLIVE-SIDED FLYCATCHER

Summary of Abundance Estimates

NORTHERN ONTARIO
Estimates from
80 of 259
10 km squares

SOUTHERN ONTARIO
Estimates from
247 of 627
10 km squares

BREEDING PAIRS

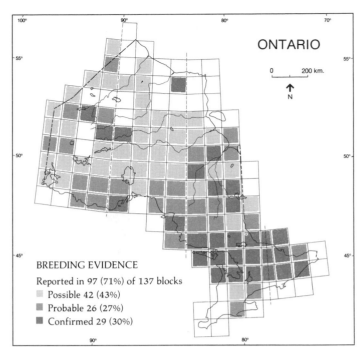

ONTARIO

0   200 km.

N

BREEDING EVIDENCE
Reported in 97 (71%) of 137 blocks

Possible 42 (43%)
Probable 26 (27%)
Confirmed 29 (30%)

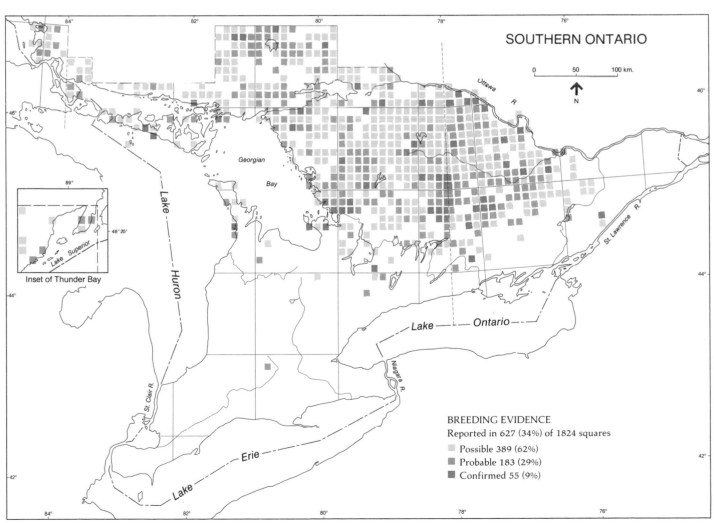

SOUTHERN ONTARIO

0   50   100 km.

N

Georgian
Bay

Lake
Huron

Lake Superior

Inset of Thunder Bay

Ottawa R.

St. Lawrence R.

Lake — Ontario

Niagara R.

St. Clair R.

Lake — Erie

BREEDING EVIDENCE
Reported in 627 (34%) of 1824 squares

Possible 389 (62%)
Probable 183 (29%)
Confirmed 55 (9%)

EASTERN WOOD-PEWEE
Pioui de l'Est
Contopus virens

The Eastern Wood-Pewee is one of the more common fly-catchers of the open deciduous and mixed forests of southern Ontario. Its plaintive, clearly-whistled 'pee-a-wee' song is one of the characteristic sounds of summer, persisting throughout the hottest period of the day and often even at night (Bent 1942).

The Eastern Wood-Pewee breeds from southeastern Saskatchewan eastward across southern Canada to Nova Scotia, and south to Texas and Florida (Godfrey 1986). It is most numerous in the wooded upland areas of the southeastern US (Robbins and Van Velzen 1974). The species winters in northwestern South America. Throughout its range, the Eastern Wood-Pewee breeds in deciduous or mixed woodlands - rarely coniferous woods - characteristically where the understory is sparse. It forages by darting out after flying insects in clearings among the trees, or along the woodland edge, and can often be seen sitting on an exposed branch.

The clearing of dense woods has increased the habitat suitable for this species, and the Pewee has become common in large urban parks and rural villages. In Ontario, it has benefitted from the thinning of the trees and clearing of understory that have occurred extensively in 'cottage country' (Clark *et al.* 1983). South of there, a number of early accounts refer to the Eastern Wood-Pewee as a common bird. Quilliam (1965) cites 4 such accounts, and Macoun and Macoun (1909) quote W.E. Saunders as saying that it is "the most common flycatcher in the country regions of southwestern Ontario with the exception of the kingbird." On Manitoulin Island, the species was thought to be "fairly common in 1912" (Nicholson 1981).

The Eastern Wood-Pewee has inconspicuous plumage, and, like most other flycatchers, returns in spring after the foliage has emerged. It is therefore difficult to see in a canopy of deciduous leaves. Hence, most of the breeding evidence for this species is based on singing (34%) or territorial (28%) males. The nest is even more difficult to see. It is small, tightly affixed (Headstrom 1970), and resembles a knob on a branch. It is usually placed in the fork of a horizontal (frequently dead) branch, 4.5 to 10 m high (or higher), and well away from the trunk of the tree (Peck and James 1987). Even though the egg-laying period, early June through mid-August, coincides with the period when most atlas field work took place, a mere 3% of the records (52 squares) consisted of nests with eggs or young. An additional 4% of all records were of nests whose contents could not be determined. Parasitism by Brown-headed Cowbirds affected 5% of the nests in the Ontario Nest Records Scheme (Peck and James 1987). Since both parents help to feed the young (Bent 1942), it is not surprising that the majority of confirmed breeding records represent fledged young (34%) or parents carrying food or faecal sacs (28%).

Except at the northern limits of the range of this species, the maps contain few surprises. They show that the Eastern Wood-Pewee is ubiquitous in most of southern Ontario and that it becomes progressively less evident, and less frequently confirmed as a breeding species, towards the northern limit of its range. It is likely that the bird breeds in all squares where breeding evidence was reported. It is of interest that one observer found 20 or 30 birds of this species in June 1985 at the southern end of Timiskaming District, east of Temagami, an area where she had not seen it at all in previous years (S. Welsh pers. comm.). In northwestern Ontario, there were records of the Eastern Wood-Pewee in every block north to the Berens River, far beyond its previously known northern range limit (Godfrey 1986).

The Eastern Wood-Pewee is common in much of southern Ontario south of the French and Mattawa Rivers. Speirs (1985) reports densities in ideal habitat of 0.5 to 1 bird/ha in southcentral Ontario, considerably higher than most of the estimates provided by atlassers. The highest densities on Breeding Bird Surveys are found south and west of Algonquin Prov. Park.-- *J.D. Rising*

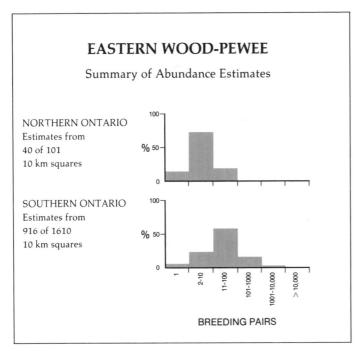

EASTERN WOOD-PEWEE

Summary of Abundance Estimates

NORTHERN ONTARIO
Estimates from
40 of 101
10 km squares

SOUTHERN ONTARIO
Estimates from
916 of 1610
10 km squares

%

BREEDING PAIRS

1 2-10 11-100 101-1000 1001-10,000 >10,000

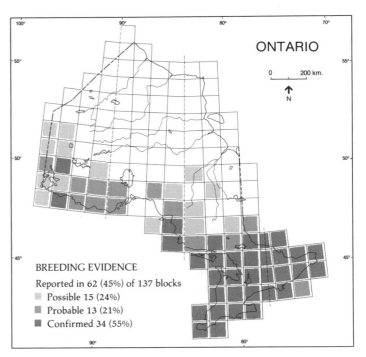

ONTARIO

0   200 km.

N

BREEDING EVIDENCE

Reported in 62 (45%) of 137 blocks

Possible 15 (24%)
Probable 13 (21%)
Confirmed 34 (55%)

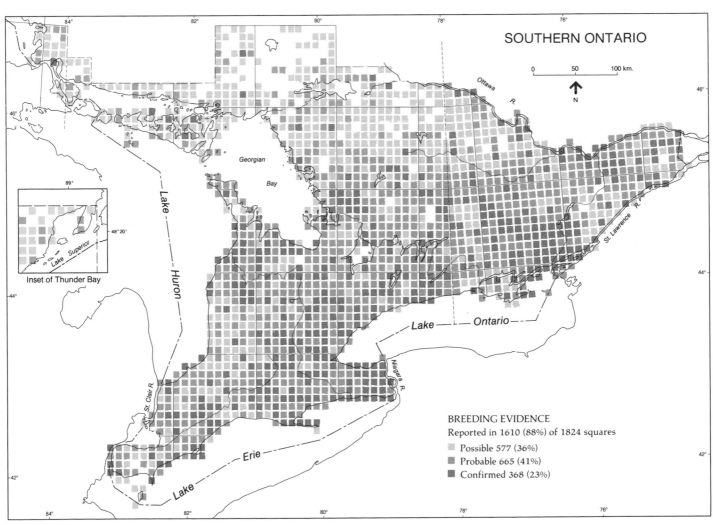

SOUTHERN ONTARIO

0   50   100 km.

N

Georgian
Bay

Lake
Huron

Lake Superior
48°20'

Inset of Thunder Bay

St. Clair R.

Lake   Erie

Niagara R.

Lake — Ontario

Ottawa R.

St. Lawrence R.

BREEDING EVIDENCE

Reported in 1610 (88%) of 1824 squares

Possible 577 (36%)
Probable 665 (41%)
Confirmed 368 (23%)

YELLOW-BELLIED FLYCATCHER
Moucherolle à ventre jaune
Empidonax flaviventris

The Yellow-bellied Flycatcher breeds throughout the Boreal Forest region of Canada, from the Maritime Provinces to the Yukon. Its range also extends into the extreme northern areas of the eastern US. In Ontario, it breeds mainly in the coniferous swamps and bogs that characterize the poorly drained areas of the Canadian Shield region. Drier shady pine and spruce forests with dense shrubs are also used. It breeds regularly, although atypically, on talus slopes along certain river valleys on the eastern side of Algonquin Prov. Park (Brunton and Crins 1975). Similar talus slope habitat is also frequented on the eastern face of the Niagara Escarpment on the Bruce Peninsula. Despite the abundance of suitable habitat in the province, nowhere is the species commonly encountered, being known to most naturalists as an irregularly common migrant along the Great Lakes shoreline, particularly during the autumn. The winter range extends from Mexico to Panama (Godfrey 1986).

It is not likely that the traditional range of the Yellow-bellied Flycatcher, unlike that of many other species, has changed greatly as a result of agricultural and urban development. Suitable coniferous habitat has always been sparse to the south of the Canadian Shield, and there is no evidence that the species was ever a common breeder south of its present range. However, records of breeding in Wellington Co (Skales 1906, Brewer 1977) and on Amherst Island near Kingston (Baillie and Harrington 1937), as well as several areas to the south of Ontario (Walkinshaw and Henry 1957) suggest that this flycatcher could potentially breed in all areas of the province where suitable habitat exists.

As shown on the maps, the Yellow-bellied Flycatcher is widely distributed throughout northern Ontario to the tree-line and in the Canadian Shield portion of southern Ontario.

The many gaps in its distribution likely reflect the difficulty in locating this species, rather than its absence, as the dense conifer swamps where it is found are not easily penetrated by atlassers. Male Yellow-bellied Flycatchers tend to be less vocal and sing less frequently from conspicuous perches than males of other *Empidonax* flycatchers that are native to Ontario. Furthermore, the 'che-lek' song of the Yellow-bellied is often quite similar to the 'che-bec' song of the Least Flycatcher, which is also found throughout the Boreal Forest region of the province (although typically in more deciduous uplands). This similarity in song might lead to some confusion in discriminating between the two species, particularly as the Yellow-bellied Flycatcher is often difficult to see in its habitat. Males are noticeably less vocal after the eggs have hatched (Walkinshaw and Henry 1957); hence, breeding birds may be completely overlooked during the latter stages of the reproductive season. Moreover, the adults are among the last to arrive on territory in spring and among the first to leave when breeding is completed (Hussell 1982); consequently, there is a very narrow period of time during which breeding can be detected by atlassers.

Predictably, most breeding evidence for this species in the province is for singing males in suitable habitat (58% of all squares, 44% of all blocks). Confirmed breeding records represent only 5% of the squares and 19% of the blocks in which it was reported; these are among the lowest proportions of any species in the province. This low confirmation rate reflects the difficulty of finding nests, which are typically well concealed in mosses on the ground, or sometimes in the roots of upturned trees (Bent 1942). Females tend to sit tightly on the nest, and both parents are extremely secretive while feeding their young. Only two active nests (both containing eggs) were found in the province during the 5 year atlas period, although the species probably breeds in most squares throughout the Boreal Forest. One nest, containing 5 eggs, was observed in mid-June 1982 near Owen Sound, an area in which Yellow-bellied Flycatcher breeding had not previously been documented, but which evidently contains a limited breeding population.

Abundance estimates suggest that the Yellow-bellied Flycatcher is fairly common in much of its breeding range in Ontario. Abundance seems to be higher, on average, in the northern parts of the province, where continuous expanses of suitable habitat predominate.-- *D.R.C. Prescott*

YELLOW-BELLIED FLYCATCHER

Summary of Abundance Estimates

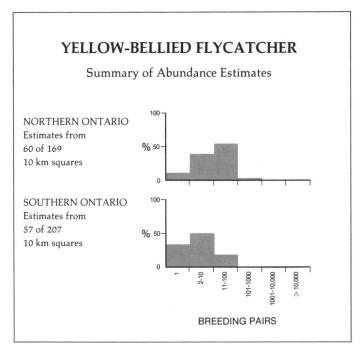

NORTHERN ONTARIO
Estimates from
60 of 169
10 km squares

SOUTHERN ONTARIO
Estimates from
57 of 207
10 km squares

%

100
50

%

100
50

1 2-10 11-100 101-1000 1001-10,000 >10,000

BREEDING PAIRS

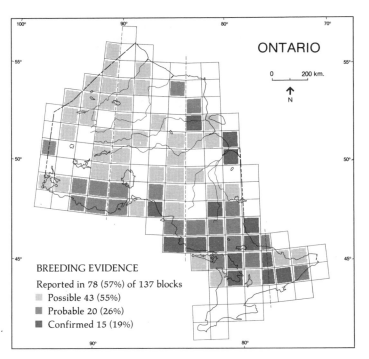

ONTARIO

0   200 km.

N

BREEDING EVIDENCE

Reported in 78 (57%) of 137 blocks

Possible 43 (55%)
Probable 20 (26%)
Confirmed 15 (19%)

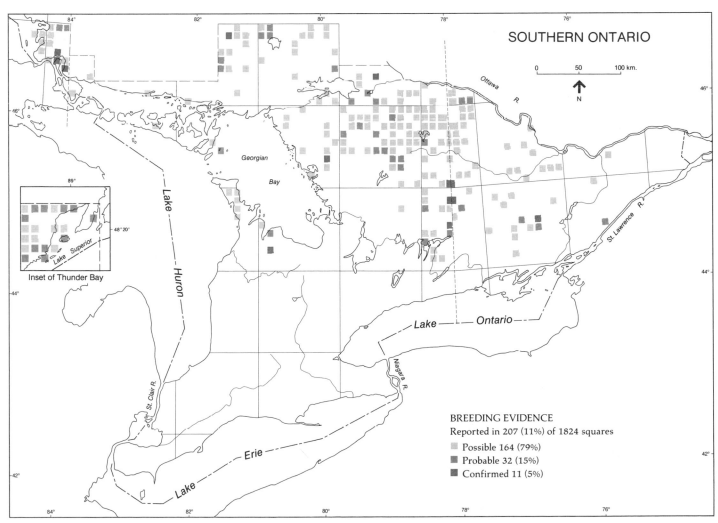

SOUTHERN ONTARIO

0   50   100 km.

N

Georgian Bay

Ottawa R.

St. Lawrence R.

Lake Huron

Lake Superior

89°

48°20'

Inset of Thunder Bay

St. Clair R.

Niagara R.

Lake — Ontario

Lake — Erie

BREEDING EVIDENCE

Reported in 207 (11%) of 1824 squares

Possible 164 (79%)
Probable 32 (15%)
Confirmed 11 (5%)

ACADIAN FLYCATCHER
Moucherolle vert
Empidonax virescens

The Acadian Flycatcher is an inconspicuous, elusive species that has a wide range in North America east of the Mississippi and is locally distributed as far north as extreme southwestern Ontario. Its winter range extends from Costa Rica to northern South America (Godfrey 1986). Its preferred breeding habitat has been described as river swamps, creek bottoms, well-wooded ravines, black ash swamps, and deep-shaded forests (Christy 1942, Godfrey 1986). In Ontario, fairly mature, deciduous forests or occasionally the edges of wooded ravines are favoured. Nests are almost always in a beech or maple tree with pendant lower branches (Christy 1942). An open space of at least 2 m beneath these branches is necessary to give room for the birds to sally for insects.

The Acadian Flycatcher reaches the northern limit of its range in Ontario. Because of the extensive clearing of mature forests for farmland and cutting for commercial or fuelwood purposes, good quality habitat has been severely reduced. It is likely, therefore, that numbers of breeding birds and perhaps even its breeding range have been reduced since the turn of the century. The species was first discovered in Canada in 1884, when a nest with 3 eggs was found in Haldimand Co in June (Baillie and Harrington 1937). In the 1960's and 1970's, nests were found as far north and east as Uxbridge and Pickering (Speirs 1985), but no records of confirmed breeding outside of extreme southwestern Ontario were obtained during the atlas period. Rondeau Prov. Park has historically been the best-known breeding area in Canada, although occasional records have been reported elsewhere in forests along the Lake Erie shoreline (*e.g.*, at Wheatley Prov. Park and Point Abino). Numbers of birds at Rondeau have apparently diminished in the past decade.

The Acadian Flycatcher is a difficult bird to find, and even more difficult to confirm as breeding. It is thinly spread throughout its available habitat, which is sparsely distributed within its range in Ontario. The various species of *Empidonax* flycatchers are difficult or impossible to distinguish from each other if the birds are not vocalizing. When they first arrive in spring, they are notoriously quiet, and the Acadian Flycatcher is no exception. Being sparsely distributed in Ontario, it probably has little territorial pressure from neighbouring birds and therefore not much singing or other territorial behaviour is exhibited during the nesting cycle. The song, when given, is explosive and distinctive, but does not carry far, and often only a partial song is given.

Partly because of its preference for the denser, mosquito-infested portions of mature woodlots, the breeding distribution and habits of the Acadian Flycatcher in Ontario have hitherto been poorly known. However, atlassers examined many potential sites, and their records have produced a more accurate picture of its distribution. The possible breeding records shown on the map may indicate unmated or transient birds. Because the Acadian Flycatcher is rare in Ontario, it seems likely that atlassers would have made a point of attempting to find a higher level of breeding evidence, had it been present. Hence, the map is probably a fairly complete portrayal of confirmed breeding sites in the province. Six of the 8 confirmed breeding records represent actual nests, 2 with eggs and 4 with young. When an observer is near the nest site, one or both adults give clear, distinctive alarm notes. If the observer is patient and quiet, one of the adults will soon return to the nest, which is a fairly visible structure usually less than 4 m above the ground near the outer part of the lower branches. It is so poorly made that the contents are easily visible from below, and it usually has several bits of nesting material dangling up to 0.5 m below the actual nest.

The atlas has improved our knowledge of the distribution of the Acadian Flycatcher in Ontario. Historically, it was known to breed in only 7 counties, primarily those along the north shore of Lake Erie (Peck and James 1987). The breeding evidence presented here represents 10 counties or regional municipalities, all within the Carolinian Forest zone. It includes locations well away from the Lake Erie shore. A nest was found at Denfield (west of London), and singing birds were observed northeast and northwest of Brantford and even within the city of Toronto. As expected for a species at the limit of its range, the available habitat is not saturated with populations of the Acadian Flycatcher, nor is the species always found in the same area each year. Although it was until recently an uncommon but regular breeding bird at Rondeau Prov. Park, it is now rare anywhere in Ontario.-- *P.A. Woodliffe*

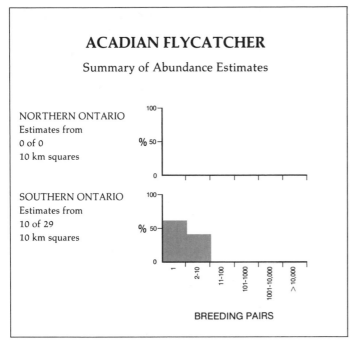

ACADIAN FLYCATCHER

Summary of Abundance Estimates

NORTHERN ONTARIO
Estimates from
0 of 0
10 km squares

SOUTHERN ONTARIO
Estimates from
10 of 29
10 km squares

%

BREEDING PAIRS

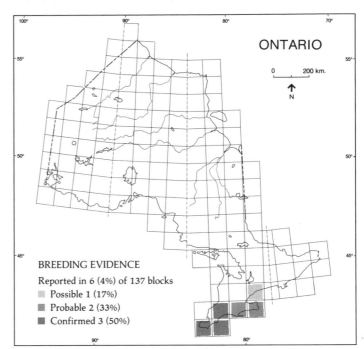

ONTARIO

0   200 km.

N

BREEDING EVIDENCE
Reported in 6 (4%) of 137 blocks
Possible 1 (17%)
Probable 2 (33%)
Confirmed 3 (50%)

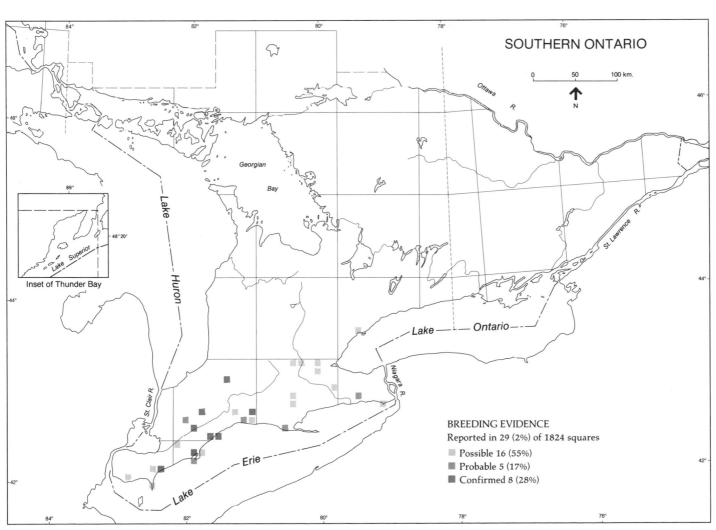

SOUTHERN ONTARIO

0   50   100 km.

N

Inset of Thunder Bay

BREEDING EVIDENCE
Reported in 29 (2%) of 1824 squares
Possible 16 (55%)
Probable 5 (17%)
Confirmed 8 (28%)

ALDER FLYCATCHER
Moucherolle des aulnes
Empidonax alnorum

The Alder Flycatcher breeds across northern North America in a band extending from Alaska and southcentral British Columbia to southern Labrador, Newfoundland, and the northeastern US. Its winter range is poorly known but is "presumably in South America" as far south as northern Argentina (AOU 1983). The species is relatively specific in its habitat requirements, preferring alder or willow thickets that border lakes and streams. As a result, it is locally very common over all of Ontario, wherever such conditions exist, although decidedly less common along the shores of Hudson Bay, at the northern limit of its range in the province. It is also largely absent as a breeding bird from the extreme southwest of the province, where agricultural development has limited the availability of suitable habitat. There is some evidence that Alder Flycatcher populations in the latter area may be declining as a consequence of the northward expansion of the closely related Willow Flycatcher (Stein 1963). These species now share many habitats where their breeding ranges overlap (Zink and Fall 1981, Barlow and McGillivray 1983). However, historic records of the breeding distributions of the Alder and Willow Flycatchers are unavailable, as these morphologically very similar species were, until 1973, treated as a single species, the Traill's Flycatcher (AOU 1973).

Because of its small size, cryptic colouration, and simple 3-syllable song ('fee-bee-o'), the Alder Flycatcher can easily be overlooked in the field. However, the male is conspicuous early in the breeding season as it calls from prominent perches within its territory. Such advertisements are usually more frequent during the morning and evening periods. At other times, however, it tends to perch low down in dense shrubbery, but may frequently be observed darting from cover in pursuit of the aerial insects on which it feeds. This is particularly true later in the breeding season, when the male, as well as the normally reclusive female, forage frequently and conspicuously to supply food for nestlings. Hence, the efficiency with which the breeding status of the Alder Flycatcher may be determined varies both diurnally and seasonally, but may also differ among areas. Males tend to call more frequently in habitats which contain other breeding pairs, presumably because of the increased need for territorial defence. The Alder Flycatcher also tends to be less vocal in habitats where it coexists with the Willow Flycatcher (Prescott unpubl. data), and may therefore be overlooked by observers. The use of tape playbacks is a valuable tool in detecting the presence of flycatchers during periods of reclusiveness; 'pishing' is somewhat less productive in detecting these birds.

The maps show the Alder Flycatcher to be at least possibly breeding over most of Ontario (86% of all blocks). However, confirmed breeding records are relatively few (7% of squares in the south; 35% of all blocks in Ontario), for at least two reasons. First, nests are frequently well concealed in dense foliage, often over standing water. Second, parent birds tend to be secretive around the nest (Bent 1942), although this behaviour is not consistent among pairs. Only 2% of all records were of active nests. Breeding was most often confirmed by observing fledged young (33%) or adults carrying food or faecal sacs (27%).

Although abundance estimates in the range of 11 to 100 pairs per square were provided more frequently (47%) than higher estimates, it is likely that most squares in southern Ontario contain far fewer than 100 pairs of Alder Flycatchers. Only 6% of abundance estimates in the south were of this magnitude, and most of these occurred in a band extending from the Kingston area to Sudbury. The species appears to be more common in northern Ontario, where suitable breeding areas are more frequent and extensive. In the northernmost part of Ontario, however, it is restricted to a very few locations and is not at all common.-- *D.R.C. Prescott*

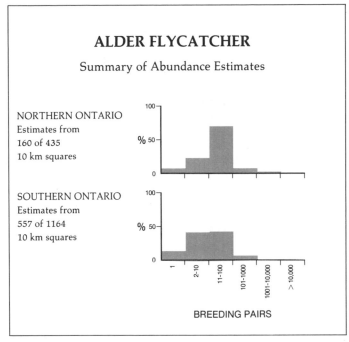

ALDER FLYCATCHER

Summary of Abundance Estimates

NORTHERN ONTARIO
Estimates from
160 of 435
10 km squares

SOUTHERN ONTARIO
Estimates from
557 of 1164
10 km squares

BREEDING PAIRS

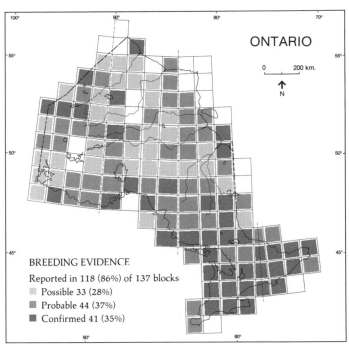

ONTARIO

0   200 km.

N

BREEDING EVIDENCE

Reported in 118 (86%) of 137 blocks

Possible 33 (28%)
Probable 44 (37%)
Confirmed 41 (35%)

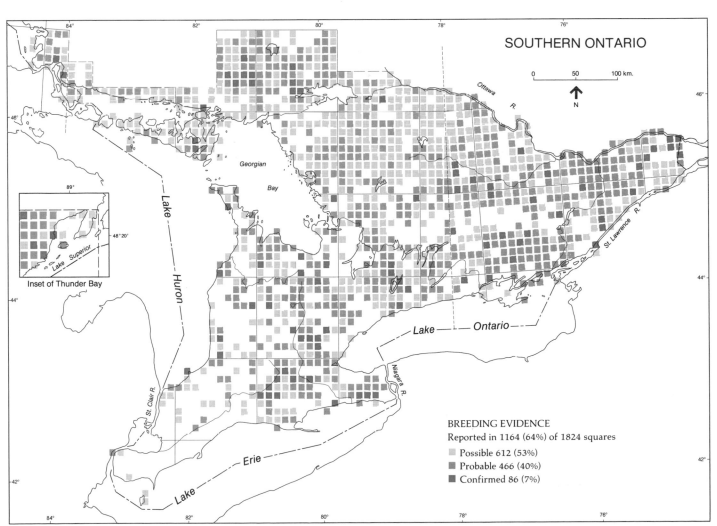

SOUTHERN ONTARIO

0   50   100 km.

N

Ottawa R.

St. Lawrence R.

Georgian Bay

Lake Huron

Lake Superior

Inset of Thunder Bay

St. Clair R.

Lake — Ontario

Niagara R.

Lake — Erie

Lake

BREEDING EVIDENCE

Reported in 1164 (64%) of 1824 squares

Possible 612 (53%)
Probable 466 (40%)
Confirmed 86 (7%)

WILLOW FLYCATCHER
Moucherolle des saules
Empidonax traillii

The Willow Flycatcher is a locally common breeding bird across southern Canada and the northern US. It inhabits secondary shrub growth in open areas (Godfrey 1986), preferring habitats dominated by willow, red osier dogwood, and, to a lesser degree, hawthorn. In Ontario, this preferred habitat is common in abandoned farmland and low-lying areas throughout the southern parts of the province, as far north as Muskoka and Parry Sound. To the north, the Willow Flycatcher is replaced by the morphologically and ecologically similar Alder Flycatcher, with which, until 1973, it was grouped as a single species, the Traill's Flycatcher (AOU 1973). There is evidence that the range of the Willow Flycatcher is pushing northward in eastern North America at the expense of its close relative (Stein 1963); this trend may now also be occurring in certain parts of southern Ontario as well (Weir 1985c), particularly where small trees and shrubs are colonizing former pastures, and where the wetlands preferred by the Alder Flycatcher have been drained for agricultural purposes. The Willow Flycatcher winters in Central and South America (AOU 1983).

Although the male is generally vocal and conspicuous in its breeding habitat, the Willow Flycatcher, like other members of the *Empidonax* genus, has proven to be a difficult species to census accurately. Its habitat in southern Ontario is often shared with the Alder Flycatcher (Barlow and McGillivray 1983, Prescott in press) and the songs of these two species are sufficiently similar that they may not be distinguishable from each other by inexperienced observers. Moreover, the habitat of the Willow Flycatcher is invariably shared with more boisterous species such as the Yellow Warbler, Common Yellowthroat, Swamp Sparrow, and Red-winged Blackbird, such that the relatively quiet 'fitz-bew' song of the male flycatcher might easily be overlooked in the field. Neverthe-

less, listening for singing males is the most profitable method of detecting the presence of the species, with the best periods being during the morning and evening, especially early in the season (late May and early June). Call notes ('whit'), which are given by both sexes during the reproductive period, are also useful detection aids, but may be confused with those of the Alder Flycatcher ('wheep') in habitats where the species coexist.

Breeding evidence gathered during the atlas period reflects the difficulty in confirming the breeding of this species. Most records (43%) represent possible breeders, with breeding confirmations comprising only 19%. Nests, which are generally located in dense willow or dogwood shrubs (Walkinshaw 1966), proved difficult to locate, and comprise only 8% of the records.

The Willow Flycatcher is widely distributed in southern Ontario, as far north as Muskoka and Ottawa. It is found in almost all squares south of the Canadian Shield, but is more sporadically distributed to the north. A sizable population has been discovered in the Sudbury area - well north of the most northerly nest records contained in the ONRS (Grey, Simcoe, Peterborough and Frontenac Counties; Peck and James 1987). Breeding Bird Surveys have previously reported the Willow Flycatcher in the Atikokan area, far from its normal range (Speirs 1985), but the species was not reported from this area during the atlas period. If it continues its recent northward expansion, the Willow Flycatcher may be expected to establish itself in this area, and perhaps also in Algoma and Timiskaming Districts. At present, the breeding ranges of both the Willow and Alder Flycatchers appear to be in a dynamic state in the province. Further observations will be required to clarify the extent of these trends in Ontario, and elsewhere.

In those squares where Willow Flycatchers were observed, densities of 2 to 10 pairs per square were most commonly estimated (55%), although more than 10 pairs per square were frequently encountered (35% of all squares with estimates). A general consensus among Regional Coordinators is that the abundance of Willow Flycatchers was underestimated in most areas because of the unfamiliarity of many atlassers with its song. Thus, higher levels of abundance may be more frequent than indicated here. The species is certainly not uncommon in a large part of its range.-- *D.R.C. Prescott*

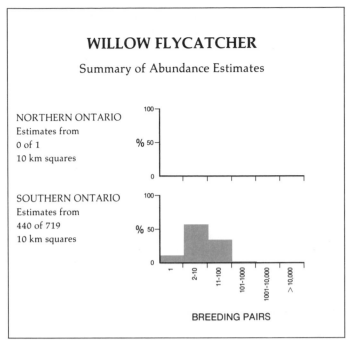

WILLOW FLYCATCHER

Summary of Abundance Estimates

NORTHERN ONTARIO
Estimates from
0 of 1
10 km squares

SOUTHERN ONTARIO
Estimates from
440 of 719
10 km squares

BREEDING PAIRS

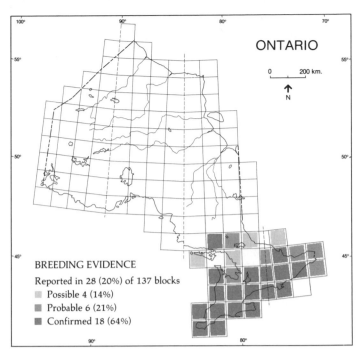

ONTARIO

0   200 km.

N

BREEDING EVIDENCE

Reported in 28 (20%) of 137 blocks

Possible 4 (14%)

Probable 6 (21%)

Confirmed 18 (64%)

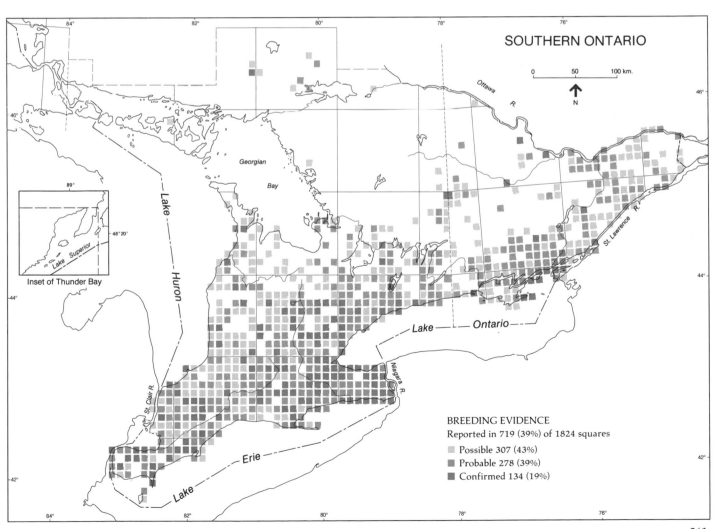

SOUTHERN ONTARIO

0   50   100 km.

N

Inset of Thunder Bay

BREEDING EVIDENCE

Reported in 719 (39%) of 1824 squares

Possible 307 (43%)

Probable 278 (39%)

Confirmed 134 (19%)

LEAST FLYCATCHER
Moucherolle tchébec
Empidonax minimus

The Least Flycatcher is the commonest and most widely distributed member of its genus, breeding from the Atlantic coast to the Cordillera and from the Appalachian Mountains north almost to the tree-line. It winters in Mexico and Central America (Godfrey 1986). The species breeds in most parts of Ontario, where it frequents open deciduous woodland and forest edge habitat (Breckenridge 1956), often in close association with human habitation (Bent 1942). The Least Flycatcher occupies a broader ecological niche than that of other eastern flycatchers in the *Empidonax* genus, which accounts for its more widespread distribution.

Historical accounts of Ontario's avifauna suggest that this species has always been a common and ubiquitous summer resident. According to McIlwraith (1894), the Least Flycatcher was "very common throughout Ontario" and was reported as being "common" from Moose Factory on James Bay south to London (Macoun and Macoun 1909). At the latter locale, the Least Flycatcher was actually considered to be more common within the city than in the surrounding countryside. Nash (1908) considered the Least Flycatcher to be a "common summer resident" and Baillie and Harrington (1937) assigned it the status of "fairly common breeding summer resident", at least in the southern two-thirds of the province. Prior to human settlement, the Least Flycatcher may have been more widely distributed south of the Canadian Shield than at present. Although the clearing of land throughout the Carolinian and southern Great Lakes-St. Lawrence Forest regions may have caused the species to disappear from some areas, forest thinning and fragmentation may have provided habitat that was actually more suitable for the species than that which existed before.

The Least Flycatcher is common but difficult to confirm as a breeding species in many regions of the province.

Although there are records in over 2100 squares, only 15% of them have confirmed status. There are several aspects of the breeding biology of the species which contribute to this low percentage. The adults establish their territory early in the season and then spend a relatively short time on the breeding grounds. In a Minnesota study, MacQueen (1950) estimated that approximately 50 days elapse between the initiation of nest-building and final feeding of the fledged young. A slightly longer average of 64 days on the breeding range has been suggested for Ontario Least Flycatchers (Hussell 1981). In addition, the Least Flycatcher occupies a relatively small territory, particularly in the southern portions of its range. In Virginia, Davis (1959) calculated an average breeding density of 1 pair/0.18 ha, significantly higher than the 1 pair/0.67 ha calculated by Kendeigh (1947) in the Lake Nipigon region. Although the male defends his territory aggressively, continually uttering the emphatic and distinctive 'chebek' call, this call is not easily detected at a distance, and males cease singing relatively early in the breeding season. Consequently, by July it is difficult to find the species and to distinguish it from other similar flycatchers unless they are singing. Yet another problem is that family groups remain intact for only a short period. MacQueen (1950) observed that, on average, the young are fed by their parents for only 10 days after fledging. Adults move south immediately upon completion of nesting and by late July most have departed the province (Hussell 1981).

The maps show that the Least Flycatcher breeds over most of Ontario. In the south, it is absent from most of Essex Co and other smaller areas where intensive agricultural practices have removed much of the forest. Other small gaps in the southern portion of its range, such as those around Toronto and along the north shore of the Niagara Peninsula, are likely due to a lack of suitable nesting habitat. The maps also suggest that the Least Flycatcher is sporadic across the northern half of the Hudson Bay Lowland and that it does not breed at all in the extreme northeastern corner of the province. The concentration of possible breeding records in the extreme northwest of the province is largely beyond the range given in Godfrey (1986). The majority of these records were of birds in deciduous areas on river banks.

The Least Flycatcher is common in much of its range in Ontario, and is particularly abundant in those parts of the Great Lakes-St. Lawrence Forest region where agriculture is not widespread. Estimates of over 1,000 pairs were made in 29 squares.-- *D.M. Fraser*

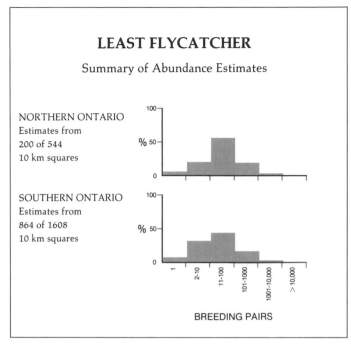

LEAST FLYCATCHER

Summary of Abundance Estimates

NORTHERN ONTARIO
Estimates from
200 of 544
10 km squares

SOUTHERN ONTARIO
Estimates from
864 of 1608
10 km squares

BREEDING PAIRS

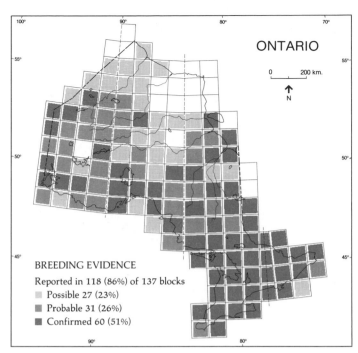

ONTARIO

0   200 km.

N

BREEDING EVIDENCE
Reported in 118 (86%) of 137 blocks

Possible 27 (23%)
Probable 31 (26%)
Confirmed 60 (51%)

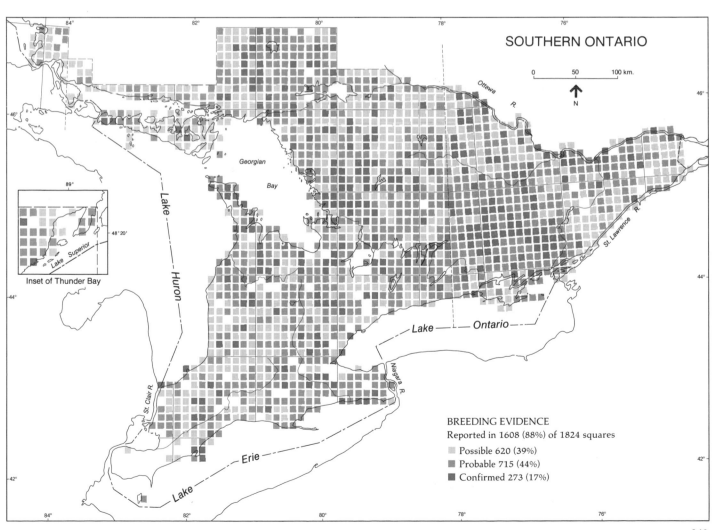

SOUTHERN ONTARIO

0   50   100 km.

N

Georgian
Bay

Lake

Huron

Ottawa R.

St. Lawrence R.

Lake — Ontario

Niagara R.

Erie

Lake

St. Clair R.

Inset of Thunder Bay

Lake Superior

BREEDING EVIDENCE
Reported in 1608 (88%) of 1824 squares

Possible 620 (39%)
Probable 715 (44%)
Confirmed 273 (17%)

EASTERN PHOEBE
Moucherolle phébi
Sayornis phoebe

The Eastern Phoebe is the only member of the tyrant fly-catcher family that has become adapted to human habitations and structures in rural Ontario. On many farmsteads, it is one of the harbingers every spring of the insectivorous birds that return later. For these reasons, despite its drab appearance, it is better known than most other members of the family.

The Eastern Phoebe is a North American bird, restricted to the area east of the Rockies. In Ontario, it breeds in almost all of the inhabited areas in or adjacent to the Carolinian and Great Lakes-St. Lawrence Forest regions. The winter range of the Eastern Phoebe is in the southern US and Mexico. Although it once nested on cliffs and fallen trees beside streams, and still does so in some more remote areas, the species today nests largely on man-made structures, typically on the beams on the underside of a bridge or beneath the overhanging eaves of a shed or cottage or in a semi-open barn. It is a common bird throughout those parts of southern Ontario where such habitat is found.

At the turn of the century, Macoun and Macoun (1909) used the words "very common" and "abundant" to describe the status of the Eastern Phoebe in Ontario. Such terms would seem a bit exaggerated today. Its apparent decline in abundance, while not affecting the extent of its distribution, may be related to the removal of derelict farmhouses, which were commonly used as nest sites, and the replacement of suitable bridge structures with more modern ones. Speirs (1985) tentatively attributed its decline to temporary climatic conditions or the widespread use of chemical insecticides.

The Eastern Phoebe is one of the easiest species to find and also to confirm as a nesting bird. Because of its fondness for human surroundings, one needs to look no further than farms, cottages, and roadsides to find it. Moreover, the spec-

ies has a persistent call note that is both penetrating and distinctive. Once the bird has been discovered, it is usually a simple matter to locate its nest; the selection of the nest sites is very predictable. Seldom is it screened from view by vegetation. However, as it is often underneath a bridge or above eye level, the nest may be somewhat inaccessible and its contents difficult to determine. Old nests of the Eastern Phoebe are fairly distinctive, though not unlike those of Barn Swallows. Individual birds normally return to the same site, frequently repairing and re-using old nests in successive years. A relatively high proportion (6%) of the records are based on the discovery of used nests. Nests of this species are most unlikely to have escaped the notice of farmers or cottagers whose buildings are being used as nest sites. Nesting is still in progress, at least for the second brood, throughout the period of the year when most atlas field work was carried out.

For reasons already mentioned, the data show a very high proportion of confirmed nesting records of the Eastern Phoebe. It nested in virtually every block south of the Boreal Forest region. Within the breeding range, however, there are extensive sections where the species was not found at all. For example, flowing streams and associated bridges are scarce, as is the Eastern Phoebe, in some of the flat glacial lake bottoms southwest of London and in limestone plains like Manitoulin Island. The uninhabited and forested interior of Algonquin Prov. Park, being largely devoid of buildings, represents another gap in the range of this species. Where it does occur there, the species frequently nests on rocky cliffs. The absence of confirmed breeding evidence from the Clay Belt or elsewhere in northeastern Ontario provides a sharply defined northern limit for its range, although there are several published sight records from beyond that point (Todd 1963). The northernmost confirmed breeding record on the Ontario map was of a used nest under a bridge 60 km northeast of Pickle Lake, and is an exceptional record. Though Godfrey (1986) gives its range north to Sandy Lake, no birds were reported there during atlas field work. The most surprising record was of a single bird seen and heard on 5 July 1983 in and around an abandoned building near the mouth of the Winisk River, hundreds of kilometres north of its normal range.

Abundance estimates for the Eastern Phoebe in southern Ontario suggest that there are usually fewer than 100 pairs per square, and that it can no longer be called a common bird.-- *F.M. Helleiner*

EASTERN PHOEBE

Summary of Abundance Estimates

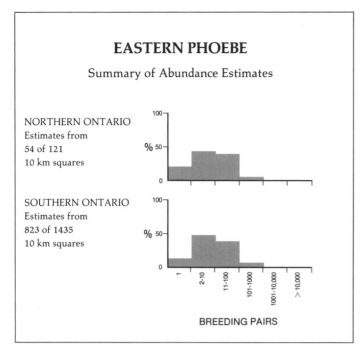

NORTHERN ONTARIO
Estimates from
54 of 121
10 km squares

SOUTHERN ONTARIO
Estimates from
823 of 1435
10 km squares

BREEDING PAIRS

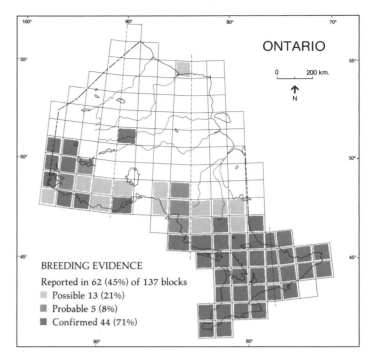

ONTARIO

0   200 km.

N

BREEDING EVIDENCE
Reported in 62 (45%) of 137 blocks
Possible 13 (21%)
Probable 5 (8%)
Confirmed 44 (71%)

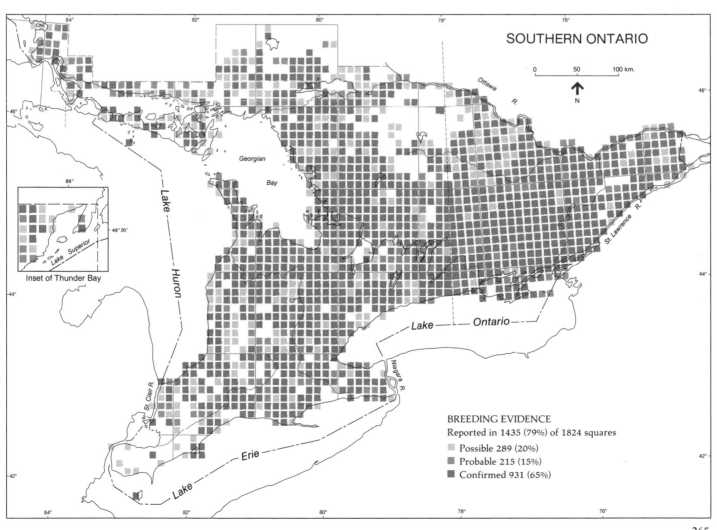

SOUTHERN ONTARIO

0   50   100 km.

N

Georgian
Bay

Ottawa
R.

St. Lawrence R.

Lake   Ontario

Niagara R.

Lake   Huron

St. Clair R.

Lake   Erie

Inset of Thunder Bay
Lake Superior
89°
48°20'

BREEDING EVIDENCE
Reported in 1435 (79%) of 1824 squares
Possible 289 (20%)
Probable 215 (15%)
Confirmed 931 (65%)

GREAT CRESTED FLYCATCHER
Tyran huppé
Myiarchus crinitus

The Great Crested Flycatcher is a common bird in eastern North America wherever broad-leaved trees occur. It breeds from Texas and the Great Plains north to southcentral Canada, and east to the Atlantic coast. Its winter range extends from Florida and Mexico south to Colombia (Godfrey 1986). In southern Ontario, the Great Crested Flycatcher occupies almost all of the Great Lakes-St. Lawrence Forest and the Carolinian Forest regions. The species is also well established in the western section of the Great Lakes-St. Lawrence Forest, between Thunder Bay and the Manitoba border. The Great Crested Flycatcher is a woodland bird, but appears to shun the deepest recesses of the forest, preferring to be near edges and clearings (Bent 1942).

As a bird of the forest edge, the Great Crested Flycatcher probably benefitted from the creation of openings in the continuous forest. The present situation, with mostly small woodlots throughout the area south of the Shield and numerous openings in the forests on the Shield, means that suitable habitat is widespread. In 1901, Young (in Macoun and Macoun 1909) commented that it was "within the last fifteen years . . . certainly becoming commoner in Ontario." BBS data show increasing numbers in the Great Lakes area between 1965 and 1979 (Robbins *et al.* 1986).

The Great Crested Flycatcher usually forages high in the canopy, where it is more often heard than seen; this may account for the fact that many of the atlas records are of singing (20%) or territorial (16%) birds, whose presence can be registered without an actual sighting of the bird. It is a cavity nester, but does not excavate its own sites. Most nests are in natural cavities or woodpecker holes, but the bird has adapted to bird houses, hollow logs, open slits in dead trees, open pipes, rotted out ends of fence rails - and many more

similar sites (Bent 1942). The height of cavities selected is variable, but the majority are from 3 to 7 m above the ground. It is therefore not surprising that 69% of the occupied nests recorded by atlassers were too inaccessible to permit viewing of the contents.

Like many other breeding insectivorous birds, the Great Crested Flycatcher, rather than being thought of as a breeding bird that migrates south for the winter, may more realistically be considered a tropical or subtropical species visiting Ontario only to nest. It spends less than one-third of the year in Ontario. Most individuals do not arrive until mid-May and are gone by early September. For successful first-try breeders the breeding season is brief. After a few days of gathering nest material and laying eggs, about 2 weeks of incubation time, another 2 week nestling period and several more days while the parents care for the fledglings, the nesting cycle is finished for another year. Unless an observer is in the vicinity during that brief period, it is not likely that breeding can be confirmed.

The maps give a good indication of the Ontario range of the Great Crested Flycatcher. Indeed, the overall extent of the area shown coincides well with the bird's known habitat and previously documented range limits. It likely breeds in all blocks in which it was reported. The species is widespread in southern Ontario. North of Sudbury, the forest has more of a coniferous content and is apparently unsuitable for the Great Crested Flycatcher. It also occurs west of Lake Superior from Thunder Bay to the Rainy River District, and there is unconfirmed breeding evidence as far north as the Red Lake block, about 200 km north of the previously documented range (Peck and James 1987). The gaps in the extreme southwest may be the result of a lack of suitable wooded land, rendering the species rare or absent in some areas. The continued destruction of woodlands in the southwest, where income from that extra bushel of corn continues to be more important than the conservation of natural areas, would probably reduce numbers further in that area.

Abundance estimates show that the Great Crested Flycatcher is generally common where it occurs in Ontario. The highest estimates are concentrated near the southern fringe of the Shield, especially to the north of Kingston. The few estimates from the north indicate lower abundance near the northern edge of the range than in the south.-- *G. Bennett*

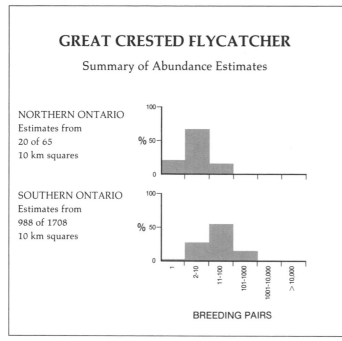

GREAT CRESTED FLYCATCHER

Summary of Abundance Estimates

NORTHERN ONTARIO
Estimates from
20 of 65
10 km squares

SOUTHERN ONTARIO
Estimates from
988 of 1708
10 km squares

BREEDING PAIRS

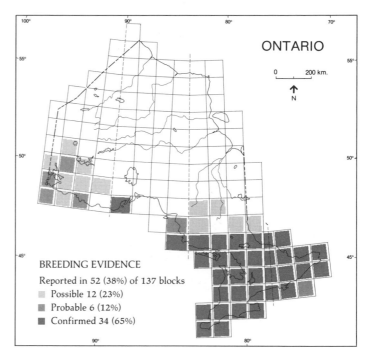

ONTARIO

0   200 km.

N

BREEDING EVIDENCE
Reported in 52 (38%) of 137 blocks
  Possible 12 (23%)
  Probable 6 (12%)
  Confirmed 34 (65%)

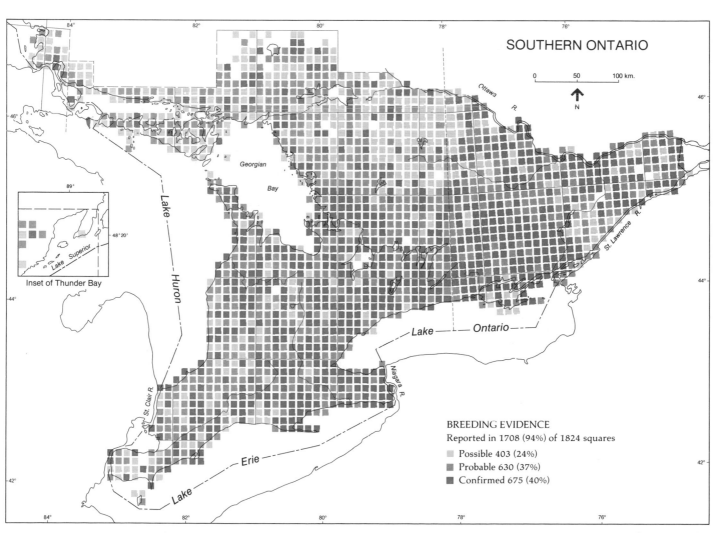

SOUTHERN ONTARIO

0   50   100 km.

N

Inset of Thunder Bay

Georgian Bay

Lake Huron

Lake Superior

Lake Ontario

Lake Erie

St. Clair R.

Niagara R.

Ottawa R.

St. Lawrence R.

BREEDING EVIDENCE
Reported in 1708 (94%) of 1824 squares
  Possible 403 (24%)
  Probable 630 (37%)
  Confirmed 675 (40%)

EASTERN KINGBIRD
Tyran tritri
Tyrannus tyrannus

The Eastern Kingbird is undoubtedly the best known and the most conspicuous of the Ontario members of the tyrant flycatcher family. It is well adapted to many rural habitats, both natural and developed, throughout at least the southern two-thirds of the province, and occasionally much further north.

The summer range of the Eastern Kingbird encompasses all but the west coast of North America, north to the Mackenzie Valley and the Maritime Provinces (Godfrey 1986). Its wintering ground is in northwestern South America (AOU 1983). In Ontario, it is a very common breeding bird throughout the agricultural lands of the Great Lakes-St. Lawrence Forest region, where it frequently uses fences or telephone wires as observation perches. It forages over the fields, and nests in hedgerows. In the forested north, it is restricted to clearings such as those along roadsides, in flooded beaver meadows, in extensive burned-over lands, or near settlements. It also nests in stumps and upended roots at the edges of lakes and rivers. The Eastern Kingbird is fairly common in the open country of the Little Clay Belt, from New Liskeard to Englehart, but less so in the Great Clay Belt around Cochrane (pers. obs.). In the more northerly parts of Ontario beyond the Albany River, it becomes progressively scarcer, but even on the shores of James and Hudson Bays the species is occasionally seen.

Because the Eastern Kingbird is never found in unbroken forest, it is likely that, when most of southern Ontario was still forested, its numbers were much lower than in recent years, when population densities of up to 20 birds per 40 ha (100 acres) have been reported (Speirs 1985).

There are very few passerine birds in Ontario which are as easy to find as the Eastern Kingbird. It is reasonably large, and habitually sits on exposed perches for prolonged periods. Intermittently, it may sally after an insect or attack whatever larger bird comes into view; even crows and hawks are vigorously pursued. Both of these habits further enhance its visibility. Apparently relying on its aggressiveness to ward off predators, the Eastern Kingbird may build its nest in places that are poorly concealed, and often draws attention to its presence by vocally confronting any intruder. The nest itself is usually less than 6 m off the ground, on a branch of a tree or bush or in some more exposed place (Godfrey 1986). Where there is a river or lake present, the nest will often be directly over the water and thus easily spotted from a canoe. It is not surprising that the species is confirmed as breeding in an exceptionally high proportion (68%) of the southern Ontario squares in which it was found.

There are few distribution maps in this atlas which show as solid a pattern south of the Canadian Shield as that of the Eastern Kingbird. Those few squares in which it was not recorded at all are mostly open water. The area in southwestern Ontario in which confirmed breeding is discontinuous is an area of intensive agriculture. Hedgerows have been almost completely removed in that part of the province. On the Shield itself, the Eastern Kingbird is somewhat more sporadic, perhaps partly a reflection of less thorough coverage of the area by field workers, but primarily because there is less suitable habitat there. Nevertheless, it is a confirmed breeding species not only throughout the agricultural lands but also in almost every 100 km block in Ontario north to the latitude of Cochrane and Thunder Bay, except for a few blocks containing only slivers of land in the province; only 6 species occur in more squares in southern Ontario than the Eastern Kingbird. There is even one block on the northern James Bay coast where the species nested in 1984: a pair was located between radar sites 415 and 416 (40 km south of Cape Henrietta Maria) on 1 July, and by 14 July they had a nest with 4 eggs. That record, and 2 others near the Hudson Bay coast (2 individuals at the north end of Kiruna Lake in June 1981 and a pair near the east shore of the Winisk River on 9 June 1983) are hundreds of kilometres north of most previously published records (Godfrey 1986).

In much of southern Ontario, the Eastern Kingbird is a common or abundant bird. The Regional Coordinator for the Kingston region has estimated that about 5,800 pairs breed in the 60 squares in his region (R. Weir pers. comm.). This is consistent with the abundance estimates offered by atlassers throughout southern Ontario, though the species is quite uncommon in the forested areas. The highest estimates, of more than 1,000 pairs per square, are from the farmland of Manitoulin Island, and between Lake Huron and the Lake Simcoe area.-- *F.M. Helleiner*

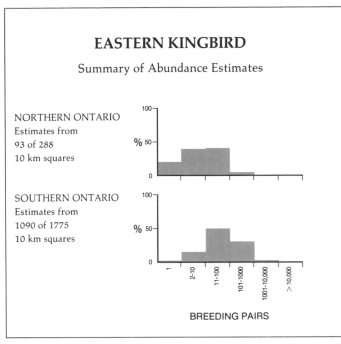

EASTERN KINGBIRD

Summary of Abundance Estimates

NORTHERN ONTARIO
Estimates from
93 of 288
10 km squares

SOUTHERN ONTARIO
Estimates from
1090 of 1775
10 km squares

BREEDING PAIRS

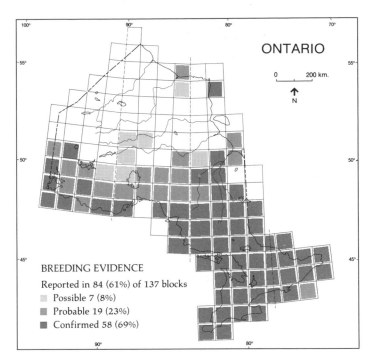

ONTARIO

0   200 km.

N

BREEDING EVIDENCE

Reported in 84 (61%) of 137 blocks

Possible 7 (8%)
Probable 19 (23%)
Confirmed 58 (69%)

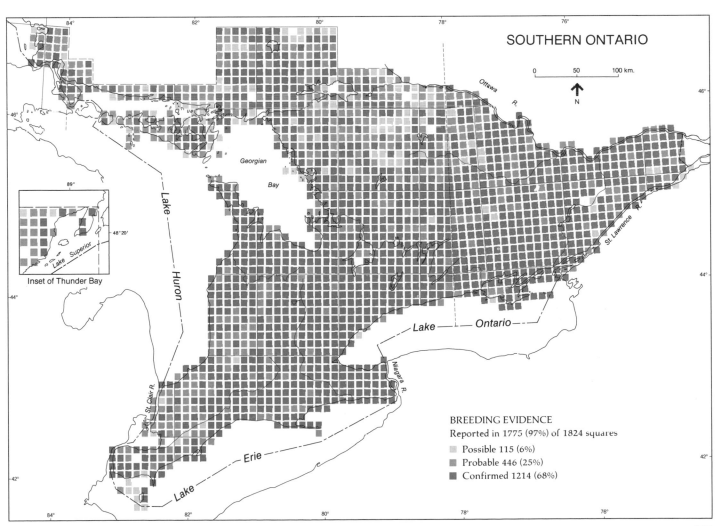

SOUTHERN ONTARIO

0   50   100 km.

N

Inset of Thunder Bay

BREEDING EVIDENCE

Reported in 1775 (97%) of 1824 squares

Possible 115 (6%)
Probable 446 (25%)
Confirmed 1214 (68%)

HORNED LARK
Alouette cornue
Eremophila alpestris

The Horned Lark is widely distributed in hot and cold deserts, on prairies, and on agricultural land in the northern hemisphere. It is the only member of the lark family which is native to the Americas. In the eastern hemisphere it breeds from North Africa and southeastern Russia, north to the arctic coasts of Norway and Siberia. In the Americas, its breeding range extends from 75^{0}N in the Canadian arctic to 5^{0}N in the mountains of Colombia, but it vacates all but the southernmost parts of Canada in winter. As many as 40 subspecies have been distinguished, of which 21 are in North America (AOU 1957, Vaurie 1959). The colour of its plumage varies geographically, with the dorsal surface tending to match that of its nesting substrate. As in many other species, the large subspecies occur in colder climates.

The breeding habitat of the Horned Lark in Ontario includes pastures and arable land, sparsely vegetated fields, and tundra. It is common in suitable open country but is absent from heavily forested regions, as can be clearly seen from the maps. Isolated occurrences within such regions are confined to airfields and other extensive man-made clearings.

Prior to settlement, the Horned Lark was probably absent as a breeding bird from Ontario, except along the James and Hudson Bay coasts. Pickwell (1931, 1942) has described the spread of the Prairie Horned Lark (*E. a. praticola*) into eastern North America. It first appeared in Ontario, probably from Michigan (Pickwell 1942), about 1868 (McIlwraith 1894) and subsequently spread throughout the cleared parts of southern and northeastern Ontario. In recent times, its range has probably contracted in parts of southcentral Ontario following abandonment of agriculture. Fleming (1901) described it as "an abundant breeding resident" in Muskoka and Parry Sound, but now there are only scattered records for those regions.

The Horned Lark is an early nester in the south, with eggs often reported in April. By mid-summer, the species becomes inconspicuous and may easily be missed by a casual observer. The song, given from the ground or in flight, is distinctive and repeated often. The species is frequently found along roadsides, so it is unlikely to be missed in squares that were covered early in the year. Confirmation of breeding is more difficult. Nests are in a depression on the ground and are surprisingly inconspicuous, even when situated on unvegetated ground such as ploughed fields. Nests are most easily found by watching females during the nest-building stage or by following either sex when they are carrying food to the nestlings. The young leave the nest before they can fly and are fed by their parents for several more days.

Confirmed breeding represents 40% of the records, and of these, 62% consist of recently fledged young and 22% of adults feeding young. Only 13% of confirmations are of nests with eggs or young, which attests to the difficulty of finding nests. The maps give a very accurate indication of distribution, but it is probable that the species breeds in most of the squares for which there are records, even though the evidence is, in many cases, circumstantial.

Four subspecies of the Horned Lark occur in Ontario. The smallest race is the Prairie Horned Lark, which breeds in southern Ontario and the Clay Belts. It is widespread throughout the agricultural areas of the south, but in forested regions it is sparsely distributed and confined to cleared sites, particularly airfields. For example, the records at Sudbury, North Bay, Emsdale, and Muskoka airports are all isolated from any other records nearby. The records for northwestern Ontario, perhaps including the rather surprising observation of a single bird in July on the Pikangikum airstrip, almost 100 km north of Red Lake, presumably refer to *E. a. enthymia*, a race which primarily inhabits the central US and is found in the adjacent parts of Manitoba. The Northern Horned Lark (*E. a. alpestris*) breeds only along the coasts of James and Hudson Bays from Moosonee to Cape Henrietta Maria and westward, where it is a common breeder in June and July on treeless raised beach ridges paralleling the shores. Somewhere along the strip of tundra between Winisk and Fort Severn, it is replaced by, or intergrades with, a fourth subspecies, Hoyt's Horned Lark (*E. a. hoyti*) (Snyder 1957a).

The abundance estimates indicate that the Horned Lark is a common bird wherever it occurs, but especially in the agricultural lands of southwestern Ontario, where it is very common or abundant. It is almost totally absent from vast areas of northern Ontario.-- *D.J.T. Hussell*

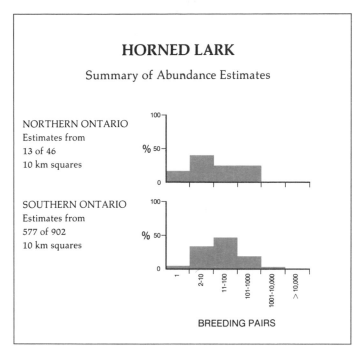

HORNED LARK

Summary of Abundance Estimates

NORTHERN ONTARIO
Estimates from
13 of 46
10 km squares

SOUTHERN ONTARIO
Estimates from
577 of 902
10 km squares

BREEDING PAIRS

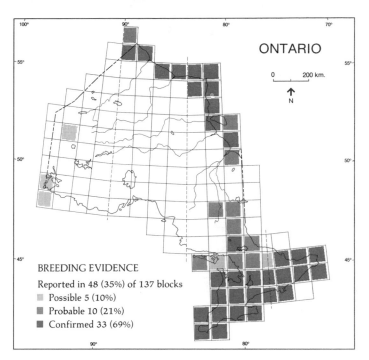

ONTARIO

0   200 km.

N

BREEDING EVIDENCE
Reported in 48 (35%) of 137 blocks

Possible 5 (10%)

Probable 10 (21%)

Confirmed 33 (69%)

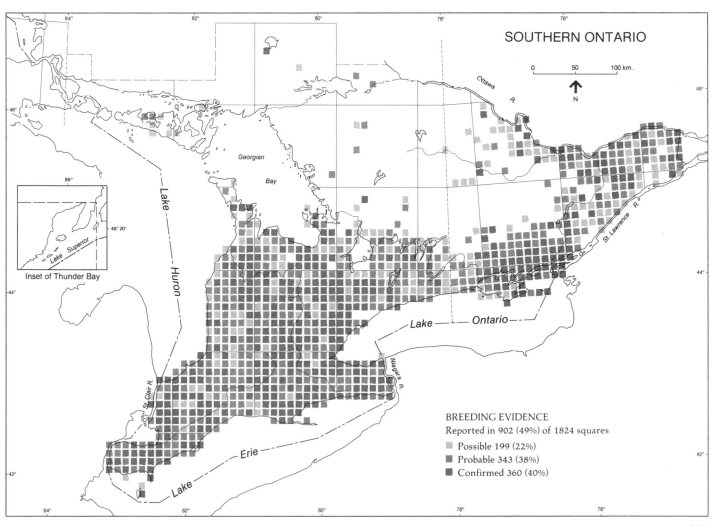

SOUTHERN ONTARIO

0   50   100 km.

N

Inset of Thunder Bay

BREEDING EVIDENCE
Reported in 902 (49%) of 1824 squares

Possible 199 (22%)

Probable 343 (38%)

Confirmed 360 (40%)

PURPLE MARTIN
Hirondelle noire
Progne subis

R. TUCKERMAN 1986

The Purple Martin is a familiar bird, thanks to the many people who erect nesting houses for it. The species is absent in the mid-west US, but breeds locally across much of the rest of North America from south of the Boreal Forest region in southern Canada to Mexico. It winters in South America. In Ontario, it is restricted to the south and the northwest (from Thunder Bay to the Manitoba border). It nests almost exclusively in colonial bird houses, often near large bodies of water. It forages for winged insects, its major source of food (Bent 1942), over the open land and water around houses and cottages.

Before European settlement, the Purple Martin nested in old woodpecker holes, in hollow trees, and on cliff ledges. The early history of the species in Ontario is poorly known. McIlwraith (1894) stated that is was "generally distributed but nowhere abundant." In the early 1900's, it was declining in parts of Ontario, *e.g.*, Toronto, and Leeds, Lanark, and Renfrew Counties (Macoun and Macoun 1909), perhaps because of nest site competition from the House Sparrow, which had recently invaded Ontario. At the same time, however, it was increasing in Bracebridge and "the settled parts of Muskoka" (Macoun and Macoun 1909). Since then, the species has become more abundant and more widely distributed. Its present range is similar to that described by Baillie and Harrington (1937).

The Purple Martin is a noisy, colonial species, that almost always nests near human habitation, where it is highly visible, even to a casual observer. Where nesting houses have been provided, it is a simple matter to confirm that it is breeding. Early May until mid-July, when most atlas field work took place, is also the period when breeding occurs (James *et al.* 1976). Of all atlas records, 72% were of confirmed breeding, and 87% of these were nests (including

those that were observed while being built and those no longer in use). As it is difficult to see inside the nest, most confirmed records were observations of adults entering or leaving the nest. Many of the possible breeding records may be of individuals scouting for new nesting sites, or individuals foraging away from the nest site. Natural nest sites are almost completely unknown in Ontario. Two reported colonies were found in holes in a limestone cliff and inside an old tree (Peck and James 1987).

Atlas field work has provided an excellent picture of Purple Martin distribution in Ontario. The species is widespread along the shores of Lakes Huron, Erie, Nipissing and Ontario, along the edge of the Canadian Shield, and north along the shore of Georgian Bay, thinning out towards the North Channel, where cottage development, with its abundance of Purple Martin houses, is more sporadic. The population west of Thunder Bay is in an area that contains habitat similar to that found along the southern edge of the Canadian Shield: Great Lakes-St. Lawrence Forest, with more deciduous than coniferous trees, and developed areas. Throughout the areas where the species occurs, there are large lakes and rivers and many dwellings where people have erected houses for the species to nest in. Such locations are scarce in northern Ontario and on the Algonquin Highlands, and the species is almost completely absent from these areas. However, it is unclear whether the Purple Martin would expand its breeding range if nest boxes were provided north of its present range and in the Algonquin Highlands, or whether there are other factors, perhaps climatic, that limit the species distribution.

BBS data indicate that the Purple Martin is most abundant in southern Ontario along the edge of the Great Lakes, especially in the extreme southwest and in the Niagara and Bruce Peninsula areas (Speirs 1985). This is consistent with the concentration of atlas records. Inland, and farther north, the species is found much more locally, and is less common overall.-- *P.D. Taylor*

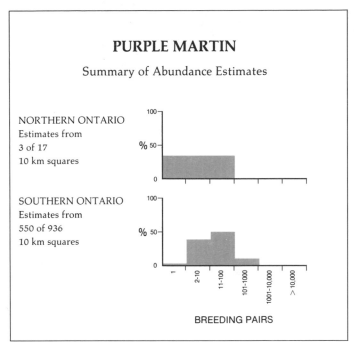

PURPLE MARTIN

Summary of Abundance Estimates

NORTHERN ONTARIO
Estimates from
3 of 17
10 km squares

SOUTHERN ONTARIO
Estimates from
550 of 936
10 km squares

BREEDING PAIRS

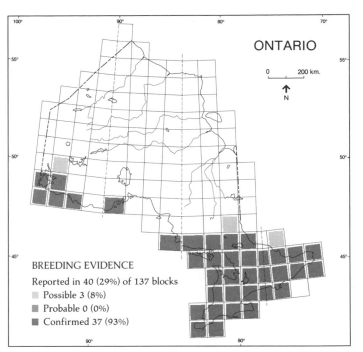

ONTARIO

0   200 km.

N

BREEDING EVIDENCE
Reported in 40 (29%) of 137 blocks
Possible 3 (8%)
Probable 0 (0%)
Confirmed 37 (93%)

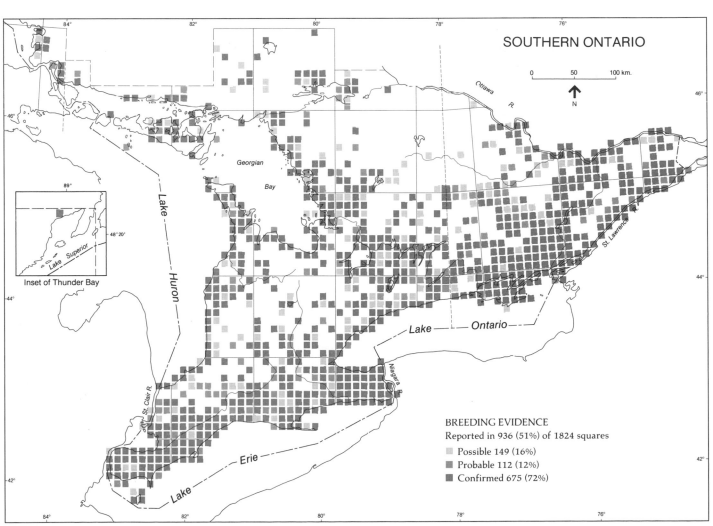

SOUTHERN ONTARIO

0   50   100 km.

N

Inset of Thunder Bay

BREEDING EVIDENCE
Reported in 936 (51%) of 1824 squares
Possible 149 (16%)
Probable 112 (12%)
Confirmed 675 (72%)

TREE SWALLOW
Hirondelle bicolore
Tachycineta bicolor

During the breeding season, the beautiful, bright, iridescent plumage and graceful flight of the Tree Swallow are familiar sights. It breeds from coast to coast from near the northern limit of trees south to the southern US, and winters from the southern states south to Central America and the West Indies.

Widespread in North America, the Tree Swallow is a common and widely distributed bird in Ontario. Its preferred habitat for feeding is open areas near ponds, lakes, and wet areas, although the species is often seen far from water. Nests are found in the vicinity of suitable feeding habitat, in tree cavities such as abandoned woodpecker holes, or in artificial nest-boxes. Tree Swallow populations have undoubtedly benefitted from the increasing popularity of nest-boxes (Holroyd 1975, Lumsden 1986). However, Erskine (1979) estimated that on the order of only 2% of the Canadian population used nest-boxes and concluded that any increase effected by their use was almost certainly insufficient to counterbalance loss of natural cavities in the settled parts of Canada. Nevertheless, BBS data indicate significant increases in Tree Swallow populations in the eastern and central regions of the continent in the 1966-79 period (Robbins *et al.* 1986).

Although nest-boxes are often put up in Ontario primarily for Eastern Bluebirds, they are readily accepted by the Tree Swallow. Although nest-box owners frequently discourage less attractive species such as the European Starling and the House Sparrow, they usually welcome the Tree Swallow (and, in fact, often mistakenly identify it as the Eastern Bluebird). The birds evidently prospect widely, as they will occupy nest-boxes erected far from trees, including those located beyond the tree-line. The Tree Swallow prefers boxes of particular dimensions if offered a choice, but many types of cavities are accepted, including those with a wide range of entrance sizes (Lumsden 1986). Other cavity-nesters (such as the House Wren) will take over nesting holes once surrounding shrub has reached the height of the nest entrance, making it less suitable for this species. Although it is normally a solitary nester because of the distribution of natural sites, the Tree Swallow will tolerate neighbours in nest-boxes 10 m away and sometimes much closer.

Although the Tree Swallow is among the first migrant species to arrive on the breeding grounds in spring, its nesting in Ontario does not begin until early May (James *et al.* 1976). Tree Swallows often feed in the vicinity of their breeding sites and are not particularly shy about visiting the nest when human observers are in the area. While artificial nest-boxes are easily examined, many natural nesting sites are out of reach. When eggs are in the nest, parents enter the nest-hole readily, and the male often perches near the nest entrance for long periods. The birds usually approach intruding humans in a display which can vary from silent circling to vigorous diving accompanied by loud chattering. Adults feed the nestlings once every 2 to 3 minutes (Kuerzi 1941, Quinney 1986), and can readily be observed carrying food to the nest-hole. It is not surprising, therefore, that two-thirds of the records (79% in the south) are of confirmed breeding, and that 72% of these are of nests with young or adults entering or leaving a nest. Normally only a single brood is raised per season, and the birds disperse from the breeding sites shortly after the young fledge. Early and late in the season, prospecting birds may enter holes without a nest being present, but such visits are infrequent and should not have caused false reports of confirmed breeding by atlassers.

The map shows the Tree Swallow to be ubiquitous in southern Ontario and widespread throughout the province where its breeding and feeding habitat requirements are met, *i.e.*, almost everywhere except on the tundra near Cape Henrietta Maria, where there are no trees large enough to allow nesting. In southern Ontario, the proportion of squares with probable or confirmed breeding records (92%) is exceptionally high. The Tree Swallow is among the most widespread breeding species in Ontario, having been reported in 128 blocks.

The concentration of possible and probable breeding records in the extreme southwest suggests that Tree Swallow numbers are lower in that area. This appears to be generally in agreement with BBS data, which show relatively low concentrations of the species in that part of Ontario (Speirs 1985). Abundance estimates suggest that the Tree Swallow is common throughout the province, and is especially common in the south. Abundance estimates of over 1,000 pairs per square were provided for 61 squares, with somewhat of a concentration of these high estimates in the Kawartha Lakes area. Its wide distribution and large population size in the heavily altered agricultural areas will continue to be enhanced as long as people provide nest-boxes in suitable habitat.-- *T.E. Quinney and E.H. Dunn*

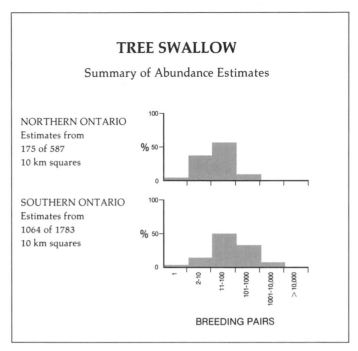

TREE SWALLOW

Summary of Abundance Estimates

NORTHERN ONTARIO
Estimates from
175 of 587
10 km squares

SOUTHERN ONTARIO
Estimates from
1064 of 1783
10 km squares

BREEDING PAIRS

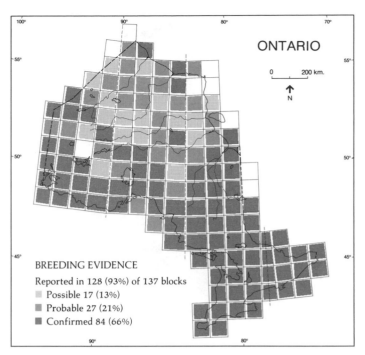

ONTARIO

0   200 km.

N

BREEDING EVIDENCE
Reported in 128 (93%) of 137 blocks

Possible 17 (13%)
Probable 27 (21%)
Confirmed 84 (66%)

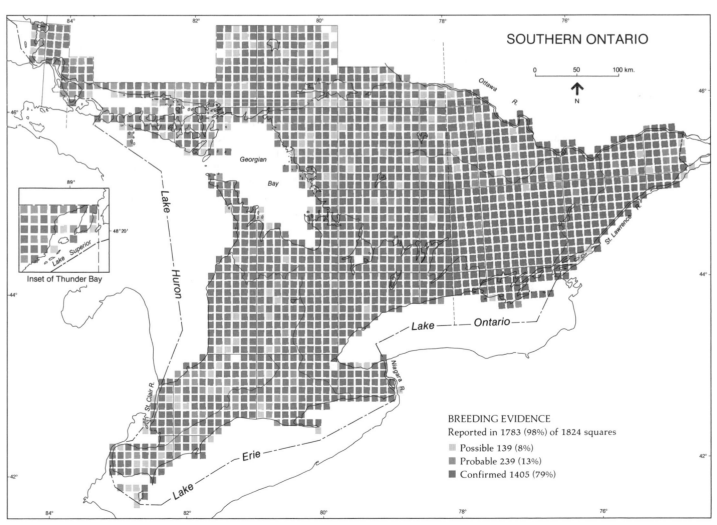

SOUTHERN ONTARIO

0   50   100 km.

N

Georgian
Bay

Lake Huron

Lake Superior

Inset of Thunder Bay

St. Clair R.

Lake   Erie

Niagara R.

Lake — Ontario

Ottawa R.

St. Lawrence R.

BREEDING EVIDENCE
Reported in 1783 (98%) of 1824 squares

Possible 139 (8%)
Probable 239 (13%)
Confirmed 1405 (79%)

NORTHERN ROUGH-WINGED SWALLOW
Hirondelle à ailes hérissées
Stelgidopteryx serripennis

The Northern Rough-winged Swallow is an inconspicuous, often overlooked, but nonetheless fascinating member of Ontario's avifauna. It is widely distributed in North and Central America, from the southern edge of the Boreal Forest zone south to Costa Rica. It breeds throughout southern Ontario and in the Rainy River District of northwestern Ontario in a wide variety of open and semi-open habitats. Winged insects (mostly Diptera and Hymenoptera) comprise the bulk of its diet (Lunk 1962). The species is common in southern Ontario, but becomes rarer as one travels north onto the Canadian Shield.

As with other swallow species (*e.g.*, Purple Martin, Cliff Swallow, Barn Swallow), the Northern Rough-winged Swallow has apparently expanded its range and increased in abundance in eastern North America since European settlement. In 1894, McIlwraith (1894) found it breeding only at London, Ontario, and in 1920, Fleming and Lloyd (in Lunk 1962) suggested that a "general extension of range" was occurring in Ontario. The apparent range expansion and increase in abundance are at least partly due to an increase in the numbers and competence of observers. How much the range has actually changed is very difficult to determine. The present range, as portrayed on the atlas maps, is essentially unchanged from that described by Baillie and Harrington (1937). BBS data indicate a continent-wide increase in Northern Rough-winged Swallow populations in the 1966-79 period (Robbins *et al.* 1986).

In Ontario, as elsewhere, the Northern Rough-winged Swallow nests frequently in situations such as drainage holes in bridges, sandy roadbanks, and gravel pits (Peck and James 1987). These sites are accessible to atlassers, so the species is not difficult to find. Nesting also occurs in natural locations, such as river banks (where it frequently nests on the fringes of Bank Swallow colonies), and these localities can be more difficult to access in areas with few roads. The species is therefore probably somewhat underrepresented away from road-accessible areas. The adults are vocal and aggressive around the nest site (Lunk 1962), making it easy to confirm nesting when individuals are discovered. Fifty-eight percent of all atlas records confirm breeding, and most of these (61%) were of adults entering or leaving a nest hole.

It is interesting to note that the Belted Kingfisher and Bank Swallow range generally throughout the province, while the Northern Rough-winged Swallow, which often shares their nesting habitat and even uses their old nests in the southern half of the province, usually extends no farther north than the Great Lakes-St. Lawrence Forest zone and the Clay Belt. The two records of possible breeding farther north indicate that it occurs even there, though it is certainly thinly distributed in that area.

The Northern Rough-winged Swallow is widespread south of the Shield, where it was reported in most squares. The lower proportion of squares with records in southeastern Ontario suggests that habitat is less suitable there than it is to the southwest, although there is no obvious explanation as to what factors are responsible. There are records from most squares on the south end of the Canadian Shield. Farther north, there are very few reports from the Algonquin Highlands, although the east-west Highway 60 corridor from Barry's Bay to the western edge of Algonquin Prov. Park is discernible, suggesting the importance of artificial structures to the local occurrence of the species. There is also a concentration of records along the Highway 17 corridor between Espanola and Blind River; except for these squares, records north of Parry Sound are scattered. As is true of the Purple Martin, the Northern Rough-winged Swallow extends its range only slightly beyond the boundary of southern Ontario in the northeast, the northernmost atlas record in that area being on the Mattagami River, south of Smooth Rock Falls. As is also true of the Purple Martin, the range of the Northern Rough-winged Swallow extends into the Great Lakes-St. Lawrence Forest zone in northwestern Ontario.

Abundance estimates show that the Northern Rough-winged Swallow is common south of the Canadian Shield and becomes increasingly uncommon and more difficult to find on the Canadian Shield. The species is less common than the similar Bank Swallow, most likely because its solitary habits restrict the population that can be accommodated in the limited number of suitable nesting sites. These nesting habits are poorly understood; further field work could reveal a larger population than is known at present.-- *P.D. Taylor*

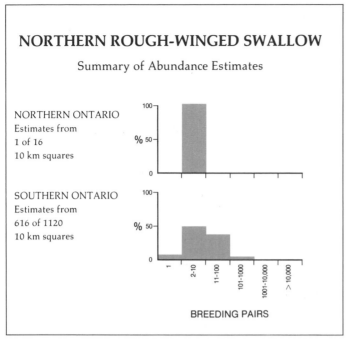

NORTHERN ROUGH-WINGED SWALLOW

Summary of Abundance Estimates

NORTHERN ONTARIO
Estimates from
1 of 16
10 km squares

SOUTHERN ONTARIO
Estimates from
616 of 1120
10 km squares

%

BREEDING PAIRS

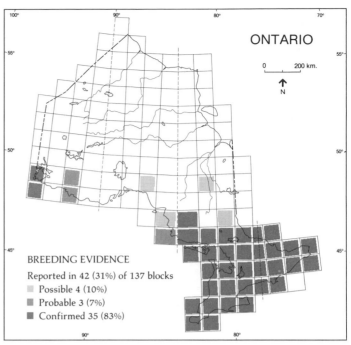

ONTARIO

0   200 km.

N

BREEDING EVIDENCE
Reported in 42 (31%) of 137 blocks

Possible 4 (10%)
Probable 3 (7%)
Confirmed 35 (83%)

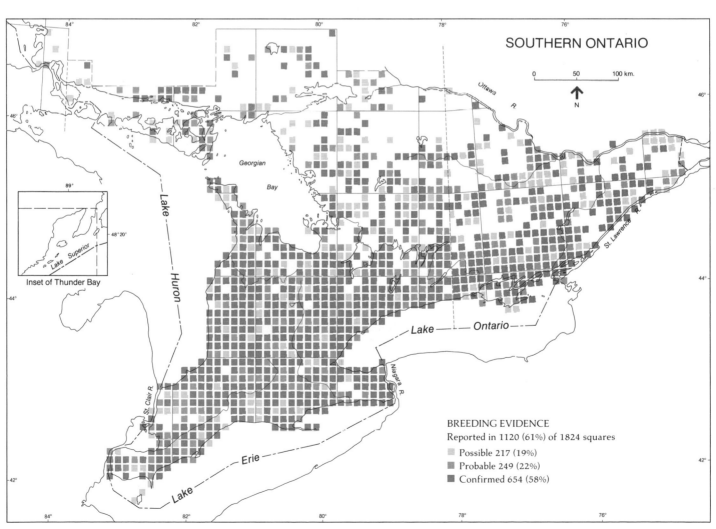

SOUTHERN ONTARIO

0   50   100 km.

N

Inset of Thunder Bay

Lake Superior

Lake Huron

Georgian Bay

Ottawa R.

St. Lawrence R.

Lake Ontario

Niagara R.

St. Clair R.

Lake Erie

BREEDING EVIDENCE
Reported in 1120 (61%) of 1824 squares

Possible 217 (19%)
Probable 249 (22%)
Confirmed 654 (58%)

BANK SWALLOW
Hirondelle de rivage
Riparia riparia

The Bank Swallow is one of the more widely distributed passerines in the world. It breeds over much of Eurasia and North America, with isolated populations in Africa, and winters in Africa and South America. In North America, it ranges from western Newfoundland across Canada to Alaska and the Bering Sea, extending north of the Arctic Circle in the Mackenzie Valley, and south to about latitude 35°N in the US.

Bank Swallows are among the most social members of their family. Breeding almost invariably occurs in a colony, often of substantial size, and birds usually feed and roost in flocks as well. Unlike the similar Northern Rough-winged Swallow, the Bank Swallow only rarely nests in man-made holes such as drain-pipes in walls. The nest is an untidy collection of straw and feathers, situated at the end of a 0.6 m long burrow, which is excavated by both sexes. Nest burrows can be made in reasonably friable sand or gravel banks; clay or very stony soils, for example, make tunnelling too difficult. Nevertheless, the species will sometimes nest in rather bizarre circumstances, such as in piles of sawdust. Like all swallows in North America, the Bank Swallow has benefitted substantially from human activities which provide nesting sites. In this case, sand-pits, railway embankments, and road-cuttings which pass through suitable soil all provide good colony sites. The species also nests, however, in river banks and lakeshore bluffs, and a much larger proportion of Bank Swallow nests are in genuinely natural settings than is the case for the Barn, Cliff, and Northern Rough-winged Swallows, or the Purple Martin.

The breeding distribution of the Bank Swallow in Ontario is apparently determined largely by two factors: the suitability of foraging areas and the availability of nesting sites. More than most swallows, this species seems to feed in close proximity to water. However, there are large areas of Ontario which are abundantly supplied with lakes and rivers where the species is sparse or absent. Presumably, the absence of nesting sites is a major factor in these cases. Although the Bank Swallow obtains its food in flight, there is little evidence that birds habitually commute long distances between the colony and foraging areas. Hence, the presence of suitable banks near water is crucial.

Bank Swallows make no attempt whatsoever to conceal their nests. Furthermore, their nesting locations in pits and cuttings are easily visited. In the north, the rivers, which seem to be their preferred habitat, were also the main routes of access for atlas coverage. Thus, the high proportion of confirmed breeding and of nests (65% of confirmed records) is easily understood.

The maps probably portray quite accurately the range of the Bank Swallow in the province. The Bank Swallow is widely distributed over the agricultural section of southern Ontario. In fact, between Windsor and Kingston, the only major gaps are in Essex Co and in parts of Haldimand-Norfolk and Niagara RMs. In both these areas, the predominant soil type is a heavy clay or clay loam, which may be unsuitable for excavation by Bank Swallows. Furthermore, the almost total absence of relief in the flatlands east of Windsor provides few of the bluffs and banks necessary for successful nesting. On the Canadian Shield, the frequency of occurrence drops off sharply. There are major areas without Bank Swallows on the southern fringe of the Shield in Hastings and Frontenac Counties, for example, and records become much sparser north of a line from Sudbury to Pembroke. In the Canadian Shield areas of northern Ontario, it is absent over large sections of the Districts of Kenora and Cochrane. North of the Shield, however, it reappears. Whereas the rivers of the Shield most often have rocky banks, the many rivers of the Lowland are lined with sand banks and cliffs in many areas, providing numerous suitable sites for colonies. For example, on a 260 km stretch of the Fawn River across the Lowland in 1983, 6 colonies were discovered, with 19, 11, 31, 117, 40, and 13 nest holes, respectively (M. Cadman pers. comm.). The rivers in far northern Ontario are important not merely as nesting sites, but also as feeding areas; birds are rarely seen on muskeg away from the rivers.

It is clear from the abundance estimates that the Bank Swallow is a common bird wherever it occurs, especially in the south. There are 18 squares (all of them in the south) with estimates of over 1,000 pairs.-- *D. Brewer*

BANK SWALLOW

Summary of Abundance Estimates

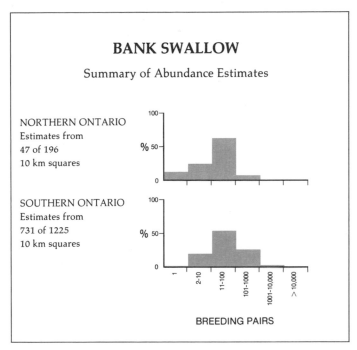

NORTHERN ONTARIO
Estimates from
47 of 196
10 km squares

SOUTHERN ONTARIO
Estimates from
731 of 1225
10 km squares

BREEDING PAIRS

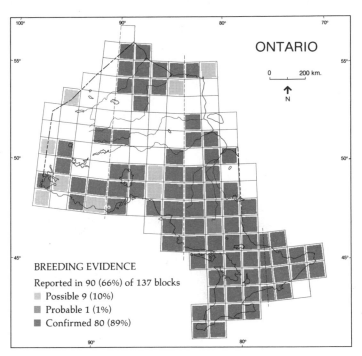

ONTARIO

0   200 km.

N

BREEDING EVIDENCE

Reported in 90 (66%) of 137 blocks

Possible 9 (10%)
Probable 1 (1%)
Confirmed 80 (89%)

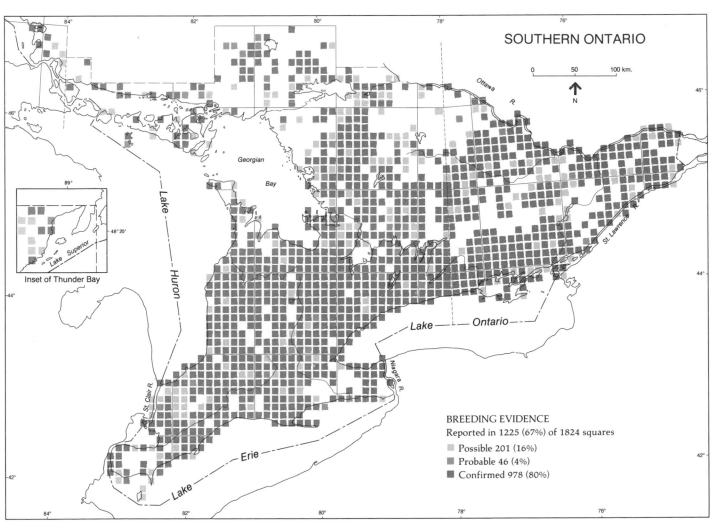

SOUTHERN ONTARIO

0   50   100 km.

N

Inset of Thunder Bay

BREEDING EVIDENCE

Reported in 1225 (67%) of 1824 squares

Possible 201 (16%)
Probable 46 (4%)
Confirmed 978 (80%)

CLIFF SWALLOW
Hirondelle à front blanc
Hirundo pyrrhonota

R. TUCKERMAN 1986

The Cliff Swallow is a widespread breeding bird in North America outside of the southeastern US, even occurring locally well north of the tree-line. It winters in South America. In Ontario, the species is widely distributed south of the Hudson Bay Lowland, and it is associated almost exclusively with areas of human settlement, nesting under the eaves of buildings or under concrete bridges. It forages in nearby open areas, pasture, and agricultural land. Winged insects are its major source of food (Bent 1942). The species is very concentrated in its distribution. Thus, it is often thought to be the rarest swallow in Ontario, but its colonial nesting habits can make it locally abundant.

Before land was settled and cleared for agriculture, the Cliff Swallow was probably rare and very locally distributed across eastern North America, nesting only on cliffs and bluffs. After settlement, the species began nesting on man-made structures and using the newly-cleared land for foraging. It rapidly increased in abundance and became more widespread (Bent 1942). By the turn of the century, the species was well established in Ontario, occupying much the same range as today. McIlwraith (1894) called it "... not equally abundant at all points, but still ... very numerous throughout the province." Since that time, the Cliff Swallow has undergone numerous and sometimes major changes in local distribution and abundance while apparently maintaining its overall population size. BBS data indicate a continent-wide increase in the 1966-79 period, and that the species is expanding its range in the southwest (Robbins *et al.* 1986).

Because of its association with man-made structures, its preferences for open, settled areas, and its usual colonial nature, the Cliff Swallow is an easy bird to find. Three-quarters of all records are of confirmed breeding, and of these, 90% are nests. Looking inside the jug-shaped nests is diffi-

cult, so most (52%) confirmations were recorded as adults entering inaccessible nests. Natural nesting sites are difficult to discover and appear to be quite rare in Ontario (Peck and James 1987). They are mentioned more frequently in older distributional accounts (*e.g.*, McIlwraith 1894), and their apparent rarity may be attributable to the ease with which atlassers or the birds can find nest sites associated with man.

The species has a tendency to abandon traditional colonies and to appear at locations where it has been absent for some time. The species is genuinely scarce or absent in 5 areas of the province. In extreme southwestern Ontario, it may be intensive agriculture and the architectural character of the associated buildings that restrict the species to a few scattered squares. It is completely absent from the Niagara Peninsula, although it formerly nested there in small numbers (H. Lancaster pers. comm.), and suitable habitat exists there even today. Across the Niagara River, in New York State, the species is also locally absent (NYBBA preliminary maps) but was formerly common (Beardslee and Mitchell 1965). Similarly, the Cliff Swallow is also largely absent from Leeds Co, and from adjoining areas of New York State, just across the St. Lawrence River (NYBBA preliminary maps). Agricultural practices, lack of suitable nesting sites, competition from House Sparrows, and migration patterns have all been postulated as possible reasons for its absence from these parts of Ontario. Recent research (Brown and Brown 1986) suggests that the species may be assessing the degree of parasitism in former nest-sites and avoiding heavily infested colonies. This kind of behaviour could explain the patchy distribution of the species in Ontario.

In the Algonquin Highlands, there are Cliff Swallow nesting records only in the few areas of permanent human settlement. There have never been any nests reported there on natural sites. In far northern Ontario, the species was found nesting on man-made structures in or near the communities of Pickle Lake, Moosonee, Fort Albany, and Big Trout Lake. Two observers, paddling down a particularly swift section of the Attawapiskat River about 100 km from James Bay, passed a colony of about a dozen active Cliff Swallow nests on a cliff, apparently the only such colony discovered in all of northern Ontario. There are older nest records from Winisk (Peck and James 1987). In the rest of far northern Ontario, the Cliff Swallow appears to be rare or absent, possibly because there are few suitable nesting sites.

In Ontario, BBS data indicate that the Cliff Swallow is most abundant in a belt from the Ottawa Valley to Manitoulin Island, in the Rainy River District, and, to a lesser extent, in the Waterloo and Wellington Co areas (Speirs 1985). Atlas abundance estimates suggest that most squares (93%) contain 100 or fewer pairs of birds, and that it is more common where it does occur in northern Ontario than in the south.-- *P.D. Taylor*

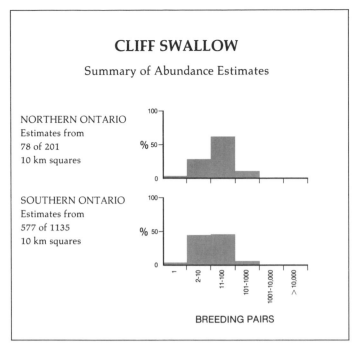

CLIFF SWALLOW

Summary of Abundance Estimates

NORTHERN ONTARIO
Estimates from
78 of 201
10 km squares

SOUTHERN ONTARIO
Estimates from
577 of 1135
10 km squares

%

%

BREEDING PAIRS

1 2-10 11-100 101-1000 1001-10,000 >10,000

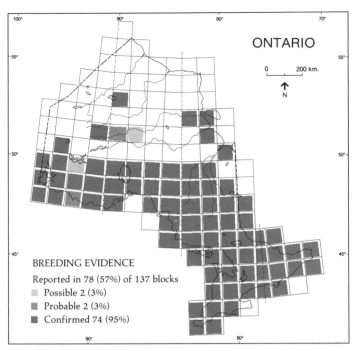

ONTARIO

0   200 km.

N

BREEDING EVIDENCE
Reported in 78 (57%) of 137 blocks

Possible 2 (3%)

Probable 2 (3%)

Confirmed 74 (95%)

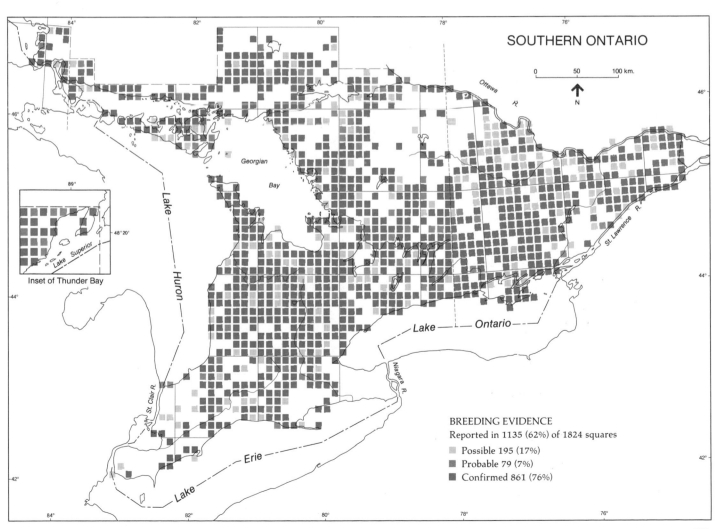

SOUTHERN ONTARIO

0   50   100 km.

N

Ottawa R.

Georgian Bay

Lake Huron

Lake Superior

Inset of Thunder Bay

89°

48°20'

St. Clair R.

Lake Erie

Niagara R.

Lake Ontario

St. Lawrence R.

BREEDING EVIDENCE
Reported in 1135 (62%) of 1824 squares

Possible 195 (17%)

Probable 79 (7%)

Confirmed 861 (76%)

BARN SWALLOW
Hirondelle des granges
Hirundo rustica

"Symbol of summer and talisman of good fortune," Ingram (1974) calls the swift and graceful Barn Swallow. Known as the Swallow in Britain, it breeds throughout the Holarctic region, and is among the most widely distributed of all bird species.

In North America, the Barn Swallow breeds north to central Alaska and Hudson Bay, and winters in South America. There is at least one breeding record from the early 1800's as far north as 67.5°N, just beyond the Arctic Circle (Baird *et al.* 1874). By this century, the northern limit in Ontario may only have been around Huntsville. In 1901, a Barn Swallow was collected at nearby Port Sydney; in 1904, Thompson (1922) reported no swallows of any kind at North Bay, then a remote 13 year-old frontier town just 150 km to the north. During the 1930's, nesting was discovered in towns along the Canadian Pacific Railway line from Lake Superior to Quebec; by the 1950's, the known range extended to the Canadian National Railway northern route right across Ontario (Speirs 1985); and, since the mid-1970's, breeding evidence has been found near Pickle Lake and along James and Hudson Bays to Winisk (Peck and James 1987). A Barn Swallow reported killed in 1924 at Moose Factory (Baillie and Harrington 1937) and a 1953 nest record for Fort Severn (Peck and James 1987) are isolated early records from the far north. BBS data show that the Barn Swallow increased throughout the continental US and Canada in the 1966-79 period (Robbins *et al.* 1986).

The Barn Swallow is usually found in open country near bodies of water such as lakes, rivers and marshes, where it feeds on the wing upon flying insects. Closed forest, barren lands, and marine environments tend to be avoided. It breeds in single pairs or in small, loose colonies and it is well adapted to, even dependent upon, humans. Its mud-and-straw nests are often on beams or ledges, or plastered on sheltered vertical surfaces inside or outside buildings or on other man-made structures such as bridges and culverts (Campbell and Lack 1985, Godfrey 1986). There are even reports of mobile nests located on boats, trains, and recently on the giant rotating receiving dish of the Algonquin Radio-Observatory. Not surprisingly, therefore, in Ontario the Barn Swallow's distribution and abundance coincide with man's.

There are a few recent records of the Barn Swallow nesting in natural habitats in Ontario: on rock cliff faces bordering water, in rocky caves, and - rare indeed - on a white birch at Presqu'ile Prov. Park. It may also nest in cutbanks (Bent 1942). Erskine's (1979) exhaustive study of nearly 5000 Canadian nest records found that only 1% of nests were on cliffs or in caves, the rest being inside (54%) or on the outside (33%) of buildings, and on other structures (12%).

The Barn Swallow is confirmed as breeding in almost every block in which it was recorded. In 86% of squares in which breeding was confirmed, the nest itself was discovered. This easy atlassing reflects the tendency of the species to choose obvious nest sites close to humans and its conspicuous behaviour, especially around the nest. Often it returns faithfully to the same location year after year, sometimes to the same nest (Austin 1961). High fecundity means atlassers get a second chance, too, for it often rears 2, and occasionally 3, broods per season; egg dates range from 15 May to 10 August in Ontario (James *et al.* 1976). The adults and fledged young eventually gather to roost in noticeable, and sometimes spectacular, numbers (*e.g.*, Clark 1984) on marsh vegetation, on telephone wires, or in trees near water, prior to migrating south.

The map indicates that the Barn Swallow is ubiquitous within the settled areas of Ontario. It was reported in 93% of the southern squares and 69% of all Ontario blocks. The northern limit at Fort Severn is consistent with extant Manitoba breeding records at Thompson and Churchill (Godfrey 1986). There are 2 large areas where records of the Barn Swallow are missing. The first is in a remote part of Algonquin Prov. Park, where it was not reported from a contiguous section of over 900 km². This void is comparable to ones that atlassers have discovered in upper New York State in the mountainous interior of Adirondack Park (NYBBA preliminary maps). The second, and by far the larger area, is in extreme northern Ontario inland from the coast. Both areas are largely free of human habitations, but do probably offer a few good natural nesting sites.

This bird is common in the south, locally common to the northern transcontinental railway, and almost absent in the far north.-- *W.R. Clark and C.A. Clark*

BARN SWALLOW

Summary of Abundance Estimates

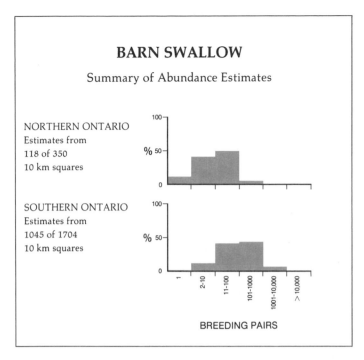

NORTHERN ONTARIO
Estimates from
118 of 350
10 km squares

SOUTHERN ONTARIO
Estimates from
1045 of 1704
10 km squares

BREEDING PAIRS

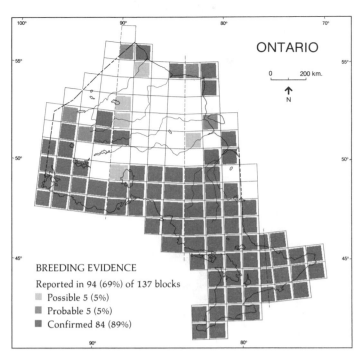

ONTARIO

0   200 km.

N

BREEDING EVIDENCE
Reported in 94 (69%) of 137 blocks
- Possible 5 (5%)
- Probable 5 (5%)
- Confirmed 84 (89%)

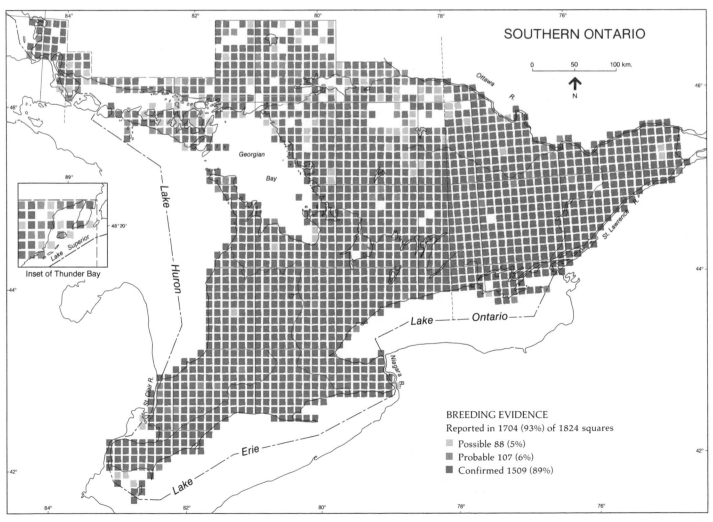

SOUTHERN ONTARIO

0   50   100 km.

N

Inset of Thunder Bay

Georgian Bay

Lake Superior

Lake Huron

Ottawa R.

St. Lawrence R.

Lake Ontario

Niagara R.

St. Clair R.

Lake Erie

BREEDING EVIDENCE
Reported in 1704 (93%) of 1824 squares
- Possible 88 (5%)
- Probable 107 (6%)
- Confirmed 1509 (89%)

283

GRAY JAY
Geai du Canada
Perisoreus canadensis

The Gray Jay is a characteristic permanent resident of Ontario's boreal forests of spruce and fir. Famous for its extreme tamability and soft, appealing plumage, it is one of the best known and most popular of our northern birds.

The range of the Gray Jay stretches across North America from Newfoundland to Alaska. It is resident from the tree-line south to the limit of significant conifer stands in the east and, in the alpine and subalpine forests of the Rocky Mountains, as far as New Mexico and Arizona (Bent 1946). Outlying populations occur in the Black Hills of South Dakota (Smith 1966) and in the Adirondacks (NYBBA preliminary maps). In Ontario, the Gray Jay occurs from the coasts of Hudson and James Bays south to the Algonquin Highlands and surrounding areas of the southern Canadian Shield. There is no indication that this range has changed in historic times (see Baillie and Harrington 1937, Peck and James 1987).

Despite the reputation of the Gray Jay for tameness, even the most confiding of individual birds are very inconspicuous in the warm months of May through September (probably because they are then less interested in humans as a potential source of storable food). Normal atlassing activity, therefore, tended to miss or underestimate this species. Moreover, it has an extremely early breeding season (March to April), which precedes the period when most of the field work took place. Ninety-six percent of all records of confirmed breeding were the result of the observation of fledged young. Fledglings remain with their parents until mid-June, when inter-sibling aggression leads to the expulsion from the natal territory of all but the dominant juvenile (Strickland unpubl.). The expelled birds sometimes succeed in joining other pairs which have no young of their own and, by October, about 30% of surviving juveniles (by then in adult plumage) are accompanying unrelated adults away from their natal territo-

ries. Therefore, the observation of single juveniles in summer with adults does not prove that those adults bred successfully or that the juvenile was produced at that particular location. It is likely, however, that, even in these cases, the young birds have been raised within a few kilometres of their adopted territory, and no serious error in the map should have resulted from the partial June dispersal of juvenile Gray Jays described here.

The maps are probably quite accurate representations of the breeding range of the Gray Jay in Ontario. In northern Ontario, it appears to be an established breeder everywhere. It was confirmed as a breeder in more blocks than any other species. Except along the Hudson and James Bay coasts, northern blocks showing less than confirmed breeding status are almost certainly a consequence of difficult access and the relatively elusive summer behaviour of this species, and are unlikely to represent areas in which the bird in really absent or a non-breeder. In southern Ontario, the stronghold of the Gray Jay is the highlands of Algonquin Prov. Park. There, and especially in the lower regions surrounding the park, where there are fewer conifers, the patchy distribution of records is probably a true reflection of a discontinuous, sparse population inhabiting less than ideal habitat. The Gray Jay is largely or completely absent as a breeding bird from areas of lower elevation (and little spruce/fir cover) on the southern Canadian Shield, including the Frontenac axis, southern Muskoka, and the Lake Nipissing-Upper French River area. The isolation of the Gray Jays of the Algonquin Highlands from those of northern Ontario is a distributional pattern also seen in the maps of several other boreal bird species. The southernmost area where the Gray Jay was detected as a breeding bird during the atlas field work is northern Peterborough Co (see also Sadler 1983). The map also shows 2 outlying Gray Jay records off the Shield - on the south shore of Manitoulin Island and in the Alfred Bog in extreme eastern Ontario. The former is apparently a self-sustaining, though small, population (Nicholson 1981), but it is not known if the Alfred Bog birds have such permanence. A single pair was observed in 1984 and 1985. The Gray Jay nested in a similar bog (Mer Bleue) near Ottawa in 1974 (Ouellet *et al.* 1976) but was not recorded there during atlas field work.

The Gray Jay was recorded in only 16% of the squares in southern Ontario and was deemed to be uncommon even there. This may be accurate in truly peripheral areas of the breeding range, but in Algonquin Park, it breeds at densities in the order of 60 pairs per square (Strickland unpubl.). Other data from La Vérendrye Park, Quebec, where the habitat is similar to that in northern Ontario, give an estimate (up to 140 pairs per 10 km square) that would probably apply to most of Ontario's boreal forest (Strickland 1968). Estimates of that magnitude were made for 16 squares in various parts of northern Ontario, though the majority of the estimates were of fewer than 100 pairs per square.-- *D. Strickland*

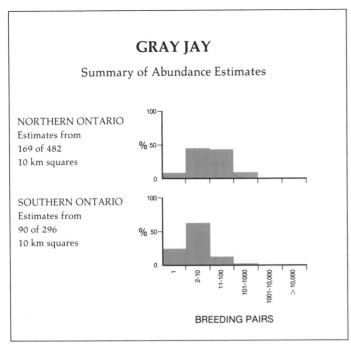

GRAY JAY

Summary of Abundance Estimates

NORTHERN ONTARIO
Estimates from
169 of 482
10 km squares

SOUTHERN ONTARIO
Estimates from
90 of 296
10 km squares

BREEDING PAIRS

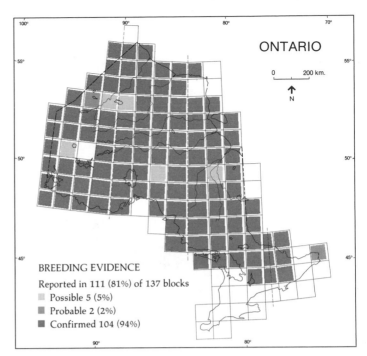

ONTARIO

0   200 km.

N

BREEDING EVIDENCE

Reported in 111 (81%) of 137 blocks

Possible 5 (5%)

Probable 2 (2%)

Confirmed 104 (94%)

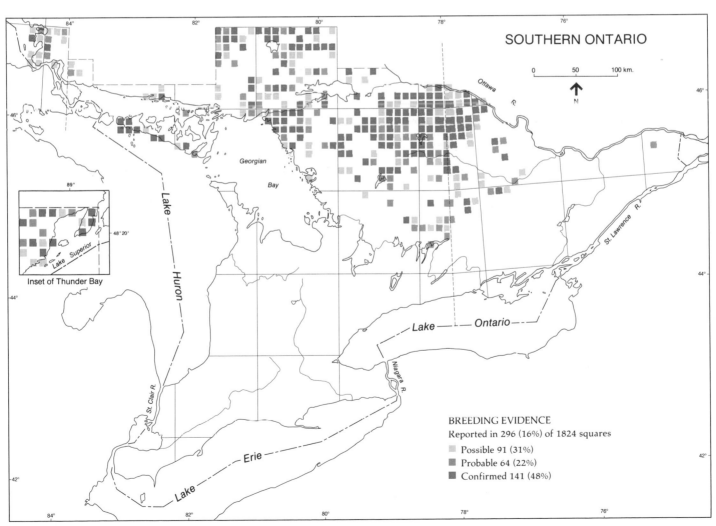

SOUTHERN ONTARIO

0   50   100 km.

N

Inset of Thunder Bay

BREEDING EVIDENCE

Reported in 296 (16%) of 1824 squares

Possible 91 (31%)

Probable 64 (22%)

Confirmed 141 (48%)

BLUE JAY
Geai bleu
Cyanocitta cristata

The Blue Jay is one of southern Ontario's best known and most conspicuous birds. It breeds only in North America, where it is a widespread permanent resident east of the Rocky Mountains, from the Gulf of Mexico to southern Canada. South of the US, its status is that of a rare fall visitor to northeastern Mexico. In Ontario, the breeding range of this species extends north to within the Boreal Forest region, but there its numbers decline sharply. It is absent only in the northern third of the province.

Within its range, the Blue Jay is at home in coniferous, deciduous, or mixed forests. It also frequents thinly wooded river bottoms, hawthorn scrub meadows, and regenerating burn-scarred hillsides. Its habitat requirements are flexible enough to allow it to adapt easily to urban environments (Bent 1946). Bushy ornamental trees and shrubs, widely used in residential landscaping, are much to this bird's liking.

As far back as historical data are available, the Blue Jay has been a common species within its Ontario range (McIlwraith 1894). Its ability to adjust quickly to whatever habitat is available has, no doubt, helped it to maintain its numbers, even in areas where the natural environment has been radically altered.

Although the Blue Jay is gregarious and raucous by nature during most of the year, its nesting habits are those of a solitary, secretive bird. Whereas the vocal activities of many species of passerine birds increase during the nesting season, those of the Blue Jay do not. The boisterous, screaming bird that dominated feeders and woodlots all winter becomes quiet and unobtrusive once nesting time arrives in mid-April. Were the species not so large and colourful, it could easily be overlooked at that time. The nest itself is also well concealed, in a variety of situations, but most commonly in white pines, hemlocks, or other evergreen trees (Bent

1946). Records of actual nests form a remarkably low proportion (27%) of all the confirmed breeding records of the Blue Jay, the majority of confirmed records being observations of fledged young (55%). At this stage, 17 to 21 days after hatching, fledglings already have a voice almost as loud as that of their parents, and are easily detected (Bent 1946). Parents seen carrying food or faecal sacs account for an additional 18% of the confirmed breeding records.

Previously published information about the distribution of the Blue Jay in Ontario consistently indicates that its northern limit is fairly well defined and extends from Sioux Lookout to Kapuskasing, with scattered sightings as far north as Pickle Lake and Fraserdale (*e.g.*, Speirs 1985, Godfrey 1986). In general, the species is confined to the Carolinian and the Great Lakes-St. Lawrence Forest zones in Ontario. As shown on the map, there are confirmed breeding records in northwestern Ontario as far north as the Red Lake area (a nest), and a possible breeding record from a location 90 km west of Pickle Lake, which is well within the Boreal Forest zone but in an area of mixed forest with hardwoods on the esker ridges. Although almost half of the Blue Jay records in southern Ontario are of possible or probable breeding, it seems likely that breeding actually does take place in almost all of those squares, because the habits of the species make it difficult to obtain evidence of confirmed breeding. In general, the map gives an accurate indication of the extent of the Blue Jay's range but probably not of its breeding status within that range.

The abundance estimates suggest that the Blue Jay is a common bird wherever it occurs in southern Ontario, as well as in parts of its northern Ontario range. It was most frequently estimated at densities of 10 to 100 pairs per square, but in 14 squares the estimates were of over 1,000 pairs. The average density in Ontario Co (now Durham RM) forests reported by Speirs (1985) was 15 birds/40 ha. That figure is consistent with an estimate of 1,000 to 10,000 pairs in a completely wooded 10 km square. However, most southern squares lack continuous woodland and would be expected to have fewer Blue Jays.-- *G. Bennett*

BLUE JAY

Summary of Abundance Estimates

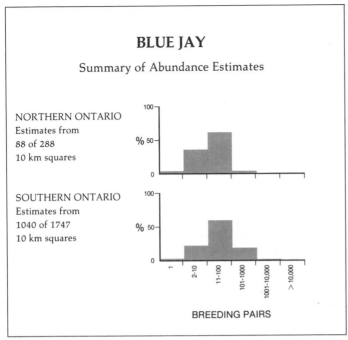

NORTHERN ONTARIO
Estimates from
88 of 288
10 km squares

SOUTHERN ONTARIO
Estimates from
1040 of 1747
10 km squares

BREEDING PAIRS

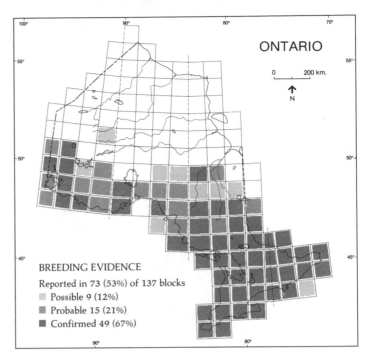

ONTARIO

0   200 km.

N

BREEDING EVIDENCE
Reported in 73 (53%) of 137 blocks

Possible 9 (12%)
Probable 15 (21%)
Confirmed 49 (67%)

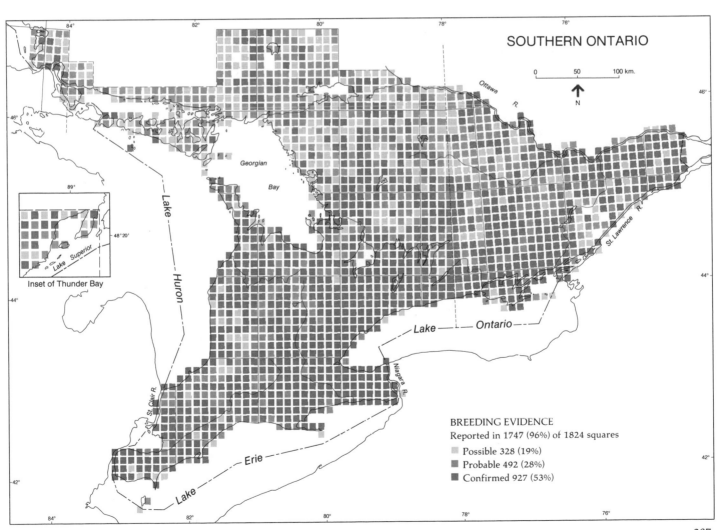

SOUTHERN ONTARIO

0   50   100 km.

N

Georgian
Bay

Lake
Huron

Lake
Superior

Inset of Thunder Bay

St. Clair R.

Niagara R.

Lake — Ontario

Lake — Erie

Ottawa R.

St. Lawrence R.

BREEDING EVIDENCE
Reported in 1747 (96%) of 1824 squares

Possible 328 (19%)
Probable 492 (28%)
Confirmed 927 (53%)

AMERICAN CROW
Corneille d'Amérique
Corvus brachyrhynchos

The American Crow is one of the few birds recognized by almost everyone. This comparatively intelligent and adaptable bird is confined to North America, but similar ecological counterparts can be found on most other continents.

The species breeds from the east coast almost to the west and from the Gulf of Mexico to the northern borders of the Prairie Provinces, including all regions of Ontario, but its winter range barely includes southernmost Canada. Its habitat preference is for open fields with scattered woods and copses, but it also occurs in reduced numbers throughout the forested parts of the province, especially near marshes, lakes, rivers, and other open areas (Godfrey 1986).

The scant evidence indicates that, before the settlement of Ontario by Europeans, the American Crow was widespread and fairly common. As the forest was cleared, the land was able to support increasing numbers of the species (McIlwraith 1894). Corn cultivation has provided a new food supply for the American Crow, and has become particularly important as a ready source of food in the winter, reducing the need for migration. Indeed, corn represents 36% of the diet of the average American Crow, the actual proportion varying with the season (Bent 1946). For this and other economic reasons, great efforts were made in the early part of this century to eliminate the species. This persecution appears to have had no long-term adverse effect on its population.

The American Crow has a loud and distinctive call note, usually rendered as 'caw', which makes the bird much easier to locate than less vocal species of the same size, even when it is well hidden in foliage. It is also a large bird, which can easily be seen, as well as heard, from a considerable distance. More often than not, the bird is seen flying from woodlot to woodlot or feeding in open fields. Usually, it keeps well out of range of direct human interference, but it also has successfully moved into urban centres.

Except in the forested northern parts of the province, the American Crow frequents agricultural land during all seasons. It nests during the period from early March to late June (Peck and James 1987). The nest is usually built at a height of above 5 m in an evergreen tree, often well out of sight among the inner branches, but some nests are located in deciduous trees, and are easily found before the leaves emerge. In Ontario, unlike some other areas, the American Crow is a solitary nester. Although the general nesting area is easy to find, the bird's secretive behaviour in the vicinity of the nest makes the actual structure difficult to locate, unless the noisy older young are present, and only 29% of the records of confirmed breeding resulted from the finding of a nest, compared to 58% from observing fledged young.

There are very few breeding birds which are more widely distributed in Ontario than the American Crow. In southern Ontario, the only extensive area where it is absent is in the western interior of Algonquin Prov. Park. That area is extensively forested with mature hardwoods that form a continuous canopy, broken only by a limited number of lakes. In the Boreal Forest region, and especially on the Hudson Bay Lowland, there are much larger areas with no records of the American Crow. These areas had enough coverage to warrant stating that the species is much less common there, but perhaps had insufficient coverage to prove that it is altogether absent. It is certainly not as prominent a species in the north as the closely related Common Raven.

The American Crow is a common bird in this province, although, as indicated by the abundance estimates, it occurs at higher densities in southern Ontario than it does in the north. The difference is likely a result of the higher proportion of agricultual land in the south.-- *R. John*

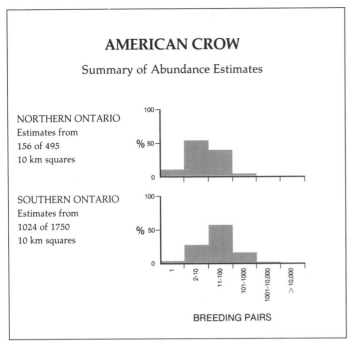

AMERICAN CROW

Summary of Abundance Estimates

NORTHERN ONTARIO
Estimates from
156 of 495
10 km squares

SOUTHERN ONTARIO
Estimates from
1024 of 1750
10 km squares

% axis with values 100, 50, 0

BREEDING PAIRS
(1, 2-10, 11-100, 101-1000, 1001-10,000, >10,000)

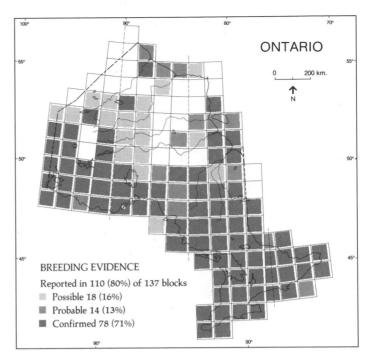

ONTARIO

0   200 km.

N

BREEDING EVIDENCE
Reported in 110 (80%) of 137 blocks

Possible 18 (16%)
Probable 14 (13%)
Confirmed 78 (71%)

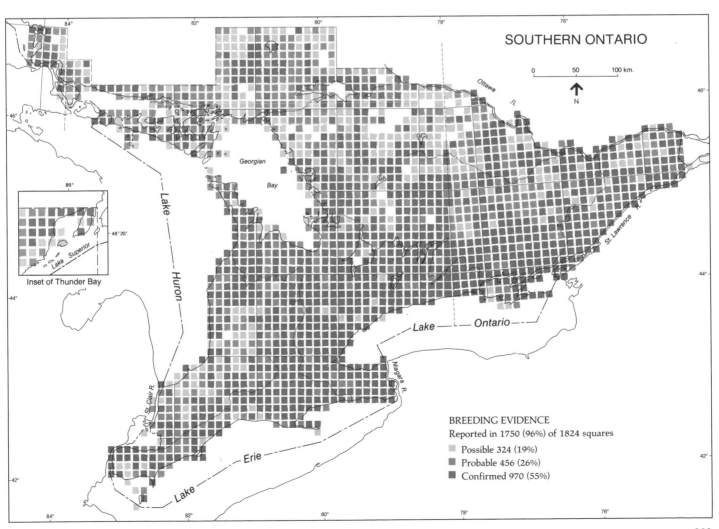

SOUTHERN ONTARIO

0   50   100 km.

N

Lake Superior
Inset of Thunder Bay

Lake Huron
Georgian Bay
St. Clair R.
Lake Erie
Niagara R.
Lake Ontario
Ottawa R.
St. Lawrence R.

BREEDING EVIDENCE
Reported in 1750 (96%) of 1824 squares

Possible 324 (19%)
Probable 456 (26%)
Confirmed 970 (55%)

COMMON RAVEN
Grand Corbeau
Corvus corax

The Common Raven, largest of the corvids, is frequently seen patrolling roads in wooded areas for road-killed animals, and has a fine-tuned judgement of when to move out of the way of approaching vehicles. It is primarily holarctic in its distribution. In North America, the species is found from coast to coast in Canada north of the prairies, southward in the mountains of the west to Central America and in the Appalachians. In Ontario, it is a year-round resident north of the southern edge of the Canadian Shield, from southcentral Ontario northward to the Hudson Bay coast. It is associated with relatively undisturbed habitats of the boreal and mixed forests of Ontario. Steep cliffs and tall trees, especially conifers, are chosen nesting sites for the species. Artificial structures such as hydro towers and utility structures are also used when natural nesting locations are lacking.

The range of the Common Raven has changed dramatically since the beginning of the 19th century. Alexander Wilson (1814) reported it to be particularly abundant along the Niagara River, and saw the species every day on a trip along the shores of Lakes Erie and Ontario in August and September. The use of poison baits and trapping to reduce wolf numbers, habitat destruction, and direct human persecution of Common Ravens contributed to its disappearance as a breeding bird from much of southern Ontario, and probably also near populated areas of the north. More recently, its range has begun to extend southward again, and the population in southern Ontario appears to be increasing fairly rapidly. In 1938, it was considered rare even in the wilder parts of Algonquin Prov. Park (MacLulich 1938), where it is now a fairly common sight, and Snyder (1953) reported an increase in the Kenora area, where it had been rare in 1930. The Common Raven is now considered a common resident on Manitoulin Island (Nicholson 1981), but was not mentioned in an earlier account of the fauna of the region (Williams 1942). Its breeding range extended onto the Bruce Peninsula in the 1970's, and the population there was approximately 10 to 20 pairs during the atlas period (M. Parker pers. comm.). One pair of birds was reported near Kolapore, in Grey Co, during the atlas period. Suitable habitat along the Niagara Escarpment may mean that the range of the Common Raven can continue to expand to the south.

The species is known for its early nesting. Courtship interactions, including spectacular aerial displays, occur in February, when snow is still on the ground. The Common Raven lays 3 to 5 eggs as early as April (James *et al.* 1976, Godfrey 1986), which partially explains why only 3 of the over 1,500 records of this species are of nests with eggs. Fortunately, nests are bulky structures, often observed on cliffs or other open sites. The species has long incubation and fledging periods, so birds on nests, nests with young, and newly fledged young are much more readily reported. Though vocal during the courtship period, the Common Raven becomes quiet and retiring during incubation and early feeding of the young. However, it is a large bird, not likely to be overlooked in those squares where it occurs. Nests can be found by following the begging cries of young, but if the adults detect a human intruder, their alarm signals induce the young to become silent. Once the young are fledged, family parties roam extensively, and it is impossible to conclude from a sighting of a family that the species bred in the immediate vicinity.

The maps clearly show the precisely defined range limits of the Common Raven in Ontario. It occurs on the Hudson Bay Lowland, the Canadian Shield to its southern boundary, and on the Bruce Peninsula and Manitoulin Island, but is absent in southwestern and much of southeastern and southcentral Ontario. The lack of extensive forest tracts may keep the species out of these regions, but recent range expansion eastward nearly to Ottawa is evident from the maps. The Common Raven is wary of humans but can also prove highly adaptable: the absence of suitable nesting sites in the Sudbury basin has resulted in some unusual nesting locations such as on towers within the city limits and 43 m below ground level in an open-pit mine.

Abundance estimates of over 10 pairs per square were made in a slightly higher proportion of northern than southern squares (30%, versus 21%), suggesting that the Common Raven maintains a greater affinity for the north despite its expansion into more southerly areas.-- *C. Blomme*

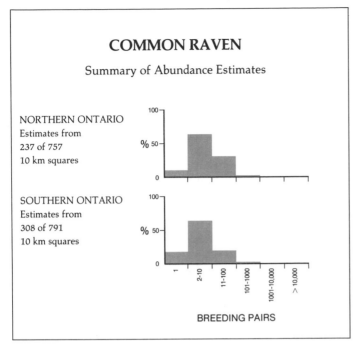

COMMON RAVEN

Summary of Abundance Estimates

NORTHERN ONTARIO
Estimates from
237 of 757
10 km squares

SOUTHERN ONTARIO
Estimates from
308 of 791
10 km squares

BREEDING PAIRS

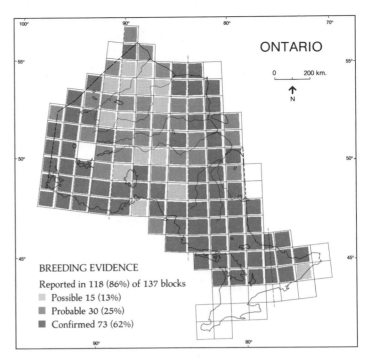

ONTARIO

0   200 km.

N

BREEDING EVIDENCE

Reported in 118 (86%) of 137 blocks

Possible 15 (13%)
Probable 30 (25%)
Confirmed 73 (62%)

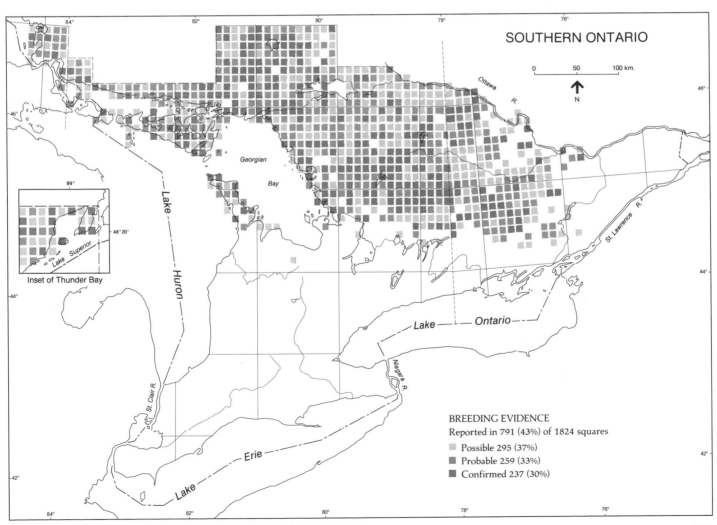

SOUTHERN ONTARIO

0   50   100 km.

N

Inset of Thunder Bay

Georgian Bay

Lake Huron

Lake Superior

Ottawa R.

St. Lawrence R.

Lake — Ontario

Niagara R.

St. Clair R.

Lake — Erie

BREEDING EVIDENCE

Reported in 791 (43%) of 1824 squares

Possible 295 (37%)
Probable 259 (33%)
Confirmed 237 (30%)

BLACK-CAPPED CHICKADEE
Mésange à tête noire
Parus atricapillus

Sprightly, energetic, inquisitive, and gregarious, the Black-capped Chickadee is among the best known and best liked of our birds. Though common and widespread all year, it is mostly during the winter that it comes to feeders, where its boldness and acrobatics are most familiar, and it is in the winter woods, often seemingly devoid of life and warmth, that its cheery 'chick-a-dee-dee' is most appreciated.

The species is a permanent resident across North America from the Boreal Forest zone south to the central US. It is most common in deciduous and mixed woods but also occurs in semi-open areas such as wooded city parks and wooded stream valleys in farming areas. In spring, the dominant one or 2 males from the winter flock establish breeding territories in the former winter flock territory, often with their mates of the previous season (Odum 1941, Smith 1976, Ficken *et al.* 1981). Other members of the winter flock disperse to territories in areas unused by chickadees in winter or, in some cases, may not breed (Smith 1967).

The distribution of the Black-capped Chickadee in Ontario has probably changed only slightly since settlement. The ability of the species to use partly open areas, including areas in towns and cities, has enabled it to maintain populations in places where other deciduous and mixed-woods species are now uncommon. Nevertheless, loss of habitat resulting from clearing of the land for agriculture has left local gaps in distribution and probably has reduced the overall population. On the other hand, the availability of food at winter feeders may have increased survival rates in some areas.

The whistled 'fee-bee' song of the Black-capped Chickadee (often translated by optimists, who occasionally hear it in mid-winter, as 'spring's here') is a familiar sound in the early springtime woods. By mid-April or even earlier, pairs are travelling together and beginning to excavate nest holes in rotten stubs and branches or to explore other potential nesting cavities. The pair spends several periods each day for a week or more excavating the nest hole, often calling softly to each other all the while. However, between excavation bouts the pair rarely approach the nest site. From egg-laying through fledging, the species is quiet and secretive. Even the 'chick-a-dee-dee' call is infrequently heard during these periods. After fledging, the family group travels together, and the noisy begging calls of the young during the first week or two after fledging are conspicuous. These characteristics are reflected in the breeding evidence for this species. Breeding was confirmed in 56% of Ontario squares in which it was reported, but over 50% of these records are of recently fledged young. In northern Ontario, over 70% of those records are of recently fledged young. Only 8% of all confirmed records are of nests with eggs.

Atlas data confirm that the Black-capped Chickadee is a widespread breeding bird in Ontario. It occurs in almost every block south of the Hudson Bay Lowland. In most of the far northern blocks, either the species was not reported at all or breeding was not confirmed, but in the southern portion of the Lowland, probable or confirmed breeding was reported in several blocks. In southern Ontario, the species was present in 95% of the squares and breeding was probable or confirmed in 88% of these. It was absent from many squares in Essex Co and parts of Kent and Lambton Counties in extreme southwestern Ontario. This probably reflects the very extensive agricultural clearing and lack of woodlots in this area. The lack of Black-capped Chickadees in some parts of this area seems to occur even in winter. Results of Christmas Bird Counts for 1968 to 1977 suggest that its numbers in southwestern Ontario may be lower than in other parts of southern Ontario at that season, probably for the same reason (Speirs 1985).

Abundance estimates for the Black-capped Chickadee suggest populations of between 11 and 1,000 pairs in 76% of the squares in southern Ontario. As might be expected, towards the northern edge of the range there is a smaller proportion of estimates in the categories of over 100 pairs than there is in the south, but even there it is a common bird. The average Black-capped Chickadee territory size of 5 ha (Odum 1941, Smith 1967) suggests that the maximum number in any one square, if its habitat is uniformly suitable, would rarely exceed 2,000 pairs.-- *M. McLaren*

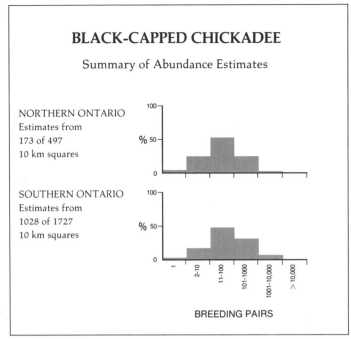

BLACK-CAPPED CHICKADEE

Summary of Abundance Estimates

NORTHERN ONTARIO
Estimates from
173 of 497
10 km squares

SOUTHERN ONTARIO
Estimates from
1028 of 1727
10 km squares

BREEDING PAIRS

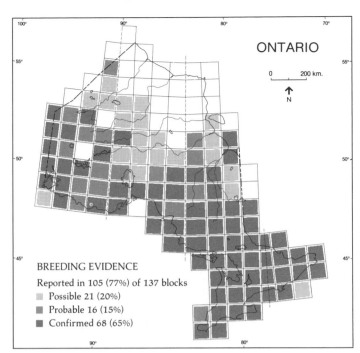

ONTARIO

0   200 km.

↑
N

BREEDING EVIDENCE
Reported in 105 (77%) of 137 blocks

Possible 21 (20%)
Probable 16 (15%)
Confirmed 68 (65%)

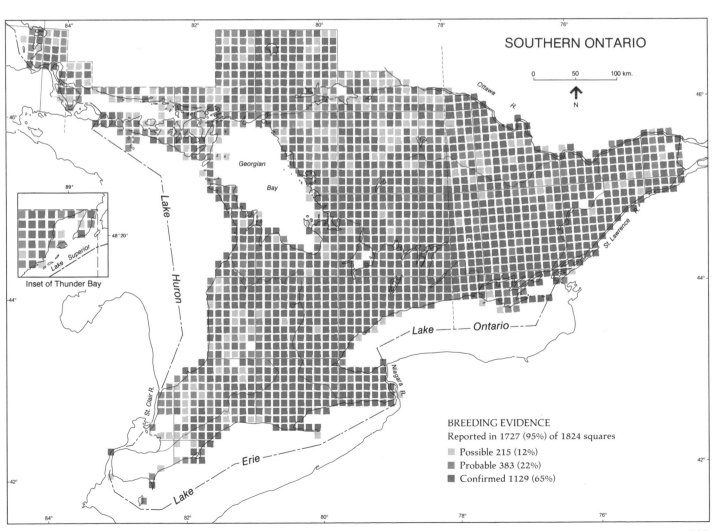

SOUTHERN ONTARIO

0   50   100 km.

↑
N

Inset of Thunder Bay

BREEDING EVIDENCE
Reported in 1727 (95%) of 1824 squares

Possible 215 (12%)
Probable 383 (22%)
Confirmed 1129 (65%)

BOREAL CHICKADEE
Mésange à tête brune
Parus hudsonicus

The Boreal Chickadee occurs across North America in northern coniferous forests. Its range extends into the northern US in the mountains of both the west and the east as well as in southern extensions of Boreal Forest near Lake Superior. In Ontario, the species occurs across the northern part of the province and south to Algonquin Prov. Park. The Boreal Chickadee is primarily a species of spruce forests, although it also occurs in coniferous forests of mixed species and, to a lesser extent, in areas with a limited deciduous component to the forest.

The northern distribution of the Boreal Chickadee, as well as its preference for forests growing on poor soils, has protected it from habitat loss associated with extensive clearing for agriculture. Although logging has undoubtedly removed some habitat, this has probably not been a significant factor in changing the range of the species. Wildfires in both presettlement and historic times have always been a factor in changing the local distribution of appropriate habitat. Despite such considerations, the numbers and overall distribution of the Boreal Chickadee, in broad overview, are probably little different from what they were when Europeans first arrived in northern Ontario.

The Boreal Chickadee is not an easy species to locate or to confirm as a breeding species. In early spring, as the snow is melting, pairs can be found travelling together, often feeding on the ground where melting snow has revealed material dropped from the trees over the winter (McLaren 1975). Although the species gives various territorial calls in early spring (McLaren 1976), it is generally quiet during the nesting season. The Boreal Chickadee excavates its nest hole in early May, a time of year when few atlassers were afield, and, unlike the Black-capped Chickadee, which excavates in the open before the leaves are on the trees, the Boreal Chickadee usually nests in deep woods. It is very inconspicuous during egg-laying and incubation but somewhat less so while feeding young in the nest. After fledging, the begging calls of the young are fairly conspicuous but probably not familiar to many atlassers.

The inconspicuous behaviour of breeding Boreal Chickadees is reflected in the atlas data. Unlike records of the Black-capped Chickadee or of many other songbirds, records of singing males are outnumbered by records of birds in suitable habitat. Probable or confirmed breeding records were obtained for only 39% of the squares where this species was recorded. Most of the probable breeding records consisted of sightings of pairs, and most (73%) of the confirmed records were of fledged young. Nests were reported in only 3 squares.

The Boreal Chickadee is widely distributed in northern Ontario. The scarcity of records along some parts of the coast of Hudson Bay may be a result of lack of appropriate habitat, but elsewhere in the north the lack of records more likely reflects the inconspicuous nature of the species. Nevertheless, atlas breeding evidence adds considerably to the knowledge of specific breeding sites in the north (Peck and James 1987). The range of the Boreal Chickadee extends into southern Ontario mainly in the area north of Sudbury and in the Algonquin Highlands. A few scattered records occur as far south as northern Muskoka and southern Renfrew or northern Hastings Counties.

The atlas data probably represent the distribution of the Boreal Chickadee in southern Ontario fairly accurately. Although southern Ontario lies in the Great Lakes-St. Lawrence Forest region, south of the main range of Boreal Chickadee in the Boreal Forest region, a cluster of records falls in the Algonquin Highlands, representing a disjunct population in an area which shows greater affinities to the Boreal Forest region than other adjacent sections of the Great Lakes-St. Lawrence Forest region.

Abundance estimates for the Boreal Chickadee are generally higher in the north than in the south. Most estimates in the south are in the 2 to 10 pairs category with a higher estimate in only one square. These low estimates probably reflect the inconspicuous behaviour of breeding Boreal Chickadees as well as their scarcity in that part of the province. In the north, 52% of the estimates are of over 10 pairs per square, and 6% are of over 100. Erskine (1977) found densities of Boreal Chickadees ranging from 2 to 11 per km^2 (200 to 1,100 per square) in various parts of Canada including northern Ontario. Typical densities were about 4 to 5 per km^2, and the highest occurred during spruce budworm outbreaks (Gage *et al.* 1970). The density of Boreal Chickadees in parts of Algonquin Prov. Park (M. McLaren unpubl. data) suggests that at least some squares would support well over 100 pairs even at the southern edge of the range, and therefore the range of abundance estimates shown here may well be too low.-- *M. McLaren*

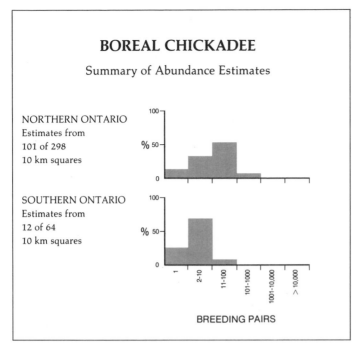

BOREAL CHICKADEE

Summary of Abundance Estimates

NORTHERN ONTARIO
Estimates from
101 of 298
10 km squares

SOUTHERN ONTARIO
Estimates from
12 of 64
10 km squares

BREEDING PAIRS

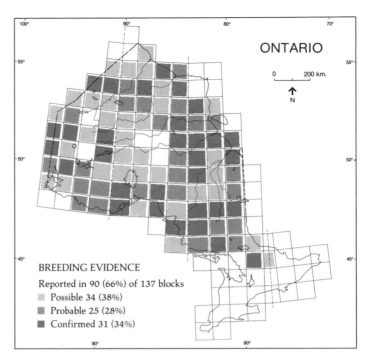

ONTARIO

0   200 km.

N

BREEDING EVIDENCE

Reported in 90 (66%) of 137 blocks

Possible 34 (38%)

Probable 25 (28%)

Confirmed 31 (34%)

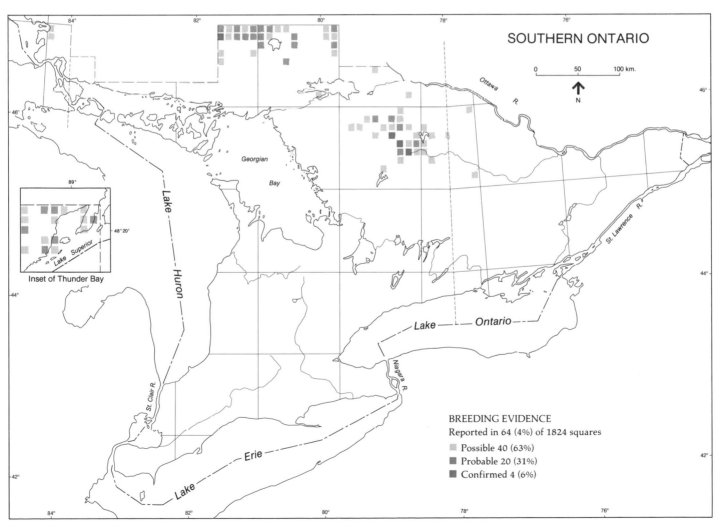

SOUTHERN ONTARIO

0   50   100 km.

N

Georgian
Bay

Lake Huron

Lake Superior

Inset of Thunder Bay

Ottawa R.

St. Lawrence R.

Lake — Ontario

Niagara R.

Lake — Erie

Lake

St. Clair R.

BREEDING EVIDENCE

Reported in 64 (4%) of 1824 squares

Possible 40 (63%)

Probable 20 (31%)

Confirmed 4 (6%)

TUFTED TITMOUSE
Mésange bicolore
Parus bicolor

The Tufted Titmouse is found through much of the eastern US and eastern Mexico. Its range extends into extreme southern Ontario, where it is rare but slowly increasing in numbers.

In Ontario, there are breeding records from Sarnia, London, Welland, Hamilton, and Lorne Park near Toronto (Godfrey 1986), although wanderers may occasionally be seen a little farther north. Within this breeding range, it may be found anywhere that tall deciduous trees are present, but it is nowhere common. Suitable locations may occur in urban, agricultural, or forested areas, with a preponderence in the vicinity of bird feeders, especially those which are stocked throughout the spring.

The Tufted Titmouse was first reported in Ontario (at Point Pelee) on 2 May 1914, at about the same time as the first Northern Cardinal, but the Titmouse has spread much more slowly. The first summer record (at Fisher's Glen in Haldimand-Norfolk RM) occurred on 7 July 1932, and the first indication of breeding was provided on 6 September 1936 by a sighting of an adult with 2 full-sized young at Bull's Point near Hamilton (Woodford 1962). However, it was 1955 before a nest was reported: it contained 4 eggs and was found at Sarnia in exactly the same site (a knot-hole in an old apple tree) where nesting had been reported in 1953 (Baillie 1955). Unfortunately, the 1955 nest remained undocumented, and it was not until 1977 that a fully acceptable breeding record was reported. This involved a nest (actually found in 1971) in a pipe in a Port Colborne backyard, for which photo-documentation was secured (James 1984b).

There was a marked influx in the late fall and winter of 1961-62. The species was then found in 15 new locations in Ontario, including sites as far north and east as Owen Sound, Toronto, and Kingston (Woodford 1962). Subsequent population levels remained somewhat elevated. In extreme southwestern Ontario, there were only 3 records before 1955, but since 1966 up to 3 birds have been reported annually (Kelley 1978). In that area, numbers were again low during the atlas period. The number of reports during the atlas period indicates that there have been minor population increases in the Niagara and Kitchener regions.

The overall pattern of slow increase is in agreement with range expansion occurring in the northeastern US. Expansion is probably slow because the Tufted Titmouse is extremely sedentary; US banding statistics show that less than 1% of birds travel more than 30 km from the banding site (Elder 1985). However, one factor which may be assisting the species is the greater number of bird feeders in recent years. Near St. Catharines, a feeder which is stocked year-round with sunflower seeds has been attracting Tufted Titmice for 20 years.

Nests are usually built in small natural cavities or woodpecker holes in deciduous trees, but other sites include rotten stumps and nest boxes. However, only 8 nests have ever been reported in Ontario (ONRS), and none were found during the atlas period. In fact, the Tufted Titmouse is infrequently observed during the nesting season and seems to be most vocal in March and April. Once the birds have left the vicinity of feeders to nest in wooded areas, they are less easily observed, especially when foraging silently in the upper canopy of large deciduous trees. Six of 7 confirmed breeding records of the Tufted Titmouse concerned observations of fledged young, perhaps because they tend to remain in family groups for at least a month or two after fledging. The other confirmed record was of an adult bird carrying food or a faecal sac. The records of possible breeding may reflect the atlassing difficulties mentioned above, but may also indicate, as might be expected, that unpaired birds occur sporadically in the Carolinian Forest zone, which is the northern extreme of the species' breeding distribution.

The map indicates the essentially Carolinian range of this species, and shows a concentration of records on the Niagara Peninsula. The paucity of records in extreme southwestern Ontario may be due to the extensive clearing for agriculture which has occurred there, but one might have expected more records in the Long Point region, where some of the best remnant Carolinian forest occurs. The isolated observation of a single singing bird between Barrie and Orillia on 18 June 1985 would be totally unexpected were it not for a handful of other Ontario sightings of the species at that latitude (Woodford 1962). That record, unlike the others, was in the heart of the breeding season.

Abundance estimates were provided for 9 squares; most are of one pair, but estimates from 2 squares near Niagara-on-the-Lake are of 2 to 10 pairs. One of the latter squares had reports from 4 sites and the other had 3. Thus, the Tufted Titmouse must be considered a rare bird anywhere else in Ontario and uncommon even in the Niagara Region.-- *H.G. Currie*

TUFTED TITMOUSE

Summary of Abundance Estimates

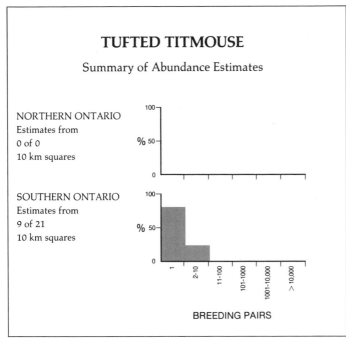

NORTHERN ONTARIO
Estimates from
0 of 0
10 km squares

SOUTHERN ONTARIO
Estimates from
9 of 21
10 km squares

% 50 / 100 / 0

BREEDING PAIRS

1   2-10   11-100   101-1000   1001-10,000   >10,000

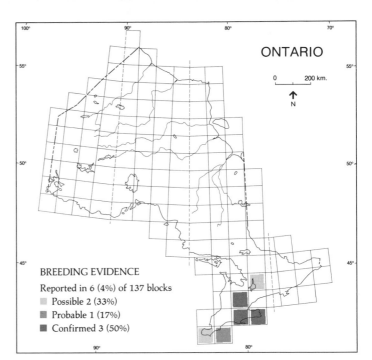

ONTARIO

0   200 km.
↑ N

BREEDING EVIDENCE
Reported in 6 (4%) of 137 blocks
☐ Possible 2 (33%)
☐ Probable 1 (17%)
■ Confirmed 3 (50%)

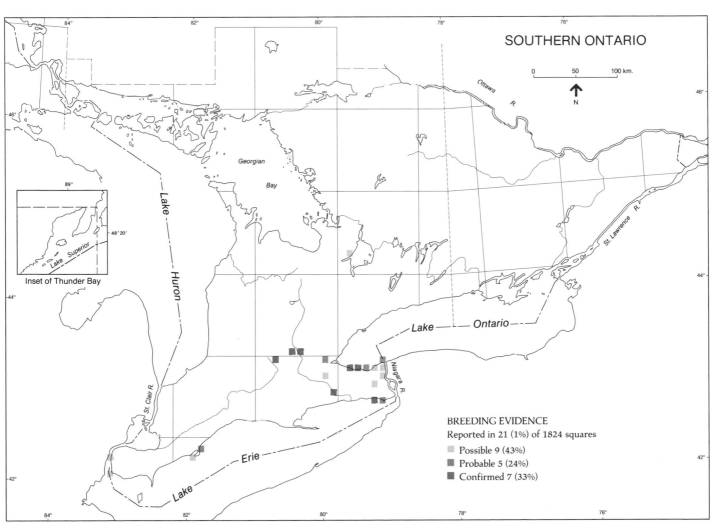

SOUTHERN ONTARIO

0   50   100 km.
↑ N

Inset of Thunder Bay

Georgian Bay

Lake Huron

Lake Erie

Lake Ontario

St. Clair R.

Niagara R.

Ottawa R.

St. Lawrence R.

Lake Superior

BREEDING EVIDENCE
Reported in 21 (1%) of 1824 squares
☐ Possible 9 (43%)
■ Probable 5 (24%)
■ Confirmed 7 (33%)

297

RED-BREASTED NUTHATCH
Sittelle à poitrine rousse
Sitta canadensis

Ontario's two nuthatches assort themselves rather predictably between two general habitat types. Reflecting their scientific names, the Red-breasted definitely prefers coniferous forests, which are typically Canadian, while the White-breasted is usually found in hardwoods, which characterize the Carolinian Forest region. Although both species are well known for creeping about on tree trunks and main branches, often upside down, the smaller Red-breasted Nuthatch spends considerable time foraging in the thick growth of branch ends and in the dense crowns of spruces and other evergreens with thick foliage. As a breeding bird, the Red-breasted Nuthatch is found from Alaska to Newfoundland and southward; except for certain northern states, in the US it extends south only in the Appalachians and the various western mountain ranges.

McIlwraith (1894) claimed that the Red-breasted Nuthatch was found in southern Ontario only during the spring and fall. Macoun and Macoun (1909) indicated that it was found as a breeder at least as far south as Muskoka and the Bruce Peninsula, and at least as far north as the Moose River, where it was reportedly common. More recently, Baillie and Harrington (1937) documented breeding as far south as Peel Co, as well as for Thunder Bay and Algoma in the north. In the south, that limit stood until 1974 when nesting was reported in pine plantations at St. Williams in Haldimand-Norfolk RM (Dunn *et al.* 1975). This apparent absence or scarcity in the southern part of Ontario around the turn of the century and the apparent range extension since then can be partly attributed to changes in the numbers of field observers. However, given that it is at present fairly widespread in the south, it seems safe to say that, although its breeding distribution in the north seems unlikely to have changed, the Red-breasted Nuthatch has penetrated farther south in Ontario in historic times, and breeding densities there have doubtless increased as well. This is almost certainly due to its colonization of conifer plantations south of the Canadian Shield. As more abandoned farmland in southern Ontario is converted to conifer plantation, we can expect Red-breasted Nuthatch populations there to thrive.

The Red-breasted Nuthatch can be an unobtrusive bird. The often dense forests in which it breeds, its muted notes never strung into song, and its small size make it less than conspicuous. For these reasons, several Regional Coordinators consider that the species is underrepresented on the maps. The nest is usually a cavity in a dead conifer, at heights of from 2 to 12 m (Harrison 1975), where it is easy to find. Only 22% of the confirmed breeding records consist of nests. The bird, however, is not difficult to identify, and when it does vocalize, its notes are distinctive. Its presence can be overlooked, but there is little chance of misidentification.

A comparison of the maps of the 2 Ontario species of nuthatches shows that, as White-breasted Nuthatch density begins to thin in the southern part of the Canadian Shield, the Red-breasted Nuthatch numbers increase. Although breeders and presumed breeders are scattered as far south as Kent and Elgin Counties, a fairly definite change in density and status occurs along the southern margin of the Shield. North of this line, most squares host birds, while south of it, most do not. This is in keeping with the habitat requirements, specifically the coniferous forest component. The absence of the Red-breasted Nuthatch from the Hudson Bay Lowland is difficult to explain, as spruce forests are extensive there. It may be that trees of sufficient size for foraging are too few.

In some winters, usually in alternate years, there is a major exodus of Red-breasted Nuthatches to the eastern US (Bock and Lepthien 1972). In the intervening years, however, large numbers spend the winter in southern Ontario, especially when an excellent cone crop has developed. During the atlas period, the winter of 1984-85 was such a year. As with certain finches that were present in southern Ontario in high numbers during that winter, the following nesting season found an increase in the numbers of Red-breasted Nuthatches in the south. Several Regional Coordinators in that part of the province noted these increases, and, were it not for the 1985 season, the distribution and status of the Red-breasted Nuthatch south of the Shield may have appeared somewhat different. Notably, in the southwest, atlassers confirmed breeding in Kent, Elgin, Middlesex, Huron and Oxford Counties, all of which are outside of the range mapped by Peck and James (1987).

It is evident from the abundance estimates that the Red-breasted Nuthatch is a common breeding bird where it occurs in northern Ontario, but tends to be less common in the south. Almost two-thirds of the estimates from southern Ontario squares in excess of 100 pairs per square were made in 1985, the season after the irruption mentioned above.-- *A. Mills*

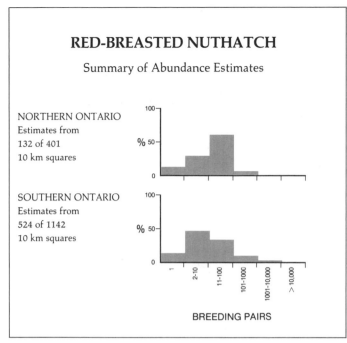

RED-BREASTED NUTHATCH

Summary of Abundance Estimates

NORTHERN ONTARIO
Estimates from
132 of 401
10 km squares

SOUTHERN ONTARIO
Estimates from
524 of 1142
10 km squares

BREEDING PAIRS

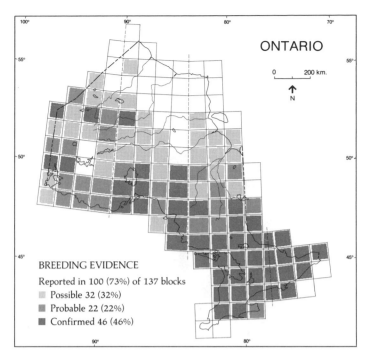

ONTARIO

0   200 km.

N

BREEDING EVIDENCE

Reported in 100 (73%) of 137 blocks

- Possible 32 (32%)
- Probable 22 (22%)
- Confirmed 46 (46%)

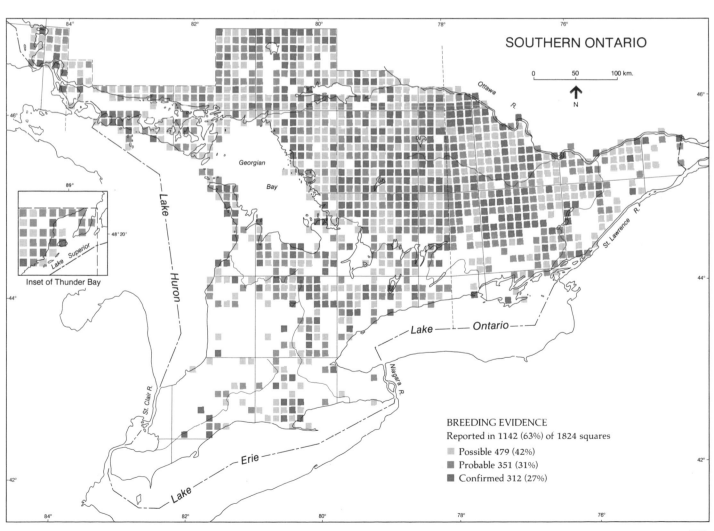

SOUTHERN ONTARIO

0   50   100 km.

N

Inset of Thunder Bay

Georgian Bay

Lake Huron

Lake Superior

Lake Erie

Lake Ontario

St. Lawrence R.

Ottawa R.

Niagara R.

St. Clair R.

BREEDING EVIDENCE

Reported in 1142 (63%) of 1824 squares

- Possible 479 (42%)
- Probable 351 (31%)
- Confirmed 312 (27%)

WHITE-BREASTED NUTHATCH
Sittelle à poitrine blanche
Sitta carolinensis

Just as the Red-breasted Nuthatch is generally a species of coniferous forests, the White-breasted Nuthatch is generally a species of mature broad-leaved woodlands. It is tolerant of mixed forests, however, and the shade trees of suburbs and other residential areas also provide suitable habitat. It is a permanent resident or irregular partial migrant wherever it breeds, and its distribution extends from southern Canada to Mexico and from the Atlantic to the Pacific Oceans, with some significant gaps in its range in mid-continent. In Ontario, it occurs widely throughout the Carolinian Forest zone and throughout the less coniferous portions of the Great Lakes-St. Lawrence Forest region. Familiar even to those who are not keen bird-watchers, its distinctive nasal calls and especially its habit of 'walking' upside down on tree trunks make it noteworthy.

There is no evidence to suggest that the status of this species at present in Ontario is different from what it was in earlier times; neither its numbers nor its distribution appear to have changed, although the loss of most of the deciduous forests in agricultural areas undoubtedly reduced its numbers. Early accounts such as McIlwraith (1894) and Macoun and Macoun (1909) indicate that it was a common resident in the province south of the French River. Since then, the northern limit has been extended somewhat to Temagami and Kenora (Baillie and Harrington 1937), but this almost certainly reflects an improvement in ornithological coverage rather than a recent geographic range expansion. Today, the White-breasted Nuthatch is fairly common north to the latitude of Parry Sound and Algonquin Prov. Park and relatively scarce in the hardwood or mixed forests north of Lake Huron and east of Lake Superior.

In the opinion of a number of Regional Coordinators, the White-breasted Nuthatch was found less often than expected. Consequently, the number of records of confirmed breeding and the abundance estimates are also lower than expected. This situation is probably attributable to several factors. Although it is often a noisy and conspicuous bird, especially in the winter and early spring, the White-breasted Nuthatch becomes extremely quiet and inconspicuous during nesting. It ceases delivery of its song of repeated nasal notes and creeps about unobtrusively among the higher tree trunks and branches. It does not resume its more noticeable and vocal forms of behaviour until later in the summer, after breeding is over. It tends to nest high in trees, usually in an inconspicuous cavity. Furthermore, it is an early nester, with no known Ontario egg dates beyond the end of May (Peck and James 1987). Relatively few nests were found during atlas field work, and almost half (46%) of all records of confirmed breeding are based on observations of newly fledged young.

The distribution map derived from atlas field work confirms previous reports of White-breasted Nuthatch distribution. Except for Essex and Kent Counties, which have little extensive forest, the species is generally found throughout the south, having been reported in 73% of the squares in that part of the province. Its frequency of occurrence begins to thin at the latitude of Lake Nipissing, as previously reported, and the distribution north of there is sparse and irregular as the proportion of hardwood forest decreases. There are also notable gaps southward into the Algonquin Highlands. There are records of confirmed breeding in the vicinities of North Bay, Sudbury, and Sault Ste Marie, however, and a few populations are found north of those cities: for example, to the east and northwest of Temagami. Farther west, the White-breasted Nuthatch is found in several Thunder Bay area squares, 2 of which have records of confirmed breeding. It is also found in the vicinity of Kenora, including individuals found northward along the Manitoba border to the latitude of Red Lake. In short, most Ontario White-breasted Nuthatches are found south of Lake Nipissing, but a sparse population is widely scattered north of there and in the northwest.

Abundance estimates fall into the 11 to 100 pairs per square category more often than any other in the south, but in the north only 1 of 9 estimates exceeds 10 pairs per square. The 5 highest estimates are from the southern Georgian Bay, Lindsay, and Haliburton Co areas.-- *A. Mills*

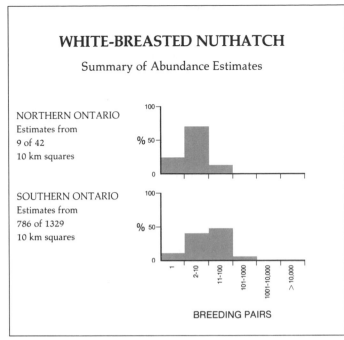

WHITE-BREASTED NUTHATCH

Summary of Abundance Estimates

NORTHERN ONTARIO
Estimates from
9 of 42
10 km squares

SOUTHERN ONTARIO
Estimates from
786 of 1329
10 km squares

BREEDING PAIRS

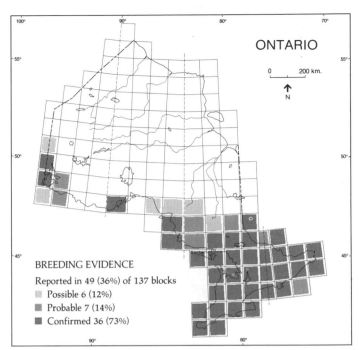

ONTARIO

0   200 km.

N

BREEDING EVIDENCE
Reported in 49 (36%) of 137 blocks

Possible 6 (12%)
Probable 7 (14%)
Confirmed 36 (73%)

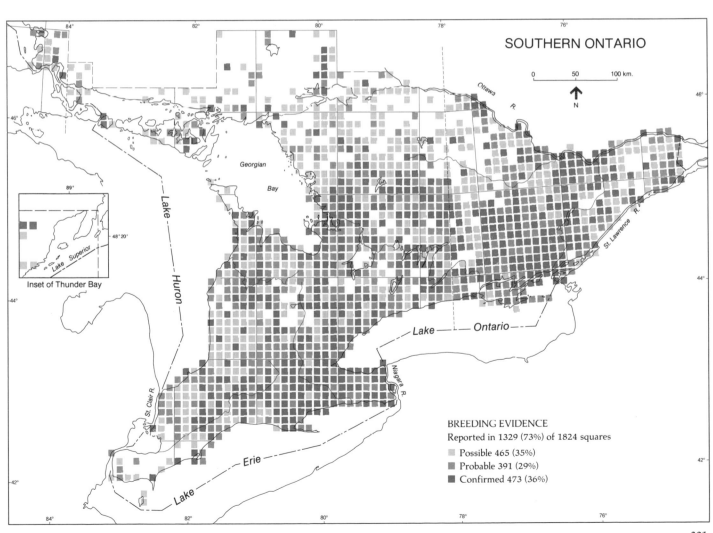

SOUTHERN ONTARIO

0   50   100 km.

N

Inset of Thunder Bay

BREEDING EVIDENCE
Reported in 1329 (73%) of 1824 squares

Possible 465 (35%)
Probable 391 (29%)
Confirmed 473 (36%)

BROWN CREEPER
Grimpereau brun
Certhia americana

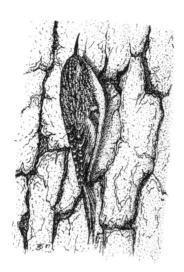

There are 7 species of creepers in the world, but the Brown Creeper is the only one which breeds in North America. Like the other members of the family, it is secretive and unobtrusive.

The Brown Creeper breeds across southern Canada from coast to coast, and south in the various mountain ranges as far as Central America in the west and North Carolina in the east. In Ontario, it breeds in almost all parts of the province, though only rarely in the extreme southwest or on the Hudson Bay Lowland. Its winter range encompasses most of the breeding range, although there is partial withdrawal from the higher latitudes (Godfrey 1986). The preferred habitat for the species is mature woodland, especially in wet areas, and often with large dead trees. Coniferous, deciduous, and mixed woods are frequented, and bogs are also used.

The historical distribution and abundance of the Brown Creeper in Ontario are difficult to assess, because it is a bird that apparently warranted only limited mention in early publications. Baillie and Harrington (1937) described it as an "inconspicuous" bird that bred "sparingly" in most of the southern half of the province. The most northerly breeding evidence that they cited consisted of fledged young near Kapuskasing and Minaki. Given the lack of obvious changes to the northern forests, it seems likely that the species has not expanded its range northward since that time, but rather that our knowledge has improved with more northern coverage.

The Brown Creeper is a rather difficult bird to detect, and consequently may easily be overlooked in its wooded habitat. It is small, usually solitary, and exceptionally well camouflaged against the trunks or limbs of trees, where it spends most of its time foraging for insects or other small arthropods located in crevices in the bark. Its upward movement along that surface or its hasty descent through the air to another nearby tree are most often the forms of behaviour that betray its presence. It often appears oblivious to human presence, even in the vicinity of its nest. Instead of engaging in conspicuous distraction displays or other aggressive or noisy activity under such circumstances, the Brown Creeper commonly adopts an absolutely motionless posture for several minutes when threatened, relying entirely on its cryptic colouration for concealment. Even its vocalizations are weak and infrequent, particularly its song, which is "rather rarely heard" (Bent 1948) and not widely recognized. It is probably for this reason, in addition to the bird's lack of sexual dimorphism, that there are as many records of Brown Creepers having been merely observed in suitable habitat in breeding season (late April to late June in Ontario) as there are of singing males. Together, however, these 2 possible breeding categories comprise 58% of all records (56% in the south, 70% in the north), and records of probable and confirmed breeding are correspondingly few: 23% and 18%, respectively.

The Brown Creeper almost always builds its nest in a well concealed location under the large, peeling flakes of bark that occur as a degenerative condition of a recently dead tree, a condition that is often the result of prolonged flooding. More rarely, the nest is built in a tree cavity. The mean nest height is 3.8 m in American elm and 2.6 m in balsam fir (Davis 1978). Except when the young begin their begging cries, the surreptitious behaviour of the adults makes the nest exceedingly difficult to find. In fact, only 6% of all records are of occupied nests.

The maps indicate a more general breeding distribution of the Brown Creeper in Ontario than a casual observer would suspect. However, even on the southern Shield, where records seem most concentrated, the range as shown is far from continuous. This is probably because of the specificity of the bird's nesting habitat requirements (and possibly in part because of difficulty in finding the bird due to the behavioural traits noted above). The flooded mature forests at Point Pelee Nat. Park and Rondeau Prov. Park provide excellent nesting habitat and are the only places southwest of Ingersoll where breeding of the Brown Creeper was confirmed. In the north, breeding was confirmed in 2 blocks north of the continuous range described by Godfrey (1986), reinforcing his observation that there is a disjunct population in the Sutton Ridges, and adding a site near Little Sachigo Lake in the northwest.

Abundance estimates suggest that the Brown Creeper is uncommon even where it does breed in southern Ontario, but is fairly common in those northern squares where it occurs. Although the modal abundance estimate is higher in the north, there are 16 estimates of over 100 pairs per square in southern Ontario. Fourteen of these 16 high estimates are from squares on the heavily wooded Canadian Shield; only 2 of them are from the primarily agricultural area farther south.-- *C. Blomme*

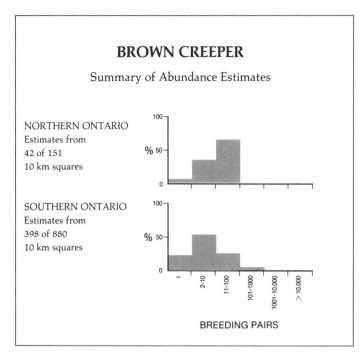

BROWN CREEPER

Summary of Abundance Estimates

NORTHERN ONTARIO
Estimates from
42 of 151
10 km squares

SOUTHERN ONTARIO
Estimates from
398 of 880
10 km squares

%

BREEDING PAIRS

1 2-10 11-100 101-1000 1001-10,000 >10,000

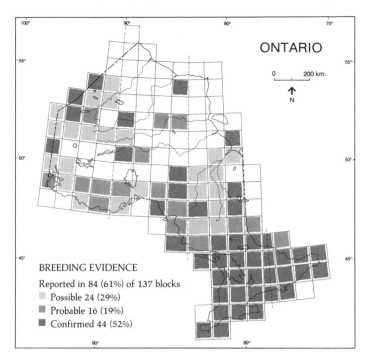

ONTARIO

0 200 km.

N

BREEDING EVIDENCE

Reported in 84 (61%) of 137 blocks

Possible 24 (29%)
Probable 16 (19%)
Confirmed 44 (52%)

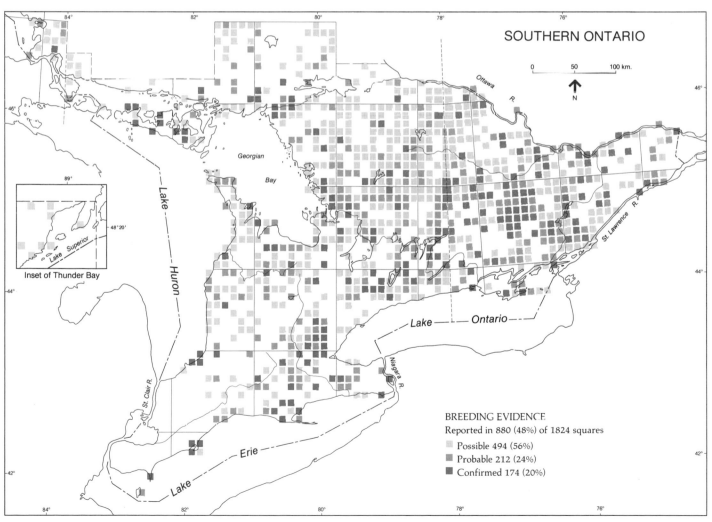

SOUTHERN ONTARIO

0 50 100 km.

N

Ottawa R.

Georgian Bay

Lake Huron

St. Clair R.

Lake Erie

Niagara R.

Lake Ontario

St. Lawrence R.

Inset of Thunder Bay

89°

48°20'

Lake Superior

BREEDING EVIDENCE

Reported in 880 (48%) of 1824 squares

Possible 494 (56%)
Probable 212 (24%)
Confirmed 174 (20%)

CAROLINA WREN
Troglodyte de Caroline
Thryothorus ludovicianus

The Carolina Wren is the largest and rarest of the wrens which breed regularly in Ontario. Within its limited range, it can be found at any time of the year.

The Deciduous Forest region, which is widespread in the eastern US but is restricted in Canada to the Carolinian Forest zone of southwestern Ontario, defines the main breeding range of the Carolina Wren. It nests in the thickets and tangles of open deciduous woodlands, and in residential areas if the dense cover it requires for protection against intruders is close by.

In the early part of the 20th century, the Carolina Wren expanded its range northward to include most of the eastern US (Bent 1948). It has been resident in Ontario since at least 1891, when a bird was collected at Mount Forest (McIlwraith 1894). Baillie and Harrington (1937) found it to be an irregular and uncommon resident in most of the southern parts of the province, north to Lambton Co and east to Victoria and Northumberland Counties. Fourteen years later, however, Snyder (1951) considered it so rare in Ontario that he did not discuss the species at all. Descriptions of the extent of its range vary from one account to another because of fluctuations in population size due to the susceptibility of the species to severely cold weather. A harsh Ontario winter often results in the depletion of the population, especially in areas north and northeast of the Carolinian Forest zone. This lack of hardiness has likely slowed its expansion farther north in Ontario. More recent data (James *et al.* 1976) indicate that, during the 1970's, it was a rare permanent resident north to Georgian Bay and east to Durham RM, and a 1975 nesting in southern Quebec (Godfrey 1986) showed that it is capable of breeding even north and east of that area. The severe winters of 1976-77 and 1977-78 caused a marked reduction in Carolina Wren numbers in eastern North America (Robbins *et al.*

1986), from which the species is still recovering.

Finding the Carolina Wren is usually not difficult, because it advertises itself well by means of its explosive, loud, and rich song, which can be heard for relatively long distances. It sings from dawn until dusk and through every month of the year, becoming more persistent during the late winter and early spring months; singing or territorial birds constitute 46% of all records of this species. In addition, the Carolina Wren is an inquisitive bird which responds well to 'pishing' and other sounds made to attract its attention. It does not linger, however. After exploding from its hiding place in woodland thickets, brushpiles, or tangles, it flits quickly out of sight to resume its normal skulking behaviour, offering little evidence to confirm that it is breeding. There are many records in the possible and probable breeding categories, partly because of its secretive behaviour around the nest, but partly, no doubt, because it is at the extreme northern limit of its range in Ontario and is so sparsely distributed that birds may not always find mates. A pair of Carolina Wrens may rear 2 broods between May and mid-July, but patience is required to find the well hidden cavity containing the nest. The best time to confirm breeding is during the noisy period after the young have left the nest and the group travels about as a family. Of 12 confirmed breeding records, 6 are of fledged young and 4 are of nests with young. While the Carolina Wren prefers dense bushy thickets, on occasion it becomes a familiar backyard bird, using nooks and crannies in buildings and other structures for its nest. In those instances the nest is more obvious and breeding is easier to confirm.

Urban areas in which the Carolina Wren was confirmed as breeding during the atlas period include Bronte, Grimsby, St. Catharines, Niagara-on-the-Lake, Port Colborne, Port Dover, and London. Other confirmed records were at Point Pelee Nat. Park and Rondeau Prov. Park. The most easterly record was at Carleton Place, with 2 other southeastern Ontario records in the town of Kingston. As is evident from the map, the great majority of records occurred in squares adjacent to the shorelines of the Great Lakes, especially Lake Erie and the northern shore of the Niagara Peninsula. Mild, sheltered conditions in protected shrubby areas, such as stream valleys, and nearby food sources, such as bird feeders, are apparently important in helping the birds survive cold winters, and maintaining the province's small breeding population.

There is no square in Ontario where abundance of the Carolina Wren was estimated at more than one pair. It is a rare and local species.-- *M.A. Richard*

CAROLINA WREN

Summary of Abundance Estimates

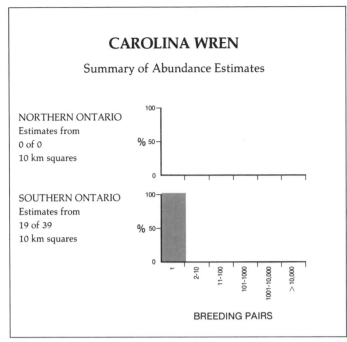

NORTHERN ONTARIO
Estimates from
0 of 0
10 km squares

SOUTHERN ONTARIO
Estimates from
19 of 39
10 km squares

% 50

% 50

100

0

100

0

1 2-10 11-100 101-1000 1001-10,000 >10,000

BREEDING PAIRS

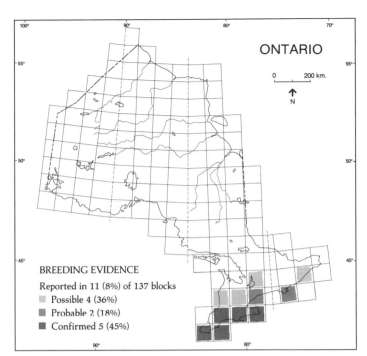

ONTARIO

0   200 km.

N

BREEDING EVIDENCE
Reported in 11 (8%) of 137 blocks
- Possible 4 (36%)
- Probable 2 (18%)
- Confirmed 5 (45%)

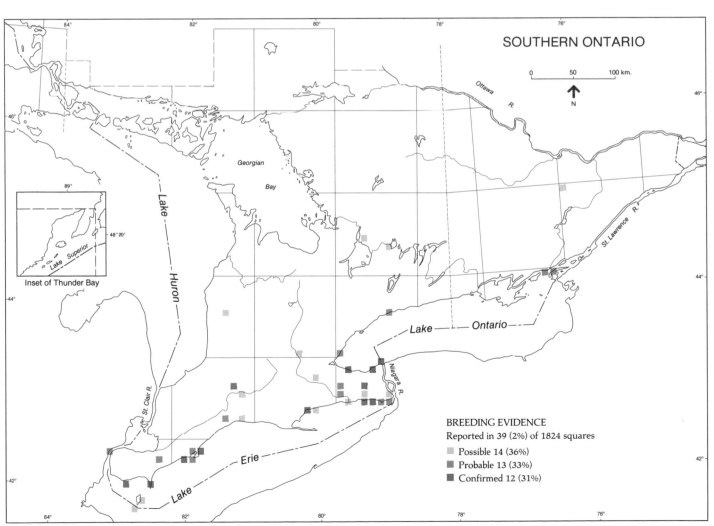

SOUTHERN ONTARIO

0   50   100 km.

N

Inset of Thunder Bay

Georgian Bay

Lake Huron

Lake Superior

St. Clair R.

Lake Erie

Niagara R.

Lake Ontario

Ottawa R.

St. Lawrence R.

BREEDING EVIDENCE
Reported in 39 (2%) of 1824 squares
- Possible 14 (36%)
- Probable 13 (33%)
- Confirmed 12 (31%)

HOUSE WREN
Troglodyte familier
Troglodytes aedon

The House Wren is the commonest member of the wren family south of the Canadian Shield in Ontario, and is known to many people in that area because of its loud bubbling song, aggressive nature, and adaptability to human habitation. The House Wren is a species of the New World, breeding in an area from southern Canada to central Chile and Argentina. In Ontario, it is fairly common in the southern portion of the province (Speirs 1985) in an area outlined, in general, by the limits of the Carolinian Forest and Great Lakes-St. Lawrence Forest regions. Its winter range extends from the southern half of the US south through the rest of its breeding range (AOU 1983). Its preferred habitats are thickets and woodland openings with shrubbery, most commonly in close proximity to human abodes. It will readily occupy nest boxes and other artificial cavities (Stokes 1979, Godfrey 1986), particularly where natural nest sites such as cavities in stumps are not available.

The historical range of the House Wren in Ontario differs little in its overall extent from that of the present day. In northern Ontario, it may well be that in earlier times, before forest clearing and settlement moved into that part of the province, the lack of suitable habitat there restricted the species to the southernmost fringes of the region. Baillie and Harrington (1937) reported the House Wren to be common in southern Ontario north to Rainy River and Lake Abitibi, but qualified that statement with the phrase "rather sparingly in the north." This same range was documented in 1951 by Snyder (1951). MacLulich (1938) described the House Wren as "a common breeding summer resident around settlements and open places" in Algonquin Prov. Park, where there were extensive burned and clear-cut areas, which have since grown over. Settlements and open places are gone from the park - and so is the House Wren. This may also be the case

elsewhere on the northern fringe of its range.

The male House Wren sings loudly and almost continuously throughout the breeding season and is difficult to overlook at that time. Singing males and birds exhibiting territorial behaviour, presumably by song, constitute over one-quarter of all records of this species. That figure would be even higher, except that it is also relatively easy to confirm breeding in the House Wren, as exemplified by the fact that confirmation was recorded in 61% of the squares in which it was found. The House Wren commonly produces 2, and rarely 3, broods per year (Bent 1948, Stokes 1979, Godfrey 1986) between the end of May and late August (James *et al.* 1976, Speirs 1985). The nests are built conspicuously in nesting boxes or in old trees or stumps near the border of woods: 72% of the confirmed breeding records (44% of all records) are of nests, primarily in situations where their contents could not be inspected. The territory is small (0.2 to 0.3 ha), and the singing male remains close to the nest site while the female incubates the eggs. Both parents feed the young. The young leave the nest after 12 to 18 days (Bent 1948), and the fledglings continue to be fed by both parents. The birds are particularly conspicuous at this time making breeding easy to confirm. Thus, 27% of the confirmed breeding records (17% of all records) are of fledged young or parents carrying food for young. The aggressive nature of the House Wren also aids in verifying breeding, as the bird frequently comes out into the open scolding vehemently with its loud alarm call, especially in response to 'pishing'.

The maps present a clear picture of the distribution of the House Wren in Ontario. It is ubiquitous in most of the area south of the Canadian Shield. Elsewhere in the province, its occurrence is much more sporadic. Possible breeding records extend north as far as the latitude of Kapuskasing, Marathon, and Red Lake, with one exceptional record of an unmated male that sang for 2 weeks at a site on the shore of James Bay, 27 km north of Moosonee.

The House Wren is common in most parts of its range in southern Ontario, and very common in the area between Lake Huron and the western end of Lake Ontario. In 13 squares, abundance estimates indicated densities of over 1,000 pairs per square.-- *J. Gould*

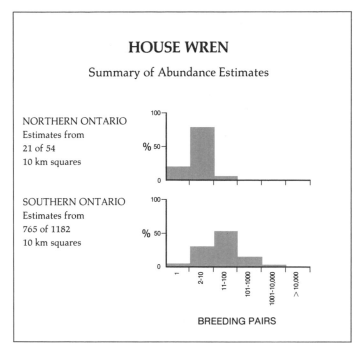

HOUSE WREN

Summary of Abundance Estimates

NORTHERN ONTARIO
Estimates from
21 of 54
10 km squares

SOUTHERN ONTARIO
Estimates from
765 of 1182
10 km squares

% — 100 / 50

% — 100 / 50

BREEDING PAIRS

1   2-10   11-100   101-1000   1001-10,000   > 10,000

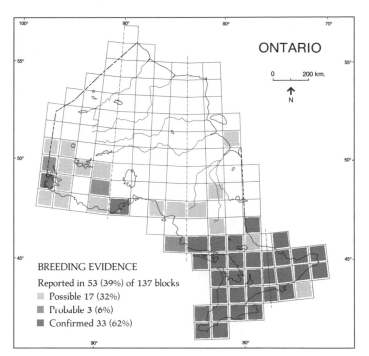

ONTARIO

0   200 km.

N

BREEDING EVIDENCE

Reported in 53 (39%) of 137 blocks

Possible 17 (32%)

Probable 3 (6%)

Confirmed 33 (62%)

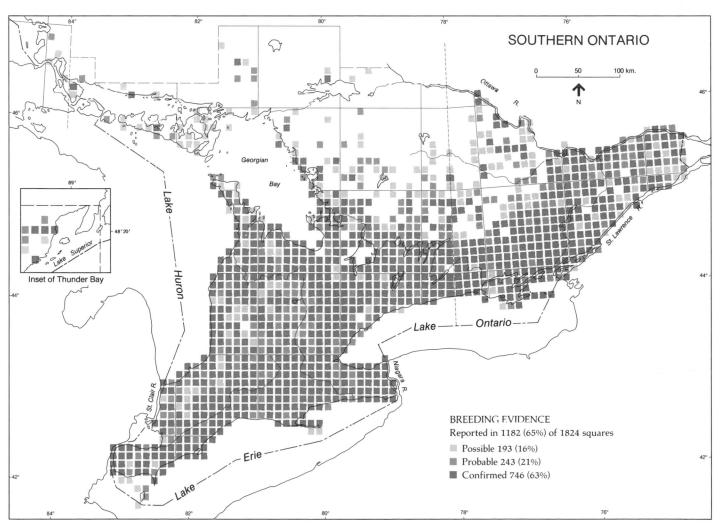

SOUTHERN ONTARIO

0   50   100 km.

N

Inset of Thunder Bay

BREEDING EVIDENCE

Reported in 1182 (65%) of 1824 squares

Possible 193 (16%)

Probable 243 (21%)

Confirmed 746 (63%)

WINTER WREN
Troglodyte des forêts
Troglodytes troglodytes

The Winter Wren is well known for its exuberant, melodious song of high, tinkling trills and tumbling warbles. It is the only North American wren to be found in Europe and Asia, where it is a common garden bird (the 'Jenny' Wren), and seems to be quite different from birds of the North American populations in disposition, habitat, and vocalizations.

In North America, the Winter Wren is primarily a bird of the coniferous forest regions. It is widespread through western conifer habitats in Canada and the US, in the Boreal Forest zone, and in the hemlock and pine communities of the Great Lakes-St. Lawrence Forest region and the Acadian Forest region of the Maritime Provinces (Erskine 1977). It is common across Ontario, where it haunts the cedar swamps, spruce bogs, and dark, moist evergreen woodlands of the Boreal Forest and similar habitats in the central and southern portions of the province. Most Winter Wrens migrate to the US in the fall, but small numbers winter in southern Ontario.

Historically, the Winter Wren has been found in suitable habitat in most of Ontario. In 1951, Snyder reported that it was found across the province, but sparingly and irregularly in agricultural areas, rarely in the western parts of the province near Manitoba, and not at all in the vast area of the Hudson Bay Lowland (Snyder 1951). It was absent from the southern fringe of Ontario on the 1968 to 1977 Breeding Bird Surveys (Speirs 1985), although road-side surveys of this sort are not well suited to sample the deep-woods habitat of the Winter Wren. By 1976, the known range of the species extended north to Sandy Lake and Fort Albany, and south to London (James *et al*. 1976). In 1977, it experienced a precipitous decline in numbers throughout North America. The unusually harsh, cold winter in the US in 1976-77 was the apparent cause (Speirs 1985). It was not until 1981 that the Winter Wren began to reach its former abundance in Ontario (Mansell 1983).

The preference of the Winter Wren for low, dense ground cover, fallen logs, and tangled roots, along with its diminutive size, dark colour, and retiring habits, make it difficult to see. It is mouselike in its movements, and is seldom seen in open flight. Once found, however, it is not really shy, but is often indifferent to human presence unless responding to 'pishing'. Its rambling song is well known and readily detected. Singing and territorial birds represent 78% of all atlas records in southern Ontario, and 87% in the north. Extreme patience is required if one is to observe the bird's entry into the well hidden nest. Not surprisingly, nests with eggs or young comprise only 12 of the 1,583 records across the province. The easiest time to confirm breeding for this species is during the final few weeks of the breeding period, when parents and fledglings travel about as a family: over half of all confirmed breeding records were obtained in this way. Nevertheless, for such a widespread species, a confirmation rate of only 8% is exceptionally low. Most records of confirmed breeding are from southern Ontario and the adjacent portion of northern Ontario, where atlassers were numerous, many return visits were possible, and access was not difficult. Abundance estimates, perhaps more accurately, indicate that the species is more abundant in the north.

Because of its song, the Winter Wren was probably overlooked in very few squares in which it bred, and its range is therefore well defined on the map. However, there is a strong likelihood that the map understates its breeding status by showing only possible or probable breeding in many squares where it actually did breed. The range defined by Snyder (1951) as north to the Hudson Bay Lowland is still largely valid, but small numbers of Winter Wrens also occur in densely wooded areas along river banks across the Lowland, as they probably did in Snyder's time and before. The boreal affinity of the species is evident from the map. In southern Ontario, pockets of coniferous habitat provide breeding areas south of the Canadian Shield: notably the areas of fairly extensive coniferous wetlands and cedar swamps of Manitoulin Island, the Bruce Peninsula, and Grey, Dufferin, and Wellington Counties, as well as numerous smaller isolated areas elsewhere. The most southerly record of confirmed breeding was in Rondeau Prov. Park, in 1981. This record suggests that the species would be more widespread in southern Ontario if more extensive woodlands remained.

Seventy-eight percent of abundance estimates were of between 2 and 100 pairs in both southern and northern Ontario; but a larger proportion of the northern estimates were of over 10 pairs (65% versus 42%), suggesting that the Winter Wren is more common in northern Ontario.-- *M.A. Richard*

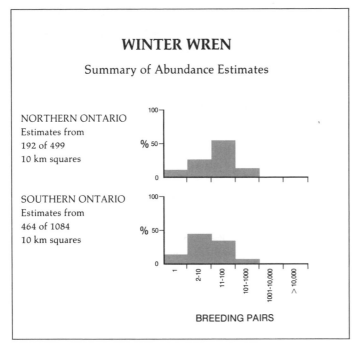

WINTER WREN

Summary of Abundance Estimates

NORTHERN ONTARIO
Estimates from
192 of 499
10 km squares

SOUTHERN ONTARIO
Estimates from
464 of 1084
10 km squares

BREEDING PAIRS

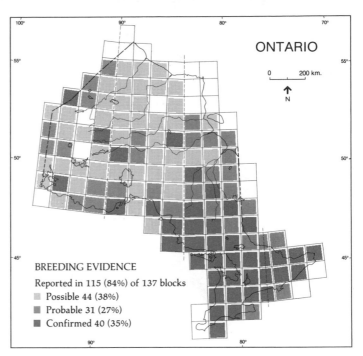

ONTARIO

0   200 km.

N

BREEDING EVIDENCE
Reported in 115 (84%) of 137 blocks

Possible 44 (38%)
Probable 31 (27%)
Confirmed 40 (35%)

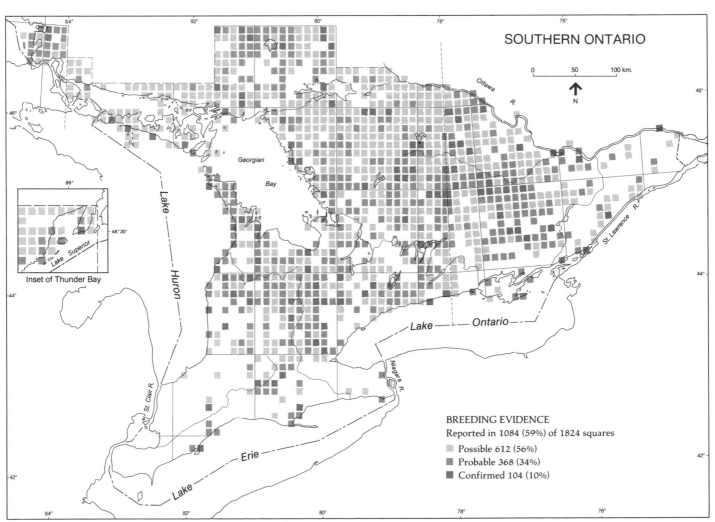

SOUTHERN ONTARIO

0   50   100 km.

N

Inset of Thunder Bay

Georgian Bay

Lake Huron

Lake Superior

St. Clair R.

Lake Erie

Lake Ontario

Niagara R.

Ottawa R.

St. Lawrence R.

BREEDING EVIDENCE
Reported in 1084 (59%) of 1824 squares

Possible 612 (56%)
Probable 368 (34%)
Confirmed 104 (10%)

SEDGE WREN
Troglodyte à bec court
Cistothorus platensis

The Sedge Wren, formerly known as Short-billed Marsh Wren, is one of the smallest and shiest members of the wren family. It is also one of the most poorly known because of its scarcity and its reclusive behaviour.

The Sedge Wren is a New World species that breeds in 3 disjunct populations, one from east-central Alberta to southern New Brunswick and south into the northeastern US, and the other 2 in Central and South America (AOU 1983). The North American population winters along the eastern seaboard and Gulf Coast of the US and Mexico. True to its name, the Sedge Wren prefers wet or moderately wet sedge meadows, bogs, grassy marshes, or old fields, particularly wet fields with low shrubby willow or alder growth (Speirs 1985, Godfrey 1986). Because of these preferences, its distribution is highly localized within its breeding range, but it is often inexplicably absent from seemingly suitable habitat.

There can be little doubt that pre-settlement southern Ontario, with its unbroken forest, contained less suitable Sedge Wren habitat than it does at present, when the species is usually found in agricultural areas. Nevertheless, in more recent years, the distribution of the Sedge Wren has become more restricted, at least in the eastern half of its range (Vickery 1983), probably as a result of the decimation of its habitat due to the development of more intensive agriculture. In southern Ontario, the present breeding range corresponds closely to its recorded historical range. Baillie and Harrington (1937) described the distribution as being from the southern portion of the province north to Kenora, southern Sudbury District, and Carleton Co, and also called it an irregular breeder throughout its range. The results of the Breeding Bird Survey, for the period 1968 to 1977, re-affirmed this distribution pattern in the southwestern and southcentral portions of the province, where it was described as being rare to uncommon (Speirs 1985). James *et al.* (1976) described it as being uncommon in Ontario north to Kenora and Lake Nipissing, and occasional to rare farther north to Kirkland Lake. The atlas maps substantiate its irregular breeding distribution throughout southern Ontario, but document the occurrence of the Sedge Wren farther north in both the northeastern and northwestern parts of the province than it was previously known.

The Sedge Wren is difficult to find and to confirm as a breeding species, both because of its small size and elusive nature and because its habitat is somewhat inaccessible. It does not flush readily and is often seen only briefly before retreating to the cover of grasses and sedges. The male sings from thickets of low shrubs and seldom uses a prominent perch. The song is somewhat muted in comparison to that of other wrens, and, like the bird itself, is not well known and can easily be overlooked. The nest of the Sedge Wren is well concealed in upright grasses or sedges (Godfrey 1986); nests with eggs or young were reported in only 6 squares. Several 'dummy' nests are built in addition to the one used for raising young. Sedge Wrens usually nest in small colonies, but these colonies are unstable and the same breeding area is not necessarily used yearly. The high proportion of possible and probable breeding records attests to the difficulty of confirming nesting of this species.

Until atlas field work was begun, the Sedge Wren was considered rare in Ontario, was generally poorly known, and was probably not looked for by inexperienced birders. To a certain extent, this is still the case, and more extensive searching of suitable habitat throughout Ontario would undoubtedly add more breeding localities to the maps. However, the scattered distribution shown does reveal some pattern. The majority of records come from southern Ontario. The intensively cultivated area of the southwest contains few records, although there are some from protected areas such as Rondeau Prov. Park and Long Point. Records increase with the percentage of neglected farmland in Wellington, Grey, and Bruce Counties and near the margin of the Canadian Shield. On the Shield, pockets of former agricultural land, marshes, and wet beaver meadows probably account for the majority of the records. Except for a few new northern records, the species is concentrated in the Great Lakes-St. Lawrence Forest region, both in northwestern and in southern Ontario.-- *J. Gould*

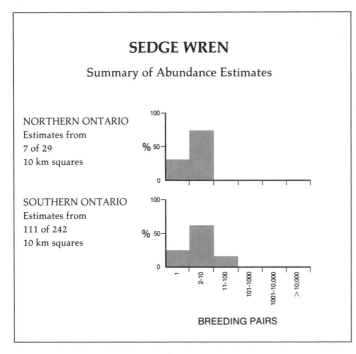

SEDGE WREN

Summary of Abundance Estimates

NORTHERN ONTARIO
Estimates from
7 of 29
10 km squares

SOUTHERN ONTARIO
Estimates from
111 of 242
10 km squares

BREEDING PAIRS

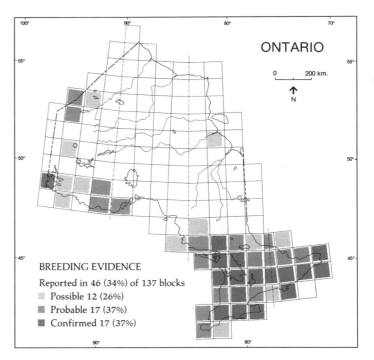

ONTARIO

0   200 km.

N

BREEDING EVIDENCE
Reported in 46 (34%) of 137 blocks
Possible 12 (26%)
Probable 17 (37%)
Confirmed 17 (37%)

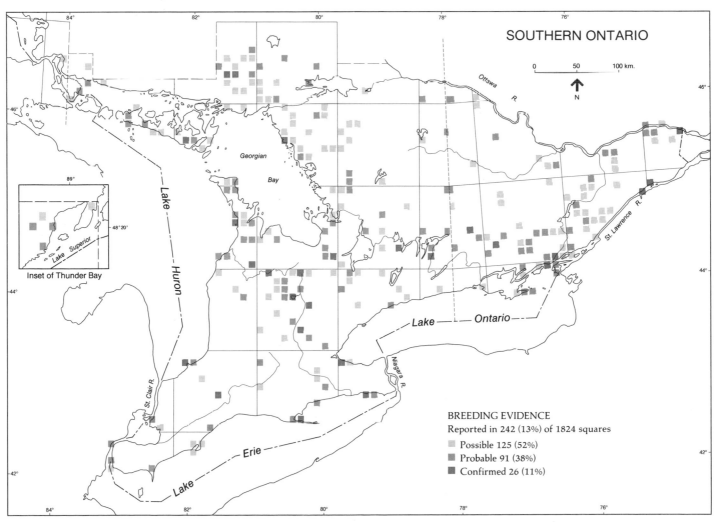

SOUTHERN ONTARIO

0   50   100 km.

N

Georgian
Bay

Lake Huron

Lake Superior

Inset of Thunder Bay

Ottawa R.

St. Lawrence R.

Lake — Ontario

Niagara R.

St. Clair R.

Lake Erie

Lake

BREEDING EVIDENCE
Reported in 242 (13%) of 1824 squares
Possible 125 (52%)
Probable 91 (38%)
Confirmed 26 (11%)

MARSH WREN
Troglodyte des marais
Cistothorus palustris

The Marsh Wren (formerly known as the Long-billed Marsh Wren) is a noisy denizen of cattail marshes that breeds in loose colonies locally across much of central Canada, through a large part of the US, and into south-central Mexico. It winters in the southern US and Mexico. In Ontario, it is fairly widespread in the south, becoming much more sparsely distributed north of the Great Lakes-St. Lawrence Forest region. It occurs west of Lake Superior and there was an isolated atlas record from the coast of James Bay. Large expanses of cattail marsh, usually interspersed with open water, provide the favoured habitat (Verner and Engelsen 1970). In some areas at least, and when given a choice, it seems to prefer nesting in narrowleaf cattail rather than the broadleaf species (Welter 1935, Bent 1948). In any case, stands of moderate density are favoured (Verner 1965). Breeding frequently occurs in marshes dominated by other tall, robust emergent types of vegetation (*e.g.*, bur-reed and *Phragmites*), and less often in tall grass/sedge marshes. Its somewhat patchy distribution in the province is primarily related to the occurrence of its preferred habitats. Even so, it is often inexplicably absent from areas which contain apparently suitable habitat.

The former and present ranges of the Marsh Wren are virtually identical (Baillie and Harrington 1937, Godfrey 1986). Although the atlas data indicate a slight northward extension, this is likely due to increased coverage of remote areas and does not indicate a range expansion *per se*. Like many other wetlands-dependent species, the Marsh Wren has probably suffered local declines in southern Ontario due to habitat loss.

Although somewhat elusive, the Marsh Wren is still very easy to detect. It is an incessant singer, both day and night, and the song ('rattle') is loud and distinctive. Furthermore, it is a quite gregarious bird and, barring habitat changes, gener-

ally occupies the same traditional breeding stations in successive years. Although nests are easily found, they are difficult to reach in their marshy locations. Not only is there a rather low proportion of confirmed breeding records (31%), but many of these (17%) are of inaccessible nests whose contents were not, or could not, be checked. The propensity of the male Marsh Wren to build several 'dummy' nests (Verner 1965) further hampers efforts at finding functional nests. It may well be that some of the 15 records of used nests, representing 11% of the confirmed breeding records, were 'dummy' nests.

Given the above, the maps give an accurate representation of the distribution of the Marsh Wren in Ontario. Except in the most remote areas where coverage was minimal, the majority of breeding sites were probably detected. There is a strong likelihood that breeding does occur even in those squares where the record shows only possible or probable breeding. Its patchy distribution is strongly related to the occurrence of its preferred habitat. The southern Ontario distribution of the Marsh Wren is very similar to that of the Black Tern, which shares the same type of habitat, except that the Marsh Wren can occur in smaller patches of marsh and is therefore more widespread away from the Great Lakes shoreline in the southwest. Records of the Marsh Wren are most abundant and most uniformly distributed in the vicinity of eastern Lake Ontario. A similar tendency was indicated by results summarized from the BBS (Speirs 1985). Its Ontario range is apparently occupied by 2 discrete populations representing 2 different subspecies, *C. p. dissaeptus* in the south and *C. p. iliacus* in the northwest (Godfrey 1986). There was no evidence from atlas field work of a continuous distribution around the north shore of Lake Superior as mapped by Godfrey (1986). A noteworthy extralimital breeding record was established on 9 July 1983, when a nest with young, surrounded by 3 dummy nests, was found at Big Piskwanish Point on James Bay -- a location approximately 600 km north of the previously known breeding range of the species (Weir 1983). For almost 4 weeks during the following summer, a male sang persistently at North Point, about 30 km farther south along the coast. Although an empty nest was found nearby, this bird apparently remained unmated.

Because the Marsh Wren is basically a gregarious nesting species, it is usually locally common in southern Ontario. Some estimates of over 100 pairs per square have been provided from marshes in various parts of the south, including Walpole Island on Lake St. Clair, Long Point, Turkey Point, and Port Maitland on Lake Erie, Matchedash Bay on Georgian Bay, Luther Lake, and Scugog, Sturgeon, and Pigeon Lakes in the Kawartha region of southcentral Ontario.-- *J.D. McCracken*

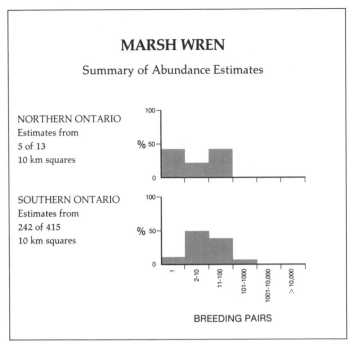

MARSH WREN

Summary of Abundance Estimates

NORTHERN ONTARIO
Estimates from
5 of 13
10 km squares

SOUTHERN ONTARIO
Estimates from
242 of 415
10 km squares

BREEDING PAIRS

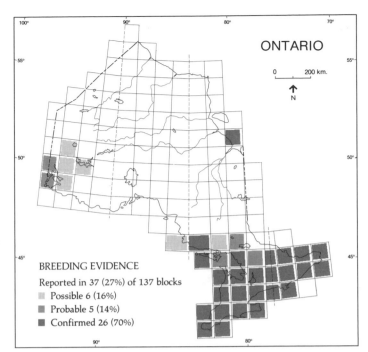

ONTARIO

0   200 km.

N

BREEDING EVIDENCE
Reported in 37 (27%) of 137 blocks
Possible 6 (16%)
Probable 5 (14%)
Confirmed 26 (70%)

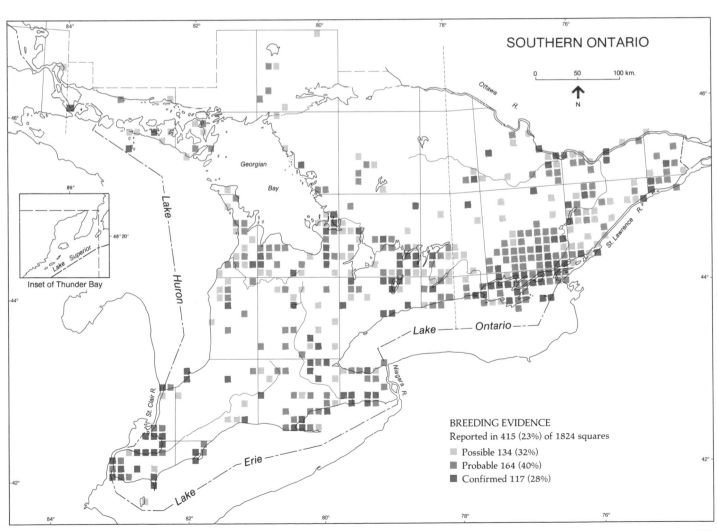

SOUTHERN ONTARIO

0   50   100 km.

N

Inset of Thunder Bay

BREEDING EVIDENCE
Reported in 415 (23%) of 1824 squares
Possible 134 (32%)
Probable 164 (40%)
Confirmed 117 (28%)

313

GOLDEN-CROWNED KINGLET
Roitelet à couronne dorée
Regulus satrapa

Kinglets are among the smallest birds in Canada, and among songbirds, the Golden-crowned Kinglet is the smallest.

The breeding range of the Golden-crowned Kinglet spans North America in a broad horseshoe pattern, extending north from Central America along the American Cordillera, across Canada north of the prairies to Newfoundland, and reaching south into the eastern states along the Appalachian Mountains. It winters as far north as southern Canada. The species breeds in areas of closed coniferous forest, its northern limit generally following the northern limit of the Boreal Forest region. The Golden-crowned Kinglet is commonly found throughout Ontario wherever mature coniferous stands occur; in particular, the greater the dominance of spruce, the greater will be the breeding density of this species (Salt 1957, Franzreb 1984). The species is also associated with mature stands of fir or hemlock, and occasionally with pines.

This species has long been known to breed in Ontario, although its breeding range has not been well documented. It may have been more numerous in southern Ontario in pre-settlement days, becoming relegated more to the north as land clearing proceeded. Subsequently, the species has likely re-occupied some of its presumed former breeding range in southern Ontario as the spruce and pine plantations scattered throughout the area have begun to mature. In the future, as these same plantations are thinned or logged, its local breeding density will likely decrease (Andrle 1971), while increasing where there are newer plantations.

Although the maps accurately reflect the distributional limits of the species, confirmed breeding is almost certainly underrepresented. The song of the Golden-crowned Kinglet is weak and could easily be lost among the other woodland noises, especially on windy days. This species prefers the middle to upper portions of dense spruce trees, and spends most of its feeding time gleaning along small branches and twigs, where it is not easily seen. Golden-crowned Kinglets seldom become agitated when intruders approach, and the small mossy nests (2 to 16 m high) blend in well with the surrounding foliage and are therefore very difficult to find: only 7 of the 934 records in all of Ontario are of nests. Because of these difficulties, 56% of the records of this species are classified as possible breeding records. Almost every case of confirmed breeding (96%) involved observations of fledged young or of adults with food for young. The species likely breeds in all blocks within the Boreal Forest zone. In areas outside of the Boreal Forest, it should be expected to breed where there is a sufficient density of tall, mature spruce and other conifers. On the Hudson Bay Lowland, such conditions occur mostly in well-drained areas: for example, the only confirmed record on the Lowland was a 1981 observation of fledged young on the Sutton Ridges. Elsewhere on the Lowland, birds were reported in low numbers along river banks in the west, but the Sutton Ridges record is the only one in the east. The Golden-crowned Kinglet's distribution is increasingly spotty the farther south one moves from the Boreal Forest zone. In particular, in southwestern Ontario its distribution is closely associated with spruce and pine plantations. Concentrations of probable and confirmed breeding records are evident near Guelph, Luther Lake, and the Turkey Point area, all areas with good atlasser coverage and extensive conifer plantations. Squares in southern Ontario which lack mature spruce plantations are unlikely to harbour breeding pairs of Golden-crowned Kinglets. Most Regional Coordinators felt that the atlas coverage underrepresented the true confirmed breeding range of the species: given more time and effort, it would have been confirmed as breeding in more squares.

Golden-crowned Kinglets are common in much of northern Ontario, but less so in the south, based on the limited number of abundance estimates made during the atlas period. The highest abundance estimates (over 100 pairs per square) were provided in squares scattered throughout the area from Huntsville north to Pickle Lake, with none south of Bracebridge or Bancroft.-- *B. Klinkenberg*

GOLDEN-CROWNED KINGLET

Summary of Abundance Estimates

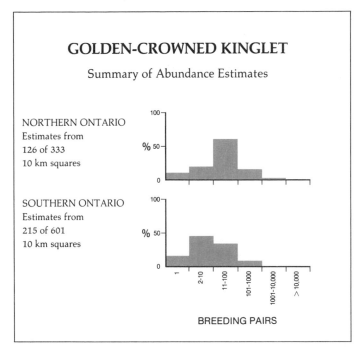

NORTHERN ONTARIO
Estimates from
126 of 333
10 km squares

SOUTHERN ONTARIO
Estimates from
215 of 601
10 km squares

BREEDING PAIRS

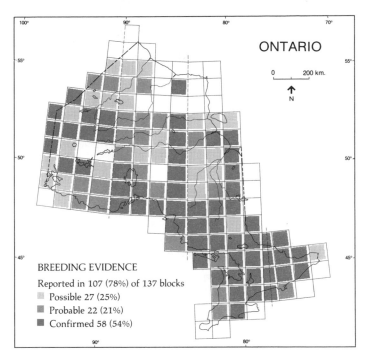

ONTARIO

0   200 km.

N

BREEDING EVIDENCE
Reported in 107 (78%) of 137 blocks

Possible 27 (25%)
Probable 22 (21%)
Confirmed 58 (54%)

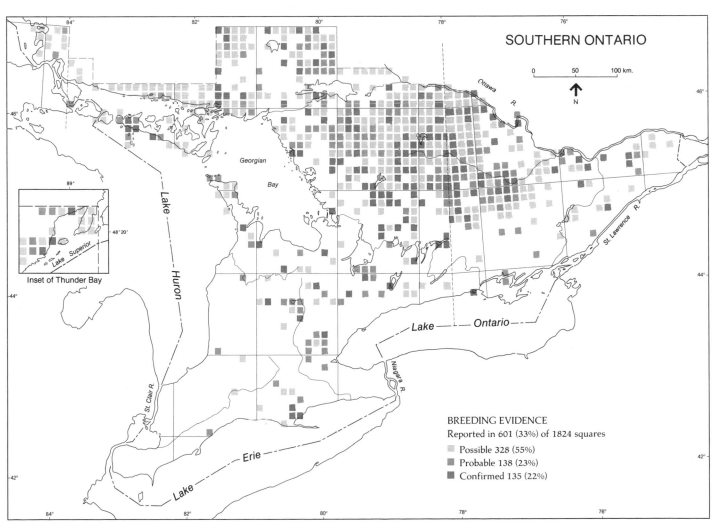

SOUTHERN ONTARIO

0   50   100 km.

N

Inset of Thunder Bay

BREEDING EVIDENCE
Reported in 601 (33%) of 1824 squares

Possible 328 (55%)
Probable 138 (23%)
Confirmed 135 (22%)

RUBY-CROWNED KINGLET
Roitelet à couronne rubis
Regulus calendula

The Ruby-crowned Kinglet breeds from Newfoundland in the east across Canada to Alaska in the west, and south in the mountains to the southwestern US. It winters as far north as southern Ontario, and in most of the US south to Central America. It is less restrictive in its habitat (and food) preferences than the Golden-crowned Kinglet (Salt 1957, Franzreb 1984), and can be found breeding in evergreen stands (preferably spruce) in a wide range of habitats, ranging from pure boreal forest to mixed woodlands. It is often found in open or edge areas within the evergreen stand.

The earliest records of bird life in Ontario confirm that the Ruby-crowned Kinglet seldom breeds south of Lake Simcoe and Georgian Bay. McIlwraith (1894) stated, "Their breeding ground is far north," and had never seen the bird in summer. Baillie and Harrington (1937), on the other hand, while citing no breeding evidence south of Muskoka and Parry Sound, ventured the opinion that, "It probably breeds throughout Ontario." During the 1970's, nests were found at Bronte (near Hamilton) and near Fort Erie (R. Curry pers. comm.).

The Ruby-crowned Kinglet prefers the middle and upper portions of tall coniferous trees, where it is more easily heard than seen. Its oft-repeated song is loud and rollicking. Not surprisingly, singing males account for half of all records of this species. Unlike the Golden-crowned Kinglet, the male of this species becomes agitated when intruders approach the nest, often restlessly flicking its wings while calling incessantly. Although such behaviour is likely to indicate the presence of a nest, the structure itself is small, obscure, and extremely difficult to locate. Only 13 of the confirmed breeding records are of nests.

The Ruby-crowned Kinglet is a slightly more 'northern' bird than the Golden-crowned in its summer distribution.

While its breeding range extends across almost all of northern Ontario, with very few blocks lacking any reported breeding evidence, there are very few confirmed breeding records south of the Shield, and even very few on the southern edge of the Shield itself. This fact is somewhat surprising and suggests that although conifer plantations are suitable for the Golden-crowned Kinglet, they are not suitable for Ruby-crowned Kinglets. Given the behaviour of the species, it is unlikely that it was merely overlooked in that part of the province. The range overlap of the 2 species of kinglets is of interest. Although the Ruby-crowned Kinglet tends to breed farther north than the Golden-crowned Kinglet, it winters farther south (Lepthien and Bock 1976). As indicated by Laurenzi *et al.* (1982), the Ruby-crowned Kinglet may be winter-limited, with its survival and abundance related more to the severity of the winter than to summer breeding successes. Away from the Canadian Shield, there are only a few locations in southern Ontario where the Ruby-crowned Kinglet breeds regularly: coniferous habitat near the western shore of the Bruce Peninsula and the southern edge of Manitoulin Island, and the Alfred Bog (east of Ottawa).

Within its range, the Ruby-crowned Kinglet is at least as abundant as most other songbirds. Twelve percent of abundance estimates were of over 100 pairs per square, with a concentration of these high estimates at the latitude of the Albany River. Previously, it was thought to be an uncommon summer resident across all of Ontario (Baillie and Harrington 1937), but the atlas data suggest that it could be classed as rare in southwestern and southeastern Ontario and common across central and northern Ontario. BBS data for central and northern Ontario and Quebec indicate that the populations of this species fluctuate widely, with its numbers decreasing by over one-half within a 12 year period (1967 to 1979), and then doubling over the next 4 year period (1979 to 1983).-- *B. Klinkenberg*

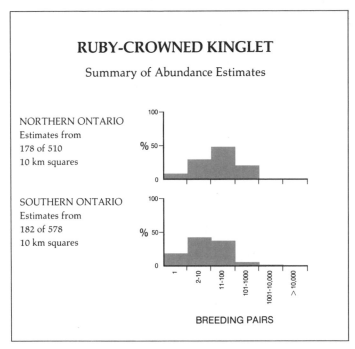

RUBY-CROWNED KINGLET

Summary of Abundance Estimates

NORTHERN ONTARIO
Estimates from
178 of 510
10 km squares

SOUTHERN ONTARIO
Estimates from
182 of 578
10 km squares

%

BREEDING PAIRS

1   2-10   11-100   101-1000   1001-10,000   >10,000

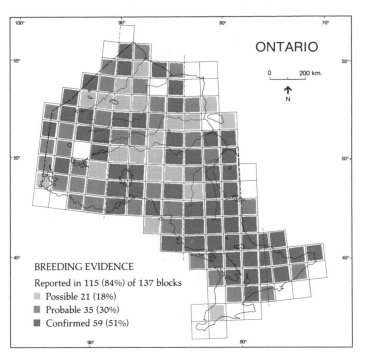

ONTARIO

0   200 km.

N

BREEDING EVIDENCE

Reported in 115 (84%) of 137 blocks

Possible 21 (18%)
Probable 35 (30%)
Confirmed 59 (51%)

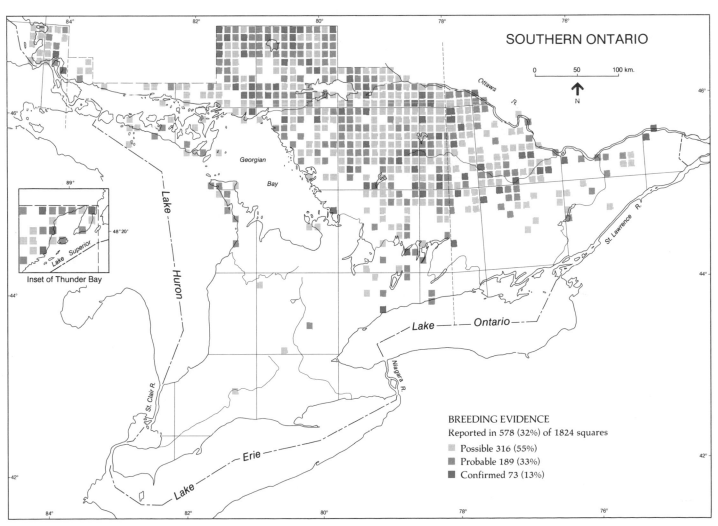

SOUTHERN ONTARIO

0   50   100 km.

N

Ottawa R.

Georgian Bay

Lake Huron

Lake Superior

89°
48°20'

Inset of Thunder Bay

St. Clair R.

Lake — Ontario —

Niagara R.

St. Lawrence R.

Lake — Erie —

BREEDING EVIDENCE

Reported in 578 (32%) of 1824 squares

Possible 316 (55%)
Probable 189 (33%)
Confirmed 73 (13%)

BLUE-GRAY GNATCATCHER
Gobe-moucherons gris-bleu
Polioptila caerulea

The Blue-gray Gnatcatcher breeds from northern Central America, throughout the US, and north to extreme southern Canada. It winters from the southern US southward (Godfrey 1986). In Ontario, it is confined to the Carolinian and Great Lakes-St. Lawrence Forest zones, where it breeds in a wide spectrum of forested and semi-forested habitats, including both primary and secondary deciduous woodlands, oak savannas, and, rarely, coniferous plantations, orchards, hawthorn savannas, and low shrubby tangles along woodland borders. However, deciduous woodlands, particularly those along watercourses and wooded swamps, are evidently preferred.

Historically, the Blue-gray Gnatcatcher was unknown as a breeding species throughout much of northeastern North America. However, in the last 50 years, its range has undergone a considerable northeastward expansion to include much of New England, southern Ontario, and extreme southwestern Quebec. Prior to 1930 in Ontario, the species was restricted almost entirely to the counties south and west of Aylmer and London, where it was described as a "not uncommon" to fairly common summer resident (Macoun and Macoun 1909, Saunders and Dale 1933). Beginning about 1940, its range underwent several marked extensions, apparently the result of major flight years in 1940, 1953, and 1963; a pattern of expansion closely mirrored in adjacent New York State (Bull 1974). In Ontario, the most dramatic extension of the Blue-gray Gnatcatcher range occurred in 1953, when it was first reported at Kingston (Quilliam 1965), at Prince Edward Co (Sprague and Weir 1984), at Collingwood (Devitt 1967), and near Tobermory, Bruce Co, where nesting was confirmed (Moore 1954). By 1964, it had apparently become established as a breeding species north to northern Lake Simcoe and east to Presqu'ile Prov. Park and, by the mid- to late 1970's, east to Charleston Lake in Leeds Co, where a nest was found in 1978 (ONRS). It also nested in the 1970's in the Minesing Swamp and Wasaga Beach Prov. Park in Simcoe Co. Since the late 1970's, however, the northward spread of the Blue-gray Gnatcatcher in Ontario has been less dramatic.

The scratchy, scarcely-audible, warbled song of the male Blue-gray Gnatcatcher is delivered for only a short time between the arrival of the species in late April and the onset of egg-laying in early to mid-May. The species is often initially detected on the basis of the almost constant nasal 'twee' call, rather than the inconspicuous song itself. Thus, singing males constitute a rather low 15% of all records. Slightly less than half (42%) of all records of confirmed breeding involve fledged young, a reflection of the fact that, in Ontario, noisy, active family groups may regularly be found well into August. Whether these groups are the result of renestings due to failed earlier attempts or, possibly, rare second broods, is not known.

It is clear from the data that the Blue-gray Gnatcatcher is no longer the rare bird in Ontario that it was half a century ago (Baillie and Harrington 1937). The Carolinian Forest remains the stronghold for the species in the province, though the species is far from ubiquitous, even there. Its absence from many squares in the extreme southwest is likely due to lack of habitat, caused by forest clearing. Elsewhere, its absence may be due to a general unfamiliarity among atlassers with the weak call of the species, and its inconspicuousness. Expansion north and east into the Great Lakes-St. Lawrence Forest has been less extensive; the species apparently avoids the forest of the Canadian Shield and, for the most part, is confined to the shores of Lake Ontario and the St. Lawrence River. Despite recent nestings in Ontario near Ottawa, and in the Gatineau Hills and at Montreal in adjacent Quebec (David 1980), the species is evidently still largely absent from the eastern counties of Ontario. Interestingly, the Blue-gray Gnatcatcher was not recorded either in extreme western Rainy River District, where it had previously been recorded in early July 1980 (Goodwin 1980), or at Sault Ste Marie, an area where it might reasonably have been expected to occur in view of recent nesting in the upper peninsula of adjacent Michigan (Pinkowski 1978). More intensive coverage of these regions might have revealed the presence of this species.

Abundance estimates suggest that, in southern Ontario, the Blue-gray Gnatcatcher is an uncommon breeding species. However, all 18 of the estimates of over 10 pairs per square are from southwestern Ontario, where it now appears to be a fairly common bird.-- *D.A. Sutherland and M.E. Gartshore*

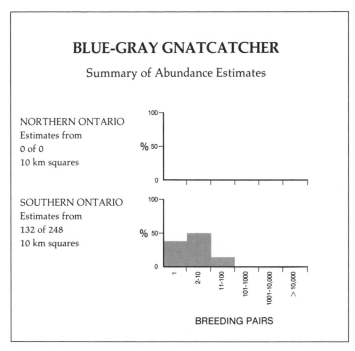

BLUE-GRAY GNATCATCHER

Summary of Abundance Estimates

NORTHERN ONTARIO
Estimates from
0 of 0
10 km squares

SOUTHERN ONTARIO
Estimates from
132 of 248
10 km squares

%

BREEDING PAIRS

1 2-10 11-100 101-1000 1001-10,000 >10,000

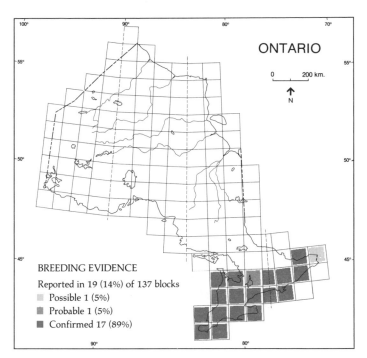

ONTARIO

0   200 km.

N

BREEDING EVIDENCE

Reported in 19 (14%) of 137 blocks

Possible 1 (5%)
Probable 1 (5%)
Confirmed 17 (89%)

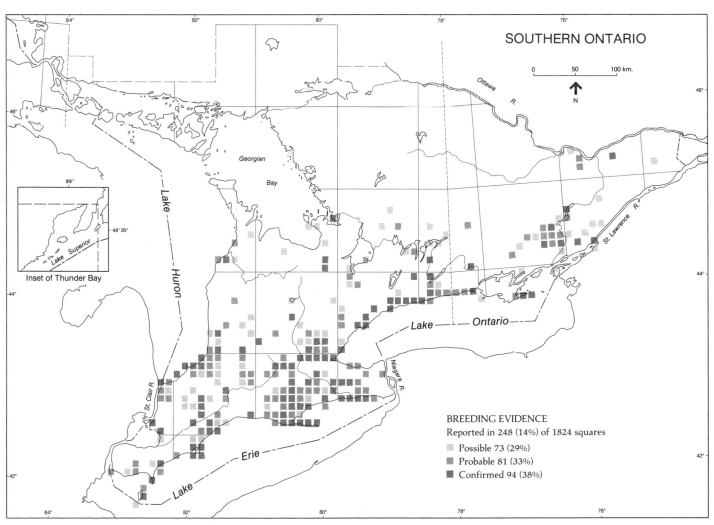

SOUTHERN ONTARIO

0   50   100 km.

N

Inset of Thunder Bay

Georgian Bay

Lake Huron

Lake Superior

St. Clair R.

Lake Erie

Lake Ontario

Niagara R.

Ottawa R.

St. Lawrence R.

BREEDING EVIDENCE

Reported in 248 (14%) of 1824 squares

Possible 73 (29%)
Probable 81 (33%)
Confirmed 94 (38%)

EASTERN BLUEBIRD
Merle-bleu de l'Est
Sialia sialis

The Eastern Bluebird is a popular summer resident of eastern North America, well liked for its colourful plumage and cheery song. In Ontario, it breeds throughout the province in agricultural areas and clearings south of the continuous boreal forest. Within these areas, the species inhabits cultivated lands, fallow fields, grazed pastures, old fields, large lawns, country cemeteries, golf courses, and forest clearings. Although the list of habitats is long, a particular area is suitable only if the immediate area provides a cavity (natural or man-made) for nesting and room-sized patches of bare ground or short grass for feeding. Because of these special habitat needs, it is not present in all habitats and is often absent from intensively farmed land and forest openings, but it can be encouraged to nest in these areas if nest boxes are erected. Individuals throughout the range of the Eastern Bluebird have established nest box trails in hopes of increasing its population (Zeleny 1976, 1977, Walton 1986).

During the past century, the Eastern Bluebird has undergone a general decline in numbers as well as several population fluctuations of shorter duration. Early observers in Ontario reported the Bluebird to be common or even abundant (Quilliam 1973, Risley 1981). Nash reported seeing "many thousands" passing from west to east over Toronto during a spring in the early 1900's (Barry 1974). Following these accounts, Eastern Bluebirds were reported less commonly around the towns and cities, possibly because of competition with the House Sparrow and loss of natural cavities. Along with these observations were reports that the species was subject to precipitous declines in numbers following severe winters on their wintering areas in the southeastern US. The winters of 1895-1896 and 1911-1912 were particularly devastating, but apparently the population recovered soon afterwards (Taverner 1922b, Bent 1949). During the 1930's

and 1940's, the Eastern Bluebird was still considered as regular or fairly common, though less so than before (Baillie and Harrington 1937, Snyder 1941). More recently, severe winters and the gradual loss of suitable habitat around urban areas have continued to plague the Bluebird. The winters of 1957-1958, 1976-1977, and 1977-1978 caused an estimated 60 to 90% decline in numbers in Ontario and across its range (Bystrak 1980, Risley 1981). At present, winter conditions have ameliorated, and Bluebird trail operators and monitoring schemes such as the BBS suggest that the population is gradually recovering.

Several habits of the Eastern Bluebird make it an easy bird to record and then to confirm as breeding. Its habit of perching on wires or dead branches while scanning for food makes it conspicuous, as does its cheerful, easily recognized call. Its cavity-nesting habit ensured that atlassers who saw a nearby nest box would check it. The species also has a long breeding season, with egg dates ranging from 27 March to 11 August (Risley 1981). Finally, because the young remain with the adults following fledging, confirmed breeding can be established even late in the season. Of the over 30 cavity-nesting species in Ontario, only a handful, mostly swallows, surpass the Eastern Bluebird in percentage of confirmed breeding records. Most records of confirmed breeding of this species are of nests with young (35%), fledged young (20%), or adults entering nest holes (20%).

The distribution of the Eastern Bluebird as shown here is generally an accurate picture of its historical distribution. The records cover all of its known range except for occasional previous breeding records from farther north, *e.g.*, Favourable Lake (Hope 1938) and Moosonee (Goodwin 1975). Major concentrations of the species are along the Niagara Escarpment and the southern edge of the Canadian Shield. This distribution closely follows the distribution of limestone plains and glacial till moraines. These are areas of poor rocky soil and sparse vegetation, which are chiefly used for grazing. In the intensively agricultural areas of southwestern Ontario, the species has a more scattered distribution. In these areas, its presence is linked to availability of suitable nestboxes or cavities. At the northern edge of its range, the Eastern Bluebird becomes sparsely distributed because forest openings and agricultural areas are fewer, and the bird could be overlooked if the natural cavities in these areas are not as obvious as boxes. Thus, in northern Ontario, the map may slightly underrepresent its true distribution.

The Eastern Bluebird was declared to be rare by COSEWIC and MNR based on its low population following the severe winters of 1976 to 1978. At present, the population is recovering and the species might now be considered uncommon rather than rare. Abundance estimates suggest that most occupied squares have from 2 to 10 pairs of Bluebirds in southern Ontario, but only one pair in the north. With the help of nest boxes erected by the Bluebird enthusiasts, the species may continue to frequent our countryside.--
C. Risley

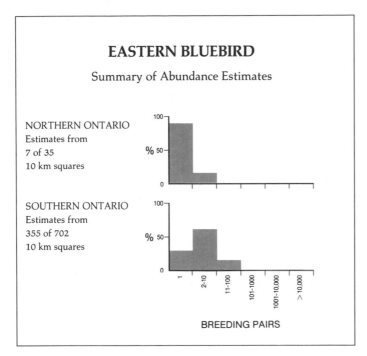

EASTERN BLUEBIRD

Summary of Abundance Estimates

NORTHERN ONTARIO
Estimates from
7 of 35
10 km squares

SOUTHERN ONTARIO
Estimates from
355 of 702
10 km squares

%

BREEDING PAIRS

1 2-10 11-100 101-1000 1001-10,000 >10,000

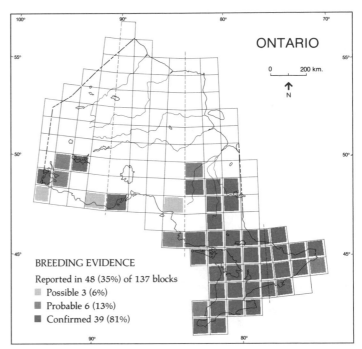

ONTARIO

0   200 km.

N

BREEDING EVIDENCE

Reported in 48 (35%) of 137 blocks

- Possible 3 (6%)
- Probable 6 (13%)
- Confirmed 39 (81%)

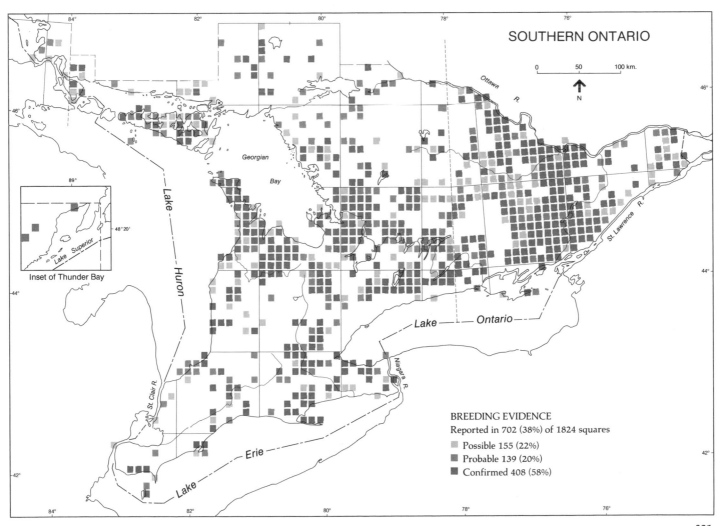

SOUTHERN ONTARIO

Ottawa R.

0   50   100 km.

N

Georgian Bay

Lake Huron

Lake Superior

Inset of Thunder Bay

48°20'

89°

St. Clair R.

Niagara R.

Lake Ontario

St. Lawrence R.

Lake Erie

BREEDING EVIDENCE

Reported in 702 (38%) of 1824 squares

- Possible 155 (22%)
- Probable 139 (20%)
- Confirmed 408 (58%)

VEERY
Grive fauve
Catharus fuscescens

The Veery, with its warm russet colouration and its pleasant song, is well known over most of the settled parts of Ontario.

The breeding range of the Veery extends in a band across the continent from New England and southern Newfoundland to central British Columbia, and south through the mountain areas of the western US. It winters in South America. In Ontario, it is associated with cool, damp deciduous or mixed forests, with a fair amount of bushy undergrowth and ferns. These conditions may be found at the edge of the forest, but this species is more frequently found in young or disturbed forests, where other required conditions are met. Its range covers all parts of the province where Carolinian or Great Lakes-St. Lawrence Forests are found, but it also extends somewhat farther north into the Boreal Forest region in deciduous growth such as poplar, birch, cherry, and alder; such conditions commonly occur in wetlands or follow forest fires or timber cutting.

Judging from accounts written 50 to 100 years ago (McIlwraith 1894, Baillie and Harrington 1937), the status of the Veery in Ontario has changed little in the last century. The influence of forestry operations, or clearance of woods for agriculture in the more settled regions of the southwest may have eliminated the Veery from a few areas of Ontario, and reduced its numbers in others in historical times. Even the mere thinning or clean-up of woodlots in such areas could affect moisture level and understory density and so inhibit their use for nesting. The overall geographic extent of its range in Ontario has remained largely unchanged, however.

Like our other thrushes, the Veery is mostly a ground-foraging bird, but its niche is different from that of the hardwood- and moisture-loving Wood Thrush in that it uses

perches from which to hunt for food items in the litter. It also does a fair amount of aerial flycatching, during which it is more conspicuous. The nest is less easily found than the bird. It is built on or near the ground, often on a stump, and is quite well hidden. The incubating bird does not flush easily. Therefore, it is not surprising that only 6% of the Veery records consist of nests with eggs or young. However, the song is distinctive and persistent, even though the bird does not sing for about 2 weeks after arrival on its territory (Bent 1949), making it harder to find at that crucial time. It is heard mainly in the evening and morning, as with other thrushes, but sometimes in the daytime. Birds are aggressive in defence of the nest and territory. They will come out in response to human 'pishing' or tapes. Atlas records include a high percentage of singing and territorial birds (50%) and of birds exhibiting agitated behaviour or distraction displays (16%).

The maps show the range of the Veery extending somewhat farther north than that given in Godfrey (1986), even as far north as the Albany River, near Fort Hope. It should be kept in mind that the occurrence of any species thins out towards the outer limits of its range and usually shows some outliers in pockets of suitable habitat. Some of these stations may prove to be temporary. However, there is confirmed breeding evidence from several locations as much as 100 km or more to the north of the range limits stated by Godfrey (Timmins, Kirkland Lake, Matachewan, Virginiatown, Schreiber, and northwest of Nipigon); it is likely that these represent a long-term range expansion. In southern Ontario, the Veery is almost totally absent as a breeding bird from the intensively cultivated farmland of Essex Co, the fruit lands of the Niagara Peninsula below the Escarpment, and the urban environs of Toronto, all of which lack suitable Veery habitat.

Abundance estimates provided by atlassers show generally high densities, indicating that the Veery is common in both southern and northern Ontario. Ontario had the highest counts of Veerys of any state or province in the BBS (Robbins *et al.* 1986), with particularly high counts concentrated in the Rainy River District and in a belt running east from Georgian Bay to the Quebec border (Speirs 1985). BBS data from 1967 to 1983 indicate a decrease in abundance in the region of Ontario south of a line from Tobermory to Pembroke (although not north of there) at a rate of 50% per 25 years, and that this rate of fall-off has been accelerating: at the rate of decline for the period 1979 to 1983, a 50% population decrease would occur in 7 years. In Maine, a similar decline has been associated with an increase in the Wood Thrush (Morse 1971).-- *D.C. Sadler*

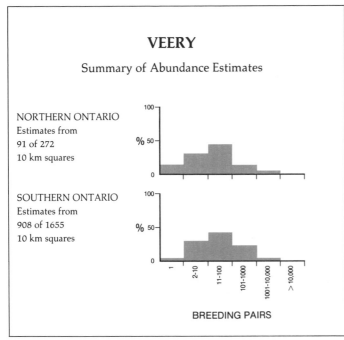

VEERY

Summary of Abundance Estimates

NORTHERN ONTARIO
Estimates from
91 of 272
10 km squares

SOUTHERN ONTARIO
Estimates from
908 of 1655
10 km squares

BREEDING PAIRS

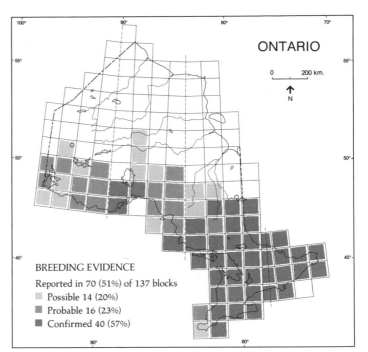

ONTARIO

0   200 km.

N

BREEDING EVIDENCE
Reported in 70 (51%) of 137 blocks
  Possible 14 (20%)
  Probable 16 (23%)
  Confirmed 40 (57%)

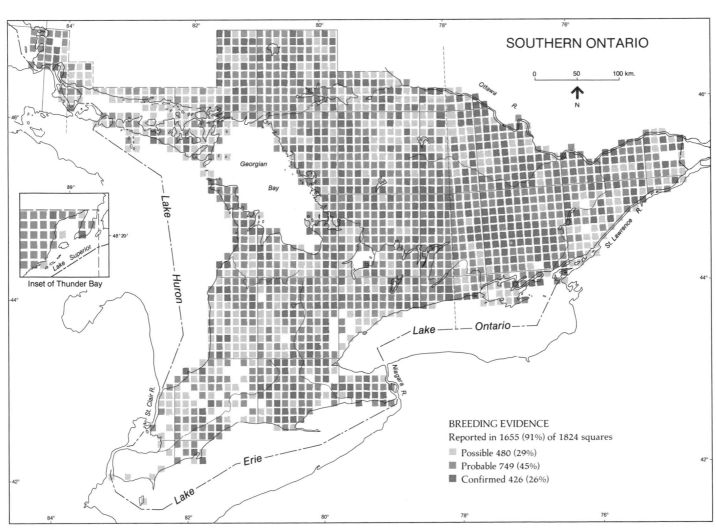

SOUTHERN ONTARIO

0   50   100 km.

N

Inset of Thunder Bay

BREEDING EVIDENCE
Reported in 1655 (91%) of 1824 squares
  Possible 480 (29%)
  Probable 749 (45%)
  Confirmed 426 (26%)

SWAINSON'S THRUSH
Grive à dos olive
Catharus ustulatus

The Swainson's Thrush, formerly known as the Olive-backed Thrush, is typical of the northern woods. Its breeding range extends in a belt across the continent from the tree-line south to southern Canada, with extensions in the western mountains to California and Colorado, and in the Appalachians of the east. In New York State, it is found in the Adirondacks, the Catskills, and even the hill country south of Lake Ontario (NYBBA preliminary maps). It winters in Mexico and Central and South America. In Ontario, the southern edge of its breeding range coincides fairly well with the edge of the Canadian Shield and occurs at an equivalent latitude on the Bruce Peninsula. It continues north in forested country to near the coast of Hudson Bay. In the vast muskeg areas of the Lowland, it is found mainly along river banks and old beach lines, where more densely-treed boreal forest habitat occurs. It also occurs in a few islands of suitable habitat south of the Shield. The Swainson's Thrush prefers conifers, especially spruce and fir, but has also been found to frequent deciduous shrubs in some parts of its range (McLaren and McLaren 1981a). This flexibility of habitat choice presumably accounts for its widespread range in Ontario.

Earlier observers, such as McIlwraith (1894), knew the Swainson's Thrush only on migration. Devitt (1967) reported a nest with eggs found at Barrie in 1915, and the species has recently been reported from the Alfred Bog near Ottawa. By the time that Baillie and Harrington (1937) were writing, it had been commonly found as far north as Moose Factory, where a nest was found in 1930, and even farther north along the Manitoba border at 53[0] to 54[0]N. McLaren and McLaren (1981a) found the species in all forest and scrub habitats sampled at selected sites along a transect from Pickle Lake northwest to the Manitoba border in June 1977. Schueler *et al.* (1974) give records at Hawley Lake and Win-

isk. A nest was found during the atlas period near Fort Severn, and there were reports from much of the rest of the Hudson Bay Lowland. It is clear that our knowledge has gradually filled in as a result of more coverage of northern areas.

It is difficult to confirm that the Swainson's Thrush is breeding. It inhabits dense growth, and usually conceals its nest at a height of less than 2 m above the ground (Bent 1949). Only the female builds the nest and incubates the eggs, and she is quiet and wary, but both parents bring food to the young. This species does more aerial flycatching than the other thrushes, but is a difficult bird to see; the main clue to its presence is its song. Two-thirds of the records of this species are of singing males or birds on territory. Yet, like other thrushes, it sings mainly in low levels of light (from 3:15 to 9:00 a.m. and from 4:30 to 7:30 p.m. at the latitude of southern Ontario (Bent 1949)) so observers who visited a square at other times could easily have missed this species.

The maps show a range that includes all parts of the province except for the far northeast and the southwest. In general, there are almost no records south of the latitude of Lake Simcoe, and few south of Muskoka. There are, however, isolated records in Dufferin and Grey Counties (on the slopes of the Niagara Escarpment), in the reforested conifer plantations on the interlobate moraine of Northumberland Co, and in scattered patches of forest in the western part of Prescott Co. The preference of the Swainson's Thrush for coniferous habitat is evident in the south; the species is widespread in and around the Algonquin Highlands, the Temagami area north of Sudbury, the south shore of Manitoulin Island, and the north end of the Bruce Peninsula, but is only scattered throughout the rest of the south. The species was not mapped on the Bruce Peninsula by Godfrey (1986), but appears to be well established there. Gaps showing in the core of the range on the map may have breeding Swainson's Thrushes if requisite forest cover exists. However, the large gap between Sudbury and Lake Nipissing suggests that habitat is not suitable in much of that area, perhaps because agricultural land-use and airborne pollution have reduced the extent of forests there.

Abundance estimates suggest that the Swainson's Thrush is common throughout its Ontario range, with somewhat higher breeding densities in the north. In his analysis of 142 northern breeding bird plot studies from 1970 to 1979, Erskine (1980) found this bird to be among the commonest species in 46 of 142 plots. It was most abundant in conifers, especially those with budworm infestation, where densities of up to 140 males per km^2 were noted. Other estimates are of 5.6 pairs per km^2 at Pickle Lake (James 1980) and a maximum of 14 per km^2 from there northward (McLaren and McLaren 1981a). Erskine (1978) listed it as fourth in abundance out of 62 species on BBS routes run in central Ontario and Quebec from 1969 to 1975. BBS data indicated an increase in Ontario in the 1965 to 1979 period (Robbins *et al.* 1986).-- *D.C. Sadler*

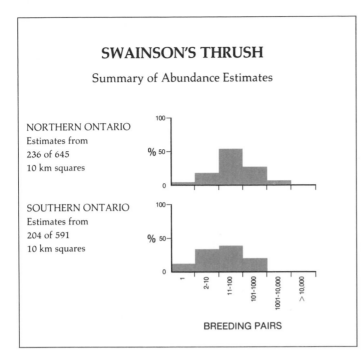

SWAINSON'S THRUSH

Summary of Abundance Estimates

NORTHERN ONTARIO
Estimates from
236 of 645
10 km squares

SOUTHERN ONTARIO
Estimates from
204 of 591
10 km squares

%

BREEDING PAIRS

1   2-10   11-100   101-1000   1001-10,000   >10,000

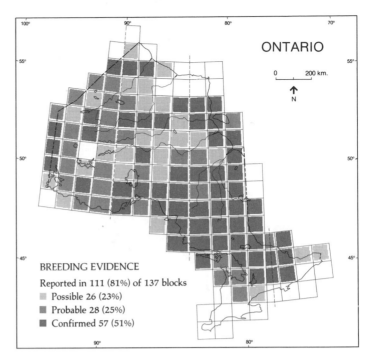

ONTARIO

0   200 km.

N

BREEDING EVIDENCE
Reported in 111 (81%) of 137 blocks

Possible 26 (23%)
Probable 28 (25%)
Confirmed 57 (51%)

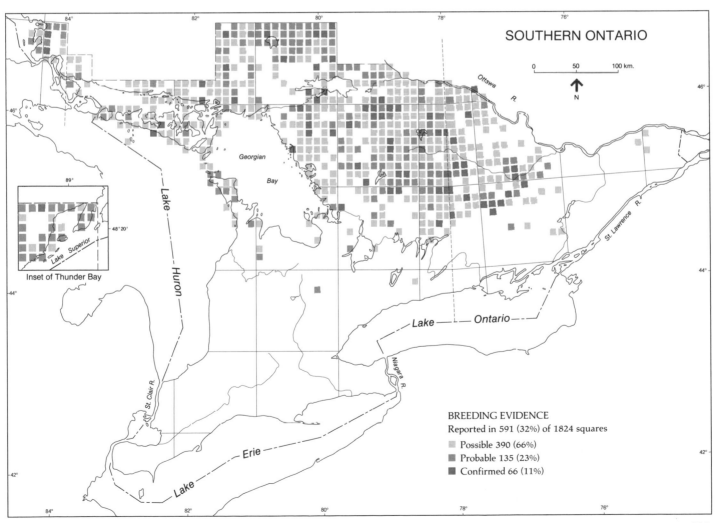

SOUTHERN ONTARIO

0   50   100 km.

N

Inset of Thunder Bay

Georgian Bay

Lake Huron

Lake Superior

Ottawa R.

St. Lawrence R.

Lake Ontario

Niagara R.

St. Clair R.

Lake Erie

BREEDING EVIDENCE
Reported in 591 (32%) of 1824 squares

Possible 390 (66%)
Probable 135 (23%)
Confirmed 66 (11%)

HERMIT THRUSH
Grive solitaire
Catharus guttatus

The Hermit Thrush has the distinction of being one of the most gifted singers in North America. Its song has been variously described as exquisite, ethereal, beautiful, bell-like, and flute-like. To those familiar with the song, it is likely to evoke recollections of cottage country or of wilderness.

The Hermit Thrush breeds in woodlands across the northern half of North America, extending farther south in the forests of the Rocky Mountains and the Appalachians. It winters in much of the US, Mexico, and Central America. In Ontario, the species occupies the Boreal Forest and Great Lakes-St. Lawrence Forest zones north to the tree-line along the Hudson Bay coast. Within this range, it occurs in a variety of woodland types: rocky and dry jack pine stands, uniform black spruce bogs, mixed deciduous-coniferous woodlands, and dry, sandy deciduous woods, usually regenerating with trembling aspen. Such a broad range of habitats allows the species to occupy much of central and northern Ontario, as indicated by the maps.

The present range of the Hermit Thrush appears to extend beyond its former limits. Baillie and Harrington (1937) reported that it bred north only to the transcontinental line of the Canadian National Railway and probably to James Bay, while Speirs (1985), Peck and James (1987), and particularly the present atlas indicate breeding north to Hudson Bay. This range extension is likely more apparent than real, reflecting more detailed information that formerly was lacking.

Confirmation of breeding for the Hermit Thrush is difficult to obtain. Only 11% of the records are of confirmed breeding, mostly representing fledged young or adults carrying food or faecal sacs. Nests are usually placed on the ground and are well hidden. Although the species breeds throughout most of the months when field work was most active, it can be elusive and unobtrusive on its nesting grounds. The Hermit Thrush is most readily detected by its song, which is easily recognized and carries long distances; 73% of all records of breeding evidence were of singing or territorial males.

Within southern Ontario, the Hermit Thrush is generally absent as a breeding bird in the southwestern part of the province from southern Bruce Co to the north shore of Lake Ontario. It is distinctly local and thinly distributed south and east of the Canadian Shield, from the Bruce Peninsula to the St. Lawrence River. Isolated populations occur along the north shores of Lakes Ontario and Erie in reforested areas of the Oak Ridges Moraine and the Norfolk Sand Plains (where birds have been known to occur in summer in pine plantations since the early 1970's, and where a nest was first found in 1976 (ONRS)). The widespread reforestation program across southern Ontario appears to have benefitted the Hermit Thrush in several areas. Stands of red and jack pine on dry sandy soils have certainly increased available habitat in the York and Durham RMs and in the Peterborough and Kingston areas, all of which have Hermit Thrush records. To the north, the species becomes progressively more common and widespread, at least into the central part of the province, as indicated by atlas data and the results of Breeding Bird Surveys (Speirs 1985). There are some squares and blocks in the Boreal Forest zone and the Shield section of the Great Lakes-St. Lawrence Forest zone in which there are no records. Either, as in northern Ontario, these areas were difficult to gain access to, or the Hermit Thrush occurred in sufficiently low numbers to escape detection. The species is probably absent from few squares or blocks from the Muskoka, Haliburton, and Renfrew districts north to the tree-line.

The maps accurately reflect the distribution of the Hermit Thrush within the province. BBS counts indicate no change in the density of Hermit Thrushes during the period 1967 to 1983, and most Regional Coordinators indicate no obvious change in numbers. It is evident from the abundance estimates in both southern and northern Ontario that the Hermit Thrush is a common breeding bird in suitable habitat.-- *R. Knapton*

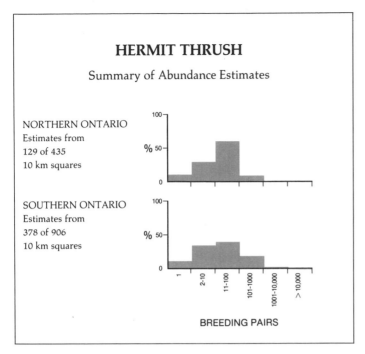

HERMIT THRUSH

Summary of Abundance Estimates

NORTHERN ONTARIO
Estimates from
129 of 435
10 km squares

SOUTHERN ONTARIO
Estimates from
378 of 906
10 km squares

BREEDING PAIRS

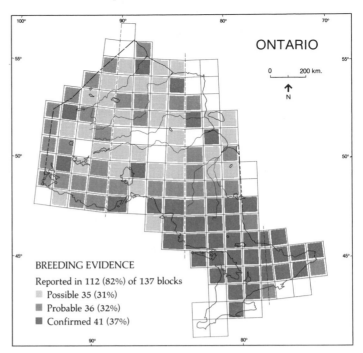

ONTARIO

0   200 km.

N

BREEDING EVIDENCE
Reported in 112 (82%) of 137 blocks

Possible 35 (31%)
Probable 36 (32%)
Confirmed 41 (37%)

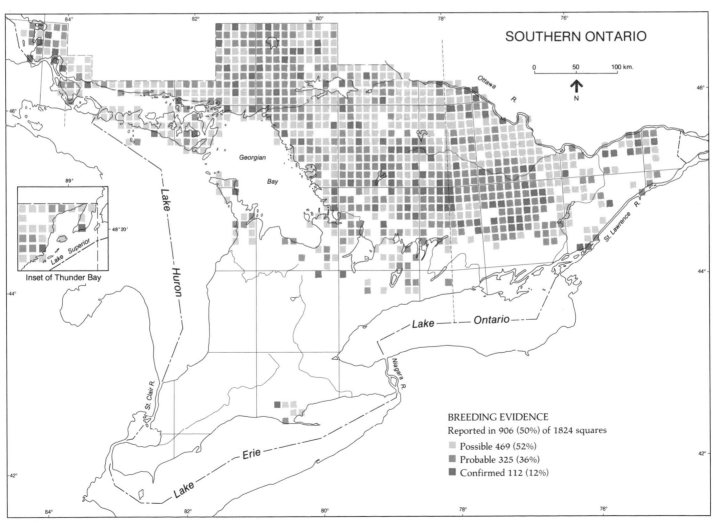

SOUTHERN ONTARIO

0   50   100 km.

N

Inset of Thunder Bay

BREEDING EVIDENCE
Reported in 906 (50%) of 1824 squares

Possible 469 (52%)
Probable 325 (36%)
Confirmed 112 (12%)

WOOD THRUSH
Grive des bois
Hylocichla mustelina

This, the largest and most vocal (Dilger 1956a,b) of the brown forest thrushes, also has the most southerly breeding range. It is best known for its loud but musical song.

The breeding range of the Wood Thrush is confined to the eastern half of the continent, as far south as the Gulf of Mexico. Its range covers all of southern Ontario, following closely the Great Lakes-St. Lawrence Forest and Carolinian Forest zones. Its winter range is in Mexico and Central America. The optimum nesting habitat is moist, mature hardwood or mixed forest, preferably undisturbed, with some trees greater than 12 m tall and an abundance of deciduous sapling growth, in which the nest is usually situated. The minimum required size of a nesting woodlot is fairly small, perhaps 5 ha (P. Eagles pers. comm.); Bent (1949) gives territory size as 0.08 to 0.8 ha. In Connecticut, it is most abundant in hardwood forest edges (Bertin 1977). Its widespread range in highly agricultural southwestern Ontario reflects its use of relict woodland stands, including those along rivers and lakeshores. Although McIlwraith (1894) noted, "They avoid the dwellings of man," this bird has more recently shown tolerance even of wooded areas within cities (Brackbill 1958).

Early records (*e.g.*, Macoun and Macoun 1909) reported a more restricted southern range for the Wood Thrush than at present, while Baillie and Harrington (1937) stated, "[it] cannot be considered a common summer resident of Ontario, excepting in the extreme south-western peninsula." They gave records from Kingston in 1905 and Algonquin Prov. Park in 1934. In eastern Canada and the New England states, the Wood Thrush has expanded its breeding range in recent years (Bent 1949, Godfrey 1966, 1986), perhaps because it is more adaptable to human settlement than the Hermit Thrush and Veery, whose ranges have contracted during the same

period (Morse 1971). BBS data from the period 1968 to 1975 (Erskine 1978) show an appreciable population increase for the Wood Thrush in southern Ontario and Quebec. Previous northern Ontario records worth noting are occurrences in 1957 at Dorion (Speirs 1985) and in 1974 at Wawa and north of Rainy River (Goodwin 1974).

Like most of our thrushes, the Wood Thrush is a fairly secretive bird. The male spends much of its time in the forest floor litter while the female incubates. He makes only brief visits to the nest. Song is given mainly in the early morning (often before dawn) and in the evening. The male also sings nearby when the female is off the nest. It is not surprising, then, that 62% of atlas records are of singing or territorial males, the 2 most frequently used codes for this species. The song is recognized by most atlassers, although some may have difficulty in telling it from the song of the Hermit Thrush, especially at a distance (Erskine 1978). These 2 species may at times respond to each other's songs, though at rather a low level. Both parents feed the young. The nest is among the easiest of all thrush nests to find, being at a convenient height (1.5 to 4.5 m) above ground in a low, or perhaps the lowest, fork of a tree (Brackbill 1958, Godfrey 1986). However, several Regional Coordinators have noted difficulty in confirming breeding, presumably because of the extensive and densely vegetated woodland that it generally inhabits. This is reflected in the southern Ontario map and statistics, which show a high proportion of possible and probable breeding records, even though breeding likely does occur in most squares for which there are records.

The maps show confirmed and possible breeding records north of the previously known breeding range (Godfrey 1986). Apparently, breeding occurs in pockets of suitable habitat in what is mainly boreal forest. The most northeasterly record is of a singing male found in 1983 near Kapuskasing. If northward expansion continues, one might expect the Wood Thrush to breed in the Great Lakes-St. Lawrence Forest region in northwestern Ontario. An isolated breeding locality, far north of the normal range, has been reported in southwestern Manitoba (Godfrey 1986). Farther south, the Wood Thrush occurs in almost every square north to Manitoulin Island and the French River, even in areas where many squares have only very small, isolated woodlots. It is absent, however, from many of the small islands on the eastern shore of Georgian Bay.

Squares in the southern sector of the province were most often (52%) thought to contain 11 to 100 breeding pairs of Wood Thrushes, suggesting that this species is fairly common in this area. It is less abundant in northern squares where estimates were made, with 2 to 10 pairs being most frequent (67%).-- *D.C. Sadler*

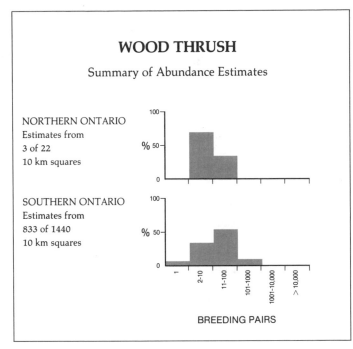

WOOD THRUSH

Summary of Abundance Estimates

NORTHERN ONTARIO
Estimates from
3 of 22
10 km squares

SOUTHERN ONTARIO
Estimates from
833 of 1440
10 km squares

BREEDING PAIRS

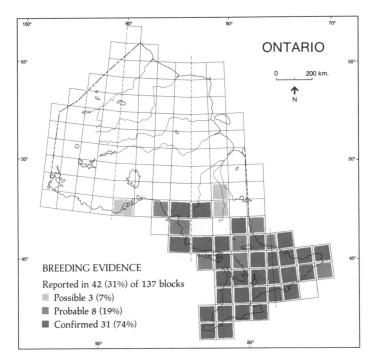

ONTARIO

0   200 km.

N

BREEDING EVIDENCE

Reported in 42 (31%) of 137 blocks

Possible 3 (7%)
Probable 8 (19%)
Confirmed 31 (74%)

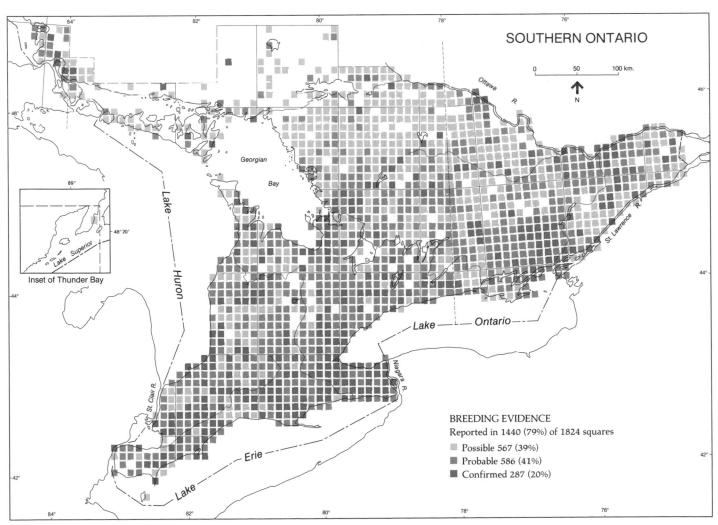

SOUTHERN ONTARIO

0   50   100 km.

N

Georgian
Bay

Lake
Huron

Lake
Superior

Inset of Thunder Bay

St. Clair R.

Ottawa
R.

St. Lawrence R.

Lake — Ontario —

Niagara
R.

Lake — Erie

BREEDING EVIDENCE

Reported in 1440 (79%) of 1824 squares

Possible 567 (39%)
Probable 586 (41%)
Confirmed 287 (20%)

AMERICAN ROBIN
Merle d'Amérique
Turdus migratorius

The American Robin is probably best known to most people as the harbinger of spring. Although small numbers winter in sheltered areas in extreme southern Canada, American Robins become most evident when they begin to appear on lawns and in gardens soon after the snow leaves the ground. Some other species return earlier, but they are not as closely associated with human dwellings, are not as readily identified, or do not have the fine and well known singing voice of the American Robin.

It is found virtually everywhere across the continent in summer except southern Florida, the arctic tundra, and Central America south of Mexico. Robins can be seen from sea level to alpine meadows, and from deserts to humid woodlands, and from the tropics of southern Mexico to the edges of the tundra in northern Alaska and Quebec. Few other species are found in such a range of areas or climates. In winter, it is found throughout most of the US and Central America south to Guatemala.

As might be expected of such a ubiquitous species, the American Robin has adapted to a wide variety of habitats. It may be found even in areas of low or sparse plant growth; in residential areas, lawns with ornamental trees and shrubbery provide not only sufficient habitat, but also ideal conditions for ground foraging. It is much more common in built up areas than in more remote situations. In forested regions, it prefers forest edges and openings, burned or cut-over areas, fens and bogs, and even lake and river shores. It is absent only from extensive unbroken forests or from some very rocky areas with little soil or ground vegetation.

The American Robin is found throughout Ontario, with the exception of extensive areas of tundra, and even there it may occasionally appear. It has no doubt occupied this range for centuries. However, the species has probably greatly increased in numbers in the last couple of hundred years as forest clearing and urbanization of much of the continent have taken place. Human activities have altered the environment very favourably for the American Robin. Although pesticide contamination adversely affected it, at least locally, Breeding Bird Surveys reveal virtually no change in its abundance since 1967 in Ontario.

Robins must be among the easiest of birds to find and to confirm as breeding in urban and agricultural settings. Birds hop about in the open, nests are often placed in conspicuous places, even on or in buildings, and, when an observer is near a nest with young, the parent birds usually scold loudly. Old nests built of mud often last for months after being occupied, and, in southern Ontario at least, each pair is likely to build 2 or 3 nests each season. Used nests should be conspicuous and readily identifiable. Thus, it is not surprising that only one regular southern Ontario species (the European Starling) has a higher rate of breeding confirmation. Despite the close association of the American Robin and people in settled areas of the province, there is another segment of the population that inhabits woodlands, and these birds are as secretive and unapproachable as any other wild species. It is as if there were 2 species occupying the same province, one accustomed to living in close proximity to people and the other wild and remote. The nests of the forest dwellers are much more difficult to find, but the birds themselves are still readily visible. They often forage in open areas and they carol conspicuously in both morning and evening. While they cannot be as easily watched, it is possible to find them and to confirm breeding with relative ease.

As expected, the American Robin was found and confirmed breeding in virtually all squares of southern Ontario. A few squares with no records may represent those with much more water than land area, and/or very rocky areas. In the northern part of southern Ontario, a few squares are recorded as having only possible or probable breeding. Although the species may be scarce in those squares, it is more likely that they did not receive sufficient coverage. The same is probably true for northern blocks with records of possible or probable breeding, with the exception of some that contain only tundra habitat. Blank areas on the maps are probably the result of insufficient field work. In many instances, the birds are probably present in very low numbers, and perhaps clumped locally in suitable habitat, so that finding them would require more time and travel than atlassers had available.

The most frequently given abundance estimate in southern Ontario was 101 to 1,000 pairs per square, and this seems reasonable for most southern squares in urban and agricultural areas. Higher abundance is certainly possible in many areas. Across the forested parts of the province, abundance is probably lower and this is reflected in the generally lower estimates from northern Ontario, where 11 to 100 pairs were most frequently reported.-- *R.D. James and R.C. Long*

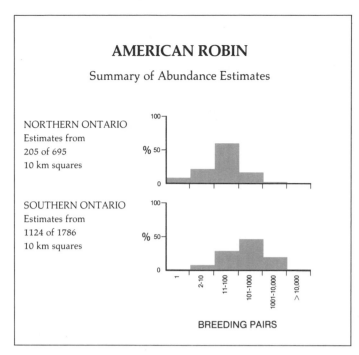

AMERICAN ROBIN

Summary of Abundance Estimates

NORTHERN ONTARIO
Estimates from
205 of 695
10 km squares

SOUTHERN ONTARIO
Estimates from
1124 of 1786
10 km squares

BREEDING PAIRS

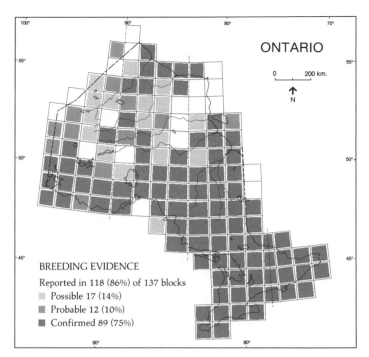

ONTARIO

0   200 km.

N

BREEDING EVIDENCE
Reported in 118 (86%) of 137 blocks
Possible 17 (14%)
Probable 12 (10%)
Confirmed 89 (75%)

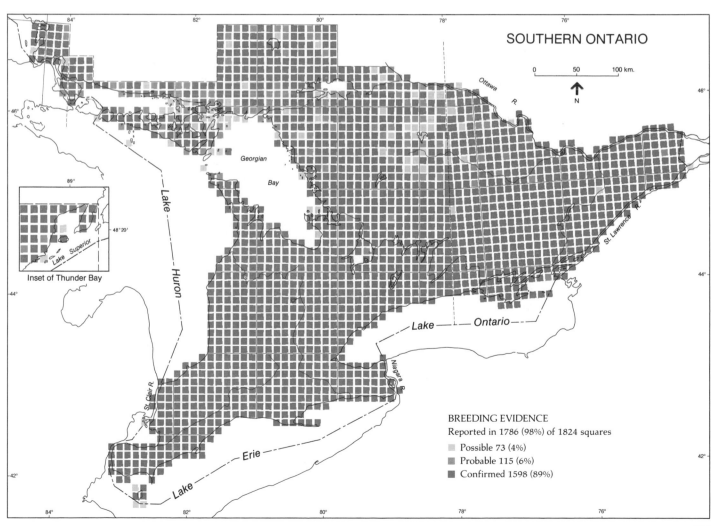

SOUTHERN ONTARIO

0   50   100 km.

N

Georgian Bay

Ottawa R.

St. Lawrence

Lake — Ontario

Lake Huron

Niagara R.

St. Clair R.

Lake Erie

Inset of Thunder Bay
Lake Superior

BREEDING EVIDENCE
Reported in 1786 (98%) of 1824 squares
Possible 73 (4%)
Probable 115 (6%)
Confirmed 1598 (89%)

GRAY CATBIRD
Moqueur chat
Dumetella carolinensis

The Gray Catbird surely must rank as one of the most familiar of southern Ontario's summer birds. Its attractive, intricate
song may be heard in country lane or suburban garden so
long as there is a patch of shrubs, vines, or other thick, tangled vegetation.

The species breeds widely throughout North America
from southern Canada to the southern US (AOU 1983). It
winters from the southeastern US to Central America and the
Caribbean. Any brush or scrubby habitat, such as is found in
ravines and valleys in farming country, at the edges of
woodlots, or in hedgerows around gardens or fields, is likely
to support Gray Catbirds.

There is no evidence to suggest that this species has
changed much in status in the last 100 years. McIlwraith
(1894) described it as a very common resident in southern
Ontario although, prior to this, it undoubtedly benefitted
from the clearing of the forests and creation of scrub and
edge habitat which accompanied the settling of the province
by Europeans. Baillie and Harrington (1937) likewise called it
a common summer resident north to southern Kenora and
Sudbury Districts.

The Gray Catbird is a relatively easy species to confirm as
a breeding bird. It often raises 2 broods each summer (Bent
1948), and nesting activity is still in progress in July or even
August. Egg dates in Ontario range from 2 May to 18
August, but most are in the first 3 weeks of June (Peck and
James 1987). The Gray Catbird often sings from dawn until
late morning and again in the evening. Moreover, upon close
approach to the nest or young, both parents scold the
intruder with the well known snarl or a low, sharp 'qwerk'.
Most nests are less than 3 m above the ground in low shrubs
and vines (Bent 1948), the ideal height for atlassers in search
of confirmation of breeding, although nests can be difficult to
find in the heart of tangles or thorn scrub. A relatively high
52% of records are of confirmed breeding, and 33% of those
are of nests with eggs or young. However, the species likely
breeds in many of the squares in which it was reported but
not confirmed, especially those south of the Canadian Shield.
Generally, the map is a good indication of the breeding distribution of the Gray Catbird in Ontario.

The northern limit of the range coincides quite neatly
with the northern edge of the mixed deciduous-coniferous
Great Lakes-St. Lawrence Forest zone, scarcely penetrating
the Boreal Forest zone at all. Although the Gray Catbird will
nest in conifers (Bent 1948), its preferred habitat of deciduous
shrubs and brambles is found mostly south of this ecological
boundary. In the Algonquin Highlands, where there is a high
content of coniferous forest, there is at best a spotty distribution of this species. In fact, it is probably less common there
now than 50 years ago, when even more of Algonquin Prov.
Park was subjected to clear-cut logging. Regeneration after
clear-cutting creates extensive shrubby areas, highly suitable
for the Gray Catbird. The scarcity of the species in the most
northerly squares in southern Ontario is also probably a
result of the more coniferous habitat in that area. To the
west of Lake Superior, in an area of modified mixed forest
and cleared habitat, the species occurs in the Rainy River and
Kenora Districts as a scarce and probable breeder. It has previously been confirmed as breeding in that area (Speirs 1985,
Peck and James 1987). It can occur as an isolated bird in the
nesting season almost anywhere throughout the Boreal Forest
zone, as evidenced by the singing bird at Big Trout Lake in
1975 (Lee 1978), and one or more at North Point north of
Moosonee in 1978, 1983, and 1984.

It is in settled southern Ontario that the Gray Catbird
reaches its greatest abundance. Seventy-six percent of all
abundance estimates from southern Ontario are of more than
10 pairs per square, and 28% are of more than 100 pairs. A
measure of just how abundant the species can be in optimum habitat was the count of 27 birds per 40 ha in a hawthorn-apple plot in Pickering (Speirs 1985). Although this
converts to more than 6,000 birds per square, suitable Gray
Catbird habitat would seldom exceed one-third of the area of
any square. Nevertheless, it is likely that numerous squares
containing abandoned farmland near the southern edge of
the Shield support more than 1,000 pairs.-- *R. Curry*

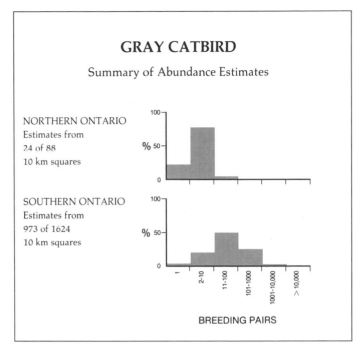

GRAY CATBIRD

Summary of Abundance Estimates

NORTHERN ONTARIO
Estimates from
24 of 88
10 km squares

SOUTHERN ONTARIO
Estimates from
973 of 1624
10 km squares

BREEDING PAIRS

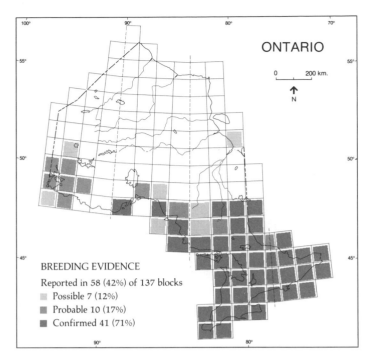

ONTARIO

0   200 km.

N

BREEDING EVIDENCE

Reported in 58 (42%) of 137 blocks

Possible 7 (12%)

Probable 10 (17%)

Confirmed 41 (71%)

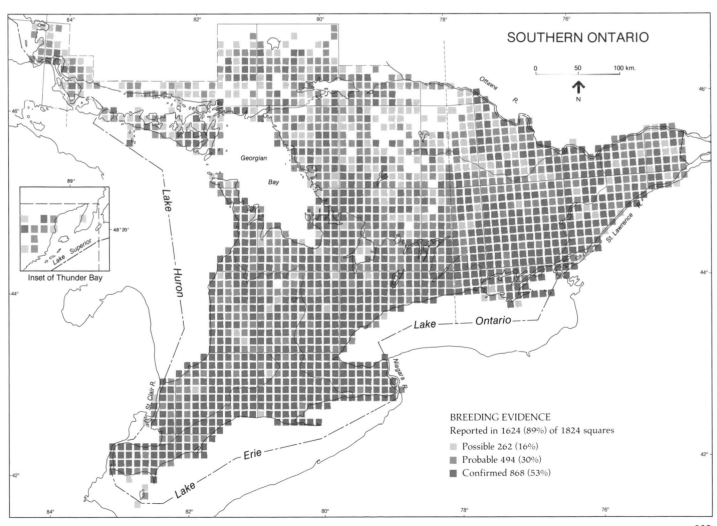

SOUTHERN ONTARIO

0   50   100 km.

N

Inset of Thunder Bay

BREEDING EVIDENCE

Reported in 1624 (89%) of 1824 squares

Possible 262 (16%)

Probable 494 (30%)

Confirmed 868 (53%)

NORTHERN MOCKINGBIRD
Moqueur polyglotte
Mimus polyglottos

The Northern Mockingbird is resident across North America north to southern Canada in the Prairie Provinces, Ontario, Quebec, and the Maritime Provinces, with the northern populations being partially migratory. In Ontario, it occurs irregularly to Lake Nipissing and occasionally farther north. The species favours suburban gardens, orchard margins, hedges, and artificial plantings. The presence of fruit such as that of the multiflora rose, planted as hedgerows, is especially important to its survival in winter.

The Northern Mockingbird is a relative newcomer to Ontario. A pair spent the summer and presumably nested at Hamilton in 1883 (McIlwraith 1894). Baillie (1967) reported the first nesting about 1906 in Amherstburg, 20 km south of Windsor, and Baillie and Harrington (1937) referred to a nesting at Point Pelee in 1909. As recently as 1965, it was still considered to be an irregular, very rare summer resident in western New York State and the adjacent Niagara Peninsula (Beardslee and Mitchell 1965). In the 19th century, the species was commonly kept as a cage bird (McIlwraith 1894, Bent 1948), and it is possible that the first occurrences recorded in Ontario were of escaped individuals. Today, however, it is still expanding its range northward.

The Northern Mockingbird is justifiably renowned as a loud, brilliant, and indefatigable singer. This, and its affinity for human settlement, suggest that the maps present a fairly complete picture of its occurrence and range. The 42% of records involving possible breeding may in part represent pioneering individuals. Confirmation of breeding (24% of all records) is relatively easy as the species is remarkably vigorous and aggressive in defence of its nest or young. That this percentage is not higher reflects the low numbers of birds in most squares and the propensity of the Northern Mockingbird to stray into areas near the fringe of its range where it may not find a mate. To a certain extent, it may also reflect the lack of observer time invested in looking for confirmation where the bird is scarce.

It is clear from the maps that the Niagara Peninsula is the heart of its range in Ontario: the largest contiguous group of occupied squares (29) and most of the breeding confirmations came from this area. On the Lake Ontario plain, from the Niagara River west to Hamilton, it is a fairly common bird in orchards and gardens. The northern half of the Peninsula experiences the mildest winters in the province, and deep snow seldom lingers for more than a few days. For an essentially non-migratory species, these benign conditions have permitted successful overwintering and a consequent build-up of the population. Elsewhere, no simple pattern is to be found on the maps. The words "irregular" and "erratic" have been used by some Regional Coordinators to describe its status. There has been a slow, steady northward progression since the 1960's. It appears that individual birds and pairs strike out in search of suitable breeding habitat. If they find it and can survive the winters, then a new nucleus is begun. This seems to have been the case at Ottawa (where it first nested in 1971), and at Sudbury (where it first nested in 1972). Further evidence of pioneering is apparent from the confirmed breeding record from Cochrane, the probable breeding record from Moosonee, and the scattered possible breeding records north and west of Lake Superior. If present trends continue, there is likely to be infilling along the most densely populated and urbanized parts of the Windsor-Ottawa corridor, especially where snowfall is relatively low and artificial plantings provide winter food and cover. To the north and west, the erratic pattern should continue, with nesting occurring almost anywhere but winter die-off limiting any systematic advance. The extent to which birds withdraw in fall from the summer range and return to the same territories in spring is unclear. A check of the 1984-85 Christmas Bird Count revealed only 6 individuals away from the Niagara Peninsula. If there is a tendency for the Northern Mockingbird to adapt by seasonally migrating from the harsher parts of its range, though the evidence for such a trend is scanty, then there is little to prevent a more widespread expansion in Ontario.

The abundance estimates also demarcate the Niagara Peninsula as the core of the Ontario range of the species. Twenty-three of 25 estimates of more than a single pair per square are from the Niagara Peninsula, and all 9 estimates of more than 10 pairs are from that area.-- *R. Curry*

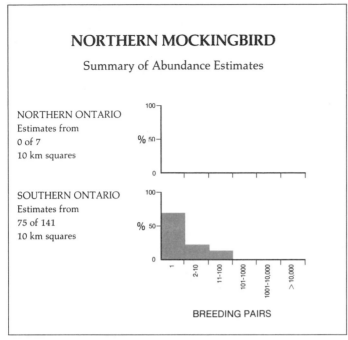

NORTHERN MOCKINGBIRD

Summary of Abundance Estimates

NORTHERN ONTARIO
Estimates from
0 of 7
10 km squares

SOUTHERN ONTARIO
Estimates from
75 of 141
10 km squares

BREEDING PAIRS

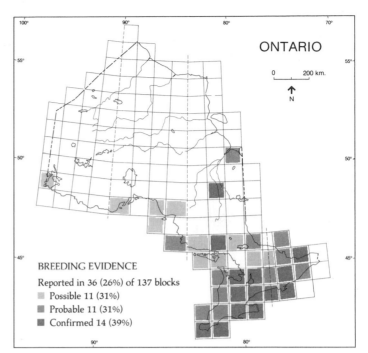

ONTARIO

0   200 km.

N

BREEDING EVIDENCE

Reported in 36 (26%) of 137 blocks

Possible 11 (31%)
Probable 11 (31%)
Confirmed 14 (39%)

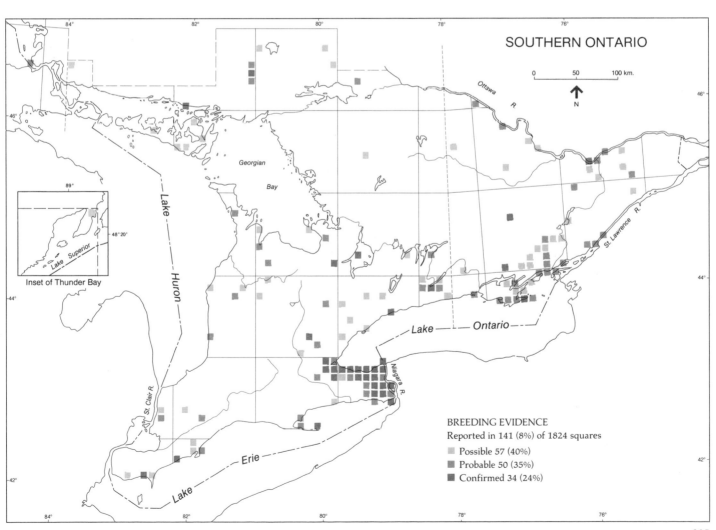

SOUTHERN ONTARIO

0   50   100 km.

N

Inset of Thunder Bay

BREEDING EVIDENCE

Reported in 141 (8%) of 1824 squares

Possible 57 (40%)
Probable 50 (35%)
Confirmed 34 (24%)

BROWN THRASHER
Moqueur roux
Toxostoma rufum

The Brown Thrasher is a North American species that breeds from the Boreal Forest zone to the Gulf of Mexico and west to the foothills of the Rocky Mountains. Its winter range includes much of the southeastern US. In Ontario, it inhabits brushy hillsides, overgrown hawthorn pastures, woodland edges and especially areas of marginal farmland within the Great Lakes-St. Lawrence and Carolinian Forest zones. Because the Brown Thrasher shows a strong preference for early successional habitats, its occurrence is often quite localized. Unlike the closely related Gray Catbird, it tends to shun close proximity to human settlement (Bent 1948).

Like many other species, the Brown Thrasher has benefitted from the clearing of the forests for cultivation and settlement. It was described as "common" near Hamilton in the late 19th century, though less common than the Gray Catbird (McIlwraith 1894), and "fairly common" earlier in this century (Baillie and Harrington 1937). The comment by McIlwraith (1894) that it was unaccountably absent from certain areas is still true today, but this is likely explained by subtle factors in its habitat requirements not obvious to bird-watchers. Since the 1930's, when there were no definite nesting records for Manitoulin Island and the distribution of the species lay entirely south of Lake Nipissing, it has expanded its range northward following farm abandonment and concomitant growth of scrub. In contrast, it is now probably declining in some of the same general areas, where reforestation, either natural or artificial, has progressed further. Nevertheless, no significant population trend was detected from BBS data in Ontario in the 1965 to 1979 period (Robbins *et al.* 1986).

The Brown Thrasher sings loudly from a treetop perch in semi-open country along roadsides and country lanes. As a result, it is easily detected. However, when it has eggs or young, the species can be very secretive (Bent 1948). The nest

is fairly large, but often concealed in the thickest of tangles (or on the ground underneath), especially in tangles with thorns; nevertheless, nests with eggs or young were found in 32% of the squares in which breeding was confirmed. Thrashers are aggressive and noisy in defence of their young, in or out of the nest. The loud alarm call, a 'smack' or 'tchink', no doubt helped to confirm breeding by leading atlassers to fledged young or adults carrying food. These categories make up almost 60% of all confirmed breeding records. The Brown Thrasher usually rears 2 broods per season (Bent 1948), providing a substantial period of time in which evidence of breeding can be found. Egg dates in Ontario range from 29 April to 20 July, with a peak period from 25 May to 9 June (Peck and James 1987).

The map depicts quite accurately the breeding distribution of the species. Discontinuities within its range represent more heavily forested regions. It is noticeably absent from the Algonquin Highlands, with their extensive areas of uninterrupted mature forests and lack of agricultural land. The Brown Thrasher occurs in the mixed farming Sudbury Basin and along the north shore of Lake Huron, but is scarce in the uncleared mixed Great Lakes-St. Lawrence Forest immediately to the south and limited by the Boreal Forest to the north. In northwestern Ontario, there are a number of blocks where possible and probable breeding was reported. It is described as being common in summer in southeastern Manitoba (Cleveland *et al.* 1980), so nesting in northwestern Ontario should be expected. A record of 2 males singing at Pickle Lake in the summer of 1976 or 1977 (Speirs 1985) was at about the same latitude as the persistently singing male in breeding habitat that was caught and banded in 1983 at North Point, 27 km north of Moosonee.

Abundance estimates are available for 56% of the squares where some evidence of breeding was obtained, 93% of which are of 100 or fewer pairs per square. In optimum habitat in Ontario (hawthorn-apple plot in Durham RM), the species was found at a density of 16 birds per 40 ha (Speirs 1985). Judging from the distribution of suitable habitat, the Brown Thrasher probably reaches its greatest abundance at the southern edge of the Canadian Shield, declining fairly rapidly northward and more gradually southward.-- *R. Curry*

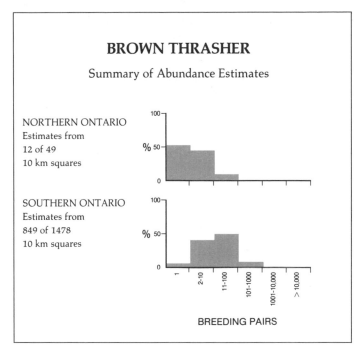

BROWN THRASHER

Summary of Abundance Estimates

NORTHERN ONTARIO
Estimates from
12 of 49
10 km squares

SOUTHERN ONTARIO
Estimates from
849 of 1478
10 km squares

BREEDING PAIRS

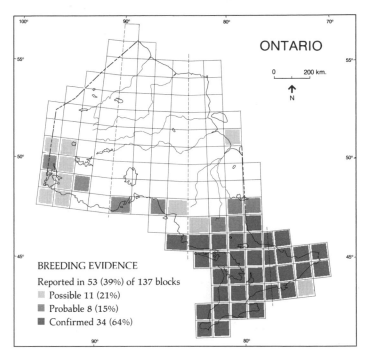

ONTARIO

0   200 km.

N

BREEDING EVIDENCE
Reported in 53 (39%) of 137 blocks
Possible 11 (21%)
Probable 8 (15%)
Confirmed 34 (64%)

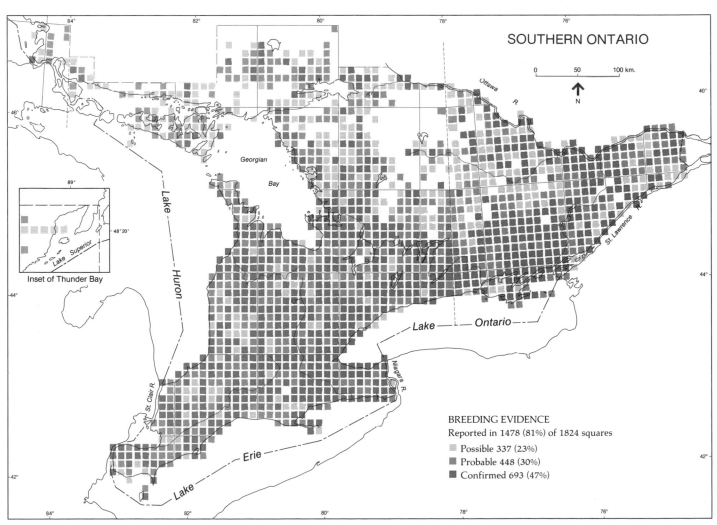

SOUTHERN ONTARIO

Ottawa R.

0   50   100 km.

N

Georgian Bay

Lake Huron

Lake Superior
48°20'
Inset of Thunder Bay

St. Lawrence R.

Lake — Ontario —

Niagara R.

St. Clair R.

Erie

Lake

BREEDING EVIDENCE
Reported in 1478 (81%) of 1824 squares
Possible 337 (23%)
Probable 448 (30%)
Confirmed 693 (47%)

CEDAR WAXWING
Jaseur des cèdres
Bombycilla cedrorum

The Cedar Waxwing is a popular species because of its attractive and elegant plumage and its highly social nature. The waxwings are unusual among temperate zone songbirds because of their reliance upon fruit as their major food for most of the year.

The Cedar Waxwing breeds only in North America. Its nesting range is primarily in southern Canada and the northern and central US (AOU 1983), but it may be somewhat erratic around the fringes of this range. Its winter range extends from extreme southern Canada to Central America. Within this area, the Cedar Waxwing occupies forest edges, open woodland, second-growth, residential areas, and other similar places.

The distribution of this species has probably changed little in historical time. Because of its reliance on edge habitats, it may have benefitted in some places from the opening of the forest by modern man, and may be lower in numbers in some areas where agriculture has cleared extensive areas, but overall the range has probably not changed much. McIlwraith (1894) stated, "The Cedar Bird is generally distributed throughout Ontario," and Macoun and Macoun (1909) mention observations from a number of widely separated locations as far apart as Moose Factory and London.

The Cedar Waxwing is usually conspicuous and is not difficult to find during the breeding season. It often perches in obvious locations and is active and visible throughout the day. Its calls are weak and unobtrusive, but are given frequently and are fairly distinctive, allowing observant birders to detect easily the presence of the species. Although the Cedar Waxwing is relatively easy to find and observe, it nests later than most birds (largely from late June to late August) and may not yet have been breeding at the time when many atlassers were active. Leck and Cantor (1979)

suggest that this late breeding date is a recent change and that it may be a response to parasitic pressure from the Brown-headed Cowbird, but it may simply reflect the date of peak berry production. The adult birds visit the nest only infrequently and this may also make breeding difficult to confirm. If disturbed near the nest, Cedar Waxwings give a high, thin distress call but do not usually perform a distraction display or show conspicuous agitation near the nest. The observation of fledged young represents 33%, and nests represent 41%, of confirmed breeding records. The use of some of the probable breeding codes for the Cedar Waxwing calls for a certain amount of judgement. The species may guard a small area around the nest in the early stages of the breeding cycle (Putnam 1949) but does not defend a true territory as most other songbirds do. Moreover, male and female Cedar Waxwings are very difficult to distinguish in the field, and birds of either sex forage together in small groups, even during the breeding season. Thus, a loose interpretation has to be put on records in the 'pair' breeding category, which constitute 29% of all the records, and the 'territorial behaviour' breeding category, which constitutes another 5%. Also, unlike most songbirds, this species flies up to 2 km from the nest to search for food (Mountjoy unpubl. data); hence, some birds recorded as breeding were undoubtedly nesting in nearby squares rather than the ones in which they were recorded. Fortunately, however, because the species is common and not very specific in its habitat requirements, it very likely does nest in most squares in southern Ontario, and none of these problems is likely to have had much effect on the accuracy of the maps.

The Cedar Waxwing is one of Ontario's most widespread birds. Breeding evidence has been secured for all parts of the province, extending from extreme southwestern Ontario to the tree-line near Hudson Bay. The Cedar Waxwing is widely distributed in southern Ontario, where it was reported in 98% of the squares. To the north, the species was reported in almost every block north to the northern section of the Hudson Bay Lowland, where it was reported in only about half of the blocks. This suggests that suitable habitat is less widespread in the far north. It was, however, recorded as a probable breeder even as far north as the Fort Severn area. This generally concurs with previously published reports. For example, Schueler *et al.* (1974) recorded Cedar Waxwings as far north as Winisk and had evidence of breeding from Aquatuk Lake (100 km to the southeast).

It is evident from the abundance estimates that, in addition to being widely distributed, the Cedar Waxwing is also a common bird in almost all areas of Ontario where it breeds.-- *D.J. Mountjoy and R. John*

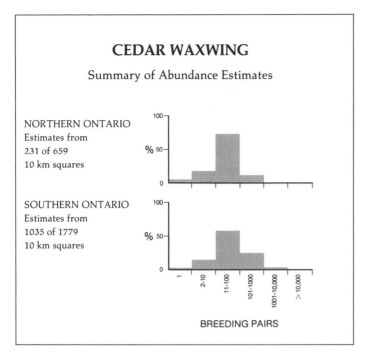

CEDAR WAXWING

Summary of Abundance Estimates

NORTHERN ONTARIO
Estimates from
231 of 659
10 km squares

SOUTHERN ONTARIO
Estimates from
1035 of 1779
10 km squares

%50

100

%50

100

1   2-10   11-100   101-1000   1001-10,000   >10,000

BREEDING PAIRS

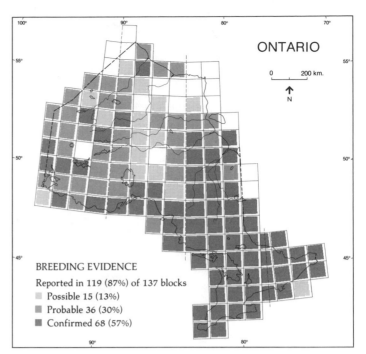

ONTARIO

0   200 km.

N

BREEDING EVIDENCE
Reported in 119 (87%) of 137 blocks

Possible 15 (13%)
Probable 36 (30%)
Confirmed 68 (57%)

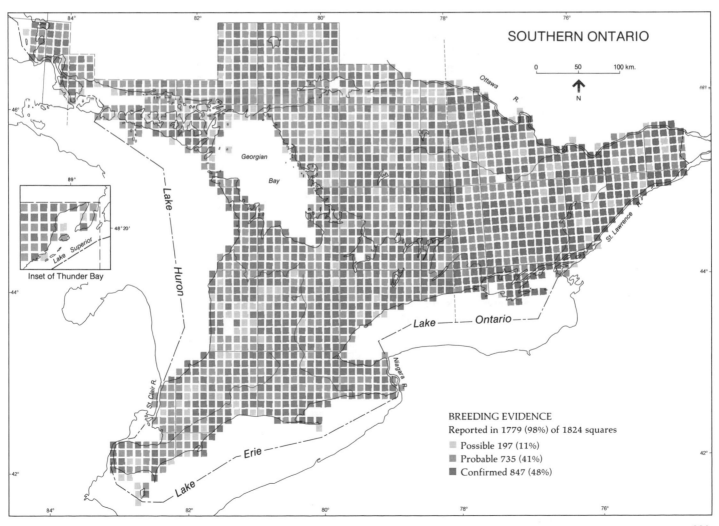

SOUTHERN ONTARIO

0   50   100 km.

N

Georgian Bay

Ottawa R.

St. Lawrence R.

Lake Ontario

Niagara R.

Lake Erie

Lake Huron

Lake Superior

St. Clair R.

89°   48°20'

Inset of Thunder Bay

BREEDING EVIDENCE
Reported in 1779 (98%) of 1824 squares

Possible 197 (11%)
Probable 735 (41%)
Confirmed 847 (48%)

LOGGERHEAD SHRIKE
Pie-grièche migratrice
Lanius ludovicianus

The Loggerhead Shrike is widespread in North America south of the Canadian Shield and the Boreal Forest region. In Ontario, the species is primarily restricted to cleared areas in the Great Lakes-St. Lawrence Forest ecosystem. Grazed pasture, and marginal farmland with scattered hawthorn shrubs, fenceposts, wires, and an associated low-lying wetland, are the preferred breeding habitats, and provide the open, short-grass habitat needed for the location and capture of insect and small vertebrate prey (Miller 1931). This habitat preference has resulted in a concentration of breeding pairs in the poorer farming areas near the northern margin of southern Ontario's cultivated land.

The species was not originally a resident of northeastern North America, but expanded its range north and east with the clearing of forests for settlement and agriculture. It was first reported in Ontario in 1860 near Hamilton (McIlwraith 1886). The eastward range expansion continued until early in the 20th century, by which time the species had also reached Quebec and the Maritime Provinces. At the turn of the century, it was common along the St. Lawrence River in eastern Ontario and in Muskoka, had extended to the Bruce Peninsula, and had been reported near Ottawa (Campbell 1975). The northward extension of the range apparently continued into the 1940's, by which time the species had been reported near Lake Nipissing, Sault Ste Marie, Thunder Bay, and the Rainy River area.

Generally, since the 1940's the range of the Loggerhead Shrike has gradually contracted in eastern Canada and across the northeastern US. The species probably no longer breeds in the Maritime Provinces, is a rare breeder in Quebec, and, as shown by BBS data, has declined significantly in Ontario. Currently, it is rare in southwestern Ontario and is seldom reported near Sault Ste Marie, Thunder Bay, or Lake of the Woods. It is now listed as threatened or endangered in most northeastern states, and is showing signs of decrease in almost all parts of its range. The causes of its decline are not well understood, although habitat loss due to intensive agriculture, and pesticide accumulation have been suggested (Cadman 1986). Many birds, especially fledglings, are killed by cars because of the species' habit of hunting near roadsides.

Although the song of the Loggerhead Shrike is not well known, there are other aspects of the behaviour of the species which make it easy to locate: its frequent use of roadside habitat, the open country it inhabits, and the choice of conspicuous hunting perches. Only 13% of the possible breeding records are of singing males. Once a pair is located, the bulky nest is relatively easy to find, usually in a hawthorn; 15% of all records are of nests. However, confirmation of breeding was most often achieved by locating the active and noisy family groups. Egg-laying is early, often in April, so young are out of the nest when most atlassers are in the field in late spring.

Atlas data have improved greatly our knowledge of the breeding range and population level of this species in Ontario. Although breeding records were known previously from immediately south and east of the Shield, the concentration of records in that area, as revealed by atlas coverage, was somewhat surprising. The existence of agricultural land has permitted a few birds to move up the Ottawa Valley to near Pembroke, and onto the Bruce Peninsula and Manitoulin Island. However, intensive agricultural practices, whereby hedgerows are removed and most land is cultivated, are the most likely reason for the withdrawal of the breeding population from the southwest. A pair of birds with unfledged young near London was reported in both 1982 and 1983; this is the only confirmed breeding evidence to the south and west of Guelph. More concentrated field work might reveal that the species still breeds, at least occasionally, in the Rainy River area, because sightings have been made there recently and there is a breeding population in nearby southeastern Manitoba (Cleveland *et al.* 1980).

Abundance estimates were provided with about 50% of Loggerhead Shrike records. The estimated numbers of pairs per square suggest that it is rare in most areas of Ontario. One square in Carden and Eldon Townships, just east of Lake Simcoe, where the species has traditionally been reported, was estimated to contain more than 10 breeding pairs. During the atlas period, the province's breeding population was probably 50 to 100 breeding pairs in any one year, with all but a few of those in southern Ontario. The lower value is probably more accurate in the later years of the project as there was a noticeable decline in some well-covered areas (*e.g.*, near Ottawa, C. Hanrahan pers. comm.) during the atlas period. COSEWIC has designated the species as 'threatened' in Canada.-- *M.D. Cadman*

LOGGERHEAD SHRIKE

Summary of Abundance Estimates

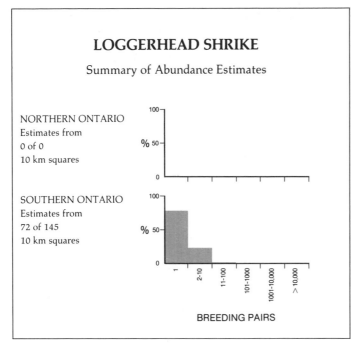

NORTHERN ONTARIO
Estimates from
0 of 0
10 km squares

SOUTHERN ONTARIO
Estimates from
72 of 145
10 km squares

BREEDING PAIRS

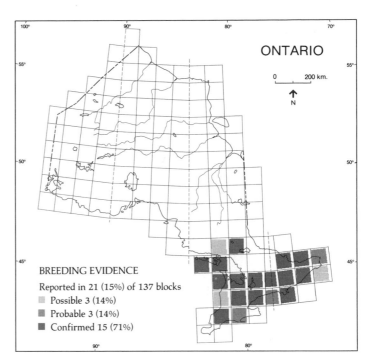

ONTARIO

0   200 km.

N

BREEDING EVIDENCE

Reported in 21 (15%) of 137 blocks

Possible 3 (14%)

Probable 3 (14%)

Confirmed 15 (71%)

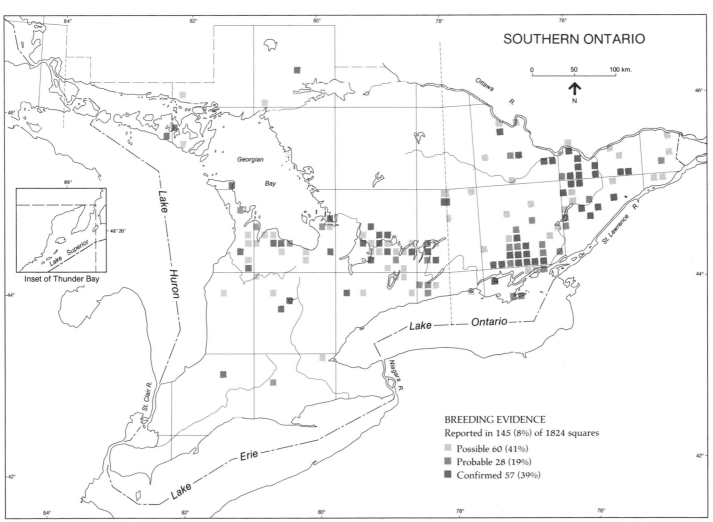

SOUTHERN ONTARIO

0   50   100 km.

N

Inset of Thunder Bay

Georgian Bay

Lake Huron

Lake Superior

St. Clair R.

Niagara R.

Lake Erie

Lake Ontario

Ottawa R.

St. Lawrence R.

BREEDING EVIDENCE

Reported in 145 (8%) of 1824 squares

Possible 60 (41%)

Probable 28 (19%)

Confirmed 57 (39%)

EUROPEAN STARLING
Etourneau sansonnet
Sturnus vulgaris

The European Starling is a temperate grassland species with a wide distribution in the Eastern Hemisphere, extending from Iceland to northern Russia, south to northeastern China and north Africa, and west to the Azores. The species is a relative newcomer to North America, where it has been introduced.

As one of the most common birds in North America, the European Starling is well known and has been recorded in a variety of habitats. During the breeding season, it is found in cities, towns, farmlands, and wooded areas, and may nest in building recesses, eavespipes, mail boxes, bird houses, cavities in trees or fence posts, and holes in embankments. In forested areas in Ontario, it nests only within foraging distance of cleared land (Lumsden 1980). It is thought to compete for nesting cavities with native species such as the Eastern Bluebird, Tree Swallow and Red-headed Woodpecker (Snyder 1951, Godfrey 1986). The European Starling is a common winter resident in southern Ontario, although part of the population migrates farther south.

Several attempts were made to introduce the European Starling to North America, but it was the March 1890 release of about 60 birds in Central Park, New York City, that gave the species its foothold on this continent. Within 80 years, its breeding range extended from the Canadian arctic to Mexico (Feare 1984). Its colonization of North America coincided with a period of rapid habitat modification. In Europe, this species had expanded its range into habitats opened up by agriculture and settlement. Judging from its success in North America, the European Starling adapted well to similar land-use practices here.

In Ontario, the European Starling was first reported breeding at Burlington, Halton Co, in 1922, but had been sighted as early as 1914 near Niagara Falls (MacCrimmon

1977). By the mid-1930's, breeding had been reported in the southern Ontario counties of Halton, Frontenac, Lambton, Victoria, York, Carleton, Essex, Northumberland, and Ontario (now Durham Region) (Lewis 1928, Baillie and Harrington 1937). In northern Ontario, a specimen was found dead at York Factory on Hudson Bay in 1931 (Lloyd 1934), and there were reports of nestings in Sudbury, Algoma, and Thunder Bay Districts by 1937 (Baillie and Harrington 1937). The first nesting on the Hudson Bay coast was reported at Winisk in 1967 (Lumsden 1980).

The European Starling is easy to find since it frequents areas of human land-use and often selects nest cavities in open areas. Its squat-bodied, delta-winged silhouette is also distinctive in flight. As in the case of most cavity-nesters, few records (3%) are of nests with eggs. However, noisy young birds in the nest are easy to locate, and account for the highest percentage of records (24%) in any of the categories used to confirm breeding. Food, carried to the young, is very obvious in the adult's long, thin bill, and this cue was used frequently to identify breeding activity.

Away from human settlements and fields, the European Starling is less common. The map shows the sparsity of records in Algonquin Prov. Park and parts of the Parry Sound and Sudbury Districts. North of paved roads, records are scattered and correspond to areas of human settlement: Red Lake, Pickle Lake, Big Trout Lake, Webeque, Fort Hope, Winisk, Attawapiskat, Fort Albany, and Moosonee. European Starlings were reported in most blocks in which settlements are present, and were even reported in an area of tundra near Cape Henrietta Maria where buildings provided nesting sites. They were not reported in a few small towns, most of which were located in the far northwestern part of Ontario. These include: Big Beaver House, Wunnummin Lake, Kasabonika, Fort Severn, Favourable Lake, Deer Lake, Indian Village, Sandy Lake, Opasquia, Sachigo Lake, and Bearskin Lake.

Evidence for the beginning of a decline in abundance of this species comes from Snyder (1951), who stated, "The Starling is now somewhat less numerous in southern Ontario than it was in the early years of the 1930's." Trend analysis of the Breeding Bird Survey data has indicated a 54% decline in European Starling populations from 1967 through 1983 in central and northern Ontario and Quebec, and a 29% decline in the same period in southern Ontario and Quebec (B. Collins pers. comm.). In spite of declines, the European Starling remains a very widespread and abundant species in southern Ontario, as indicated by the fact that 40% of abundance estimates in that area were of 101 to 1,000 pairs, and 23% were of 1001 to 10,000 pairs. It seems likely that any decline in the last several decades represents a stabilization of the population, and it is certain that the European Starling is here to stay.-- *I. Bowman*

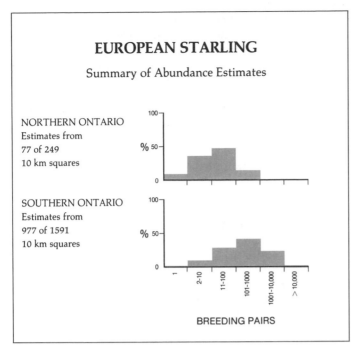

EUROPEAN STARLING

Summary of Abundance Estimates

NORTHERN ONTARIO
Estimates from
77 of 249
10 km squares

SOUTHERN ONTARIO
Estimates from
977 of 1591
10 km squares

BREEDING PAIRS

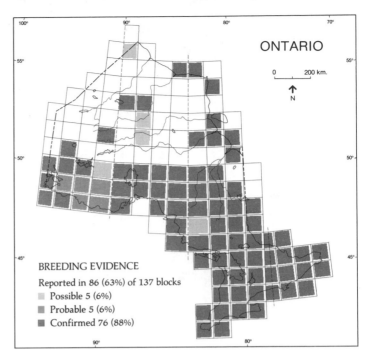

ONTARIO

0   200 km.

N

BREEDING EVIDENCE
Reported in 86 (63%) of 137 blocks

- Possible 5 (6%)
- Probable 5 (6%)
- Confirmed 76 (88%)

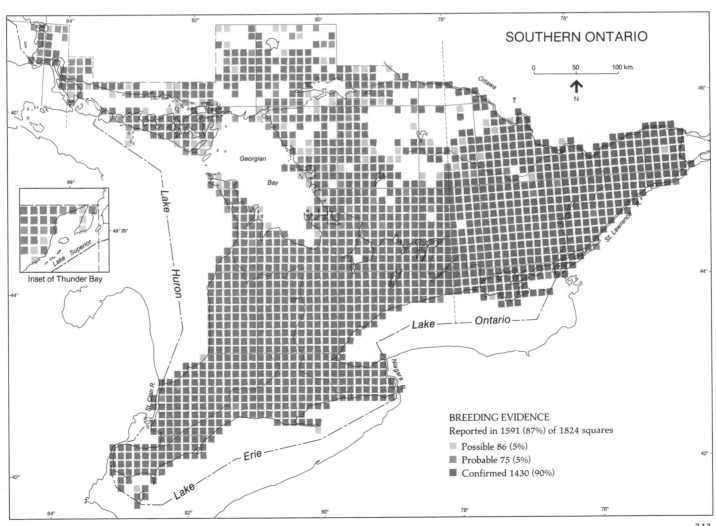

SOUTHERN ONTARIO

0   50   100 km.

N

Inset of Thunder Bay

BREEDING EVIDENCE
Reported in 1591 (87%) of 1824 squares

- Possible 86 (5%)
- Probable 75 (5%)
- Confirmed 1430 (90%)

WHITE-EYED VIREO
Viréo aux yeux blancs
Vireo griseus

The White-eyed Vireo occurs in summer over most of the eastern half of the US as far west as central Texas and eastern Nebraska and as far north as southern Michigan and Massachusetts, including the southern part of Ontario. In winter, it resides in the states along the coast of the Gulf of Mexico and in much of Central America.

The White-eyed Vireo occupies both wet and dry habitat types, generally where dense shrubbery is available in such places as overgrown fields, young second growth woodlands, woodlot edges and clearings, and open swampy woodlands. Many such areas result from human activity and therefore the amount of suitable habitat has probably increased over the last century although the overall amount of forest cover has decreased. The White-eyed Vireo has been expanding its range northward, very noticeably in the past 3 or 4 decades, likely taking advantage of habitats altered by human activity.

There are 2 unconfirmed Ontario nesting records for this species, dating back almost a century: one in Toronto (Macoun and Macoun 1909), and a second near Niagara (Beardslee and Mitchell 1965). However, considerable doubt has been expressed about their validity because there is no other indication that the White-eyed Vireo ever occurred regularly in or near Ontario at that time. The species was first confirmed as nesting in Ontario in 1971 (Peck 1976), in the vicinity of Rondeau Prov. Park. Since then, small numbers of White-eyed Vireos have appeared annually and with continued expansion it has become a regular part of our avifauna. Small populations occur in and near Point Pelee Nat. Park, and Rondeau. Near Lake Erie, it is likely to be regular every summer in several places, and single pairs could be expected to nest almost anywhere in the Carolinian Forest region. While some White-eyed Vireos wander in spring somewhat north of the Carolinian area, it is unlikely that the species

has nested north of the latitude of Hamilton. If the northward expansion continues, and there is every indication that it will, evidence of breeding should be looked for elsewhere, particularly in the Kingston area where summer sightings are now occurring.

The feature most likely to be noticed first about the White-eyed Vireo is its unusual song. Beginning and ending with 'chicks' or 'chucks', the song is a series of 6 to 9 notes of various warbles, slurs, and scolding sounds, often including phrases that are similar to those of numerous other songbirds (Adkisson and Connor 1978). Compared to that of our other vireos, it is a very distinctive song which is of paramount importance in the location and identification of this species.

Although locating White-eyed Vireos should be easy because of the species' loud and distinctive song, confirmation of breeding is likely to be more difficult. The dense shrubbery used for nesting inhibits observation of both birds and their nests. The birds, however, apparently allow rather close approach without alarm, which should be helpful in determining breeding status. At least the behaviour of the birds can be seen, even when a bird on a nest may go unnoticed. The relatively small number of atlas records probably reflects the small population in Ontario, as much as the difficulty in locating birds that are present.

The map indicates clearly that the breeding population is confined mainly to the counties bordering Lake Erie, and that the species perhaps occurs most frequently in the western third of this range. Even there, a few of the atlas records are likely to be of unmated males. The record along the north shore of Lake Ontario probably represents an unmated male also. Any of these birds could have bred if they had found mates, but with the very small population there is little chance of each male finding a partner. The record from the Bruce Peninsula must be considered a vagrant bird with virtually no chance of nesting successfully during the atlas years.

Although it may be considered rare in Ontario, the White-eyed Vireo is expanding its range northward, and may become more abundant in the future. At present, an abundance of less than 10 pairs per square is to be expected everywhere except for the 2 previously mentioned parks where it could be classified as uncommon, and estimates of slightly over 10 pairs are appropriate. The atlas project has come at a very good time to help document the northward spread of the White-eyed Vireo.-- *R.D. James*

WHITE-EYED VIREO

Summary of Abundance Estimates

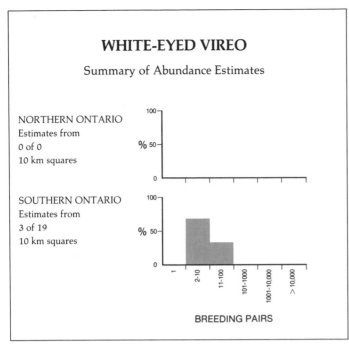

NORTHERN ONTARIO
Estimates from
0 of 0
10 km squares

SOUTHERN ONTARIO
Estimates from
3 of 19
10 km squares

BREEDING PAIRS

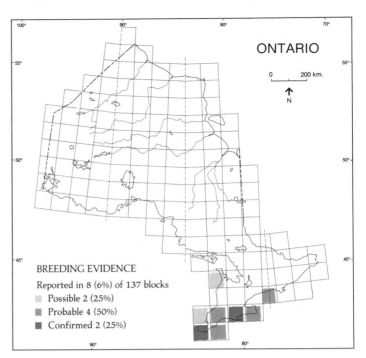

ONTARIO

0   200 km.

N

BREEDING EVIDENCE
Reported in 8 (6%) of 137 blocks
▢ Possible 2 (25%)
▨ Probable 4 (50%)
▧ Confirmed 2 (25%)

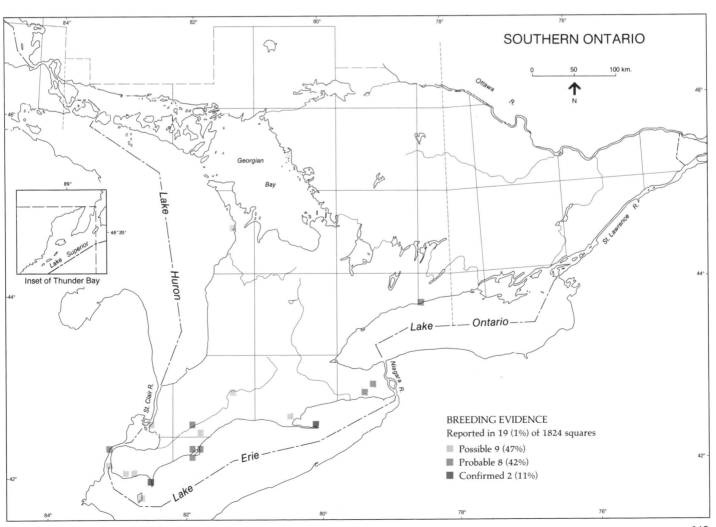

SOUTHERN ONTARIO

0   50   100 km.

N

Georgian Bay

Lake Huron

Lake — Ontario

St. Lawrence R.

Ottawa R.

Niagara R.

St. Clair R.

Lake — Erie

Inset of Thunder Bay
Lake Superior
48°20'
89°

BREEDING EVIDENCE
Reported in 19 (1%) of 1824 squares
▢ Possible 9 (47%)
▨ Probable 8 (42%)
▧ Confirmed 2 (11%)

SOLITARY VIREO
Viréo à tête bleue
Vireo solitarius

The Solitary Vireo is partial to larger forest tracts, usually well removed from human habitation. In Ontario, pairs are typically well spaced and often isolated from the sound of others. The Solitary Vireo breeds throughout much of the forested parts of North and Central America, usually at higher latitudes or higher elevations, where conifers form part of the tree cover. It winters in the southern US and Central America.

In Ontario, the Solitary Vireo occupies a wide band across the centre of the province, becoming rarer in the north as tree size diminishes, and rarer in the south where forests are few. The species prefers mixed coniferous/deciduous or pure coniferous forest in Ontario, although elsewhere it may occupy essentially pure deciduous forest (James 1979). It is the only vireo in the province making extensive use of coniferous trees for feeding and nesting. The Solitary Vireo nests in relatively mature forests, where there is typically a nearly continuous canopy of trees overhead, but it may rarely be in younger growth adjacent to taller trees. In southern Ontario, it is confined largely to areas on the Canadian Shield where extensive forests still prevail, with isolated birds occurring in remnant patches of woodland or in pine plantations in agricultural regions south of the Shield. In the north, it is likely to be scarce or absent from much of the northern Hudson Bay Lowland, where tree cover is scattered.

Its occupancy of more remote forests and its relatively low density make this species somewhat difficult to locate, unless one is familiar with its loud song. Unfortunately, it is the unmated males that sing most persistently (James 1978a). The song is also difficult for some to distinguish from that of the more ubiquitous Red-eyed Vireo. When incubating, the male Solitary Vireo may remain silent for periods of half to three-quarters of an hour and could easily be passed unnoticed. Since incubation is likely to have begun before the end of May or shortly thereafter, some atlassers were probably not in the field until after the birds became rather silent. When off the nest, the male often wanders considerable distances, while singing only slowly and sporadically. This makes him difficult to follow when attempting to locate a nest. However, both parents feed the young, so that if birds are found, confirmation of breeding should be readily determined at the nestling stage. Nests are often low in trees and shrubs, where they can usually be seen without great effort.

The Solitary Vireo may have occurred throughout southern Ontario in presettlement days, when forests occupied almost all of that area, but, if so, it has retreated northward following forest clearing. Most remaining forest patches south of the Canadian Shield are too small to maintain permanent populations today. Some larger and more mature plantations of pines may be the only places where the species regularly occurs in small numbers. These isolated areas have only recently become mature enough to attract Solitary Vireos. For example, the group of records north of Long Point on Lake Erie reveals the maturity of the conifer plantations on and near the St. Williams Forestry Station.

The relatively small number of squares on the Canadian Shield with records of the Solitary Vireo suggests a scattered distribution, but perhaps also that this species was difficult to detect. The small number of breeding confirmations suggest that the bird's behaviour made it difficult to confirm. The maps, however, do seem to reflect rather well the distribution known before the atlas period. In southern Ontario, records are concentrated in the Algonquin Highlands and near Bon Echo Prov. Park on the Canadian Shield. However, habitat is apparently suitable from central Georgian Bay and Ottawa northward, so almost every square might be expected to contain at least one nesting pair. Elsewhere, sightings are more likely of isolated pairs. Suitable habitat is also widespread in northern Ontario, north to at least Big Trout Lake and Moosonee. It seems likely that with sufficient effort breeding could have been confirmed in most, if not all, blocks with records of only possible or probable breeding, and in many blocks south of the Lowland in which the species was not reported. Close to Hudson Bay, there would ordinarily not be any Solitary Vireos, as the map indicates.

The Solitary Vireo is a widespread species and not rare, despite populations that never seem to be as dense as we might like to see them. Although most observers indicated an abundance of less than 10 pairs per square (55% of all squares with estimates), this may be too low an estimate for most squares on the Canadian Shield. Between 11 and 100 pairs would likely be more appropriate for most of these squares and between 101 and 1,000 pairs would be attained in many. Declines in numbers, if happening at all today, are more likely the result of habitat changes in wintering areas.-- *R.D. James*

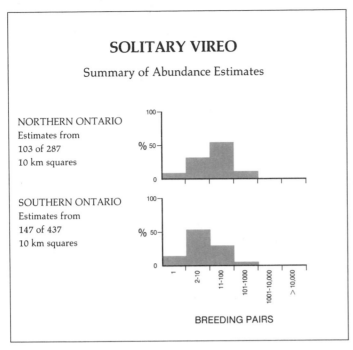

SOLITARY VIREO

Summary of Abundance Estimates

NORTHERN ONTARIO
Estimates from
103 of 287
10 km squares

SOUTHERN ONTARIO
Estimates from
147 of 437
10 km squares

BREEDING PAIRS

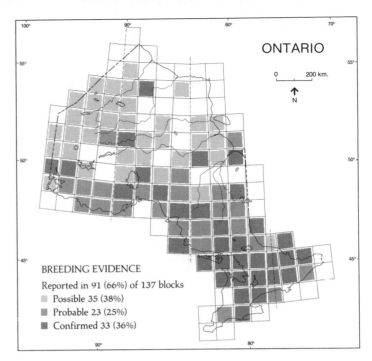

ONTARIO

0   200 km.

N

BREEDING EVIDENCE
Reported in 91 (66%) of 137 blocks

Possible 35 (38%)
Probable 23 (25%)
Confirmed 33 (36%)

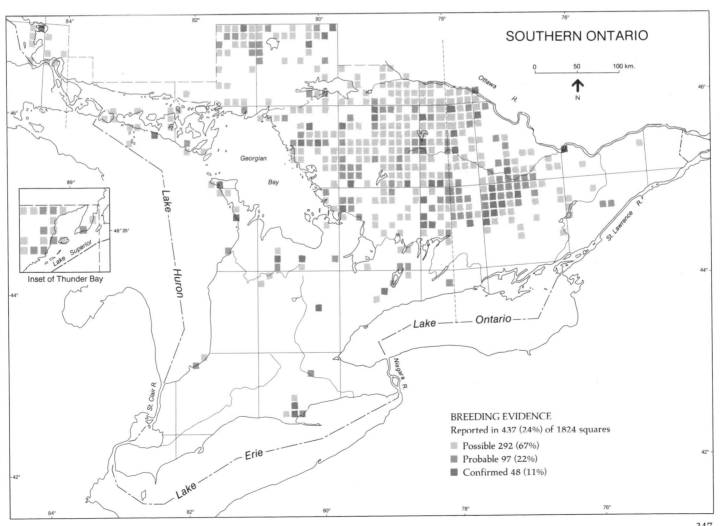

SOUTHERN ONTARIO

0   50   100 km.

N

Inset of Thunder Bay

BREEDING EVIDENCE
Reported in 437 (24%) of 1824 squares

Possible 292 (67%)
Probable 97 (22%)
Confirmed 48 (11%)

347

YELLOW-THROATED VIREO
Viréo à gorge jaune
Vireo flavifrons

The Yellow-throated Vireo is found almost exclusively in deciduous woodlands. It often occurs in the same forests as the Red-eyed Vireo, and in such situations generally stays higher in the trees.

The Yellow-throated Vireo may be found in summer throughout eastern North America, from southern Manitoba, Ontario, and Quebec south to eastern Texas, the Gulf of Mexico, and central Florida. In Ontario, it is found north to Kenora, Sault Ste Marie, and Ottawa, but in areas of coniferous forest on the Canadian Shield it may be absent. It is also absent from regions dominated by extensive agriculture where forest cover is sparse or absent. It occupies woodlands that typically contain relatively mature trees with large spreading canopies. Such woods may be open, with trees well separated, or more closed where there is little undergrowth of shrubs or young trees (James 1979). Densities are highest where trees cover sufficient area that several pairs can occupy large territories with space to spare. In winter, it resides in southern Mexico, Central America, Colombia, and Venezuela.

The nesting period for the Yellow-throated Vireo usually begins earlier than for most small birds, with many individuals incubating eggs by the end of May. Once incubation is under way, the males become almost silent. When they do sing, they are likely to be some distance from their nests (James 1984a). Thus, this vireo may be easily overlooked in June, when most atlassers were searching. The Yellow-throated Vireo's song is similar to that of the Red-eyed and Solitary Vireos, but it is recognizable to the experienced observer, sounding more hoarse than the songs of the other 2 species. Nest placement and foraging occur high in trees so that neither is conspicuous. Nests may be difficult to locate during incubation unless the behaviour of the birds is understood. Both sexes incubate and the sitting bird leaves the nest as the approaching bird flies into the nest tree. Unless this exchange is noticed, it appears as if one bird passes into and out of the tree without stopping at the nest. Both adults feed the young in the nest and for a couple of weeks after they leave so confirmation of breeding should be easiest at such times (James 1978a). The small number of confirmed breeding records suggests that people had difficulty with this task.

This vireo was found in 25% of the squares in the south and virtually all of these lie south of 45°30'N latitude. The map reflects its distribution fairly well, but there are wide gaps in its occurrence. Given the apparent suitability of the habitat, there probably should have been more records in the western part of the Quetico Forest section. The greatest concentrations appear along the southern edge of the Canadian Shield, especially north and east of Kingston. This is an area of maturing second growth forests where agriculture is marginal and habitat is ideal. Only a few birds range north to Sault Ste Marie and Manitoulin Island or as far east as southern Renfrew Co. Many of these may not be successful breeders, but could nest if mates appeared. There is another substantial concentration of records north of the shore of Lake Erie, as one might expect of southern species, although forest cover is often sparse in that area. I would have expected more records along the Niagara Escarpment between Lake Ontario and southern Georgian Bay. In addition, the absence of the species from many squares from Prince Edward Co west to Lake Huron is probably related to a lack of suitable habitat, but low atlassing effort may have played a role. Overall, the number of records may not be far below the actual number of squares occupied.

The Yellow-throated Vireo is not an abundant species in Ontario. Even where comparatively common, it tends to be widely spaced in the available habitat. Individual pairs may occupy woodlots far removed from any other pair. It likely occupies much the same range as it did before settlement of this continent, but has no doubt decreased in abundance as forest clearing took place. About 50% and 22% of the squares with abundance estimates were thought to contain 2 to 10 pairs per square and 11 to 100 pairs per square, respectively. These are probably reasonable estimates for most predominantly agricultural areas. I would expect few areas to contain more than 100 pairs per square and scarcely any over 1,000 pairs per square. Land clearing and forest fragmentation have no doubt contributed to population declines in the past. Present populations appear stable, but loss of tropical forests may affect them in the future.-- *R.D. James*

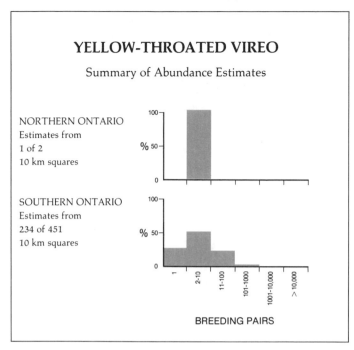

YELLOW-THROATED VIREO

Summary of Abundance Estimates

NORTHERN ONTARIO
Estimates from
1 of 2
10 km squares

SOUTHERN ONTARIO
Estimates from
234 of 451
10 km squares

%

100

50

0

%

100

50

0

1
2-10
11-100
101-1000
1001-10,000
>10,000

BREEDING PAIRS

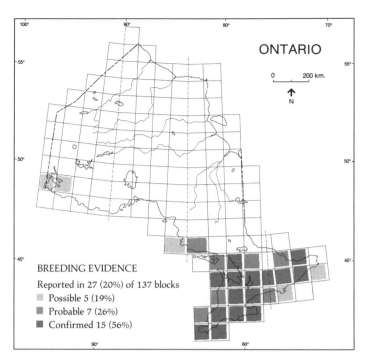

ONTARIO

0   200 km.

N

BREEDING EVIDENCE

Reported in 27 (20%) of 137 blocks

Possible 5 (19%)
Probable 7 (26%)
Confirmed 15 (56%)

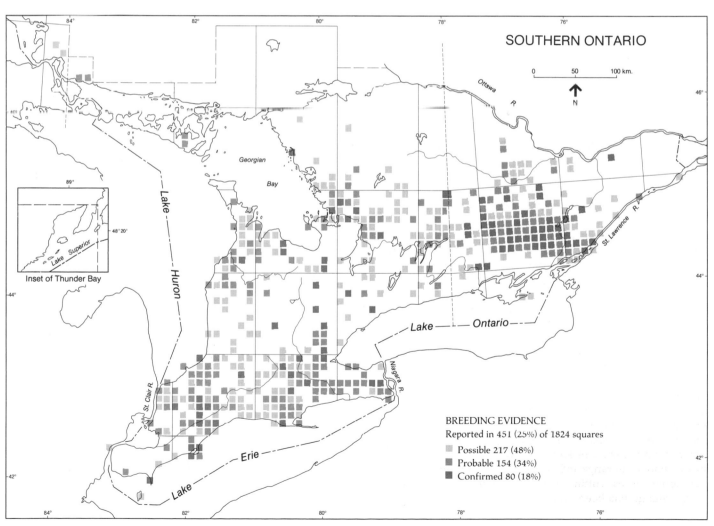

SOUTHERN ONTARIO

0   50   100 km.

N

Georgian Bay

Lake Huron

Lake Superior
Inset of Thunder Bay

Ottawa R.

St. Lawrence R.

Lake — Ontario

Niagara R.

St. Clair R.

Lake Erie

BREEDING EVIDENCE

Reported in 451 (25%) of 1824 squares

Possible 217 (48%)
Probable 154 (34%)
Confirmed 80 (18%)

WARBLING VIREO
Viréo mélodieux
Vireo gilvus

While the Warbling Vireo is among the most plain and unadorned of birds with respect to plumage, its song is one of the more elaborate and pleasing to listen to. The sedate warbling is distinctive, and is a good guide to the identification of this species.

The Warbling Vireo is found across North America in summer, but is more widespread in the west. Its range extends from northern British Columbia, southeastern Yukon, and southwestern Northwest Territories southeast to Manitoba, east to Kenora, Thunder Bay, southern Ontario, extreme southern Quebec, and New Brunswick. It also occurs south to the mountains of central Mexico. The Warbling Vireo is migratory, spending the winter in Mexico and Central America.

In Ontario, the Warbling Vireo is usually found in association with maple trees, although it also occurs in poplars. It is unusual to find this vireo in groves of broadleaved trees that are surrounded by coniferous woods; it seems to require areas where only deciduous trees dominate the landscape. Also, it does not usually occur in dense or extensive woodland, but prefers open woods, woodland edges, groves, and scattered trees along watercourses, laneways, rural roadways, parks, towns, and cities. In all situations, it is partial to older and taller trees (James 1976).

The Warbling Vireo is not a rare species in southern Ontario, but tends to be thinly scattered. Isolated pairs are often encountered, while sometimes several birds are close together. This type of distribution results from the fragmented type of habitat that the species prefers. This species has probably been a resident of Ontario for centuries, but has extended its range following human settlement and subsequent forest cutting. Even in this century, a significant range change has been noted. It was formerly unknown in

the western Rainy River District (Snyder 1938) and scarcely ranged north of southern Georgian Bay and Ottawa in the eastern part of the province (Baillie and Harrington 1937). However, it is now numerous in the Rainy River region and regular in the Kenora area. It is found throughout southern Ontario, regularly north to at least Sault Ste Marie and Sudbury, particularly where settlement has removed much of the forest cover.

The Warbling Vireo should be among the easiest of birds to find and confirm as a breeder. It is often found near human habitation and along roadways. Although it tends to remain high in trees where it is relatively well concealed among the foliage, its song is not only distinctive, but also given throughout the nesting season. The male often sings when sitting on the nest (Howes-Jones 1985). The main difficulty is in actually finding the nest, which is typically high in a tree and completely obscured from view by leaves. Both sexes feed the young, thereby making breeding easy to confirm as they carry food into the nest tree.

The large number of records on the atlas maps indicates that the Warbling Vireo is easily found and is widely distributed in southern Ontario. South of 45°N latitude, the Warbling Vireo is present in nearly every square. Its absence in several squares in Essex Co, for example, is probably due to a lack of suitable habitat. The few blank squares in the area south of Georgian Bay and Ottawa may reflect a lack of observer coverage. It is absent from much of the highlands in and about Algonquin Prov. Park. In the most northern parts of southern Ontario, the distribution becomes more patchy, as shown on the map. Especially interesting are the records in northern areas. The few records north of Kenora in the west, about Lake Superior, and in the Clay Belt area in the east represent range extensions. The atlas data show that this species is expanding its range northward into settled and disturbed regions, although at present, it is widely scattered and occasional.

Most observers estimated abundance of up to 100 pairs per square and that is probably appropriate in most of the continuous range in southern agricultural areas. A few people suggested densities of over 1,000 pairs per square, but such a level is unlikely to occur because prime breeding habitat is seldom found throughout an entire square. The BBS results from 1967 to 1983 show a significant increase in numbers in southern Ontario. Although the Warbling Vireo is never likely to become as abundant as the Red-eyed Vireo, we can look forward to a continuing increase in its number in Ontario.-- *R.D. James*

WARBLING VIREO

Summary of Abundance Estimates

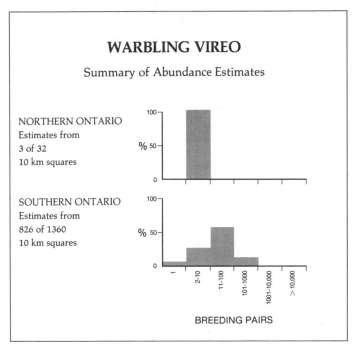

NORTHERN ONTARIO
Estimates from
3 of 32
10 km squares

SOUTHERN ONTARIO
Estimates from
826 of 1360
10 km squares

%

%

100

50

0

100

50

0

1 2-10 11-100 101-1000 1001-10,000 >10,000

BREEDING PAIRS

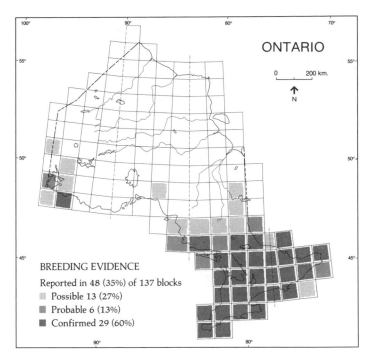

ONTARIO

0   200 km.

N

BREEDING EVIDENCE
Reported in 48 (35%) of 137 blocks

Possible 13 (27%)
Probable 6 (13%)
Confirmed 29 (60%)

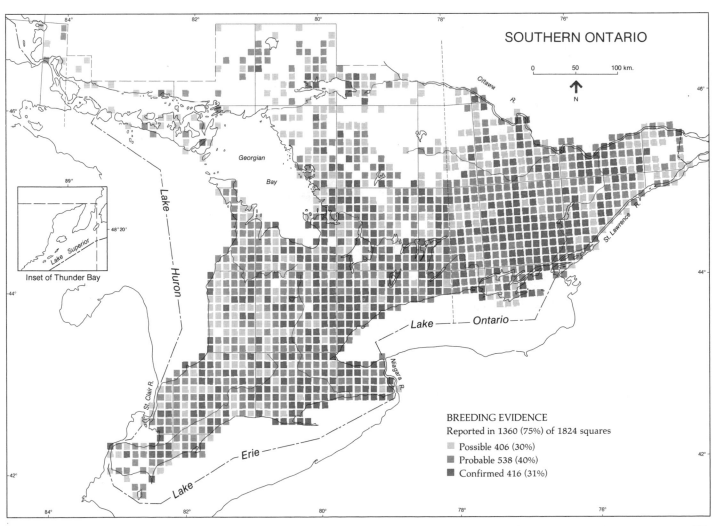

SOUTHERN ONTARIO

0   50   100 km.

N

Georgian
Bay

Lake
Huron

Lake — Ontario

Lake
Erie

Lake

St. Clair R.

Niagara R.

St. Lawrence R.

Ottawa R.

89°

48°20'

Lake Superior

Inset of Thunder Bay

BREEDING EVIDENCE
Reported in 1360 (75%) of 1824 squares

Possible 406 (30%)
Probable 538 (40%)
Confirmed 416 (31%)

PHILADELPHIA VIREO
Viréo de Philadelphie
Vireo philadelphicus

Although the Philadelphia Vireo looks somewhat like a small Red-eyed Vireo, the 2 species are easily distinguished by plumage. However, the songs of these 2 vireos are so similar that most observers are not able to distinguish them. Even experienced listeners prefer to check their identification by sight. The resemblance is so great that the Red-eyed Vireo apparently responds to a Philadelphia Vireo song as if it were that of another Red-eyed Vireo. It appears that, where the 2 species occur in the same woodlands, the Philadelphia Vireo is able to maintain a territory, at least for a few days, and largely exclude the Red-eyed Vireo using this similar song as the primary means of territorial defence. This avoids, to some extent, potentially injurious physical encounters with its larger relative (Rice 1978a), and thus the 2 species are able to coexist in some of the same habitat (Rice 1978b). This behaviour created a difficult challenge for atlassers trying to discern the presence of the Philadelphia Vireo.

The Philadelphia Vireo breeds across Canada from eastern British Columbia to western Newfoundland, but not in Prince Edward Island or Nova Scotia (Godfrey 1986). A few breed in some parts of the extreme northern US, from North Dakota to Maine. It winters in Central America and northwestern South America (Colombia). It prefers open woodlands in the breeding season, particularly where trembling aspen is the dominant cover. Growths of alders under and among or adjacent to aspen groves are also extensively used. The ranges of the Philadelphia Vireo and of trembling aspen correspond rather closely except in British Columbia. The species seldom occurs in the most northern parts of Ontario, where aspens are few, and is uncommon south of the Boreal Forest, where maples dominate the deciduous woodlands. In southern Ontario, it is a rather rare species, and those that do breed there are confined mainly to the Canadian Shield.

It is most abundant in the Boreal Forest, becoming more numerous as the Red-eyed Vireo becomes less common. In the vicinity of Moosonee, the Philadelphia Vireo outnumbers the Red-eyed Vireo by at least 2 to 1.

The Philadelphia Vireo might be overlooked, not only because of a confusing song, but also because it is not nearly as persistent a singer as the Red-eyed Vireo. Once mated, its song may be completely masked by that of its more vocal relative. However, once located, at least during the incubation period, it is easily confirmed as breeding, for the male often sings while sitting on the nest. Considering the relative rarity of the species in southern Ontario, it is encouraging to see as many records as there are. The number is higher than would have been expected before the atlas began. While some birds have probably been missed, the scattering of records on the map is representative of the species' distribution. Males are less likely to find a mate in an area of low density and hence may have sung longer and more persistently in southern Ontario than males farther north. This would have been of considerable help in finding the birds in squares where this species was not expected. The low number of breeding confirmations in this same area suggests also that many males were unmated when found. As expected, records are concentrated on or close to the Canadian Shield. Others, on the Bruce Peninsula and in the southern Georgian Bay area also occupy suitable areas of more extensive forests along the Niagara Escarpment.

Across northern Ontario, the species probably breeds in nearly every block, including those where it went unrecorded by atlassers. The northern limits appear to be from about 55^0N in the west to 53^0N in the east and conform well to those shown in Godfrey (1986), although all confirmed records lie somewhat further south within the limits defined by Peck and James (1987).

The Philadelphia Vireo is widespread and common in northern Ontario, but considerably less so in the south. The species may well be increasing in southern Ontario as second growth forests mature. It does well in such forests as long as trembling aspen is a significant component.-- *R.D. James*

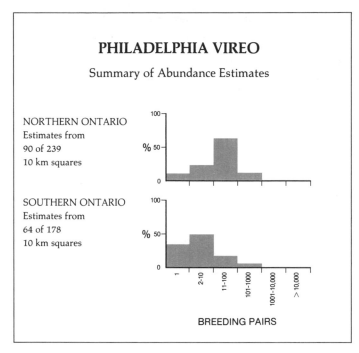

PHILADELPHIA VIREO

Summary of Abundance Estimates

NORTHERN ONTARIO
Estimates from
90 of 239
10 km squares

SOUTHERN ONTARIO
Estimates from
64 of 178
10 km squares

BREEDING PAIRS

x-axis labels: 1, 2-10, 11-100, 101-1000, 1001-10,000, >10,000

y-axis: %, 0, 50, 100

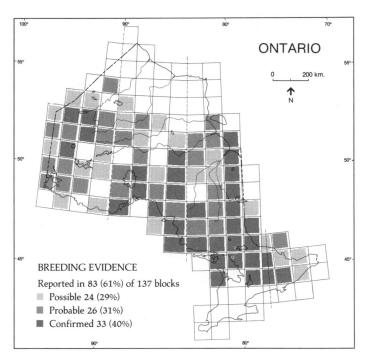

ONTARIO

0   200 km.

N

BREEDING EVIDENCE
Reported in 83 (61%) of 137 blocks

Possible 24 (29%)
Probable 26 (31%)
Confirmed 33 (40%)

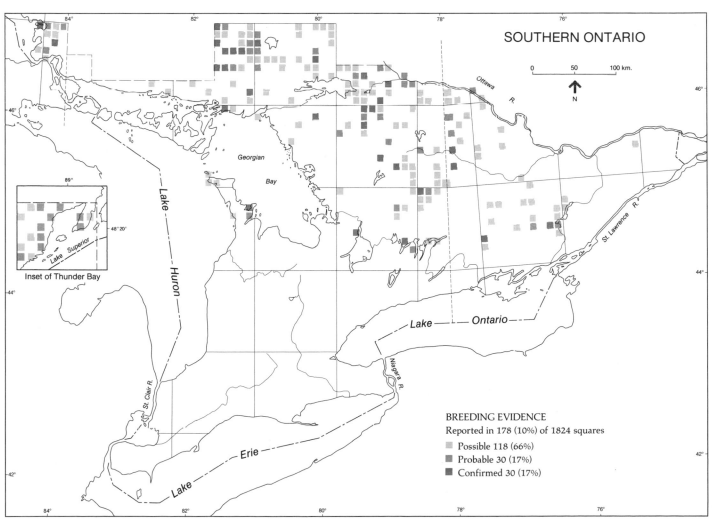

SOUTHERN ONTARIO

0   50   100 km.

N

Georgian Bay

Ottawa R.

St. Lawrence R.

Lake Huron

Lake Ontario

Lake Erie

St. Clair R.

Niagara R.

89°   48°20'

Lake Superior

Inset of Thunder Bay

BREEDING EVIDENCE
Reported in 178 (10%) of 1824 squares

Possible 118 (66%)
Probable 30 (17%)
Confirmed 30 (17%)

RED-EYED VIREO
Viréo aux yeux rouges
Vireo olivaceus

The Red-eyed Vireo is among the most numerous of songbirds in eastern North America. There is scarcely a single deciduous woods, or grove of broad-leaved trees within most of our coniferous forests, without its complement of these birds. The male is the most persistent of singers, capable of delivering more than 20,000 songs a day (Lawrence 1953b) and singing through most of the hot summer days, when other birds are more likely to be silent. Anybody familiar with its song could not fail to detect it, but catching a glimpse of the reddish iris of the adult bird may prove much more difficult as it forages amid windblown leaves, with which its colouration blends. It is found across North America and in all Canadian provinces. In the west, it is found north to the Mackenzie River region into the southern Northwest Territories, and in the east to central Quebec. Breeding populations extend south to Oregon, eastern Texas, and central Florida. In winter, it resides in northern South America.

While the Red-eyed Vireo is numerous and readily located, confirmation of breeding is not easily achieved. The noisy male does not assist with nest building or incubation, as do males of all other Ontario vireos. The female goes quietly about building, and, if incubating, she is likely to sit tight as one walks by. If flushed from the nest, she will probably not make a sound, and hence finding a nest before young are present may be little more than good luck. The male does assist with feeding the young, however, so that both may be seen carrying food. Nests are usually placed relatively close to the ground, often at or just above eye level, where birds can be seen carrying food much more easily than if nests were high in trees, where they are occasionally placed. Perhaps more importantly, the nests persist through the winter. They are easily found after the leaves have fallen in autumn. In southern Ontario at least, these nests are the

only vireo nests likely to be near the ground in deciduous trees or shrubs, so that identification is almost certain. Philadelphia Vireo nests may be placed in similar situations, particularly in more northern parts of Ontario, but that species is relatively scarce in southern Ontario. Used nests make up only 1% of the confirmed breeding records in Ontario.

The Red-eyed Vireo is most likely to be found in broadleaved forests and woodlots (James 1976). It requires a patch of at least half a hectare in size, and may occur as a single pair in an isolated patch of trees. However, it prefers larger forests, including rather short, second growth to mature trees with a good growth of understory trees and shrubs. It may also occupy mixed forests and hence is found throughout most of the Boreal and Hudson Bay Lowland forests. It is likely to be only thinly distributed in the most northern parts of its range, where conifers dominate the landscape. In southern agricultural areas, it has disappeared from many areas as land was cleared, although it may still be found wherever woodlots persist. The BBS results show a continuing slow decline in the southernmost parts of the province, likely the result of continued habitat destruction. Populations in central Ontario show virtually no change in numbers during the same period, from 1967 to 1983.

The atlas maps present a fairly accurate picture of the expected distribution of the Red-eyed Vireo. The species was recorded in 97% of the squares in southern Ontario and was missed mainly in those covered largely by water. Across northern Ontario it was found in 38% of all squares with atlas data, and occurred in most blocks, with probable and confirmed records at least as far north as Fort Albany and Big Trout Lake. In a few squares and blocks, inadequate coverage may have occurred. Inaccessibility to some areas may account for the absence of the species, but in several contiguous blocks along the north coast, the absence is real and related to unsuitable habitat.

Most observers considered an abundance of between 11 and 100 pairs per square to be most appropriate (45% of all estimates), and this may be true for most squares dominated by agriculture in the south and coniferous forest in the north. However, estimates between 101 and 10,000 pairs were made often and seem reasonable for mixed forests of the Great Lakes-St. Lawrence Forest region. Five squares in the province were thought to contain more than 10,000 pairs, or in excess of 100 pairs per square kilometre. While such a density could be reached locally, it would never extend uniformly over an entire square. These abundance estimates show that the species is abundant in Ontario.-- *R.D. James*

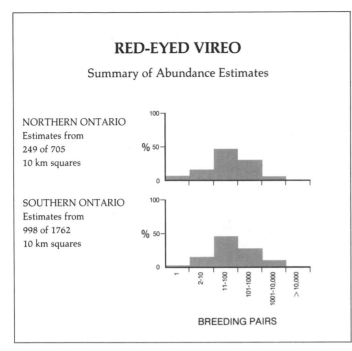

RED-EYED VIREO

Summary of Abundance Estimates

NORTHERN ONTARIO
Estimates from
249 of 705
10 km squares

SOUTHERN ONTARIO
Estimates from
998 of 1762
10 km squares

100

% 50

0

100

% 50

0

1 2-10 11-100 101-1000 1001-10,000 >10,000

BREEDING PAIRS

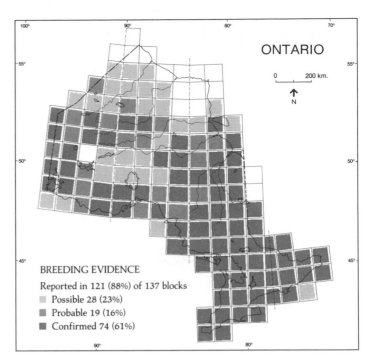

ONTARIO

0   200 km.

N

BREEDING EVIDENCE

Reported in 121 (88%) of 137 blocks

Possible 28 (23%)

Probable 19 (16%)

Confirmed 74 (61%)

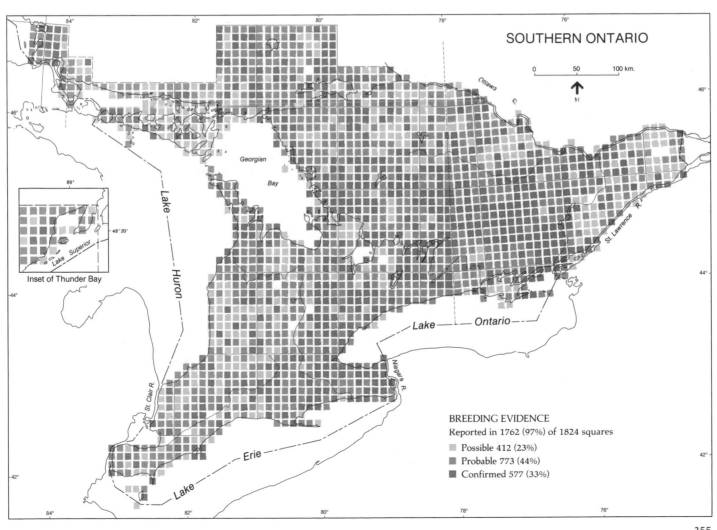

SOUTHERN ONTARIO

0   50   100 km.

N

Inset of Thunder Bay

Lake Superior

48°20'

89°

Georgian Bay

Lake Huron

St. Clair R.

Lake Erie

Lake Ontario

Niagara R.

Ottawa R.

St. Lawrence R.

BREEDING EVIDENCE

Reported in 1762 (97%) of 1824 squares

Possible 412 (23%)

Probable 773 (44%)

Confirmed 577 (33%)

BLUE-WINGED WARBLER
Paruline à ailes bleues
Vermivora pinus

With the closely related Golden-winged Warbler, the Blue-winged Warbler has attracted much attention from ornithologists in the past century. It has exhibited a dramatic range expansion in the American northeast, and it frequently hybridizes with the Golden-winged Warbler where their ranges overlap. Overall, the Blue-winged Warbler seems to be expanding and consolidating new range at the expense of the Golden-winged Warbler.

One hundred years ago, the main population of this species was located west of the Appalachians and south of the Great Lakes, with isolated populations along the Atlantic seaboard. Since that time, the Blue-winged Warbler range has spread north into Minnesota, Ontario, and the southern New England states, and south to North Carolina; previous gaps have also become incorporated into the breeding range (Gill 1980). The winter range is in Mexico and Central America. Favoured breeding habitat is similar to that of the Golden-winged, and consequently both species may inhabit the same area. Woodland openings, stream edges, willow swamps, and old fields colonized by young deciduous growth are frequently used for breeding. The Blue-winged Warbler is less of a habitat specialist than the Golden-winged, and will consequently use a wider range of habitat types (Confer and Knapp 1981).

Accounts of Ontario's bird fauna from around the turn of the century fail to mention the Blue-winged Warbler. The province's first record was of a specimen collected at Point Pelee before 1908 (Snyder 1941). Beginning in the early 1930's, the species was reported with increasing regularity in localities in the extreme south, mostly during May (Speirs 1985). These may have been disoriented migrants, but with hindsight they can now be viewed as the vanguard of Ontario's now substantial breeding population. A summering bird

was noted near Woodstock in 1953 (Berger 1958). The first breeding records for the province were reported at Ancaster and Milton in 1956. At least 3 of the first 6 Ontario breeding records were of mixed pairings between Blue-winged and Golden-winged Warblers (Baillie 1963b). In the past 3 decades, breeding Blue-winged Warblers have spread into many places in the south of Ontario, with the southern margin of the Canadian Shield from north of Kingston to Muskoka being the most recent conquest. Golden-winged Warblers have declined in the south where the Blue-winged has become increasingly common (Weir 1985).

The typical song of this warbler is quite distinctive, although hybrids may sing an identical song, as will some Golden-winged Warblers. In Ontario, where the 2 species overlap and hybrids may occur, visual contact is the safest means of identification. However, identification by song is usually correct. Because the preferred habitats are often damp and dense, Blue-winged Warblers are not easily seen. Female birds are often overlooked, as are males when they become quieter later in the season. Singing males accounted for 37% of all records, while territorial birds accounted for 19%. Breeding was confirmed in 20% of squares reporting the species, most frequently by observation of birds carrying food or faecal sacs. Being a ground-nesting warbler of dense habitats, it is not surprising that nests were found in only 5 squares during the atlas period.

The maps indicate that the stronghold of Ontario's present Blue-winged Warbler range is south of a line extending from Toronto to Grand Bend, especially in the block south and west of Hamilton. Several Regional Coordinators found that even over the course of the atlas project, numbers increased in their regions (*e.g.*, Waterloo, P. Eagles pers. comm.). These numerical increases and the recent (albeit sparse) colonization of the Rideau and Muskoka Lakes regions suggest that the range expansion is not yet complete. Many deserted farms in these regions are at a successional stage that may foster further expansion and numerical increase.

Abundance estimates were provided for 59% of the squares in which the species was recorded. Except for 5 squares with estimates of over 10 breeding pairs, abundance levels reported were of 1 pair (42%) or 2 to 10 pairs (52%), averaging lower than estimates for the Golden-winged Warbler. The 5 squares with abundance estimates of 11 to 100 pairs were all located along Lake Erie. If the Blue-winged Warbler continues to exhibit a pattern of increase and expansion, a future atlas project will doubtless produce a very different breeding status and distribution.-- *A. Mills*

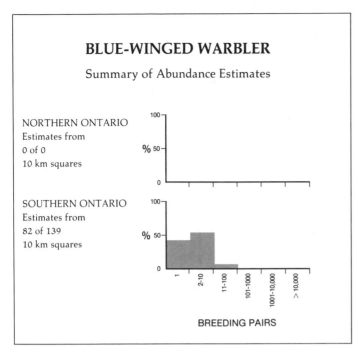

BLUE-WINGED WARBLER

Summary of Abundance Estimates

NORTHERN ONTARIO
Estimates from
0 of 0
10 km squares

SOUTHERN ONTARIO
Estimates from
82 of 139
10 km squares

%

BREEDING PAIRS

1 2-10 11-100 101-1000 1001-10,000 >10,000

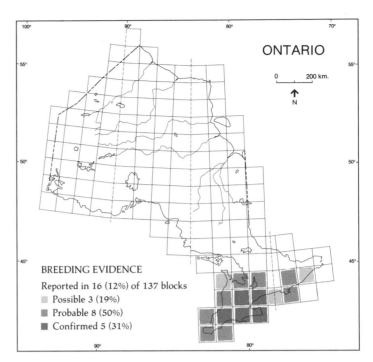

ONTARIO

0   200 km.

N

BREEDING EVIDENCE
Reported in 16 (12%) of 137 blocks
Possible 3 (19%)
Probable 8 (50%)
Confirmed 5 (31%)

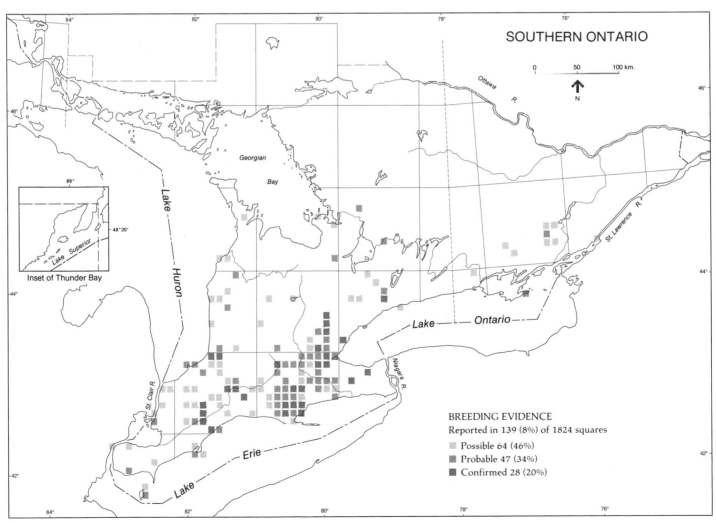

SOUTHERN ONTARIO

0   50   100 km.

N

Georgian Bay

Lake Huron

Ottawa R.

St. Lawrence R.

Lake Ontario

Niagara R.

St. Clair R.

Lake Erie

Inset of Thunder Bay
Lake Superior

BREEDING EVIDENCE
Reported in 139 (8%) of 1824 squares
Possible 64 (46%)
Probable 47 (34%)
Confirmed 28 (20%)

GOLDEN-WINGED WARBLER
Paruline à ailes dorées
Vermivora chrysoptera

The Golden-winged and Blue-winged Warblers are the subjects of considerable excitement. They hybridize extensively where their breeding ranges overlap, and are undergoing dramatic range changes, both in Ontario and elsewhere.

The Golden-winged Warbler favours damp habitats in relatively early stages of succession. In southern Ontario, this includes abandoned fields in which shrubs and small hardwood trees grow, especially those bordered by low woodland or wooded swamps, as well as hillsides with young growth. On the Canadian Shield, many Golden-winged Warblers also use alder bogs, especially when a few taller species such as black ash or even spruce or tamarack are present. At present, the breeding distribution extends from Minnesota to the New England states, and southward in the Appalachians as far as Georgia. The species winters from southern Mexico to northwestern South America.

In the first quarter of this century, the Golden-winged Warbler was restricted to the southwestern part of the province (Macoun and Macoun 1909). The first provincial breeding record occurred at London in 1912 (Baillie 1968), and the first Toronto sighting came in 1928. In the 1930's, it was increasingly reported, and by 1948 it had been found as a breeder at London, Simcoe, and Toronto. In 1953, Snyder stated that it was scarce in the south and was expanding northwards (Berger 1958). By the early 1960's, it had consolidated and expanded its range northward and eastward, and breeding had occurred near Owen Sound, Gravenhurst, Peterborough, and Kingston (Mills 1981). During the 1970's, the Golden-winged Warbler invaded the southern Canadian Shield on a broad front, ranging from the Rideau Lakes to Parry Sound and even north to Sudbury. At the same time, the species spread into previously uninhabited parts of Quebec, Wisconsin, and Minnesota (reported variously in American Birds).

During the last 20 years, Golden-winged Warbler numbers have declined in the 'older' range in the south. In some regions, where it was once one of the most common breeding warblers, it is now quite scarce. This provincial decline is part of a widespread decline through much of its eastern North American breeding range, and is coincident with, and geographically related to, an increase and range expansion of the closely related Blue-winged Warbler. For the American northeast, Gill (1980) has documented the general pattern of replacement of Golden-winged Warblers by Blue-winged Warblers within about 50 years of initial contact. The precise mechanisms are not known. Blue-winged Warblers are considered habitat generalists while Golden-winged Warblers supposedly have more specific requirements (Confer and Knapp 1981). Competition for territories between the 2 species as well as genetic introgression due to hybridization have been implicated (Gill 1980). The plight of the Golden-winged Warbler is sufficiently alarming that it was added to the Blue List in 1981. If the Blue-winged Warbler expansion continues and that species invades the northern margin of the Golden-winged Warbler range (which is now the latter's stronghold), the continued existence of the Golden-winged Warbler in Ontario is questionable.

The Golden-winged Warbler has an easily recognizable song; but hybrids, and occasionally Blue-winged Warblers, can sing an identical song. In addition, the Golden-winged Warbler occasionally sings the Blue-winged song. Except for this source of possible confusion, the song is a fairly dependable identifying character. Song is less frequent after the young appear in the latter part of June; this quietness and the often dense, damp habitat combine to make the species rather inconspicuous later in the season. Confirmation of breeding in the south was obtained in 106 of 467 squares reporting the species; half of these were observations of birds carrying food or a faecal sac, while another 27% were of fledged young. In at least one case near Long Point, a Golden-winged Warbler successfully mated with a Brewster's Warbler, the first generation hybrid of a Golden-winged/Blue-winged pairing.

The maps clearly indicate that the provincial breeding stronghold is currently located along the southern margin of the Canadian Shield. The extreme southwest is now sparsely populated. A few individuals were reported from localities further north and west, including birds near Sault Ste Marie and Lake of the Woods. These may represent a new vanguard as the species continues to invade northward. Abundance estimates (235 squares) indicate that 2 to 10 pairs per square is usual (53%).-- *A. Mills*

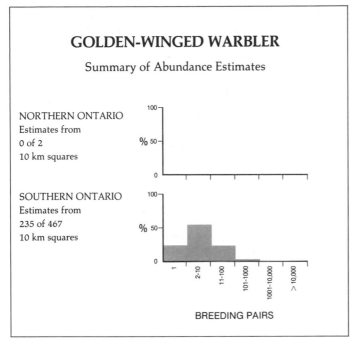

GOLDEN-WINGED WARBLER

Summary of Abundance Estimates

NORTHERN ONTARIO
Estimates from
0 of 2
10 km squares

SOUTHERN ONTARIO
Estimates from
235 of 467
10 km squares

BREEDING PAIRS

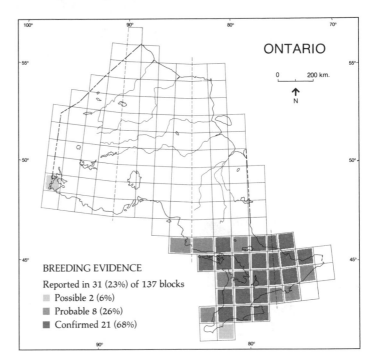

ONTARIO

0   200 km.

N

BREEDING EVIDENCE

Reported in 31 (23%) of 137 blocks

Possible 2 (6%)
Probable 8 (26%)
Confirmed 21 (68%)

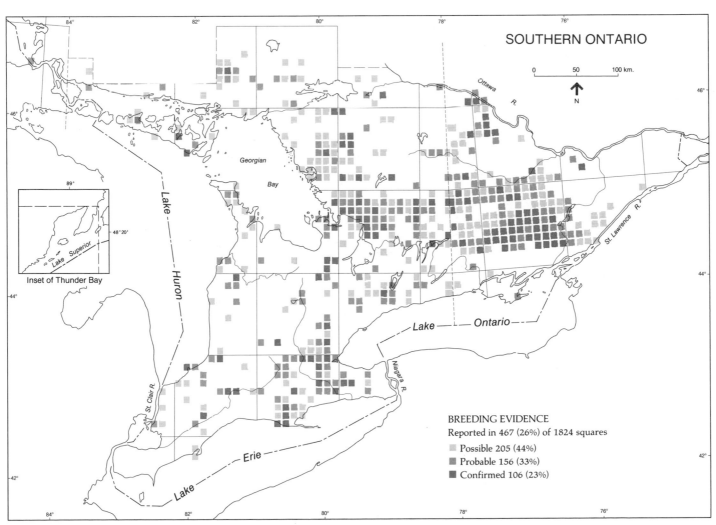

SOUTHERN ONTARIO

0   50   100 km.

N

Inset of Thunder Bay

BREEDING EVIDENCE

Reported in 467 (26%) of 1824 squares

Possible 205 (44%)
Probable 156 (33%)
Confirmed 106 (23%)

BREWSTER'S WARBLER
Paruline de Brewster
Vermivora chrysoptera x V. pinus

Where the ranges of the Golden-winged and Blue-winged Warblers overlap, they frequently hybridize to produce fertile offspring. The more frequent of the 2 hybrid types produced is the Brewster's Warbler, while the recessive type, known as the Lawrence's Warbler, occurs much less frequently. The typical Brewster's plumage results from a mating between a pure Golden-winged and pure Blue-winged. Whenever a mating involves at least one hybrid individual, a variety of plumages can result, one of which is the Lawrence's Warbler (Parkes 1951, Gill 1980).

Differences in breeding habitat between the Blue-winged and Golden-winged Warblers are apparently slight. Not surprisingly, the hybrids inhabit similar environments - generally damp, dense, early successional stages of deciduous forests. The breeding distribution of the Brewster's Warbler is more or less equivalent to the range of overlap between the breeding distributions of the parent species. Because the ranges of the parent species have changed markedly in the past century, the range of overlap and consequently the distribution of the hybrids have changed accordingly. At some point in the recent past, perhaps the mid-19th century, the Blue-winged and Golden-winged Warblers presumably had stable and more or less isolated ranges. At that time, hybrids were probably extremely rare or even non-existent: indeed, they were not described until 1874.

Brewster's Warblers begin showing up in an area more or less simultaneously with the first appearance of Blue-winged Warblers. This is probably because many of the Blue-winged Warblers at the range frontier cannot find a conspecific mate and so pair with Golden-winged Warblers. In 1965, a male Brewster's Warbler was found paired with a female Golden-winged Warbler near Oshawa; they were feeding a young Brown-headed Cowbird (Tozer and Richards 1974). Speirs

(1985) cites records indicating that Brewster's Warblers began showing up in southern Ontario with some frequency in the 1950's, the decade in which the Blue-winged Warbler was first found breeding in the province. With the expansion of the Blue-winged Warbler into southeastern Ontario in the 1970's, Brewster's Warbler sightings were noted there also (*e.g.*, Weir and Quilliam 1980).

Brewster's Warblers could be easily overlooked by atlassers for a variety of reasons. Song enhances observer success in finding most warblers, but this hybrid is known to sing perfect Golden-winged or Blue-winged Warbler patterns. It also sings a pattern incorporating elements of both parent songs, one with which few people are familiar. Despite this, about half the atlas records are of singing birds. The plumage is not very striking and, without thorough observation, many people would not be able to make a quick identification. The dense, often damp habitats it favours do not improve the likelihood of finding it, nor do they make confirmation of breeding easy, especially when an observer assumes that a breeding incident involving either a Golden-winged or a Blue-winged Warbler will automatically involve a conspecific mate.

During atlas field work, Brewster's Warblers were found in 24 squares. Most were located in the block south and west of Hamilton (where Ontario's present Blue-winged Warbler population is most dense), but individuals were found as far east as Chaffey's Locks (40 km north of Kingston) and as far north as Parry Sound. Rare Brewster's/Brewster's pairings presumably do occur; Carpentier (1983) documented such a mating near Havelock (40 km northeast of Peterborough) in 1982 and P. Eagles (pers. comm.) observed 2 Brewster's Warbler south of Cambridge in 1984. Near Long Point, a Brewster's/Golden-winged mating was confirmed in 1982, as was a Brewster's/Blue-winged mating in 1983.

Only 8 squares reporting the Brewster's Warbler included abundance estimates: 5 estimated one pair per square and 3 estimated 2 to 10 pairs per square. The 3 highest estimates came from the Dundas Valley near Hamilton, the Brantford Indian Reserve and the Beverly Sparrow Field south of Cambridge, presumably areas with a high frequency of hybridization. Gill (1980) has summarized the stages a region of Blue-winged/Golden-winged Warbler interaction undergoes in the approximately 50 years it takes for the Blue-winged Warbler to replace the Golden-winged Warbler. The Brewster's Warbler plumage dominates the hybrids in the early stages of interaction. Once a population is composed almost entirely of Blue-winged Warblers, hybrids are scarce and are usually in the Lawrence's plumage.-- *A. Mills*

BREWSTER'S WARBLER

Summary of Abundance Estimates

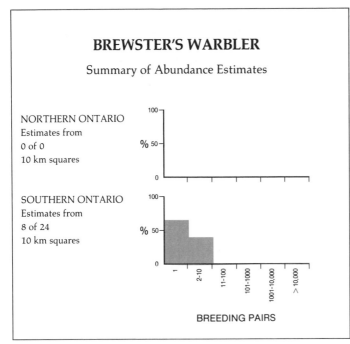

NORTHERN ONTARIO
Estimates from
0 of 0
10 km squares

SOUTHERN ONTARIO
Estimates from
8 of 24
10 km squares

BREEDING PAIRS

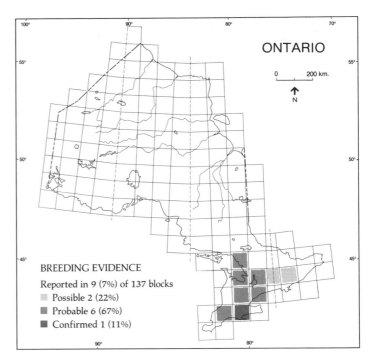

ONTARIO

0   200 km.

BREEDING EVIDENCE
Reported in 9 (7%) of 137 blocks
Possible 2 (22%)
Probable 6 (67%)
Confirmed 1 (11%)

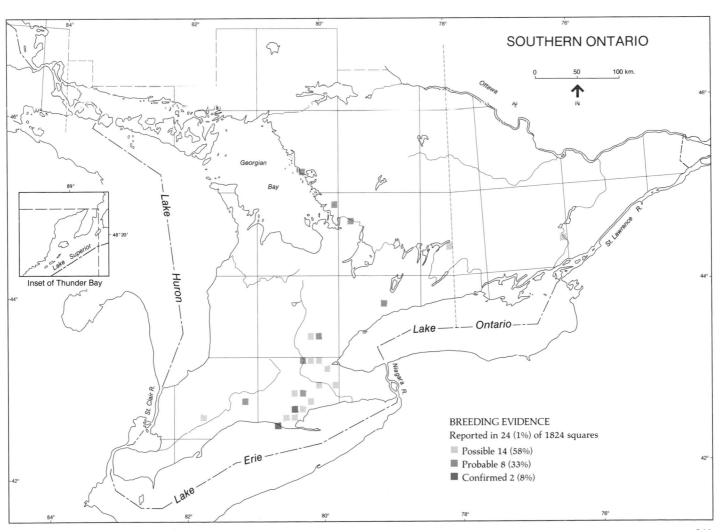

SOUTHERN ONTARIO

0   50   100 km.

Georgian
Bay

Lake Huron

Lake — Ontario

Lake Erie

Lake

St. Clair R.

Niagara R.

St. Lawrence R.

Ottawa R.

Inset of Thunder Bay

Lake Superior

BREEDING EVIDENCE
Reported in 24 (1%) of 1824 squares
Possible 14 (58%)
Probable 8 (33%)
Confirmed 2 (8%)

LAWRENCE'S WARBLER
Paruline de Lawrence
Vermivora chyrosptera x V. pinus

The Lawrence's Warbler was first described in 1874, and for many years it, and the less rare but still uncommon Brewster's Warbler, posed a problem for those trying to ascertain their status in the North American bird fauna. We now know that Golden-winged and Blue-winged Warblers frequently hybridize where their breeding ranges overlap, and that 2 hybrid types that differ in several plumage characteristics are produced. The typical Brewster's Warbler plumage results from a pure Blue-winged/Golden-winged mating, while the distinctive Lawrence's Warbler, as well as variations of the Brewster's plumage and parental plumages, result from the back crossing of a hybrid with one of the parent species, or from a rare mating of 2 hybrids. Parkes (1951) and Gill (1980) provide a more detailed discussion of the genetics of this complex.

The breeding distribution of the Lawrence's Warbler has followed the changes in distribution of the parent species. It is generally found in those regions where both parent species are found, although it also shows up in areas where the establishment of the Blue-winged Warbler and the disappearance of the Golden-winged Warbler are recent events (Gill 1980). Like the parent species, it is found in old fields reclaimed by secondary deciduous growth, along stream edges, and in other dense young habitats, especially if moist.

The Lawrence's Warbler is evidently as rare in Ontario as it is elsewhere in its range. Speirs (1985) cites only 7 Ontario sight records. All but the first individual near Kingston in July 1947 were noted in May, and 4 of the sightings were made at Point Pelee Nat. Park. There were 2 additional migration records at Long Point in 1985.

During the last 2 years of atlas field work, the Lawrence's Warbler was found on 4 occasions at 3 widely separated localities. The bird north of Kingston and the Muskoka indi-

viduals were at the edge of the Blue-winged Warbler's range, while the bird at Cayuga (35 km south of Hamilton) was in Ontario's Blue-winged Warbler stronghold. A few individuals may have been overlooked because the Lawrence's Warbler can sing a song identical to either parent species, or because of impenetrable habitats or the unfamiliarity of the species to most observers. As Blue-winged and Golden-winged Warblers continue to undergo range changes and to hybridize in Ontario, we can expect to witness accompanying changes in the status and distribution of the Lawrence's Warbler.-- *A. Mills*

LAWRENCE'S WARBLER

Summary of Abundance Estimates

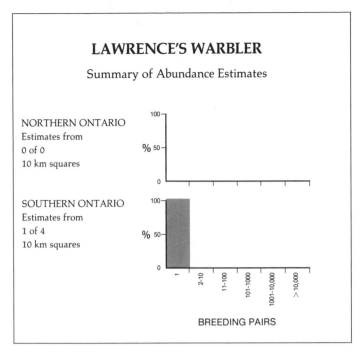

NORTHERN ONTARIO
Estimates from
0 of 0
10 km squares

SOUTHERN ONTARIO
Estimates from
1 of 4
10 km squares

BREEDING PAIRS

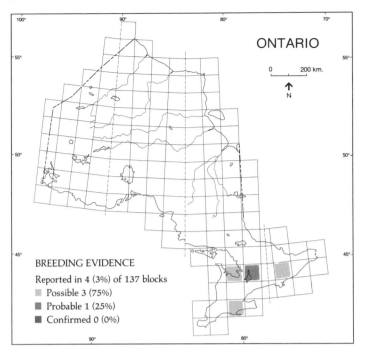

ONTARIO

0   200 km.

N

BREEDING EVIDENCE
Reported in 4 (3%) of 137 blocks
Possible 3 (75%)
Probable 1 (25%)
Confirmed 0 (0%)

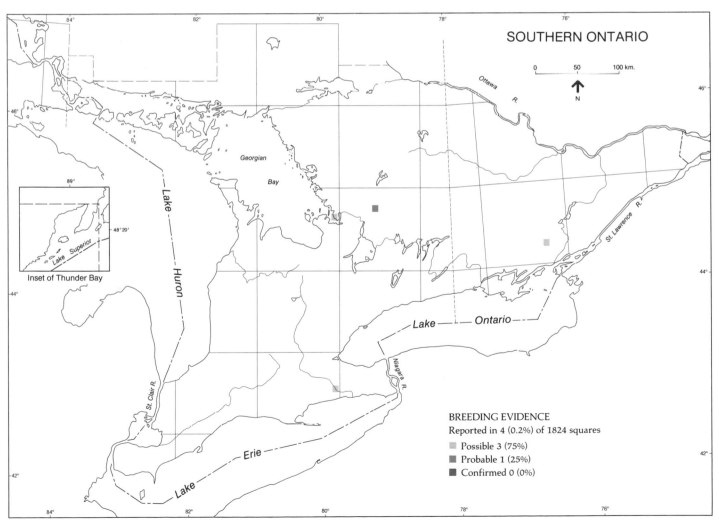

SOUTHERN ONTARIO

0   50   100 km.

N

Georgian
Bay

Ottawa R.

St. Lawrence R.

Lake Huron

Lake

Lake   Ontario

Niagara R.

St. Clair R.

Lake   Erie

Lake

89°
48°20'
Lake Superior
Inset of Thunder Bay

BREEDING EVIDENCE
Reported in 4 (0.2%) of 1824 squares
Possible 3 (75%)
Probable 1 (25%)
Confirmed 0 (0%)

TENNESSEE WARBLER
Paruline obscure
Vermivora peregrina

The plainly marked Tennessee Warbler is one of the more common wood warblers of the forested regions of Ontario. Occurring in coniferous, deciduous, and mixed forests, alder and willow swales, and cutovers, the species has a breeding range that is almost exclusively Canadian, from central Newfoundland to the Pacific coast, excluding the southern portions of all provinces from Quebec westward. In Ontario, it breeds regularly in central and northern areas of the Canadian Shield and on the Hudson Bay Lowland. The Tennessee Warbler is adept at exploiting spruce budworm epidemics, to which it responds both functionally through immigration, and by true numeric increase by enlarging clutch size by 50% or more. It winters from southern Mexico to Venezuela (Godfrey 1986).

J. Hughes-Samuel, quoted by Macoun and Macoun (1909), disagreed with the general opinion of the day that the Tennessee Warbler was rare. He believed it was common but often overlooked because of its dull colouration. The species was first confirmed as a breeder from Michipicoten before 1874, according to Baillie and Harrington (1937), who reported it as a common breeder north of 47°N. Manning (1952) found it along the Shagamu River on the Hudson Bay coast and it has also been reported from Winisk (Schueler *et al*. 1974) and Fort Severn (Hope 1940).

In spite of the Tennessee Warbler's drab colouration, the prominent eye stripe, thin bill, and active, warbler-like behaviour make the species readily identifiable, and its curious nature makes it readily observable. It has a persistent, loud, and monotonous 3-part song that is readily learned, and the species occurs so widely in the north that one never forgets it. It arrives on the breeding grounds in late May but seems to take a while to settle down to active territorial advertisement. Singing often occurs from dawn to mid-day or later, and frequently persists into early July. It is a ground nester and can readily be found feeding young from mid-June through the first 2 weeks of July.

Although most common in the mixed forest, it is found in many habitats. Welsh and Fillman (1980) recorded it in successional stages ranging from 3 year old lowland cutovers to mature lowland and upland forests. In the James Bay area, Welsh and Ross (unpubl.) found it common to locally abundant in all forested habitats from beach ridges (400 pairs/km^2) to rich, treed fens. In all habitats selected, there is a strong association with shrubs, particularly speckled alder (Welsh unpubl.).

The atlas maps support the historical range and are in close accord with the maps of Griscom and Sprunt (1979) and Godfrey (1986). The effect of the Tennessee Warbler's ubiquitous and vociferous nature is reflected in its recorded occurrence in 79% of the blocks in the province, of which more than half are confirmed records. The occasional gap in the north probably reflects inadequate coverage, with the exception of the 4 blocks in the Cape Henrietta Maria area, and one at West Pen Island on the Manitoba border, where it seems to be absent. Most individuals in the south presumably occurred in patches of coniferous forest within the Great Lakes-St. Lawrence Forest zone, as suggested by one Regional Coordinator. Some of the more southerly birds may be non-breeders, The low level of confirmation in southern Ontario supports the suggestion that some southern birds may be unattached males. Although the carrying of food was important (68% of all block data), confirmation of breeding was achieved by all other means as well.

The abundance records in the south reflect the uncommon status of the species south of the Boreal Forest transition zone. James *et al*. (1976) considered the species to be a common summer resident south to Algonquin Prov. Park.

The irregular abundance of the species and its rapid fluctuations in density can be seen in the changes that occurred from 1979 to 1983 at Manitouwadge, 70 km north of Pukaskwa Nat. Park (Welsh unpubl.). Overall, the population on all research plots increased by a factor of 9 in 4 years, and then decreased by 30% the next year. In mature forest, density ranged from 75 pairs/km^2 to a high of 610 pairs/km^2, while, during the same period on cutovers less than 10 years of age, the population went from 0 to 336 pairs/km^2.

The very high densities encountered both with and without budworm suggest that more than 10,000 pairs per square is probable in prime areas and most northern squares would be above 100 pairs. However, no estimates of more than 10,000 pairs were reported, and only 18 are of more than 1,000 pairs, which suggests that most atlassers underestimated the abundance of this species.-- *D.A. Welsh*

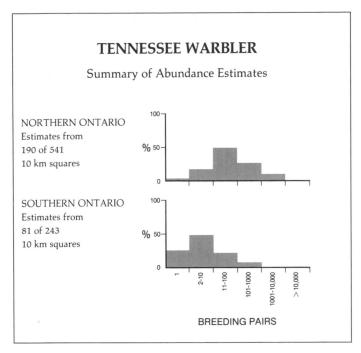

TENNESSEE WARBLER

Summary of Abundance Estimates

NORTHERN ONTARIO
Estimates from
190 of 541
10 km squares

SOUTHERN ONTARIO
Estimates from
81 of 243
10 km squares

BREEDING PAIRS

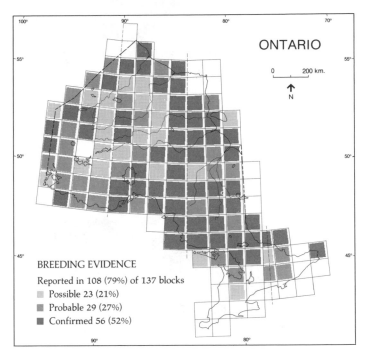

ONTARIO

0   200 km.

N

BREEDING EVIDENCE
Reported in 108 (79%) of 137 blocks
Possible 23 (21%)
Probable 29 (27%)
Confirmed 56 (52%)

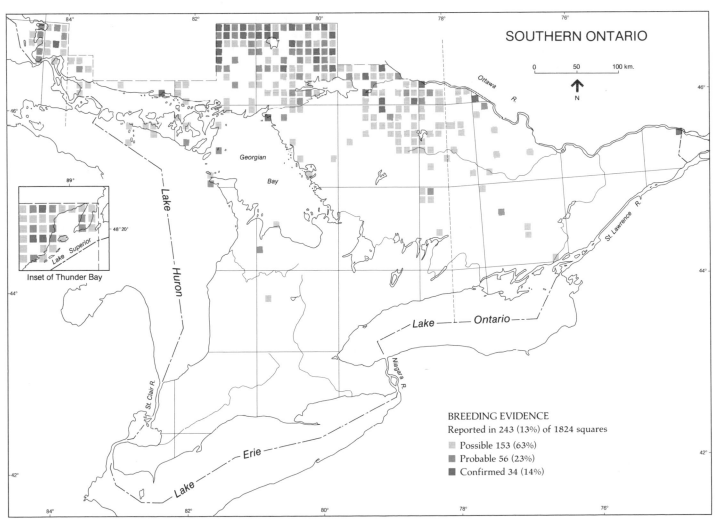

SOUTHERN ONTARIO

0   50   100 km.

N

Ottawa R.

Lake Huron

Georgian Bay

St. Lawrence R.

Inset of Thunder Bay

Lake Superior

48°20'

89°

St. Clair R.

Niagara R.

Lake — Ontario

Lake — Erie

BREEDING EVIDENCE
Reported in 243 (13%) of 1824 squares
Possible 153 (63%)
Probable 56 (23%)
Confirmed 34 (14%)

NASHVILLE WARBLER
Paruline à joues grises
Vermivora ruficapilla

The Nashville Warbler breeds across much of southern Canada and the northern US. It is a common summer resident throughout much of the northern and central portions of Ontario. Migrating from its wintering areas in Mexico and Guatemala, it is one of the earliest warblers to arrive in Ontario in spring. Breeding evidence was obtained from 71% of the squares in the south, and from 71% of the blocks in all of Ontario. This widespread distribution is undoubtedly due to its ability to use a wide variety of habitats. Although second-growth, mixed woods appear to be the preferred habitat, the Nashville Warbler has also been recorded in old field habitat, cedar-spruce swamps, and wet coniferous woods (Speirs 1985).

Like other species that are commonly found in second-growth mixed forest and old field habitats, it seems likely that the Nashville Warbler may be considerably more common at present than in former years, at least in central Ontario, where it has benefitted from extensive clearing and lumbering operations over the past century. On the other hand, the extensive development of farmland in the southwestern portion of the province may have caused the local extirpation of the breeding population in that region. The Nashville Warbler is a common nesting species in areas of suitable habitat in Michigan and New York adjacent to southwestern Ontario.

The song of the Nashville Warbler is characteristic, carries well, and is easily identifiable. Like other warbler species, males are generally very conspicuous during courtship and the early stages of nesting, moving constantly through their territories and singing regularly. Females, on the other hand, are generally inconspicuous, remaining in low vegetation near their nests. Nests are built on the ground and are generally well concealed under a shrub or fern. The difficulty of finding nests (before the eggs hatch) is evidenced by the fact that atlassers found nests with eggs in only 7 squares in the south and in 2 northern squares. After hatching, both parents participate in feeding the nestlings and adults carrying food can commonly be seen. Confirmed breeding was reported in 27% of the squares in which the species was found. Most of the confirmed evidence was of parents carrying food, or of fledged young, rather than nest-associated events. Singing or territorial males constituted the highest breeding evidence in 48% of southern Ontario squares, and 53% of all of the squares in the province from which the species was reported.

James *et al.* (1976) indicated that the species occurred south to the London area and north to Moosonee. The atlas data agree with this, with some evidence of breeding beyond both extremes. In the Carolinian Forest zone of southwestern Ontario, nesting is local and perhaps occasional. Of the 32 squares in the 3 most southerly blocks where some evidence of breeding was obtained, breeding was confirmed in only 5 squares (16%), a substantially lower value than the 32% (410 of 1289) in all of southern Ontario; this probably reflects the low breeding density of the species in this area. Much of the breeding in the southernmost areas occurs in cedar swamps, possibly a reflection of the lack of second-growth, open, mixed forest that the species apparently prefers. Records of singing males and agitated behaviour in the Pinery Prov. Park are noteworthy.

Possible breeding evidence, primarily in the form of singing males, in some far northern blocks north of the Canadian Shield, suggests that the nesting distribution of this species extends farther north than previously thought. McLaren and McLaren (1981a) noted that Nashville Warblers were commonly heard singing in the Wetiko Hills in the far northwest, but Godfrey (1986) considers the area north of the Albany River to be almost entirely outside the range of this species.

Atlassers provided estimates of breeding abundance for 47% and 34% of the squares in the south and north where some evidence of breeding by Nashville Warblers was obtained. In 98% of these squares, the abundance estimates were between 1 and 1,000 pairs and in 73% of them the estimates were between 2 and 100 pairs. All of the highest values, those over 1,000 pairs per square, were from the Canadian Shield north of a line from Parry Sound to Perth, through Algonquin Prov. Park to Timmins. This is consistent with BBS data, which show particularly high densities in the Lake of the Woods area and in a band running east from Lake Superior to the Quebec border (Speirs 1985). James (1980) recorded a density of 10.4 pairs/100 ha near Pickle Lake; densities of up to 206 territories/100 ha have been recorded in northeastern Ontario (Welsh and Fillman 1980). Martin (1960) calculated that the average size of a territory of the Nashville Warbler in Algonquin Prov. Park was about 0.4 ha.-- *P.L. McLaren*

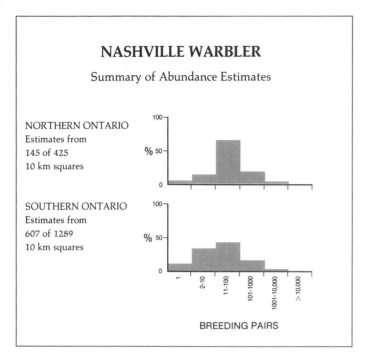

NASHVILLE WARBLER

Summary of Abundance Estimates

NORTHERN ONTARIO
Estimates from
145 of 425
10 km squares

SOUTHERN ONTARIO
Estimates from
607 of 1289
10 km squares

BREEDING PAIRS

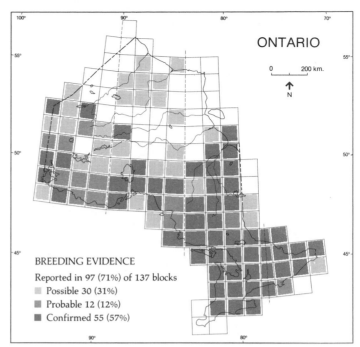

ONTARIO

0   200 km.

N

BREEDING EVIDENCE
Reported in 97 (71%) of 137 blocks
Possible 30 (31%)
Probable 12 (12%)
Confirmed 55 (57%)

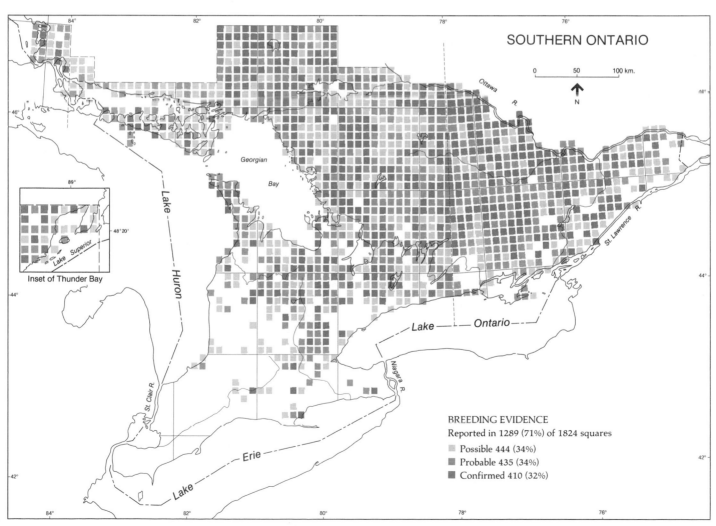

SOUTHERN ONTARIO

0   50   100 km.

N

Inset of Thunder Bay

BREEDING EVIDENCE
Reported in 1289 (71%) of 1824 squares
Possible 444 (34%)
Probable 435 (34%)
Confirmed 410 (32%)

NORTHERN PARULA
Paruline à collier
Parula americana

This diminutive but colourful member of the wood warbler family is one of Ontario's least-known warbler species, both because it is not at all common and because its habitat and behaviour make it difficult to find. For many years, the Northern Parula had a certain notoriety in Ontario birding circles for having successfully eluded all attempts to find its nest in the province (Baillie and Harrington 1937).

The Northern Parula spends the summer throughout eastern North America from southern Canada to the Gulf of Mexico and the winter in Central America and the West Indies (AOU 1983). There is a concentration of the species in the Adirondack region of New York State (NYBBA preliminary maps). In Ontario, the breeding range extends from the southern edge of the Canadian Shield and adjacent areas well into the Boreal Forest zone, but the species is almost exclusively a migrant in the rest of southern Ontario. The distribution during the breeding season is partially governed by the distribution of *Usnea* lichen, in which the Northern Parula builds its nest (Godfrey 1986). This type of vegetation commonly dangles from the branches of conifers in Ontario. However, where *Usnea* is scarce, a hanging cluster of hemlock or spruce may be used and some lichen added (Godfrey 1986).

There are no definite historical records of the Northern Parula which suggest any change in its range over the years. Vague terms in the early literature, like "abundant" or "moderately common" (Macoun and Macoun 1909), seem to indicate a subsequent decline in numbers, but the summer distribution records from the past are so few that no change in breeding range can be determined. Regional Coordinators disagree on whether the species has been declining or increasing in numbers in recent years. No nest of this species had been found in Ontario before 1943, largely because the nest is exceptionally well concealed.

Finding the Northern Parula on its breeding grounds can be a challenge. Its song is fairly distinctive but is rather weak and can easily be overlooked in a chorus of other bird songs. To see the tiny songster is even more difficult, because it frequents higher branches and often remains deep in a spruce swamp, shunning the 'edge' habitats in which birds are easier to locate. The majority (58%) of atlas records of this species are of singing males. Breeding was confirmed in only 6 squares in southern Ontario and 7 squares in the north. This is not surprising, since the nest is extremely difficult to find. It is usually "5 to 15 feet (1.5 to 4.5 m) above the ground," though a Quebec study found them to be somewhat higher (Bent 1953), and is constructed so as to resemble a hanging clump of moss or lichen. Nests were found in only 3 squares in all of Ontario.

The maps document particularly well the relative scarcity of the Northern Parula in Ontario. The southern Ontario map shows a scattered distribution, with the highest concentration of records in the Algonquin Highlands, but other records are thinly spread across much of the southern Shield. The species is apparently rare in agricultural areas south of the Shield, and was not confirmed as breeding in that area. Other concentrations are evident on the Bruce Peninsula, the southern edge of Manitoulin Island, and north of Sault Ste Marie. Even in the Algonquin region, which has long been considered the core of its Ontario breeding range, it was found in only 33% of the squares, but the species probably occurs in 50% or more of the squares in that region (R. Tozer pers. comm.). It was found in only about 10% of the squares visited within its breeding range. The actual observations probably coincide very closely with the existence of extensive spruce forests festooned with *Usnea*. In much of the Boreal Forest zone, however, some other unknown limiting factor seems to come into play, for, as one Regional Co-ordinator observed, there is plenty of *Usnea* on the spruces in the northern third of Ontario, but records of the Northern Parula are scarce at that latitude. Moreover, there are few records from that part of the province adjacent to the Manitoba border, which also has apparently suitable habitat. Several sightings shown on the map, notably one at Garret Lake (about 54°N) near the edge of the Hudson Bay Lowland, represent significant northward range extensions. There can be little doubt that the species breeds in many of the areas where it is shown as only possibly or probably breeding, even in those northern areas in which it had not previously been found at all.

Atlassers who attempted to estimate the abundance of Northern Parulas generally estimated from 2 to 10 pairs in their squares. In a few cases, estimates range in the hundreds. The highest estimates, those of over 100 pairs per square, are from the Haliburton and Nipissing areas. The small number of atlas records and the generally low abundance estimates indicate that the Northern Parula is at best an uncommon bird in Ontario.-- *F.M. Helleiner*

NORTHERN PARULA

Summary of Abundance Estimates

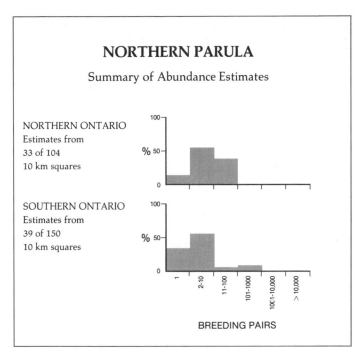

NORTHERN ONTARIO
Estimates from
33 of 104
10 km squares

SOUTHERN ONTARIO
Estimates from
39 of 150
10 km squares

BREEDING PAIRS

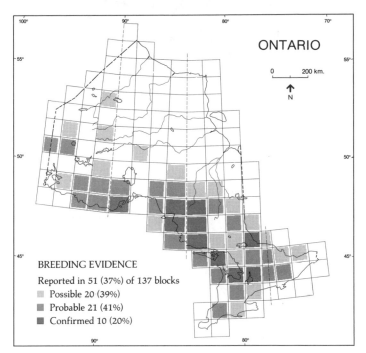

ONTARIO

0   200 km.

N

BREEDING EVIDENCE

Reported in 51 (37%) of 137 blocks

Possible 20 (39%)
Probable 21 (41%)
Confirmed 10 (20%)

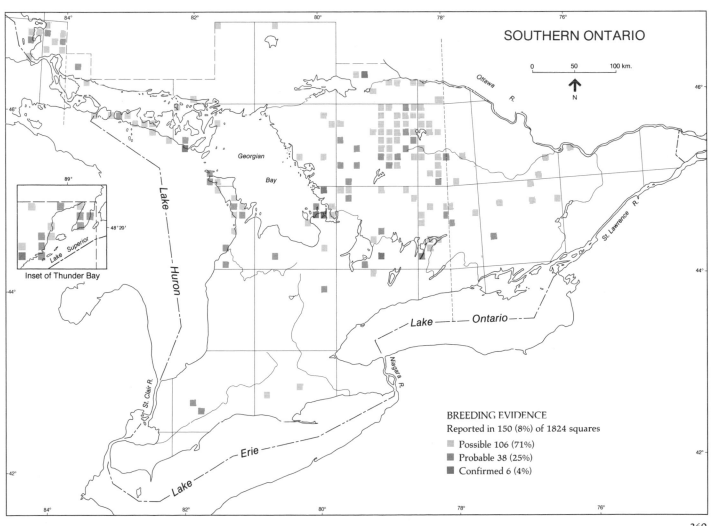

SOUTHERN ONTARIO

0   50   100 km.

N

Inset of Thunder Bay

BREEDING EVIDENCE

Reported in 150 (8%) of 1824 squares

Possible 106 (71%)
Probable 38 (25%)
Confirmed 6 (4%)

YELLOW WARBLER
Paruline jaune
Dendroica petechia

This colourful little songster is perhaps one of the commonest warblers throughout North America. Easily identified by sight, and, with a little practice, by its song, its presence is acknowledged and appreciated by most observers.

In the breeding season, the Yellow Warbler ranges across much of North America from Alaska to Newfoundland, south to Baja California and the southeastern US, and to the West Indies and northern South America (AOU 1983, Godfrey 1986). It breeds throughout Ontario. Favouring moist habitats (Bent 1953), it is often a bird of suburban yards and relatively open areas with dense scrub. Dense forested stands are avoided by the species.

The Yellow Warbler unquestionably benefitted as land was logged and cleared for agriculture, giving rise to more of the scrubby areas that are its preferred habitat. Taverner (1919) described it as the commonest breeding warbler in southern Canada. The eastward spread of the Brown-headed Cowbird from the central plains, following the clearing of forest lands in the last century, has been extensively documented. However, the impact of this brood parasite on the Yellow Warbler is unclear. Some parasitic attempts are successful, but perhaps not as frequently as might be expected. Bent (1953) and Taverner (1919) point out an intriguing behavioural adaptation of the species to offset this pressure. Yellow Warblers frequently abandon nests or build two-storey ones if parasitized. If 2 or 3 warbler eggs have been laid before the cowbird egg is laid, incubation will probably proceed. Interestingly, H.F. Perkins (in Bent 1953) noted that of the 43 Yellow Warbler nests found in areas frequented by Red-winged Blackbirds, only one was parasitized by cowbirds. He attributed this low proportion to an intolerance by Red-winged Blackbirds of cowbirds near their nests.

Highly visible, vocal and common, the Yellow Warbler was easily found by atlassers, as evidenced by its being recorded in 92% of the blocks in Ontario and 93% of the squares in the south. The similarity of its song to one of the songs of the Chestnut-sided Warbler, which occupies similar habitats, may lead to its being overlooked in areas where its population is low. The birds aggressively defend their territories against most intruders. Breeding is easily confirmed because the scrubby areas where the Yellow Warbler breeds are often open, making the birds easy to observe and follow. Spectacular distraction displays are performed close to the nest. Both parents make frequent trips to feed the young in the nest and for a period after the young fledge. An analysis of the southern Ontario data reveals that 12% of the records involve nests with eggs, 8% are of nests with young, and 22% represent birds carrying food or faecal sacs. Breeding was confirmed in over half (51%) of the squares in which the Yellow Warbler was reported in the province.

Atlas data show that the Yellow Warbler is one of Ontario's most widespread species. The few gaps in the Ontario map, with the exception of some northern tundra locations, probably represent areas where the species was overlooked. Breeding could probably have been confirmed in almost all 100 km blocks. The Yellow Warbler is often found in wetland shrubby areas, such as along river banks, across northern Ontario. In southern Ontario, the species was found in almost every square, except for the heavily wooded interior of Algonquin Prov. Park and marginal squares along the shores of the Great Lakes.

Abundance estimates indicate that the Yellow Warbler is generally common throughout Ontario. The estimates show a somewhat higher population density in the south, probably reflecting the fact that human activity tends to create good Yellow Warbler habitat.-- *G. Carpentier*

YELLOW WARBLER

Summary of Abundance Estimates

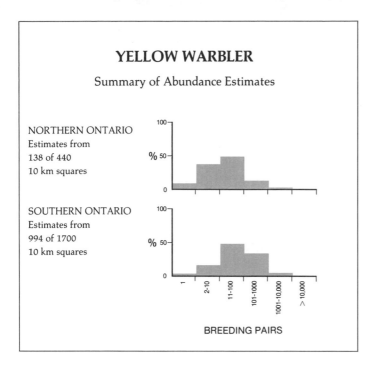

NORTHERN ONTARIO
Estimates from
138 of 440
10 km squares

SOUTHERN ONTARIO
Estimates from
994 of 1700
10 km squares

BREEDING PAIRS

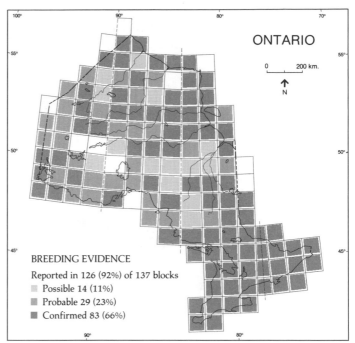

ONTARIO

0   200 km.

N

BREEDING EVIDENCE

Reported in 126 (92%) of 137 blocks

- Possible 14 (11%)
- Probable 29 (23%)
- Confirmed 83 (66%)

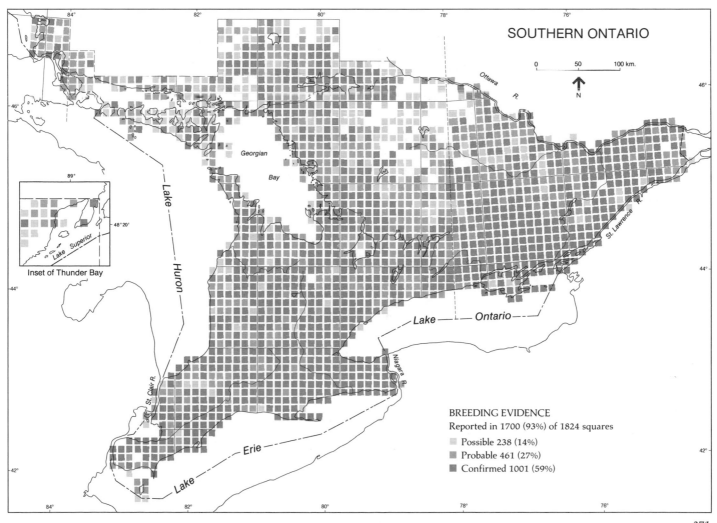

SOUTHERN ONTARIO

0   50   100 km.

N

Georgian Bay

Ottawa R.

St. Lawrence R.

Lake Huron

Lake Ontario

Niagara R.

St. Clair R.

Lake Erie

Lake Superior

Inset of Thunder Bay

BREEDING EVIDENCE

Reported in 1700 (93%) of 1824 squares

- Possible 238 (14%)
- Probable 461 (27%)
- Confirmed 1001 (59%)

CHESTNUT-SIDED WARBLER
Paruline à flancs marron
Dendroica pensylvanica

If you had been a contemporary of Audubon or Wilson, you might never have had the pleasure of seeing the Chestnut-sided Warbler. In a lifetime of exploring and collecting, Audubon saw but one and Wilson only a handful (Bent 1953). The present widespread distribution and abundance of this species are a product, for the most part, of the land-use patterns that have occurred since the arrival of European settlers.

At present, the Chestnut-sided Warbler is common in northeastern North America wherever suitable habitat occurs. The species spends the winter in Central America. The preferred breeding habitat is shrubby second-growth deciduous forest found adjacent to stands of mature forest where regeneration is well-advanced. The Chestnut-sided Warbler can be found in a variety of situations - both man-made and natural. While forest edges, abandoned orchards, small clearings, and hedgerows all provide attractive and suitable habitat, this warbler is most abundant in areas where large tracts of agricultural land have been abandoned and where the forest is in the process of regeneration after fire or logging. The Chestnut-sided Warbler can be found throughout Ontario in any of these situations.

Because Ontario was mostly forested prior to the arrival of European settlers, the Chestnut-sided Warbler must have been a rarer breeder in the province than it is today and may have been common only in areas that were regenerating after natural fires. Population increases probably kept pace with the clearing of the land as forest edges were created. However, it was only when marginal land was abandoned and logged areas began to regenerate that a population explosion occurred. In 1901, it was labelled as an abundant summer resident in southcentral Ontario by Fleming (1901) - an area which had been cleared in the mid- to late 1800's. Today, its

numbers may be declining in some areas. In Algonquin Prov. Park it was "even more common in the past when burns and young forests were more widespread" (R. Tozer pers. comm.). The continued presence of this species as a common breeder in the province seems likely for some time, considering present land-use patterns.

To the unpractised ear, the song of the Chestnut-sided Warbler can often resemble that of the Yellow Warbler, perhaps causing it to be overlooked in a few squares, especially south of the Shield where the Yellow Warbler is much more abundant and widespread. Habitat preferences of these 2 species are also similar. In a Minnesota study of the habitat relationships of 16 species of wood warblers, the Chestnut-sided Warbler was found to have very similar habitat preferences to those of the Yellow Warbler (Collins *et al.* 1982). Confirmed evidence of breeding was primarily in the form of adults carrying food or of fledged young. The number of nests found in the south (3% of the total breeding evidence) is low considering that the species generally nests within 1 m of the ground (McLaren 1975, Harrison 1984), and that these nests are quite easy to locate.

The northern limit of confirmed and probable breeding is at about 51^0N. Two confirmed breeding records in blocks containing the Moose and Albany Rivers east of Moosonee are on the southern section of the Hudson Bay Lowland and are more than 100 km north of the northern range limit mapped by Godfrey (1986) and Peck and James (1987). Scattered possible records indicate that the breeding range may extend still farther north, with one exceptional outlying record of a singing male on the east bank of the Fawn River in the northern section of the Hudson Bay Lowland. Nevertheless, the Chestnut-sided Warbler is primarily a bird of the Great Lakes-St. Lawrence Forest zone and its penetration of the Boreal Forest is limited by the presence of deciduous tress in suitable openings. BBS data indicate that it reaches it greatest abundance on the Canadian Shield (Speirs 1985). South of the Shield, its distribution becomes progressively more scattered and there are large unexplained gaps in Prince Edward Co and in the vicinity of Lakes Simcoe and Scugog. In the extreme southwest its absence is presumably related to a lack of suitable habitat in areas of intensive agriculture. The prevalence of possible records in the southwest is consistent with the sparse population indicated by BBS counts (Speirs 1985).

Abundance estimates suggest that the Chestnut-sided Warbler is a common species, being most abundant in the central portions of the province. The higher estimates, those of over 1,000 pairs per square, came from the Canadian Shield: from Muskoka and Haliburton in the south to Lake Temagami in the north. The single abundance estimate of over 10,000 pairs came from a square 20 km east of North Bay, presumably in an extensive area of highly suitable habitat. Ontario has the highest average BBS counts for this species of any state or province (Robbins *et al.* 1986).-- *D. Martin*

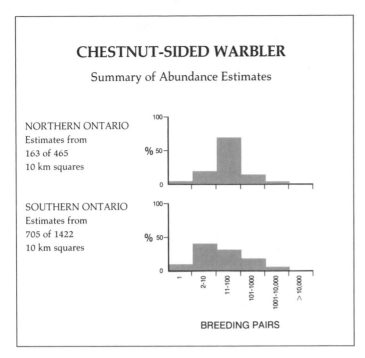

CHESTNUT-SIDED WARBLER

Summary of Abundance Estimates

NORTHERN ONTARIO
Estimates from
163 of 465
10 km squares

SOUTHERN ONTARIO
Estimates from
705 of 1422
10 km squares

BREEDING PAIRS

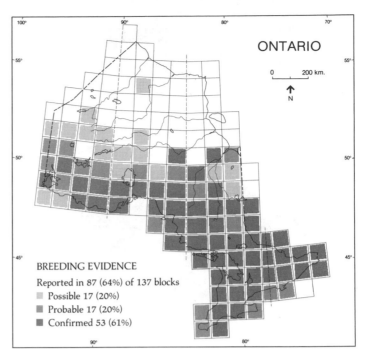

ONTARIO

0   200 km.

N

BREEDING EVIDENCE
Reported in 87 (64%) of 137 blocks
Possible 17 (20%)
Probable 17 (20%)
Confirmed 53 (61%)

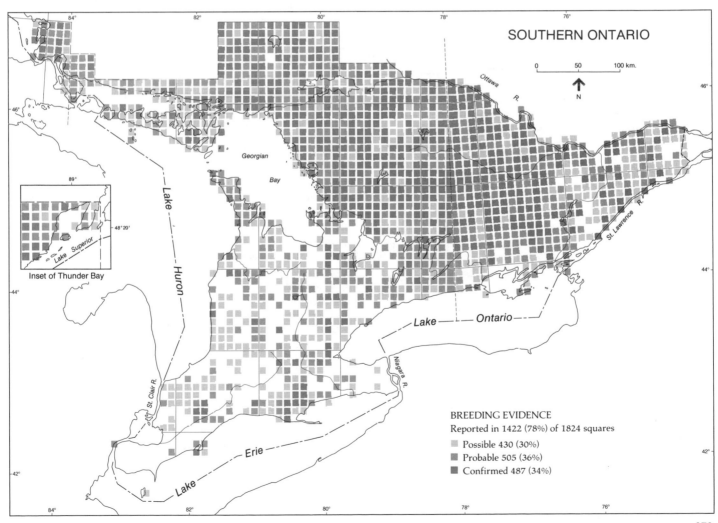

SOUTHERN ONTARIO

0   50   100 km.

N

Inset of Thunder Bay

BREEDING EVIDENCE
Reported in 1422 (78%) of 1824 squares

Possible 430 (30%)
Probable 505 (36%)
Confirmed 487 (34%)

MAGNOLIA WARBLER
Paruline à tête cendrée
Dendroica magnolia

One of the most common and widespread breeding warblers of the province is the Magnolia Warbler. The species is broadly distributed in Canada and in the northern US east of Michigan. Its Canadian breeding distribution stretches from Newfoundland to British Columbia and the western Mackenzie District. In Ontario, this brilliant 'Black-and-Yellow Warbler', as it was once called, breeds throughout the Canadian Shield, the Hudson Bay Lowland, the Boreal Forest, and northern portions of the Great Lakes-St. Lawrence Forest zones. Evidence of breeding was found in 81% of the atlas blocks, significant areas of absence occurring only in the far north along Hudson Bay and in the south along Lakes Erie and Ontario.

In 1909, Macoun and Macoun reported that the Magnolia Warbler bred from Moose Factory on Hudson Bay south to the Bruce Peninsula and Ottawa. James *et al.* (1976) considered it common in Ontario from the tree-line south to Guelph. The species was common in upper New York State during their atlas period (1980 to 1986) (NYBBA preliminary maps).

The Magnolia Warbler is found in mixed and coniferous forests, particularly along edges and in natural openings. Its single strongest habitat association is with small white spruce and balsam fir. A listing of preferred habitat in Ontario would include natural canopy breaks caused by rock outcrops, streams, windfalls, and roadcuts within mature spruce-fir and mixed forests. Edges of woodlands, pastures and old fields with young spruce and balsam fir, powerline corridors, and disturbed woodland are all potential breeding sites. The species quickly replaces the Chestnut-sided Warbler in cutover forest and mixed woods when young balsam fir and spruce grow through the shrub layer. All of these habitats can be readily sampled by atlassers. The association of the species with roads, trails, and river edges makes it even more likely to be found.

Early in the season, male Magnolia Warblers are particularly conspicuous as they sing incessantly and often display at eye level with their tail fanned and their wings spread to show off their brilliant colouration. The Magnolia Warbler tends to concentrate its activities, particularly foraging, within 6 m of the ground (McLaren 1975) and is consequently readily detectable by eye. Additionally, it sings a loud, distinctive, easily identifiable song from conspicuous perches. The only potentially confusing attribute of the species is a tendency to alter its song as the season progresses from an emphatic and definite song to a much less definite and more tentative rendition. Nests are usually low (0.25 to 2 m) in small conifers and have been recorded in almost all parts of the province. The species is noisy and obvious for the first few days after the young leave the nest, as both parents attend to the new fledglings.

The maps appear to reflect accurately the expected widespread range. The data are remarkable in the high level of breeding confirmation throughout Ontario and the exceptional frequency of records in the Great Lakes-St. Lawrence Forest zone. The general distribution pattern to the northern sections of the Hudson Bay Lowland Forest zone supports recently published records and indicates the regular occurrence of the species nearer to the tree-line than is shown on previous maps like that of Godfrey (1986), which indicate a more limited range. With more time it seems certain that all of the northern blocks within the mapped range would have had confirmed breeding records. An overall pattern of increasing density from south to north is clear in the southern Ontario map. Of particular interest are the records, including numerous breeding confirmations, south of the Canadian Shield. The southernmost records are near Long Point (McCracken 1987). Two Regional Coordinators suggested that suitable habitat for the species in the south is increasing as red and scotch pine plantations mature. However, it is difficult to assess whether or not there has been a corresponding southward range expansion, because documentation of the previous range is relatively incomplete; moreover, there is some evidence that the species may formerly have been more widespread in the southwest (Sutherland 1986).

The abundance estimates reflect the widespread occurrence of this species in the northern half of southern Ontario, and in northern Ontario. Martin (1960) recorded a density equivalent to 10,250 pairs per 100 km^2 (extrapolated from much smaller units) in Algonquin Prov. Park. On the Canadian Shield section of the Boreal Forest zone, 1,000 pairs per square would be usual, and densities of at least several thousand pairs per 100 km^2 would certainly occur in cutover and disturbed areas. The Magnolia Warbler was one of the most commonly recorded breeding birds in cutover boreal forest in northeastern Ontario (Welsh and Fillman 1980). Abundance levels are undoubtedly lower in the undisturbed and flatter regions of the Hudson Bay Lowland.-- *D.A. Welsh*

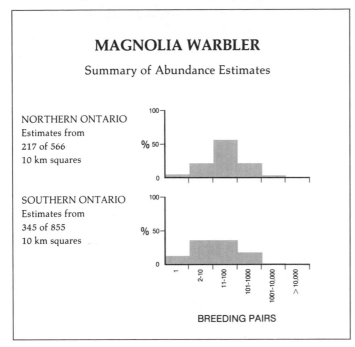

MAGNOLIA WARBLER

Summary of Abundance Estimates

NORTHERN ONTARIO
Estimates from
217 of 566
10 km squares

SOUTHERN ONTARIO
Estimates from
345 of 855
10 km squares

%

BREEDING PAIRS

1 2-10 11-100 101-1000 1001-10,000 >10,000

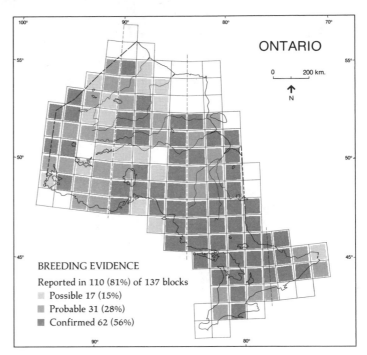

ONTARIO

0   200 km.

N

BREEDING EVIDENCE
Reported in 110 (81%) of 137 blocks
Possible 17 (15%)
Probable 31 (28%)
Confirmed 62 (56%)

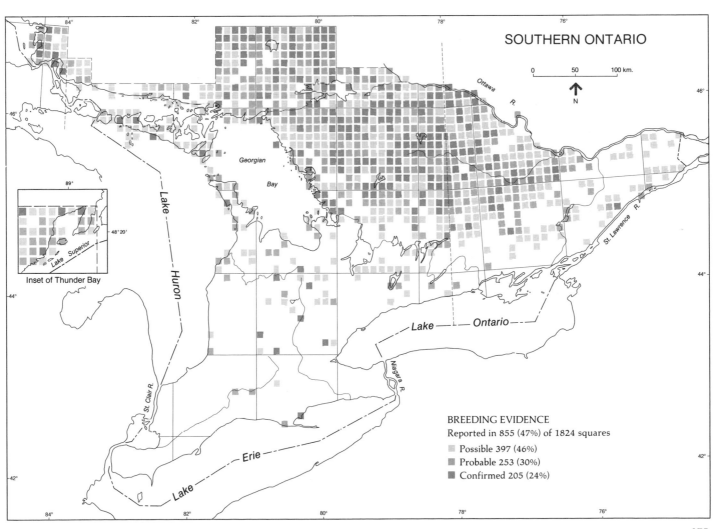

SOUTHERN ONTARIO

0   50   100 km.

N

Georgian Bay

Ottawa R.

Lake Huron

Lake Superior

89°
48°20'

Inset of Thunder Bay

St. Clair R.

Niagara R.

Lake — Ontario

St. Lawrence R.

Lake — Erie

BREEDING EVIDENCE
Reported in 855 (47%) of 1824 squares
Possible 397 (46%)
Probable 253 (30%)
Confirmed 205 (24%)

CAPE MAY WARBLER
Paruline tigrée
Dendroica tigrina

The Cape May Warbler, a tree top singer and nester of the coniferous forest, is a spruce budworm specialist like the Bay-breasted and Tennessee Warblers. The range of the species extends from eastern British Columbia to eastern Nova Scotia but it is most common in Ontario, Quebec, and the Maritimes, where it reaches peak densities coincident with spruce budworm outbreaks. It occurs across Ontario from Algonquin Prov. Park in the south to Moosonee and Sachigo Lake in the north. Its range in the province rather closely matches the Canadian Shield.

Historically, the species has been considered rare in the province (Macoun and Macoun 1909, Baillie and Harrington 1937). Early breeding records are from Bruce Co (1934) and Sudbury District (1937) (Baillie and Harrington 1937). Smith (1957) considered it rare in the Clay Belt, and MacLulich (1938) and Martin (1960) recorded it uncommonly in Algonquin Prov. Park. Historical records exist for the Ottawa area, and Baillie and Harrington (1937) claimed it was common in certain places on the Bruce Peninsula.

Welsh and Ross (unpubl.) found it on 8 of 32 James Bay coastal zone sites in 1978, where it occurred exclusively in mature conifer stands. Kendeigh's (1947) study of bird populations during a spruce budworm outbreak was the first to report high breeding densities: 70 pairs/km^2. Sanders' (1970) follow-up study 20 years later in the same area near Black Sturgeon Lake did not record the species at all. Welsh (unpubl.) found 148 pairs/km^2 in 1983 on Sanders' study plots. In another area, the density increased from 34 pairs/km^2 in 1979 to 370 pairs/km^2 4 years later, but in 1983 the species had declined to 243 pairs/km^2, and in 1986, during a visit in late June, the species was almost impossible to locate. Several Regional Coordinators commented on declines in their regions during the atlas period.

Of the boreal forest warblers, the Cape May is certainly one of the more specialized in its habitat selection. It normally breeds only in mature coniferous and mixed forests, selecting the tallest trees to use as song posts. The song is somewhat thin, is low in volume, and is normally sung from the top of a tree at least 10 m high. Nests have been found in both black and white spruce and balsam fir from 10 m to over 20 m in height, usually near the top of the tree in heavy foliage. In a 5 year study at Manitouwadge, Welsh (unpubl.) found the species only in stands greater than 50 years of age, and greater than 15 m in height.

This particularly elegant species sings actively for only the first few weeks after its arrival on territory. By early June in Maine, it is often silent for long periods (Griscom 1938). The song could be confused with that of other species, and the Cape May is normally quiet when the late-arriving, and similar sounding Bay-breasted Warbler is singing most actively. Most Regional Coordinators believe that the species is underrepresented in the atlas data for their region because of difficulties in detection, and several commented on problems with confirmation of breeding. The easiest method is to locate parents carrying food to young. Food-carrying accounted for 86% of all breeding confirmations, and, as might be expected, only one nest-related encounter occurred.

The maps of Cape May Warbler distribution accurately reflect the range in the province, but it seems likely that it was missed in a number of blocks in the north where it actually bred. A detailed examination of the map, along with accessory habitat information and known records, suggests that it may occur in as many as 27 blocks where no records were obtained. The fact that most northern atlassing was conducted relatively late in the season in 1984 and 1985, at a time well after active singing, and during years when the population may have been declining, could explain the apparent gaps in the mapped distribution. Like the Bay-breasted Warbler, the distribution of the species in the south coincides with the high relief rocky areas of the Great Lakes-St. Lawrence Forest zone which abound in spruce and fir. The map is somewhat remarkable in the rather high frequency of occurrence in Algonquin Prov. Park and the northern part of southern Ontario. One might have expected a similar distribution with less frequent occurrence. The scarcity of records from the Bruce Peninsula and the Ottawa area is noteworthy.

The abundance estimates reasonably reflect density for the southern distribution of the species. As expected, estimates from the north average higher. In its principal range in northern Ontario, the studies cited above indicate that most squares would be expected to contain an average of 100 to 1,000 pairs, with occasional squares containing between 1,000 and 10,000 pairs.-- *D.A. Welsh*

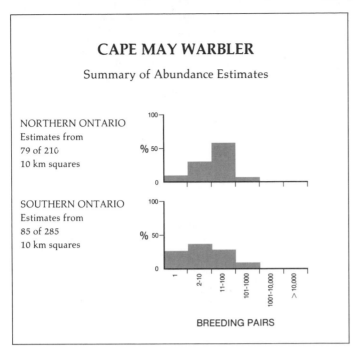

CAPE MAY WARBLER

Summary of Abundance Estimates

NORTHERN ONTARIO
Estimates from
79 of 210
10 km squares

SOUTHERN ONTARIO
Estimates from
85 of 285
10 km squares

BREEDING PAIRS

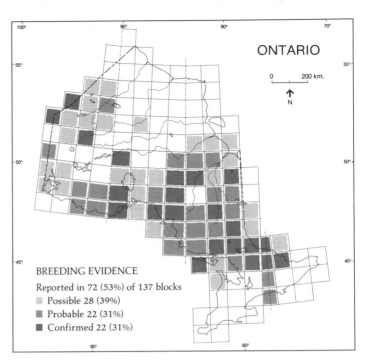

ONTARIO

0   200 km.

N

BREEDING EVIDENCE
Reported in 72 (53%) of 137 blocks
Possible 28 (39%)
Probable 22 (31%)
Confirmed 22 (31%)

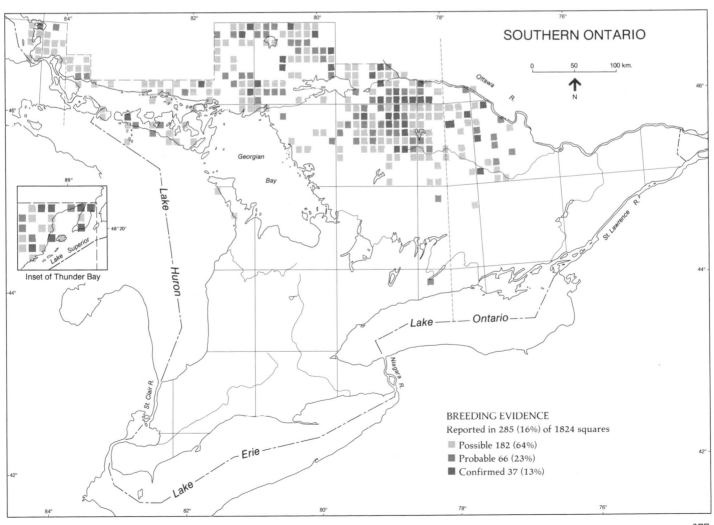

SOUTHERN ONTARIO

0   50   100 km.

N

Inset of Thunder Bay

BREEDING EVIDENCE
Reported in 285 (16%) of 1824 squares
Possible 182 (64%)
Probable 66 (23%)
Confirmed 37 (13%)

BLACK-THROATED BLUE WARBLER
Paruline bleue à gorge noire
Dendroica caerulescens

The breeding range of the Black-throated Blue Warbler extends from southeastern Manitoba to Nova Scotia, and south through the Appalachian Mountains to northeastern Georgia. Its distribution in Ontario is largely confined to the Canadian Shield. Upland mixed and deciduous forests with a dense understory of shrubs and saplings are the favoured breeding habitats of the Black-throated Blue Warbler. It also nests in brushy clearings, second growth woodland, and selectively logged areas. In Vermont, Laughlin and Kibbe (1985) found that these habitats are used for nesting provided that the canopy layer remains intact. The species winters mainly in the West Indies (Godfrey 1986).

The breeding distribution of the Black-throated Warbler in Ontario appears to have remained largely unchanged over the last century. The earliest published accounts, however, suggest that, in the southern parts of the province, this species was once more common than today. McIlwraith (1894) cites an 1886 nest from Listowel, Perth Co, and Macoun and Macoun (1909) mention an 1894 nest from Wildwood, Wellington Co, two areas from which the species is now apparently absent. Although McIlwraith (1894) acknowledged that Black-throated Blue Warblers occasionally breed in southern Ontario, he contended that "the majority unquestionably go farther north." According to Baillie and Harrington (1937), the species "occurs sparingly" in open deciduous woodland throughout its range.

The virtual disappearance of the Black-throated Blue Warbler as a breeding species in those parts of southern Ontario away from the Canadian Shield is largely attributable to widespread clearing of forests for agriculture. Its general absence from agricultural areas is also apparent from preliminary New York State atlas data.

The Black-throated Blue Warbler builds its nest close to the ground in dense undergrowth, usually at a height of less than 1 m (Harding 1931). The nest is difficult to find, accounting for the fact that only 2% of all southern Ontario records involve nests containing eggs or young. These factors may have contributed to the low level of breeding confirmation from the 703 squares in southern Ontario in which it was recorded. Several Regional Coordinators consider the Black-throated Blue Warbler to be underconfirmed in their regions.

Like many of the wood warblers, the Black-throated Blue Warbler spends a relatively short time on the breeding grounds. Nice (1930a) found that, on average, the period from initiation of nest construction to fledging of the young covers only 26 days. Adult birds, however, feed their young on territory for up to 10 days after fledging (Harding 1931). In 86% of all southern Ontario squares in which breeding was confirmed, the evidence involved fledged young or adults carrying food or faecal sacs.

Male Black-throated Blue Warblers are very vocal when advertising their territories, and, unlike most wood warblers, continue to sing for up to several weeks following completion of the fledgling feeding period (Black 1975). It is not surprising, therefore, that singing or other territorial behaviour accounted for 64% of all Ontario square records.

The northern limit of the Black-thoated Blue Warbler's breeding distribution in Ontario closely conforms to the northern boundary of the Great Lakes-St. Lawrence Forest region. The species is widely distributed within a zone extending from the southern edge of the Canadian Shield north to Kapuskasing, occurring in virtually every block, and in every square over large areas. It is apparently less widespread in a narrow band along the north shore of Lake Superior west to Thunder Bay. In the Rainy River District and the northwest generally there were very few records, all of possible breeding, and its range appears to be more restricted than indicated by Godfrey (1986).

Many of the squares in which breeding of the Black-throated Blue Warbler was confirmed are located in the Algonquin Highlands. The affinity of the species for areas of higher elevation is apparent in Vermont (Laughlin and Kibbe 1985) and New York (NYBBA preliminary maps). The species is absent as a breeding bird from the southwestern and southeastern parts of the province, where the land-use is predominantly agricultural.

Abundance estimates of over 1,000 pairs per square were all from the Haliburton Highlands and Algonquin Prov. Park areas. Estimates of 101 to 1,000 pairs per square were also from these areas, as well as from the Muskoka, Parry Sound, and Nipissing East atlas regions.-- *D.M. Fraser*

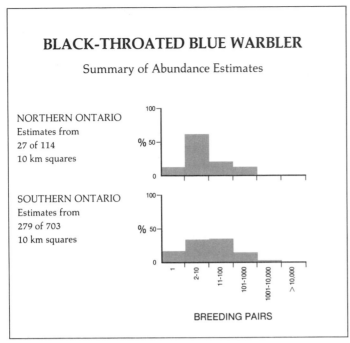

BLACK-THROATED BLUE WARBLER

Summary of Abundance Estimates

NORTHERN ONTARIO
Estimates from
27 of 114
10 km squares

SOUTHERN ONTARIO
Estimates from
279 of 703
10 km squares

% 100
    50
    0

% 100
    50
    0

1 2-10 11-100 101-1000 1001-10,000 > 10,000

BREEDING PAIRS

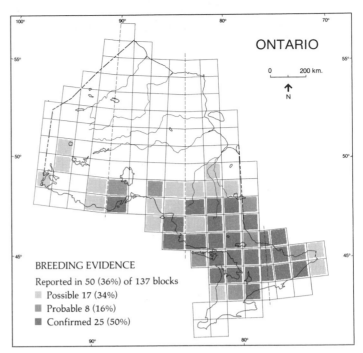

ONTARIO

0 200 km.

N

BREEDING EVIDENCE

Reported in 50 (36%) of 137 blocks

Possible 17 (34%)
Probable 8 (16%)
Confirmed 25 (50%)

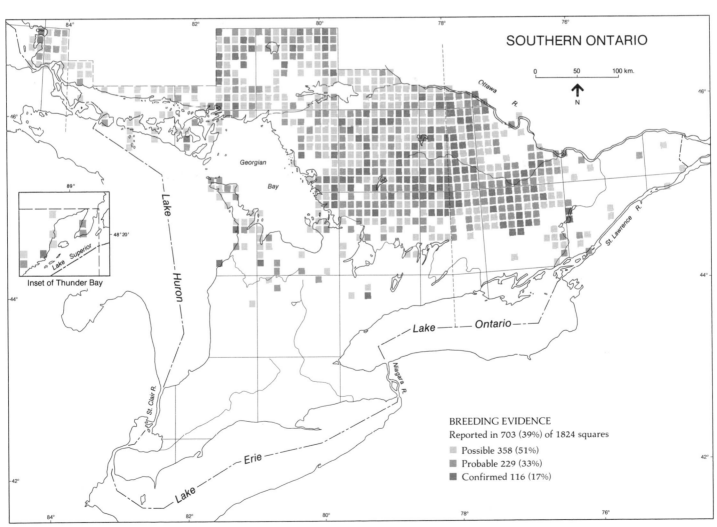

SOUTHERN ONTARIO

0 50 100 km.

N

Georgian Bay

Lake Huron

Lake Superior

89°
48°20'

Inset of Thunder Bay

St. Clair R.

Niagara R.

Lake — Ontario —

Lake — Erie —

Ottawa R.

St. Lawrence R.

BREEDING EVIDENCE

Reported in 703 (39%) of 1824 squares

Possible 358 (51%)
Probable 229 (33%)
Confirmed 116 (17%)

YELLOW-RUMPED WARBLER
Paruline à croupion jaune
Dendroica coronata

The Yellow-rumped Warbler is one of the most common, widespread, and familiar of all of the wood warblers. It is the earliest warbler to return in the spring, and the latest to leave in the autumn. Several often linger into the early winter in the extreme southern parts of the province, though the normal winter range is in the US, Mexico, Central America, and the West Indies. This account deals with the eastern subspecies of the Yellow-rumped Warbler, formerly known as the Myrtle Warbler. Its western relative, the Audubon's Warbler, is not known to breed in Ontario.

The Yellow-rumped Warbler is widespread as a breeding species throughout the wooded parts of Canada, and extends southward into the northern states. Although typically a Boreal Forest species, it is found in coniferous or mixed forests from the tree-line on the Hudson Bay Lowland south through the Great Lakes-St. Lawrence Forest region. It has also been reported as a breeding species in scattered localities within the Carolinian Forest region, in cool areas with remnant conifer woodland, and in pine plantations. It prefers dry coniferous woodlands dominated by balsam fir and various species of spruce, but white and jack pine, hemlock, and white cedar woodlands are also suitable, as are wetter habitats where black spruce or tamarack occur. Its habitat often includes scattered openings caused by various forms of disturbance (logging, fire, or abandoned agricultural lands).

This species has probably been common and widespread in the Boreal Forest since the end of the Pleistocene glaciations, when suitable habitats became available. It is an adaptable, opportunistic species, being capable of colonizing edges produced by power-lines or logging, and its population densities and territory size are little affected by such disturbances (Apfelbaum and Haney 1981, Clark *et al.* 1983, Niemi and Hanowski 1984). There is little evidence to suggest that it is undergoing (or has undergone) any major change in its distribution in Ontario, with the possible exception of a limited expansion in the south. Several Regional Coordinators commented on the colonization of mature pine plantations by this species. Breeding Bird Survey results from central and northern Ontario suggest a slight general trend toward an overall increase, during the period from 1967 to 1983. Hall (1984) suspects that it is extending its breeding range southward in Pennsylvania and West Virginia.

The Yellow-rumped Warbler is very vocal, both during the onset of breeding activity, when territories are being established, and after fledging of the young (Nice 1930b). In addition, it responds well to 'pishing', making it easy to detect. Although the birds are relatively quiet during the incubation and nestling periods (12 to 13 and 12 to 14 days, respectively), the adults become quite vocal after the young have fledged and are easily agitated, performing elaborate distraction displays (Bent 1953). Feeding of the young is performed by both sexes (Nice 1930b). As the breeding behaviour of the species would suggest, the majority of breeding confirmations (86%) in Ontario were made by observing fledged young, or adults carrying food or faecal sacs.

The Yellow-rumped Warbler is one of the most widespread species of warbler in Ontario. It occurs from the northernmost blocks along Hudson Bay, through all of the north, to the forests of the southern Canadian Shield, and spottily into remnant forests in built-up southern Ontario. Confirmed breeding was reported as far south as Turkey Point on Lake Erie, Woodstock, and Kettle Point on Lake Huron. Apart from the few exclusively tundra blocks on the shores of Hudson and James Bays, those blocks in the north which lack records of this species are not likely to represent real distributional gaps, but probably reflect difficulty of access and reduced atlasser activity.

In the central part of its range in Ontario, the Yellow-rumped Warbler is common to abundant. Population densities generally range from 0.25 to 1.0 pairs per hectare (Martin 1960). However, it is rare and localized in the south. It is also uncommon in apparently suitable habitat in parts of the north (Peterson 1985). Abundance estimates are predominantly in the 2 to 100 pairs per square range, but estimates of more than 1,000 pairs per square were also reported from southern Ontario. The majority of these higher abundance estimates, those above 1,000 pairs per square, were from the Algonquin, Parry Sound, and Manitoulin atlas regions, in the Great Lakes-St. Lawrence Forest region.-- *W.J. Crins*

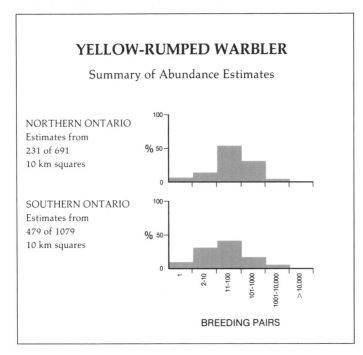

YELLOW-RUMPED WARBLER

Summary of Abundance Estimates

NORTHERN ONTARIO
Estimates from
231 of 691
10 km squares

SOUTHERN ONTARIO
Estimates from
479 of 1079
10 km squares

BREEDING PAIRS

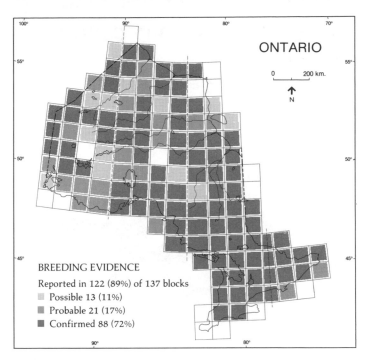

ONTARIO

0   200 km.

N

BREEDING EVIDENCE
Reported in 122 (89%) of 137 blocks
☐ Possible 13 (11%)
☐ Probable 21 (17%)
■ Confirmed 88 (72%)

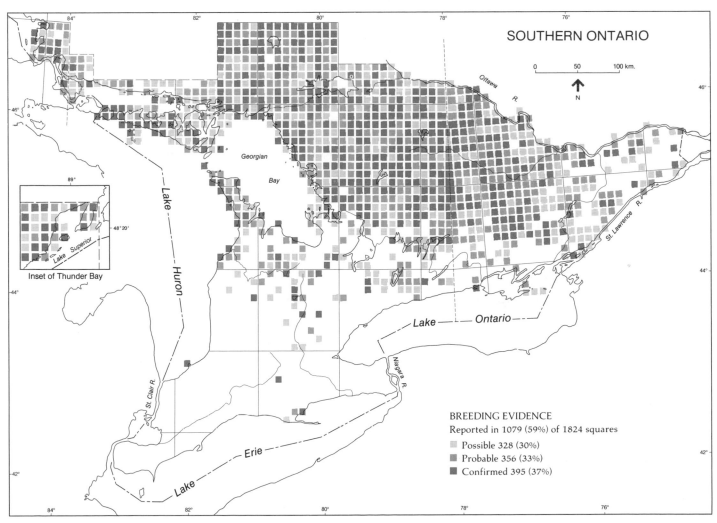

SOUTHERN ONTARIO

0   50   100 km.

N

Georgian
Bay

Lake
Huron

Lake

Ottawa
R.

St. Lawrence R.

Lake — Ontario

Niagara R.

Inset of Thunder Bay

Lake Superior

St. Clair R.

Lake — Erie

Lake

BREEDING EVIDENCE
Reported in 1079 (59%) of 1824 squares
☐ Possible 328 (30%)
☐ Probable 356 (33%)
■ Confirmed 395 (37%)

BLACK-THROATED GREEN WARBLER
Paruline verte à gorge noire
Dendroica virens

The Black-throated Green Warbler is a widespread and often common breeding bird throughout the coniferous forest regions east of the Rocky Mountains in North America. It winters in the extreme southern US, Mexico, and Central America. In Ontario, it is found in the southern two-thirds of the province, mainly within the Great Lakes-St. Lawrence and the southern part of the Boreal Forest regions. It is generally a species of coniferous or mixed forests in which pines, spruces, white cedar, hemlock, and/or balsam fir may be dominant, but it may also occur in deciduous forests composed of beech, maples, or birches. The important aspect of the habitat for this species appears to be three-dimensional structure, not species composition. Since the Black-throated Green Warbler is a foliage-gleaner, that is, it picks insects from the surfaces of leaves, a multilayered leaf canopy (provided by many conifers) is an important component of its habitat (Collins 1983). The forests in which it occurs usually have a well-developed shrub layer, and may contain small clearings. At the southern edge of its range in Ontario (along Lakes Erie and Ontario), it inhabits hemlock forests in ravines, wet cedar swamps, and conifer plantations.

There may have been some decline in the abundance of this species in southern Ontario since the turn of the century. Clark *et al.* (1983) have suggested that the Black-throated Green Warbler is limited to undisturbed forests in central Ontario. In the author's experience, this is an overstatement, but it is likely that increasing development has reduced population densities somewhat. The southern areas which still support populations of this species (sites along Lake Erie, in particular) are potentially in danger of being cleared for agricultural and recreational purposes.

The Black-throated Green Warbler is one of the most

aggressive species of wood warbler in eastern North America (Morse 1976). It actively chases conspecifics and other species, making it highly visible. In its foraging activity, it is also one of the most active warblers, moving quickly from site to site, and chipping continuously while doing so (MacArthur 1958). It sings frequently while establishing and defending its territory, and song activity continues well into the summer (until at least mid-July in central Ontario). Over half (51%) of the atlas records are of singing males. From the perspective of someone trying to find this species, its propensity for foraging and nesting at high levels in trees is offset by its high activity level. The nest itself is difficult to find. It is usually situated more than 2 m above ground level, on a horizontal branch or near the top of a conifer. The incubation period lasts for 12 days, and the young remain in the nest for an additional 8 to 10 days (Pitelka 1940). After the eggs have hatched, it is a relatively simple task to confirm the Black-throated Green Warbler as a breeding bird, because both adults actively collect food for the young, and remove faecal sacs from the nest (88% of southern Ontario confirmations were of adults carrying food, or fledged young). Parental care continues after the young have fledged. The species is under-confirmed in the atlas data base (17% and 9% of southern and northern Ontario records, respectively), perhaps because atlassing effort tapered off after the end of June, when the birds were feeding young.

This species is widespread and common in much of southcentral Ontario, wherever coniferous or mixed forests occur. Outlier populations persist where heavily wooded areas remain in the south. The patchy nature of its occurrence in the south is expected, and is a function of the shortage of suitable habitat. It was almost certainly more common on a local scale prior to the clearing of land in this part of the province. The paucity of confirmations of breeding in the north may be a function of less intensive atlassing effort due to difficulty of access. Perhaps there are lower population densities in northern Ontario. Nevertheless, atlas data have extended the known range of the species north as far as Wunnummin Lake. In the south, the majority of squares near the southern edge of the Canadian Shield (Muskoka, Haliburton, Nipissing, Parry Sound, Sudbury) have at least possible breeding evidence for this species.

Abundance estimates were provided for 41% and 35% of the squares where this species was reported in southern and northern Ontario, respectively. In the north, 67% of estimates are of between 11 and 1,000 pairs. Virtually all of the estimates over 1,000 pairs per square are from the Haliburton, Parry Sound, and Manitoulin atlas regions.-- *W.J. Crins*

BLACK-THROATED GREEN WARBLER

Summary of Abundance Estimates

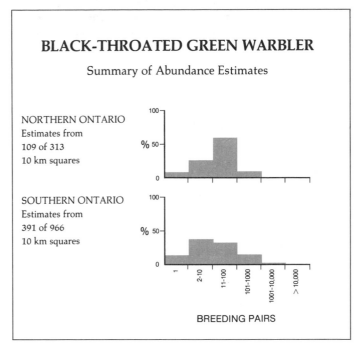

NORTHERN ONTARIO
Estimates from
109 of 313
10 km squares

SOUTHERN ONTARIO
Estimates from
391 of 966
10 km squares

%

BREEDING PAIRS

1   2-10   11-100   101-1000   1001-10,000   >10,000

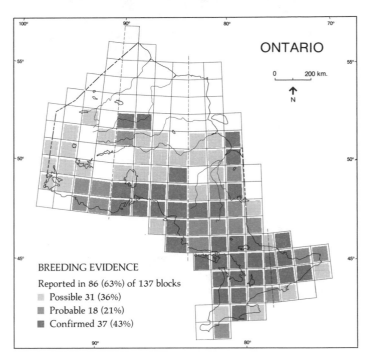

ONTARIO

0   200 km.

N

BREEDING EVIDENCE

Reported in 86 (63%) of 137 blocks

Possible 31 (36%)

Probable 18 (21%)

Confirmed 37 (43%)

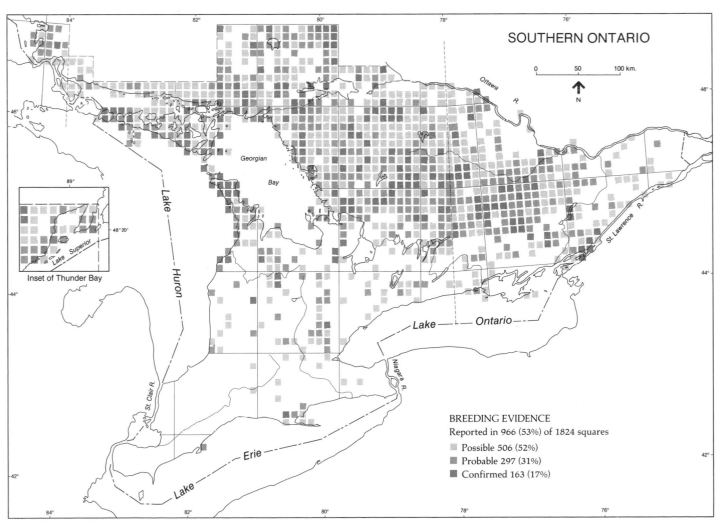

SOUTHERN ONTARIO

0   50   100 km.

N

Georgian Bay

Lake Huron

Lake Superior

Inset of Thunder Bay

Ottawa R.

St. Lawrence R.

Lake Ontario

Niagara R.

St. Clair R.

Lake Erie

BREEDING EVIDENCE

Reported in 966 (53%) of 1824 squares

Possible 506 (52%)

Probable 297 (31%)

Confirmed 163 (17%)

BLACKBURNIAN WARBLER
Paruline à gorge orangée
Dendroica fusca

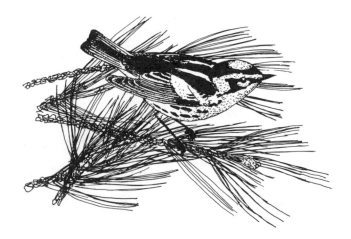

The Blackburnian Warbler is a widespread breeding bird throughout northeastern North America. It winters in central and northwestern South America. It nests in the southern two-thirds of Ontario, mainly within the Great Lakes-St. Lawrence and southern Boreal Forest regions, and is most common near the southern limit of the Canadian Shield. It is characteristic of moist to dry hemlock forests, but also occurs in other types of conifer-dominated woodland (including white cedar, pine, and spruce forests), and in some types of hardwood forest, especially those that contained native chestnut (Nice 1932, Bent 1953, Lawrence 1953a). Particularly in southern parts of the province, it has also adapted to mature conifer plantations. It forages and nests in the upper levels of the forest canopy.

The distribution of the Blackburnian Warbler in Ontario prior to the arrival of Europeans is unknown, but it may have been common in the south, when native chestnut was more common in the forests of that part of the province. In the main part of its present range, it is certainly subject to local changes in abundance. These changes are due in part to its requirement for mature forest. It is intolerant of disturbances that affect forest structure (Apfelbaum and Haney 1981, Clark *et al.* 1983).

The Blackburnian Warbler is conspicuous and vocal early in the breeding season. It often selects high, exposed perches as singing posts. Thus, its presence early in the breeding season is easy to detect. Forty-eight percent of all records in Ontario were of singing males. However, the nest is difficult to locate. It is situated 2 m or more above the ground (often more than 10 m), in a crotch or fork of branches or on a horizontal limb, often over-arched by other branches (Lawrence 1953a). The female does not flush easily during the incubation stage. After the eggs have hatched, and during the nes-

tling stage, both parents are very attentive. The young are also quite vocal (Nice 1932, Lawrence 1953a), making confirmation of this species as a breeding bird relatively simple from late June to mid-July. In 90% of the Ontario squares in which Blackburnian Warblers were confirmed as breeding, the evidence consisted of fledged young, or adults carrying food or faecal sacs.

The Blackburnian Warbler is widespread on the Canadian Shield, but is spottily distributed to the north and south, much like the Black-throated Green Warbler. Both species occur in remnant mature coniferous forests and in plantations in southern areas, and may have been more widespread in this area prior to settlement. The atlas maps show confirmed breeding south to the St. Williams Forestry Station near Long Point. The species also extends north into the Boreal Forest region to the Sandy Lake, Lake Nipigon, and Cochrane areas. The northwestern records confirm and extend the known distribution of this species northward and westward in Ontario. A number of blocks between Red Lake and Cochrane lack records of this species. These may reflect real gaps in distribution, but it is also possible that the species was overlooked there, because of low population densities or limited access to those blocks.

Territory size appears to be more variable in the Blackburnian Warbler than in other southern Shield species of wood warblers (Kendeigh 1947, Martin 1960). In response to increasing spruce budworm populations, nesting densities of Blackburnian Warblers increase, and territory sizes decrease (Morris *et al.* 1958). The variability of territory size may also be due to the patchy distribution of mature forest. It is also near the bottom of the social dominance hierarchy among spruce-woods warblers (Morse 1976), which might tend to reduce its population densities in areas where other species, such as the aggressive Black-throated Green Warbler or Magnolia Warbler, occur in high densities. Accordingly, abundance estimates vary widely, ranging from 1 to more than 1,000 pairs per square where this species occurs. However, most squares where estimates were provided were thought to contain between 2 and 100 pairs (75%). All 9 abundance estimates of 1,000 pairs per square or more, were from the Haliburton, Algonquin, Nipissing East, and Kirkland Lake atlas regions. The abundance estimates of 101 to 1,000 pairs were distributed more widely throughout the range of the species in Ontario.-- *W.J. Crins*

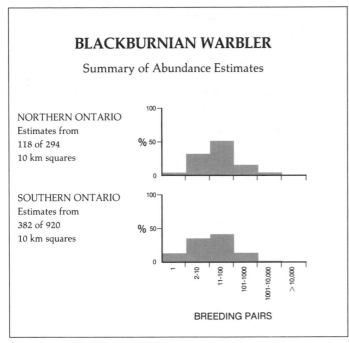

BLACKBURNIAN WARBLER

Summary of Abundance Estimates

NORTHERN ONTARIO
Estimates from
118 of 294
10 km squares

SOUTHERN ONTARIO
Estimates from
382 of 920
10 km squares

%

BREEDING PAIRS

1   2-10   11-100   101-1000   1001-10,000   >10,000

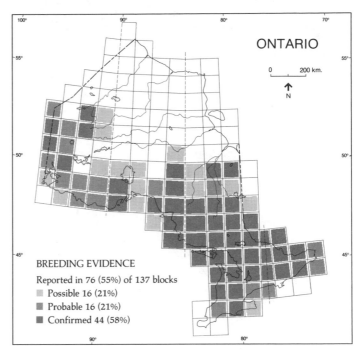

ONTARIO

0   200 km.

N

BREEDING EVIDENCE

Reported in 76 (55%) of 137 blocks

Possible 16 (21%)
Probable 16 (21%)
Confirmed 44 (58%)

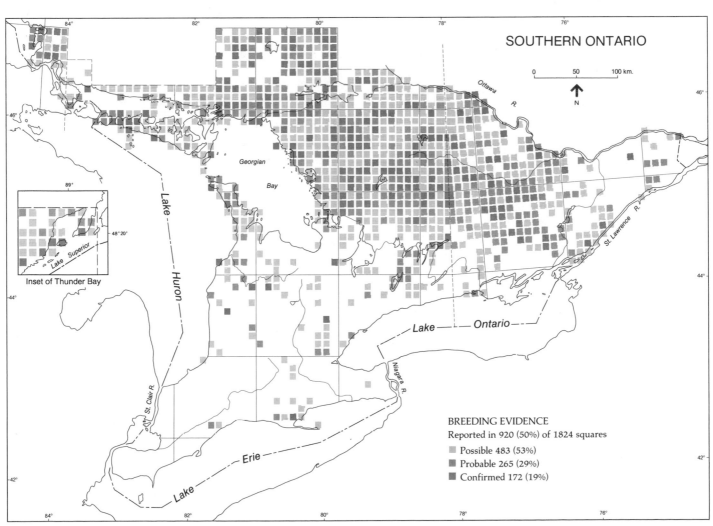

SOUTHERN ONTARIO

0   50   100 km.

N

Inset of Thunder Bay

Georgian Bay

Lake Huron

Lake Superior

Ottawa R.

St. Lawrence R.

St. Clair R.

Niagara R.

Lake Ontario

Lake Erie

BREEDING EVIDENCE

Reported in 920 (50%) of 1824 squares

Possible 483 (53%)
Probable 265 (29%)
Confirmed 172 (19%)

PINE WARBLER
Paruline des pins
Dendroica pinus

The Pine Warbler is a breeding bird of eastern North America from the Gulf coast to Ontario. In Ontario, it breeds across the south and as far north as the northern shore of Lake Superior. Its range is closely tied to the distribution of pine forests, which comprise the preferred habitat for this secretive warbler. The species, therefore, varies in numbers considerably according to the number and frequency of these forests. Once on the breeding grounds, the Pine Warbler spends its time high in the tallest pines, with white pines usually preferred but red pines used occasionally. It winters in the southeastern US.

In Ontario, the historical range of this species has varied according to the fate of the host forests. It is probable that, before European settlers arrived, the species was abundant in the extensive groves of pines that had grown in the clearings created by Indian agriculture and by fire. As the ravages of the square timber loggers swept across Ontario in the first half of the 19th century, the population of Pine Warblers is assumed to have dropped as the pines were systematically removed. By 1886, McIlwraith indicated that the Pine Warbler had a spotty distribution, being unknown at London, but "rather a common species" at Hamilton. Baillie and Harrington (1936) stated that the species was found locally from the "southern border north to the northern Algoma district and Renfrew Co." As the pine populations have recovered somewhat, it is reasonable to speculate that the Pine Warbler populations may have experienced a modest increase since the 1800's. There is evidence that the bird is now moving into the pine plantations in southwestern Ontario that are reaching a suitable size after 40 to 50 years of growth.

The Pine Warbler is not easy to find. It is drab in colour, feeds inconspicuously in the tops of tall trees and has a song that is easily confused with that of the much more abundant Chipping Sparrow. McIlwraith (1886) used the older name for this species, the "Pine-creeping Warbler", a name which aptly describes its propensity to creep on the branches and trunks of pines searching for insects. It is probable that these characteristics resulted in somewhat of an underrecording of the species by some atlassers. On the other hand, its habitat preference is so specific that atlassers familiar with the Pine Warbler would have no difficulty in knowing where to look for it, such habitat being relatively scarce in southern Ontario.

The difficulty in actually seeing the species is reflected in the high percentage (63%) of all southern Ontario records that are of singing males recorded once or on territory. Correspondingly, only 19% of the southern records are of confirmed breeding and nests with eggs or young were found in only 6 squares.

The maps show the wide distribution of the species across the southern third of the province. The heaviest concentration of records comes from the Great Lakes-St. Lawrence Forest region on the Canadian Shield. It is here that the tall pines are found most frequently, either as small clumps of trees in old clearings or as extensive blocks of forest growing on old fire sites. Off the Shield, there are pockets of birds, closely correlated with white and red pine populations in parks or areas of rough land. In the southwest, the populations in Pinery and Rondeau Prov. Parks, on Lakes Huron and Erie respectively, show the importance of these protected lands in providing suitable habitat in the otherwise uniform agricultural landscape. The pines in the remaining forested habitat on the Norfolk Sand Plain and on the morainic hills between Cambridge and Hamilton are reflected clearly on the map. In the north, the map shows a distributional limit that corresponds to the northern limit of white pine (Elias 1980). The atlas adds to the distributional knowledge of this species with the discovery of breeding populations between the previously known disjunct ranges in the Lake of the Woods and Lake Superior areas (Godfrey 1986).

Fifty-four percent of the abundance estimates were of between 2 and 10 pairs in a square, but the estimates varied from 1 to over 1,000 pairs, depending largely on the available habitat in any particular square. Those squares in the southwest with only small remnants of the forest remaining often had low abundance estimates, while squares in the forest country of southcentral Ontario had higher estimates. The highest abundance estimates (greater than 100 pairs per square) were all recorded from 2 distinct areas, east of Georgian Bay from Parry Sound to North Bay and north of Lake Ontario from Kingston to Brockville.-- *P.F.J. Eagles*

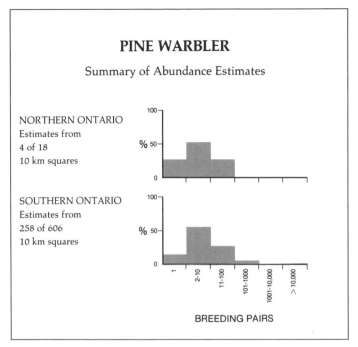

PINE WARBLER

Summary of Abundance Estimates

NORTHERN ONTARIO
Estimates from
4 of 18
10 km squares

SOUTHERN ONTARIO
Estimates from
258 of 606
10 km squares

BREEDING PAIRS

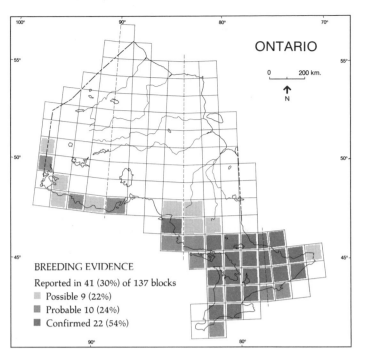

ONTARIO

0   200 km.

N

BREEDING EVIDENCE
Reported in 41 (30%) of 137 blocks
☐ Possible 9 (22%)
☐ Probable 10 (24%)
■ Confirmed 22 (54%)

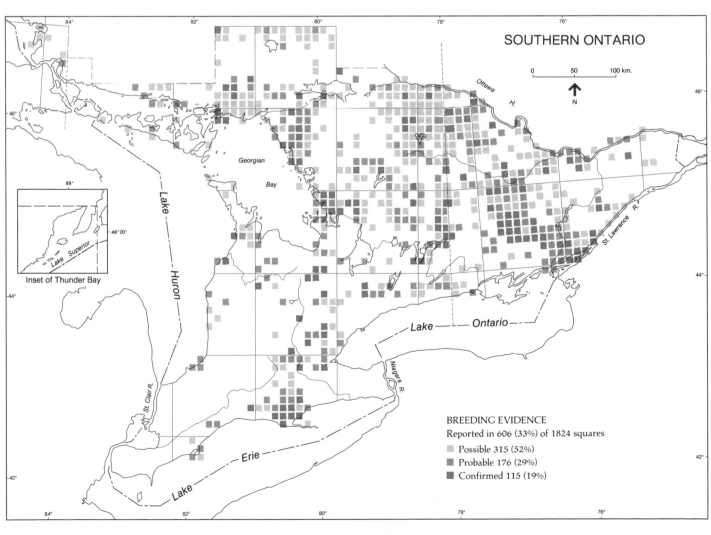

SOUTHERN ONTARIO

0   50   100 km.

N

Inset of Thunder Bay

BREEDING EVIDENCE
Reported in 606 (33%) of 1824 squares
☐ Possible 315 (52%)
☐ Probable 176 (29%)
■ Confirmed 115 (19%)

KIRTLAND'S WARBLER
Paruline de Kirtland
Dendroica kirtlandii

The Kirtland's Warbler, which may once have been a regular breeder in Canada, is now a rare visitor. A total of 37 records have been documented in Canada since 1900. The western limit of its range in Canada may be defined by a sighting at Long Point on Lake Winnipegosis, Manitoba, by P.A. Taverner in 1921, and the eastern limit by a possible sighting in Mount Royal Park, Montreal, by E. Klukas in 1977. All 37 sightings were of single birds, males, or birds of unknown sex, except for a pair seen with at least one juvenile near Barrie in 1945 (Speirs 1984). This incidence is Ontario's only confirmed breeding record. Verified historical sightings in Ontario range from Point Pelee in the west to the Kingston area in the east, and to McVicar on the Bruce Peninsula, Pointe au Baril (northwest of Parry Sound), and Petawawa in the north.

At present, the only known breeding grounds are in young jack pine stands in central Michigan. The estimates of breeding pairs obtained from the annual June census of singing males in Michigan have ranged from 200 to 243 in the last decade. The 1986 census team reported 210 breeding pairs, all in Michigan. The winter range for the species includes the Bahama Islands, the adjacent Grand Turk and Caicos Islands, Hispaniola, and perhaps the east coast of Mexico (Mayfield 1960, Huber 1982, Walkinshaw 1983, Byelich *et al.* 1986).

The only breeding season record in Ontario during the atlas period was of a single male which maintained a territory for at least 10 days in early July 1985. The bird was singing in its typical habitat, a jack pine forest. Specific locational details are being withheld for conservation purposes, even though the bird did not reappear in this location in 1986.

In 1986, an intensive survey of potential breeding habitat across Ontario was completed by the Ministry of Natural Resources. No Kirtland's Warblers were found. The species is listed as endangered throughout North America, including Ontario. Though extremely rare in Ontario, it is always found in relatively pure jack pine stands or plantations of more than 20 ha on well-drained sands or on shallow soils covering bedrock. Trees 2 to 7 m tall are preferred, though taller ones are also acceptable, if young trees are present in scattered openings. Kirtland's Warbler males usually respond well to audio-projections of the species' song. They sing throughout the day, from about mid-May to the beginning of July, and are loud, mellifluous singers.-- *P. Aird and D. Pope*

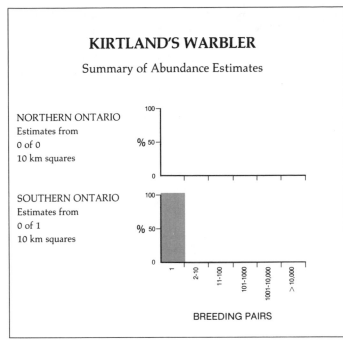

KIRTLAND'S WARBLER

Summary of Abundance Estimates

NORTHERN ONTARIO
Estimates from
0 of 0
10 km squares

SOUTHERN ONTARIO
Estimates from
0 of 1
10 km squares

BREEDING PAIRS

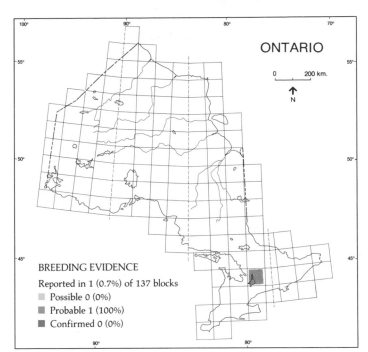

ONTARIO

0   200 km.

N

BREEDING EVIDENCE
Reported in 1 (0.7%) of 137 blocks
  Possible 0 (0%)
  Probable 1 (100%)
  Confirmed 0 (0%)

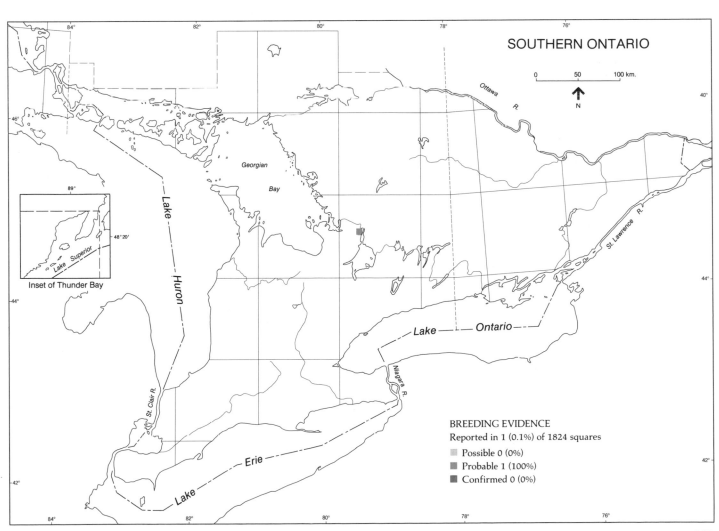

SOUTHERN ONTARIO

0   50   100 km.

N

Georgian Bay

Lake Huron

Lake Superior
Inset of Thunder Bay

St. Clair R.

Lake Erie

Niagara R.

Lake Ontario

Ottawa R.

St. Lawrence R.

BREEDING EVIDENCE
Reported in 1 (0.1%) of 1824 squares
  Possible 0 (0%)
  Probable 1 (100%)
  Confirmed 0 (0%)

PRAIRIE WARBLER
Paruline des prés
Dendroica discolor

The Prairie Warbler breeds in eastern North America and winters largely in the West Indies and Bahamas. Although fairly common in much of its US range, it is rare and localized in Ontario, the only Canadian province in which it is known to breed. Preferred habitats in Ontario are characterized by open, scrubby vegetation and, usually, dry conditions. The most important habitat is the rocky pine-oak-juniper scrub found along the southeastern Georgian Bay shoreline and at scattered locations along the southern edge of the Canadian Shield, notably in Peterborough and Frontenac Counties. Ground-Juniper is an important component of this habitat (Lord 1955) and is present in sand dune habitats used along parts of the Lake Huron shoreline. There is also limited use of early successional habitats in Ontario, including forest-edge deciduous scrub and young pine plantations with deciduous regeneration.

Before European settlement of North America, the Prairie Warbler was probably rare and local, with most populations scattered around the margins of its present range (Nolan 1978). A large increase in the US population has occurred in this century as the species has exploited newly available scrublands, regenerating old fields and young conifer plantations. During the last 20 years, range expansions have occurred in these habitats along the northern edge of the US range in Ohio, New York, and Maine. A similar expansion does not seem to have taken place in Ontario, where isolated populations have probably been in existence for centuries. However, since the 1920's, substantial population declines have been documented in two counties. Habitat loss caused by cottage development was largely responsible for the disappearance of the Wasaga Beach 'colony' in Simcoe Co, which once numbered at least 150 pairs (Devitt 1967). The same pressures seem to have affected the Pinery-Ipperwash

population in Lambton Co, which has gradually declined since the 1930's, so that today only a remnant population of less than 10 pairs persists in Pinery Prov. Park. On the other hand, regular breeding in Haldimand-Norfolk RM was probably initiated in the 1930's, as young pine plantations became established in the St. Williams/Turkey Point area.

This species does not present any special atlassing problems, except for difficulties of access to the remoter parts of its range. The Prairie Warbler is best located by its highly distinctive song, but since breeding sites bordering lakes and rivers are often accessible only by boat, opportunities for obtaining higher categories of breeding evidence may have been limited in some cases. Observers in areas where breeding densities are low have found that Prairie Warblers tend to sing only in the early morning and hence may be missed later in the day. This could have been a problem in Peterborough Co and parts of Muskoka and Parry Sound Districts, where the species was not recorded in some squares despite pre-atlas period records.

Although some isolated birds were undoubtedly missed, the map provides a good representation of the actual distribution, which was fairly well known prior to the atlas period (Lambert and Smith 1984a). Gaps in the Georgian Bay area very likely reflect problems with coverage of the extensive shoreline habitats rather than complete absence of the birds. Atlas field work turned up some new sites within the previously known range, particularly in northern Frontenac Co, and produced the first breeding season records for Haliburton, Hastings, Lennox and Addington, and Lanark Counties. Also of interest was the finding that hydro rights-of-way are being used as breeding habitat in Frontenac Co. In addition, a few new sites in successional habitat were found south of the Shield, with a record of a nest with eggs in Waterloo RM, and single records of territorial males in Middlesex Co and York RM. In Parry Sound District, the extent of the breeding range was poorly known before the atlas; however, it now seems that Pointe au Baril may mark the northern limit of the range. The absence of atlas records from Bruce Co is also worth noting, in view of historic records suggesting scattered breeding, and confirmed breeding at one site in the 1950's.

Abundance estimates from the atlas reflect the small size of the Ontario population, which was previously estimated as probably not exceeding 500 pairs (Lambert and Smith 1984b). This estimate was based on a known breeding population in 1983-84 of between 321 and 336 pairs, and, taking account of 1985 field work, it seems that an estimate of somewhat less than 500 pairs is reasonable. The Prairie Warbler was designated as "rare" by COSEWIC in 1985, and is also listed as a "rare" species by MNR.-- *A.B. Lambert*

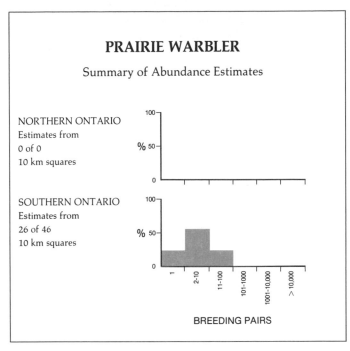

PRAIRIE WARBLER

Summary of Abundance Estimates

NORTHERN ONTARIO
Estimates from
0 of 0
10 km squares

SOUTHERN ONTARIO
Estimates from
26 of 46
10 km squares

BREEDING PAIRS

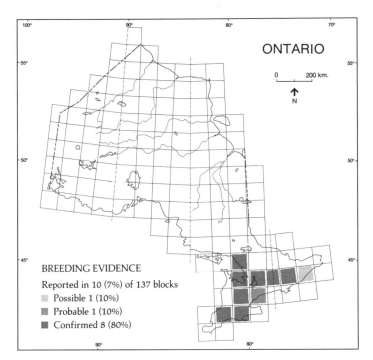

ONTARIO

0   200 km.

N

BREEDING EVIDENCE

Reported in 10 (7%) of 137 blocks

Possible 1 (10%)
Probable 1 (10%)
Confirmed 8 (80%)

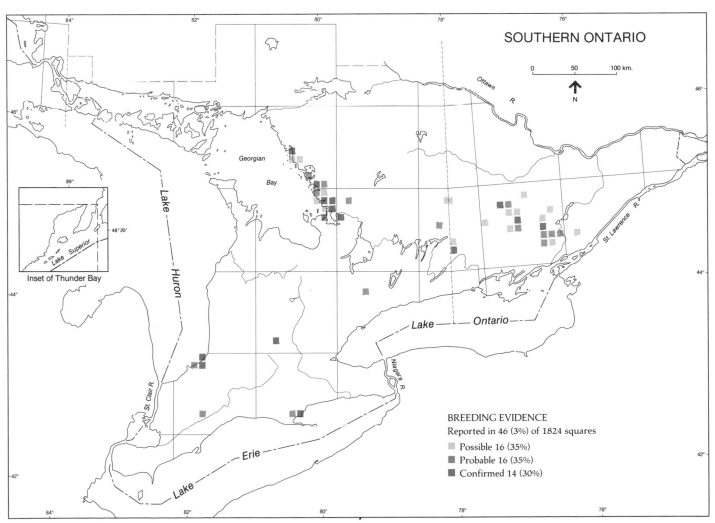

SOUTHERN ONTARIO

0   50   100 km.

N

Inset of Thunder Bay

BREEDING EVIDENCE

Reported in 46 (3%) of 1824 squares

Possible 16 (35%)
Probable 16 (35%)
Confirmed 14 (30%)

PALM WARBLER
Paruline à couronne rousse
Dendroica palmarum

The Palm Warbler is a regular, but uncommon, broadly scattered breeding species throughout most of Canada. It has very specific habitat requirements, normally breeding in mature bogs and bog heaths with scattered black spruce. It occasionally settles in regenerating cutover spruce forests with clearings, and the critical habitat components seem to be openings with sphagnum moss and shrub cover adjacent to spruce or tamarack.

Although originally described from Hispaniola, most Palm Warblers would rarely see a palm tree. The species breeds from Newfoundland and Labrador to the western Mackenzie District and northern Alberta. It winters mostly in the southeastern US, but some birds go as far as the southern parts of Central America and the Caribbean (AOU 1983). Its Canadian range approximately coincides with the Boreal Forest region but does not extend much west of Great Bear Lake. It is rare in southern Ontario because of scarcity of suitable habitat but it occurs widely in the north.

Macoun and Macoun (1909) provide mostly scattered records of migrants but note a specimen collected by Drexler at Moosonee in July 1860. Early breeding records from Mer Bleue near Ottawa by Saunders in 1901, and Young in 1908, are mentioned in Baillie and Harrington (1937) and Speirs (1985), respectively. James *et al.* (1983) report a nesting Palm Warbler from the Sutton Ridge area of the northern Hudson Bay Lowland. McLaren and McLaren (1981a) recorded the species at Little Sachigo Lake and North Caribou Lake in the northwest and at Wetiko Hills near the Manitoba border. Although normally at low densities, it can be locally abundant. Elder *et al.* (1971) discovered 15 territories on their 10 ha breeding bird census plot (150 pairs/km^2), and Welsh and Ross (unpubl.) found the species at 3 of 32 sites in the James Bay section of the Hudson Bay Lowland, one of which con-

tained more than 100 territories with an estimated average density of 272 pairs/km^2. In spite of its uncommon breeding status, the species is well known by sight to most birdwatchers.

The bright yellow under tail coverts, olive-coloured rump, and white corner tail spots make good field marks, as does the rusty chestnut cap during breeding season. Its most conspicuous field characteristic is its habit of constant, vigorous tail-wagging. The song is appropriately described as a Junco-like or Chipping Sparrow-like trill, but can usually be recognized as being shorter and less sibilant or forceful than these songs, often rising somewhat in the middle. The very specific choice of habitat, usually bogs, bog heaths, and barrens with scattered trees, provides the males with apparently essential song posts (Welsh 1971) while retaining open areas for foraging. The species is primarily insectivorous, and most insects are caught from the ground or low shrubs in the more open sections of its territory. Several authors (Knight 1904, Welsh 1971, Peterson 1980) have recognized the importance of the wooded margin or scattered clumps of trees as integral components of the territory. In large open bogs, Palm Warblers normally occur only in embayments around the edge. Nests are normally on or near the ground and are usually well concealed but can often be located with persistent observation. Both parents feed the young. Regional Coordinators did not note specific problems with atlassing the species, except difficulty in finding it. It seems that the restricted habitat of the species often resulted in its being missed.

Examination of breeding evidence shows that the most common means of confirmation was the observation of fledged young or of adults carrying food or faecal sacs. Breeding was confirmed in almost half of the blocks in which the species was found.

The maps reflect the overall range of the Palm Warbler but clearly contain gaps. For example, the species was a locally abundant breeder in the blocks south and southeast of Moosonee in 1978 (pers. obs.), and presumably still breeds there. Even though uncommon and scattered, it should be expected in most blocks in the north, particularly in the Clay Belt and Hudson Bay Lowland. The 3 records in the south seem reasonable, and their scarcity reflects a general absence of suitable habitat.

Atlas abundance estimates appear to be generally too low. The species certainly breeds in low density on the Canadian Shield because of the limited size of habitat units, and there would rarely be more than 10 pairs per square. In the extensive interior of the Hudson Bay Lowland, 11 to 100 pairs would be more expected as an average and densities of over 100 pairs per square undoubtedly occur regularly.-- *D.A. Welsh*

PALM WARBLER

Summary of Abundance Estimates

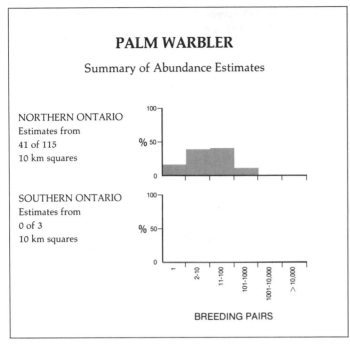

NORTHERN ONTARIO
Estimates from
41 of 115
10 km squares

SOUTHERN ONTARIO
Estimates from
0 of 3
10 km squares

BREEDING PAIRS

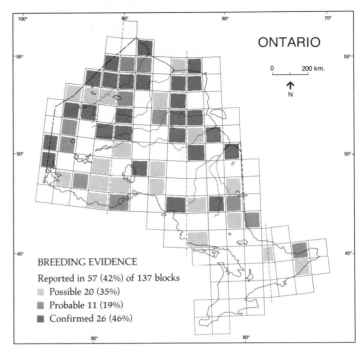

ONTARIO

0   200 km.

N

BREEDING EVIDENCE
Reported in 57 (42%) of 137 blocks
Possible 20 (35%)
Probable 11 (19%)
Confirmed 26 (46%)

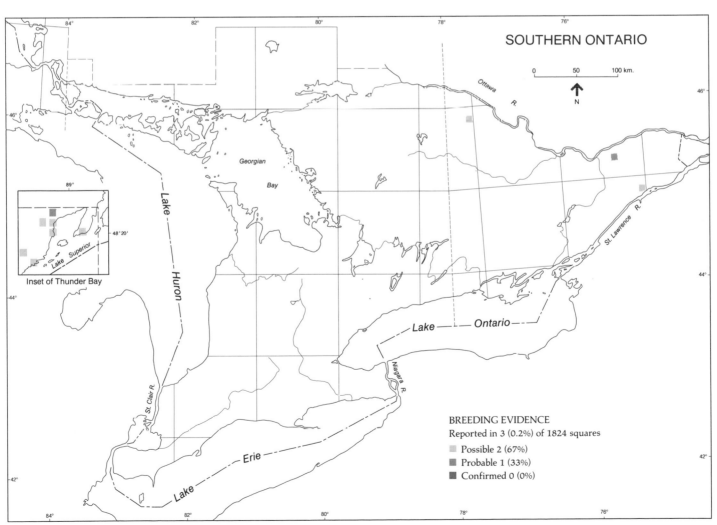

SOUTHERN ONTARIO

0   50   100 km.

N

Georgian
Bay

Lake
Huron

Lake — Superior
Inset of Thunder Bay

Ottawa R.

St. Lawrence R.

Lake — Ontario

Niagara R.

St. Clair R.

Lake — Erie

Lake

BREEDING EVIDENCE
Reported in 3 (0.2%) of 1824 squares
Possible 2 (67%)
Probable 1 (33%)
Confirmed 0 (0%)

BAY-BREASTED WARBLER
Paruline à poitrine baie
Dendroica castanea

'From rags to riches' aptly reflects the recurring fate of the Bay-breasted Warbler. The Bay-breasted Warbler is primarily a spruce budworm specialist and as such its population size is closely linked to the cyclic nature of the budworm population. As a regular, low-density resident of mature coniferous and mixed Boreal Forest stands, the species breeds over much of Canada from northeastern British Columbia and the Yukon to eastern Newfoundland. In Ontario, the primary range coincides with the Canadian Shield and Hudson Bay Lowland regions including the Boreal Forest and the northern sections of the Great Lakes-St. Lawrence forest zone. The Bay-breasted Warbler's distribution could easily be used as a map of spruce and balsam fir distribution in the province.

Macoun and Macoun (1909) reported a nest found in 1899 north of Waterloo. Baillie and Harrington (1937) considered the species as a not uncommon, although unconfirmed, breeding species in the Boreal Forest. Early records indicate that the species occurred from the northern Bruce Peninsula to Moose Factory. Most ornithologists considered it uncommon until Kendeigh (1947) reported 230 pairs/km^2 during a 1946 study of a spruce budworm outbreak at Black Sturgeon Lake, near Lake Nipigon. Speirs (1949) subsequently reported 120 pairs/km^2 in the same area in 1946. Several subsequent studies have confirmed that Bay-breasted Warblers often exhibit a strong breeding density increase in response to budworm population increases. From 1966 to 1968, Sanders (1970) found no Bay-breasted Warblers on plots at Black Sturgeon Lake during a period of low budworm density. In continued study on those same plots, Welsh (unpubl.) found densities of 190 pairs/km^2 in 1983 at the start of a new budworm peak in the area, confirming the cyclic response of the species to budworm.

The Bay-breasted Warbler inhabits spruce and balsam fir forests almost exclusively. It is well known to ecologists as the inner mid-canopy specialist in MacArthur's (1958) studies of habitat niche partitioning in wood warblers. In Ontario, in mature forests higher than 15 m, it regularly forages in the mid and inner upper portion of the tree, leaving the tree tops for the Cape May and occasional Blackburnian and Yellow-rumped Warbler, with which it regularly co-occurs. When budworm densities are high, it readily expands into younger stands and takes over the upper canopy as well, often forcing other species like the Golden-crowned Kinglet to lower strata. Large changes in density and habitat dramatically affect the probability of discovering the species. At high density, the species sings incessantly, often for long periods from the same perch, and the sibilant song is unlikely to be missed by the experienced atlasser, but the species is much less conspicuous at lower densities in mature stands. After incubation, both parents actively feed the nestlings, and young fledglings can readily be seen foraging in the mid-canopy. Breeding of the species was most often confirmed as adults carrying food or faecal sacs, and frequently the best evidence was of singing males.

The atlas maps contain few surprises but provide documentation of what was previously a mostly inferred range. Gaps in the central north undoubtedly reflect inadequate coverage rather than absence, whereas the species is certainly uncommon north of Attawapiskat in the northeast. Although the species has been previously recorded north of the Severn River (McLaren and McLaren 1981a, b), the atlas map reflects a presence in the northwest in contrast to the published maps (Robbins *et al.* 1966, Griscom and Sprunt 1979, Godfrey 1986) in which the range swings to more southern latitudes from east to west across the province. The atlas maps closely coincide with the habitat distribution. The range in the south is as expected from published records.

Spruce budworm densities reached their highest recorded levels in 1975. The resultant Bay-breasted Warbler population was likely to be several times higher than in recorded history. As the outbreak continued into the atlas period, the range shown on the atlas maps is certainly maximal and we can expect the frequency of occurrence in the south to decrease and the density to drop throughout. As densities of more than 100 pairs/km^2 in good habitat are common throughout the north, it is certain that the Ontario population was several million during the atlas period, and abundance estimates of more than 1,000 pairs per square would have been appropriate for the Boreal Forest zone, but not for the northern Lowlands. The abundance levels for southern Ontario reflect density well, while those for the north are generally severely underestimated.-- *D.A. Welsh*

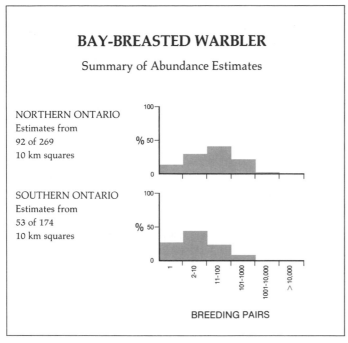

BAY-BREASTED WARBLER

Summary of Abundance Estimates

NORTHERN ONTARIO
Estimates from
92 of 269
10 km squares

SOUTHERN ONTARIO
Estimates from
53 of 174
10 km squares

BREEDING PAIRS

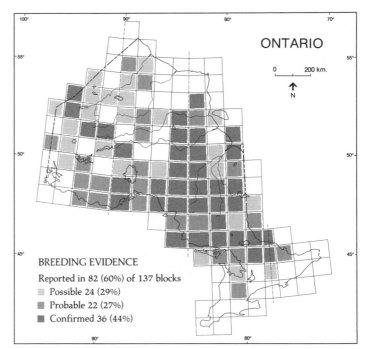

ONTARIO

0   200 km.

N

BREEDING EVIDENCE
Reported in 82 (60%) of 137 blocks

Possible 24 (29%)
Probable 22 (27%)
Confirmed 36 (44%)

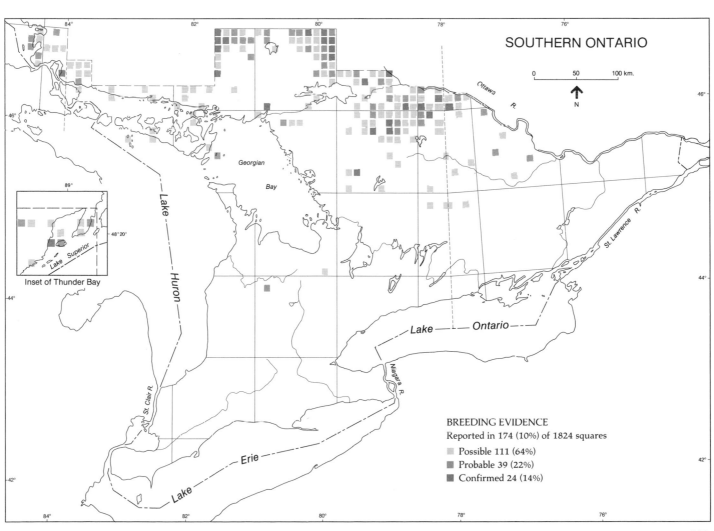

SOUTHERN ONTARIO

0   50   100 km.

N

Inset of Thunder Bay

BREEDING EVIDENCE
Reported in 174 (10%) of 1824 squares

Possible 111 (64%)
Probable 39 (22%)
Confirmed 24 (14%)

CERULEAN WARBLER
Paruline azurée
Dendroica cerulea

The Cerulean Warbler is a neotropical migrant that spends the winter months in central and northern South America and the summer months in extreme southern Canada and the eastern US. Within Canada, it is found only in southern Ontario and extreme southern Quebec. The Cerulean Warbler is uncommon and breeds only in the broad-leaved forests of the southern Great Lakes-St. Lawrence and Carolinian Forest zones. The species frequents the upper canopy of large deciduous trees within extensive blocks of woodland.

In 1886, McIlwraith reported that in southern Ontario the species was a "regular summer resident" and "local in its distribution." Baillie and Harrington (1937) stated that the Cerulean Warbler bred in only 6 of the southwestern Ontario counties but that records from Ottawa and Leeds Co suggested a broader range. It is probable that the species had declined in numbers earlier, during the early 1800's, as the extensive deciduous forests were cut down by the European settlers to make way for agriculture. The entire range of the species in Canada was undoubtedly affected by the significant decline of forest cover. Today the species could be described as being local in distribution and uncommon.

The Cerulean Warbler is small, has a weak voice, and frequents the tops of tall deciduous trees. The species is not easy to find, and even harder to confirm as a breeding bird. The nests are situated high in the trees, at a height of between 7.5 and 18 m (Godfrey 1986), making breeding confirmation difficult. The nests are so difficult to find that it took 50 years for the first nest to be reported after the species was first discovered in the early 1800's (Bent 1953). In Ontario, 48% of all atlas records were of singing males and a further 26% were of birds on territory. Confirmations of breeding were rare, with only 11% of all records being in this category, and nests being reported from only 3 squares.

In its Ontario range, the species is found in areas that still contain substantial blocks of forest. In southwestern Ontario, it occurs in river valleys, large parks, and other areas with large trees. Farther north, it occurs in the more extensive blocks of deciduous forest along the southern margin of the Canadian Shield.

The maps show an unusual distribution comprised of 2 apparent bands of occurrence, which is similar to that of the more numerous and widespread Yellow-throated Vireo. The southern band is in the Carolinian Forest zone, an area where the Cerulean Warbler has been known to occur for over 100 years. In New York State, atlas records reveal that the majority of the population is found along the southern shore of Lake Ontario, an area directly east of Ontario's Carolinian Forest region (NYBBA preliminary maps). The more northerly band of Cerulean Warbler distribution extends from the Bruce Peninsula to eastern Ontario along the edge of the Canadian Shield, with a notable concentration of records on the Frontenac Axis north of Kingston. This northern distribution comes as a surprise to many of Ontario's naturalists, as it is north of the range described by Godfrey (1986), but it validates the suggestion of a northern population made by Baillie and Harrington in 1937. Nevertheless, northward expansion, as in the eastern US (Laughlin and Kibbe 1985), may also be responsible for this newly discovered pattern. The northern band of distribution occurs in the Great Lakes-St. Lawrence Forest zone. This forest type is predominantly deciduous trees but some conifers, such as white pine, occur frequently. The northern band contains a higher proportion of forested land than does the more southerly area, and the presence of the Cerulean Warbler is possibly a reflection of this fact. Records in that part of Ontario are more sporadic than the extent of suitable habitat would suggest. An inconspicuous species like this one could very well be present, and even breeding, in more squares throughout that heavily wooded area than are shown here.

In Ontario, the unexpected extent of this northern range may be an indication of a similar trend; there are probable breeding records as far north as northern Georgian Bay and the western outskirts of Ottawa.

This species is uncommon in Ontario. It was recorded from only 108 atlas squares. Eighty percent of the abundance estimates are of fewer than 11 pairs per square.-- *P.F.J. Eagles*

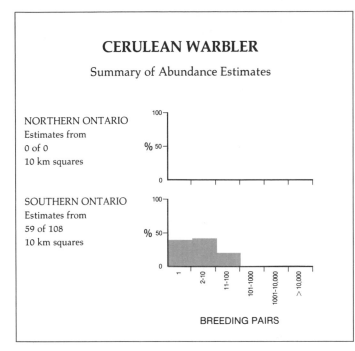

CERULEAN WARBLER

Summary of Abundance Estimates

NORTHERN ONTARIO
Estimates from
0 of 0
10 km squares

SOUTHERN ONTARIO
Estimates from
59 of 108
10 km squares

BREEDING PAIRS

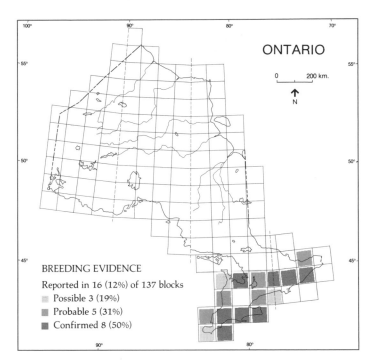

ONTARIO

BREEDING EVIDENCE
Reported in 16 (12%) of 137 blocks
Possible 3 (19%)
Probable 5 (31%)
Confirmed 8 (50%)

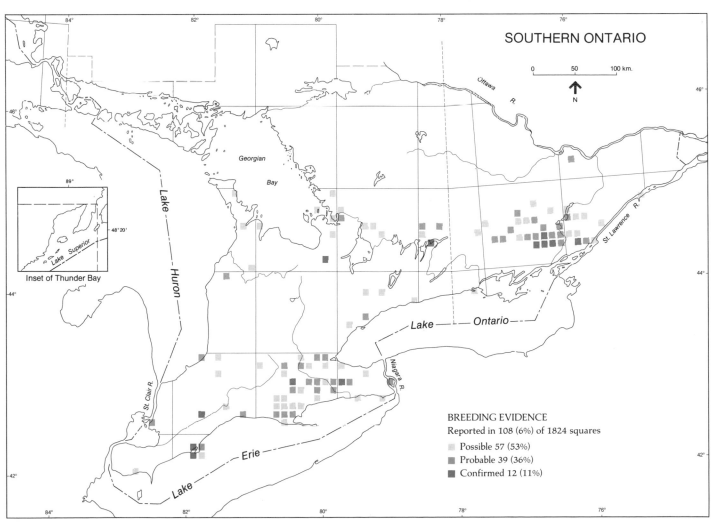

SOUTHERN ONTARIO

Inset of Thunder Bay

BREEDING EVIDENCE
Reported in 108 (6%) of 1824 squares
Possible 57 (53%)
Probable 39 (36%)
Confirmed 12 (11%)

BLACK-AND-WHITE WARBLER
Paruline noir et blanc
Mniotilta varia

Unobtrusive but widespread, the Black-and-White Warbler breeds in deciduous and mixed woods from Newfoundland and the Mackenzie River valley south to Georgia and eastern Texas. It winters from the extreme southern US to northern South America. In Ontario, it has been found nesting from near London north to James Bay, although its main range is in the Great Lakes-St. Lawrence Forest zone from Rainy River and Kenora east and south to Ottawa and Guelph.

The literature suggests that the Black-and-White Warbler was formerly more numerous in southwestern Ontario, and that it was scarcely known in far northern Ontario. Saunders and Dale (1933), for example, noted that, in Middlesex Co, "every cedar swamp [has] its quota," while Snyder mentioned several summer records north of Long Point in the 1920's (Speirs 1985). Atlassers recorded this species in only 3 dozen squares from Hamilton and Stratford southward. Baillie and Harrington (1937) documented that the Black-and-White Warbler reached northward only to the Canadian National transcontinental railroad line, while Godfrey (1986) shows its range extending well north of this line, at least as far as a line from Sandy Lake to Fort Albany. Speirs (1985) cites several records as far north as 55^{0}N on the Manitoba border. Atlassers obtained breeding evidence considerably further north than the range described by Godfrey (1986). Likely these northern records reflect more thorough recent coverage of the far north, rather than an actual extension of the breeding range. In the north, it probably occurs in small numbers wherever there are substantial numbers of deciduous trees mixed in with the conifers. In the southwest, however, it is almost certain that this species really is as scarce as the records indicate, mainly because of the loss of sufficiently large woodlands, and, in particular, of the cedar swamps.

The Black-and-White Warbler normally places its nest on the ground, at the base of a tree, stump, or large rock, typically concealed within a drift of dead leaves; occasionally it will nest in the top of a stump (Bent 1953, Harrison 1984). The nest normally contains 5 eggs, and this number seldom varies (Bent 1953). Its sibilant and frequently repeated song readily reveals its presence on breeding territory. The species has a propensity to reduce the song level in June, but it does sing through July and as late as August. It is not shy, and will go to its nest even when closely observed. It may perform a distraction display when disturbed at the nest (Harrison 1984). However, it is rather thinly distributed over its desired habitat, and conceals its nest well, so that finding nests calls for a long, patient search, frequently through difficult terrain. Despite these difficulties, the breeding confirmation rate of 25% of southern squares, and 10% of northern squares in which it was recorded, is not noticeably different from that of other warbler species.

The southern Ontario map shows the species as occurring in the majority of squares where woods remain in southern Ontario, and north into the Boreal Forest of northern Ontario. The scattered locations both in far northern and in southwestern Ontario reflect the scarcity of the species in these areas. The lack of confirmed records in northwestern Ontario is likely a reflection of the difficulty of finding a small, quiet forest bird of inconspicuous habits and the fact that much of the northern atlassing took place in late June and July when singing is reduced. A number of atlas Regional Coordinators noted that it likely breeds wherever suitable habitat occurs, although some insisted that this habitat could include smaller woodlots as well as the larger continuous stands which most often seem to attract it.

The Black-and-White Warbler appears to be relatively abundant where it occurs. It ranks along with the Yellow-rumped, Black-throated Green, and Chestnut-sided Warblers in the frequency of its occurrence. The atlas abundance data show that the species was estimated to have higher population numbers in northern Ontario. Of 46 BBS routes through southcentral and northern Ontario, only one had more than 10 Black-and-White Warblers per 50 stops, 5 had more than 6 birds per 50 stops, and 35 averaged one bird per 50 stops over the years 1968 to 1977 (Speirs 1985). It is, in short, a widespread and common species.-- *J. Cartwright*

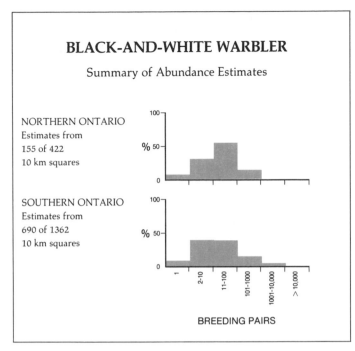

BLACK-AND-WHITE WARBLER

Summary of Abundance Estimates

NORTHERN ONTARIO
Estimates from
155 of 422
10 km squares

SOUTHERN ONTARIO
Estimates from
690 of 1362
10 km squares

%

100

50

%

100

50

1　2-10　11-100　101-1000　1001-10,000　>10,000

BREEDING PAIRS

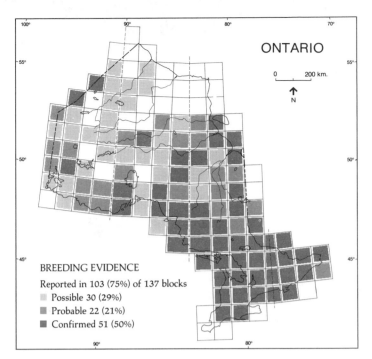

ONTARIO

0　200 km.

N

BREEDING EVIDENCE

Reported in 103 (75%) of 137 blocks

Possible 30 (29%)

Probable 22 (21%)

Confirmed 51 (50%)

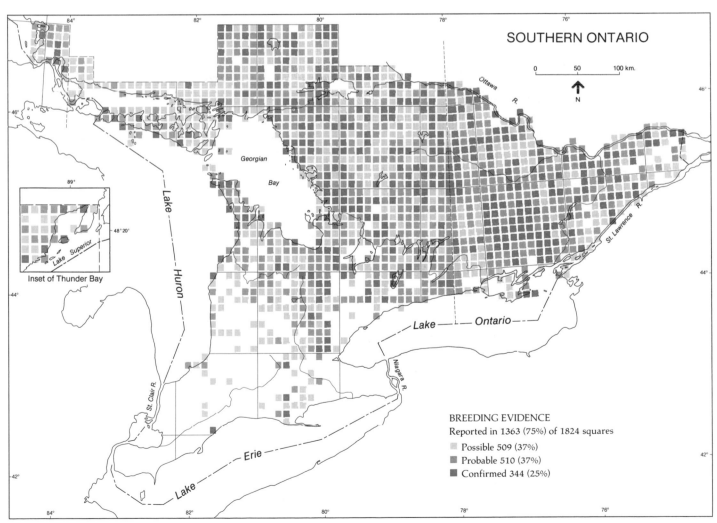

SOUTHERN ONTARIO

0　50　100 km.

N

Lake Huron

Georgian Bay

Lake Superior

Inset of Thunder Bay

St. Clair R.

Lake Erie

Niagara R.

Lake Ontario

Ottawa R.

St. Lawrence R.

BREEDING EVIDENCE

Reported in 1363 (75%) of 1824 squares

Possible 509 (37%)

Probable 510 (37%)

Confirmed 344 (25%)

AMERICAN REDSTART
Paruline flamboyante
Setophaga ruticilla

Long a favourite among birders for its sprightliness and colour, the American Redstart has one of the most widespread breeding ranges of any wood warbler, breeding across North America from Newfoundland west to Alaska, and south to the Gulf of Mexico. Its Ontario range centres on the Great Lakes-St. Lawrence Forest, although it has been found breeding north to 53°N, from James Bay to the Manitoba border. Primarily a bird of the deciduous understory and edges of woodlands, it feeds largely by flycatching among the understory, and nests in the crotch of a young tree between 1 and 7 m above the ground; thus it needs woods in which there is considerable regeneration of young growth. Its preferred breeding habitat includes open and semi-open deciduous and mixed forests (Collins *et al.* 1982). Its winter range extends from the extreme south of the US to northern South America.

While the American Redstart occupies essentially the range indicated by Godfrey (1986), its numbers seem to have declined in recent years. Ontario BBS data from 1967 to 1983 show a slight, but fairly consistent decline of about 2.5% a year. Since the American Redstart is a bird of maturing rather than fully mature woods, decreases in numbers may simply reflect the fact that specific woods do become less suitable as the trees mature. However, one Regional Coordinator suggested that there might be a drop because more control of forest fires may mean fewer regenerating stands (R. Tozer pers. comm.), while some observers have warned that massive deforestation of passerine wintering habitat in Central America and the West Indies may result in a reduction in the populations of many of the neotropical migrants, including the American Redstart (Terborgh 1980). However, in the core of its Ontario range, the American Redstart is still a common and conspicuous summer resident. The fact that

males sing well into July, the species' practice of foraging in the understory with frequent flycatching sallies, its lack of shyness around humans, and its ability to live in relatively small woodlots, all ensure that it will be readily noted wherever it occurs.

The maps show that the American Redstart is widespread in those portions of Ontario with deciduous forests. The gaps in the distribution map in the south reflect a shortage of suitable habitat, rather than any failure by observers to detect the species. In Essex Co and other intensively farmed areas in southwestern Ontario, the absence of American Redstarts is due to a lack of forest habitat. In the northern Boreal Forest, the few small pockets of aspen and birch provide limited habitat for the species, as the somewhat scattered records from more northern blocks suggest. However, there were a number of atlas records in places north of the previously known range, including the lower and middle portions of the Attawapiskat River and its tributaries, and a scattering of records along the Manitoba border south from Lake Opasquia. Quite possibly there are a few other breeding pairs in the pockets of deciduous trees along this fringe of the species' range, but it is not a species one would expect in predominantly coniferous forest.

Within the mixed forest of the Great Lakes-St. Lawrence region, the American Redstart is a common species, according to atlas data. Abundance estimates for southern Ontario suggest that it is more abundant than such deciduous forest species as the Black-throated Green and Black-throated Blue Warblers, although less common than the Ovenbird. Similarly, the BBS data compiled by Speirs (1985) for the years 1968 to 1977 suggest a common, generally distributed species, except in the heavily farmed portions of southern Ontario and parts of far western Ontario. It was most common on Breeding Bird Surveys in parts of the mixed Great Lakes-St. Lawrence Forest zone, with peaks of 20.4 individuals per 50 stops in the Huntsville area, and 18 per 50 stops in the Ottawa Valley east of North Bay. While the apparent decline in numbers of this species gives some cause for concern, the American Redstart will likely be with us for some time yet.-- J. Cartwright

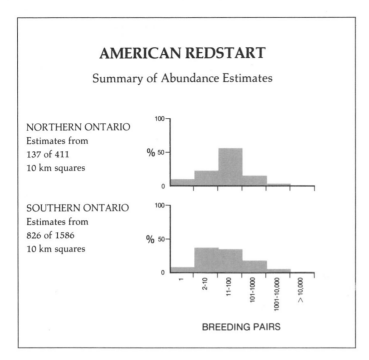

AMERICAN REDSTART

Summary of Abundance Estimates

NORTHERN ONTARIO
Estimates from
137 of 411
10 km squares

SOUTHERN ONTARIO
Estimates from
826 of 1586
10 km squares

% 100

% 50

% 100

% 50

1 2-10 11-100 101-1000 1001-10,000 >10,000

BREEDING PAIRS

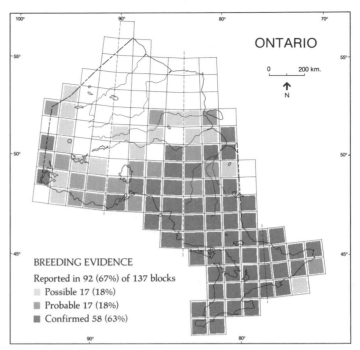

ONTARIO

0   200 km.

N

BREEDING EVIDENCE

Reported in 92 (67%) of 137 blocks

Possible 17 (18%)

Probable 17 (18%)

Confirmed 58 (63%)

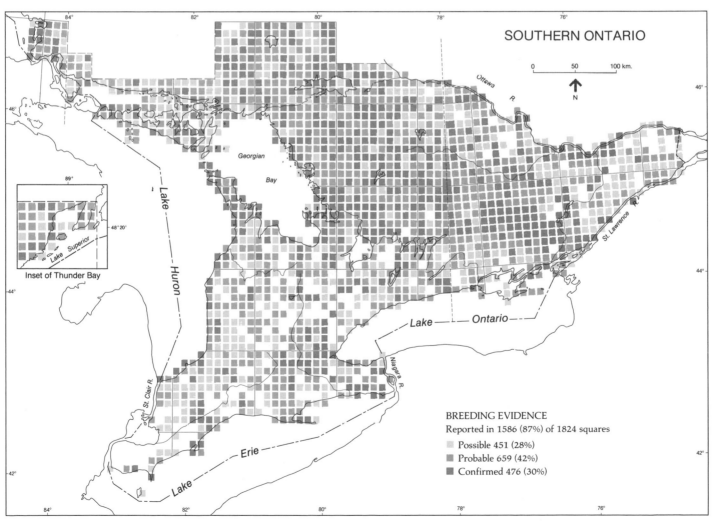

SOUTHERN ONTARIO

Ottawa R.

0   50   100 km.

N

Georgian Bay

Lake Huron

St. Lawrence

Lake Superior

Inset of Thunder Bay

St. Clair R.

Niagara R.

Lake — Ontario —

Lake — Erie

BREEDING EVIDENCE

Reported in 1586 (87%) of 1824 squares

Possible 451 (28%)

Probable 659 (42%)

Confirmed 476 (30%)

PROTHONOTARY WARBLER
Paruline orangée
Protonotaria citrea

In Ontario, the Prothonotary Warbler is a denizen of the southern swamps, where its brilliant plumage and ringing song contrast strikingly with the sombre dark waters.

The breeding range of the Prothonotary Warbler includes most parts of the eastern US. Its winter range extends from southern Mexico to northwestern South America. Ontario's largest and most regular breeding populations occur along the north shore of Lake Erie, where the species is particularly associated with sandspits (Point Pelee, Rondeau, and Long Point). It nests singly at times, but more commonly in groups, no doubt because of the localized nature of its habitat. The Prothonotary Warbler occurs in a variety of deciduous swampy conditions, most typically in fairly open circumstances afforded by extensively flooded tracts of silver maple-red maple swamp, buttonbush swamp, and black willow-maple-ash floodplains. It is nearly always closely associated with an open pool of standing or flowing water. Tree cavities are used as nest sites; thus, dead trees are usually an important component of the habitat. Nests are invariably situated low, generally from near water level up to 3 m in height. Mosses account for the majority of the nest material, and are a critical habitat requirement. The shrub and herbaceous cover is always well interspersed with open water and is seldom dense (McCracken 1981).

Although the Prothonotary Warbler could be regarded as having expanded its range into Ontario during this century (Snyder 1957b, Baillie 1967), the evidence for this is not strong. Its patchy and inaccessible habitat, combined with the scant early ornithological coverage of the province, could very well have left its breeding populations undiscovered prior to the 1930's. It is hard to believe that the Rondeau population, first documented in 1929 and later reported to contain as many as 100 pairs in 1933 (Baillie and Harrington 1937), was of recent origin. Although this population estimate may have been too high (A. Woodliffe pers. comm.), the Prothonotary Warbler had obviously been well established at that site long before its discovery. Evidence for long-term population trends for the species in Ontario is difficult to assess (McCracken 1981). The Rondeau population was said to have crashed because of the removal of dead timber (Baillie 1967), or the very low water levels of the 1930's (A. Woodliffe pers. comm.). Recent population increases have been reported at a few sites, notably Long Point (McCracken 1987). Despite the discovery of several new sites where nesting may have taken place during the atlas period, only one additional confirmed breeding station was found (Pinery Prov. Park). In 1986, birds were rediscovered at 2 former known breeding locations: Point Abino and Hamilton (D. Sutherland and R. Curry pers. comm.). Prothonotary Warblers were apparently absent from these sites during the atlas period, as it is doubtful that they could have been overlooked by altassers. Recolonization of formerly held habitat, rather than colonization of new areas, seems to attest to the rather narrow habitat tolerance of the species and its general inability to make widespread expansions in Ontario. All things considered, its population is relatively stable.

Although its habitat is difficult to access, the Prothonotary Warbler is fairly easy to detect. The song of the male is persistent, loud, and easily recognizable. Confirming breeding is also a fairly simple task; the male frequently feeds the female on the nest and both parents care for the young. The low, cavity nest is usually easily found and inspected. In addition, the male often builds one or more rudimentary 'dummy' nests early in the breeding season, making discovery of the functional nest that much simpler. The species displays remarkable site-tenacity and usually returns to breed in the same area, often at the same nest site, year after year (Walkinshaw 1941, McCracken 1981).

The maps conform well to other published accounts of the range of this warbler in Ontario (Godfrey 1986), and they effectively portray the localized distribution of the species.

Although the Prothonotary Warbler is rare and extremely localized in Ontario, it can be considered locally uncommon at Rondeau and Long Point. Weir (1985c) reported that up to 30 singing males have been counted at Rondeau, and A. Woodliffe (pers. comm.) estimates that the area may support from 40 to 50 pairs. Populations in the Long Point region currently number about 20-25 pairs (McCracken 1987). The entire known Canadian breeding population is estimated to consist of fewer than 80 pairs. As such, COSEWIC has designated the species as rare in Canada (McCracken 1981).-- *J.D. McCracken*

PROTHONOTARY WARBLER

Summary of Abundance Estimates

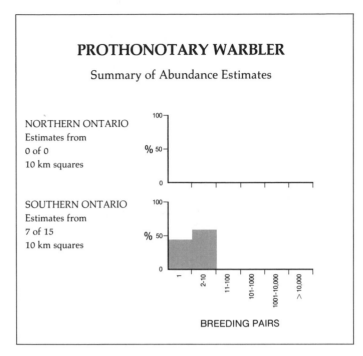

NORTHERN ONTARIO
Estimates from
0 of 0
10 km squares

SOUTHERN ONTARIO
Estimates from
7 of 15
10 km squares

BREEDING PAIRS

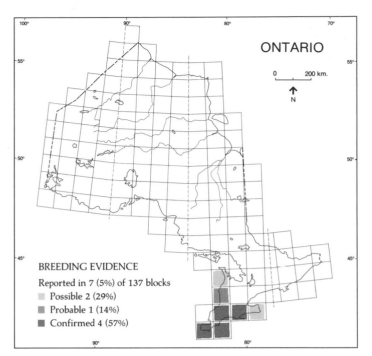

ONTARIO

BREEDING EVIDENCE
Reported in 7 (5%) of 137 blocks

Possible 2 (29%)
Probable 1 (14%)
Confirmed 4 (57%)

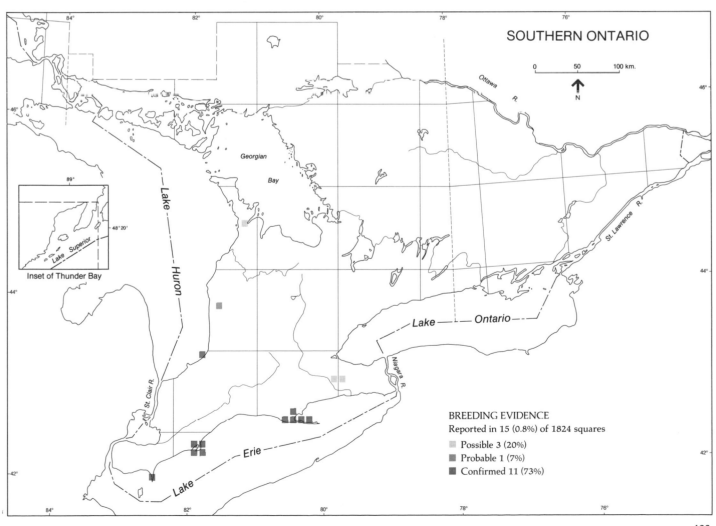

SOUTHERN ONTARIO

Inset of Thunder Bay

BREEDING EVIDENCE
Reported in 15 (0.8%) of 1824 squares

Possible 3 (20%)
Probable 1 (7%)
Confirmed 11 (73%)

OVENBIRD
Paruline couronnée
Seiurus aurocapillus

The Ovenbird is widely distributed across the forested areas of Canada and the eastern US, occurring in all Canadian provinces and territories. It is predominantly a bird of undisturbed mature forests, both deciduous and mixed. Typical nesting habitat consists of closed canopy forest with relatively little ground vegetation. The Ovenbird breeds throughout much of Ontario where habitat conditions are suitable. Prior to the atlas project, nesting had been reported everywhere in the province except the extremities to the southeast, southwest, and north (Peck and James 1987). It winters from the southern US to Venezuela (Godfrey 1986).

There is little evidence to indicate that the distribution of the Ovenbird has changed greatly in recent times. It was described by Baillie and Harrington (1937) as a "common summer resident throughout most of Ontario." The gradual depletion of forested land in southern Ontario in the past no doubt progressively reduced Ovenbird populations; however, the species is still found in woodlots, forested ravines, and other areas with residual timber. It is not often found in urban areas, even where apparently suitable habitat occurs (J.B. Falls pers. comm.). In this respect, it is similar to other ground nesting bird species which suffer from domestic cat and dog predation in such areas.

The presence of Ovenbirds is readily detected. Its characteristic 'teacher-teacher-teacher' song is loud and unmistakable, thus making possible (*i.e.*, singing males) or probable (*i.e.*, territorial birds) breeding records relatively easy to obtain. Territorial males account for over one-half (57%) of the probable breeding reports in southern Ontario. Ovenbird nests are among the most difficult nests to find, as they are covered on the top (hence the bird's name) and well camouflaged on the forest floor. Adults rarely flush unless the observer is almost on top of the nest. Consequently, nests were rarely found, accounting for less than 5% of all breeding records, and 22% of the confirmed breeding reports in southern Ontario.

The atlas data show that the Ovenbird is distributed over a large portion of Ontario, including virtually all of southern Ontario and most of the forested parts of northern Ontario. The lack of confirmed records in several blocks within the species range in northern Ontario suggests low sampling intensity as well as lower densities of Ovenbirds in predominantly coniferous habitat. The most northerly records, which occurred along the Fawn and Sachigo Rivers, were from aspen woodlands on well-drained riverbanks. The open coniferous muskeg habitat away from the northern rivers is unsuitable for Ovenbirds. The species is absent from a few squares in southwestern Ontario where the removal of forests for agriculture has greatly reduced available habitat. The same is true, to a somewhat lesser extent, in the heavily urbanized area around Toronto. The Ovenbird likely nests in a large percentage of the squares in which it was recorded. The ease with which the Ovenbird is located has facilitated a very thorough documentation of its Ontario range.

The species is best described as common and widespread throughout Ontario, except on the northern section of the Hudson Bay Lowland where it is absent or scarce, generally echoing Baillie and Harrington's (1937) assessment. Abundance estimates from southern and northern Ontario do not differ greatly, but average somewhat higher in the south. The Ovenbird is most abundant in the Great Lakes-St. Lawrence Forest region, where mixed and hardwood forests predominate. The highest reported abundance estimates, those of over 1,000 pairs per square, were from squares on the Canadian Shield in central Ontario. BBS data show that Ontario is among the 4 states or provinces with the highest counts of Ovenbirds (Robbins *et al.* 1986) and indicate a significant increase in breeding populations in the province from 1967 to 1983.-- *E.R. Armstrong*

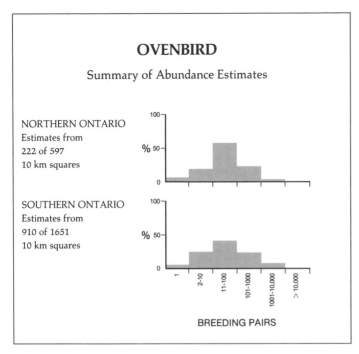

OVENBIRD

Summary of Abundance Estimates

NORTHERN ONTARIO
Estimates from
222 of 597
10 km squares

SOUTHERN ONTARIO
Estimates from
910 of 1651
10 km squares

BREEDING PAIRS

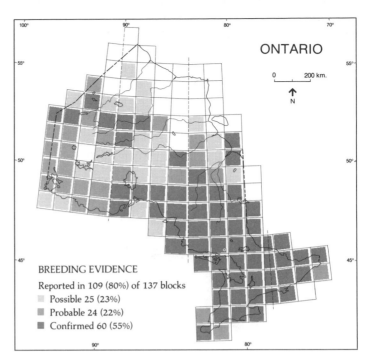

ONTARIO

0   200 km.

N

BREEDING EVIDENCE

Reported in 109 (80%) of 137 blocks

Possible 25 (23%)
Probable 24 (22%)
Confirmed 60 (55%)

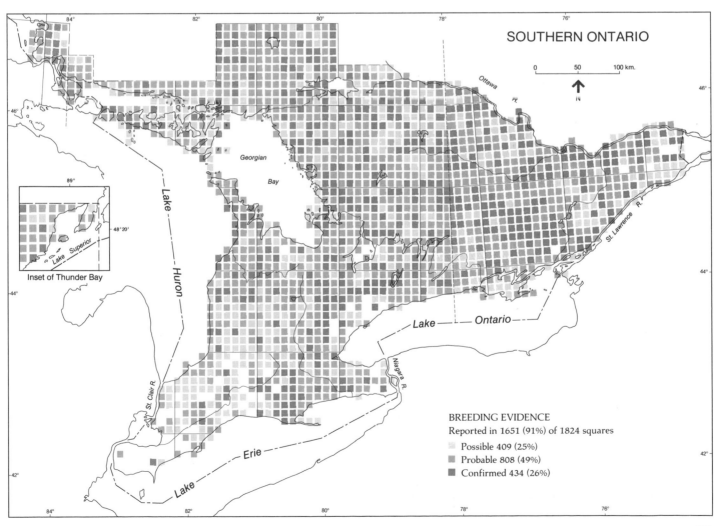

SOUTHERN ONTARIO

0   50   100 km.

N

Georgian

Bay

Lake

Huron

Lake

Erie

Lake — Ontario

Niagara R.

St. Clair R.

St. Lawrence R.

Ottawa R.

Inset of Thunder Bay

Lake Superior

48° 20'

89°

BREEDING EVIDENCE

Reported in 1651 (91%) of 1824 squares

Possible 409 (25%)
Probable 808 (49%)
Confirmed 434 (26%)

NORTHERN WATERTHRUSH
Paruline des ruisseaux
Seiurus noveboracensis

The Northern Waterthrush is a spring migrant to Ontario from its winter range that extends from the southern US to northern South America. On the breeding grounds, its loud, distinctive song can be heard from the northeastern US across all of Canada as far north as the tree-line. In Ontario, it has been known to breed from southern Ontario to the salt water coast in the north.

The Northern Waterthrush is a bird of forested wetlands, a habitat that is abundant in Ontario. It prefers swamps but can also be found on lake edges, along river banks, and in wooded ravines. As early as 1886, McIlwraith stated that this bird "was found in all suitable places throughout the country." Baillie and Harrington (1936) stated that the species breeds "throughout the province." The number and distribution of the species in Ontario do not appear to have changed noticeably since these reports.

The Northern Waterthrush is highly vocal: in 59% of all Ontario squares in which it was recorded, the highest level of breeding evidence consisted of birds recorded singing or on territory. The species is reasonably easy to find once the atlasser gains access to its habitat, which can be a somewhat daunting proposition. The bird responds aggressively to 'pishing', thereby making it easy to record. Access to the wet forest, full of tangles and fallen logs and home to hordes of mosquitoes, requires determination on the part of the atlasser. The large number of records in the province shows both the high numbers of the species and the perseverance of the volunteer field workers. Laughlin and Kibbe (1985) report that confirmation of breeding was relatively easy to document during the nesting period in Vermont. Many Ontario Regional Coordinators have a different opinion, and report that the species is underconfirmed. The percentage of records in the confirmed breeding category in Ontario, 17%, is low

for such a common bird and shows the difficulty in finding the well-hidden nests in banks and upturned tree roots (Bent 1953) as well as the difficulty in accessing the wetland habitat. Fifty-six percent of confirmed records are of adults carrying food or faecal sacs. Only 1% of 1,574 records is of nests with eggs or young. Fledged young were easier to find, with 4% of all records in this category.

The maps show the wide distribution of the Northern Waterthrush from the warm shores of Lake Erie to the cool areas near the coast of Hudson Bay. The atlas field work revealed a considerable number of records in southwestern Ontario, an area lying south of the previously published range of the species (Godfrey 1986). The Rondeau Prov. Park records are the southernmost of the atlas and are over 100 km south of Godfrey's (1986) range map. It is probable that this southward range extension is just an increase in our knowledge of the species, rather than a real change in its breeding range.

The absence of the bird in some of the apparently suitable habitat in southern Ontario is mystifying. A number of Regional Coordinators commented on this fact, with no possible reasons given. For example, the Regional Coordinator for the Hamilton region stated that it was quite curious that the species was not found in the wet woods on the Niagara Peninsula since it was abundant west and north of there (R. Curry pers. comm.).

Where it occurs, this species is common in its preferred habitat in Ontario and was recorded in 63% of southern Ontario squares and 48% of the squares visited in northern Ontario. The majority of abundance estimates in southern Ontario are of between 1 and 100 pairs per square, with 2 to 10 per square being most frequently reported. In the north, atlassers estimated the abundance to be somewhat higher, with 11 to 100 pairs per square being most frequent.

The highest abundance estimates in southern Ontario, those above 100 pairs per square, are in the atlas regions of Peterborough, Haliburton, Algonquin, and Nipissing East. In northern Ontario, the region containing Big Trout Lake, has higher estimates of abundance than elsewhere.-- *P.F.J. Eagles*

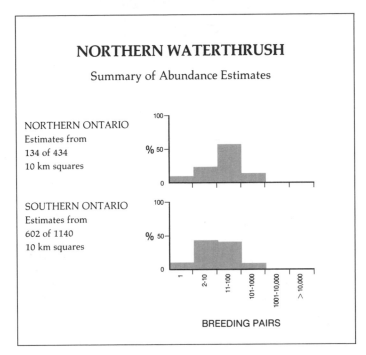

NORTHERN WATERTHRUSH

Summary of Abundance Estimates

NORTHERN ONTARIO
Estimates from
134 of 434
10 km squares

SOUTHERN ONTARIO
Estimates from
602 of 1140
10 km squares

BREEDING PAIRS

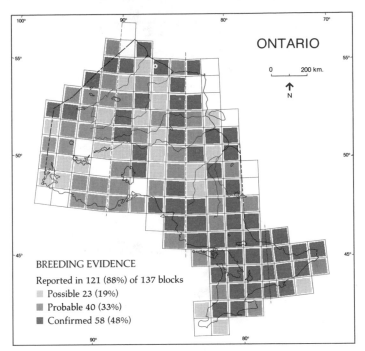

ONTARIO

0   200 km.

N

BREEDING EVIDENCE
Reported in 121 (88%) of 137 blocks

Possible 23 (19%)
Probable 40 (33%)
Confirmed 58 (48%)

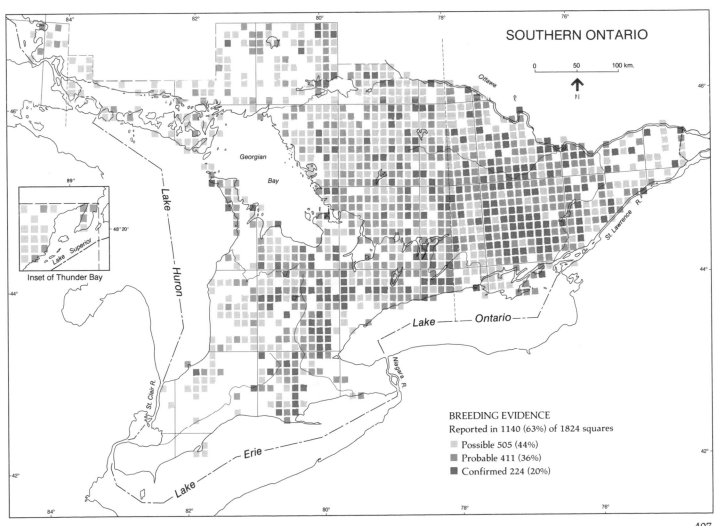

SOUTHERN ONTARIO

0   50   100 km.

N

Georgian Bay

Lake Huron

Lake Superior

Inset of Thunder Bay

Ottawa

St. Lawrence R.

Lake — Ontario

Niagara R.

Lake Erie

St. Clair R.

BREEDING EVIDENCE
Reported in 1140 (63%) of 1824 squares

Possible 505 (44%)
Probable 411 (36%)
Confirmed 224 (20%)

LOUISIANA WATERTHRUSH
Paruline hochequeue
Seiurus motacilla

The Louisiana Waterthrush is one of the many southern birds that are common in their range across the eastern US but rare in Canada, with southern Ontario being its only foothold in the country. This bird's winter range is from Mexico to northern South America. The Louisiana Waterthrush is similar to its close relative, the Northern Waterthrush, but is by far the rarer of the 2 species in Ontario.

The Louisiana Waterthrush is eagerly sought by birders because of its rarity and because of its unique habitat: deep forested ravines that contain tumbling waters, moss-covered boulders, and often a waterfall. As early as 1886, McIlwraith captured the essence of the bird when he stated that when it is found in "a rocky ravine, its loud, clear notes are almost sure to be heard in the spring, mingling with the sound of the falling water." Less commonly, this species occupies forested swamps, the preferred habitat of the Northern Waterthrush.

This species seems always to have been rare in Ontario. McIlwraith (1886) stated that it was "not generally distributed" and when compared to the Northern Waterthrush was "by no means so common a bird." He reported finding the species "in a deep ravine, which there cut through the mountain wall" west of Hamilton. Today, the species still breeds in some of the forested ravines where rivers cut through the Niagara Escarpment in the Hamilton area, suggesting a high site tenacity of the bird. Baillie and Harrington (1937) reported that the Louisiana Waterthrush "reaches the lower edge of Ontario in small numbers, occurring north to Middlesex and east to Frontenac Counties."

The species is so rare in Canada that it is not well known to the majority of birders in Ontario. It is very similar in appearance to the Northern Waterthrush; hence, careful observation is required to discriminate between the two. For-

tunately, it has a unique song and alarm call that permit easy discrimination from the Northern Waterthrush by those atlassers who are familiar with the 2 species. Nevertheless, its unfamiliarity to many may have resulted in underrecording of the species.

The maps show the limited distribution of the species in Ontario. It was confirmed as a breeding bird in 12 of the 40 squares where it was found during the atlas period. The most frequently reported breeding category was that of singing male, which provided 33% of all records. Field work in 1986 is not included in the atlas data base, but in the Haldimand-Norfolk RM the species was found in 4 additional squares (J. McCracken pers. comm.) and in Elgin Co in 4 additional squares (W. Lamond pers. comm.). It is probable that the species occurred in these additional 8 squares during the atlas period but was not recorded until the detailed field work associated with Environmentally Sensitive Area inventories was conducted in 1986. Therefore the Louisiana Waterthrush was recorded in 48 squares between 1981 and 1986.

Atlas data show several extensions north of the previously known range of the Louisiana Waterthrush. Godfrey (1986) stated that the bird was found only in extreme southern Ontario. The map shows a much more northerly and easterly distribution, including confirmed breeding in the Bayfield River valley along Lake Huron, probable breeding in ravines in the Niagara Escarpment in Grey Co along Georgian Bay, and confirmed breeding north of Kingston. Probable breeding in several locations between Cambridge and Long Point was recorded in sites where the species was previously unknown. The southeastern Ontario records are not very far north of the previously known breeding range in New York State. R. Weir (pers. comm.) feels that the records in the Kingston region are the result of a range expansion that has taken place during the last 10 years. Much of the range change, and especially that along the northern reaches of the Niagara Escarpment, may be a result of the increase in field work associated with the atlas, and not an actual change in the distribution of the species.

The Louisiana Waterthrush is rare in Ontario. Most of the abundance estimates (82%) are for only one pair per square. Four squares (18%) have abundance estimates of between 2 and 10 pairs, with atlassers indicating that the actual abundance is in the low end of this range. Therefore, it is possible to estimate that the Ontario population of this species is between 50 and 100 pairs. The population is stable, but to the author's knowledge only 3 of the recorded sites are located in any form of protected area. All of the other sites are on private land. It is to be hoped that the difficult terrain on the majority of the sites will protect the birds from extensive land development, but the over-harvesting of trees is a distinct threat.-- *P.F.J. Eagles*

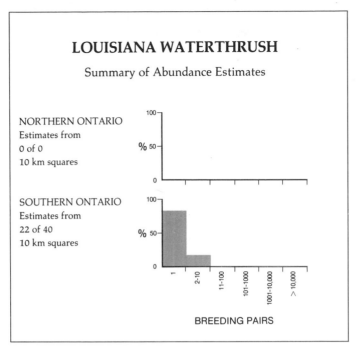

LOUISIANA WATERTHRUSH

Summary of Abundance Estimates

NORTHERN ONTARIO
Estimates from
0 of 0
10 km squares

SOUTHERN ONTARIO
Estimates from
22 of 40
10 km squares

BREEDING PAIRS

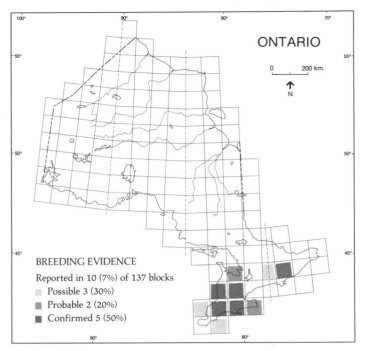

ONTARIO

0   200 km.

N

BREEDING EVIDENCE

Reported in 10 (7%) of 137 blocks

Possible 3 (30%)

Probable 2 (20%)

Confirmed 5 (50%)

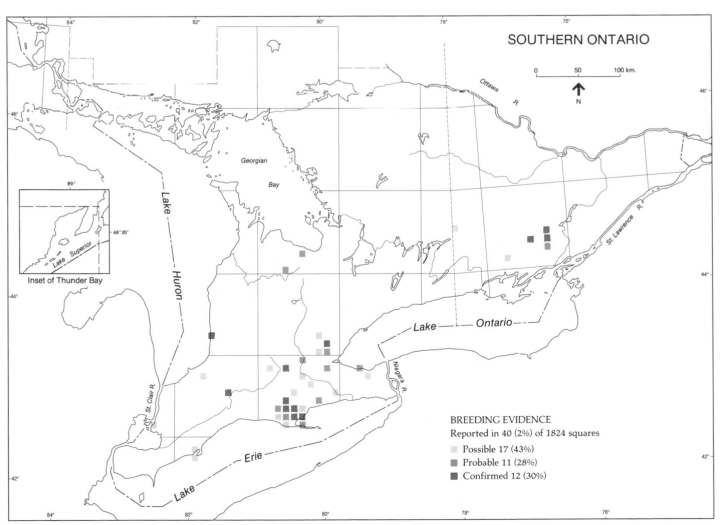

SOUTHERN ONTARIO

0   50   100 km.

N

Inset of Thunder Bay

BREEDING EVIDENCE

Reported in 40 (2%) of 1824 squares

Possible 17 (43%)

Probable 11 (28%)

Confirmed 12 (30%)

KENTUCKY WARBLER *
Paruline du Kentucky
Oporornis formosus

This species is locally common in the southeastern US, and is a regular breeder as far north as central Ohio. It winters from the West Indies to northern South America.

The Kentucky Warbler likely nests very rarely, sporadically, and locally in the southwestern part of the province (especially the Rondeau, Long Point, and Kettle Point regions), north to the eastern Lake Ontario region. However, no nests have ever been found in Canada and breeding was not confirmed during the atlas period. Including the atlas data, there is barely even a handful of summer records for the province. Moist deciduous or mixed woodlands, which support very dense herbaceous and shrub covers, provide suitable habitat. Such features are often associated with ravines, creek bottomlands, swamp edges, and blow-downs.

Because of the extreme rarity of the Kentucky Warbler in Canada, its historical status is difficult to assess. It was considered "accidental" in most early Canadian references (*e.g.*, Stirrett 1945). Although it is now considered to be a likely breeding species in Ontario, this status probably reflects heightened ornithological coverage of the province rather than increased numbers of birds.

Although it is a colourful species and a loud and persistent singer, the Kentucky Warbler is elusive and is seldom seen. Its song could be confused with that of a Carolina Wren, an Ovenbird, a Mourning Warbler, or even a Northern Cardinal. The nest is extraordinarily difficult to find (Griscom and Sprunt 1957, Bent 1953, Chapman 1968). The dense and impenetrable nature of the Kentucky Warbler's habitat, together with a general unfamiliarity of atlassers with its song, probably also contributes to a paucity of records. In total, there were breeding records in 9 atlas squares in Ontario. Seven were in the Carolinian zone of Ontario in the southwest and 2 were in eastern Ontario. Despite its recog-

nized rarity, it is entirely likely that the Kentucky Warbler was slightly underrecorded during the atlas field work.

The atlas records were scattered, with distinct concentrations in Rondeau Prov. Park and eastern Haldimand-Norfolk RM. Of the 2 probable records reported, one concerns an agitated male found near Rockport, Leeds Co on 6 June 1982 by Tom Hince. The other involved a territorial and agitated male noted by the author, near Vanessa, Haldimand-Norfolk R.M. on several dates from 11 to 19 July 1985. Remarkably, what was presumably the same bird maintained a territory at precisely the same location again throughout the 1986 breeding season (McCracken 1987). No breeding evidence higher than the atlas code for agitated behaviour was recorded. As intriguing as these records are, a mated pair has apparently yet to be documented in Ontario. Perhaps this is because females are very secretive, particularly around the nest (Chapman 1968).

No abundance estimates were submitted by atlassers, but data from the USRF's show that only one male was observed in each square.-- *J.D. McCracken*

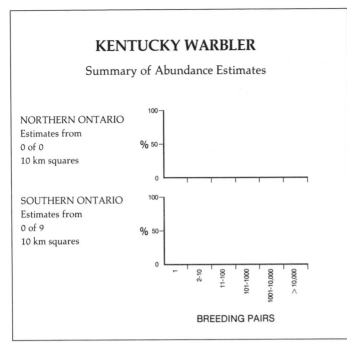

KENTUCKY WARBLER

Summary of Abundance Estimates

NORTHERN ONTARIO
Estimates from
0 of 0
10 km squares

SOUTHERN ONTARIO
Estimates from
0 of 9
10 km squares

BREEDING PAIRS

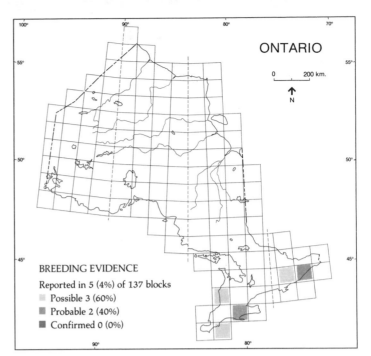

ONTARIO

0   200 km.

N

BREEDING EVIDENCE

Reported in 5 (4%) of 137 blocks

Possible 3 (60%)
Probable 2 (40%)
Confirmed 0 (0%)

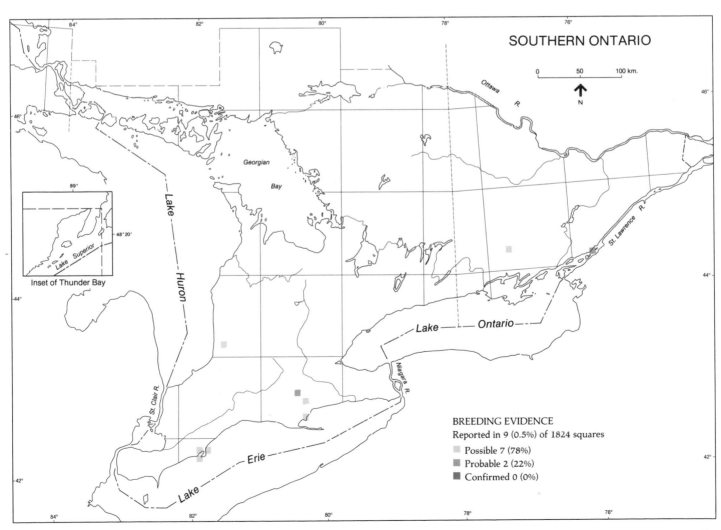

SOUTHERN ONTARIO

0   50   100 km.

N

Georgian Bay

Lake Huron

Lake Superior
Inset of Thunder Bay

St. Clair R.

Lake Erie

Niagara R.

Lake Ontario

Ottawa R.

St. Lawrence R.

BREEDING EVIDENCE

Reported in 9 (0.5%) of 1824 squares

Possible 7 (78%)
Probable 2 (22%)
Confirmed 0 (0%)

CONNECTICUT WARBLER
Paruline à gorge grise
Oporornis agilis

The Connecticut Warbler, while not rare in Ontario, is one of the most elusive of all the wood warblers, sought after by many a bird-watcher and familiar to very few. It has the distinction of being one of only 5 species accepted as breeding in Ontario by the Ontario Bird Records Committee for which material evidence of breeding has not been preserved (James 1984b).

The breeding range of the Connecticut Warbler lies almost entirely in Canada, in a narrow band extending from the Peace River district eastward through central Saskatchewan, Manitoba, and Ontario to the westernmost parts of Quebec (Godfrey 1986). Its winter range is primarily in northwestern South America (AOU 1983). The Connecticut Warbler can frequently be found in loose colonies, with several singing males being audible from a single point. Typically, this species occurs in extensive, fairly open, spruce bogs and tamarack fens, with a well developed understory (D. Welsh pers. comm.). Other open forests such as young jack pine stands or open poplar woods are also frequented in certain areas, suggesting that dry sites are sometimes as suitable as boggy ones, provided that the forest is not too dense. The Connecticut Warbler is not ubiquitous within its range, even in suitable habitat. For this reason, because it seldom passes through the densely populated parts of Ontario in migration, hides exceptionally well from view, and is a late spring migrant, the species is commonly regarded as being rare in Ontario, although it is actually quite common in some places within its breeding range.

Early descriptions of the status of the Connecticut Warbler in Ontario conform closely to its current status. Inasmuch as its breeding range has seen little ecological alteration over the years, this is hardly surprising.

Most descriptions of Connecticut Warbler behaviour stress that it is a 'skulker' and very difficult to see. In contrast, its song is powerful and almost unmistakable. It is therefore not surprising that 65% of records of this species are of singing males. Few birds proved to be as difficult as this one to upgrade from the status of possible breeding to that of probable or confirmed breeding. A strong singer like the Connecticut Warbler would be easily upgraded to 'territory' status by repeated visits by atlassers, but many of the squares in which this species was reported were remote and were visited only once. Almost 75% of the blocks shown on the map are at the possible breeding level, the highest percentage of possible breeding records among widespread passerine species.

Connecticut Warbler nests had completely eluded naturalists until 1883, when the first nest was discovered (Setton 1884). The only nest reported in Ontario, at Sibley Prov. Park, Thunder Bay District, on 24 July 1971, unfortunately was not adequately documented (Peck 1976, Peck and James 1987). However, there have been several sightings of adults feeding fledged young, including one at Sesikinika Lake, near Kirkland Lake, in 1985. The other 2 blocks in which confirmed breeding was established were based on sightings of adults carrying faecal sacs or food for young. The known range of this species in Ontario has been extended considerably northward as a result of the atlas project. It is now thought to be at least possibly breeding at several locations in the Severn River basin, including along the lower Sachigo River, 300 km northeast of Sandy Lake, which was the previously known northern limit of its range (James *et al.* 1976). As these records were all collected in June or July, it is highly likely that they represent breeding birds. The gaps in the atlas maps within the area northward from Lakes Huron and Superior may well contain breeding Connecticut Warblers.

The abundance estimates from southern and northern Ontario were similar. The highest estimates, those over 10 pairs per square, were scattered in a band diagonally across the Ontario range from near Blind River in the east to North Spirit Lake in the northwest.-- *F.M. Helleiner*

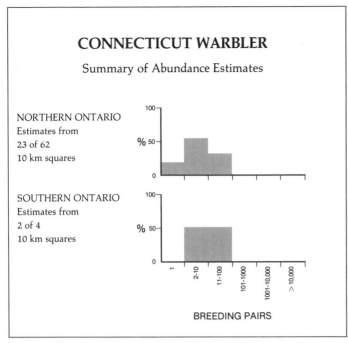

CONNECTICUT WARBLER

Summary of Abundance Estimates

NORTHERN ONTARIO
Estimates from
23 of 62
10 km squares

SOUTHERN ONTARIO
Estimates from
2 of 4
10 km squares

%

BREEDING PAIRS

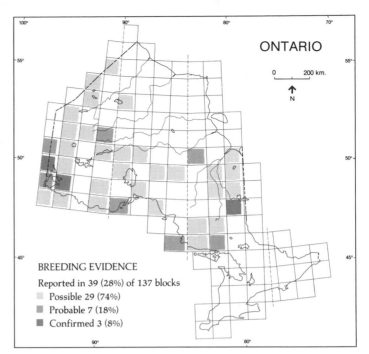

ONTARIO

0   200 km.

N

BREEDING EVIDENCE

Reported in 39 (28%) of 137 blocks

Possible 29 (74%)

Probable 7 (18%)

Confirmed 3 (8%)

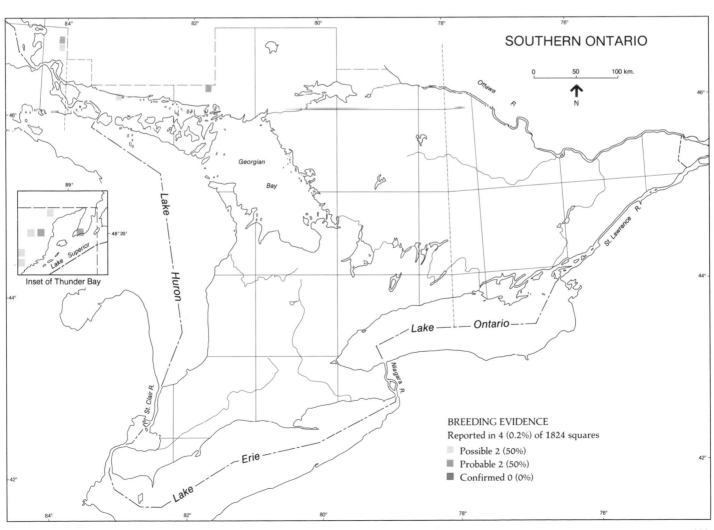

SOUTHERN ONTARIO

0   50   100 km.

N

Georgian
Bay

Lake
Huron

Lake

Superior

Inset of Thunder Bay

Ottawa R.

St. Lawrence R.

Lake — Ontario

Niagara R.

St. Clair R.

Lake

Erie

Lake

BREEDING EVIDENCE

Reported in 4 (0.2%) of 1824 squares

Possible 2 (50%)

Probable 2 (50%)

Confirmed 0 (0%)

413

MOURNING WARBLER
Paruline triste
Oporornis philadelphia

The Mourning Warbler breeds from northern Alberta east to Newfoundland, and south through the Appalachian Mountains to northwestern Virginia. The species is widely distributed throughout most of Ontario, where it typically frequents forest clearings, burned and cut-over areas, and the margins of swamps and watercourses. Although the Mourning Warbler occupies a wide variety of habitat niches, it exhibits a decided preference for mesic areas with dense shrubby undergrowth such as raspberry thickets, nettles, and brier patches. It winters in central America and northern South America.

Very little was known about the Ontario status and distribution of the Mourning Warbler prior to the turn of the century. McIlwraith (1894) provides no information concerning the nesting ecology of the species, while Nash (1908) considered it "a rather rare summer resident." Macoun and Macoun (1909) cite only a single Ontario nest, from Leeds Co. Two breeding records were obtained from Toronto in the 1890's (Fleming 1907). Baillie and Harrington (1937) considered the Mourning Warbler to be "a fairly common summer resident," breeding north to the Algoma, Cochrane, and Kenora Districts.

It is difficult to ascertain whether the early Ontario naturalists were unfamiliar with the habits of this somewhat elusive species, or whether the Mourning Warbler was once less common than today. On the Canadian Shield, its present status as one of the more common breeding warblers has probably undergone little change since the days of pre-European settlement. Several Regional Coordinators suggested that, as a breeder, the Mourning Warbler is currently more numerous and widespread than previously suspected, possibly because of the presence of brushy edges and openings in abundant regenerating, second-growth woodland.

Confirmed breeding status was attained in only 20% of the 1579 squares in which this species was recorded. A late spring migrant, the Mourning Warbler typically arrives in southern Ontario during the third week of May. By the time it has reached its breeding grounds, much of the vegetation is in full foliage. The Mourning Warbler's habit of skulking low to the ground in dense leafy underbrush makes it extremely difficult to observe. In spite of its inconspicuous nature, the presence of a territorial male is easily detected by its distinctive rolling song. It is not surprising, therefore, that 69% of all atlas records involved singing or territorial males.

Mourning Warblers build bulky nests on or near the ground, rarely at a height greater than 1 m (Bent 1953). Nests are typically well concealed in brier tangles, in tussocks of grass, or at the base of fern fronds (Bent 1953). Not only are the nests extremely difficult to locate, but their whereabouts are rarely betrayed by the adult birds. When incubating, females sit tightly on the nest and, if disturbed, walk away from the nest stealthily rather than taking flight. Nests with eggs or young were reported from only 8 squares in southern Ontario, and from 6 northern squares.

Both the incubation and fledging periods of the Mourning Warbler are short. Eggs usually hatch 12 days after laying (Hofslund 1954). Nestlings may fledge as early as 7 days after hatching (Cox 1960), often before they are capable of flying. Family groups may remain together for several weeks following fledging (Cox 1960). Adult Mourning Warblers readily become agitated while attending their young and are much more conspicuous during this period. This behaviour is reflected by the fact that 63% of all breeding confirmations in the province were of adults carrying food for young.

The Mourning Warbler breeds throughout most of Ontario, as far north as Sandy Lake and the Attawapiskat River, though the frequency of records drops off near the northern edge of the range. Along the north shore of Lake Erie, it is very sparsely distributed because of a lack of suitable habitat in this highly agricultural landscape. The absence from Prince Edward and Frontenac Counties is difficult to explain since apparently suitable habitat is present in these areas.

Abundance estimates show that the Mourning Warbler is common in Ontario. The highest estimates, those over 100 pairs per square, come from the centre of the Ontario range, an area from Haliburton north to Timmins and west to Lake of the Woods.-- *D.M. Fraser*

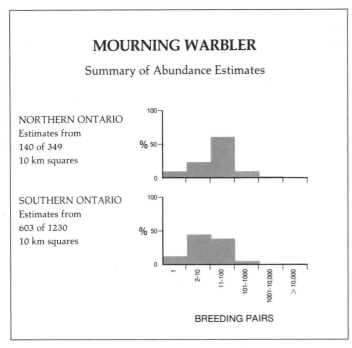

MOURNING WARBLER

Summary of Abundance Estimates

NORTHERN ONTARIO
Estimates from
140 of 349
10 km squares

SOUTHERN ONTARIO
Estimates from
603 of 1230
10 km squares

BREEDING PAIRS

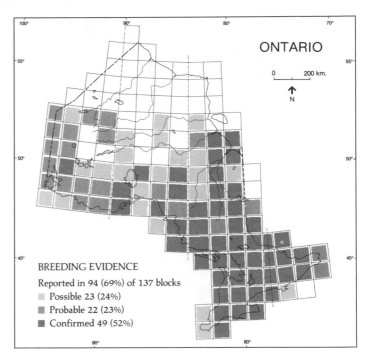

ONTARIO

0   200 km.

↑
N

BREEDING EVIDENCE
Reported in 94 (69%) of 137 blocks

☐ Possible 23 (24%)
▨ Probable 22 (23%)
■ Confirmed 49 (52%)

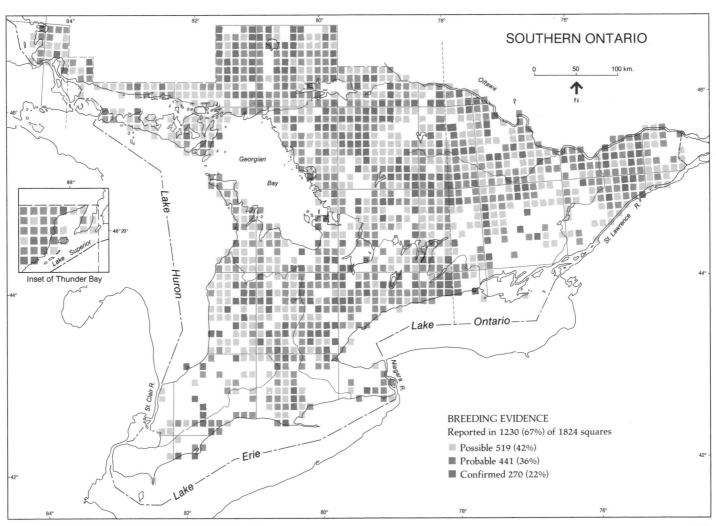

SOUTHERN ONTARIO

0   50   100 km.

↑
N

Georgian
Bay

Lake
Huron

Lake Superior

Inset of Thunder Bay

St. Clair R.

Lake   Erie

Niagara R.

Lake — Ontario

Ottawa

St. Lawrence R.

BREEDING EVIDENCE
Reported in 1230 (67%) of 1824 squares

☐ Possible 519 (42%)
▨ Probable 441 (36%)
■ Confirmed 270 (22%)

COMMON YELLOWTHROAT
Paruline masquée
Geothlypis trichas

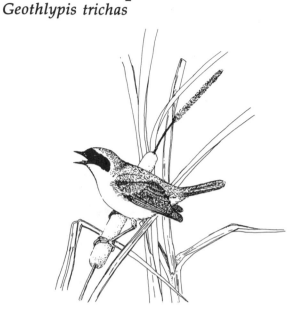

The Common Yellowthroat is a summer resident across North America from Alaska to Newfoundland and throughout the US and much of Mexico. It winters in the southern US, the West Indies, Mexico, and Central America. In Ontario, the Common Yellowthroat is a widespread and common nesting species. It occurs in many habitats that have open areas and low vegetation, from dry second-growth old fields to bogs, marshes, and swamps to black spruce forest (Erskine 1977), but it is primarily a bird of wetlands.

Given its ability to use a wide variety of habitats, the Common Yellowthroat has probably always been a widespread species in Ontario. The extensive clearing of land over the past century has provided additional suitable habitat throughout its range, and, as a result, the species may be even more widespread now than in earlier years.

The song of the Common Yellowthroat is loud, carries well, and is easily identifiable. During the nesting season, the male can be seen singing from prominent locations, such as shrubs in cattail marshes, but spends most of its time, like the female, moving through the lower levels of vegetation. Other than for singing, Common Yellowthroats do not regularly use the upper levels of vegetation for their daily activities. Males are aggressive and can often be observed chasing each other. The species also reacts to the presence of observers by loud scolding and visible approaches. The data reflect these activities, with 46% of all probable records being of agitated behaviour.

The nest is usually constructed on, or a few centimetres above, the ground, often in open areas. Although it is generally well hidden, the openness of the vegetation facilitates the discovery of the nest, usually by flushing the female or by following the adults. Both parents participate in feeding the nestlings, and adults carrying food are commonly seen.

These characteristics are reflected in the breeding evidence obtained for this species. Evidence of confirmed breeding was obtained in 53% of southern Ontario squares and 54% of the blocks in the province with records of the species. These are remarkably high breeding confirmation values for a warbler, exceeded only by the Yellow and Prothonotary Warblers. Nests with eggs or young were located in 91 squares in southern Ontario. However, the 2 most common categories of confirmed breeding in the south are adults with food (57%) and fledged young (19%).

Breeding evidence was obtained from 97% of the squares in the south and from 78% of the blocks in all of Ontario. The maps accurately depict the overall distribution of the Common Yellowthroat in Ontario. Although James *et al.* (1976) indicated that the species occurred north only to Fort Albany and Sioux Lookout, atlassers found evidence of breeding much further north, at some inland locations near Hudson Bay. In the south, the species was absent from very few squares, such as those along Great Lakes shorelines which have only small amounts of land.

Atlassers provided estimates of breeding abundance for 53% of the squares in the province where some evidence of breeding by the Common Yellowthroat was obtained. In 73% of these squares, the abundance estimates are between 11 and 1,000 pairs. All of the abundance estimates of over 1,000 pairs per square come from south of a line from Mattawa to Sault Ste Marie. Notably, no such high estimates come from southwestern Ontario, south and west of Stratford, presumably because of a lack of extensive suitable habitat in such a highly agricultural area.

Atlassers considered this species to be the most abundant of all warbler species in Ontario. Other researchers have also recorded very high densities of Common Yellowthroats. Martin (1960) estimated a density of 335 pairs per 100 ha in dry bog habitat in Algonquin Prov. Park, and Welsh and Fillman (1980) recorded a density of 244 pairs per 100 ha in a black spruce swamp in northeastern Ontario. The territory of the Common Yellowthroat is small: Martin (1960) estimated an average size of 0.14 ha in Algonquin Prov. Park, and Stewart (1953) estimated the average size to be 0.5 ha in southern Michigan. Hess (in Bent 1953) found 17 Common Yellowthroat nests in a 0.2 ha swamp in Illinois.-- *P.L. McLaren*

COMMON YELLOWTHROAT

Summary of Abundance Estimates

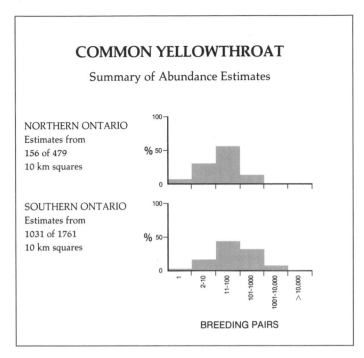

NORTHERN ONTARIO
Estimates from
156 of 479
10 km squares

SOUTHERN ONTARIO
Estimates from
1031 of 1761
10 km squares

BREEDING PAIRS

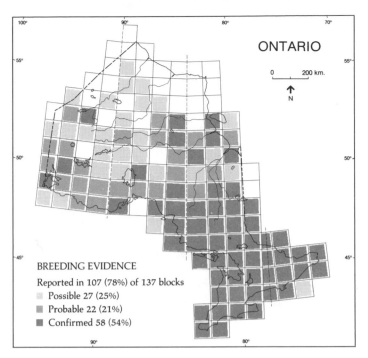

ONTARIO

0   200 km.

N

BREEDING EVIDENCE
Reported in 107 (78%) of 137 blocks
☐ Possible 27 (25%)
▨ Probable 22 (21%)
■ Confirmed 58 (54%)

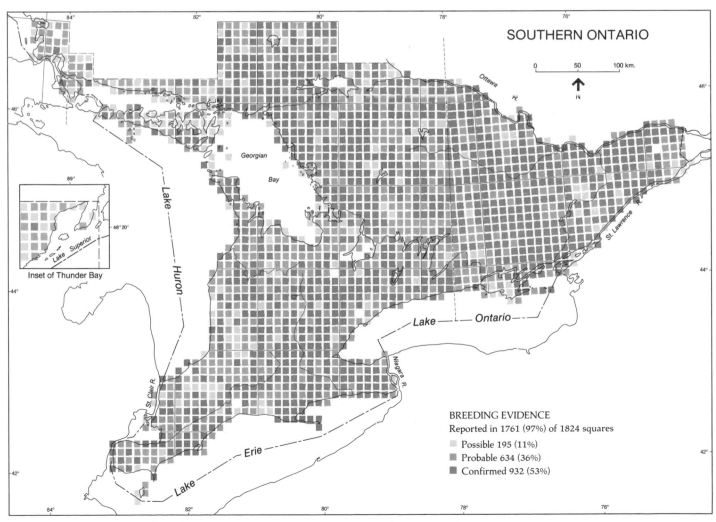

SOUTHERN ONTARIO

0   50   100 km.

N

Georgian
Bay

Lake
Huron

Lake

Ottawa R.

St. Lawrence R.

Lake — Ontario

Niagara R.

Erie

Lake

St. Clair R.

89°
48° 20'
Lake Superior
Inset of Thunder Bay

BREEDING EVIDENCE
Reported in 1761 (97%) of 1824 squares
☐ Possible 195 (11%)
▨ Probable 634 (36%)
■ Confirmed 932 (53%)

417

HOODED WARBLER
Paruline à capuchon
Wilsonia citrina

The Hooded Warbler is a widespread breeder in the eastern US but rare in southern Ontario. It winters in Central America. In Ontario, the Hooded Warbler occurs in mature, upland deciduous or mixed forest, where clearings have been created naturally or by logging. It occupies small clearings with low, dense, shrubby vegetation less than 2 m in height. In Ontario, it is restricted to larger forest tracts in the Carolinian Forest region dominated by white oak, red maple, white pine, and American beech. It is less common in the ecotones of sugar maple-silver maple swamp, eastern hemlock-yellow birch ravines, and mature white pine plantations with a dense deciduous shrub layer.

Although the Hooded Warbler was reported in Ontario perhaps as early as 1878, it was evidently regarded as an overshoot on migration (Baillie 1925, Baillie and Harrington 1937). The first nest was discovered in 1949 in Springwater Forest (Elgin Co); three nests were reported from the same locality in 1950 and 1952 (Baillie 1962). Nests were subsequently found near Newbury (Middlesex Co), at Mansewood (Halton RM), and southeast of Tillsonburg (Haldimand-Norfolk RM) (ONRS). The Hooded Warbler probably always bred in southern Ontario in small numbers but generally escaped the notice of naturalists unfamiliar with its habits.

It is probable that the Hooded Warbler is underrepresented on the maps. For instance, prior to 1985 only one atlas record was forthcoming from the Norfolk Sand Plain, where many records of the species were established during a detailed 1985 and 1986 field study. Birds singing in May in suitable habitat may have been dismissed as migrants by atlassers, but the authors found a male, later known to be breeding, on territory on 11 May. In some cases, higher breeding evidence may not have been sought because the presence of a male, without evidence of a female, may have led some observers to believe that the individual was unpaired. Another difficulty is that older females may closely resemble males and consequently be confused with them (Lynch *et al.* 1985). Most females seem to be remarkably cryptic until feeding large, fledged young, and they do not always flush when the nest is approached. Paired males may wander up to 750 m from their usual singing post and may then have been dismissed as unattached wanderers. Recognition of the alarm call (loud chip) is very useful for finding and confirming the species. No specific distraction displays have been noted, and some birds, even with a nest nearby, show little sign of agitation. However, adults may feed nestlings in plain view of an observer and give away the location of the nest. Territories are occupied in successive years: on the Norfolk Sand Plain, of 19 territories occupied in 1985, 47% were reoccupied in 1986.

The maps show the range limits of the Hooded Warbler in Ontario. However, the density of observations is somewhat less than expected, given the suitability of many areas. Most breeding season records during and prior to the atlas period were from the Carolinian Forest region, particularly on sandy soil such as the Norfolk, Bothwell, and Caradoc Sand Plains and near Port Franks (Lambton Co). The Rondeau Prov. Park record may refer to a late-arriving migrant or an overshoot returning southward, as the habitat at this site is, for the most part, unsuitable for breeding.

Nesting areas are few and widely spaced, but, where a suitable patch of forest exists, the density of birds can be quite high. Near Walsingham (Haldimand-Norfolk RM), at least 9 pairs were found in 75 ha, and in Springwater Forest, Speirs and Frank (1970) found 10 pairs in approximately 40 ha. Atlas results and 1986 observations suggest that this species is probably uncommon on the Norfolk Sand Plain in eastern Elgin Co, southern Oxford Co, and Haldimand-Norfolk RM but is rare elsewhere in southern Ontario. In 1986, more than 27 pairs were found in the vicinity of Big Creek in Haldimand-Norfolk RM. A further 9 pairs were found during 1986 at 6 sites in Kent and Elgin Counties (A. Wormington pers. comm.). Given that many apparently suitable woodlands have not been investigated for this species, the population in Ontario may well approach 100 pairs. The Hooded Warbler has never been listed as rare in Ontario by the MNR or COSEWIC because it was considered to be a straggler that only occasionally bred in Ontario. The Hooded Warbler should now be added to these lists as a rare but regular breeding species.-- *D.A. Sutherland and M.E. Gartshore*

HOODED WARBLER

Summary of Abundance Estimates

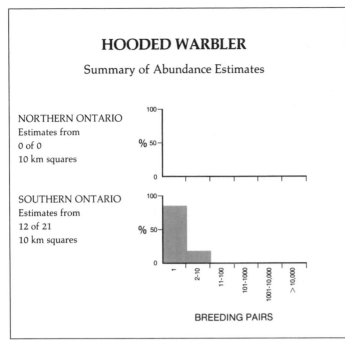

NORTHERN ONTARIO
Estimates from
0 of 0
10 km squares

SOUTHERN ONTARIO
Estimates from
12 of 21
10 km squares

BREEDING PAIRS

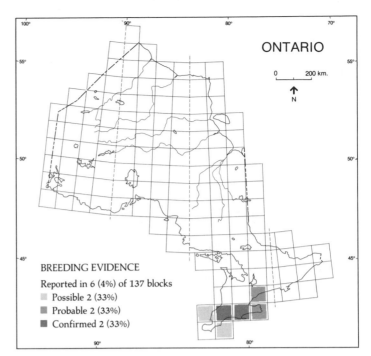

ONTARIO

0   200 km.

N

BREEDING EVIDENCE

Reported in 6 (4%) of 137 blocks

Possible 2 (33%)
Probable 2 (33%)
Confirmed 2 (33%)

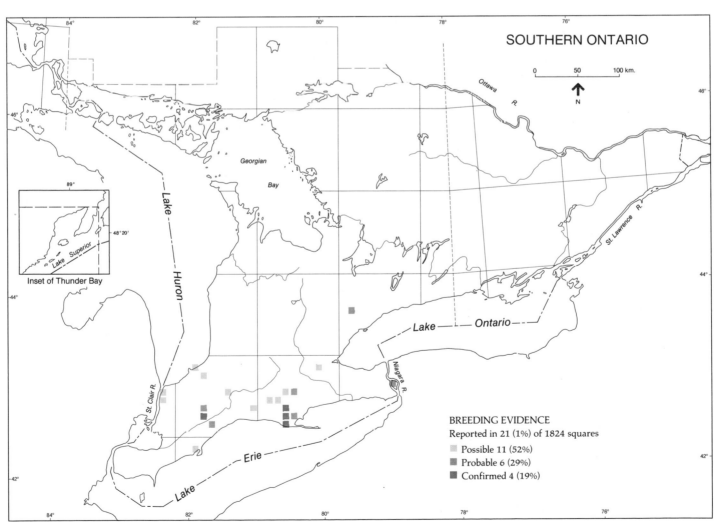

SOUTHERN ONTARIO

0   50   100 km.

N

Inset of Thunder Bay

BREEDING EVIDENCE

Reported in 21 (1%) of 1824 squares

Possible 11 (52%)
Probable 6 (29%)
Confirmed 4 (19%)

WILSON'S WARBLER
Paruline à calotte noire
Wilsonia pusilla

This brightly coloured species breeds from coast to coast in the northern two-thirds of North America, north to the treeline, and in the mountains of the west, south to southern California. Most of its winter range is in Central America, north to the southern US (AOU 1983). The Wilson's Warbler is a spring and fall migrant in most of southern Ontario, but breeds throughout almost all of the rest of the province. It frequents wet, wooded habitat, where there are high shrubs or low deciduous trees, commonly alder or willow. Boggy areas surrounded by cedar, tamarack, and spruce trees are prime nesting habitat for this warbler (M. Austen pers. comm.). Such environments are common along streams throughout northern Ontario. Although the Wilson's Warbler is widespread in Ontario during the breeding season, having been found in 63% of the blocks, it is nowhere an abundant bird.

Most of the range of the Wilson's Warbler in Ontario was not occupied by human settlement until the present century. Thus, it is not surprising that early breeding records of this species are scarce and are confined to southern Ontario. McIlwraith (1894) refers to a surprising, perhaps even questionable, record of a nest in an Ottawa garden. Two other equally doubtful nests, one in a hemlock tree near Ottawa and the other, a more typical one, in Leeds Co, are described by Macoun and Macoun (1909). Apart from these, the most southerly nesting evidence found in Ontario prior to the atlas period was from Sudbury (Baillie and Harrington 1937). At the higher elevations of New York State, a few records of possible or probable breeding in recent years have been noted by atlas workers (NYBBA preliminary maps).

Nests of Wilson's Warblers are very difficult to locate. Peck and James (1987) note only 5 locations in the entire province where nests have been found. The atlas field work revealed nests in only 3 blocks. This lack of nest records is probably due to the species' preference for habitats that are difficult to access and, more importantly, to its habit of being a 'very close sitter' while incubating, thereby not revealing its presence (M. Austen pers. comm.). An atlasser drifting in a canoe or standing at the edge of a road can often find this species foraging or singing at eye level. It is easily attracted by 'pishing', and has a weak song that is usually the easiest way of finding the bird in the dense alder thickets that it inhabits. In 75% of the 37 squares in which breeding was confirmed, the evidence consisted of adults carrying faecal sacs or food for young. Where possible and probable breeding was established, the single male and territory codes were by far the most frequently used ones, no doubt because it is much easier to detect a Wilson's Warbler song than to pursue the bird in its wet, dense habitat.

This species is one of four typically northern wood warblers in Ontario, the others being Palm, Orange-crowned, and Blackpoll. To a greater extent than any of the other three, it was discovered by atlassers to be possibly or probably breeding in southern Ontario. Because the Wilson's Warbler is a late migrant in spring, some of these southern records may actually have been migrant birds, but there is sufficient information in the records to warrant a re-interpretation of the previously known breeding range of the species. It apparently breeds sporadically south as far as the 20^0C July isotherm. The patchiness of the distribution indicated by the atlas maps is somewhat surprising. To be sure, it is not a common bird anywhere within its Ontario range, but it almost certainly is present in most of the northern blocks where it was not found. Possibly its song is not well enough known to some atlassers to have enabled them to recognize it.

Anyone looking for Wilson's Warblers in southern Ontario during the breeding season is likely to be disappointed. During the 5 summers of atlassing, it was found in only 46 squares in that part of the province, and in only 10 squares south of the French and Mattawa Rivers; most southern Ontario records are from the Sudbury and Sault Ste Marie areas. Abundance estimates in the south, are also low, generally in the range of 2 to 10 pairs. There are probably fewer than 1,000 pairs breeding in all of southern Ontario, and most of those are in the northern fringe of the area. Abundance estimates were provided from 54 of the 186 squares in which the species was recorded in northern Ontario. These verify that the Wilson's Warbler is more common in the north. In 4 out of the 54 squares, the estimated abundance was greater than 100 pairs per square.-- *F.M. Helleiner*

420

WILSON'S WARBLER

Summary of Abundance Estimates

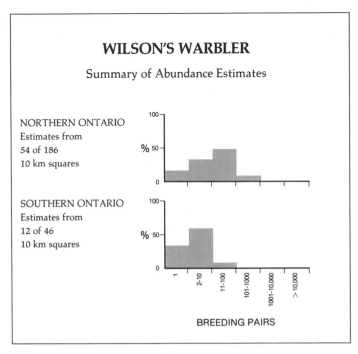

NORTHERN ONTARIO
Estimates from
54 of 186
10 km squares

SOUTHERN ONTARIO
Estimates from
12 of 46
10 km squares

BREEDING PAIRS

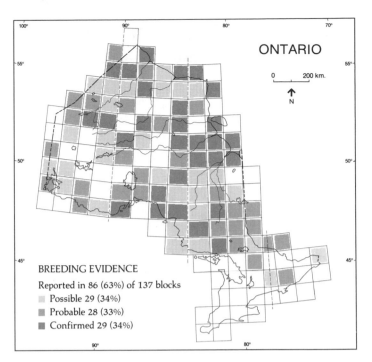

ONTARIO

0   200 km.

N

BREEDING EVIDENCE
Reported in 86 (63%) of 137 blocks
Possible 29 (34%)
Probable 28 (33%)
Confirmed 29 (34%)

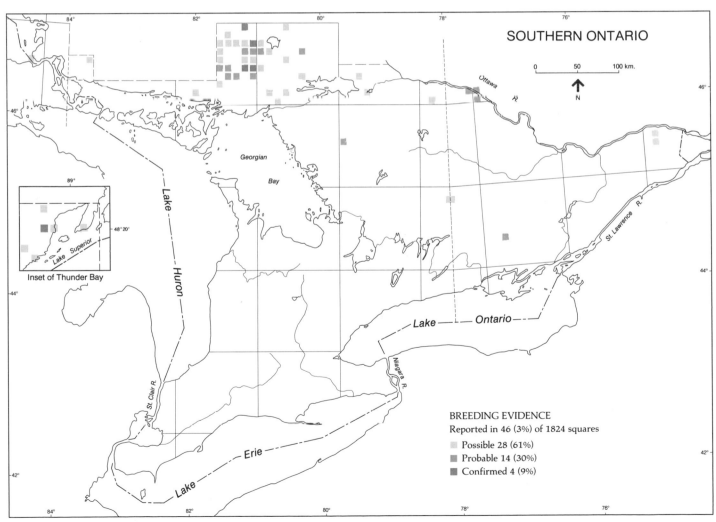

SOUTHERN ONTARIO

0   50   100 km.

N

Inset of Thunder Bay

BREEDING EVIDENCE
Reported in 46 (3%) of 1824 squares
Possible 28 (61%)
Probable 14 (30%)
Confirmed 4 (9%)

CANADA WARBLER
Paruline du Canada
Wilsonia canadensis

The Canada Warbler is a summer resident across Canada from northern Alberta to Nova Scotia, and in the northeastern US south through the Appalachian Mountains. In its summer range, it is most common in mixed coniferous and deciduous forests, often in wet low-lying areas such as cedar woods or alder swamps. The Canada Warbler winters in South America, from Colombia to Peru. It is a widespread nesting species in Ontario; breeding evidence was obtained from 52% of the squares in the south and 61% of the blocks in all of Ontario.

Little is known concerning the historic distribution and abundance of the Canada Warbler in Ontario. It is fair to suppose, however, that numbers are much lower now in southwestern Ontario than they were before the land was cleared for farming. That it still occurs in local pockets of wet woods throughout much of southwestern Ontario and in adjacent areas of suitable habitat in Michigan and New York suggests that it formerly nested throughout this region. On the other hand, the elimination of the original pine forests in parts of southcentral Ontario and their resulting replacement by mixed coniferous and deciduous forest indicate that there may be a somewhat more widespread distribution of the Canada Warbler in that area today than in the last century.

The song of the Canada Warbler is loud, and the species sings throughout much of the nesting cycle. However, the song is difficult to describe and this suggests that the species may not always be identified when heard. Moreover, the fact that this species often occurs in relatively inaccessible habitats suggests that it may sometimes be identified, but not confirmed as a breeding species. Nests are built on or near the ground, often in stumps of fallen logs, and are apparently difficult to find. (Atlassers found nests in only 8 squares in the south). After the eggs hatch, both parents par-

ticipate in feeding the young, even after fledging. These nesting characteristics are reflected in the breeding evidence obtained for this species. Over 80% of the evidence obtained in the south was of only possible or probable breeding, and in over half of these the evidence consisted of a male seen or heard. Most of the small amount of evidence of confirmed breeding consisted of adults carrying food, or of fledged young.

In general, the atlas maps reflect quite accurately the previously known distribution of the Canada Warbler in Ontario. In southwestern and southeastern Ontario, the species may be more widespread and numerous than indicated by the maps. Although breeding was confirmed in several squares in those areas (*e.g.*, Kent, Niagara, Ottawa), several of the Regional Coordinators in that area suspected that the species was underconfirmed in many squares and unrecorded in some where they were in fact present. The centre of the Canada Warbler's range in Ontario appears to extend from the southern Georgian Bay region north to Moosonee and Kenora. Virtually all squares and blocks within this region have some evidence of breeding in them and many, especially the blocks in northcentral and northeastern Ontario, have confirmed breeding. The apparent absence of the Canada Warbler from several blocks in northcentral and far northern Ontario may be more an indication of the relative inaccessibility of those regions than of its actual distribution. However, the absence of deciduous trees and understory, which are the preferred habitat of this species, will also limit numbers.

Atlassers provided estimates of breeding abundance for 41% of the squares in the south, and 31% of northern squares, in which some evidence of breeding by the Canada Warbler was obtained. In 92% of these squares, the abundance estimates were between 1 and 100 pairs per square, which suggests that, although widespread, the species is not particularly abundant. Canada Warblers were apparently more common in the more northern parts of the province. Martin (1960) estimated the density of Canada Warblers in a hemlock-cedar forest in Algonquin Prov. Park as 5 pairs/100 ha and densities of up to 128 pairs/100 ha have been recorded in mixed-wood cutover forest in northeastern Ontario (Welsh and Fillman 1980). Martin (1960) calculated that the average size of a territory of the Canada Warbler in Algonquin Prov. Park was about 0.2 ha.-- *P.L. McLaren*

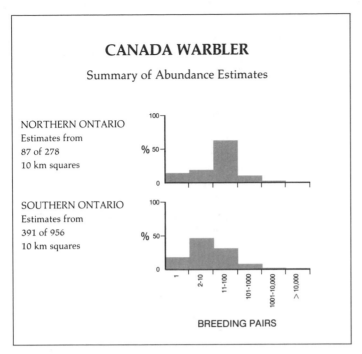

CANADA WARBLER

Summary of Abundance Estimates

NORTHERN ONTARIO
Estimates from
87 of 278
10 km squares

SOUTHERN ONTARIO
Estimates from
391 of 956
10 km squares

%

BREEDING PAIRS

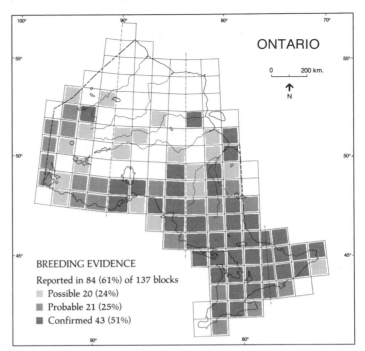

ONTARIO

0   200 km.

N

BREEDING EVIDENCE
Reported in 84 (61%) of 137 blocks
Possible 20 (24%)
Probable 21 (25%)
Confirmed 43 (51%)

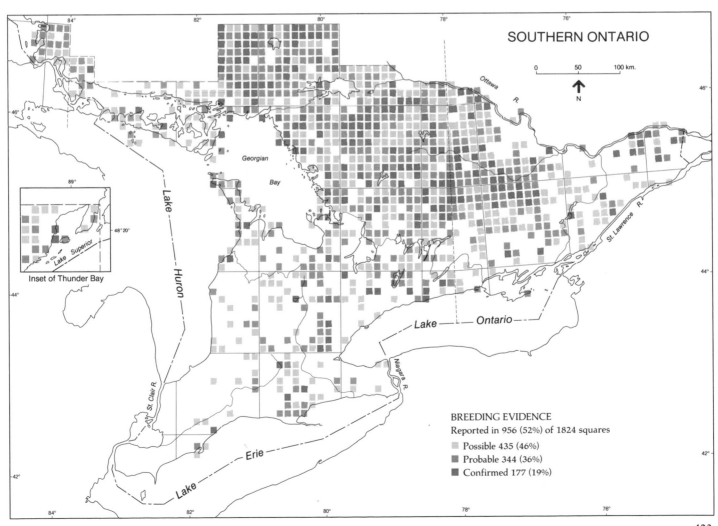

SOUTHERN ONTARIO

0   50   100 km.

N

Lake Huron

Georgian Bay

Ottawa R.

St. Lawrence R.

Lake Ontario

Niagara R.

St. Clair R.

Lake Erie

Lake Superior

89°   48°20'

Inset of Thunder Bay

BREEDING EVIDENCE
Reported in 956 (52%) of 1824 squares
Possible 435 (46%)
Probable 344 (36%)
Confirmed 177 (19%)

YELLOW-BREASTED CHAT
Paruline polyglotte
Icteria virens

The Yellow-breasted Chat is the largest of the wood warblers that occur in Canada. Its secretive habits make it difficult to observe, but when the species is found, its bright colours make it pleasant to view. The Chat has a distinctive and loud song, making it easy to record. These factors, along with the bird's rarity in Ontario, combine to increase its desirability to birders.

This species breeds across North America but comes into Canada at just a few southern spots, such as southern British Columbia, the southern parts of the Prairie Provinces, and southern Ontario. It migrates to the extreme southern US, Mexico, and Central America for the winter. Shrubby tangles and deciduous thickets are the preferred habitat.

The Yellow-breasted Chat is notorious for its varied song that contains many types of notes, whistles, and squawks (Bent 1953). Such calls are usually given from the centre of clumps of thick, tangled brambles or shrubbery. The calls are distinctive but can be confused with those of Gray Catbird, Brown Thrasher, and Northern Mockingbird. These factors may have resulted in a degree of underrecording of the species by atlassers.

This species has never been widespread in Ontario. McIlwraith (1886) reported that he knew of only 3 records of the Yellow-breasted Chat for Ontario, all at Hamilton or Point Pelee. Macoun and Macoun (1909) reported an expansion of the known range to the east to include Toronto, and stated that the species was, according to Saunders, "a constant and not very rare resident of the vicinity of Pelee point." Baillie and Harrington (1937) stated that it was known to breed only in the southwestern Counties of Essex and Elgin but occurred as far north as Middlesex Co and as far east as Peel Co. Godfrey (1986), gave a range of the entire Carolinian Forest zone. There is little evidence of significant change in the range of this species in Ontario over time, just an increase in knowledge of the range.

The Yellow-breasted Chat is highly vocal but secretive. Therefore, it is most easily found through its call, as shown by the 69% of all records in the singing male or territorial categories. Breeding of the species is very difficult to confirm because of its skulking habits and the dense nature of its preferred breeding habitat. Only 4 records (9%) are of confirmed breeding and nests were found in only 3 squares. The atlas data reveal that 45 squares contained this species at some time during the 5 year field work period.

The map suggests a distribution that is more widespread than previously known, but the confirmed records are all in the breeding range defined by Godfrey (1986). As expected, the majority of records are from the Carolinian Forest region, but a few records came from such northern and eastern locations as Shelburne and the Kingston area.

The abundance estimates provided by atlassers indicate that the species had a small population at any place of occurrence in Ontario. All estimates but one are of populations of from 1 to 10 pairs per square.

The Pelee Island atlas square is the only one that has an abundance estimate of more than 10 pairs. This square and the one including Point Pelee Nat. Park are probably the areas with the highest density in Ontario, at slightly more than 10 pairs (M. Oldham pers. comm.). However, elsewhere in Essex Co, scrub habitats are very rare, thereby limiting numbers. Land in this part of Ontario is so productive for agriculture that once it is cleared, it is seldom allowed to revert to scrub. This lack of suitable breeding habitat may be a limiting factor in many parts of the agricultural southwest (M. Oldham pers. comm.). In the more northern portions of its Ontario range, the species often does not return to the same site each year. However, Yellow-breasted Chats do return to certain southern areas each year, *i.e.*, Pelee Island, Point Pelee Nat. Park, and Rondeau Prov. Park (A. Woodliffe pers. comm.). Therefore, the 45 squares in which it was reported over the 5 year period may represent a population of around 50 pairs in Ontario in any breeding season.-- *P.F.J. Eagles*

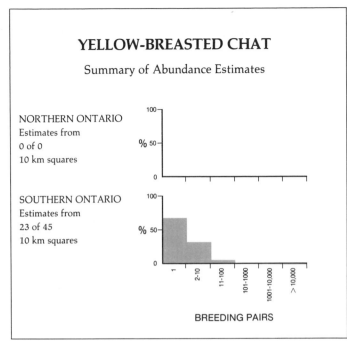

YELLOW-BREASTED CHAT

Summary of Abundance Estimates

NORTHERN ONTARIO
Estimates from
0 of 0
10 km squares

SOUTHERN ONTARIO
Estimates from
23 of 45
10 km squares

BREEDING PAIRS

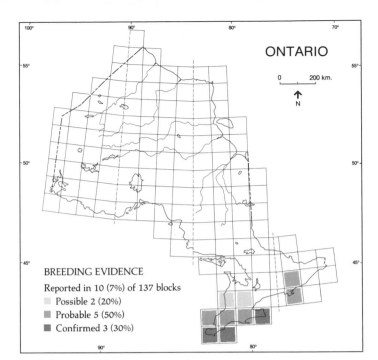

ONTARIO

0   200 km.

N

BREEDING EVIDENCE

Reported in 10 (7%) of 137 blocks

Possible 2 (20%)
Probable 5 (50%)
Confirmed 3 (30%)

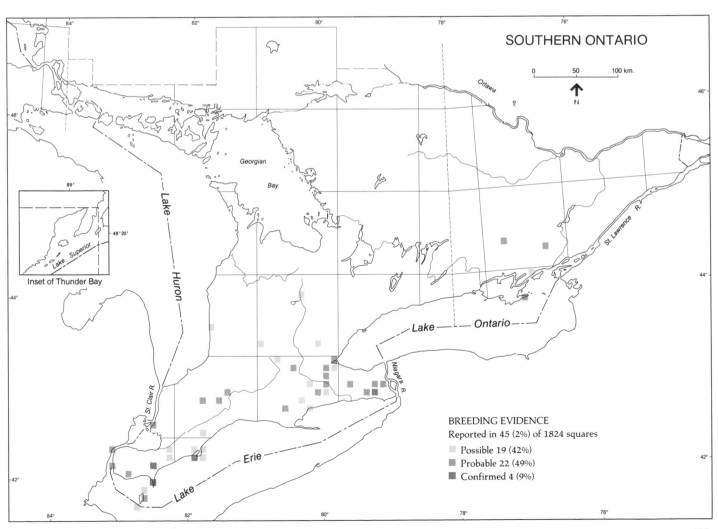

SOUTHERN ONTARIO

0   50   100 km.

N

Georgian Bay

89°
48°20'
Lake Superior
Inset of Thunder Bay

Lake Huron

St. Clair R.

Lake Erie

Niagara R.

Lake — Ontario —

Ottawa

St. Lawrence R.

BREEDING EVIDENCE

Reported in 45 (2%) of 1824 squares

Possible 19 (42%)
Probable 22 (49%)
Confirmed 4 (9%)

SUMMER TANAGER *
Tangara vermillon
Piranga rubra

The Summer Tanager is a colourful, distinctive species which always attracts attention among the birding fraternity when it puts in an appearance in Ontario.

The breeding range of this species is restricted to North America, being primarily in the southern US and northern Mexico. The species ventures somewhat farther north in the eastern US, regularly breeding as far north as southern Ohio (National Geographic Society 1983). The winter range extends from northern Mexico to central South America. In Ontario, it is known as a rare migrant in the south, with records as far north as Ottawa, Manitoulin Island, and Ouimet Canyon and Neys Prov. Parks north of Lake Superior (Speirs 1985). Most sightings are during spring or fall (James *et al.* 1976), indicating either an overshooting of its normal range during the spring migration or a post-breeding tendency to wander northward. There is a summer record in June 1965 (Speirs 1985) at Rondeau Prov. Park. Its preferred breeding habitat includes pine forests, mixed woods, upland woods, and even groves of shade trees near houses (Bent 1958, Terres 1980). Nests of this species are frequently low (up to 10 m) and often on a horizontal branch over roadways. Even so, they are difficult to find (Bent 1958).

The first record for the atlas was a singing male in late May of 1983 at the Royal Botanical Gardens in Hamilton. During May of 1985, there was a moderate influx of Summer Tanagers at Rondeau Prov. Park, with at least 8 birds arriving, including 2 females. A pair was seen together in suitable habitat on 26 May. Few birders frequented the area in June, but on 9 July the birds were recorded again. They were then seen or heard regularly until 30 July. The birds were observed copulating (R. Knapton pers. comm.). However, on several occasions when the area was checked, no sign of the birds was apparent, even though they had been seen earlier

on the same day. These individuals, being beyond the normal range of the species and with no neighbouring birds of the same species, would not likely feel territorial pressure and therefore would exhibit little territorial behaviour or singing. This quietness could be the reason why they were not recorded in June. Even when observed in July, the birds were elusive, and no nest was found. However, after the leaves of the trees had fallen, a nest resembling that of a tanager was discovered a short distance from the birds' main area of activity. It was in a location more typical of the Summer Tanager than the Scarlet Tanager, which had also been observed in the general area. Since the nests are virtually indistinguishable, the evidence for Summer Tanager nesting is inconclusive.

The atlas maps likely indicate fully the distribution of this species in Ontario during the atlas years. Because influxes of Summer Tanagers occur only occasionally in Ontario, and the normal northern limit of the breeding range of the species occurs at least 300 km to the south, it is likely that the lack of nesting records in the province is a true reflection of the species' status. Nevertheless, the Rondeau and Hamilton records indicate the possibility that occasional breeding may occur.-- *P.A. Woodliffe*

SUMMER TANAGER

Summary of Abundance Estimates

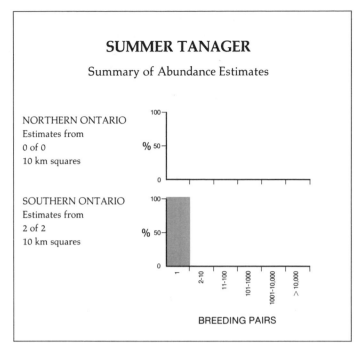

NORTHERN ONTARIO
Estimates from
0 of 0
10 km squares

SOUTHERN ONTARIO
Estimates from
2 of 2
10 km squares

BREEDING PAIRS

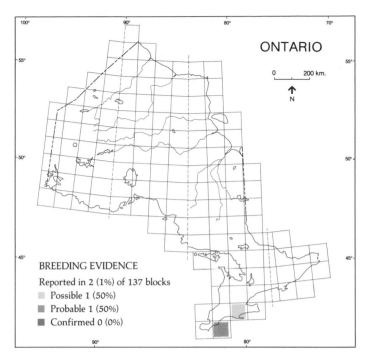

ONTARIO

BREEDING EVIDENCE

Reported in 2 (1%) of 137 blocks

Possible 1 (50%)

Probable 1 (50%)

Confirmed 0 (0%)

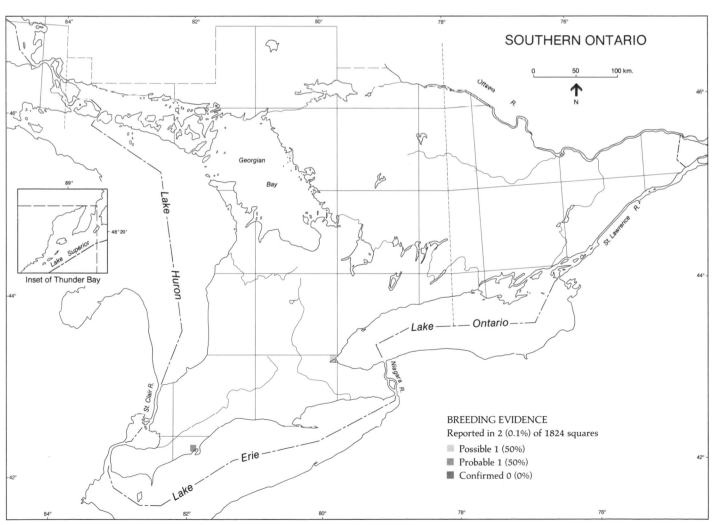

SOUTHERN ONTARIO

BREEDING EVIDENCE

Reported in 2 (0.1%) of 1824 squares

Possible 1 (50%)

Probable 1 (50%)

Confirmed 0 (0%)

Inset of Thunder Bay

427

SCARLET TANAGER
Tangara écarlate
Piranga olivacea

The 'black-winged redbird', as it is sometimes known, is surprisingly difficult to see, despite the brilliance of the adult male's plumage. Fortunately, its voice, frequently described as resembling that of a Robin with a sore throat, is quite distinctive and well known, and was undoubtedly important in helping atlassers record the species in a large number of squares and blocks.

The Scarlet Tanager breeds in eastern North America from southeastern Manitoba east to the Atlantic coast and south to central Alabama. It winters in South America from Colombia to Bolivia. In Ontario, it breeds in upland, fairly mature deciduous or mixed forests in the Carolinian and Great Lakes-St. Lawrence Forest zones. In New Jersey, Galli *et al.* (1976) consistently failed to find it in woods smaller than 10 ha; a similar preference for larger woods is also apparent in Ontario.

Historical information indicates that the breeding distribution in Ontario has changed little over time. Given the preference of the species for extensive woodland, there can, however, be little doubt that the Scarlet Tanager was far more common than it is now when southern Ontario was completely forested - before ornithological information was recorded. Within its range, Macoun and Macoun (1909) noted that it had become decidedly more abundant at Ottawa than formerly. Otherwise, published accounts tend to indicate a stable population, except for a recent decline noted in the Guelph area (Brewer 1977).

The Scarlet Tanager is quite easily overlooked, and is difficult to confirm as a breeding species. It tends to use the "retired parts of the woods" (McIlwraith 1894), is a sluggish, secretive bird, and keeps to the upper canopy of tall trees. Its nest is usually over 10 m high (Harrison 1975), quite flimsy in construction, and not easily found. Only 13% of records are of confirmed breeding, which is a low value for a fairly widespread species; most confirmed records are of adults with food for young (42%) or newly fledged young (33%). The dull plumage of the female makes it inconspicuous and adds to the difficulty of confirmation. The song and the distinctive 'chick-burr' alarm call are, however, readily identified by most atlassers.

The maps are probably a good indication of the breeding range in the province, but it is likely that Scarlet Tanagers breed in the great majority of squares from which they were reported. Those squares south of the Shield in which they were not reported probably lack the species or at best have only very low numbers; the larger woodlots used by Scarlet Tanagers are relatively scarce in this agricultural and heavily settled area and are likely to have been well covered by atlassers. The lack of records around Toronto is undoubtedly the result of the lack of suitable habitat because of urbanization, whereas most of the other areas with few or no records (*e.g.*, the extreme southwest and northern Wellington Co) are agricultural. Gaps on the southern Shield, where suitable woodland is extensive, are more likely to be squares where the birds have been overlooked. Farther north, the concentration of gaps between Sudbury and Deep River more likely represents an absence of birds or at best low breeding densities due to less suitable habitat, presumably as a result of deforestation for agriculture and by air-borne pollution. The northern range limit generally follows the northern edge of the Great Lakes-St. Lawrence Forest region, with a few records in the southern Boreal Forest region - most notably in the Clay Belt area.

Breeding Bird Survey data show that the Scarlet Tanager is at its highest density from Georgian Bay east to the Peterborough, Pembroke, and Ottawa areas (Speirs 1985). The atlas data show a similar result; the highest abundance estimates, those over 100 pairs per square, are primarily from the rich deciduous or mixed forest areas on the Canadian Shield from Kingston to Algonquin Prov. Park. Abundance estimates indicate that populations are less dense along the northern edge of the range.-- *M.D. Cadman*

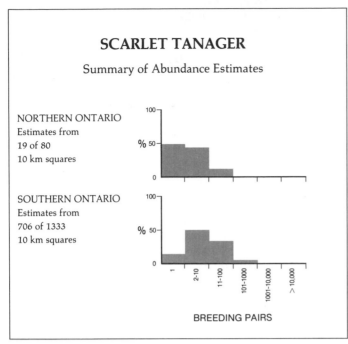

SCARLET TANAGER

Summary of Abundance Estimates

NORTHERN ONTARIO
Estimates from
19 of 80
10 km squares

SOUTHERN ONTARIO
Estimates from
706 of 1333
10 km squares

BREEDING PAIRS

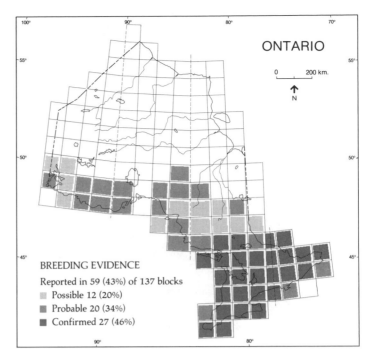

ONTARIO

0   200 km.

N

BREEDING EVIDENCE

Reported in 59 (43%) of 137 blocks

Possible 12 (20%)
Probable 20 (34%)
Confirmed 27 (46%)

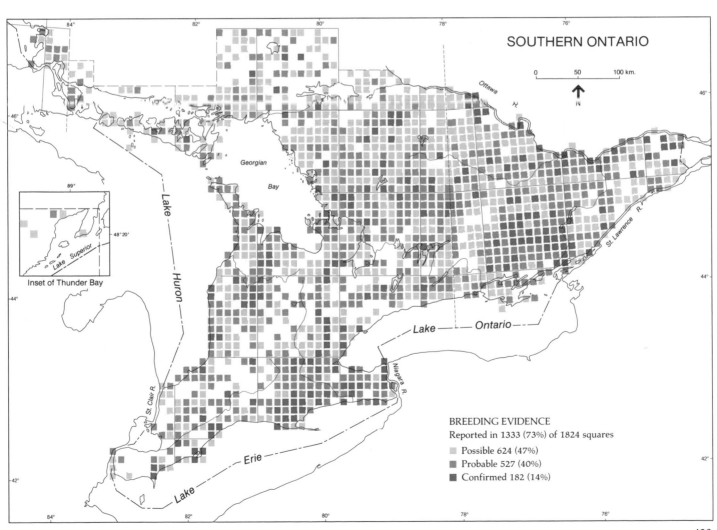

SOUTHERN ONTARIO

Georgian Bay

Lake Huron

Lake Superior

Inset of Thunder Bay

Ottawa R.

St. Lawrence R.

Lake — Ontario —

Niagara R.

Lake — Erie

St. Clair R.

0   50   100 km.

N

BREEDING EVIDENCE

Reported in 1333 (73%) of 1824 squares

Possible 624 (47%)
Probable 527 (40%)
Confirmed 182 (14%)

NORTHERN CARDINAL
Cardinal rouge
Cardinalis cardinalis

The bright red plumage of the male Northern Cardinal and the loud songs of both sexes combine to make it one of the best-known birds in southern Ontario. This finch is resident in brushy thickets and forest edge from Central America north through much of the US into southern Ontario and Quebec (AOU 1983, Godfrey 1986).

The Northern Cardinal is a newcomer to Ontario, first collected at Chatham in 1849 (Snyder 1951). McIlwraith (1894) considered the occasional wanderings of this species into Ontario to be expected on the basis of its "quite common" status in Ohio, but even then it was only a recent arrival to the Great Lakes area (Bent 1968) as part of a northward range extension (Burns 1958, Beddall 1963, Bent 1968). The first Ontario nest was found at Point Pelee in 1901, with additional nestings reported at London (1915), Brantford (1919), Toronto (1922), Orillia and Port Hope (1939), Owen Sound (1942), and Tweed (1953) (Snyder 1957b). By 1930, it was a rare permanent resident of Toronto (Fleming 1930). Baillie and Harrington (1937) considered it part of Ontario's breeding avifauna to the south and west of Toronto. A marked increase in numbers occurred at Toronto in 1938 (Snyder 1957b). Nesting was confirmed at Oshawa in 1945 (Tozer and Richards 1974), and the first nesting at Peterborough was in 1951 (Sadler 1983). Although now regular at Kingston (Quilliam 1986), nesting was not found there until 1964 (Quilliam 1965). The first confirmation of nesting at Ottawa came just over a decade ago in 1974, but it is increasing rapidly there (Hanrahan and DiLabio 1984). By 1976, its breeding range was considered to extend north to Georgian Bay and Ottawa (James *et al.* 1976), a line not substantially different from that shown by Godfrey (1986). Reasons for its range expansion are varied, but the most important factors appear to be the creation of edge type habitat by forest clear-ing, the urbanization in southern Ontario (Snyder 1957b, Burns 1958, Beddall 1963, Dance 1986), and the proliferation of bird feeders.

Although the brushy tangles that it inhabits can make the Northern Cardinal difficult to see at times (Sadler 1983), and can certainly confound nest searches, the loud whistled songs and staccato call notes make detection easy. It frequently appears in the open, often behaving conspicuously. Both sexes feed the young (Bent 1968), which often fly with loud calls into the open after their parents; 41% of all confirmed breeding records are of recently fledged young. In Ontario, the species often nests in conifers if available, but it usually sings from non-coniferous perches (Dow 1969b). The length of the nesting period in this part of its range is one of the longest, with egg dates for the general area known from 13 April to 15 August (Peck and James 1987). Adult males are more likely to disperse to new areas after nesting than are females (Dow and Scott 1971).

The conspicuous song and behaviour of this species are reflected in the fact that 40% of the records for southern Ontario are attributed to the singing male, pair, or territory categories, while another 19% involve fledged young. Nests containing eggs or young were found reasonably often, comprising 16% of southern records.

The breeding distribution shown by atlas results largely corresponds to that shown by Godfrey (1986) and is very similar to the map of documented nests shown by Peck and James (1987). Atlas data indicate more breeding in the Ottawa area than previously known (Speirs 1985), which is consistent with the recent increase of Northern Cardinals in that area (Hanrahan and DiLabio 1984). Possible nestings indicated to the north of the regular breeding range and along the northern shores of Lakes Huron and Superior are consistent with the ongoing range expansion of this species and its occasional occurrence at various sites in this area (Preece 1923, Snyder 1957b, Nicholson 1972, 1974, Baxter 1985, Speirs 1985) and even in southern Manitoba (Godfrey 1986).

Of 534 southern squares with abundance estimates, 427 (80%) had estimates from 2 to 100 pairs per square. However, because these figures include both the established and expanding ranges and because both sexes sing (Bent 1968), their meaning is difficult to interpret. Density estimates in Ontario include 15 to 20 birds per 100 ha and 1.2 birds per 100 ha (Dow 1969a, Tozer and Richards 1974). Data from BBS show abundances ranging from 0.1 to 20.1 birds per 50 stops where the species was recorded, with high values occurring primarily south and west of Toronto (Speirs 1985).-- *M.K. McNicholl*

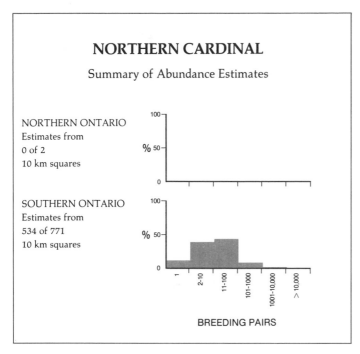

NORTHERN CARDINAL

Summary of Abundance Estimates

NORTHERN ONTARIO
Estimates from
0 of 2
10 km squares

SOUTHERN ONTARIO
Estimates from
534 of 771
10 km squares

%

100

50

0

%

100

50

0

1   2-10   11-100   101-1000   1001-10,000   > 10,000

BREEDING PAIRS

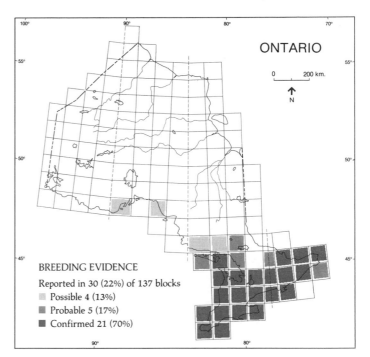

ONTARIO

0   200 km.

N

BREEDING EVIDENCE

Reported in 30 (22%) of 137 blocks

Possible 4 (13%)

Probable 5 (17%)

Confirmed 21 (70%)

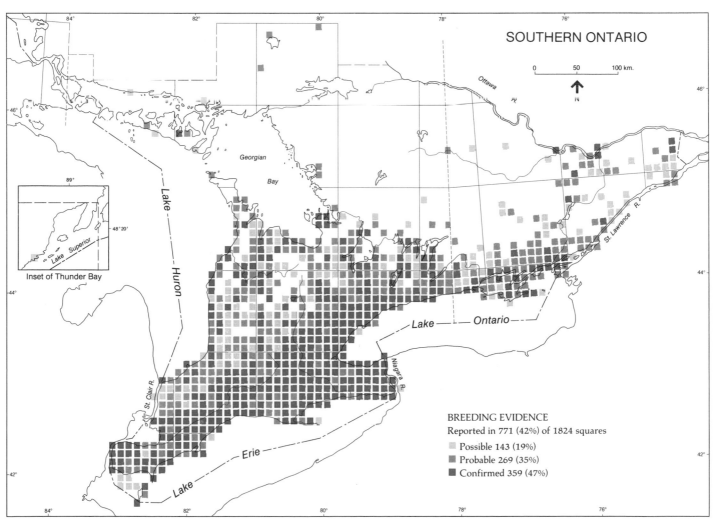

SOUTHERN ONTARIO

0   50   100 km.

N

Inset of Thunder Bay

BREEDING EVIDENCE

Reported in 771 (42%) of 1824 squares

Possible 143 (19%)

Probable 269 (35%)

Confirmed 359 (47%)

431

ROSE-BREASTED GROSBEAK
Cardinal à poitrine rose
Pheucticus ludovicianus

The Rose-breasted Grosbeak is a common neotropical migrant that is well known to all who are interested in wild birds in Ontario. It is colourful, common in suitable habitat, and has a beautiful melodic song, all of which add to its popularity. It occupies both immature and mature broad-leaved forests. It appears to be able to breed in relatively small blocks of forest such as are found in many agricultural areas.

In the US, the Rose-breasted Grosbeak breeds across most of the northern states as far west as the Dakotas. In Canada, its summer range extends from the Maritime provinces to as far west as eastern British Columbia. The bird is found throughout the southern half of Ontario in broad-leaved deciduous forests. It winters from southern Mexico to Colombia.

McIlwraith (1886) stated that the Rose-breasted Grosbeak "breeds regularly along the southern border of Ontario," implying that the bird was not found further north. By 1909, the range was seen to include all of southern Ontario (Macoun and Macoun 1909), with some indication of population increases in the previous few years. By 1937, the range was recognized as extending as far north as the transcontinental CNR line and "probably west to Lake of the Woods" (Baillie and Harrington 1937). Godfrey (1986) shows its range to be the southern half of the province, which is probably the historic range as well, with our developing knowledge gradually piecing together the real picture.

Although the female is dull brown, the male Rose-breasted Grosbeak is a conspicuous bird in song and in plumage, thereby making for relatively easy discovery by atlassers. Rose-breasted Grosbeaks respond well to 'pishing', often with both adults responding in alarm. Nests are flimsy structures, but are at a relatively low height of 1.5 to 4.5 m

(Godfrey 1986) and are among the easiest nests to find. These attributes contribute to the high proportion of probable and confirmed breeding records. Over 14% of all southern Ontario records are of nests. The species was found in 92% of the squares in southern Ontario and breeding was confirmed in 45% of the squares visited by atlassers in all of Ontario. The most frequently used breeding evidence category was that of newly fledged young.

The maps show that the Rose-breasted Grosbeak is found as far north as deciduous trees form a major portion of the forest composition. It is ubiquitous in the Great Lakes-St. Lawrence Forest zone and occurs sporadically in the southern Boreal Forest, where broad-leaved deciduous trees occur in climatically favoured spots. The bird's absence in the extreme southwestern portion of southern Ontario is indicative of the almost total clearing of forest due to agriculture in this area. The absence of records from a few squares elsewhere in southern Ontario is probably due to an inability of particular atlassers to find the species even where it actually occurs.

The Rose-breasted Grosbeak is common in suitable habitat in Ontario. The abundance codes estimated by atlassers indicate numbers of pairs per square ranging mostly from 2 to 1,000, with 11 to 100 pairs being the most frequently estimated range. All of the abundance estimates of over 1,000 pairs per square were submitted from an area of the province that stretches from Owen Sound eastward along the edge of the Canadian Shield to Kingston.

Breeding Bird Survey data for southern Ontario and Quebec show a significant increase in the numbers of the Rose-breasted Grosbeak in the 1967 to 1983 period, with a population doubling time calculated to be 17 years. In central and northern Ontario and Quebec, the data showed a significant increase in the 1967 to 1979 period, with the growth rate slowing down from 1979 to 1983 to produce an overall doubling rate of 20 years. Laughlin and Kibbe (1985) report that BBS data showed a 10% a year increase in this species in Vermont in recent years. Presumably, the regrowth of forest in previously cleared areas, in much of the northeastern US, is an important reason for the increase in numbers. It is also noteworthy that such an increase is sustained over the time of very extensive forest clearance in the winter range of the species in the tropics.-- *P.F.J. Eagles*

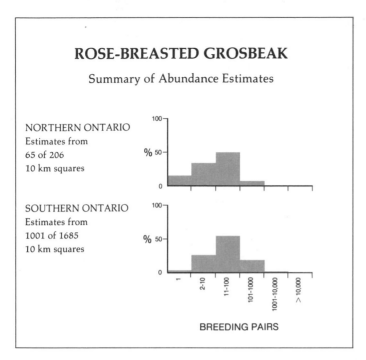

ROSE-BREASTED GROSBEAK

Summary of Abundance Estimates

NORTHERN ONTARIO
Estimates from
65 of 206
10 km squares

SOUTHERN ONTARIO
Estimates from
1001 of 1685
10 km squares

BREEDING PAIRS

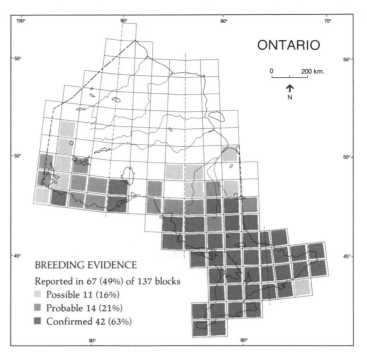

ONTARIO

0   200 km.

N

BREEDING EVIDENCE

Reported in 67 (49%) of 137 blocks

Possible 11 (16%)

Probable 14 (21%)

Confirmed 42 (63%)

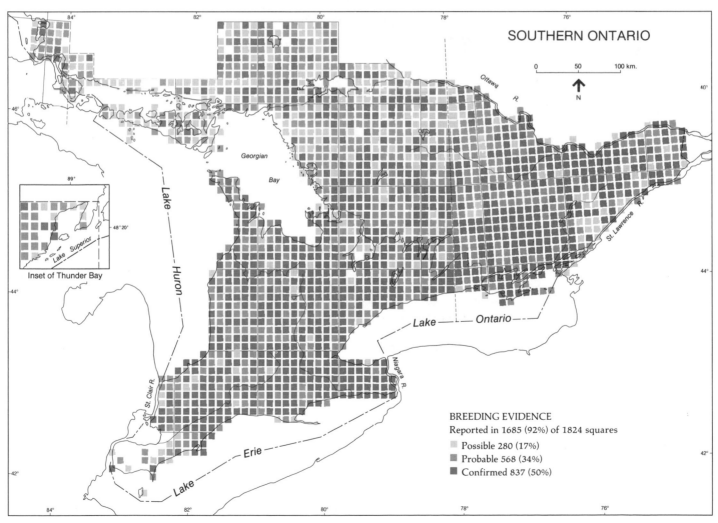

SOUTHERN ONTARIO

Ottawa R.

Georgian Bay

Lake Huron

Lake Superior

Inset of Thunder Bay

St. Lawrence R.

Lake Ontario

Niagara R.

St. Clair R.

Lake Erie

BREEDING EVIDENCE

Reported in 1685 (92%) of 1824 squares

Possible 280 (17%)

Probable 568 (34%)

Confirmed 837 (50%)

INDIGO BUNTING
Passerin indigo
Passerina cyanea

The bright blue male Indigo Bunting is one of the most strikingly coloured songbirds that breed in Ontario. The female, however, is a nondescript brownish bunting. After arriving in the spring, the males are often seen singing from a tree at the edge of a wood or hedgerow.

The Indigo Bunting breeds from southeastern Saskatchewan eastward across southern Canada to southern New Brunswick, and south to the Gulf coast, and winters mainly in Mexico and Central America. It has been increasing its range in the southwestern US in recent years, and now breeds locally as far west as California. Throughout its range, the Indigo Bunting nests in deciduous woodland edge, second-growth, and shrubby old fields. On occasion, it occurs in mixed woods. The vegetation that comes in following the cutting of deciduous woods is often ideal habitat for this species, as is abandoned pasture. Consequently, its numbers have increased in northeastern North America in recent times (Robbins *et al.* 1986).

Territorial males advertise by singing persistently from a conspicuous perch, and, because of this, the species is probably rarely overlooked. Territorial behaviour enabled atlassers to infer probable breeding in 302 squares, 19% of the squares from which the species was recorded. Such records, however, may not indicate nesting in this species because many singing males are unmated. Indigo Buntings ignore 'pishing', but males respond to song playback. Nesting of the Indigo Bunting commences in early June and continues into early August (Speirs 1985). The nests usually contain 3 or 4 eggs and females may be double-brooded. Ten to 15% of the males that have mates are polygynous (Carey and Nolan 1979, Payne 1982). The nest is a cup of dry grass and stems, lined with rootlets, hair, or feathers, placed in a crotch, 0.5 to 6 m above the ground, but usually less than 1 m high, in low bushes or vines (Harrison 1978, Peck and James 1987). In Ontario, raspberry bushes are often selected for nest sites.

Because the males are conspicuous, the species was probably rarely overlooked during the atlas field inventories. However, breeding was confirmed in only 30% of squares in southern Ontario in which evidence of breeding was found. Nests are often difficult to find; atlassers found nests with eggs or young in only 60 squares (only 4% of records). When an intruder is near to a nest with young, both parents may give an excited *chip* note, and atlassers noted this behaviour in 14% of the squares in which Indigo Buntings were reported. Apparently, the female alone cares for the young; adults carrying food or faecal sacs were observed in 11% of the squares with records, with fledged young being found in an additional 10%. Males in their first breeding season show a great deal of variation in plumage colouration, ranging from birds that are nearly indistinguishable from 'adult' males to individuals that could be confused with females. Many of these first year males do mate.

The Indigo Bunting shows an affinity for the deciduous and mixed woods characteristic of the Great Lakes-St. Lawrence Forest region, but largely avoids the Boreal Forest. Those records from areas of northern Ontario that are predominantly coniferous forest indicate the presence of pockets of suitable deciduous forest habitat. Thus, the map is an accurate reflection of the Indigo Bunting's breeding distribution in Ontario. Atlas data indicate that the species is scarce in the Nipissing and Parry Sound Districts, except perhaps in some undisturbed areas. Otherwise, it is found virtually throughout southern Ontario. The atlas maps show a number of records north of the breeding range shown by Godfrey (1986), most specifically along the north shore of Lake Superior.

Atlassers estimated 100 or more breeding Indigo Buntings in only 12% of the southern Ontario squares for which abundance estimates were available, and fewer than 11 pairs in virtually all squares where estimates were provided in northern Ontario. As a roadside species, it is probably easy to overestimate its abundance, but atlassers seem to have taken this into consideration in their estimates. On Breeding Bird Surveys, the highest densities are recorded in southwestern Ontario and in Rainy River District (Speirs 1985). Although numbers are low in northeastern and northcentral Ontario, the species probably breeds where atlassers found singing males in these regions.-- *J.D. Rising*

INDIGO BUNTING

Summary of Abundance Estimates

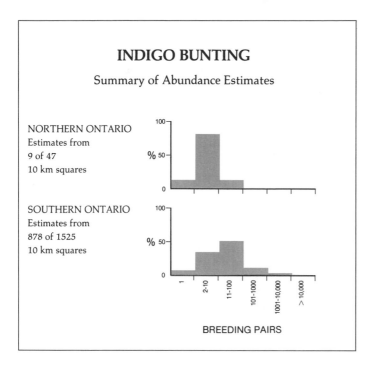

NORTHERN ONTARIO
Estimates from
9 of 47
10 km squares

SOUTHERN ONTARIO
Estimates from
878 of 1525
10 km squares

BREEDING PAIRS

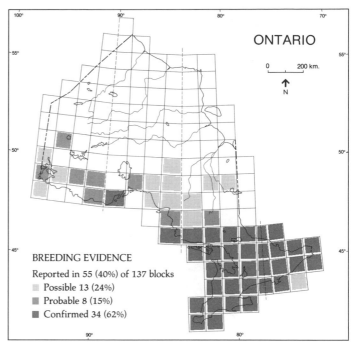

ONTARIO

BREEDING EVIDENCE
Reported in 55 (40%) of 137 blocks

Possible 13 (24%)
Probable 8 (15%)
Confirmed 34 (62%)

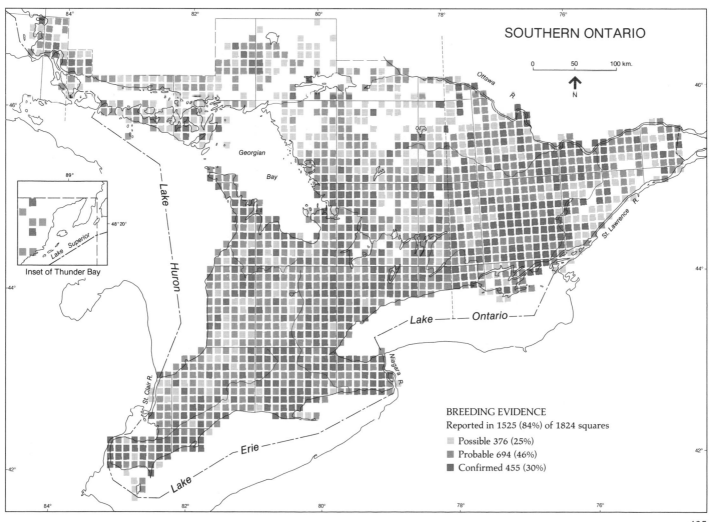

SOUTHERN ONTARIO

Inset of Thunder Bay

BREEDING EVIDENCE
Reported in 1525 (84%) of 1824 squares

Possible 376 (25%)
Probable 694 (46%)
Confirmed 455 (30%)

DICKCISSEL
Dickcissel
Spiza americana

The Dickcissel is a species of the tall and mid-grass prairies. Its main nesting range extends from the High Plains in the west to the Appalachian Mountains in the east and from the Gulf coast in the south to southern Canada in the north (Zimmerman 1971). In Canada, it is found only sporadically in the southern Prairie Provinces and southwestern Ontario. The Dickcissel winters in South America from Venezuela to Panama (Fretwell 1977).

The preferred nesting habitat is that of old fields where forbs predominate. The species is more characteristic of disturbed and agricultural habitats than true grasslands, except in certain years (Zimmerman 1971).

The Dickcissel is a distinctive bird, in both voice and appearance. The loud, staccato song is delivered incessantly from conspicuous perches through the nesting season, as is typical of polygynous species (Zimmerman 1966). Males are brightly coloured, resembling a miniature meadowlark. The males are larger than the females, weighing 30 g compared to 25 g (Fretwell 1977). The species is relatively easy to find when present in suitable breeding habitat. However, the song would be known to very few atlassers in Ontario.

Saunders and Dale (1933) reported that "the Dickcissel occurred more or less frequently and regularly in the southwestern counties up to 1895, the first record of its appearance in Middlesex was in the year 1895 when there was somewhat of a 'visitation' of them in Ontario." The species was not found again in Middlesex until 1930. A pair with young was found at Point Pelee Nat. Park in 1936 (Baillie and Harrington 1937). There are only 4 nesting records for the province, 3 of them in the late 1800's. The only nesting record in this century was in Middlesex Co in 1972 (Peck and James 1987). Godfrey (1986) gives the Ontario range as being south of a line from the Bruce Peninsula to Toronto and describes the

breeding as extremely sporadic. Fretwell (1977) states that, on the edges of its range, the Dickcissel is erratic in distribution and unpredictable in abundance.

The Dickcissel is a polygynous species with many males defending territories that contain more than one female (Zimmerman 1966). The female builds the nest in well-concealed locations deep within thick vegetation, making it difficult to locate. The male is very attentive to the female but not at all to the nest, eggs, or young (Zimmerman 1966). Nest disturbance by Brown-headed Cowbirds and by predators, most frequently snakes, reaches over 50% by early July (Zimmerman 1984), and second nesting is frequent (Harmeson 1974). The harvesting of alfalfa fields has been known to thwart nesting attempts in Manitoba (Godfrey 1986).

Dickcissel numbers are decreasing. Fretwell (1977) found that, in the northern parts of its range, the Dickcissel had an "almost uniform nesting failure" and that "females were much scarcer than males (about 1 male in 3 was mated)." Fretwell (1977) has found that, on the wintering grounds in Venezuela, the males forage on the large, rich seeds of domestic crops while the females forage on weed seeds. The abundant and rich food source may provide a selective advantage for the survival of the larger males. Therefore, by the end of the winter season, the population structure has been altered so that there are many more males than females. Back on the breeding grounds, the few remaining females have difficulty nesting because Dickcissels are not as successful in breeding at low densities, possibly because of heavy Brown-headed Cowbird nest parasitism (Fretwell 1977). Fretwell (1977) proposes that the Dickcissel be regarded as a threatened species because the species may have a low density limit and, when the number of females falls below this limit, the species will "fade into extinction."

Only 2 breeding records of this species occurred during the 1981 to 1985 atlas period in Ontario. In June 1982, Martin Parker observed a singing male Dickcissel over a 3 week period in pastureland on the Bruce Peninsula. This record is well away from the Lake Erie counties, where the species has normally been found. On 18 June 1985, Scott Connop observed a male in a clover field in Lambton Co. No records of a pair, a nest, or young were established. This is much in keeping with Fretwell's (1977) findings in the populations in the northern US, where, at low population densities, females are scarce.

This is one of the rarest breeding birds in Ontario. It is likely that no nestings occur in most years. The species is irruptive, but even during a period of higher numbers, only a few birds are found in the province, and few if any of them breed.-- *P.F.J. Eagles*

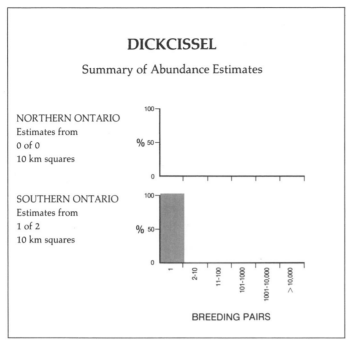

DICKCISSEL

Summary of Abundance Estimates

NORTHERN ONTARIO
Estimates from
0 of 0
10 km squares

SOUTHERN ONTARIO
Estimates from
1 of 2
10 km squares

BREEDING PAIRS

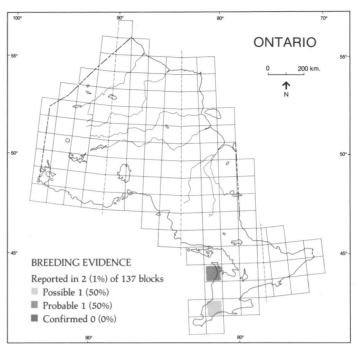

ONTARIO

0   200 km.

N

BREEDING EVIDENCE

Reported in 2 (1%) of 137 blocks

Possible 1 (50%)
Probable 1 (50%)
Confirmed 0 (0%)

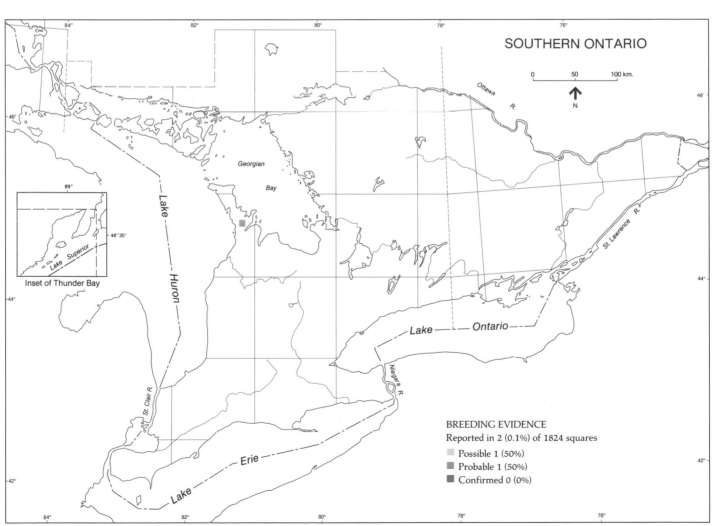

SOUTHERN ONTARIO

0   50   100 km.

N

Georgian Bay

Lake Huron

Lake Superior

Inset of Thunder Bay

Ottawa R.

St. Lawrence R.

Lake — Ontario

Niagara R.

St. Clair R.

Lake — Erie

BREEDING EVIDENCE

Reported in 2 (0.1%) of 1824 squares

Possible 1 (50%)
Probable 1 (50%)
Confirmed 0 (0%)

RUFOUS-SIDED TOWHEE
Tohi à flancs roux
Pipilo erythrophthalmus

The Rufous-sided Towhee is widespread throughout eastern North America, north to southern Ontario and the Gulf of Maine, and in western North America north to the Fraser Valley. The species is also found in central North America north to the southern Prairies. It migrates out of the northern parts of its range for the winter. In Ontario, it is widespread, but not abundant, throughout all of the south. Its preferred habitat is dense, brushy cover with an accumulation of leaf litter. Pasture lands and abandoned fields where young trees and shrubs develop comprise suitable breeding habitat, as do woodland edges and openings. Areas of intensive agriculture and dense forests are avoided by the Rufous-sided Towhee.

The range of the Rufous-sided Towhee has probably not changed substantially since McIlwraith (1894) reported that the species occurred in southern Ontario. Undoubtedly, the opening of the forests by the early settlers in the 1800's had benefitted this species by the creation of suitable brushy habitat. Godfrey (1986) gave its northern limits as "north to Sudbury, North Bay, Ottawa." However, there is some evidence that the range is now shrinking at its northern limits in Ontario. Atlas information indicates that the species breeds sparsely along that line, and the Sudbury Regional Coordinator reported that the species seems to have become extirpated in that region, even in favourable open country.

Although the Rufous-sided Towhee's call notes and song are easily recognizable, at least in eastern Canada, and also carry well, the ventriloquial quality of its voice can make the bird difficult to see in its preferred brushy habitat. As characteristic of a bird with a well-known song, over 50% of atlas records are of singing or territorial birds. It is often located by the loud rustling of leaf litter as it scratches vigorously in search of insects, seeds, and wild fruits. The female is very secretive near the well camouflaged nest, which is built on the ground or in a small shrub in dense undergrowth. Many atlassers found breeding of this species difficult to confirm. In only 21% of the squares in which this species was observed, was it confirmed as breeding.

The maps reflect the distribution of the Rufous-sided Towhee reasonably accurately and fully. Undoubtedly the clearing of woodlots and the drainage of swamps, in order to expand farmland, are reducing favourable habitat for this species. However, in the most heavily deforested area of Ontario, the Carolinian Forest zone, the species is found and is a confirmed breeder in a high proportion of squares. The Rufous-sided Towhee is largely absent from areas of dense forest, such as Algonquin Prov. Park, and areas of almost total forest clearance, such as Essex Co. However, there are some areas with seemingly ideal habitats that are blank: within central Wellington and Huron Counties and north of Peterborough, for example. Many Regional Coordinators in these areas expressed surprise that the distribution was not more general. The reasons for such absences are unknown. There are some apparent outliers of the range of this species, notably in the Rainy Lake area of northwestern Ontario, which is not far from the known range in Manitoba to the west (Godfrey 1986).

The Rufous-sided Towhee seems less abundant than was expected. Half of the squares for which abundance estimates were given were in the category of 2 to 10 pairs per square. Several Regional Coordinators indicated a decline in numbers starting in the mid-1970's. Breeding Bird Survey reports (Silieff and Finney 1981) clearly show such a decline, starting in 1976. The expansion of farmland could be a partial explanation for this trend.-- *H. and S. Inch*

RUFOUS-SIDED TOWHEE

Summary of Abundance Estimates

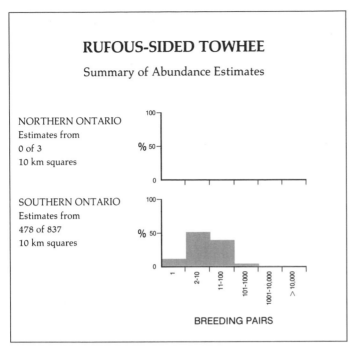

NORTHERN ONTARIO
Estimates from
0 of 3
10 km squares

SOUTHERN ONTARIO
Estimates from
478 of 837
10 km squares

BREEDING PAIRS

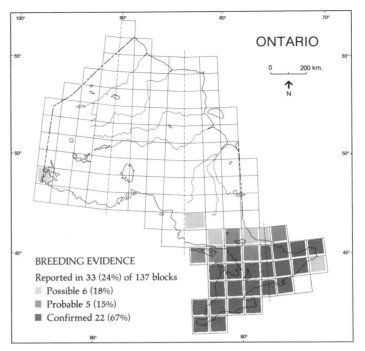

ONTARIO

0   200 km.

N

BREEDING EVIDENCE
Reported in 33 (24%) of 137 blocks
Possible 6 (18%)
Probable 5 (15%)
Confirmed 22 (67%)

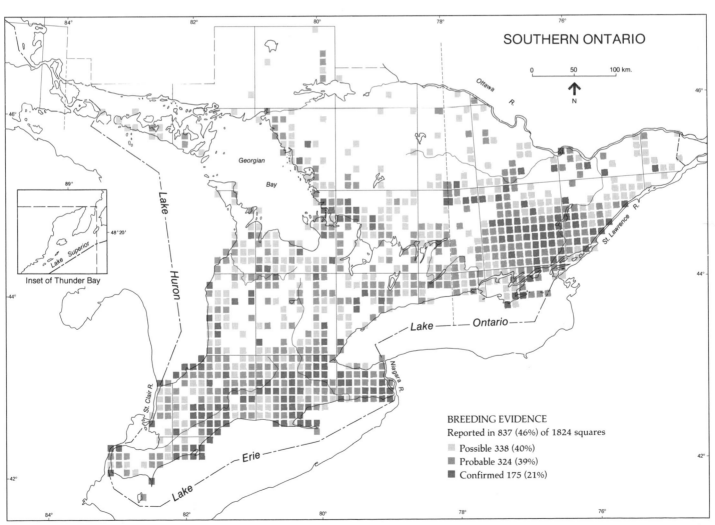

SOUTHERN ONTARIO

Georgian
Bay

Ottawa R.

0   50   100 km.

N

Lake Huron

Lake Superior

48°20'

89°

Inset of Thunder Bay

St. Clair R.

St. Lawrence R.

Lake — Ontario

Niagara R.

Lake — Erie

BREEDING EVIDENCE
Reported in 837 (46%) of 1824 squares
Possible 338 (40%)
Probable 324 (39%)
Confirmed 175 (21%)

CHIPPING SPARROW
Bruant familier
Spizella passerina

The Chipping Sparrow is a common summer resident throughout North America, where its range extends north to the limits of the Boreal Forest. In Ontario, it breeds from the US border to the Hudson Bay coast. Its preferred habitat is in open, grassy areas, bordering woodland or containing scattered thickets of trees. Because of this preference, the Chipping Sparrow has adapted well to man-made environments. It is probably now most common in settled areas characterized by dwellings surrounded by lawns, gardens, and shrubs. Such habitats are ideally suited to its nesting and feeding requirements. Though it is a common summer resident, there are few winter records for the Chipping Sparrow in Ontario (Speirs 1985). Most birds have left the province by late October and do not return until late in the following April. The main winter range is from the central US southward.

Apparently, the Chipping Sparrow was always common in eastern North America (Stull 1968). However, with the clearance of the Carolinian Forest, and its replacement with agricultural land, the numbers of this species may have initially declined. In recent times, the species has recovered in numbers and is now common and abundant throughout most parts of its range where suitable habitat exists (Speirs 1985). Its adaptation to man-made environments has undoubtedly contributed to its present success.

The Chipping Sparrow is one of our smallest sparrows. The adults are easily recognizable by their unstreaked grey breasts, rusty crowns, and black eye stripe with white line above it. The song is a simple monotone 'trill', and the call a distinct 'chip', from which the species gets its English name. However, several species produce 'trills', and, especially in areas where it is suspected to be uncommon, care must be taken to confirm identification.

Immediately after returning from the wintering grounds, the males occupy 'territories' (recent evidence suggests that the species does not occupy classical territories, Middleton unpubl.), and singing is frequent. Within days, the pairs are formed and nesting begins. At this time, singing diminishes and the birds are difficult to find. The small compact nest is most often at moderate height in a coniferous tree (Buech 1982, Reynolds and Knapton 1984) and is built entirely by the female with dry grass, rootlets, and a lining that is often of fine hair. Nest-building takes about 4 to 5 days. Nests were found in 382 squares throughout the province. The clutch of 4 light blue eggs hatches within 11 to 12 days; the young leave the nest when about 10 days old (Reynolds and Knapton 1984). It is the observation of adults with food for young, or of newly fledged young, that most often allows confirmation of nesting. For example, in southern Ontario, 49% of all records were of these 2 types of observation. Although second nesting occurs if the first nest is destroyed, few Chipping Sparrows produce 2 broods in a season. Egg-laying is complete for the season by mid-July. In contrast to the adults, the juvenile Chipping Sparrow is streaked on the breast and lacks the chestnut crown. As a result, the juveniles are sometimes confused with other sparrows.

Because the Chipping Sparrow is well known and conspicuous, the atlas data accurately reflect the breeding distribution. It is one of Ontario's most widespread species as it was encountered in 97% of all squares in southern Ontario and in 82% of all blocks in the province. In the south, it is likely that breeding occurs in every square. This is not surprising in view of its apparent preference for man-made habitats. Lack of breeding confirmation may simply have resulted from difficulty with access. In the north, the data show breeding in almost every block south of the Hudson Bay Lowland. On the Lowland, it is mostly found near the James Bay coast, and this may be further evidence that human settlements provide suitable nesting habitat. Farther west, the species occurs north to the Hudson Bay coast near Fort Severn. The most northerly confirmed record was of a nest with eggs near the Fawn River. These northern confirmations of breeding indicate an extension of the range shown by Godfrey (1986).

In addition to being a widely distributed species, the Chipping Sparrow is also one of the most abundant species in the province. Abundance estimates are somewhat higher from southern Ontario, with 68 squares having estimates of greater than 1,000 pairs. There is only one equivalent estimate from the north. The 2 highest estimates, greater than 10,000 pairs per square, came from north of Kingston in the area of Frontenac Prov. Park. The extensive areas of suitable habitat created by human activities probably explain the high estimates in the south.-- *A.L.A. Middleton*

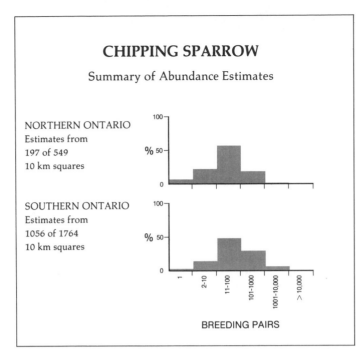

CHIPPING SPARROW

Summary of Abundance Estimates

NORTHERN ONTARIO
Estimates from
197 of 549
10 km squares

SOUTHERN ONTARIO
Estimates from
1056 of 1764
10 km squares

%

BREEDING PAIRS

1   2-10   11-100   101-1000   1001-10,000   >10,000

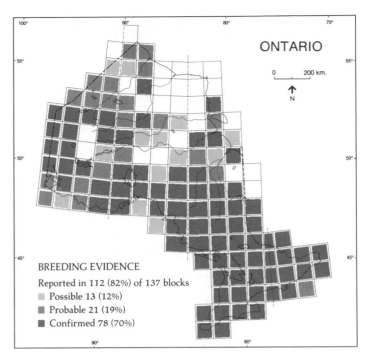

ONTARIO

0   200 km.

N

BREEDING EVIDENCE
Reported in 112 (82%) of 137 blocks
Possible 13 (12%)
Probable 21 (19%)
Confirmed 78 (70%)

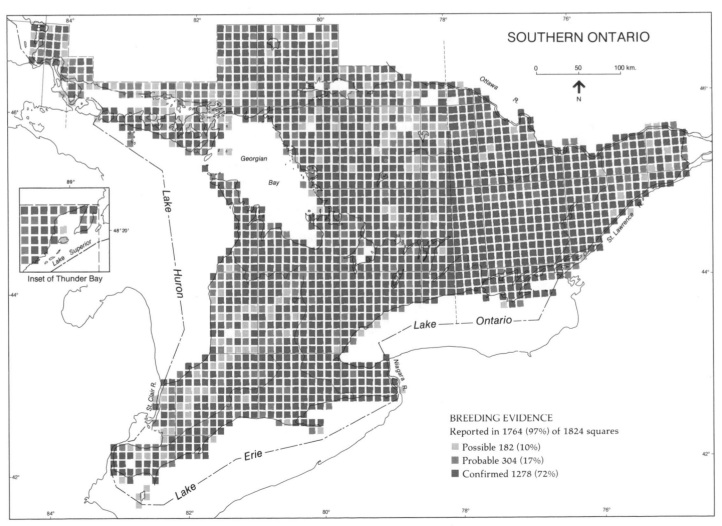

SOUTHERN ONTARIO

0   50   100 km.

N

Georgian Bay

Lake Huron

Lake Superior

Inset of Thunder Bay

89°
48°20'

St. Clair R.

Lake Erie

Ottawa R.

St. Lawrence R.

Lake — Ontario

Niagara R.

BREEDING EVIDENCE
Reported in 1764 (97%) of 1824 squares
Possible 182 (10%)
Probable 304 (17%)
Confirmed 1278 (72%)

CLAY-COLORED SPARROW
Bruant des plaines
Spizella pallida

The Clay-colored Sparrow is a widespread summer inhabitant of the North American Great Plains and Great Lakes regions. A common species of brushy open areas in the prairies, it occupies a broad range of habitat types, from young pine plantations to abandoned fields grown up to shrubs and small trees, regenerating burns, and thickets along the edges of waterways. In Ontario, the species is not confined to any particular biogeographic region, and occurs as a breeding species at suitable localities scattered throughout the province. Confirmed breeding occurs from coastal James Bay through the Boreal Forest to the Great Lakes-St. Lawrence Forest ecosystem, with possible or probable breeding occurring along the north shore of Lake Erie in the Carolinian Forest zone. The species winters in Mexico.

The Clay-colored Sparrow has expanded its range considerably eastward during the past 50 years. In the area from Lake Superior west to the Manitoba border, Baillie and Harrington (1937) listed it as "not uncommon", but elsewhere in the province the species was reported to be rare and sporadic in occurrence. It is now found in several localities across southern Ontario (Mills 1981, Speirs 1985, Peck and James 1987) east to the Quebec border. Until 1984, there were only 14 breeding records for the Clay-colored Sparrow reported in the New York State Breeding Bird Atlas (NYBBA preliminary maps). Although hybridization with the much more numerous Chipping Sparrow has been occasionally reported, this is probably not a major factor in influencing the distribution or abundance of the species within the province.

As is true of many small, brown, and relatively inconspicuous passerine birds, the Clay-colored Sparrow is most readily detected by song. The 2 to 5 low-pitched, insect-like buzzes are distinctive, and a given male will frequently sing for extended periods of time. Clay-colored Sparrows are nei-

ther shy nor retiring; they are often easy to find and approach once the song has been located. This is reflected in the fact that singing and territorial males were recorded as the highest level of breeding evidence in 61% of all squares in which the species was reported in the province. Nests are also not particularly difficult to find; the low percentage of squares (5%) that had eggs or young probably represents a shortage of time available for atlassers to search for nests, rather than well-concealed nest locations. Confirmed breeding occurred in only 38 of the 179 squares in which the species was reported in Ontario. Adults carrying faecal sacs or food was the most frequently reported breeding category, amounting to 45% of confirmed breeding records.

The maps accurately reflect the distribution of the Clay-colored Sparrow in southern Ontario. However, it is possible that there are more blocks in northern Ontario in which the species occurs but where lack of exhaustive coverage failed to discover it. Although Godfrey (1986) mentions a recent range expansion north to Moosonee, the number of atlas records from the far northern part of Ontario was unexpected and suggests a much more extensive range than was previously realized. Habitats occupied in southern Ontario are subject to successional change, and therefore birds occur for several years, then disappear. This is most evident in conifer plantations, which can hold numerous pairs for several years, but which harbour declining numbers as the trees become mature. For example, the species appears to be slowly declining in the Long Point area, probably because of pine forest maturation. On the other hand, the species is slowly expanding in southeastern Ontario (*i.e.*, the Kingston and Ottawa areas) as more suitable habitat becomes available. It remains at a low level elsewhere (*i.e.*, Parry Sound, Waterloo, Richmond Hill, Sudbury, and Peterborough).

Populations in habitats other than coniferous plantations are more likely to persist, as the habitat does not change so quickly. This is most likely the case in abandoned fields with alders and small poplars in the Sudbury RM, willow scrub and open bogs in the Ottawa area, and fields with young red cedar near Kingston.

Most abundance estimates in Ontario are of between 2 and 10 breeding pairs per square. In 30% of the remaining squares, the estimate was of only one pair, and an unexpectedly high 23% of estimates are of between 11 and 100 pairs. The species may be more numerous in northern Ontario, particularly in the Lake of the Woods area (Speirs 1985).-- *R. Knapton*

CLAY-COLORED SPARROW

Summary of Abundance Estimates

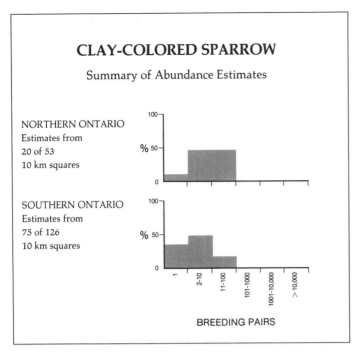

NORTHERN ONTARIO
Estimates from
20 of 53
10 km squares

SOUTHERN ONTARIO
Estimates from
75 of 126
10 km squares

%

BREEDING PAIRS

1 2-10 11-100 101-1000 1001-10,000 > 10,000

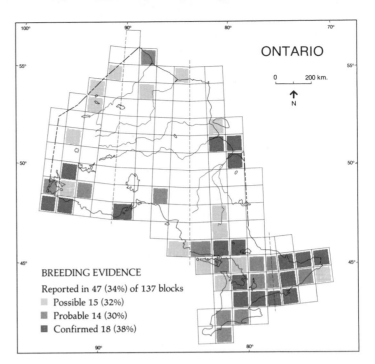

ONTARIO

0   200 km.

N

BREEDING EVIDENCE

Reported in 47 (34%) of 137 blocks

Possible 15 (32%)

Probable 14 (30%)

Confirmed 18 (38%)

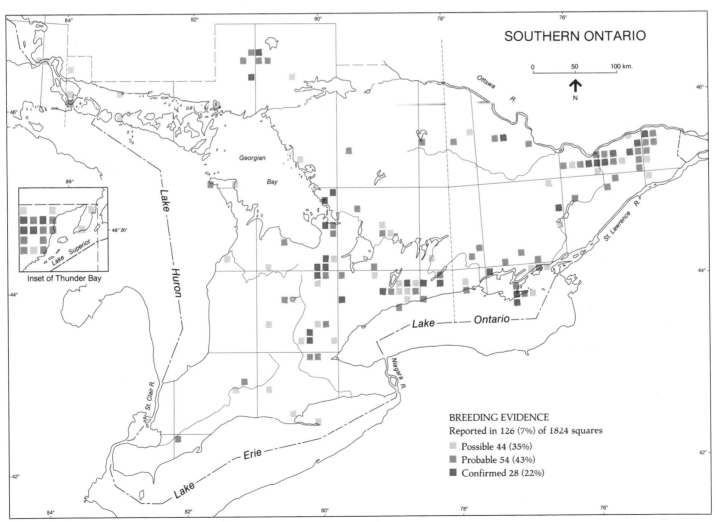

SOUTHERN ONTARIO

0   50   100 km.

N

Georgian Bay

Ottawa R.

St. Lawrence R.

Lake Huron

Lake Superior

89°

48°20'

Inset of Thunder Bay

Lake — Ontario

Niagara R.

St. Clair R.

Lake Erie

BREEDING EVIDENCE

Reported in 126 (7%) of 1824 squares

Possible 44 (35%)

Probable 54 (43%)

Confirmed 28 (22%)

FIELD SPARROW
Bruant des champs
Spizella pusilla

The Field Sparrow breeds across eastern North America from the Great Plains to the Atlantic Ocean, and south from southern latitudes of Ontario, Quebec, and New Brunswick to the Gulf of Mexico. In winter, it withdraws from the northern edge of its breeding range, although small numbers winter in southern Ontario. In Ontario, it inhabits marginal farmland, young conifer plantations, neglected pastures, and overgrown fields south of the Boreal Forest. Abandoned farm fields and powerline corridors are the typical habitat along the southern margin of the Canadian Shield.

The clearing of the original forests for agriculture and the subsequent abandonment of farms created considerable areas of suitable habitat within the province. Probably for that reason, both the range and the abundance of the Field Sparrow have increased during this century in Ontario. The limit of its distribution is now considerably beyond the northern limit stated by Baillie and Harrington (1937), at "Muskoka District, north Frontenac Co and Ottawa." Deserted farmland in the northern part of southcentral Ontario has allowed the species to spread into Parry Sound and Nipissing Districts (Mills 1981, Peck and James 1987) and onto Manitoulin Island. The species is essentially absent from northern Ontario. In recent years, suitable Field Sparrow habitat has been lost to reforestation and natural successional processes that result in unsuitable wooded areas. Urbanization, and the expansion of intensively cultivated areas, especially in southwestern Ontario, have also eliminated suitable habitat.

The Field Sparrow is not likely to be overlooked by atlassers searching in appropriate habitat. This sparrow perches conspicuously on small trees and shrubs in fairly open habitat, and its song is distinctive, loud, and far-carrying. Furthermore, it has a fairly long breeding season in southern Ontario, often into late July. Egg dates range from 4 May to 9 August, with a peak period from 26 May to 21 June (Peck and James 1987). Nests are usually well-hidden; 28% of confirmed breeding records were of nests with either eggs or young. Confirmation of breeding is mainly from observations of adults carrying food (34%) or the presence of fledglings (34%). Breeding was confirmed in 40% of the squares in which the species was reported.

The maps show an Ontario range that is similar to that portrayed by Godfrey (1986), but with few records along the northern edge of the range. They reflect an accurate picture of the distribution of the Field Sparrow in Ontario, even indicating some areas where it is conspicuously absent. It is not found in the Toronto area, presumably because of urbanization, nor in some squares in the heavily-cultivated farmland counties of southwestern Ontario such as Essex and Perth. It is curiously absent from seemingly suitable habitat in Wellington and Dufferin Counties, and in parts of southeastern Ontario where much old field habitat exists. There is plenty of apparently suitable habitat on Manitoulin Island but the species is quite scarce there. Comments from Regional Coordinators on changing abundance suggest that there has been a decline due to a decrease in the amount of abandoned farmland, parasitism by the Brown-headed Cowbird, and fewer burns in the Great Lakes-St.Lawrence Forest region.

In most squares, abundance estimates of between 2 and 10 (40%) or between 11 and 100 (45%) pairs per square were reported. Field Sparrows have been known to occur at locally high densities in some areas of Ontario, for example, over 100 birds per 100 ha in Mara Township, Durham RM (Speirs 1985). BBS data show highest counts in the southwest (Speirs 1985). Widespread decreases in the continental population of the Field Sparrow were indicated by BBS counts for the 1965 to 1979 period, but an increase was recorded in the Canadian part of the range (Robbins *et al.* 1986). The Field Sparrow is a common species in suitable habitat.-- *R. Knapton*

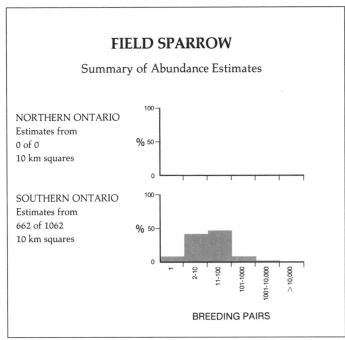

FIELD SPARROW

Summary of Abundance Estimates

NORTHERN ONTARIO
Estimates from
0 of 0
10 km squares

SOUTHERN ONTARIO
Estimates from
662 of 1062
10 km squares

%

BREEDING PAIRS

1 · 2-10 · 11-100 · 101-1000 · 1001-10,000 · > 10,000

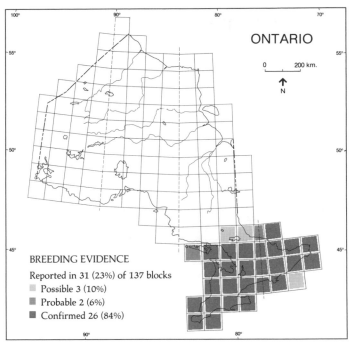

ONTARIO

0   200 km.

N

BREEDING EVIDENCE

Reported in 31 (23%) of 137 blocks

Possible 3 (10%)
Probable 2 (6%)
Confirmed 26 (84%)

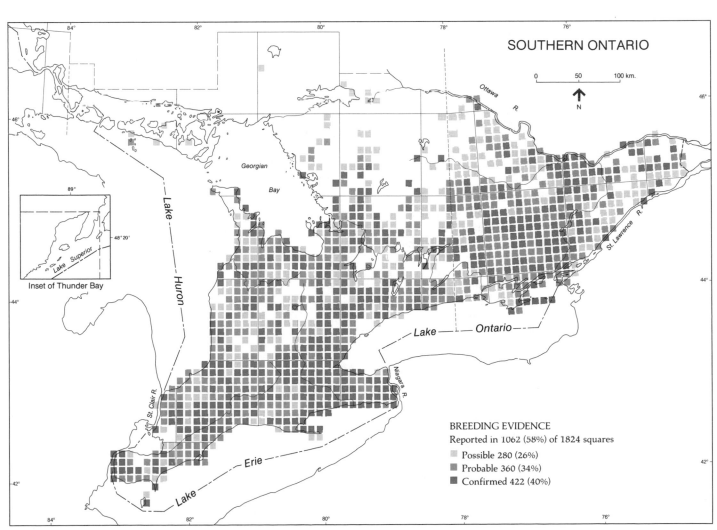

SOUTHERN ONTARIO

0   50   100 km.

N

Georgian Bay

Lake Huron

Lake Superior

Inset of Thunder Bay

Ottawa R.

St. Lawrence R.

Lake — Ontario

Niagara R.

St. Clair R.

Lake — Erie

BREEDING EVIDENCE

Reported in 1062 (58%) of 1824 squares

Possible 280 (26%)
Probable 360 (34%)
Confirmed 422 (40%)

VESPER SPARROW
Bruant vespéral
Pooecetes gramineus

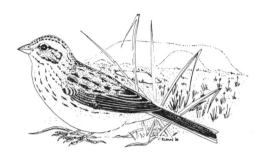

The Vesper Sparrow is a bird of open fields that tend to have wooded hedgerows or regrowth. The white lateral tail feathers (junco-like), evident when a Vesper Sparrow flushes, are a good field mark.

The Vesper Sparrow breeds from southern British Columbia eastward across the Canadian prairies through central Ontario to Nova Scotia (uncommonly), and south to the central US. Its winter range is in the southern half of the US and Mexico. It generally inhabits well-drained, often dry, grassland areas, frequently bordering woods or interspersed with trees or shrubs. In Ontario, it can be found in open, dry conifer plantations, and atlassers often found it in the scrubby regrowth area of a gravel pit, or in grain fields. Paul Eagles (pers. comm.) found a Vesper Sparrow nest that was not abandoned even after hay in the field had been cut, raked, and baled. The absence of the species from many squares on the Canadian Shield doubtless is a reflection of the scarcity of suitable habitat in this region. Regional Coordinators indicated that the Vesper Sparrow had decreased in numbers in some regions (*e.g.*, Parry Sound area) in recent years, and McIlwraith (1894) considered the Vesper Sparrow to be common throughout southern Ontario (and commoner than the Savannah Sparrow) - certainly not the case today. Changes in agricultural practices of the past generation apparently have benefitted the Savannah Sparrow at the expense of the Vesper.

The male frequently sings from a high branch in a tree bordering a field, facilitating finding territories. Atlassers indicated probable nesting in 415 squares, and territorial behaviour was cited as evidence in 52% of these. Vesper Sparrow eggs are laid from late April through early August (Peck and James 1987); females are double-brooded (Berger 1968). The nest, which is a loosely-woven cup of grasses lined with finer grasses or rootlets, is placed on the ground (Headstrom 1970). The female, if flushed from the nest, will often give a broken wing distraction display, or fan her tail so that the white feathers are conspicuous in an attempt to lead the intruder from the nest; such distraction displays were used to confirm nesting in 24 squares, and agitated behaviour was noted in 56 others. Atlassers found nests in 82 squares - 6% of those in which the species was reported. Both parents help to feed the young; the male may take primary responsibility for the first brood as the female starts a second (Berger 1968).

In southern Ontario, atlassers found evidence of breeding in 67% of the squares. The Vesper Sparrow likely occurs in most squares south of the Shield, but may be absent in heavily urbanized or intensively farmed areas, perhaps explaining the absence of records of it in the Toronto region, and in a few squares in extreme southwestern Ontario. On the Canadian Shield, the map reflects the presence of agricultural areas, with a noticeable lack of records in the dense forests of the Algonquin Highlands. The field areas adjacent to Highway 11 between Bracebridge and North Bay are clearly visible on the southern Ontario map.

In northern Ontario, Godfrey (1986) shows the northern edge of the breeding range to be the Albany River. The atlas map shows a portion of the northern range limit to be much further south. In the northeast, Lake Abitibi is the approximate northern limit, 400 km south of Fort Albany. In the northwest, the atlas data show a range similar to that of Godfrey (1986).

The Vesper Sparrow is generally uncommon in Ontario. Atlassers estimated fewer than 100 pairs from 94% of the squares from which abundance estimates were made. Breeding Bird Survey data (Speirs 1985) show the species to be most common in Essex, Lambton, and Kent Counties, but also common on the Bruce Peninsula, north and east of Toronto, and in the western Rainy River District.-- *J.D. Rising*

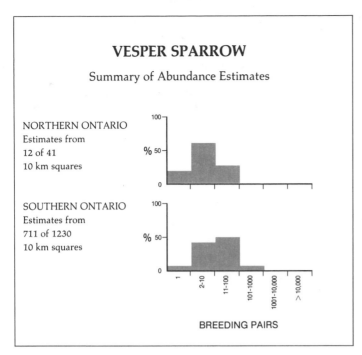

VESPER SPARROW

Summary of Abundance Estimates

NORTHERN ONTARIO
Estimates from
12 of 41
10 km squares

SOUTHERN ONTARIO
Estimates from
711 of 1230
10 km squares

BREEDING PAIRS

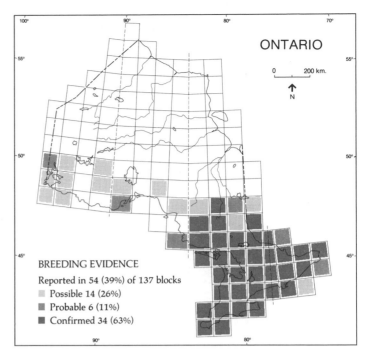

ONTARIO

0   200 km.

N

BREEDING EVIDENCE

Reported in 54 (39%) of 137 blocks

Possible 14 (26%)
Probable 6 (11%)
Confirmed 34 (63%)

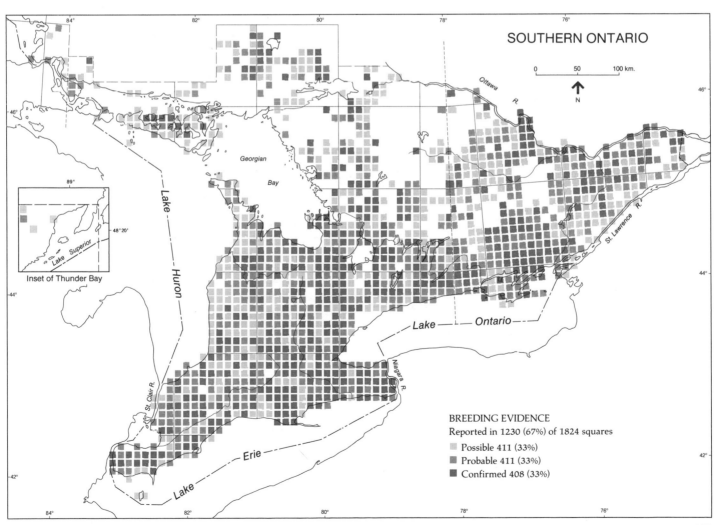

SOUTHERN ONTARIO

0   50   100 km.

N

Inset of Thunder Bay

Georgian
Bay

Lake Huron

Ottawa R.

St. Lawrence R.

Lake — Ontario

Niagara R.

St. Clair R.

Lake — Erie

BREEDING EVIDENCE

Reported in 1230 (67%) of 1824 squares

Possible 411 (33%)
Probable 411 (33%)
Confirmed 408 (33%)

SAVANNAH SPARROW
Bruant des prés
Passerculus sandwichensis

The Savannah Sparrow is the most common breeding bird in fields and meadows throughout Ontario. Although not a colourful species, it is easily seen and well-known to all students of Ontario birds.

The Savannah Sparrow is found virtually throughout Canada, south of the arctic islands, southward into the northeastern and middle prairie states, and, in the west, in mountain meadows south to southern Mexico. In the north, Savannah Sparrows nest in sedge meadows bordered by willows and sweet gale, and elsewhere, in grassy fields and pastures. Savannah Sparrows winter from Delaware and central Pennsylvania, west through Oklahoma, and south into central Mexico.

Prior to the 19th century, the Savannah Sparrow was probably restricted to large meadows, burns, and sedge bogs in Ontario. Indeed, McIlwraith (1894) writes, "... they are not very plentiful anywhere [in Ontario]," and at the time that McIlwraith wrote, the Vesper Sparrow was apparently the commonest grassland sparrow in Ontario, which certainly is not the case today. Both of these species would have benefitted from the clearing of the forests. However, the 19th and early 20th century practice of leaving one-quarter of the fields fallow each year appears to have favoured the Vesper Sparrow, whereas the widespread use of chemical fertilizers and the planting of alfalfa in pastures favours the Savannah Sparrow.

Territorial males are easily located as they sing persistently from the top of a fence post or tall weed. Males also respond to song play-back, but generally ignore 'pishing'. In 36% of the squares in which breeding was probable, territorial behaviour was noted by atlassers. Late in incubation, females sit tight on their eggs, and may run from the nest for a short distance before flushing. Occasionally, an incubating female will perform a rodent-run distraction display. Distraction displays were reported in 18 squares in southern Ontario. Both parents help feed the young. In 632 squares (75% of those in which nesting was confirmed), atlassers confirmed nesting on the basis of adults carrying food, or with fledged young. Nests are nearly always placed on the ground, and are usually well concealed. As common as the species is, atlassers reported nests with eggs or young in only 159 squares.

In Ontario, atlassers found probable or confirmed breeding in nearly every square south of the Canadian Shield, and in the blocks in the Clay Belt and Hudson Bay Lowland regions. With careful search, this species could doubtless be found in other squares or blocks. Nonetheless, the map clearly reflects its distribution and abundance; it is safe to infer that Savannah Sparrows are not common in those squares on the southern projection of the Shield (*e.g.*, Nipissing and Parry Sound Districts; Lanark and Frontenac Counties) and blocks in the forested regions of the Algonquin Highlands and northern Ontario. South of the Shield, the species is often found in conjunction with the Bobolink and Eastern Meadowlark. Along the coasts of James and Hudson Bays, the Savannah Sparrow is often the commonest species in relatively dry sedge meadows, interspersed with small willows and other shrubs.

There is slight geographic variation among populations of Savannah Sparrows in Ontario. For example, birds breeding along the coasts of James and Hudson Bays are slightly larger than those breeding in the south: 46 males breeding at Winisk, on Hudson Bay, had an average wing length of 70.8 mm, whereas 40 breeding near Wallaceburg, Kent Co had an average wing length of 68.5 mm. Northern birds also have slightly more extensive ventral spotting, and darker browns than the southern ones.

Savannah Sparrows are generally abundant where found; for example, over 30 birds/ha were found in fields near Pickering (Speirs 1985), and densities at least that high occur in large sedge meadows along the coasts of James and Hudson Bays (Schueler *et al.* 1974), and elsewhere in suitable habitat. More than 100 breeding Savannah Sparrows were estimated in 44% of the squares where abundances were estimated; over 1,000 were estimated in 12% and over 10,000 in 4 squares (0.4%)! These 4 squares were at Lebanon (south of Listowel), Mississauga, Etobicoke, and Cumberland (east of Ottawa). On Breeding Bird Surveys, the highest numbers are reported from the Bruce Peninsula and the Ottawa valley (Speirs 1985).-- *J.D. Rising*

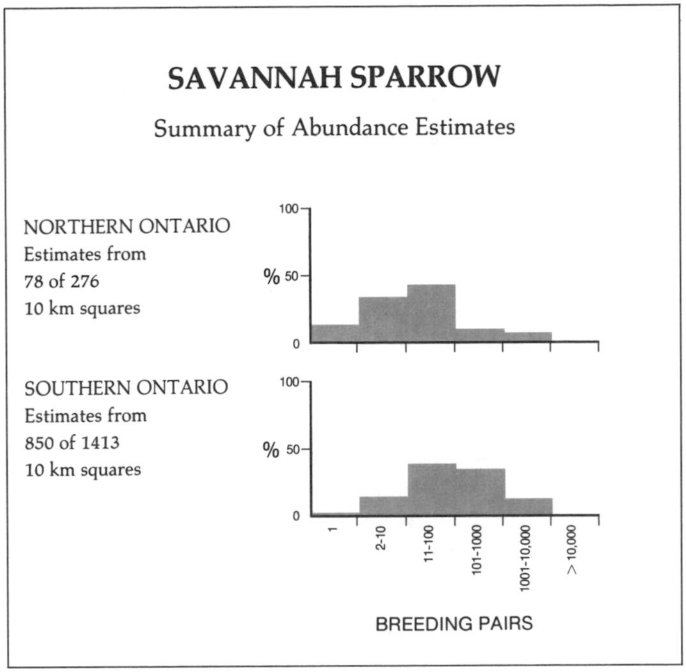

SAVANNAH SPARROW
Summary of Abundance Estimates
NORTHERN ONTARIO
Estimates from
78 of 276
10 km squares
SOUTHERN ONTARIO
Estimates from
850 of 1413
10 km squares
%
0
50
100
1
2-10
11-100
101-1000
1001-10,000
> 10,000
BREEDING PAIRS

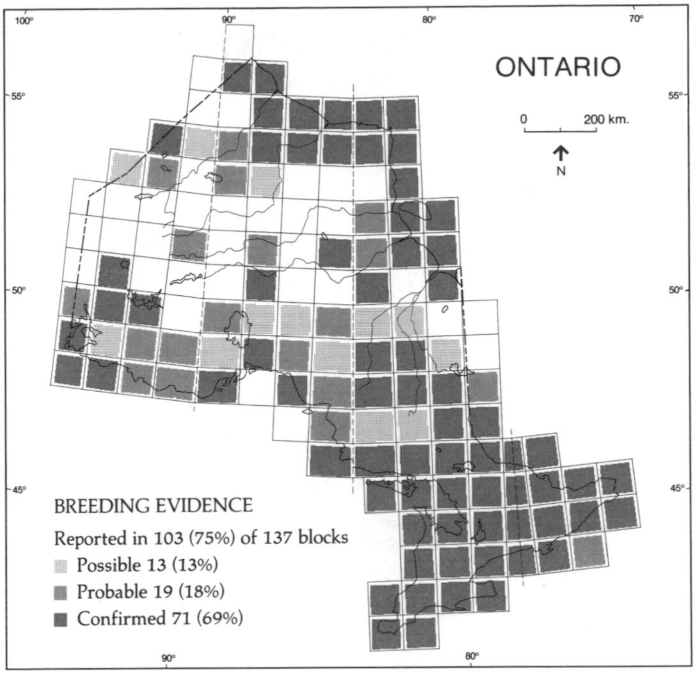

ONTARIO
0
200 km.
N
BREEDING EVIDENCE
Reported in 103 (75%) of 137 blocks
Possible 13 (13%)
Probable 19 (18%)
Confirmed 71 (69%)

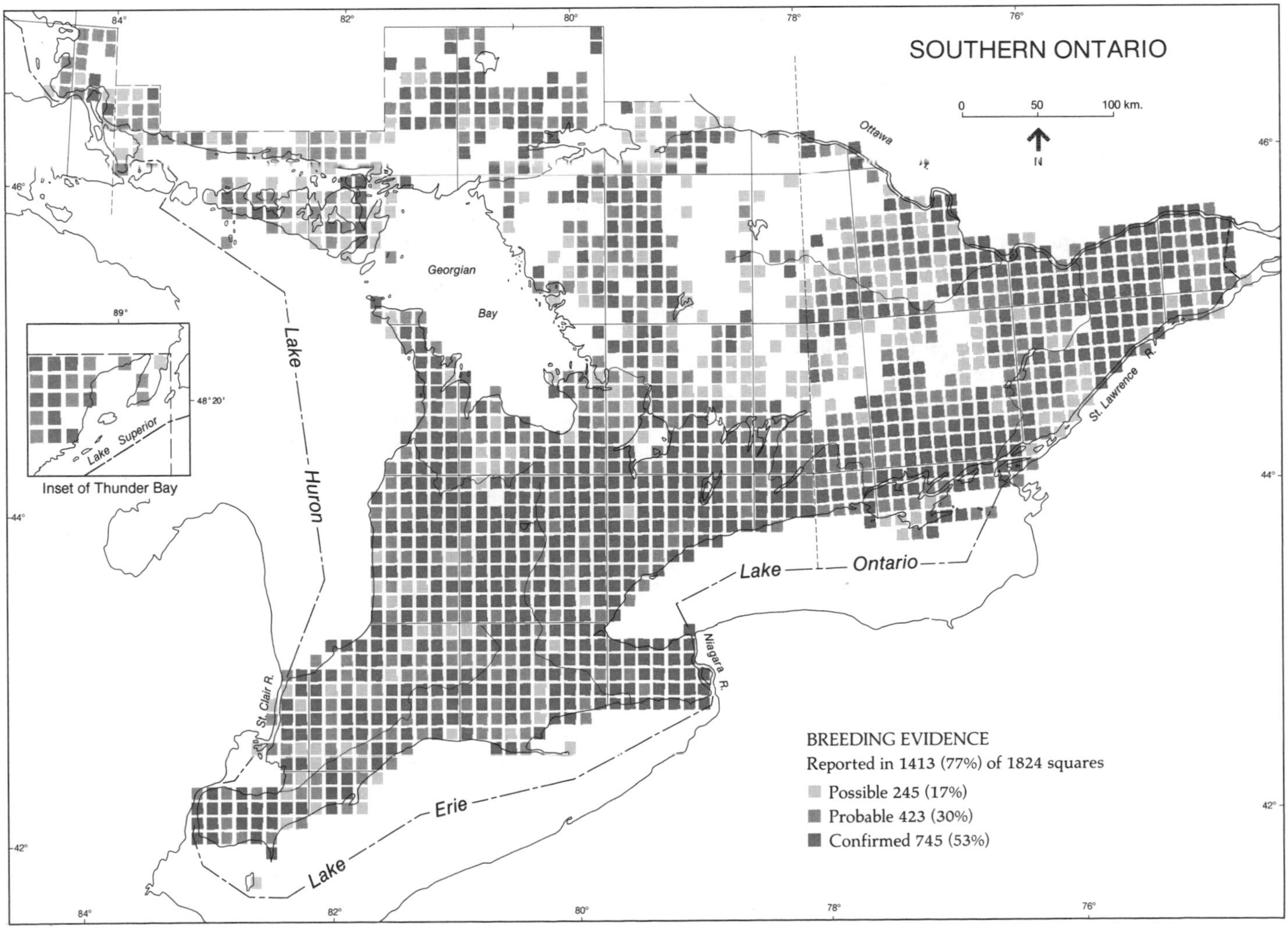

SOUTHERN ONTARIO
0
50
100 km.
N
Ottawa
Georgian
Bay
Lake
Huron
Lake
Ontario
Lake
Erie
St. Clair R.
Niagara R.
St. Lawrence R.
Lake Superior
Inset of Thunder Bay
BREEDING EVIDENCE
Reported in 1413 (77%) of 1824 squares
Possible 245 (17%)
Probable 423 (30%)
Confirmed 745 (53%)

GRASSHOPPER SPARROW
Bruant sauterelle
Ammodramus savannarum

The Grasshopper Sparrow is a grassland species. It is much more locally distributed and difficult to see than the Savannah Sparrow (with which it often occurs in Ontario).

The Grasshopper Sparrow breeds in the southern interior of British Columbia, eastward across the southern Canadian prairies and southern Ontario and Quebec, and south to the southern US. There is also a resident population from Mexico to northern South America and the Lesser Antilles. Throughout its range, the Grasshopper Sparrow shows a preference for fields that are treeless, and in which there is low cover of grasses and some taller weeds (such as mullein), which it uses for song perches. In Ontario, it is often found in well-drained grasslands growing in sandy soils. It is absent from fields with more than 35% of the area in shrubs (Smith 1968). It winters from the southern US southward to Costa Rica (AOU 1983).

In northeastern North America, the clearing of land for agriculture has permitted the species to spread. Prior to the 20th century, the Grasshopper Sparrow was quite rare in Ontario; McIlwraith (1894) knew of only two records for the province. Today, within its range, it is most frequently found in unmowed fields, and less frequently in grain and hay fields. Prior to recent times, much of the habitat in Ontario where it is found today would not have been suitable for the Grasshopper Sparrow. Even in recent times, however, the numbers in Ontario appear to have fluctuated.

Although it is a bird of sparse fields, the Grasshopper Sparrow can be difficult to locate. Territorial males, however, sing persistently, as indicated by the recording of the 'territory' category of probable breeding in 111 of the 440 southern Ontario squares where the species was located. The song is insect-like and may have been overlooked by atlassers not familiar with it. In Ontario, nesting takes place from early May through mid-August, females are probably double-brooded, and 4 to 5 eggs are generally laid (Peck and James 1987). Nests are placed on the ground, often partially covered, and are extremely difficult to find. Atlassers located nests with eggs or young in only 14 squares. The female sits closely on the nest. When leaving, she slips off the nest and runs a distance in the grass before flying. Sometimes she gives a broken-wing distraction display; this behaviour was observed by atlassers in 3 squares. The male appears to show no alarm when an intruder approaches the nest. He simply stops singing and hides in the grass (Smith 1968). Nevertheless, atlassers recorded agitated behaviour in 16 squares. Both parents feed the young (Smith 1968); 56% of all confirmed breeding was based on adults carrying food, and an additional 21% was based on reports of fledged young. In the wide-open grassland habitat of this species, the carrying of food is fairly easy to see, and this is reflected by the high numbers of these 2 levels of breeding evidence.

Atlassers found evidence of breeding most frequently in squares south of the Canadian Shield, where the Grasshopper Sparrow probably occurs more widely than the map indicates; the species likely nests in most of the 443 squares in which it was seen in Ontario. There is a concentration of records along the Niagara Escarpment, in the Guelph area, and north of Lake Ontario, particularly on the Oak Ridges Moraine and in the vicinity of Kingston. In these areas, morainic deposits are common. These, when cleared of woodlands, provide the well-drained grasslands preferred by the species. The southern edge of the Shield forms the northern limit of the main range of the species and there are only scattered records farther north. The probable and possible records from near Sudbury and northwestern Ontario in the Lake of the Woods area indicate range expansions into those regions.

Local densities can be fairly high: Speirs (1985) found densities of 1.4 to 2 birds/ha in certain fields near Whitby, and the species might be expected to be common in abandoned fields just south of the Shield. Atlassers, however, estimated fewer than 100 pairs per square in 97% of the 286 squares from which abundance estimates were made. All 10 of the abundance estimates of over 100 pairs per square came from an area of southern Ontario roughly outlined by Oshawa, Burlington, Arthur, and Barrie. It is likely that the many dry, old fields that are not extensively cultivated for crops provide ideal habitat for Grasshopper Sparrows in this area.-- *J.D. Rising*

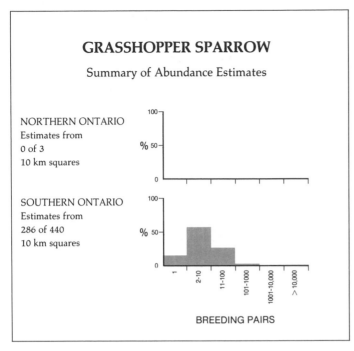

GRASSHOPPER SPARROW

Summary of Abundance Estimates

NORTHERN ONTARIO
Estimates from
0 of 3
10 km squares

SOUTHERN ONTARIO
Estimates from
286 of 440
10 km squares

BREEDING PAIRS

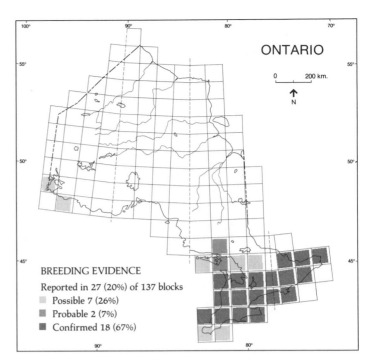

ONTARIO

0   200 km.

N

BREEDING EVIDENCE
Reported in 27 (20%) of 137 blocks
Possible 7 (26%)
Probable 2 (7%)
Confirmed 18 (67%)

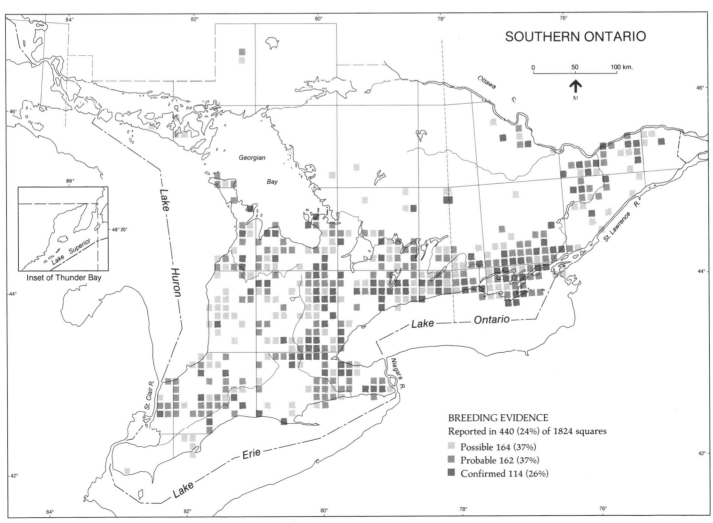

SOUTHERN ONTARIO

0   50   100 km.

N

Inset of Thunder Bay

Georgian Bay

Lake Huron

Lake Superior

Lake Ontario

Lake Erie

St. Clair R.

Niagara R.

St. Lawrence R.

Ottawa

BREEDING EVIDENCE
Reported in 440 (24%) of 1824 squares
Possible 164 (37%)
Probable 162 (37%)
Confirmed 114 (26%)

HENSLOW'S SPARROW
Bruant de Henslow
Ammodramus henslowii

The Henslow's Sparrow is an enigmatic species, not easy to locate or to census. It is sporadically distributed throughout its range, often changing breeding locations from one year to the next. The breeding range of the Henslow's Sparrow is restricted to eastern North America, extending from the central Great Plains east to the Atlantic Ocean, and south from the Canadian Shield to Kansas and North Carolina. The winter range lies along the Gulf of Mexico and southern Atlantic coastal states. In Ontario, it occurs primarily in cleared areas in the Great Lakes-St. Lawrence Forest ecosystem. Suitable habitat includes large grassy areas that have lain fallow for several years and hence have formed a ground mat of dead vegetation and have song perches. Moist grassy areas are frequently used.

The species was probably not originally found in Ontario, but expanded into the province with the clearing of forest for agriculture and the subsequent abandonment of fields. McIlwraith (1894) does not mention the Henslow's Sparrow in his review of the birds of Ontario. It was first reported in the province in 1898 in Lambton and Leeds Counties (Macoun and Macoun 1909). The species evidently became fairly common locally in Essex and Lambton Counties at the turn of the century, and soon after, records from other parts of the province indicated a gradual northward and eastward extension during the next few decades (Knapton 1984). By the mid-1930's, the northern limit of its range stretched from Sarnia to Bradford (Baillie and Harrington 1937). Since the 1930's, the species has been found north to Manitoulin Island and east to Leeds and Grenville Counties, considerably beyond Baillie and Harrington's limit (Knapton 1984). However, it now appears to be absent or very rare in much of its range in Ontario, especially in now intensively cultivated southwestern Ontario, a region which certainly had far more

suitable habitat earlier in this century (Knapton 1982).

The Henslow's Sparrow is easily overlooked because of its unobtrusive behaviour. Although not particularly shy, it is an inconspicuous bird which, when disturbed, will fly a short distance, drop to the ground, and then run through the vegetation to escape detection. Its song is the best means of finding the species, but the song itself is very short, albeit distinctive, somewhat ventriloquial, and easily drowned out by the songs of other species nesting in the same habitat, especially Bobolinks. Naturalists in Ontario are generally unfamiliar with its song, and therefore a number of birds may have been overlooked. The species sings regularly during the night, and some of the atlas records have been a result of this behaviour.

In late spring, individual males may occupy a seemingly suitable field for a short time, and then disappear. The 12 squares showing possible breeding evidence may include such wandering, prospecting males. The scattered nature of the records suggests that the species may be underrepresented and that pairs have been overlooked. Nevertheless, the map, being a composite of 5 years' records, inevitably exaggerates the picture for any one season. The concentration of records in southeastern Ontario in the Kingston area is possibly a reflection of the large area of abandoned, slightly moist fields occurring on poor soil in that part of the province.

In only 8% of the 38 squares in which the Henslow's Sparrow was reported was breeding confirmed, with fledged young in 2 squares and adults carrying food in one. Not surprisingly, in 71% of the squares, the highest level of breeding evidence consisted of singing or territorial males. There was only one group of Henslow's Sparrows greater than 10 pairs known in Ontario during the atlas period, although much larger 'colonies' have occurred in the province previously (Knapton 1984). In 12 of the 32 squares for which abundance was estimated, there was thought to be one pair per square, and in 59% the atlassers gave estimates of between 2 and 10 pairs per square. However, not all squares were occupied in each year and the number of squares occupied appears to fluctuate from year to year. Even in a year of high numbers, however, it is unlikely that the total provincial population exceeds 50 pairs. The low numbers have resulted in the Henslow's Sparrow being classed as threatened by COSEWIC and by the MNR. The species is surprisingly widespread in New York State (NYBBA preliminary maps), where presumably more suitable habitat or better environmental conditions occur.-- *R. Knapton*

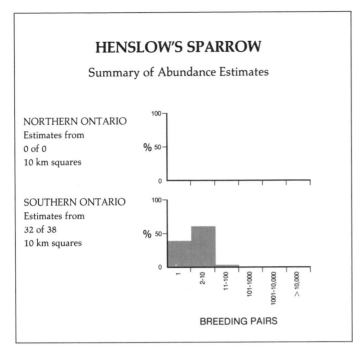

HENSLOW'S SPARROW

Summary of Abundance Estimates

NORTHERN ONTARIO
Estimates from
0 of 0
10 km squares

SOUTHERN ONTARIO
Estimates from
32 of 38
10 km squares

BREEDING PAIRS

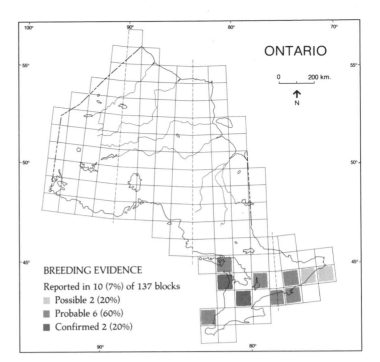

ONTARIO

0   200 km.

N

BREEDING EVIDENCE

Reported in 10 (7%) of 137 blocks

Possible 2 (20%)
Probable 6 (60%)
Confirmed 2 (20%)

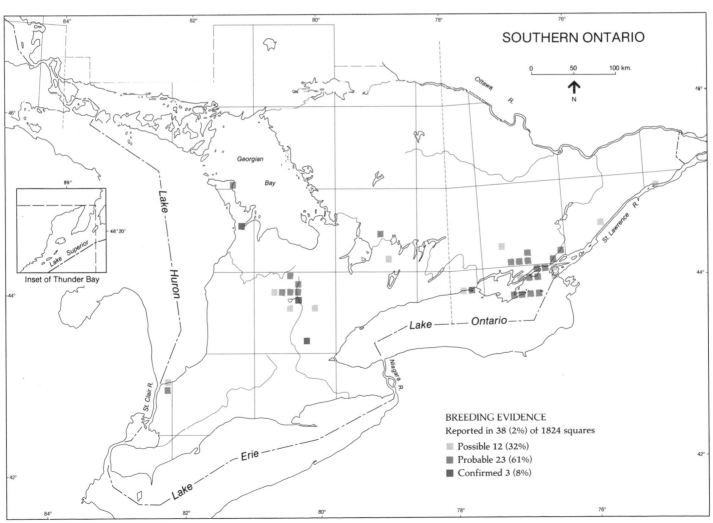

SOUTHERN ONTARIO

0   50   100 km.

N

Georgian Bay

Lake Huron

Lake Superior

Inset of Thunder Bay

Lake Ontario

Ottawa R.

St. Lawrence R.

Niagara R.

St. Clair R.

Lake Erie

BREEDING EVIDENCE

Reported in 38 (2%) of 1824 squares

Possible 12 (32%)
Probable 23 (61%)
Confirmed 3 (8%)

LE CONTE'S SPARROW
Bruant de Le Conte
Ammodramus leconteii

The Le Conte's Sparrow is confined to North America, with its main breeding range being in the Prairie Provinces and the extreme northern prairie states, extending eastward across Ontario into Quebec. In Ontario, it normally breeds in the northwest, on the Hudson Bay Lowland, and occasionally south of the Canadian Shield. The winter range is in the southern and central US.

The Le Conte's Sparrow nests in a variety of open habitats. These include prairie, and open areas within aspen parkland, eastern deciduous forest, and boreal forest. Normally, it nests in rank vegetation in dry or damp meadows, and more rarely in very wet areas. It is often found on the drier edges of marshes and wet meadows, in grasses and sedges, with some willow and alder (Murray 1969). The nest is constructed on or near the ground beneath tangles of dead vegetation. It is usually among the thickest vegetation and is extremely well hidden. Probably, only about 50 nests have ever been found (Walkinshaw 1968). One nest with eggs was found by Chris Rimmer on the James Bay coast during the atlas period.

The range of the Le Conte's Sparrow appears to have expanded in recent times, but probably only because early ornithologists were unfamiliar with it or rarely observed it. In fact, it has likely declined in southern Ontario because of habitat destruction, but has remained fairly stable in the north. McIlwraith (1886) did not list this species for Ontario, and it was first recorded in the province at Toronto in May of 1897. Baillie and Harrington (1937) reported small colonies at Baden (Waterloo Region), Toronto, and Bradford, but were unaware of the northern population. The Holland Marsh colony, near Lake Simcoe, was the best known population in southern Ontario, with up to 15 pairs being present before the habitat was converted to agricultural use (Devitt 1967).

Except for local changes in southern Ontario, it is unlikely that the Le Conte's Sparrow has undergone any significant population changes in the province.

This is one of the most elusive sparrows in North America. It was first described in 1790 and then went unrecorded for over 50 years before it was observed again (Walkinshaw 1968). There are a number of reasons for its unobtrusiveness. It rarely perches above the vegetation and prefers to run on the ground, instead of flying, when disturbed. The song is a very weak, insect-like, double-noted buzz, which lasts for only about a second. The song is easily overlooked during the day, when other birds are singing, but fortunately it also sings well into the evening and before dawn, when most other grassland species are silent. Forty percent of all atlas records are of singing males. The alarm note suggests the anxiety calls of a small mammal, and, because the bird is seldom observed, the sounds may not be recognized by those unfamiliar with them. Within its range, it tends to be locally distributed and semi-colonial, although pairs maintain definite territories throughout the nesting season (Murray 1969). Breeding was confirmed in 4 squares. The southernmost confirmed record was of a fledged young found by Craig Campbell near Luther Marsh. Other confirmations were of fledged young, a distraction display, and the nest mentioned earlier.

The Le Conte's Sparrow distribution on the maps conforms closely to that presented by Murray (1969). The broad nesting distribution shown by Godfrey (1986) is misleading as the bird occurs only in widely separated areas, with only sporadic records in the central and southern parts of the province. The greatest concentrations of the species are in central areas of the James Bay coast, and in northwestern Ontario. In this regard, at least within northern Ontario, its distribution is similar to that of other prairie species such as Marbled Godwit, Wilson's Phalarope, and Sharp-tailed and Clay-colored Sparrows. Although the atlas has provided the best information to date on the breeding distribution of the Le Conte's Sparrow in Ontario, its range and abundance have probably been underestimated. Because of its obscurity and the relatively few field workers who are familiar with it, there are undoubtedly many areas in which it has been overlooked. In particular, its habitat preference suggests that it may be present in some inland blocks containing portions of the Hudson Bay Lowland, although it is evidently more abundant in coastal areas. Its occurrence in southern Ontario may also have been underestimated. It should be looked for in marginal farmlands on the Dundalk Uplands in Grey Co, and near the edge of the Canadian Shield.

The Le Conte's Sparrow was reported from only 40 squares in all of Ontario, and only 2 of those were in the south. In northern Ontario, it was more frequent, and 7 of 12 squares with abundance estimates were thought to contain 11 to 100 pairs. Five of the 7 highest estimates come from the Thunder Bay area and the other 2 are from the coast of James Bay.-- *A. Sandilands and C. Campbell*

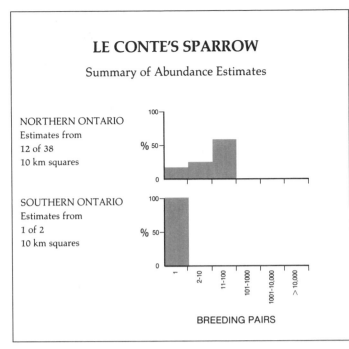

LE CONTE'S SPARROW

Summary of Abundance Estimates

NORTHERN ONTARIO
Estimates from
12 of 38
10 km squares

SOUTHERN ONTARIO
Estimates from
1 of 2
10 km squares

%

100

50

0

%

100

50

0

1   2-10   11-100   101-1000   1001-10,000   >10,000

BREEDING PAIRS

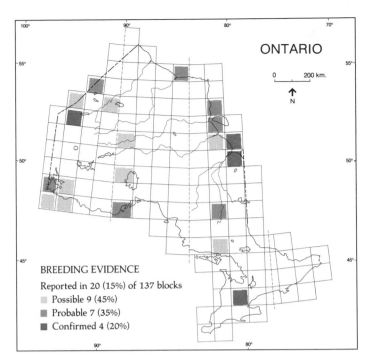

ONTARIO

0   200 km.

N

BREEDING EVIDENCE

Reported in 20 (15%) of 137 blocks

Possible 9 (45%)

Probable 7 (35%)

Confirmed 4 (20%)

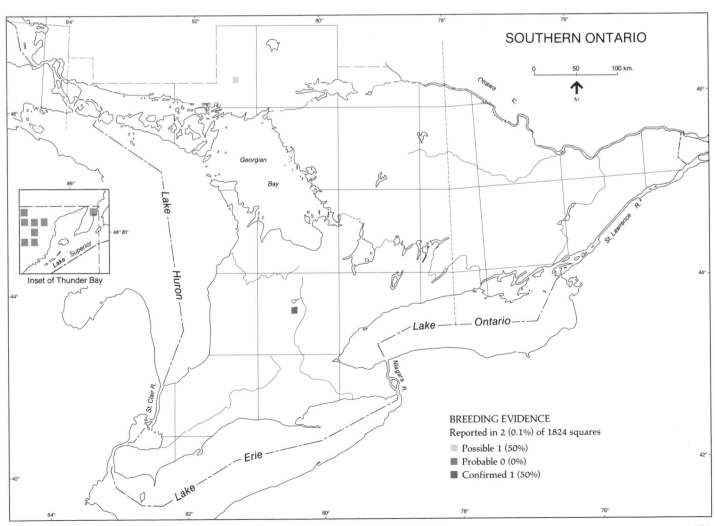

SOUTHERN ONTARIO

0   50   100 km.

N

Georgian

Bay

Ottawa

Lake

Huron

St. Clair R.

Inset of Thunder Bay

Lake   Superior

89°

48°20'

Lake — Ontario

Niagara R.

St. Lawrence R.

Erie

Lake

BREEDING EVIDENCE

Reported in 2 (0.1%) of 1824 squares

Possible 1 (50%)

Probable 0 (0%)

Confirmed 1 (50%)

SONG SPARROW
Bruant chanteur
Melospiza melodia

The Song Sparrow is one of the most widespread sparrows in Ontario. Both the call note and the song, which is often given from a conspicuous perch, are well known and distinctive.

The Song Sparrow breeds from the tree-line in Canada and Alaska south to the mid-Atlantic and northern plains states, and, in the mountains, to the central highlands of Mexico. Throughout its range, it breeds in brushy edge habitats, often near water. Along both coasts, the Song Sparrow breeds in salt marshes. The Song Sparrow breeds nearly throughout Ontario, in woodland edges, gardens, swamps, and brushy clearings and pastures, and has been found breeding on islands as small as 0.02 ha (Van Buskirk 1985). Although the Song Sparrow is one of the most geographically variable species in North America, there is only subtle geographic variation east of the Rocky Mountains (Aldrich 1984). The winter range is from southern Canada to Mexico. The majority of the birds that breed in Ontario winter in the US.

The cutting of forests and the creation of scrub and edge habitats in Ontario have created much excellent breeding habitat for the Song Sparrow, and its numbers have likely increased since forest clearance began.

The Song Sparrow begins nesting as early as mid-April. Most nests are placed on the ground, or low in bushes, but some may be as high as 4 m (Nolan 1968). Two or 3 broods are usually produced, at least in southern Ontario. This long breeding season helps atlassers find and confirm the species. Nests are not easy to locate, but territorial males sing persistently, and are usually not overlooked (Bart and Schoultz 1984). The abundance of the species, and the general conspicuousness of the adults in the vicinity of the nests enabled atlassers to locate nests in 17% of the squares in which Song Sparrows were reported. Song Sparrows generally react vigorously to 'pishing,' and males respond to song playback. The females sit tight on their eggs, and may run from the nest before flushing. Both parents repeat the call note when an intruder is near the nest; atlassers noted agitated behaviour in 51% of the squares from which breeding was only probable. Both parents help to feed the young; 43% of confirmed records are of parents carrying food, and this undoubtedly aided in finding nests in many others.

In Ohio, many males keep the same territory year after year, and a high percentage of females return to the same site as well (Nice 1968). In Ontario, Song Sparrows similarly seem to return to the same area to breed year after year.

In Ontario, atlassers found probable or confirmed breeding in nearly every square and block south of James Bay, accurately reflecting the ubiquity of the species. Breeding was confirmed in a very high 75% of the squares with records in southern Ontario. It is absent in the Cape Henrietta Maria block, which contains only tundra habitat, and was not found or confirmed as breeding in many of the blocks in the Hudson Bay Lowland. It is, at best, scarce in these areas.

The Song Sparrow is generally common or abundant south of James Bay. More than 10 pairs per square were reported from 91% of the squares in southern Ontario for which abundance estimates were made. Estimates were generally lower in northern Ontario, with 66% of the records being of 10 or more pairs per square and no estimates of over 1,000 pairs per square. All of the highest estimates, those over 10,000 pairs, came from southern Ontario. Interestingly, they were reported in an area stretching from Brantford in the west, along the north shore of Lake Ontario to near Brockville in the east. Speirs (1985) reported numbers as high as 1.6 birds/ha near Pickering. On Breeding Bird Surveys, the highest numbers are generally reported from south of the Canadian Shield, and from the western Rainy River District (Speirs 1985).-- *J.D. Rising*

SONG SPARROW

Summary of Abundance Estimates

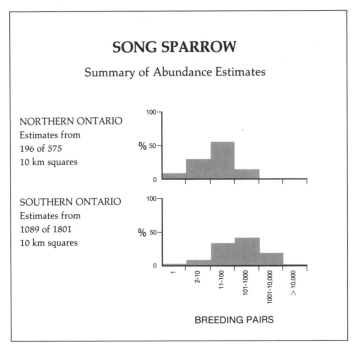

NORTHERN ONTARIO
Estimates from
196 of 575
10 km squares

SOUTHERN ONTARIO
Estimates from
1089 of 1801
10 km squares

BREEDING PAIRS

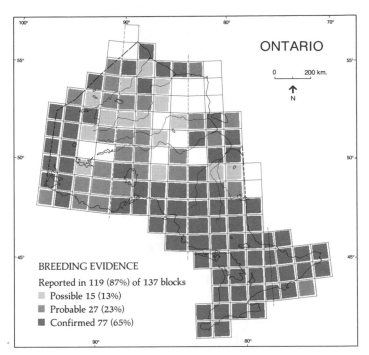

ONTARIO

0   200 km.

↑
N

BREEDING EVIDENCE
Reported in 119 (87%) of 137 blocks

Possible 15 (13%)
Probable 27 (23%)
Confirmed 77 (65%)

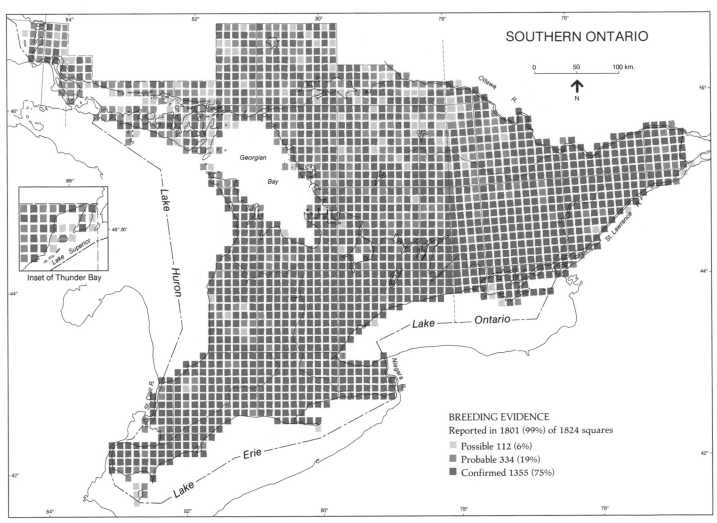

SOUTHERN ONTARIO

0   50   100 km.

↑
N

Inset of Thunder Bay

BREEDING EVIDENCE
Reported in 1801 (99%) of 1824 squares

Possible 112 (6%)
Probable 334 (19%)
Confirmed 1355 (75%)

LINCOLN'S SPARROW
Bruant de Lincoln
Melospiza lincolnii

The Lincoln's Sparrow is a fairly common breeding bird in the Boreal Forests of North America, across Canada and south in the mountains of the western US. Its winter range is from the southern US to the mountains of Central America. In Ontario, it is found throughout the Boreal Forest and muskeg of northern Ontario and in a few scattered bogs and swamps in the south. It is a bird of shrubby vegetation such as is found in moist brushy meadows, swamps and bogs, and in new growth following forest cutting and fires.

Because of its furtive, secretive habits, early ornithologists in Ontario seldom encountered the Lincoln's Sparrow, and even today most Ontario observers consider it fairly uncommon or rare. Bird banders, however, frequently encounter the species during migration, and, once the song is learned, it is readily found. Knowledgeable observers find that it is locally abundant as a breeding bird in suitable habitat in northern Ontario. The Lincoln's Sparrow has a habit of stopping singing and slinking away when approached by an observer, only to start singing again at a respectable distance (Godfrey 1986). It usually passes through southern Ontario at about the same time as the more conspicuous White-crowned Sparrow, when the Lincoln's may be found mousing around at feeders and in hedgerows.

Speirs (1985) cited summer records of Lincoln's Sparrows as far south as Big Creek, Elgin Co and breeding records in the Wainfleet Bog in Welland Co. There were summer records as far south as the Shoal Point Marsh, near Pickering, and breeding records in the Pottageville region of the Holland River Marsh. Weir and Quilliam (1980) have had summer records in the Kingston area. Numerous published records exist for the Boreal Forest region of northern Ontario, north to Fort Severn. In suitable habitat on the Hudson Bay Lowland, it is one of the more abundant breeding species

(James *et al.* 1982a). On the Canadian Shield in Ontario, this species can be found by special search south of about 47⁰N, but north of there it was encountered frequently on Breeding Bird Surveys (Speirs 1985). In New York State, the Lincoln's Sparrow has been found only in the higher elevations of the Adirondacks, where habitats resemble the northern Ontario spruce forests (NYBBA preliminary maps).

The Lincoln's Sparrow was found in 71% of the blocks covered during the field work for the atlas but in only 16% and 7% of the squares in northern and southern Ontario, respectively. Most observers were content with possible breeding records, which were the highest level of evidence found in 48% of all Ontario squares in which Lincoln's Sparrows were present. In another 31% of Ontario squares with records of Lincoln's Sparrow, probable breeding was recorded, and in 21% the species was confirmed as breeding. The nest is very difficult to find, as shown by Speirs and Speirs (1968), who found only 2 Lincoln's Sparrow nests during 2 summers of intensive study near Dorion (east of Thunder Bay), one after the young had hatched, the other by pure luck when the nest was being built. Nests with eggs or young were found in only 10% of all blocks.

The atlas field work shows the range of the Lincoln's Sparrow to be similar to that outlined by Godfrey (1986). The many records on the southern Canadian Shield from the Algonquin Highlands southeast towards Perth reveal a considerable southward extension of the known range. As expected, the species was missing from most of southwestern and southcentral Ontario, except for notable populations in large, relic wetlands such as Wainfleet Bog on the Niagara Peninsula and in the vicinity of Luther Marsh in Wellington Co. In northern Ontario, the gaps in the map are likely due to an underrecording of the species. It is probable that the Lincoln's Sparrow nests in suitable habitat all across the north.

The abundance estimates show that the species has higher populations in northern Ontario. All of the highest estimates, those over 100 pairs per square, came from northern Ontario in an area from the Pukaskwa Nat. Park north to the coasts of James and Hudson Bays. The southern Ontario abundance estimates are all of fewer than 100 pairs per square, and the majority are of only 2 to 10 pairs per square.-- *J.M. Speirs*

LINCOLN'S SPARROW

Summary of Abundance Estimates

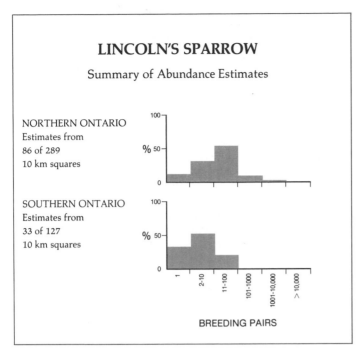

NORTHERN ONTARIO
Estimates from
86 of 289
10 km squares

SOUTHERN ONTARIO
Estimates from
33 of 127
10 km squares

BREEDING PAIRS

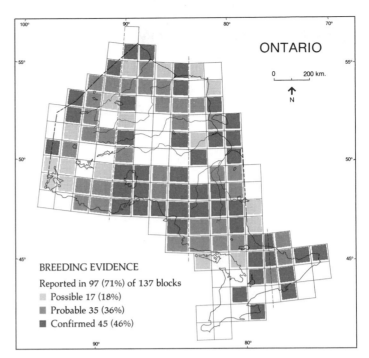

ONTARIO

0   200 km.

N

BREEDING EVIDENCE
Reported in 97 (71%) of 137 blocks
▫ Possible 17 (18%)
▨ Probable 35 (36%)
▪ Confirmed 45 (46%)

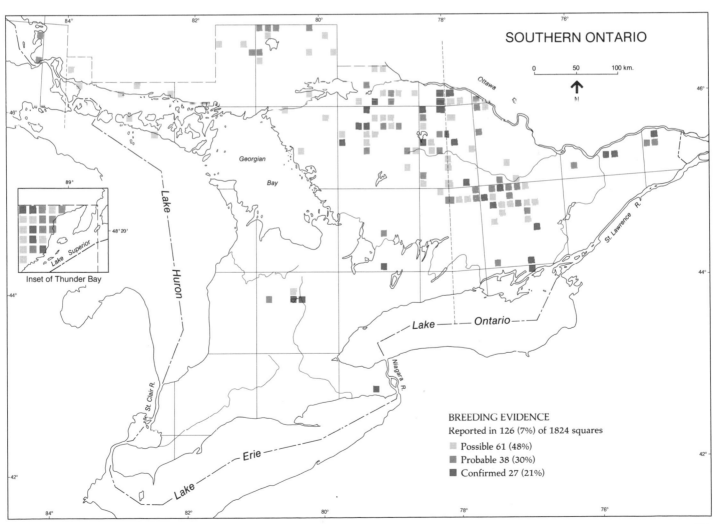

SOUTHERN ONTARIO

0   50   100 km.

N

Georgian Bay

Lake Huron

Lake Superior

Inset of Thunder Bay

Ottawa

St. Lawrence R.

Lake — Ontario

Niagara R.

St. Clair R.

Lake Erie

BREEDING EVIDENCE
Reported in 126 (7%) of 1824 squares
▫ Possible 61 (48%)
▨ Probable 38 (30%)
▪ Confirmed 27 (21%)

SWAMP SPARROW
Bruant des marais
Melospiza georgiana

As the only sparrow often found in extensive cattail marshes, the Swamp Sparrow is well-named. However, especially during migration, it is found in moist woodlands and fields as well as marshes. The chestnut cap, which shows so clearly in field guides, can at times be obscure in the field. The distinctive call and song facilitate finding the species.

The Swamp Sparrow breeds throughout Canada and the northern US east of the mountains, and as far north as the tree-line. It breeds in cattail marshes and other wetlands (such as leatherleaf-sedge marshes) that lack an overstory, in wet meadows, and in the north in open deciduous riparian thickets (willow-alder-grass).

There is no evidence to indicate any major change in the Ontario range of the Swamp Sparrow. However, the draining of wetlands has decreased the habitat available for the species in southern Ontario, especially in the southwest. The Swamp Sparrow winters regularly, but uncommonly, in the cattail marshes of southwestern Ontario (Speirs 1985). The majority of the breeding population moves southward into the US for the winter.

The song, a slow, slurred trill, and the call are both characteristic, and males sing persistently, usually from an exposed perch on a cattail, alder, or willow. It is one of the better-known sounds which emanate from Ontario's wetlands. Although many males are polygynous, both parents help to feed the young (Willson 1966); 51% of confirmed breeding records are of adults carrying food for young. Swamp Sparrow territories may vary greatly in size, and both parents forage away from their territory in nearby woods and shrubbery. However, if the nest is approached, especially early in incubation, the female quietly slips away, making it difficult to locate the nest (Wetherbee 1968). Nests are difficult to access, as they are often built over shallow water, from just above water level to 1 m above it, but usually to a height of 0.2 to 0.3 m (Peck and James 1987). Only 3% of all records are of nests with eggs or with young - a very low total for such a common and widespread bird.

In Ontario, atlassers found Swamp Sparrows virtually throughout the province. Only the Herring Gull, Mallard, and Tree Swallow were reported in more blocks. Breeding evidence was found in 55% of the squares with data province-wide, and nesting was confirmed in 31% of those. The Swamp Sparrow is doubtless scarce, if not absent, in the Cape Henrietta Maria block, which contains mainly tundra habitat, but the species could probably be found in sedge meadows and willow thickets in most of the other blocks from which no breeding evidence was found. It is absent in some squares in southwestern Ontario, around Toronto, and in Niagara and is scarce in others, undoubtedly because most of the wetlands in those regions have been drained. Otherwise, Breeding Bird Survey data indicate fairly uniform abundance, at least in the southern half of the province (Speirs 1985). Between 2 and 100 pairs were thought to be breeding in most squares for which abundance estimates were made (79% of 991 squares), and in about 1/3 of these squares fewer than 10 pairs were estimated. The Swamp Sparrow, however, does breed in some very small marshes, such as those created along roads and railways, and this in part accounts for its wide distribution. The 10 squares with the highest abundance estimates, those over 1,000 pairs per square, were scattered across the province from the Long Point Nat. Wildlife Area in the south to the Attawapiskat River in the north.-- *J.D. Rising*

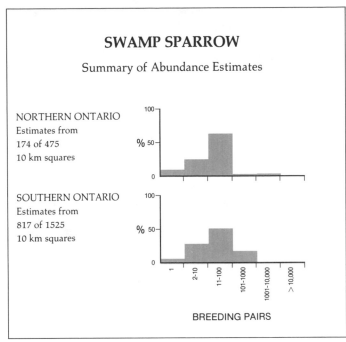

SWAMP SPARROW

Summary of Abundance Estimates

NORTHERN ONTARIO
Estimates from
174 of 475
10 km squares

SOUTHERN ONTARIO
Estimates from
817 of 1525
10 km squares

BREEDING PAIRS

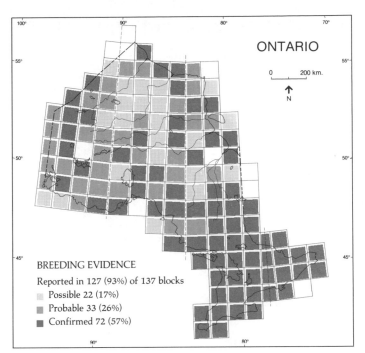

ONTARIO

0   200 km.

N

BREEDING EVIDENCE
Reported in 127 (93%) of 137 blocks
☐ Possible 22 (17%)
▨ Probable 33 (26%)
■ Confirmed 72 (57%)

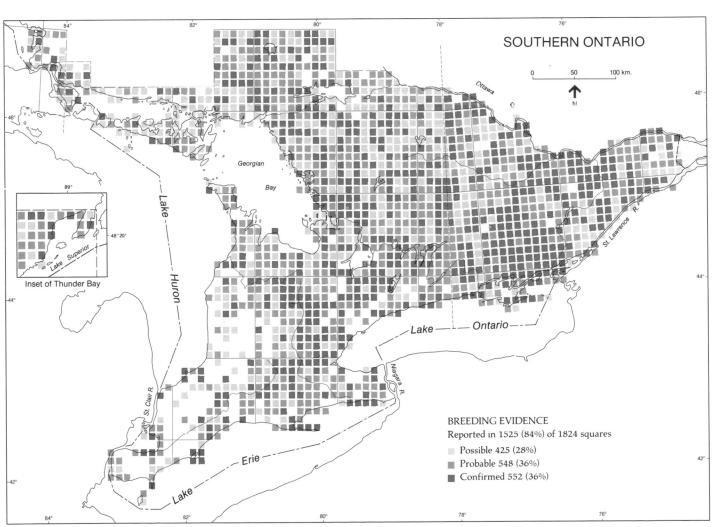

SOUTHERN ONTARIO

0   50   100 km.

N

Inset of Thunder Bay

BREEDING EVIDENCE
Reported in 1525 (84%) of 1824 squares
☐ Possible 425 (28%)
▨ Probable 548 (36%)
■ Confirmed 552 (36%)

WHITE-THROATED SPARROW
Bruant à gorge blanche
Zonotrichia albicollis

The White-throated Sparrow breeds in a broad band of coniferous and mixed forest extending north to the tree-line and from the mountains of northwestern Canada to the Atlantic Provinces and the northeastern US (Godfrey 1986). In Ontario, it is one of the most widespread land birds. In the southwest, and along the north shore of Lake Ontario, it is an abundant migrant but a sparse breeder. Further north, in southcentral and southeastern Ontario, it is a common summer resident and its distinctive whistled song is a familiar symbol of 'cottage country'. Over a vast area of northern Ontario, the White-throated Sparrow is arguably the most abundant and best known bird of the forest.

This species is most common in semi-open forests of spruces, balsam fir, or jack pine, or in mixtures of conifers with aspen and white birch. The White-throated Sparrow nests on or near the ground in openings and along forest edges, often in patches of low shrubs such as blueberry. Old burns and cutover areas with slash and forest regeneration also provide suitable habitat (Lowther and Falls 1968).

A comparison of the present distribution with that indicated in early writings (*e.g.*, McIlwraith 1894) suggests that the range of this species has not changed appreciably. Since it is essentially a bird of the forest edge, it must have decreased with the wholesale clearing of the land and drainage of coniferous wetlands, but it may have increased where fires and logging have opened areas of dense forest. Since the White-throated Sparrow does not occur in pure deciduous stands, it was probably never a common summer resident in southwestern Ontario. In the main part of its breeding range, Breeding Bird Surveys since 1967 show no consistent trends in changes in abundance. Locally, numbers respond to habitat change and have been known to increase during outbreaks of forest insects (Falls 1981).

The White-throated Sparrow occurs in 2 different breeding plumages in both sexes. Brightly marked 'white-striped' birds of either sex tend to mate with duller 'tan-striped' individuals of the opposite sex (Lowther 1961). White-striped males sing and display much more than tan-striped males; they also occur more commonly in open forests (Knapton and Falls 1982). Since the more cryptic tan-striped males may be overlooked, the abundance of the White-throated Sparrow tends to be underestimated. However, since white-striped males occur throughout the breeding range, and frequently sing their easily recognized song, it is unlikely that the species has been missed by atlassers.

White-throated Sparrows sing in trees but nest and gather food near the ground and feed their young in thickets. Incubating females are not easily flushed, and both parents are secretive when bringing food to young, making nests difficult to find. However, adults with fledged young are easily alarmed and utter loud 'chinks'. These observations may explain why, in 27% of all Ontario squares in which the species occurred, the highest level of breeding evidence was of singing males and in 29% it involved agitated parents or fledged young. Nests with eggs or young were found in only 5% of all squares in which the White-throated Sparrow was reported. Within the main breeding range, there is a record in nearly every square, but confirmations of breeding are scattered. Most Regional Coordinators in this area believe that there should have been a much higher proportion of confirmed breeding records.

The White-throated Sparrow was found breeding throughout the Boreal Forest, the Hudson Bay Lowland, and the Great Lakes-St. Lawrence Forest regions. The scarcity of confirmed breeding across Manitoulin Island may reflect the presence of deciduous forest and farmland. Near the southern edge of its breeding range and in the Carolinian Forest region, it occurs mainly in isolated coniferous wetlands or plantations.

Breeding Bird Surveys show the White-throated Sparrow to be the most abundant species in central and northern Ontario and Quebec, while it ranks 27th in southern Ontario and Quebec. Atlas abundance estimates show a similar trend, with higher populations in northern Ontario. Censuses on surveyed plots give estimated densities of males as high as $94.7/km^2$ for 11 plots at Harricanaw River near southern James Bay (James *et al.* 1982b) and 29.6 to $49.4/km^2$ for 3 plots in Algonquin Prov. Park (Lee and Speirs 1977). The highest recorded figure appears to be $152/km^2$ on a plot in Algonquin Prov. Park (Martin 1960). While these values may be exceptional, the estimates of atlassers probably are on the low side for the main breeding range. Although much less common further south, the White-throated Sparrow might breed occasionally anywhere in Ontario, where suitable habitat exists.-- *J.B. Falls*

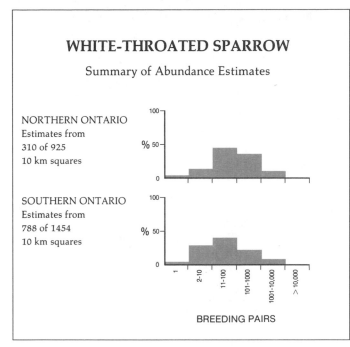

WHITE-THROATED SPARROW

Summary of Abundance Estimates

NORTHERN ONTARIO
Estimates from
310 of 925
10 km squares

SOUTHERN ONTARIO
Estimates from
788 of 1454
10 km squares

BREEDING PAIRS

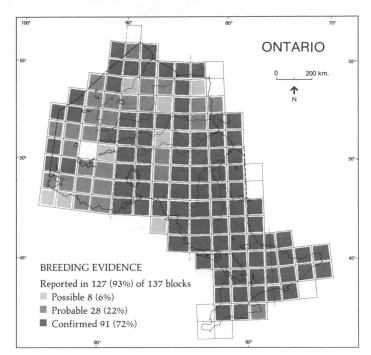

ONTARIO

0   200 km.

N

BREEDING EVIDENCE
Reported in 127 (93%) of 137 blocks

▢ Possible 8 (6%)
▨ Probable 28 (22%)
▦ Confirmed 91 (72%)

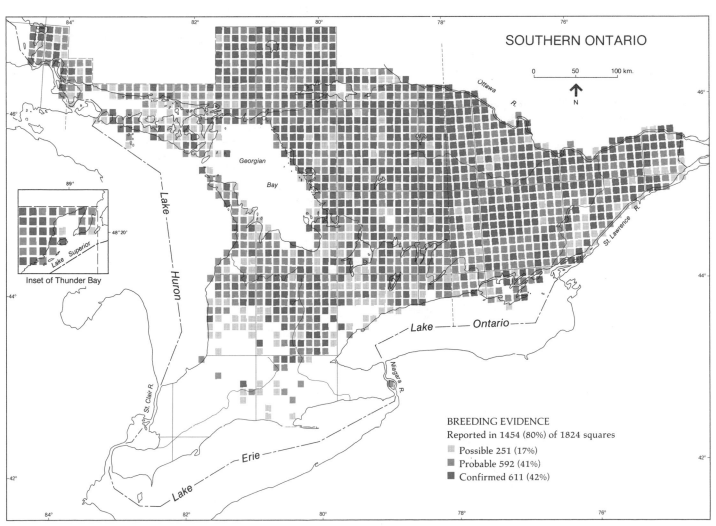

SOUTHERN ONTARIO

0   50   100 km.

N

Ottawa R.

Lake Huron

Georgian Bay

Lake Superior

Inset of Thunder Bay

St. Clair R.

Niagara R.

Lake — Ontario

St. Lawrence R.

Lake — Erie

BREEDING EVIDENCE
Reported in 1454 (80%) of 1824 squares

▢ Possible 251 (17%)
▨ Probable 592 (41%)
▦ Confirmed 611 (42%)

463

DARK-EYED JUNCO
Junco ardoisé
Junco hyemalis

The Dark-eyed Junco is perhaps the most distinctive of the common sparrows, being (in adult plumage, at least) primarily gray and white in colouration. The white lateral tail feathers are distinctive. As one of the most cold-hardy sparrows, it is most familiar around winter feeders, and many people know it as a snowbird. Dark-eyed Juncos feed on the ground, but, unlike many sparrows, they do not scratch for food.

The Dark-eyed Junco is a complex of populations formerly recognized as separate species. In Ontario, these were formerly called Slate-colored Juncos. The Slate-colored Junco breeds in the Boreal Forest region from western Alaska, central Yukon, northern Manitoba and Labrador, south to central Saskatchewan, northern Minnesota, Ontario, and New England, and in the Appalachian Mountains to northern Georgia. Throughout its range, it is common in coniferous, aspen, and birch forests, especially near clearings, where its song, a musical trill, is one of the commonest sounds. In Ontario, it is especially common in young jack pines and burned-over areas. The winter range occupies southern Canada and most of the US.

Because the Dark-eyed Junco is not a bird of the deep woods, it has perhaps benefitted in some areas from logging. Clear-cutting, however, destroys its habitat, and the widespread cutting of pine and mixed woods has doubtless reduced the number of Dark-eyed Juncos breeding in southern Ontario. Indeed, McIlwraith (1894) indicates that the species formerly bred "commonly" throughout southern Ontario, whereas today the Dark-eyed Junco is scarce or local south of the Canadian Shield, as the map indicates. Pine plantations may provide new nesting habitat for the species in this region, but to date Dark-eyed Juncos have not used the plantations to nearly the same extent as have other primarily northern species, such as Yellow-rumped or Pine Warblers.

Spring migration commences in March, with a peak, in the south, in mid-April. By early May, most Dark-eyed Juncos have departed from areas south of the Canadian Shield, and many are on breeding territories in northern Ontario. Nesting commences in mid-April, and continues through July (Peck and James 1987). The nest is built by the female, although the male may help bring nesting material, and is placed in a depression on the ground, often partially under roots or a stump or rock. It is a cup of thin twigs, grass, bark strips, and moss, lined with fine grass or hair (Eaton 1968, Harrison 1978). Although atlassers found probable or confirmed breeding in 266 southern Ontario squares, they found nests with eggs or young in only 12 (5%) of these. In northern Ontario, nests with eggs or young were reported in 10 of the 420 squares in which the species was reported. The species is apparently double-brooded (Eaton 1968, James *et al.* 1982a). Both parents feed the young (Eaton 1968). Atlassers reported adults feeding fledged young or carrying food in 78% of the southern Ontario squares, and 88% of northern Ontario squares, where breeding was confirmed. When an intruder approaches the nest, the adults become greatly agitated, flitting about, showing the white tail feathers, and rapidly giving their alarm note. Agitated behaviour was noted in 29% of all squares where probable breeding was indicated.

In Ontario, atlassers found the Dark-eyed Junco in most blocks north of the southern edge of the Canadian Shield, and it doubtless breeds throughout this region except for the Cape Henrietta Maria area, where suitable habitat may not occur. Atlassers found probable or confirmed evidence of breeding in 96 (70%) of the 137 blocks. It is regular, but not common, along the coast of Hudson Bay at Winisk and Fort Severn, though common inland at Hawley Lake and Kiruna Lake (Speirs 1985). Atlassers found it locally south of the Shield, but generally not in high density. Although atlas records are scattered over much of the northern half of southern Ontario, there is a concentration of records in the Algonquin Highlands. There are scattered records near Port Hope, Long Point, and Simcoe where breeding has been found in tracts of conifers. Breeding was confirmed in the St. Williams Forestry Station south of Simcoe in 1986.

No more than 100 Dark-eyed Juncos were estimated to be in 96% of the southern Ontario squares in which atlassers reported abundance estimates. These estimates doubtless rather accurately reflect the abundance of the species in southern Ontario, where it is really common in only a few areas. Estimates from northern Ontario suggest that it is more common there, with 12% of records being of more than 100 pairs per square. Breeding Bird Surveys show highest densities in central Ontario, from south of Timmins west to Hearst, Geraldton, and Kenora (Speirs 1985).-- *J.D. Rising*

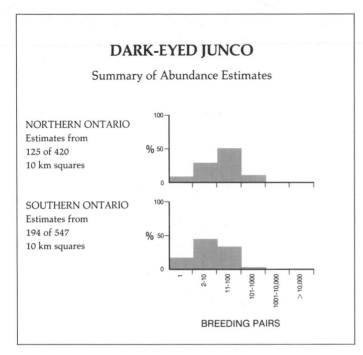

DARK-EYED JUNCO

Summary of Abundance Estimates

NORTHERN ONTARIO
Estimates from
125 of 420
10 km squares

SOUTHERN ONTARIO
Estimates from
194 of 547
10 km squares

BREEDING PAIRS

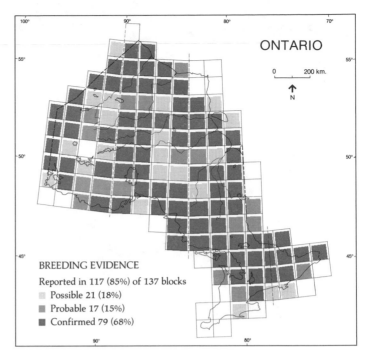

ONTARIO

0   200 km.

N

BREEDING EVIDENCE

Reported in 117 (85%) of 137 blocks

Possible 21 (18%)

Probable 17 (15%)

Confirmed 79 (68%)

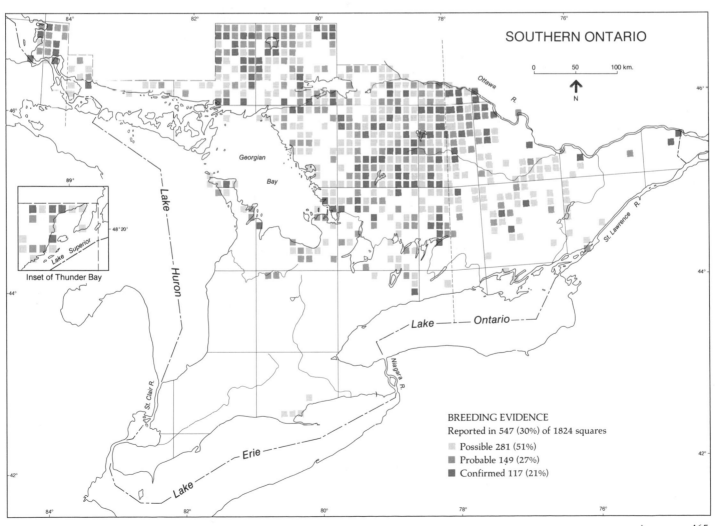

SOUTHERN ONTARIO

0   50   100 km.

N

Inset of Thunder Bay

BREEDING EVIDENCE

Reported in 547 (30%) of 1824 squares

Possible 281 (51%)

Probable 149 (27%)

Confirmed 117 (21%)

465

BOBOLINK
Goglu
Dolichonyx oryzivorus

This member of the blackbird family is perhaps best known for the impressive distances covered on its annual migration. From a breeding range in the northern US and southern Canada, it flies as far south as northern Argentina to spend the non-breeding season.

Because the Bobolink breeds in grassland habitat, its distribution throughout its range is very patchy. It varies from being locally abundant, where preferred hayfields occur, to rare or absent where the habitat is inappropriate.

The distribution of the Bobolink has changed substantially since the arrival of Europeans in North America. Because it is a grassland species, the breeding range of the Bobolink has expanded in eastern North America with the clearing of forests and the cultivation of hay. Bent (1958) cites numerous reports from around the turn of the century that document the westward spread of this species. However, Bent (1958) describes substantial population declines of the Bobolink in more recent times in the eastern part of its range. Bent considers improved hay harvesting techniques and earlier harvesting to have had some of the impact on Bobolink abundance. He also suggests that widespread hunting of Bobolinks when they were migrating through the southeastern US could have reduced northern breeding populations. Bobolinks were hunted in the past both for sport and because of the damage they did to rice crops. However, hunting of this species is no longer practised.

When Bobolinks begin breeding, in mid-May, the males are easily seen and heard. Males have very distinctive plumage, with the unusual characteristic of being black below and pale in colour above. Their song is extremely melodic and often given as part of an aerial display. Once nesting is well under way, the males are not as obvious because they sing less, but their plumage still makes them easy to see,

especially when involved in frequent chases low over grasslands. By contrast, female Bobolinks are nondescript in appearance and relatively cryptic in behaviour, particularly when nesting. Nests are constructed on the ground in hay fields, which makes them very difficult to locate. Females build the nests but never fly directly to the nest site carrying nesting material. Once incubating, females continue to provide little help in locating their nests; usually they travel some distance from the nest on the ground before flushing, and fly directly from the nest only when the observer's approach is apparently undetected. The male often has more than one mate and usually helps feed the young of the first female that nests on his territory (Martin 1974). The females carry food to the young in the nest and this behaviour resulted in almost half (42%) of all breeding confirmations. Of 1520 squares in Ontario where Bobolinks were found, only 86 (6%) had records of nests with eggs or young.

The species was found in 79% and 5% of the squares where data was collected in southern and northern Ontario, respectively. The relative ease of confirmation of Bobolink as a breeding species can be seen in its 54% confirmation rate in southern Ontario.

The distribution of the Bobolink shown on the maps reflects the association of this species with agriculture and, in particular, with hay cultivation. It is abundant throughout the agricultural zone and is found only locally (as is agriculture) as one moves into the northern, more heavily forested areas of the province. Where there is no agriculture, there are generally no Bobolinks, although they do occur in dry beaver meadows in the interior of Algonquin Prov. Park (F. Helleiner pers. comm.). The species was also found as a probable breeder in squares north of Moosonee on James Bay.

In Ontario, estimates were submitted in categories from 1 pair to over 10,000 pairs per square, suggesting a wide range of abundances, depending upon suitable habitat. BBS data summarized in Speirs (1985) show that the highest abundance of Bobolinks in Ontario occurs in an area between southern Bruce Co and southeastern Ontario. The distribution of atlas abundance estimates of over 1,000 pairs per square shows a different story. These high estimates are found in southern Ontario south and east of the Canadian Shield. Interestingly, such high estimates are lacking south and west of Chatham and in the atlas regions around Lake Ontario from Niagara to Prince Edward Co.-- *P.J. Weatherhead*

BOBOLINK

Summary of Abundance Estimates

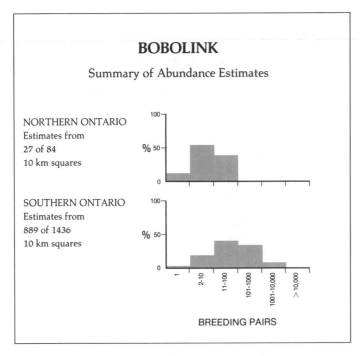

NORTHERN ONTARIO
Estimates from
27 of 84
10 km squares

SOUTHERN ONTARIO
Estimates from
889 of 1436
10 km squares

BREEDING PAIRS

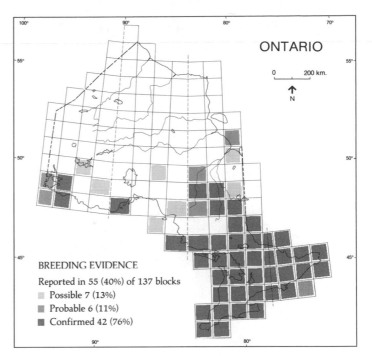

ONTARIO

0   200 km.

N

BREEDING EVIDENCE
Reported in 55 (40%) of 137 blocks
- Possible 7 (13%)
- Probable 6 (11%)
- Confirmed 42 (76%)

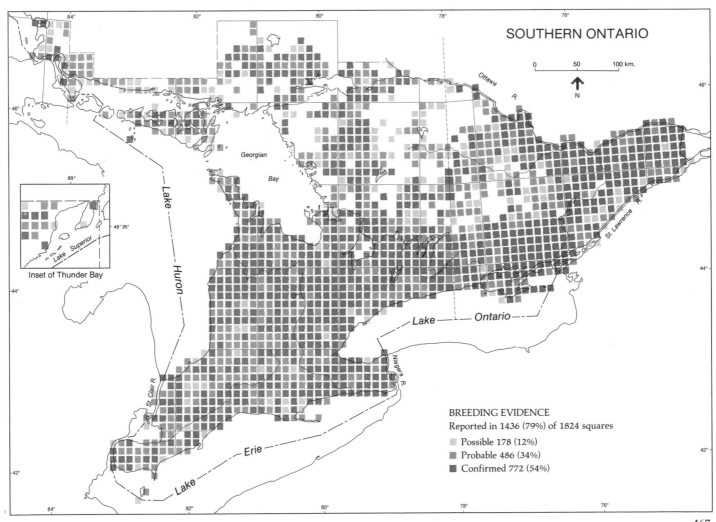

SOUTHERN ONTARIO

0   50   100 km.

N

Georgian
Bay

Lake
Huron

Lake Superior

Inset of Thunder Bay

St. Clair R.

Lake   Erie

Lake

Ottawa R.

St. Lawrence R.

Lake — Ontario

Niagara R.

BREEDING EVIDENCE
Reported in 1436 (79%) of 1824 squares
- Possible 178 (12%)
- Probable 486 (34%)
- Confirmed 772 (54%)

RED-WINGED BLACKBIRD
Carouge à épaulettes
Agelaius phoeniceus

The Red-winged Blackbird is one of the most abundant and widespread birds in North America, with a breeding distribution extending from Central America to the southern Yukon and from the Atlantic to Pacific coasts. It is also one of the most abundant species in Ontario, particularly through the agricultural zones of the eastern and southern regions of the province. Although traditionally a marsh-nesting species, in eastern North America the Red-winged Blackbird has successfully broadened its preference of nesting habitat to include hayfields, grassy roadsides, and even suburban gardens. However, the greatest breeding densities still occur in marshes. In forested parts of its range, marshes provide the only suitable breeding habitat. In the winter, it withdraws southward from Canada to the mid-latitudes of the US.

Although there have been no substantial recent changes in the distribution of the Red-winged Blackbird in Ontario, its abundance has been steadily increasing throughout the agricultural zone (Clark *et al.* 1986). This increase appears to be due to the success of the species at utilizing waste grain from agriculture, which makes up a substantial part of its diet throughout its residency in Ontario (McNicol *et al.* 1982). Ironically, it appears that the waste grain available most of the year supports the large populations responsible for serious, localized damage to standing corn crops during a brief period late each summer.

The Red-winged Blackbird is one of the easiest species to atlas. The adult male is strikingly-coloured, with scarlet epaulets contrasting with immaculate black feathers elsewhere. Its vocalizations are loud and distinctive and always made from conspicuous positions. It was found in 98% of the squares in southern Ontario and 74% of the blocks in all of Ontario. Because of the preference of the species for roadside habitat, it is probable that any census conducted along roads

would seriously overestimate its abundance (Weatherhead *et al.* 1982). Early in the breeding season (May), direct confirmation of breeding is possible because nests are usually poorly concealed, supported only by old vegetation such as standing dead cattails. Once new growth increases, nests can become much more difficult to locate, especially in grassy habitat. Indirect evidence of breeding is easily obtained later in the breeding season, either by observations of females carrying food to the nest or by the vigorous nest defence by both parents. Males defending nests often strike human intruders, a level of aggression that occasionally makes Red-winged Blackbirds unwelcome backyard residents. Worth noting is the fact that Red-winged Blackbirds usually breed polygynously, so any census of breeding birds is likely to encounter more females than males.

The species is easy to confirm as a breeding bird; 80% of the squares in southern Ontario and 75% of the blocks in all of Ontario in which it was recorded are at that level. In the province as a whole, the most frequently represented breeding categories are food carrying (21%), nests with young (17%), fledged young (15%), and nests with eggs (12%).

Because of the ease of atlassing the Red-winged Blackbird, it is likely that in all areas adequately covered, its presence would have been detected. In northern Boreal Forest regions, where this species appears to be absent, it is possible that occasional individuals breeding in isolated marshes may have been overlooked.

Within agricultural zones, the Red-winged Blackbird is unquestionably abundant. It becomes uncommon to rare north into the Boreal Forest. In southern Ontario, atlassers' estimates for abundance are from 11 to 10,000 pairs per square in 92% of the cases. The most frequently reported category was 101 to 1,000 pairs per square (44% of estimates). This clearly shows the high numbers of this species. The wide range of estimates shows habitat differences across the south. In the north, the estimates are generally of lower numbers, with 11 to 100 pairs per square being the most frequently reported category in 60% of records. Its abundance, combined with its gregarious habits and fondness of ripening corn, have made it the target of various population reduction efforts, particularly in the US. In spite of the apparent vulnerability of this species because of its use of traditional communal roost sites where populations may reach several million birds, until now attempts to control Red-winged Blackbird populations appear not to have had an impact on the population.-- *P.J. Weatherhead*

RED-WINGED BLACKBIRD

Summary of Abundance Estimates

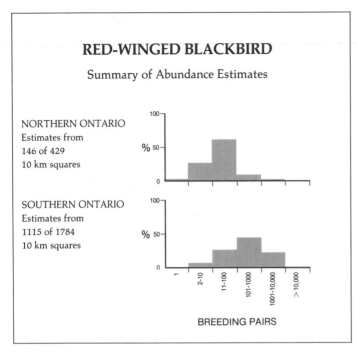

NORTHERN ONTARIO
Estimates from
146 of 429
10 km squares

SOUTHERN ONTARIO
Estimates from
1115 of 1784
10 km squares

BREEDING PAIRS

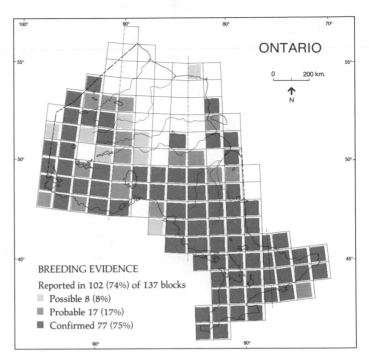

ONTARIO

0   200 km.

N

BREEDING EVIDENCE

Reported in 102 (74%) of 137 blocks

Possible 8 (8%)

Probable 17 (17%)

Confirmed 77 (75%)

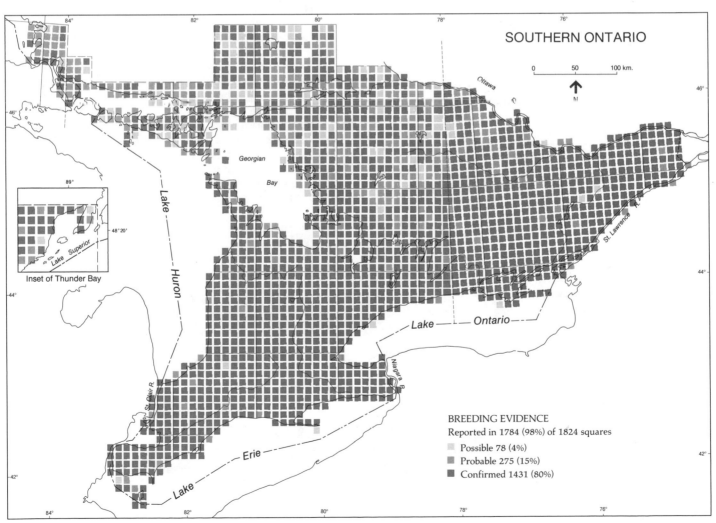

SOUTHERN ONTARIO

0   50   100 km.

N

Inset of Thunder Bay

Georgian

Bay

Lake

Huron

Lake   Superior

St. Clair R.

Lake   Erie

Niagara R.

Lake   Ontario

St. Lawrence R.

Ottawa

BREEDING EVIDENCE

Reported in 1784 (98%) of 1824 squares

Possible 78 (4%)

Probable 275 (15%)

Confirmed 1431 (80%)

469

EASTERN MEADOWLARK
Sturnelle des prés
Sturnella magna

The Eastern Meadowlark is a widely distributed breeding bird in the Americas, ranging from southern Canada to northern South America. It occupies open country in eastern North America from the Great Plains to the Atlantic seaboard, and in Ontario is found primarily south of the Canadian Shield. It withdraws southward from only the northernmost part of its range in winter. Grassy pastures and meadows are the preferred breeding habitats, providing both food for adults and young and sites for the usually well-concealed ground nest. The species also occupies cultivated land and weedy areas with trees, such as old orchards, throughout southern Ontario. It is absent from the Boreal Forest biome of northern Ontario, except along the southern edge, where cleared areas provide suitable nesting habitat.

The Eastern Meadowlark expanded its breeding range north and east in North America as a result of the clearing of the native forests for agriculture and settlement (Bent 1958). In Ontario, the species reached Muskoka by the mid-1860's and was considered common at the southern end of the District by the turn of the century (Mills 1981). Baillie and Harrington (1937) placed its northerly limit at Lake Nipissing and Englehart and across to the eastern end of Lake Superior, and commented that meadowlarks in northwestern Ontario (Thunder Bay and Kenora Districts) were more than likely Western Meadowlarks.

A few changes in the status of the Eastern Meadowlark have taken place in the past half-century. While the species is still common and widespread across most of southern Ontario, it is conspicuously absent from some squares in the intensively farmed southwest, especially in Essex and Kent Counties. Indeed, BBS counts from 1967 to 1983 indicate a long-term, significant decline in numbers in southern Ontario, probably due to the increase in intensively culti-

vated, and hence suboptimal, areas. During that period, BBS data summarized in Speirs (1985) show that the Eastern Meadowlark was most numerous within the province from Bruce and Grey Counties east to Prescott and Glengarry Counties. Both the atlas and BBS data show that the Eastern Meadowlark now occurs in northwestern Ontario, at least in small numbers.

In general, the Eastern Meadowlark is a conspicuous species of open country. The clear, slurred whistle of a territorial male carries a long distance over the countryside, and a male usually perches conspicuously while uttering its song. The distinctive shape and flight pattern of the Meadowlark mean that it can be readily identified at considerable distances. It is therefore unlikely that the species was overlooked in many squares during the atlas period. Confirmed breeding occurred in 51% of the 1279 squares in which it was reported in southern Ontario. Nests are difficult to find and account for only 17% of confirmed breeding records; most confirmations (76%) were observations of parent birds carrying food for young, and of fledglings.

The species is sporadically distributed through the Districts of Parry Sound and Haliburton, and is essentially absent from Algonquin Prov. Park, and north of Sudbury. In northern Ontario, the species was reported in only 12 squares, 11 of which had only possible breeding evidence. The most northerly records during the atlas period were made in Kenora District, although summer records of singing males have been made as far north as James Bay, for example at Attawapiskat in 1971 (Speirs 1985). An interesting future project might be to ascertain the distribution and abundance of the Eastern Meadowlark in relation to that of the Western Meadowlark in northwestern Ontario. The Eastern Meadowlark occurs in Kenora and Rainy River Districts, especially in the Lake of the Woods area, where it overlaps the distribution of the much more common Western Meadowlark in the same region. The 2 species are known to interbreed, but the offspring of such crosses are infertile (Lanyon 1979).

Forty-nine percent of squares for which there are abundance estimates give the range of breeding pairs per square as between 11 and 100 pairs; 25% of squares had estimates of between 101 and 1,000 pairs, and 2% between 1,001 and 10,000 pairs. Abundance estimates of greater than 1,000 pairs per square are generally restricted to a limited area of southern Ontario. This can best be described as the mixed farmland area bounded by Listowel in the west, Flesherton in the north, Lindsay in the east, and Brantford in the south. Four such records also occurred near Ottawa.-- *R. Knapton*

EASTERN MEADOWLARK

Summary of Abundance Estimates

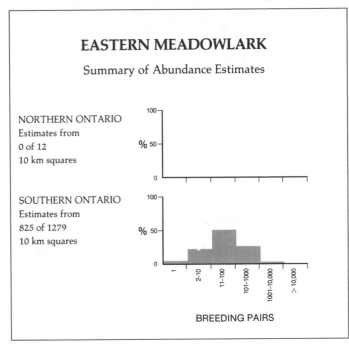

NORTHERN ONTARIO
Estimates from
0 of 12
10 km squares

SOUTHERN ONTARIO
Estimates from
825 of 1279
10 km squares

BREEDING PAIRS

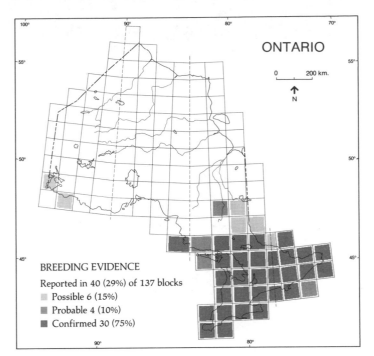

ONTARIO

0   200 km.

N

BREEDING EVIDENCE

Reported in 40 (29%) of 137 blocks

Possible 6 (15%)
Probable 4 (10%)
Confirmed 30 (75%)

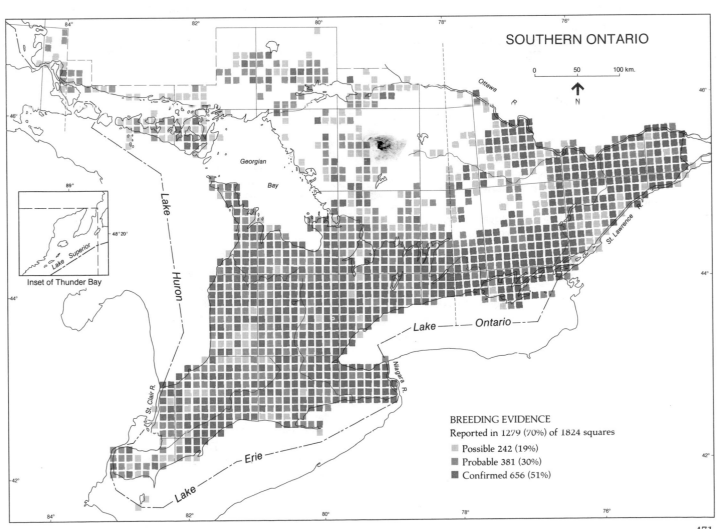

SOUTHERN ONTARIO

0   50   100 km.

N

Georgian

Bay

Lake

Huron

St. Clair R.

Lake — Erie

Lake — Ontario

Niagara R.

Ottawa R.

St. Lawrence R.

89°

48°20′

Lake Superior

Inset of Thunder Bay

BREEDING EVIDENCE

Reported in 1279 (70%) of 1824 squares

Possible 242 (19%)
Probable 381 (30%)
Confirmed 656 (51%)

471

WESTERN MEADOWLARK
Sturnelle de l'Ouest
Sturnella neglecta

The Western Meadowlark is a rare bird in Ontario fields. It looks very similar to the Eastern Meadowlark but has distinctive songs and call notes. The similarity of the 2 meadowlarks caused considerable taxonomic confusion in the early years, but, by the third edition of the AOU checklist in 1910, 2 distinct species had been recognized.

The Western Meadowlark is found abundantly on the prairies and grasslands of the central and western US and Canada. In the winter the species withdraws to areas of the US, south of its breeding range. Baillie and Harrington (1937) stated, "the Meadowlark breeding in extreme western Ontario is more than likely the Western Meadowlark," and, "There is no breeding evidence for Ontario of which we are aware, however, for the western species." However, the first Ontario nest was found in the Rainy River District in 1929 (Snyder 1938) and the first breeding evidence in the south, in 1932, was subsequently reported by Baillie (1960). Godfrey (1966) provided a range map that showed a discontinuous distribution scattered from the Manitoba border, across the southern reaches of Ontario; he extended that somewhat in his second edition (1986) to show a continuous distribution from Lake of the Woods to Thunder Bay and from Sault Ste Marie to Kingston.

The Western Meadowlark appears to have invaded southern Ontario during this century. It is probable that the northwestern Ontario populations were present much earlier as a natural extension of the abundant prairie populations to the west.

The Western Meadowlark is large and brightly coloured. It occurs in open fields where the bird is easily seen, but identification is better made by vocalizations than by appearance because of possible confusion with the very similar Eastern Meadowlark. However, most Ontario atlassers are unfamiliar with the call notes of the Western Meadowlark, making the full song the most important identifying characteristic; but even identification by song is difficult at times. Where the 2 meadowlarks overlap, individuals sometimes develop mixed repertoires of Eastern and Western songs and calls. Therefore, it is important to listen to all the vocalizations of supposed Western Meadowlarks. If they are a mixture, careful attention to details of plumage is important (J.B. Falls pers. comm.).

The bird nests on the ground with a roofed nest that is well concealed under clumps of vegetation. The nest often has a runway, made of grasses, up to 1.5 m long. Nest-building starts early in the spring, not long after the birds return to the breeding grounds (Bent 1958).

The Western Meadowlark was reported from 84 squares in the province, 69 of them in the south. The majority of the records are of the singing male (46%) and territorial (29%) categories. This suggests that the atlassers identified the bird by song and then returned at least one week later to see if it was still present. However, it is doubtful that the species bred in the majority of the squares in southern Ontario in which it was recorded. In only 6% of squares in this sector were there records of pairs of birds, and in only 10% were there records of confirmed breeding. No nests were found in the province.

The maps show a scattered distribution from the southern Manitoba border to southern Ontario. More records occurred than had been expected by many naturalists. Many Regional Coordinators described the birds as being erratic or accidental in occurrence in their area and therefore the maps probably overstate the distribution in any one year. A cluster of records in Bruce Co suggests that this area supports a regularly occurring population, despite the fact that the area is not included in the range given in Godfrey (1986).

Although the atlas data indicate that the 2 meadowlark species overlap in range in the Lake of the Woods area of northwestern Ontario, BBS data (Speirs 1985) indicate that the Western Meadowlark is by far the commoner of the 2 species there. The Coordinators from eastern Ontario and Sudbury stated that the species had largely withdrawn from their areas since the 1970's. Much the same appears to have happened throughout southern Ontario. A survey conducted by Szijj and Falls about 1960 revealed small colonies of Western Meadowlarks scattered throughout southcentral and southwestern Ontario. By the late 1970's, Falls had difficulty locating more than half a dozen birds in the same area (J.B. Falls pers. comm.).

This species is rare in southern Ontario. Atlassers in the south estimated that very low numbers of birds occurred in any one square, with one pair being recorded 64% of the time and 2 to 10 pairs, 36%. It is reasonable to suggest that under 100 pairs of birds occur in the south each year. Speirs (1985) reported 35 birds of this species on only 2 BBS routes in the Rainy River and Dryden areas. This suggests a somewhat larger number in the grasslands of the northwest, than in southern Ontario.-- *P.F.J. Eagles*

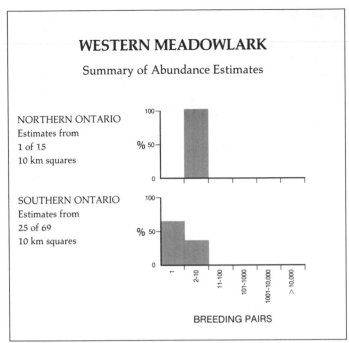

WESTERN MEADOWLARK

Summary of Abundance Estimates

NORTHERN ONTARIO
Estimates from
1 of 15
10 km squares

SOUTHERN ONTARIO
Estimates from
25 of 69
10 km squares

% 100 50 0

% 100 50 0

1 2-10 11-100 101-1000 1001-10,000 >10,000

BREEDING PAIRS

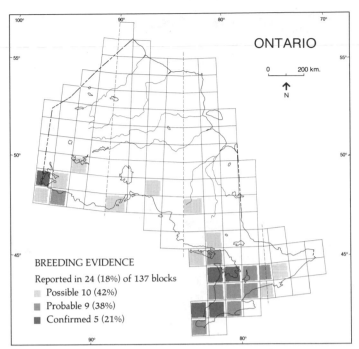

ONTARIO

0   200 km.

N

BREEDING EVIDENCE
Reported in 24 (18%) of 137 blocks

Possible 10 (42%)
Probable 9 (38%)
Confirmed 5 (21%)

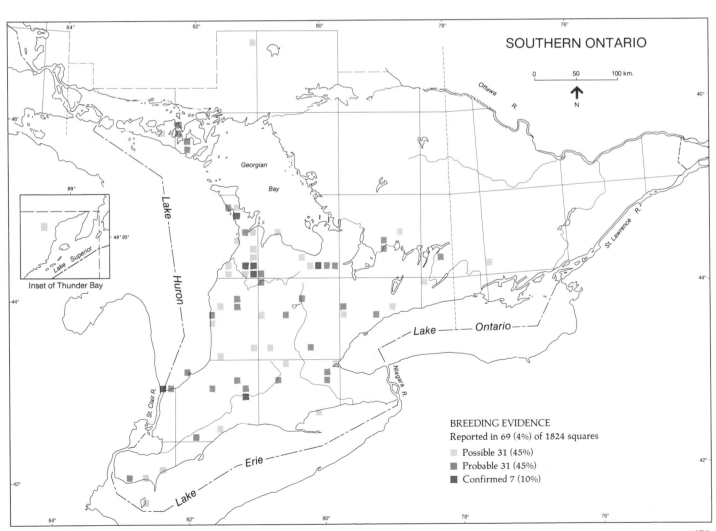

SOUTHERN ONTARIO

0   50   100 km.

N

Georgian Bay

Inset of Thunder Bay

Lake Huron

St. Clair R.

Lake Erie

Lake Ontario

Niagara R.

Ottawa R.

St. Lawrence R.

BREEDING EVIDENCE
Reported in 69 (4%) of 1824 squares

Possible 31 (45%)
Probable 31 (45%)
Confirmed 7 (10%)

YELLOW-HEADED BLACKBIRD
Carouge à tête jaune
Xanthocephalus xanthocephalus

The Yellow-headed Blackbird breeds throughout the prairies of Canada and the Great Plains of the US. East of this, breeding is only sporadic. It winters in plains habitat, in an area extending from the southern US to southern Mexico. In Ontario, it breeds only in the extreme southern portion of northwestern Ontario and in the Lake St. Clair area of southwestern Ontario. Breeding habitats of the species include permanent, deep marshes or sloughs, the marshy zone of lakes, or river impoundments, in which stands of cattails or phragmites flourish. Some marshes commonly occupied by Red-winged Blackbirds are not suitable for this species. Only water deep enough to be permanent permits overwintering of the insects the species relies on to feed its young. In marshes where they co-exist with Red-winged Blackbirds, Yellow-headed Blackbirds occupy the deeper areas (at least 0.6 to 1.2 m deep), leaving only the shoreline edges or uplands for Red-winged Blackbirds.

In 1894, McIlwraith described the Yellow-headed Blackbird as a casual visitor to Ontario, but predicted that, "it is quite probable that we may see it yet as a summer resident in the grassy meadows of Ontario." Although it may have begun breeding in Ontario as early as 1958, nesting in this province was not confirmed until 1961, when a nest was discovered near the mouth of the Rainy River (Baillie 1961, Peck and James 1987). In 1965, breeding was confirmed at a Lake St. Clair marsh, suggesting that the species was spreading into southwestern Ontario as well (Sawyer and Dyer 1968). The northwestern birds likely result from an expansion of the southern Manitoba breeding population, while those now nesting in southwestern Ontario may have originally come from the US. Although Peck and James (1987) describe the Yellow-headed Blackbird as breeding regularly only in

the Rainy River District and Kent Co, atlassers seem to have discovered breeding activity outside this limited area, specifically at Kenora and Dryden in the north and in Essex Co in the south. There is also a possible breeding record from Tiny Marsh, west of Barrie. Although there are marshes in this region that might provide suitable habitat for the species, the probability of regular breeding in this area is low.

Several features of the behaviour and appearance of the species make it a relatively easy bird to find. Second year and older males are conspicuously coloured. First year males resemble females and can be mistaken for them at a distance. Breeding males are strongly territorial and mate polygynously in colonies. Their aggressive courtship displays are relatively elaborate and conspicuous. At some nests, males assist females in feeding young. This, plus their vigorous defence of the nest against predators, makes both parents conspicuous during the nestling stage. Thus, although nests are usually placed over water, and may be inaccessible (apparently only one atlasser actually found a nest), behavioural clues can easily be used to provide evidence of breeding. The songs and calls, while probably unfamiliar to many Ontario birders, are distinctive, and a nesting colony is often audible from a distance. The young remain with their parents in the vicinity of the colony for up to 3 weeks. Breeding activity can therefore be observed in Ontario from early May through July.

Atlassers confirmed breeding in only 3 of 137 blocks covering all of Ontario, and in only 2 of 1824 squares in the south. Most atlas records lie within the previously known breeding range of the species, or describe small and expected (because of the systematic coverage afforded by the atlas program) extensions of it. Male Yellow-headed Blackbirds sometimes perform courtship and/or territorial displays during migration. They may continue to do so (and in suitable habitat) even if they fail to obtain a mate. Even accepted records outside the currently known range are unlikely to represent real and/or long term expansions of the distribution. Because Yellow-headed Blackbirds nest colonially, sightings of individuals (or even single pairs) are not usually indicative of breeding.

The Yellow-headed Blackbird has a restricted range in Ontario. Where it does breed, a colonial habit ensures that more than one pair will be present; in 2 of the 4 squares in which abundance was estimated in the south, more than 2 pairs were thought to be present. The only estimate of abundance in the north was of 2 to 10 pairs in one square in the Lake of the Woods area.-- *N. Flood*

YELLOW-HEADED BLACKBIRD

Summary of Abundance Estimates

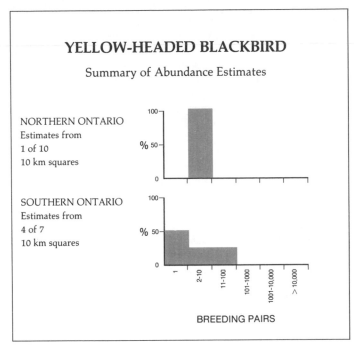

NORTHERN ONTARIO
Estimates from
1 of 10
10 km squares

SOUTHERN ONTARIO
Estimates from
4 of 7
10 km squares

BREEDING PAIRS

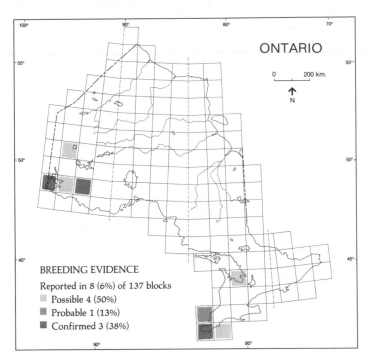

ONTARIO

0   200 km.

N

BREEDING EVIDENCE

Reported in 8 (6%) of 137 blocks

Possible 4 (50%)

Probable 1 (13%)

Confirmed 3 (38%)

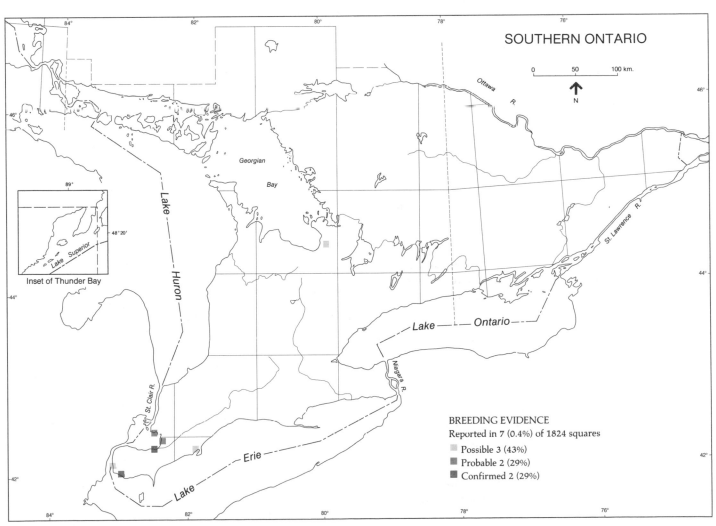

SOUTHERN ONTARIO

0   50   100 km.

N

Inset of Thunder Bay

Lake Superior

Georgian Bay

Lake Huron

Ottawa R.

St. Lawrence R.

Lake — Ontario

Niagara R.

St. Clair R.

Lake — Erie

BREEDING EVIDENCE

Reported in 7 (0.4%) of 1824 squares

Possible 3 (43%)

Probable 2 (29%)

Confirmed 2 (29%)

RUSTY BLACKBIRD
Quiscale rouilleux
Euphagus carolinus

Of all the blackbird species that breed in Ontario, the Rusty Blackbird is perhaps the least known. Its tendency to frequent remote and often inaccessible areas makes it a difficult species to study, and there has been little systematic work done on its breeding biology.

The Rusty Blackbird breeds in Alaska and across Canada from coast to coast, as well as in northern New England and New York. In Ontario, the species is widespread throughout the north but is common in the south only to Algonquin Prov. Park (James *et al.* 1976). The Rusty Blackbird is almost always found near water, on the wintering grounds as well as in summer. Its favoured breeding habitat includes bogs, fens, and other openings in coniferous woodlands bordering on lakes, rivers, swamps, or other bodies of water. Therefore, its range more or less encompasses the forest regions that contain at least some coniferous growth: in the north stopping where pure tundra begins and in the south being limited by the occurrence of pure grassland, pure deciduous forest, and/or highly developed urban or agricultural areas. Unlike many blackbirds, Rusty Blackbirds are not colonial. This, combined with their rather selective habitat preferences, means that they are rarely seen in large numbers even in the centre of their range.

It has been said that anyone "who desires to study this species on its breeding grounds must make up his mind to endure all sorts of discomforts: millions of blackflies, gnats and mosquitoes make life a burden during his stay, while the bogs and swamps through which one is compelled to flounder in search of the nest render walking anything but pleasant" (Bendire 1895). The truth of this statement is reflected in the fact that no Rusty Blackbird nests were recorded in Ontario until 1910, when one was found in Grey Co (Baillie and Harrington 1937). Subsequently, increased access to the

northern part of the province produced numerous breeding records describing a range north from Bruce, Grey, Simcoe, Muskoka, Haliburton, and Renfrew Counties to Hudson Bay. Recently, it has been suggested that the breeding range of the species is shrinking. Peck and James (1987) describe records from Bruce and Grey Counties as "historical" and indicative only of a previously larger breeding range.

Many of the habits, as well as the appearance, of the Rusty Blackbird make it relatively difficult to atlas. In summer, Rusty Blackbirds closely resemble Brewer's Blackbirds. Although their differing habitat preferences and contrasting songs can be used to distinguish them, misidentification is possible. The male Rusty Blackbird's song is short and unmelodious. Compared with other blackbird species, the Rusty is relatively quiet and inconspicuous during courtship and nesting, although it will vigorously defend its nest (Bent 1958). The breeding territory is apparently deserted almost immediately after nesting. The atlas statistics reflect these habits. Fifty percent of the squares and 30% of the blocks are represented by only possible breeding evidence. Although a few nests were found in both north and south, most confirmed breeding records consist of adults carrying food.

It is probably wise to view the atlas maps as only an approximate representation of the Rusty Blackbird's distribution. In northern Ontario, the map is conservative; the species likely nests in all blocks. In southern Ontario, on the other hand, the atlas records may overestimate the actual range. The species has been described as a common summer resident across Ontario as far south as Thunder Bay and Algonquin Prov. Park (James *et al.* 1976). Most of the 294 squares in which it was recorded are in this region. Rusty Blackbirds were also found in the Haliburton area and in other isolated locations in southeastern and southwestern Ontario, including Grey and Bruce Counties, where Peck and James (1987) suggest that the species no longer nests. Isolated areas of favoured habitat still exist in the south, and it may be that the more intense and systematic coverage afforded these areas by atlassers has produced a more accurate picture of the range of the species than was previously available. On the other hand, there have been several records of Rusty Blackbirds seen at Point Pelee in June (James *et al.* 1976), indicating that late migrants do occur in the south. In addition, as is true of many other species, non-breeders may seek out appropriate habitat and may often behave as if advertising a territory. Discovery of such individuals could lead to erroneous 'possible' or even 'probable' breeding records. The southern limit of the Rusty Blackbird's breeding range is therefore still uncertain.

Abundance estimates were made for only 77 of the 294 squares where Rusty Blackbirds were found. Three squares were estimated to have over 100 pairs, but most squares (81%) were estimated to have 10 or fewer pairs.-- *N. Flood*

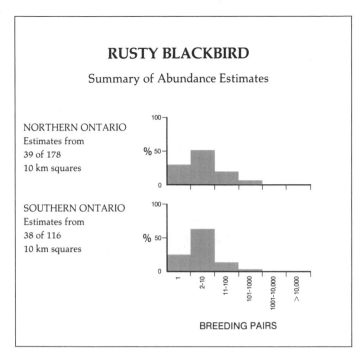

RUSTY BLACKBIRD

Summary of Abundance Estimates

NORTHERN ONTARIO
Estimates from
39 of 178
10 km squares

SOUTHERN ONTARIO
Estimates from
38 of 116
10 km squares

BREEDING PAIRS

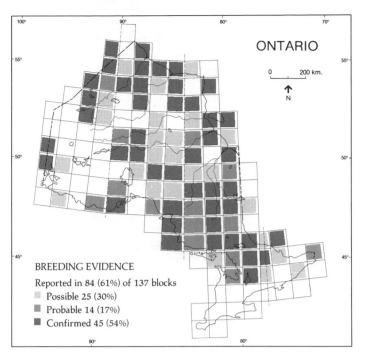

ONTARIO

0   200 km.

N

BREEDING EVIDENCE

Reported in 84 (61%) of 137 blocks

Possible 25 (30%)

Probable 14 (17%)

Confirmed 45 (54%)

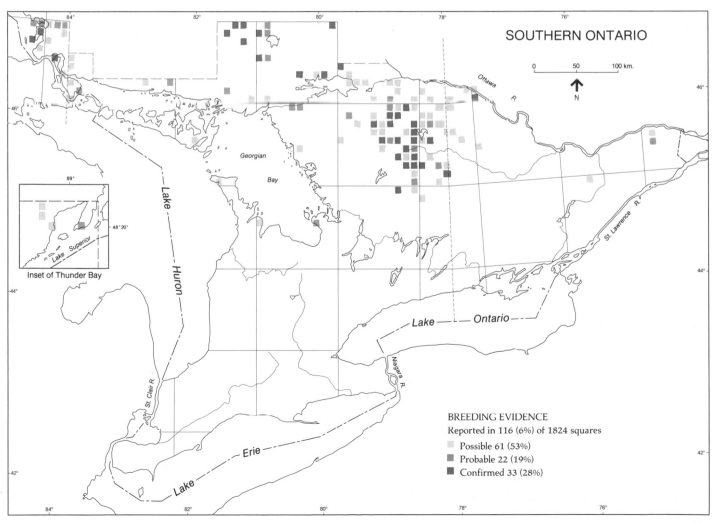

SOUTHERN ONTARIO

0   50   100 km.

N

Georgian Bay

Lake Huron

Lake Superior

Inset of Thunder Bay

Ottawa R.

St. Lawrence R.

Lake Ontario

Niagara R.

St. Clair R.

Lake Erie

BREEDING EVIDENCE

Reported in 116 (6%) of 1824 squares

Possible 61 (53%)

Probable 22 (19%)

Confirmed 33 (28%)

BREWER'S BLACKBIRD
Quiscale de Brewer
Euphagus cyanocephalus

The Brewer's Blackbird is one of the least common and lesser known blackbirds in Ontario. It is considered difficult to identify by many naturalists, at casual glance appearing somewhat intermediate between a Common Grackle and a Brown-headed Cowbird, and in the spring, a Rusty Blackbird. It can be mistaken for any of the three.

The Brewer's Blackbird is a species of western North America, with its winter range including most of Mexico. Its general habitat preference is for grassy prairies with patches of trees or shrubs or marsh edges. It is often found in the vicinity of scattered human habitation such as farm buildings. In Ontario, it is frequently found along main roads and highways with fresh-water, as opposed to stagnant, ditches.

A major expansion of the Brewer's Blackbird range occurred in this century (Bent 1958, Walkinshaw and Zimmerman 1961, Stepney 1975). It extended into both the lower and upper peninsulas of Michigan about the same time as it entered Ontario. The first breeding record in Ontario came from Thunder Bay in 1945 (Allin and Dear 1947). Prior to this, it was not known to nest east of Manitoba. It seems likely that the eastward range extension along the north shore of Lake Huron and Manitoulin Island (Speirs 1959, Devitt 1964, Baillie 1969) was initiated by expanding populations in the upper peninsula of Michigan. The first breeding records in the vicinity of Sault Ste Marie were in 1953 (Baillie 1953a,b, Speirs 1954, Wood 1955). Expansion subsequently proceeded eastward in southern Ontario to Simcoe Co and York and Durham RM's (Richards and Peck 1968, Devitt 1969, Peck and James 1987), but it did not reach southeastern Ontario.

After 1975, populations of the Brewer's Blackbird declined sharply. Only 2 squares supported the bird in southwestern Ontario during the atlas period. The species was much reduced, but persisted in the western upper Michigan peninsula, and currently occurs there in good numbers (N. Sloan pers. comm.). There are also colonies found along the north shore of Lake Huron and Georgian Bay from Sault Ste Marie to Sudbury.

Brewer's are among the most colonial of blackbirds, and colony size is a factor in locating this species. Colonies in the main parts of its range consist of 5 to 10 pairs. New colonies in Ontario are often small, from 2 to 3 pairs, and more established colonies may be up to 14 pairs (pers. obs.). During incubation, females are never conspicuous, but the location of a single male may lead to the discovery of a small colony. Foraging for food for the young birds often takes place at considerable distance from the nesting site. Single sentinel birds are left to guard a colony but if the group is small, all adult birds may be away. Unless the birds are expected and the observer waits, a colony could be missed. When the young are newly fledged, they are moved almost immediately in a family group well away from the nesting area. It is at this time that the similarity of the female and juveniles to juvenile Common Grackles is most apparent. Small colonies often re-establish in new sites in subsequent years, although usually within 1 or 2 km of the old site. Large colonies more often remain in the same site because of the extent of generally favourable habitat. Nests are often located only a few metres apart, but in atlas information only 6% of all records concern actual observations of nests with eggs or young. Adults carrying food or faecal sacs, on the other hand, represent 30% of all records and 67% of confirmed records.

The southern Ontario map reflects the distribution of this species as known prior to the atlas. Successful maintenance of breeding populations has been, for the most part, in northcentral and northwestern Ontario, at the northern limit of its summer range. This is consistent with data from Michigan, Wisconsin, and Minnesota. The current, more limited, success in southwestern Ontario is consistent with that for southern Michigan, also at the southern limit of this bird's summer range.

Abundance estimates were provided in 30% of the 145 occupied squares in the province, and agree with colony sizes discussed above. There were only 6 estimates of single pairs (14%) whereas 2 to 10 pairs were estimated in 35 squares (80%). Estimates of more than 10 pairs were provided for only 3 squares (7%), and all of these were in the Sault Ste Marie-Manitoulin Island area. Brewer's Blackbird can be considered relatively rare over much of the province but quite common locally from the Manitoba border to Thunder Bay, and from Sault Ste Marie to Sturgeon Falls, north of Lake Nipissing.-- *A.G. Gordon*

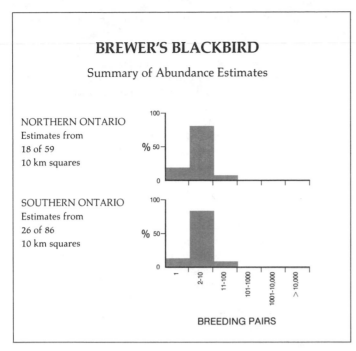

BREWER'S BLACKBIRD

Summary of Abundance Estimates

NORTHERN ONTARIO
Estimates from
18 of 59
10 km squares

SOUTHERN ONTARIO
Estimates from
26 of 86
10 km squares

BREEDING PAIRS

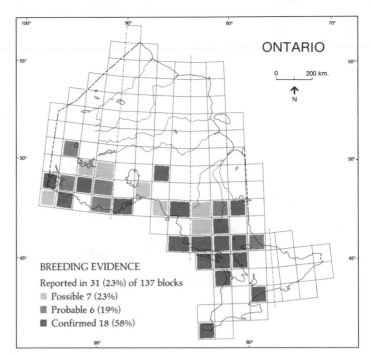

ONTARIO

0   200 km.

N

BREEDING EVIDENCE
Reported in 31 (23%) of 137 blocks
Possible 7 (23%)
Probable 6 (19%)
Confirmed 18 (58%)

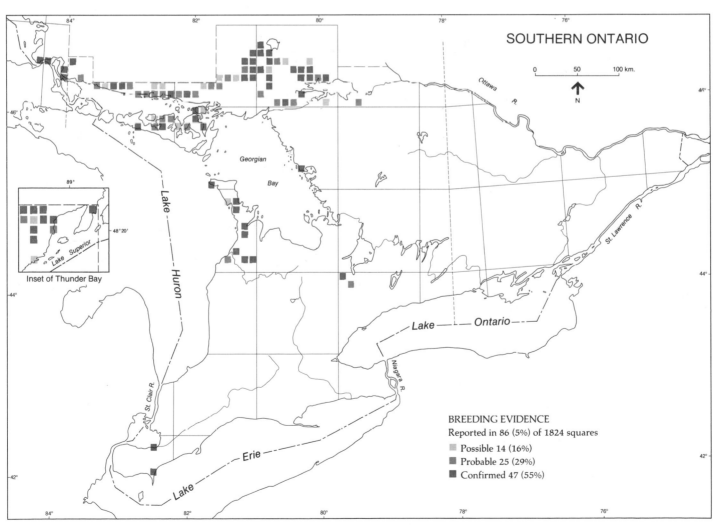

SOUTHERN ONTARIO

0   50   100 km.

N

Georgian Bay

Lake Huron

Lake Superior

89°
48°20'
Inset of Thunder Bay

Ottawa R.

St. Lawrence R.

Lake — Ontario —

Niagara R.

St. Clair R.

Lake — Erie

Lake

BREEDING EVIDENCE
Reported in 86 (5%) of 1824 squares
Possible 14 (16%)
Probable 25 (29%)
Confirmed 47 (55%)

COMMON GRACKLE
Quiscale bronzé
Quiscalus quiscula

As its name suggests, the Common Grackle is a very common species through much of its range. The arrival of this species in the spring is easily detected because the birds often associate with human settlements.

The Common Grackle breeds across most of North America east of the Rockies, primarily south of the Boreal Forest and north of Mexico. For the winter, it withdraws from only the northern parts of its range. In its choice of breeding habitat and nest sites it is among the most catholic of North American species, frequenting marshes, coniferous trees, tree stumps, hedges, and occasionally abandoned buildings (Howe 1976). Nests can be found individually or in small colonies. When not breeding, Common Grackles are very gregarious and often roost communally with Red-winged Blackbirds, European Starlings, and Brown-headed Cowbirds. The broad array of habitats used for breeding is due in part to the similarly broad diet. Outside the breeding season, Common Grackles consume large quantities of waste grain and are occasionally implicated in agricultural damage (Dolbeer 1980). In the breeding season they eat insects and earthworms, and have been observed taking the eggs and nestlings of other species of birds.

The only important historic change in the abundance of the Common Grackle in Ontario is probably associated with its success at exploiting agriculture. The availability of waste grain must have allowed this species to increase substantially in Ontario, particularly in agricultural areas. Human activity, such as land clearing for agriculture, may have been responsible for local changes in the distribution of the Common Grackle. However, because this species also breeds in areas unaffected by human activity, the overall distribution of the Common Grackle is unlikely to have been altered by man.

The Common Grackle is an easy bird to atlas. Although not colourful, the iridescent black plumage (particularly of males) is easy to see, and even breeding individuals are not secretive. In flight the wedge-shaped tail of the male is unmistakable, even at considerable distances. Furthermore, while far from melodic, the call of the Common Grackle is easily heard and recognized. The Common Grackle was found in 97% of the squares in southern Ontario and 60% of the squares with atlas data in all of Ontario. These figures show the wide distribution of the bird, its high numbers, and the ease with which it is observed.

Breeding activities continue from late April through mid-July. As a result, some aspect of reproductive activity was ongoing during the period of the year when atlas data collection took place. Females are less obvious than males during the early stages of reproduction. However, they become more obvious once they are feeding nestlings. Males do occasionally contribute substantially to rearing the young. Nests are easily observed and are usually accessible. In southern Ontario, the Common Grackle was confirmed as breeding in 1446 squares, a total exceeded by only two other species. The three most frequently used breeding categories were all in the confirmed level: food-carrying with 34%, fledged young with 19%, and nest with young with 12%.

The atlas data and maps show that the Common Grackle is widespread south of the Hudson Bay Lowland. Some records were reported on the Lowland, mostly in the southern portion adjacent to Moosonee, but one confirmed record from Attawapiskat extends the known breeding range of the species north of that given in Godfrey (1986), and two northerly possible breeding records suggest that it may breed in small numbers almost to the Hudson Bay Coast.

The abundance estimates provided by atlassers for southern Ontario reveal that the species was considered to be abundant. The most frequently represented abundance estimate is between 101 and 1,000 pairs per square, having been reported in 42% of all squares for which estimates were provided. Intrestingly, estimates were provided in all categories from 1 pair to over 10,000 pairs, possibly revealing the considerable habitat variability found in southern Ontario. The northern Ontario estimates are generally lower than those in the south. This is in agreement with the BBS data in Speirs (1985), which show a steady decline in Common Grackle numbers as one moves progressively north in Ontario.-- *P.J. Weatherhead*

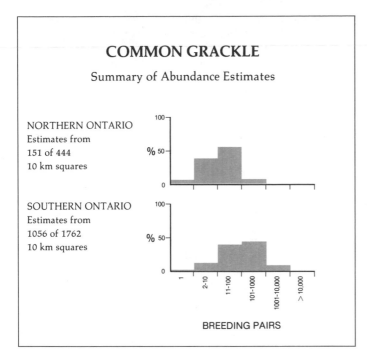

COMMON GRACKLE

Summary of Abundance Estimates

NORTHERN ONTARIO
Estimates from
151 of 444
10 km squares

SOUTHERN ONTARIO
Estimates from
1056 of 1762
10 km squares

BREEDING PAIRS

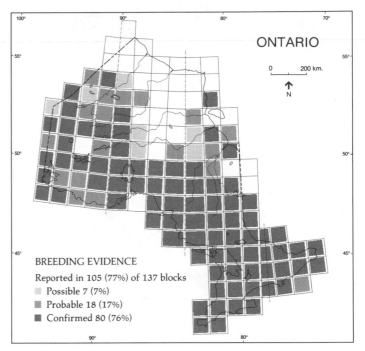

ONTARIO

0   200 km.

N

BREEDING EVIDENCE
Reported in 105 (77%) of 137 blocks
Possible 7 (7%)
Probable 18 (17%)
Confirmed 80 (76%)

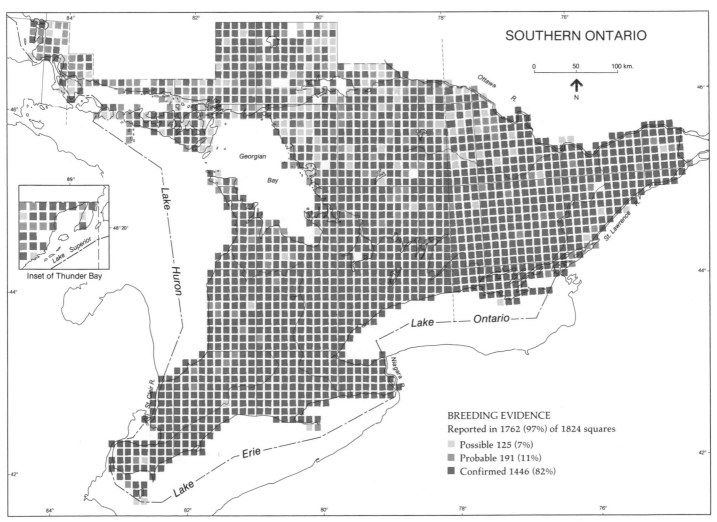

SOUTHERN ONTARIO

0   50   100 km.

N

Inset of Thunder Bay

Georgian
Bay

Lake Superior

Lake Huron

St. Clair R.

Lake Erie

Niagara R.

Lake Ontario

Ottawa R.

St. Lawrence R.

BREEDING EVIDENCE
Reported in 1762 (97%) of 1824 squares
Possible 125 (7%)
Probable 191 (11%)
Confirmed 1446 (82%)

BROWN-HEADED COWBIRD
Vacher à tête brune
Molothrus ater

The Brown-headed Cowbird breeds generally from the northern edge of agricultural land across Canada and south to the Gulf coast and central Mexico. It occurs throughout southern Ontario, but is considerably less common and more irregularly distributed in northern Ontario. This cowbird forages on the ground in short-grass areas, and thus its abundance is restricted to agricultural and residential areas. Within its preferred habitat, it is usually a very common bird. Its winter range extends from southern Canada through to Mexico.

Originally, the Brown-headed Cowbird was present only on the prairies, where it followed buffalo herds and fed on insects they disturbed. When settlers cleared much of the eastern forest for agriculture, the species spread throughout the newly created grasslands of eastern North America. By the 1880's, it was well established in southwestern Ontario (McIlwraith 1886). Records have occurred north to Big Trout Lake, Attawapiskat, and Moose Factory. In the last 100 years, it has spread northward many hundreds of kilometres. Populations west of Lake Superior were established by the mid 1940's.

The Brown-headed Cowbird never builds a nest. The female lays her eggs in the nest of another species. The cowbird egg hatches quickly, usually before that of its host, giving the young bird a head start. The host parents usually feed the nest parasite, oblivious to the fact that it is a different species.

Young Brown-headed Cowbirds make relatively loud, persistent begging calls, easily enabling atlassers to confirm the species as breeding. Fledglings account for more than half (56%) of all confirmed breeding records in Ontario for this species. The Brown-headed Cowbird parasitizes many species, including such common and widespread hosts as the Red-eyed Vireo, Yellow Warbler, Chipping Sparrow, and

Song Sparrow. Parasitized nests with eggs or young were found in 20% of all Ontario squares where the species occurred. Finding a nest of one of the host species often means also confirming the Brown-headed Cowbird as breeding. Several Regional Coordinators felt that this species was confirmed as breeding in fewer squares than it should have been. Indeed, this species likely bred in the great majority of squares in which it was reported.

Possible or probable breeding of the Brown-headed Cowbird is relatively easy to establish. Its breeding season extends from late April to mid-July. Males sing and display conspicuously from trees, rooftops, and telephone poles. Noisy feeding aggregations, composed of several males and females, occur commonly in suburban and agricultural areas.

Data from the atlas emphasize the reliance of the Brown-headed Cowbird on agricultural areas for feeding. In southern Ontario, intensively farmed areas have populations that are both widespread and abundant. In Algonquin Prov. Park and much of the French River region, agriculture is absent or limited, and the species is correspondingly absent or sporadic. However, a careful checking of the map for southern Ontario shows a line of occupied squares through Algonquin Park corresponding to Highway 60, demonstrating that only a minimum amount of forest opening is required to allow the Brown-headed Cowbird bird to populate an area. In northern Ontario, several Regional Coordinators commented that the species is confined to agricultural areas, and this is evident on the maps. It is confirmed as a breeding species as far north as Moosonee, with scattered possible breeding records in the vicinity of the towns of Pickle Lake and Fort Hope.

Within southern Ontario, the Brown-headed Cowbird was usually common, and in some squares, abundant during the atlas period. It was less common and more patchily distributed in the heavily forested areas of the French River and, particularly, Algonquin Prov. Park. In northern Ontario, it was generally rare or absent and restricted to agricultural areas. Populations in the agricultural Rainy River area have been shown to be on a par with those of southern Ontario's agricultural regions, whereas the numbers in the northeast are generally lower (Speirs 1985). Data from the BBS indicate that Brown-headed Cowbird numbers are now generally stable throughout the province.-- *D. Graham*

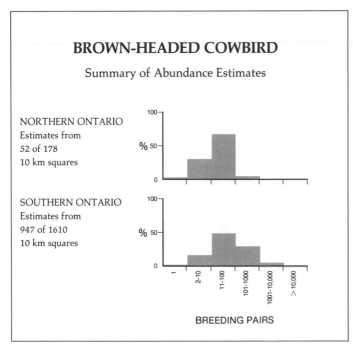

BROWN-HEADED COWBIRD

Summary of Abundance Estimates

NORTHERN ONTARIO
Estimates from
52 of 178
10 km squares

SOUTHERN ONTARIO
Estimates from
947 of 1610
10 km squares

%

%

1   2-10   11-100   101-1000   1001-10,000   >10,000

BREEDING PAIRS

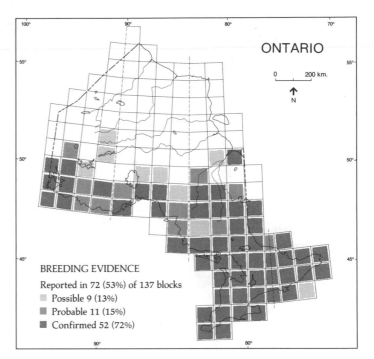

ONTARIO

0   200 km.

N

BREEDING EVIDENCE

Reported in 72 (53%) of 137 blocks

Possible 9 (13%)

Probable 11 (15%)

Confirmed 52 (72%)

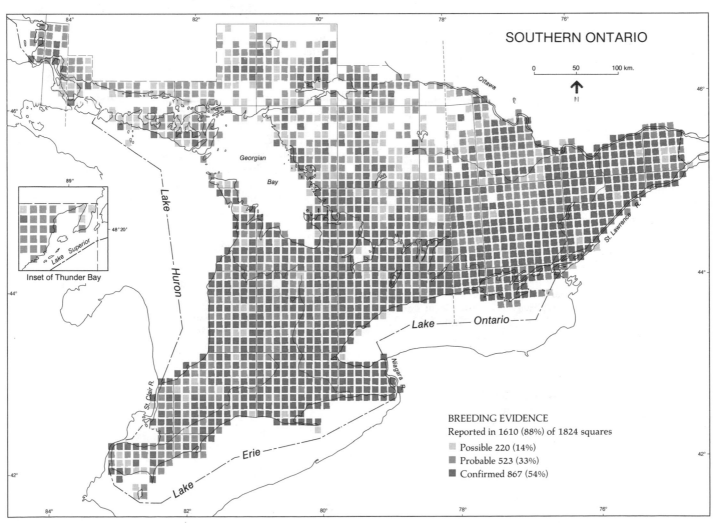

SOUTHERN ONTARIO

0   50   100 km.

N

Ottawa R.

Georgian Bay

Lake Huron

Lake Superior

Inset of Thunder Bay

St. Lawrence R.

Lake Ontario

Niagara R.

St. Clair R.

Lake Erie

BREEDING EVIDENCE

Reported in 1610 (88%) of 1824 squares

Possible 220 (14%)

Probable 523 (33%)

Confirmed 867 (54%)

ORCHARD ORIOLE
Oriole des vergers
Icterus spurius

The smaller of Ontario's breeding orioles, the Orchard Oriole is less common and more restricted in range than the Northern Oriole. The centre of the Orchard Oriole's range lies in the Mississippi River valley in the central US. It is an uncommon summer resident from southern Manitoba and Ontario through the eastern half of the US into northeastern Mexico. It winters from southern Mexico through Central America into northern South America.

As its name implies, the Orchard Oriole commonly breeds in orchards and other similar habitats. It shows a marked preference for open country, being most frequently associated with lightly wooded river courses, fields containing scattered trees, hedgerows, farmsteads, campgrounds, and residential areas. In Ontario, the Orchard Oriole essentially occurs only in the Carolinian Forest zone; here it finds the deciduous trees in which it prefers to nest (Peck and James 1987), as well as large areas of open country.

The breeding range of the Orchard Oriole in Ontario may have expanded slightly over the last century, eastward along Lake Ontario, and northward away from the Great Lakes. McIlwraith (1894) stated that the Orchard Oriole probably did not ever breed very far from the Lake Erie shore; it nested regularly west of, and in, London, but did not go as far east as Hamilton. Soon after this, a pair was found nesting in Toronto (Fleming 1907). Baillie and Harrington (1937) confirmed that the species was essentially restricted to the southern edge of the province, but noted that Orchard Orioles had been seen as far east as Durham Co and as far north as Goderich. Speirs (1985) described the species as being relatively common locally along the Lake Erie shore, but rare elsewhere in Ontario. He noted, however, that pairs had been seen during summer in different years at locations such as Prince Edward Co, Parry Sound, and Simcoe Co.

However, McCracken (1987) provided evidence that it was formerly more common in the heart of its present range in Haldimand-Norfolk RM. The Orchard Oriole is expanding its range in southern Manitoba and in Saskatchewan (Sealy 1980).

There was considerable disagreement among Regional Coordinators with respect to the degree of site tenacity displayed by Orchard Orioles, as well as with respect to apparent fluctuations in population numbers over the atlas period. This makes interpretation of atlas results somewhat difficult for this species, and suggests that range limits might better be considered fluctuating zones rather than rigid lines. Often the pioneering individuals in such zones engage in activity suggestive of breeding, but do not mate or raise young. Over three-quarters of all Orchard Oriole records in the possible breeding category are of singing males, and 42% of those in the probable breeding category are described as being territorial. Undoubtedly, many of these males remained unmated.

Apart from the above mentioned problems, the Orchard Oriole is not a difficult species to census. Orchard Orioles display behaviour patterns (foraging in the open, defending the nest vigorously against predators, and frequent singing during the courtship period) which make them easy to spot. Thirty-nine percent of all records in the confirmed breeding category involved the actual discovery of occupied nests.

In only 110 squares was any evidence of Orchard Orioles reported. Atlas records now show confirmed breeding as far east as Kingston, and as far north as Lindsay. Although records are still concentrated in a narrow band north of Lake Erie and along the western half of Lake Ontario, atlassers seem to have discovered more Orchard Orioles outside this zone than had been observed in the past. Some of this is likely due to the more extensive and systematic approach to observing and recording undertaken by atlassers. At least in part, however, these data may demonstrate a range expansion. It is probably wise to view the map of the Orchard Oriole as approximate and subject to change. Many of the more northerly records are no doubt the result of exploratory behaviour. It is doubtful that the habitat preferences of the species will permit it to move very far out of the Carolinian Forest zone.

At present, the Orchard Oriole can be described as an uncommon summer resident in Ontario. Most (68%) of the squares in which abundance was estimated appeared to support only one pair. In 18 squares, the abundance of the species was estimated to be 2 to 10 pairs, and only in the square containing Point Pelee Nat. Park were over 10 pairs considered to be present.-- *N. Flood*

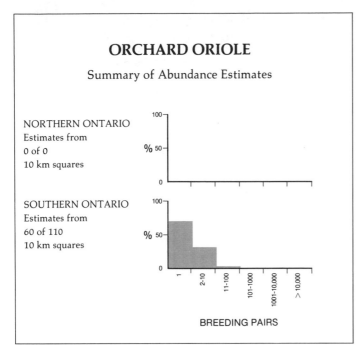

ORCHARD ORIOLE

Summary of Abundance Estimates

NORTHERN ONTARIO
Estimates from
0 of 0
10 km squares

SOUTHERN ONTARIO
Estimates from
60 of 110
10 km squares

%

100

50

0

%

100

50

0

1 2-10 11-100 101-1000 1001-10,000 >10,000

BREEDING PAIRS

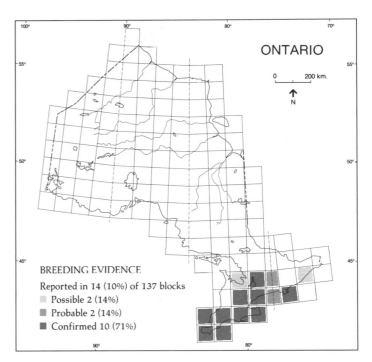

ONTARIO

0   200 km.

↑
N

BREEDING EVIDENCE
Reported in 14 (10%) of 137 blocks
☐ Possible 2 (14%)
▨ Probable 2 (14%)
■ Confirmed 10 (71%)

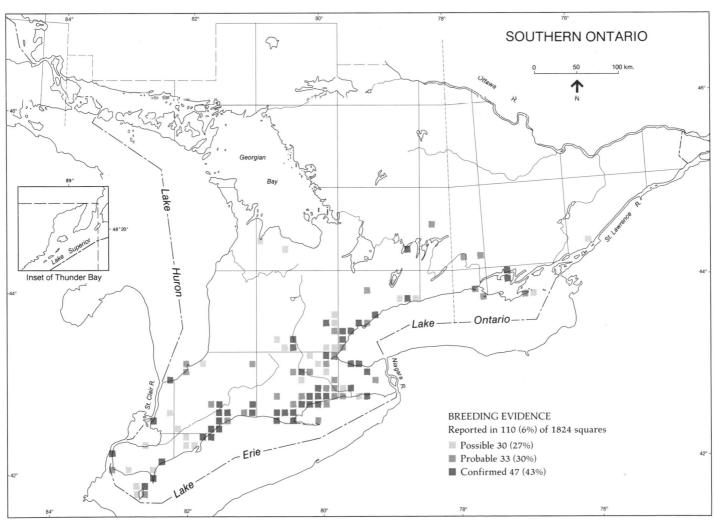

SOUTHERN ONTARIO

0   50   100 km.

↑
N

Inset of Thunder Bay

BREEDING EVIDENCE
Reported in 110 (6%) of 1824 squares
☐ Possible 30 (27%)
▨ Probable 33 (30%)
■ Confirmed 47 (43%)

NORTHERN ORIOLE
Oriole du Nord
Icterus galbula

The Baltimore Oriole, the eastern subspecies of the Northern Oriole, breeds in southern and central Canada and the eastern US, west to the prairies and Great Plains, where it meets, and interbreeds with, its western counterpart, the Bullock's Oriole. It winters from southern Mexico through Central America and into northern South America.

The breeding range of the Northern Oriole is limited by the distribution of deciduous trees, for which the species shows a marked preference when selecting nest sites. Of 401 Northern Oriole nests in Ontario reported to the ONRS, 378 (94%) were located in deciduous trees (Peck and James 1987). The species is therefore confined to the Carolinian and Great Lakes-St. Lawrence Forest regions as well as in portions of the Boreal Forest where certain deciduous tree species exist. The Northern Oriole typically nests in edge habitat: in wooded areas containing natural openings, along shorelines or roadsides, in fields containing shade trees or bordered by hedgerows, and in gardens, orchards, parks, and even backyards. Northern Orioles commonly forage in agricultural fields or other open areas for the large volume of insects they need to feed their young.

The range and abundance of the species in Ontario appears to have increased slightly over the last century. It was known to breed in the west in the Kenora area and as far north in southern Ontario as Parry Sound and Ottawa, but not between these points (McIlwraith 1894, Baillie and Harrington 1937). Increased development in the area between Kenora and Parry Sound in recent decades may have increased the amount of suitable Northern Oriole breeding habitat. Although no Northern Orioles apparently bred in the Thunder Bay area in the early part of this century, they have been known to occur there for at least a decade (James *et al.* 1976). At least some of this apparent range extension may

simply be due to an increase in the number of observers.

The Northern Oriole is relatively conspicuous and easy to atlas. The adult male is brightly coloured and, until he obtains a mate, is a loud, melodious, and frequent singer. The pensile nest is usually attached to the outermost branches of deciduous trees and thus is easily located. Parent birds generally forage in the open and are therefore easily detected carrying food or faecal sacs during the nestling period. The species does not perform distraction displays, but either or both parent(s) will mob predators that approach the nest. These types of behaviour make the Northern Oriole relatively conspicuous at or near its nest. Nests containing eggs or young were found in 19% of all squares where the species was found in Ontario. Used nests comprised an additional 4%.

Once paired, males become almost silent. Unmated males, however, continue to sing throughout the breeding season; such birds (in most instances any male singing after May) might be recorded as singing males, or on territory. In the province, 73% of records in the possible breeding category are of singing males and 31% of records in the probable breeding category are of territorial birds. The fact that males in their first breeding season are female-like in general appearance, but are reproductively mature, and therefore sing and obtain a mate, is a possible source of confusion (Flood 1984). Although on average duller in colouration than either first year or adult males, female Northern Orioles are extremely variable in appearance, and it is impossible, without extensive behavioural observations, to be certain of the sex of birds other than adult males. Finally, the relatively brief tenure of families in the nesting area might reduce atlas accuracy as both parents and young leave the area of the nest soon after fledging.

The map shows a similar distribution to that known before the atlas period. The species likely does breed in most of the southern squares in which breeding is now shown as only possible or probable. A similar situation may be true in squares surrounding Kenora. The more localized nature of the distribution in Muskoka, Haliburton, Nipissing, Parry Sound, Sudbury, and Algoma areas, however, is realistic and likely reflects the low amount of cleared land in these areas. Breeding records from Lake of the Woods to Thunder Bay are not surprising, because a band of Great Lakes-St. Lawrence Forest in that area provides suitable deciduous trees for nesting.

The species is particularly abundant in southern Ontario, where large areas of land have been cleared, producing extensive lengths of edge habitat. Almost 75% of the southern squares in which abundance was estimated were thought to contain from 11 to 1,000 pairs of Northern Orioles, suggesting that this species is one of the most common breeding birds in southern Ontario. Where it occurs in the north, it is much less abundant; all estimates are of 10 pairs or fewer per square.-- *N. Flood*

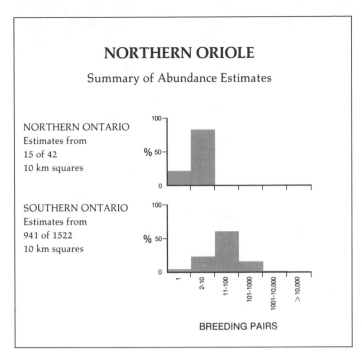

NORTHERN ORIOLE

Summary of Abundance Estimates

NORTHERN ONTARIO
Estimates from
15 of 42
10 km squares

SOUTHERN ONTARIO
Estimates from
941 of 1522
10 km squares

BREEDING PAIRS

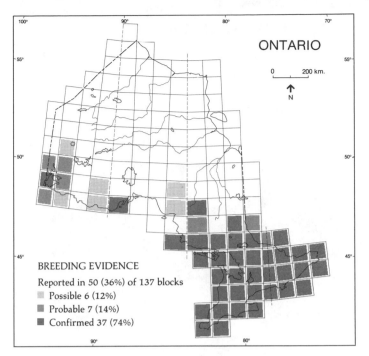

ONTARIO

0   200 km.

N

BREEDING EVIDENCE
Reported in 50 (36%) of 137 blocks
▢ Possible 6 (12%)
▨ Probable 7 (14%)
■ Confirmed 37 (74%)

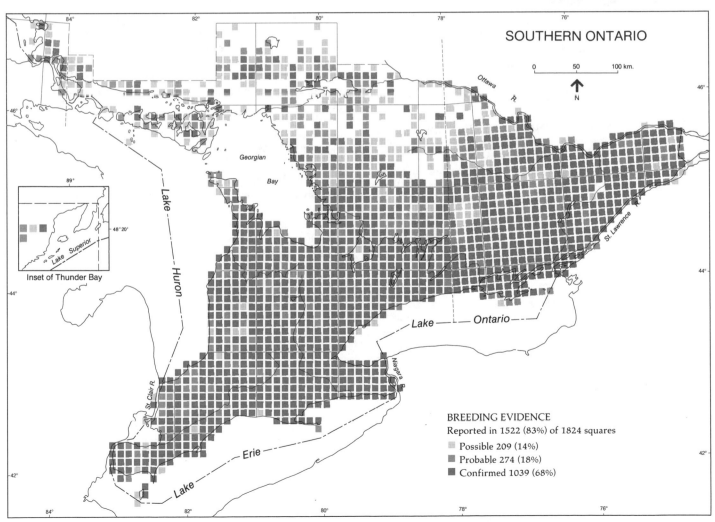

SOUTHERN ONTARIO

0   50   100 km.

N

Georgian
Bay

Lake

Huron

St. Clair R.

Ottawa R.

St. Lawrence R.

Lake — Ontario

Niagara

Erie

Lake

Inset of Thunder Bay
89°
48°20'
Lake   Superior

BREEDING EVIDENCE
Reported in 1522 (83%) of 1824 squares
▢ Possible 209 (14%)
▨ Probable 274 (18%)
■ Confirmed 1039 (68%)

487

PINE GROSBEAK
Dur-bec des pins
Pinicola enucleator

The Pine Grosbeak is one of several species of colourful 'winter finches' that are far better known at that time of year than they are during the breeding season. In some winters, the species is seen frequently in small flocks throughout Ontario. Its irregular presence in the south, and the scarcity of information about its breeding distribution, make it a bit of a mystery bird.

The Pine Grosbeak is holarctic in its range, breeding in almost all of Canada and Alaska from the tree-line to the southern parts of most of the provinces. It prefers open coniferous woods for nesting (Godfrey 1986). Information about its Ontario breeding range is scarce, but there are old (1940) records of nests from as far south as Sundridge and Temagami (Baillie 1960). It breeds in the spruce forests of northern Ontario, north to the tree-line, but is absent in the Rainy River area (Godfrey 1986). The Pine Grosbeak is seldom seen in summer and is nowhere a common bird during that season. In winter, it wanders south to the central, and occasionally southern, US.

There is little reason to suspect that the range of the Pine Grosbeak has changed over the years. Relatively little of its preferred breeding habitat has been cleared for agriculture, but major areas are now logged in clear-cut fashion. At the southern edge of its range, Speirs (1985) mentions an early July record from Rice Lake, near Peterborough, but that appears to have been an isolated occurrence south of the present breeding range. Peck and James (1987) do not accept this species as a breeder in Ontario because of the lack of material evidence of nesting; but it has been accepted by the Ontario Bird Records Committee (James 1984b, Wormington and James 1984).

The habits of the Pine Grosbeak are not conducive to finding breeding evidence. The species is secretive in summer and tends to remain well hidden in spruce forests. It seldom sings and, when it does, the song is brief, fairly weak, and unfamiliar to most Ontario naturalists. Hence, there are more records in the suitable habitat category than in the singing male category, which is unusual among passerines. Most of the breeding records (62% of the blocks; 75% of the squares) are in the possible category, a reflection of the secretive habits of the species. Most of the probable breeding records are of pairs. However, for a flocking species that moves about erratically, it may be inappropriate to interpret the presence of a pair as probable breeding in many circumstances. No nests were found during the atlas period. The one confirmed record, in 1985, was of fledged young being fed by their parents near the headwaters of the Black Duck River, approximately 10 km from the Manitoba border. The atlasser, David Shepherd, reported that many family groups were present in an area of small spruce trees along a wooded ridge between a lake and a river. Over 50 individuals were observed.

Although the maps show that the Pine Grosbeak occurs over much of Ontario, the records are sparse and there are surprising and unexplained gaps, especially in northwestern and northcentral Ontario. There is apparently suitable habitat in these areas, and Speirs (1985) cited a few summer observations from some places lacking atlas records. The Regional Coordinator for the Sault Ste Marie region expressed the opinion that the species is fairly common along the Lake Superior shore and is probably 'underatlassed'. Perhaps the lack of atlas records merely reflects the fact that the Pine Grosbeak is an uncommon species with secretive habits that occurs in northern areas where coverage, in some instances, was low. However, the large gap in northwestern Ontario may indicate that the range shown by Godfrey (1986) in the Lake of the Woods area is less extensive than he indicates. Elsewhere, the concentration of observations in the southern part of its breeding range is probably a function of more intensive coverage there, rather than an actual concentration of birds.

The atlas data substantiate the view that the Pine Grosbeak is an uncommon breeding bird in Ontario. In southern Ontario, there are very few squares (16) in which Pine Grosbeaks were recorded at all. In only 13 squares in the province did the atlassers offer abundance estimates. These estimates suggest that, at the most, there is an average of 2 to 10 pairs per square.-- *F.M. Helleiner*

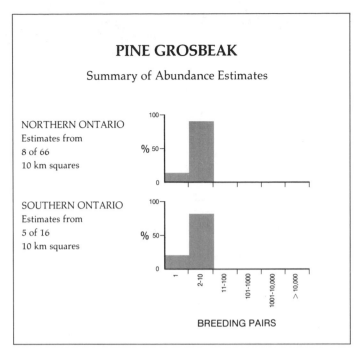

PINE GROSBEAK

Summary of Abundance Estimates

NORTHERN ONTARIO
Estimates from
8 of 66
10 km squares

SOUTHERN ONTARIO
Estimates from
5 of 16
10 km squares

% 100
50

% 100
50

1 2-10 11-100 101-1000 1001-10,000 >10,000

BREEDING PAIRS

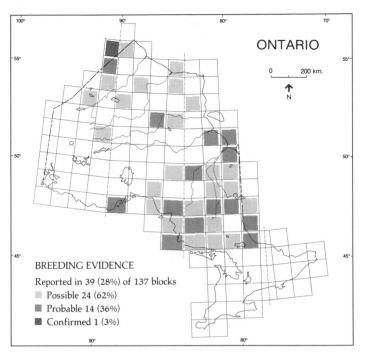

ONTARIO

0   200 km.

N

BREEDING EVIDENCE

Reported in 39 (28%) of 137 blocks

Possible 24 (62%)
Probable 14 (36%)
Confirmed 1 (3%)

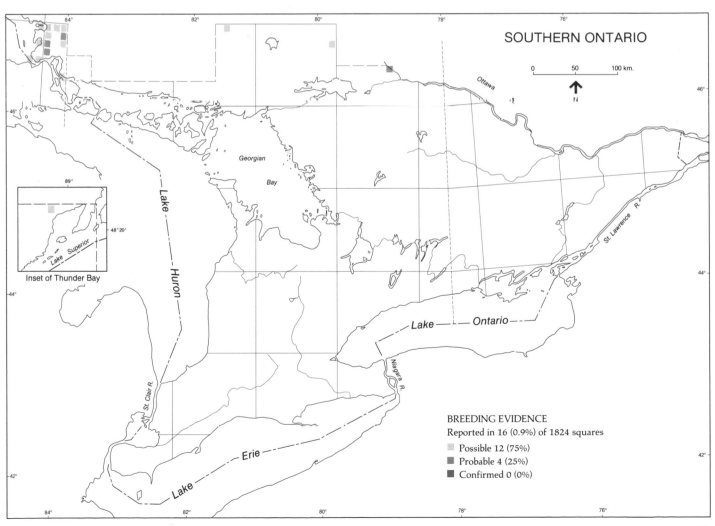

SOUTHERN ONTARIO

Ottawa

Georgian

Bay

Lake

Huron

Lake   Superior

Inset of Thunder Bay

St. Clair R.

Lake — Ontario

Niagara R.

St. Lawrence R.

Erie

Lake

BREEDING EVIDENCE

Reported in 16 (0.9%) of 1824 squares

Possible 12 (75%)
Probable 4 (25%)
Confirmed 0 (0%)

PURPLE FINCH
Roselin pourpré
Carpodacus purpureus

With the recent appearance of the House Finch in Ontario, care must be taken not to confuse it with the Purple Finch, the long-time, natural resident of the province. The Purple Finch is larger than its relative. The males are more uniformly red in colour (often described as being dipped in raspberry juice) and lack distinct striping on the flank and belly. The females, however, are very similar in appearance, but the Purple Finch is heavily striped and has an obvious black jaw stripe (Peterson 1980).

McIlwraith (1886) commented that the Purple Finch is "generally distributed across the country." By 1937, Baillie and Harrington were able to state that the species "is a fairly common summer resident of Ontario (particularly in the central and northern parts) from our southern border north to Lake of the Woods and James bay." Godfrey (1986) shows the Ontario breeding range to include all of the province north to approximately 52^0N.

The Purple Finch breeds across the continent from the Boreal Forest south to the northern US (AOU 1983). In Ontario, it is a common summer resident as far north as Moosonee, but it is infrequent in the agricultural southern parts of the province (James *et al.* 1976, Speirs 1985). This distribution is regulated by the occurrence of coniferous forest (AOU 1983), which is preferred for nesting (Blake 1968, Godfrey 1986). However, deciduous trees in open conditions, such as orchards, are also used for nesting (Godfrey 1986). In winter, however, the Purple Finch strays southward to the area from southern Canada to the Gulf coast. Here, it frequents deciduous forest and open areas with shrubbery around human habitation, and is readily attracted to feeding tables. Its winter movements are irregular.

Breeding of the Purple Finch begins about mid-May and is signalled by increased frequency of singing and a colourful mating display. The nest is usually built close to the top of a coniferous tree where the cover is thickest. Consequently, few nests are reported: only 5% of confirmed breeding records are of nests with eggs or young. The nest is formed by a cup of fine twigs, grasses, and rootlets, lined with fine moss, hair, or wool. The eggs are pale blue, speckled with dark spots, and the normal clutch is 4 or 5 eggs. Incubation is entirely by the female, who is fed on the nest by her mate. Young hatch in about 13 days, at which time both parents assume a supportive role. The young, which resemble the female, leave the nest when about 14 days old and remain dependent on the parents for a further 3 weeks. Most confirmed breeding records are obtained in this period; 65% of confirmations are of newly fledged young. Most commonly, the species is single-brooded, but double broods have been recorded.

The breeding distribution of the Purple Finch, as indicated by the maps, roughly corresponds to the previously known range of the species and shows its preference for coniferous forest. However, there are many atlas records from well north of the breeding limit shown by Godfrey (1986). There is a suggestion that the species has increased in numbers in recent years in the south. Initially, one might suspect that the confirmed nestings in southwestern Ontario could support this view. Probably, however, the species has always been a sparse, but usually unrecognized, nesting bird in parts of the south. Saunders and Dale (1933) reported a 1910 nest in Victoria Park in London. The current records probably reflect diligence of observers, which may also have given the impression of increased numbers. By contrast, the number of confirmed nestings in the north of southern Ontario seems low. This probably results from the difficulty of locating nests in thick cover. Almost certainly, many squares where the highest level of breeding evidence was in the probable category contain nesting Purple Finches. Despite these comments, the maps accurately show the species' general avoidance of the agricultural land of southern Ontario and its preference for the northern coniferous forest, a distribution that has been long recognized.

The abundance estimates suggest that there are more birds per square in the northern part of the range in Ontario. The higher estimates, those over 100 pairs per square, are scattered across the range of the Purple Finch in the province. A concentration of higher estimates is apparent in the area from Parry Sound and Haliburton in the south to the Nipissing East atlas region in the north.-- *A.L.A. Middleton*

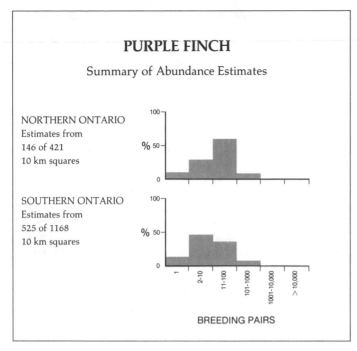

PURPLE FINCH

Summary of Abundance Estimates

NORTHERN ONTARIO
Estimates from
146 of 421
10 km squares

SOUTHERN ONTARIO
Estimates from
525 of 1168
10 km squares

BREEDING PAIRS

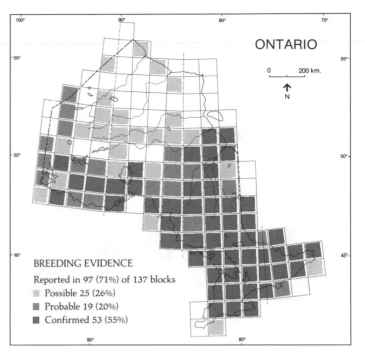

ONTARIO

0   200 km.

N

BREEDING EVIDENCE
Reported in 97 (71%) of 137 blocks
Possible 25 (26%)
Probable 19 (20%)
Confirmed 53 (55%)

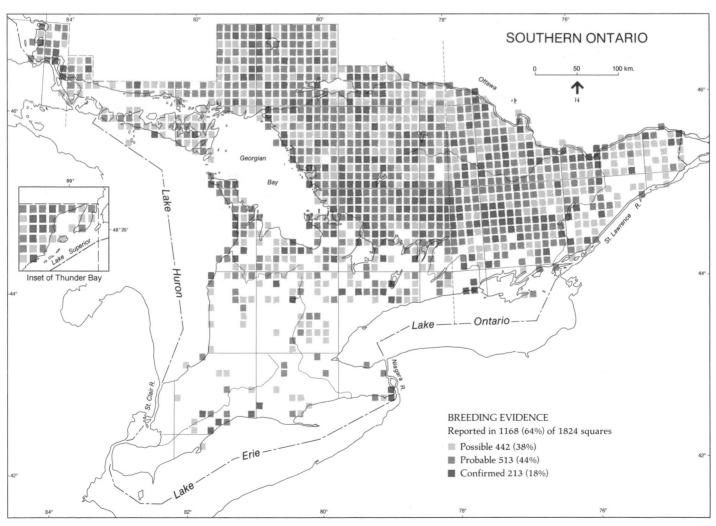

SOUTHERN ONTARIO

0   50   100 km.

N

Georgian Bay

Lake Huron

Lake Superior
89°
48°20'
Inset of Thunder Bay

St. Clair R.

Niagara R.

Lake — Ontario

Lake — Erie

Lake

Ottawa

St. Lawrence R.

BREEDING EVIDENCE
Reported in 1168 (64%) of 1824 squares
Possible 442 (38%)
Probable 513 (44%)
Confirmed 213 (18%)

HOUSE FINCH
Roselin familier
Carpodacus mexicanus

The House Finch is a recently established breeding bird in Ontario. Since its first appearance in the province in the early 1970's, the species has undergone a phenomenal increase in numbers and is rapidly becoming a common resident in the south.

This North American finch is widespread in the west from British Columbia to southern Mexico and in the east from the St. Lawrence River valley south to the Carolinas. In Ontario, it breeds across the southern margin of the province. The species is most abundant in cities, towns, and cultivated areas, which provide a variety of nesting sites (Woods 1968); its nesting requirements are similar in many ways to those of the House Sparrow. Such a commensal habit has undoubtedly assisted with successful colonization and with the recording of breeding evidence.

The House Finch is not native to eastern North America, but became established after cage-birds were liberated on Long Island, New York in 1940 (Elliott and Arbib 1953). Subsequently, the species flourished and rapidly expanded its range primarily to the west and south. By 1980, the eastern population had expanded north to Maine, south to Virginia and inland as far as the upper Ohio River valley (Bystrak 1981). The first Ontario record was from Prince Edward Co on 27 August 1972 (Sprague and Weir 1984), but not until 1978 was evidence of breeding obtained, when 2 nests were discovered in Niagara-on-the-Lake (James 1978b). In Ontario, range expansion continued mainly along the northern shores of Lakes Ontario and Erie.

The Ontario breeding distribution of this species, as depicted by the atlas maps, has essentially arisen during the atlas period. In 1981, breeding was confirmed only at Kingston, although scattered pairs may have bred all around Lake Ontario. Breeding evidence in 1982 and 1983 came mainly

from the Niagara Peninsula - Toronto area, and all along the southern edge of the province. Between 1984 and 1985, the breeding range changed little, but the number of confirmed breeding records increased twofold.

The House Finch often nests in ornamental shrubs and evergreens near dwellings. Furthermore, males are highly vocal and sing a lengthy, warbling song from conspicuous perches, so atlas volunteers should have had no difficulty in locating this species. Two broods are usually raised during a single breeding season, and nests with eggs may be found as early as 26 April and as late as 1 August. Breeding is most often confirmed by observation of recently fledged young (about 29% of all records). Approximately 20% of all records involved the discovery of nests with eggs or nestlings. On occasion, nests may be located under aluminum window awnings, and in hanging, potted plants. The conspicuous nesting activities and long breeding season of this species allowed atlassers to confirm breeding in 59% of squares where the species was found.

Despite the species' recent establishment in the province, atlas data reveal a strong trend in House Finch numbers and distribution. It is clearly a breeding bird of urban centres, being most abundant in the highly developed region of southwestern Ontario. However, because the population is still in a phase of expansion, the breeding distribution is probably not fully shown by atlas data. Agricultural practices in the east presumably facilitated the northward spread into the Ottawa Valley, which began in 1983. A nest with young was found in Pembroke in 1983. However, more northern sightings came from the northeast shore of Lake Superior at Marathon in 1976, and at Lively near Sudbury in 1979 and 1980 (Speirs 1985).

Estimates of abundance were provided for over one-half of all squares with data. A majority (54%) contained 2 to 10 breeding pairs per square, with estimates ranging from one (18%) to 101 to 1,000 (2%) breeding pairs per square. The Niagara region reported the highest abundance estimates. However, House Finch numbers have increased substantially in recent years and abundance estimates would greatly depend on when a square was censused during the atlas period. Based on BBS data, the eastern North American population has been increasing by an average of 21% per year (Robbins *et al.* 1986), which indicates an average population doubling time of only 3.3 years. Consequently, abundance estimates made during the atlas period probably underestimate the breeding population in the province.-- *D.R. Kozlovic*

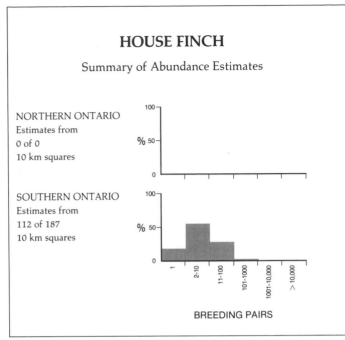

HOUSE FINCH

Summary of Abundance Estimates

NORTHERN ONTARIO
Estimates from
0 of 0
10 km squares

SOUTHERN ONTARIO
Estimates from
112 of 187
10 km squares

BREEDING PAIRS

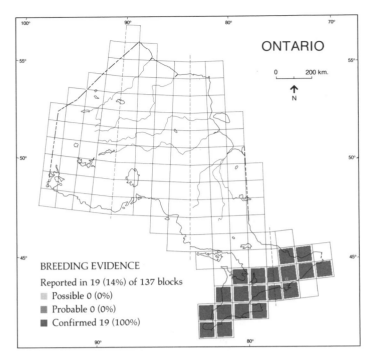

ONTARIO

0   200 km.

N

BREEDING EVIDENCE
Reported in 19 (14%) of 137 blocks
Possible 0 (0%)
Probable 0 (0%)
Confirmed 19 (100%)

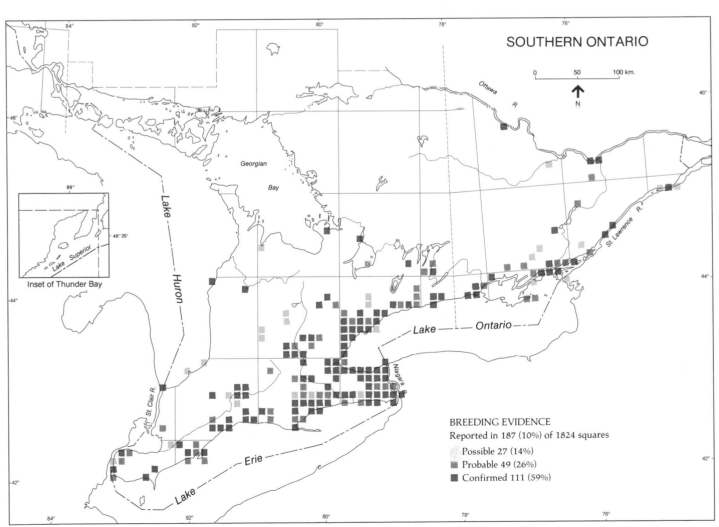

SOUTHERN ONTARIO

0   50   100 km.

N

Inset of Thunder Bay

BREEDING EVIDENCE
Reported in 187 (10%) of 1824 squares
Possible 27 (14%)
Probable 49 (26%)
Confirmed 111 (59%)

RED CROSSBILL
Bec-croisé rouge
Loxia curvirostra

These unusual finches, with their uniquely crossed mandibles, have become adapted to separating the scales of cones to extract seeds with the tongue. They can thus exploit a frequently abundant food source, but when cone crops fail over a wide area, the birds are forced to move in search of new food supplies. The resulting eruption may take them far beyond their 'normal' range, perhaps to breed in areas where they are not usually found.

The Red Crossbill has an extensive world distribution, ranging across Eurasia from northwest Africa to Japan and the Philippines (Vaurie 1959). In America, it occurs south to Nicaragua in montane pine forests, but is mainly found in southern boreal and mixed forests from Alaska to Newfoundland (AOU 1983). Six subspecies are currently recognized north of Mexico, and Dickerman (1986a) has provided a preliminary review; the race normally expected in the northeast, including Ontario, is *L. c. neogaea*, but the northwest coast form, *L. c. minor*, is also present during irruptions. In fact, 6 of 7 Ontario specimens received by the National Museums of Canada from the 1984-85 irruption, and 6 of 9 in 1950 to 1951 were *minor*, the others being *neogaea* (H. Ouellet pers. comm.). Another race, *bendirei* from the Rocky Mountains, may also have bred in Ontario (C. Benkman pers. comm.). For New York State, Dickerman (1986b) listed 17 flight years since 1872 documented by specimens of *minor*, and others in which *bendirei* occurred.

According to Godfrey (1986), the Red Crossbill's breeding range in Ontario lies within the Great Lakes-St. Lawrence and southern Boreal Forest regions. It corresponds closely with the distribution of red and white pines (Hosie 1969). The former range, prior to logging of virgin pine stands, is unknown, but since the 1920's, reforestation with both native and exotic conifers has created some new habitat south of

the Shield. However, there was little documentation of breeding anywhere in Ontario prior to the atlas, and only 5 nests had ever been reported (Peck and James 1987). They were all found in April 1948: 4 at Pimisi Bay, Nipissing District (Lawrence 1949), the fifth in Lanark Co.

Initial location of Red Crossbills is not difficult because their characteristic 'jip...jip' contact call allows identification of distant birds in flight. Furthermore, their sometimes fatal attraction to salted sand on highways in winter renders them very conspicuous. When feeding, Red Crossbills tend to be quiet and not easily seen, but courting males sing from a treetop or perform a song flight, circling the female while uttering loud whistled notes (Lawrence 1949). Nests are usually well concealed on a side branch of a conifer at a height of 7 to 18 m (Peck and James 1987), and are difficult to find. Most cases of confirmed breeding involved the observation of dependent young, which can remain in family groups for 33 days or longer (Newton 1972).

Atlas records have provided a more comprehensive picture of Red Crossbill nesting in Ontario than was previously known. When breeding occurred, it was concentrated from mid-January to May and from late June to October. No nests with eggs or young were seen, but 2 inaccessible nests were found in York RM on 18 and 23 February 1985. Nest-building was reported in one Algonquin square in 1983: on 30 January and from 30 April to 3 May (R. Tozer pers. comm.). There were also 4 reports of nest building in 1985 (16 and 21 February, 2 March and 14 May); and at least 18 records of dependent young, seen from late July to 22 October 1984, and from 2 March to early July in all other years. There were also at least 18 records of dependent young, seen from late July to 22 October (R. Tozer pers. comm.) in 1984, and from 2 March to early July in all other years. Late summer dates conform to the period when conifer seed is ripening, while early spring dates coincide with the carry-over of seed, particularly of red pine (Horton and Bedell 1960).

Data were received for 223 squares throughout Ontario (compared with 303 for White-winged Crossbill), but only 24 records (11%) involved confirmed breeding, reflecting the difficulty of confirming this species. The 1985 irruption accounted for 60% of all confirmed records: all were from Shield-edge sites or points farther south. Thus the 1981 to 1985 map generally confirms the previously known range, but shows new evidence of nesting south of the Shield. As a composite of 5 years' data, the map overstates the distribution of this species; hence the 1981 to 1983 map on page 573 is intended to show a more typical distribution in southern Ontario.

During an irruption, numbers can increase dramatically (*e.g.*, 73 atlas records from south of the Shield in 1985 compared with only 7 in the period 1981 to 1984), but at other times the background population can be so low as to be almost undetectable. Most abundance estimates (75%) indicated between 2 and 10 pairs per square, with only 4 exceeding 10 pairs. These low estimates suggest that many irrupting birds were believed not to be involved in breeding attempts.-- *H.G. Lumsden and R.B.H. Smith*

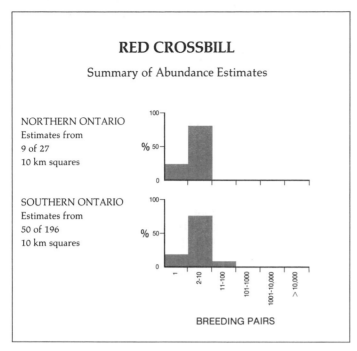

RED CROSSBILL

Summary of Abundance Estimates

NORTHERN ONTARIO
Estimates from
9 of 27
10 km squares

SOUTHERN ONTARIO
Estimates from
50 of 196
10 km squares

BREEDING PAIRS

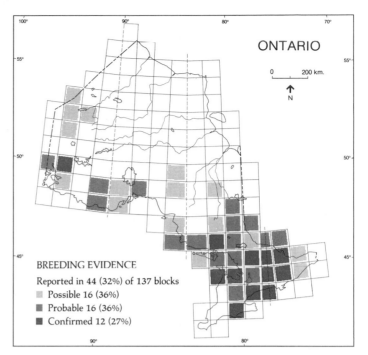

ONTARIO

0   200 km.

N

BREEDING EVIDENCE

Reported in 44 (32%) of 137 blocks

■ Possible 16 (36%)
■ Probable 16 (36%)
■ Confirmed 12 (27%)

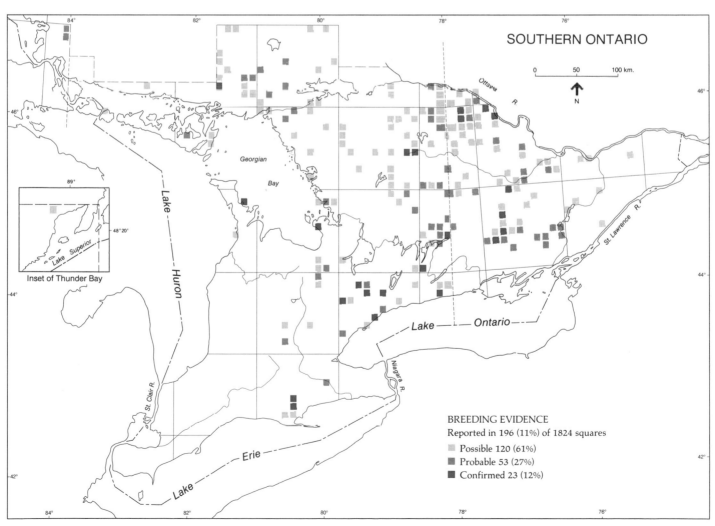

SOUTHERN ONTARIO

0   50   100 km.

N

Inset of Thunder Bay

BREEDING EVIDENCE

Reported in 196 (11%) of 1824 squares

■ Possible 120 (61%)
■ Probable 53 (27%)
■ Confirmed 23 (12%)

WHITE-WINGED CROSSBILL
Bec-croisé à ailes blanches
Loxia leucoptera

The White-winged Crossbill is found throughout the boreal forests of North America and Eurasia, with an isolated subspecies in Hispaniola. In Canada, it is present in coniferous forests from British Columbia to Newfoundland, while its Ontario range was mapped by Godfrey (1986) as extending from the southern edge of the Shield almost to the treeline. A record of 2 stubby-tailed, dependent young seen on 28 July 1969 in willows near the Brant River mouth (Lumsden unpubl.), suggests that the Ontario range extends to the Hudson Bay coast. Thus, the White-winged Crossbill occupies a substantially more northerly range than the Red Crossbill.

In contrast to the Red Crossbill's association with pines, the White-winged Crossbill exploits the cones of tamarack, spruce, and fir (Newton 1972), and eastern hemlock in Ontario (R. Tozer pers. comm.). In Siberia, a particular association with *Larix* spp. has been noted (Dement'ev and Gladkov 1952), while in southeastern Manitoba a preference for tamarack over black spruce seeds was observed during the winter of 1975-76 (Sealy *et al.* 1980). These requirements may account for its presence on the Hudson Bay Lowland, where tamarack and black spruce predominate. It is typically uncommon and unpredictable in Ontario, but during irruptions it can sometimes become locally numerous or even abundant. Major irruptions into southern Ontario were recorded in 1950-51, 1955-56, 1960-61, 1963-64, and 1971-72 (Speirs 1985), and another massive influx occurred in 1984-85.

Historically, the distribution and abundance of White-winged Crossbills has probably changed little in Ontario. Logging is unlikely to have had other than local effects on the population. Prior to 1956, a widespread die-off of tamarack occurred on the Hudson Bay Lowland (Lumsden unpubl.), but any effects on the White-winged Crossbill remain unknown. The breeding status of the White-winged Crossbill in Ontario was poorly known prior to the atlas, and only 2 nests had ever been documented (Peck and James 1987). One in Victoria Co on 19 August 1926 contained eggs (Speirs 1985), while the other, found on 20 August 1928 in Algoma District contained 3 young (Fargo and Trautman 1930). The possibility of winter nesting in Ontario was indicated by Lawrence (1949), who recorded singing males in January 1948.

At the start of its breeding cycle, the White-winged Crossbill performs song flights or sings from the tops of conifers. At this time it responds aggressively to tape recordings (Messineo 1985), and may be readily found. Its 'cheet...cheet' flight calls are somewhat similar to those of the Red Crossbill. The nest, which is difficult to find, is usually built in a thick spruce at heights ranging from 1.5 to 21 m (Harrison 1975). The only 2 reported during the atlas period were still under construction on 30 January and in early February 1983, when found near Achray in Algonquin Prov. Park. In the same area, 5 other females were observed carrying nest material, and 10 to 15 pairs were thought to be breeding nearby. They were foraging on an abundant crop of white spruce cones, but when an unusual warm spell in late February and early March caused the cones to open, the birds deserted their nests and left the area (Benkman 1985). This type of semi-colonial nesting has also been reported from New Brunwsick (Bent 1968), and from Chenango Co, New York in February 1985 (Messineo 1985). However, most atlas records occurred in late summer.

There were few records in 1981 or 1982, but numbers started to build up in 1983 and peaked in 1984, when 45% of all atlas records occurred. From mid-July 1984 onwards, there were reports of breeding activity over large areas of northern and southcentral Ontario, including 6 records of dependent young from the Algonquin region in late August and early September. Subsequently, White-winged Crossbills invaded areas south of the Shield, but unlike the Red Crossbill, there was little evidence of nesting in those areas in early 1985. Only one confirmed record was reported: a sighting of dependent young at Presqu'ile Prov. Park on 25 May 1985. However, breeding occurred in both New York State and Vermont around that time (Kibbe and Boise 1985, Messineo 1985).

The White-winged Crossbill proved difficult to atlas, since neither late winter nor late summer nesting coincided with peak atlassing times. It could have been missed in northern blocks not visited during 1984. This may explain some apparent gaps in the north. Nonetheless, the map confirms the previously known range, while the 1981-83 map on page 573 shows the sparse distribution prior to the irruption.

The majority of records (68%) were in the 'possible' category, and only 17 squares province-wide (6% of squares with records) reported confirmed breeding, mostly involving dependent young. Only 27% of squares with records provided abundance estimates, and none exceeded 100 pairs.-- *R.B.H. Smith and H.G. Lumsden*

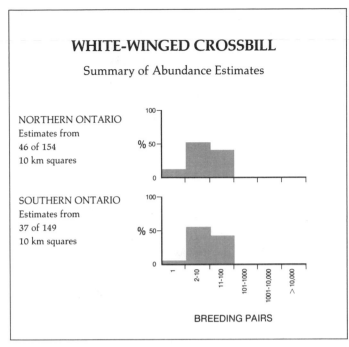

WHITE-WINGED CROSSBILL

Summary of Abundance Estimates

NORTHERN ONTARIO
Estimates from
46 of 154
10 km squares

SOUTHERN ONTARIO
Estimates from
37 of 149
10 km squares

% 50
% 50

1   2-10   11-100   101-1000   1001-10,000   > 10,000

BREEDING PAIRS

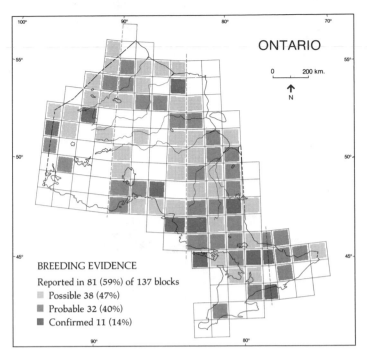

ONTARIO

0   200 km.

N

BREEDING EVIDENCE
Reported in 81 (59%) of 137 blocks
Possible 38 (47%)
Probable 32 (40%)
Confirmed 11 (14%)

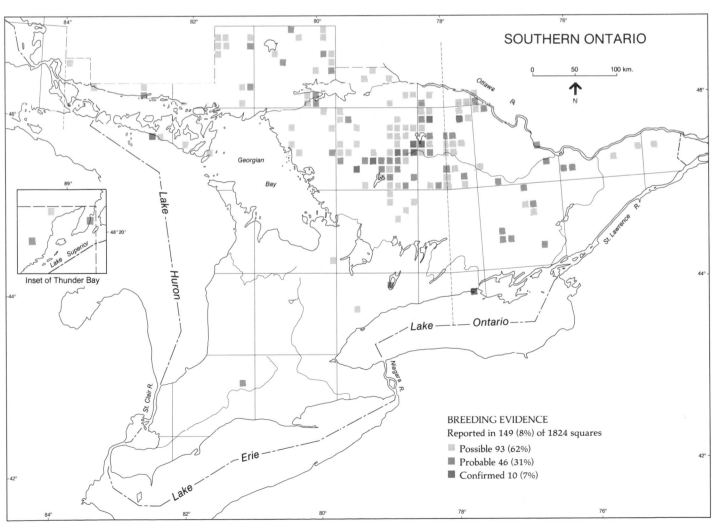

SOUTHERN ONTARIO

0   50   100 km.

N

Georgian Bay

Lake Huron

Lake Superior

Inset of Thunder Bay

St. Clair R.

Niagara R.

Lake Ontario

Ottawa R.

St. Lawrence R.

Lake Erie

BREEDING EVIDENCE
Reported in 149 (8%) of 1824 squares
Possible 93 (62%)
Probable 46 (31%)
Confirmed 10 (7%)

PINE SISKIN
Chardonneret des pins
Carduelis pinus

Because of its dark, streaked plumage, the Pine Siskin is sometimes thought of as a 'sparrow'. However, its sharply pointed bill and feeding habits identify it as a true finch. Its presence in an area is most easily recognized by its harsh, buzzy call, which is frequently described as "zwee-e-e-e-e-t" (Godfrey 1986). This species is social at all times of year and can be found in flocks even during the breeding season (Palmer 1968). The sexes are similar and cannot be easily identified in the field.

The precise breeding range of the Pine Siskin is difficult to define because it is typically a nomadic species and occurs irregularly throughout its range. Nevertheless, it breeds from coast to coast in North America in a band that roughly coincides with the mixed and Boreal Forest regions. With the exception of southwestern Ontario, the species has been previously reported to breed throughout most of the province north to a line running from Sandy Lake to Moosonee (James *et al.* 1976, Speirs 1985). The Pine Siskin breeds in coniferous and mixed woods of various densities. It also nests in urban areas in ornamental or shade trees (Godfrey 1986). Its distribution in winter is even less predictable, but it winters as far south as the Gulf states, northern Mexico and southern California (AOU 1983). In winter, it occurs erratically throughout the province, and, though abundant one year, it may be absent the next.

The nomadic, irruptive behaviour of this species makes its breeding status difficult to ascertain, because, even though it is common throughout the province at various times, its presence cannot be guaranteed. Pair formation is thought to occur within the flock. Nesting is heralded by courtship feeding in which the male offers food to the female. As with other Cardueline finches, nest-building and incubation are carried out entirely by the female. Nesting begins in late March or early April and breeding is usually complete by the end of June (Palmer 1968). The well concealed nest is usually at a medium height in the outer needles of a conifer, but deciduous trees are used from time to time. The nest is a fairly large, untidy cup of twigs, rootlets, and grass with a finer lining of hair, fur, fine rootlets, and fibres (Palmer 1968). Although single nests are occasionally encountered, nesting is usually semi-colonial. The clutch of 3 to 5 pale blue eggs hatches in about 13 days, and the young remain in the nest for a further 14 to 15 days. Fledglings remain dependent on the parents for about 3 weeks thereafter. Nests are difficult to find, and it is noteworthy that nests with eggs or young were reported from only 8 atlas squares, and confirmation of breeding from only 70 squares throughout the province.

The Pine Siskin breeds primarily on the Canadian Shield in Ontario. North of the Shield, on the Hudson Bay Lowland, there is only one confirmed breeding record. The records there are suggestive that breeding may extend further north than previously reported (Godfrey 1986), but, because those reports were from late June and July, they may reflect post-breeding dispersal. South of the Shield, breeding evidence, including several confirmed records, were reported from the Oak Ridges Moraine area north of Toronto, south and west through Waterloo. The most southerly confirmed breeding records are from Hamilton and St. Catharines. The cluster of records adhering to the outline of the north end of the Kingston region probably reflects more intensive atlas work there early in the season rather than more Pine Siskins than in surrounding areas. Confirmed breeding was also reported from Presqu'ile Prov. Park, and from several squares around the city of Ottawa, and along the coniferous western shore of the Bruce Peninsula. Southern breeding incursions are believed to follow winters of high abundance and were most frequent in the spring of 1985. The comparatively low confirmation rate for northern squares (4%) seems surprising. Almost certainly, many of the squares in which probable nesting was recorded should have contained nesting Pine Siskins. One can only presume that nesting was not confirmed because of the difficult habitat.

Although the Pine Siskin generally nests in a semi-colonial fashion, in some areas solitary nesting occurs (Middleton unpubl.). This may help to explain why 21% (a comparatively high figure) of squares in the south have estimates of only 1 pair per square. By contrast, in 57% of southern squares, density was estimated at between 2 and 10 pairs per square, which would appear to be a more typical estimate. These differing estimates point once more to the enigmatic biology of this species. Estimates suggest that the species is more common in northern Ontario, a predictable fact considering the abundance of suitable coniferous nesting habitat.-- *A.L.A. Middleton*

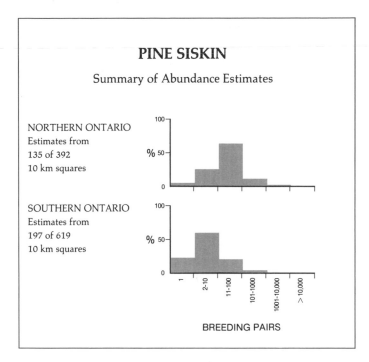

PINE SISKIN

Summary of Abundance Estimates

NORTHERN ONTARIO
Estimates from
135 of 392
10 km squares

SOUTHERN ONTARIO
Estimates from
197 of 619
10 km squares

%50

%50

BREEDING PAIRS

1 2-10 11-100 101-1000 1001-10,000 >10,000

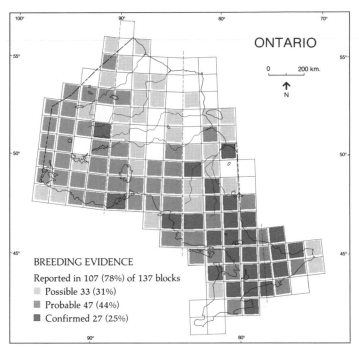

ONTARIO

0   200 km.

N

BREEDING EVIDENCE
Reported in 107 (78%) of 137 blocks

Possible 33 (31%)
Probable 47 (44%)
Confirmed 27 (25%)

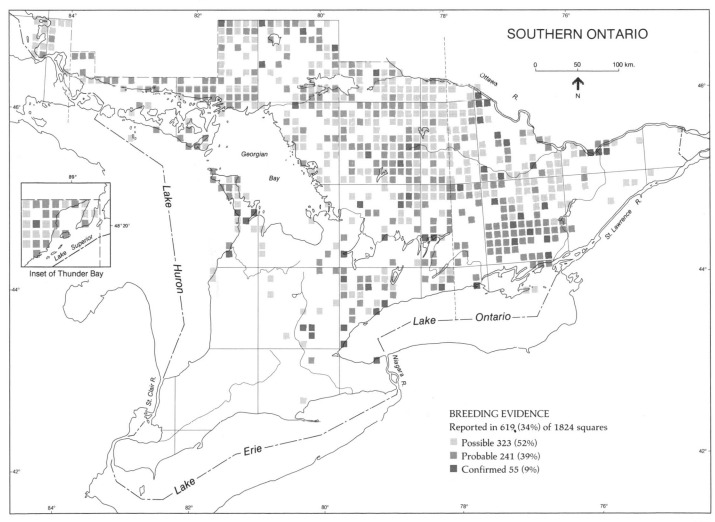

SOUTHERN ONTARIO

0   50   100 km.

N

Inset of Thunder Bay

Georgian Bay

Lake Huron

Lake Superior

Lake Ontario

Lake Erie

St. Clair R.

Niagara R.

Ottawa R.

St. Lawrence R.

BREEDING EVIDENCE
Reported in 619 (34%) of 1824 squares

Possible 323 (52%)
Probable 241 (39%)
Confirmed 55 (9%)

AMERICAN GOLDFINCH
Chardonneret jaune
Carduelis tristis

The American Goldfinch is unique because its plumage is the most seasonally and sexually dimorphic of all the members of its subfamily (Carduelinae). This results from the occurrence of 2 body moults during the year (Middleton 1977a). Because of the male's predominantly yellow summer plumage, the species has frequently been called the wild canary.

The American Goldfinch occurs from the Atlantic to Pacific Oceans northwards into the Boreal Forest zone and south to sub-tropical regions of North America (AOU 1983). In Ontario, the species is a common summer resident throughout southern parts of the province and in agricultural areas north to Sioux Lookout and Moosonee (Godfrey 1986). In winter, the American Goldfinch occurs south of North Bay (46^{0}N) and becomes increasingly abundant with decreasing latitude. The species occurs in varied habitats, but shows a preference for wood edges, weedy fields, and river bottomlands vegetated by serviceberry and hawthorns. In some areas, the species has become a common nester in suburbs, where plantings of immature maples and the seeds produced by garden plants provide good habitat and abundant food (Middleton 1979).

Breeding Bird Survey data suggest that the American Goldfinch is declining as a breeding species in parts of its range. However, in Ontario, Christmas Bird Count data show a significant increase in numbers over the past 60 years, probably in response to winter feeding by man (Middleton 1977b). It is likely that the American Goldfinch has become more widely distributed in historic times (Snyder 1957b).

The birds themselves are fairly conspicuous, especially the males, and their song and flight calls are well known. Generally, nesting does not begin until late June and eggs do not appear until July, at a time when most song birds have finished breeding for the year. Goldfinch nests can be fre-

quently overlooked because observers do not expect nesting at such a late date; nevertheless, nests with eggs or young were reported from 201 squares in Ontario. Late nesting is probably an adaptation to the annual seed crop of plants, such as thistles, upon which the young are almost exclusively fed. The compact nest of plant fibres and down is built entirely by the female, and is usually located in the terminal forks of deciduous shrubs or trees such as hawthorn, serviceberry, or maple. The nest is quite distinctive, and used nests were reported from 30 squares. Although the species is not territorial (only the immediate vicinity of the nest is defended), evidence of breeding can be quickly identified when the males undertake their circling, loose flapping (butterfly) flights over the breeding areas. Incubation is entirely by the female. Once begun, there is little activity around the nest as she incubates for long unbroken spells. The female is fed on the nest, by her mate, who may appear with food at irregular intervals of between 45 minutes and over 2 hours. However, once the eggs hatch, both parents participate in feeding the young (Stokes 1950, Nickell 1951). The species is not aggressive at the nest, but when disturbed will utter its distinct alarm call while showing agitated behaviour. These types of behaviour are useful clues when attempting to locate the nest.

Because the American Goldfinch is a bird of open, weedy areas with early shrub growth, it is not found in mature, undisturbed woodland. Thus the distribution, as shown in the maps, is realistic. It is common in squares in which agriculture is practised but diminishes as a breeding species in the Boreal Forest zone. In southern Ontario, it is most noticeably absent from northern Algonquin Prov. Park, and the islands along the Georgian Bay shoreline. These areas are characterized by forest and rocky outcrops with an absence of suitable seed plants. The high percentage of squares in which breeding is probable can best be explained by the late nesting of the species, which occurs after many atlassers would have completed field work for the season. It is almost certain that most squares that show possible or probable breeding in southern Ontario support nesting American Goldfinches. In the north, it is clear that the species is generally absent from the Hudson Bay Lowland and the northwestern section of the Boreal Forest region. Despite intensive atlas coverage of the Moosonee area, there was no indication that the American Goldfinch breeds at or near the southern shore of James Bay, contrary to the northern limit of the range shown in Godfrey (1986) for that area.

When present, the American Goldfinch often occurs in good numbers. This may reflect its gregarious nature at all times of the year. Abundance estimates indicate that this species is more common in southern Ontario than in the north. This is probably due to the wider availability of suitable habitat in the south.-- *A.L.A. Middleton*

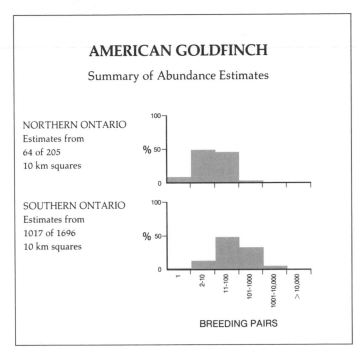

AMERICAN GOLDFINCH

Summary of Abundance Estimates

NORTHERN ONTARIO
Estimates from
64 of 205
10 km squares

SOUTHERN ONTARIO
Estimates from
1017 of 1696
10 km squares

%

BREEDING PAIRS

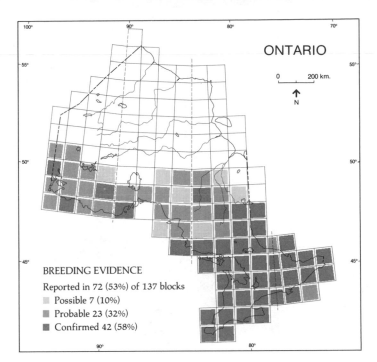

ONTARIO

0   200 km.

N

BREEDING EVIDENCE
Reported in 72 (53%) of 137 blocks

Possible 7 (10%)
Probable 23 (32%)
Confirmed 42 (58%)

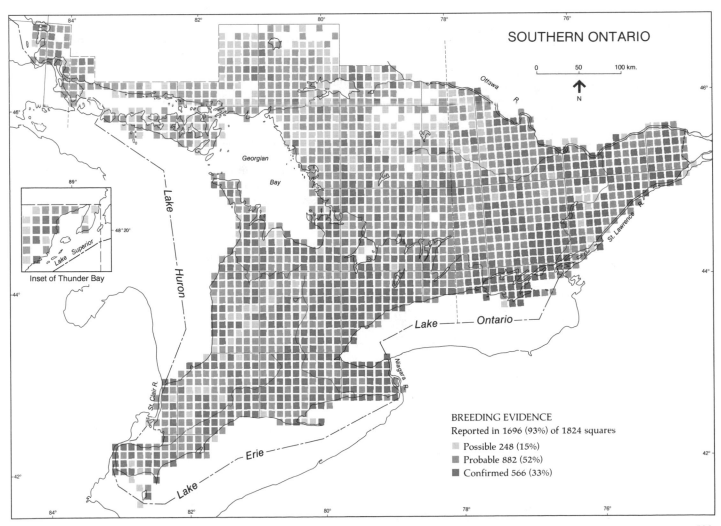

SOUTHERN ONTARIO

0   50   100 km.

N

Georgian Bay

Lake Huron

Inset of Thunder Bay

Lake Superior

89°

48°20'

St. Clair R.

Niagara R.

Lake Ontario

Lake Erie

Ottawa R.

St. Lawrence R.

BREEDING EVIDENCE
Reported in 1696 (93%) of 1824 squares

Possible 248 (15%)
Probable 882 (52%)
Confirmed 566 (33%)

EVENING GROSBEAK
Gros-bec errant
Coccothraustes vespertinus

In central Ontario particularly, the Evening Grosbeak is a familiar bird, especially at winter bird feeders. However, there was a lapse of 75 years from its first description as a species to the finding of the first nest, and a thorough knowledge of its breeding range and nesting biology is still surprisingly lacking.

The Evening Grosbeak breeds only in North America, from coast to coast across Canada, south in the east to the New England states, and in the western mountains south into Mexico. It winters throughout its breeding range and sporadically wanders south along the Atlantic seaboard to Florida and the Gulf of Mexico. In Ontario, it breeds mainly on the Canadian Shield, but wanders widely in other seasons throughout central Ontario and some southern portions of the province. The breeding habitat is primarily coniferous and mixed woodlands, and less often deciduous tree stands, parks, and orchards.

The Evening Grosbeak was originally described as a species in 1823. It was first reported in Ontario in August of the same year, in Thunder Bay District (Cooper 1825). It was next reported in Ontario at Toronto 31 years later, in 1854. The eastern population of this species has expanded during the last century. A total of 32 summer occurrence records from Ontario between 1913 and 1939 were listed by Baillie (1940), who chronicled its eastward extension in summer in Canada and the US. All but one of these records occurred in the 1920's and 1930's. This eastward range extension was in part due to the widespread planting of the Manitoba maple as a shade tree (Allen 1919), whose seeds are eaten by Evening Grosbeaks. Taverner (1921) described the situation as a "baited highway" along which the birds travelled eastward. Baillie (1940) considered the Evening Grosbeak to be an irregular breeder and summer resident along its breeding range belt, and an uncommon and locally distributed summer resident of Ontario in a narrow band extending from Algonquin Prov. Park to Lake of the Woods (Baillie and Harrington 1937). Godfrey (1986) stated that the breeding range of the species in Canada is based largely on summer occurrences, as definite breeding data are relatively few. Only 10 documented nests of the Evening Grosbeak have been reported to the ONRS (Peck and James 1987). The first nest was discovered at Clear Lake, Haliburton Co in 1944 (Peck 1973). Others were reported in the Nipissing and Thunder Bay Districts in 1945, 2 in Algonquin Prov. Park in 1946, in Nipissing District in 1973 and 1976, in Muskoka District in 1977, in Grey Co at Owen Sound in 1983, and in Timiskaming District in 1985. In all, nests with eggs or young were reported in 7 squares during the atlas period, representing 4% of all confirmations. Apparently, nests are infrequently found because of the considerable height at which they are typically situated, often in the dense foliage of conifers (Peck 1973).

Atlassing difficulties arise because the species is wide-ranging, sporadic in occurrence, often travels in pairs, and indulges in courtship feeding, display, and even copulation, in areas where no nests can be found. Thus, it is uncertain whether or not the birds nested in all atlas squares in which possible or probable breeding evidence was reported. Nevertheless, the atlas maps show a clearly defined summer range, and the location of known nests suggests that if any inappropriate codes were used, they have not affected the maps significantly.

The maps reflect data received from 1114 squares across the width of central Ontario, north to Moosonee in the east and Pickle Lake in the west, and in that part of southern Ontario on the Canadian Shield. Probable breeding records indicate that the breeding range may extend up to 200 km north of that shown by Godfrey (1986). Records are scattered south of the Shield, but reports of nests at Owen Sound and Lake Scugog during the atlas period indicate that the species may be extending its breeding range even farther south.

Abundance estimates provide evidence that, in suitable wooded habitat, the Evening Grosbeak is a fairly common breeding species.-- *G.K. Peck*

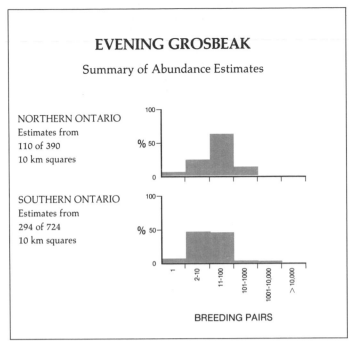

EVENING GROSBEAK

Summary of Abundance Estimates

NORTHERN ONTARIO
Estimates from
110 of 390
10 km squares

SOUTHERN ONTARIO
Estimates from
294 of 724
10 km squares

BREEDING PAIRS

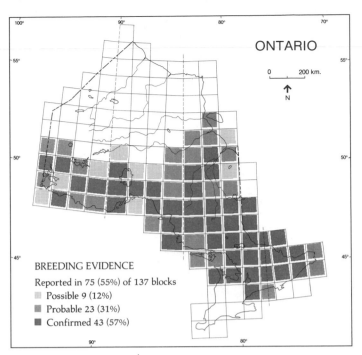

ONTARIO

0   200 km.

N

BREEDING EVIDENCE

Reported in 75 (55%) of 137 blocks

Possible 9 (12%)
Probable 23 (31%)
Confirmed 43 (57%)

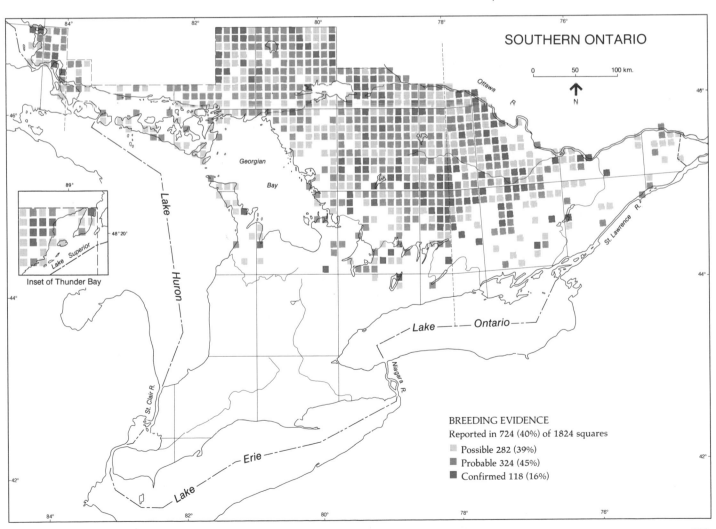

SOUTHERN ONTARIO

0   50   100 km.

N

Lake Huron

Georgian Bay

Lake Superior

Inset of Thunder Bay

St. Clair R.

Lake Erie

Lake — Ontario

Niagara R.

Ottawa R.

St. Lawrence R.

BREEDING EVIDENCE

Reported in 724 (40%) of 1824 squares

Possible 282 (39%)
Probable 324 (45%)
Confirmed 118 (16%)

HOUSE SPARROW
Moineau domestique
Passer domesticus

The House Sparrow is native to most of Eurasia south of tundra zones, where it frequents urban areas and farmlands. It has been introduced to North America, and to the Caribbean, southern South America, Australia, New Zealand, and southern Africa (Long 1981). Currently, it is resident and breeding in urban areas and farmland in virtually every square in southwestern and southeastern Ontario and in the southern half of southcentral Ontario. From the southern edge of the Canadian Shield northward, its distribution is patchy, centering on scattered communities. In the north, it is confined to a few urban areas and farms.

The House Sparrow was introduced from Great Britain and Germany into North America in metropolitan New York in the early 1850's. It spread from there by natural dispersal and by repeated introduction, being first imported into Ontario at Ottawa in 1870 (Barrows 1889). Since the 1880's, it has been widely regarded as an agricultural and aesthetic pest in North America. Its numbers increased dramatically into the 1920's, with the species achieving its present coast-to-coast range by 1940 (Long 1981). Subsequently, the overall abundance of the House Sparrow has declined in northeastern North America, which has been attributed in part to die-off in severe winters (Robbins *et al.* 1986). Atlas data show no decrease in range in southern Ontario. Except for the record in the Moosonee area, no additional breeding sites were confirmed in far northern Ontario than were previously reported (Barlow 1966). Today, the species continues to spread into remote areas around human habitation in the southwest deserts of the US.

The House Sparrow breeds from April to August in most inhabited parts of Ontario (Peck and James 1987), and is readily distinguishable from all other urban species by its non-musical chattering song. Because the House Sparrow occupies urban and agricultural habitats, and is noisy and conspicuous, it is easy to find and confirm as breeding. House Sparrows are largely sedentary, but may disperse from one site to another, which in the north may result in their absence for several years from previously occupied sites (Barlow 1966). Nests are mostly placed in man-made structures, and occasionally in natural cavities, as well as in trees - usually planted ornamentals. Once House Sparrows are detected in the breeding season, activities such as nest building, the carrying of food to nestlings or fledglings, or the presence of small flocks of birds in juvenal plumage are easily observed as the birds make little attempt at concealment. Nest-site search is facilitated by the loose colonial nesting habit of this species - with several nests placed close together in nearby cavities (Godfrey 1986). Atlassers were able to confirm breeding in 83% of 1,229 southern squares and 45% of 99 northern squares with breeding evidence, indicating the relative ease with which breeding can be authenticated.

Because of its sedentary nature, the presence of a pair of adult House Sparrows in the appropriate habitat is highly (at least 80%) predictive of breeding. Where birds have not been found in seemingly suitable urban areas on the Canadian Shield of the south and in large areas in northern Ontario, the likelihood that any breeding birds were overlooked must be regarded as very low. Given that the usual distances moved daily are not great - less than 4 km (North 1973, Will 1973) - and that the greatest distances travelled in North America are no more than a few dozen kilometres (Will 1973), coupled with the patchy distribution of suitable urban habitat, it is not surprising that House Sparrows have not occupied all potential breeding sites in the north. Distance, expanses of inhospitable forest, basic intolerance of extreme cold (Blem 1973), and the fact that dispersal is primarily by small numbers of immature birds in autumn (Will 1973) better explain the distribution of House Sparrow records in northern Ontario than any lapses on the part of atlassers.

In southern Ontario, the House Sparrow is a common to abundant permanent resident with 11 to more than 10,000 breeding pairs in evidence in 89% of squares for which abundance estimates are available. In all parts of northern Ontario, it appears to be absent or rare to locally common, depending on remoteness, severity of winter climate, abundance of food resources, shelter in winter, and chance.-- *J.C. Barlow*

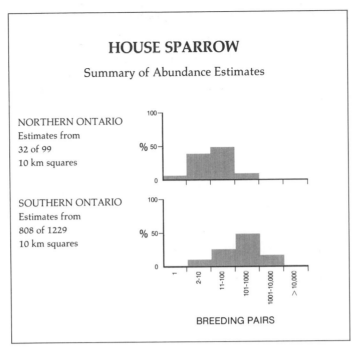

HOUSE SPARROW

Summary of Abundance Estimates

NORTHERN ONTARIO
Estimates from
32 of 99
10 km squares

SOUTHERN ONTARIO
Estimates from
808 of 1229
10 km squares

%

100

50

0

%

100

50

0

1   2-10   11-100   101-1000   1001-10,000   > 10,000

BREEDING PAIRS

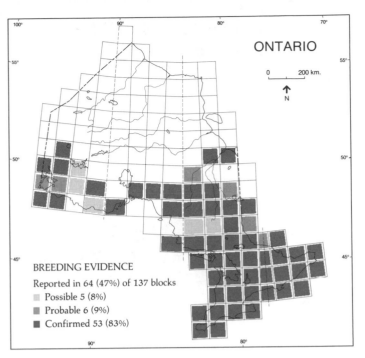

ONTARIO

0   200 km.

N

BREEDING EVIDENCE
Reported in 64 (47%) of 137 blocks
Possible 5 (8%)
Probable 6 (9%)
Confirmed 53 (83%)

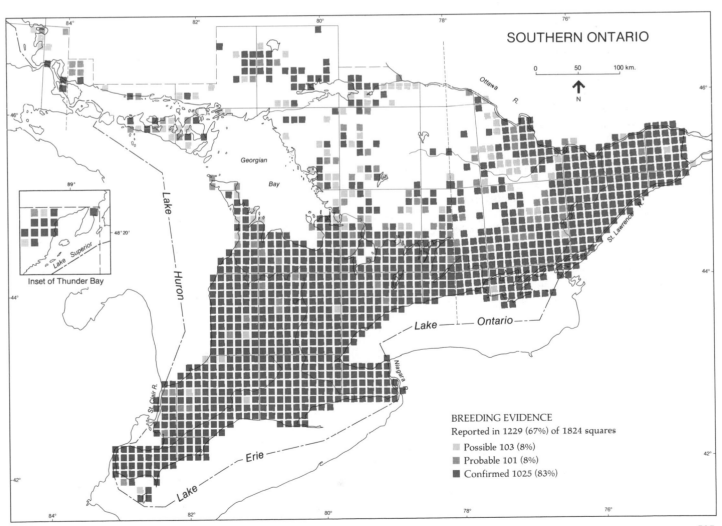

SOUTHERN ONTARIO

0   50   100 km.

N

Inset of Thunder Bay

BREEDING EVIDENCE
Reported in 1229 (67%) of 1824 squares
Possible 103 (8%)
Probable 101 (8%)
Confirmed 1025 (83%)

RED-THROATED LOON
Huart à gorge rousse
Gavia stellata

The Red-throated Loon is the most northerly of our loons, breeding across the northern part of the Nearctic and Palearctic regions, north to Ellesmere Island. It winters south of its breeding range, mainly at sea along both the Atlantic and Pacific coasts. It nests on freshwater lakes and ponds, most often in coastal tundra, but also south of the Tundra region on coastal flats, and less frequently on interior lakes in both treeless and forested areas. In Ontario, it breeds primarily along our northern sea coast, with 2 undocumented and isolated breeding records reported along the north shore of Lake Superior, south of the more normal range.

Breeding evidence for the Red-throated Loon in Ontario was first reported on 1 July 1912, with the undocumented finding of a nest with two eggs near Thunder Cape on Lake Superior (Baillie and Harrington 1936). A later sighting of adults with 2 young at Rossport on Lake Superior, in July 1941, also went undocumented (Baillie and Hope 1943). A flightless young attended by its parents was seen on Cape Henrietta Maria by R.H. Smith on 22 August 1944 (Smith 1944). A nest with 2 eggs was found on 3 August 1962 near Hudson Bay, west of Cape Henrietta Maria, by D.W. Simkin, who later established the first documented provincial breeding, by photographing adults and 2 downy young, on 29 July 1966, also west of Cape Henrietta Maria (Simkin 1968).

The Red-throated Loon generally prefers more arctic conditions than the Pacific Loon, and this undoubtedly accounts for the fewer breeding records in the province. The Red-throated Loon is a solitary nester, usually on smaller, shallower ponds than the Pacific Loon. It feeds customarily, if not always, away from the nest pond, often at sea, and birds passing overhead are easily detected by their quacking flight calls. On the nest pond, it usually allows close approach before taking flight, so the presence of breeders is unlikely to

have gone undetected by atlassers who were in the vicinity of a nesting territory.

There were records from 12 squares in 9 blocks during the atlas period. Records came from 6 squares in the vicinity of Cape Henrietta Maria, where breeding was confirmed in 2 squares. One nest from which observed young had recently departed was found on 9 July 1984, near James Bay, 40 km south of Cape Henrietta Maria, and 2 nests, each containing eggs, were recorded on Cape Henrietta Maria on 30 June 1985 (Peck and James 1987, atlas data). The remaining 5 records of possible and probable breeding were scattered in 5 blocks on the Hudson Bay Lowland, and along the Hudson Bay coast, suggesting the possibility of at least occasional breeding in those areas.

The Red-throated Loon is rare to uncommon in its northern Ontario breeding habitat, even in the Cape Henrietta Maria area where it has been reported most frequently. In 1944, R.H. Smith observed only one young and 6 adult Red-throated Loons in the tundra area of Cape Henrietta Maria, compared with 4 young and at least 35 adult Pacific Loons (Smith 1944). These observations are consistent with recent records.-- *G.K. Peck*

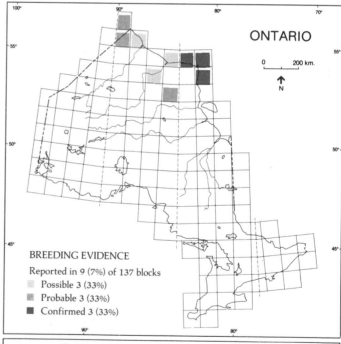

ONTARIO

0   200 km.

N

BREEDING EVIDENCE

Reported in 9 (7%) of 137 blocks

Possible 3 (33%)
Probable 3 (33%)
Confirmed 3 (33%)

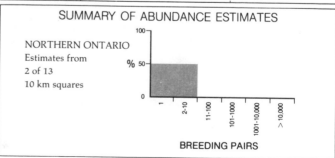

SUMMARY OF ABUNDANCE ESTIMATES

NORTHERN ONTARIO
Estimates from
2 of 13
10 km squares

BREEDING PAIRS

PACIFIC LOON
Huart du Pacifique
Gavia pacifica

breed.

During the atlas period, breeding was confirmed in 10 squares west of Winisk, extending the known Ontario breeding range of the species to within 20 km of the Manitoba border. A nest with young was reported on East Pen Island in 1985, and flightless young were reported on the adjacent mainland. Breeding was confirmed in 3 coastal squares in another block west of Fort Severn. Between Winisk and Fort Severn, breeding was confirmed in 5 squares, and possible and probable records were reported in 2 more, all in the same 100 km block: all squares were coastal. Breeding was confirmed in 8 squares east of Winisk, including nests near radar sites 415 and 416 in 1984, and on Cape Henrietta Maria in 1985; probable breeding was reported in 6 other squares east of Winisk. The most surprising record is a probable breeding record for a Pacific Loon over 100 km inland near the Black Duck River along the Manitoba border.

The single abundance estimate of more than 10 pairs came from the Cape Henrietta Maria square. Elsewhere, 2 to 10 pairs were estimated to occur in 13 squares, 4 of which are west of Winisk.-- *G.K. Peck*

The Pacific Loon breeds in areas of tundra and muskeg across the Nearctic and the eastern portion of the Palearctic, and winters south of its breeding range, mainly off the Pacific coast. In Ontario, it breeds only along our northern coast; the limited extent of the tundra/tree-line area of Ontario is the major factor limiting the range and abundance of this species in the province. In suitable habitat, such as the Cape Henrietta Maria region in Polar Bear Prov. Park, many of the larger tundra sloughs have a breeding pair (Peck 1970, 1972).

The first evidence of breeding of the Pacific Loon in Ontario was the observation of 3 broods of flightless young at and near Cape Henrietta Maria, 16 to 21 August 1944 (Smith 1944, Baillie 1961). The first nests (6 in total) were found on the tundra near the abandoned radar site 416 in the Cape Henrietta Maria region, between 23 June and 2 July 1970 (Peck 1970, 1972). Prior to the atlas project, summer occurrences were also noted at Long Ridge Point between Moosonee and Fort Albany, and as far west as Fort Severn, while breeding was confirmed as far west as Winisk (Peck and James 1983).

The Pacific Loon nests usually on larger sloughs and lakes than does the Red-throated Loon. Nesting is solitary. Pacific Loons are often conspicuous and can be observed from a great distance in open tundra habitat. Upon close approach by an intruding human or an aircraft, breeding birds perform a surprising and spectacular distraction display consisting of a "splash dive" preceded by a yelp (Palmer 1962, Cramp and Simmons 1977); this display was used to confirm breeding in 6 squares. Family groups are easily observed and remain together until autumn. Non-nesting birds are present in the breeding range, and it is conceivable that some possible breeding records are of birds that did not

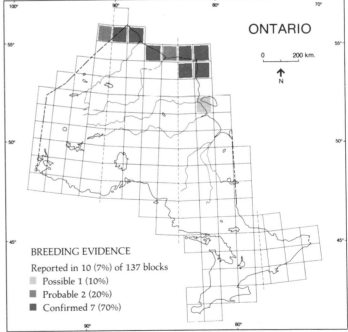

ONTARIO

BREEDING EVIDENCE

Reported in 10 (7%) of 137 blocks

- Possible 1 (10%)
- Probable 2 (20%)
- Confirmed 7 (70%)

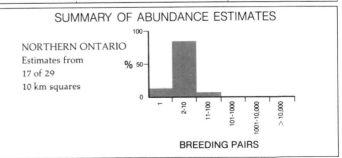

SUMMARY OF ABUNDANCE ESTIMATES

NORTHERN ONTARIO
Estimates from
17 of 29
10 km squares

BREEDING PAIRS

AMERICAN WHITE PELICAN
Pélican blanc d'Amérique
Pelecanus erythrorhynchos

few fish of economic importance. Any continuing disturbance at a breeding colony, where its ground nests are very vulnerable, could lead to nest desertion and ultimately to the breakup of the colony. Its wandering tendencies at any time of the year, together with the fact that it often forages at long distances from the breeding colony (Macins 1977), are reasons for reports of the species in locations where it is not known to breed; breeding season reports were received from 3 blocks adjacent to that containing the Lake of the Woods colony. All were thought to be foraging birds and are not shown on the map.

The Canadian populations of the American White Pelican were thought to be undergoing a long-term decline by the 1970's, and the species was designated as being 'threatened' in this country by COSEWIC (Markham 1978). However, the national populations have since stabilized or increased, and the species has recently been delisted (Koonz 1987). Despite a continuing increase in the Ontario population, the breeding areas in this province are highly localized, and the species is still considered to be 'endangered' in Ontario by the MNR.-- *G. K. Peck*

The American White Pelican is one of North America's biggest and most conspicuous birds, and its 3 m wingspread is the largest of any Canadian bird. It usually breeds on remote islands in fresh water lakes, but occasionally in brackish or salt water. The breeding range includes the western Canadian provinces and some western states between latitudes 27^0N and 60^0N. In Ontario, the American White Pelican is known to breed at only one location, at the southern end of Lake of the Woods in Kenora District. It generally winters south of its breeding range along the Pacific and Gulf coasts, as far south as Guatemala (Palmer 1962).

Although the American White Pelican was reported from Lake of the Woods as long ago as 1775 (Henry 1809), it was first reported nesting there in June 1938, at which time about 8 pairs were found with nests containing eggs, on Dream Island (Baillie 1939). Despite periodic visits to this area, no nesting White Pelicans were reported there from 1940 to 1953. From 1954 on, an increasing number of Pelicans (up to 500) nested on 2 small islands south of Dream Island. These sites were abandoned in 1968 and all the nesters moved to 2 of the Three Sisters Islands, 23 km south of Dream Island. The breeding population there has increased dramatically, from about 235 individuals in 1968, to about 1,684 pairs by the start of the atlas period (1981), and to over 6,500 pairs by 1986 (Macins 1977, pers. comm.). A new colony has recently (1982) been reported from Burton Island, several kilometres southeast of the Three Sisters Islands (Ryder *et al.* 1983).

The species is colonial and breeds on small, relatively remote islands on permanent water, which are usually free from disturbance by man. In Ontario, it often nests in company with Double-crested Cormorants and Herring Gulls. In the past, it was persecuted by fishermen, and without adequate protection might still be, even though its diet includes

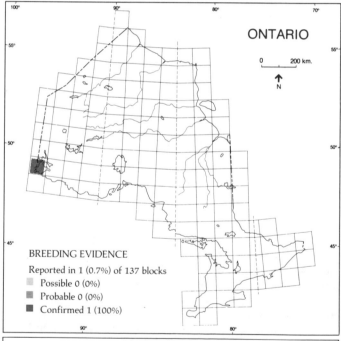

ONTARIO

0   200 km.

BREEDING EVIDENCE

Reported in 1 (0.7%) of 137 blocks

Possible 0 (0%)
Probable 0 (0%)
Confirmed 1 (100%)

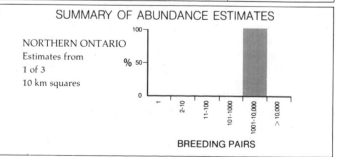

SUMMARY OF ABUNDANCE ESTIMATES

NORTHERN ONTARIO
Estimates from
1 of 3
10 km squares

BREEDING PAIRS

TUNDRA SWAN
Cygne siffleur
Cygnus columbianus

The Tundra Swan (formerly known as the Whistling Swan in North America) has a circumpolar breeding distribution. In North America, it breeds from Baffin Island to Alaska, and winters on the Atlantic and Pacific coasts and formerly on the Gulf coast, where small numbers are again returning. The Eurasian race is known as the Bewick's Swan in Britain.

The Ontario coast of Hudson Bay supports the southernmost breeding population of Tundra Swans in the world. The species breeds within 40 km of the Hudson Bay coast between James Bay and the Kinusheo River, 100 km to the west. Breeding west of the Kinusheo to the Manitoba border is seldom more than 5 km from the coast. The species migrates across most of Ontario, particularly through the southern Great Lakes. The Tundra Swan breeds on rich tundras, and rarely in open areas south of the tree-line. Nests are placed on islands and peninsulas in tundra ponds and lakes, and near sluggish rivers, but sometimes may be up to 200 m from water. They are built on dry sites, but in swampy tundra they are on hummocks which may be surrounded by melt water at breakup (M. McLaren pers. comm.).

The former abundance of the Tundra Swan is unknown, but it was reduced to low numbers in the early years of this century. Recovery started with the protection provided by the Migratory Bird Treaty of 1916. Winter inventories throughout the continent were started in 1949; the lowest number recorded was 49,000 in 1950. A substantial upward trend, with some setbacks, has been recorded since. The winter inventory in 1984-85 indicated a total of 143,000, of which 92,000 were on the Atlantic and 49,000 on the Pacific coast (USFWS and CWS 1985). Breeding Tundra Swans were extirpated in Ontario and the southern Hudson Bay region during the early days of the fur trade. The first recent breeding

record was in 1973, but in 1984 there were 20 to 25 breeding pairs in northern Ontario (Lumsden 1984b).

In Ontario, eggs are laid in late May and hatch after about 32 days, in late June or early July. The fledging period varies with latitude and day length, being about 45 days at 70°N (Scott 1972). The nesting period therefore gives an atlasser almost 3 months during which breeding can be confirmed. From the air, the Tundra Swan on the breeding grounds is conspicuous and easy to see from a distance of up to 1.1 km (Lumsden 1984). Those with broods do not flush from the aircraft. On the ground, it is visible from considerable distances over the flat tundra. It may be heard as far away as 6.5 km under favourable conditions (Palmer 1976a).

Breeding evidence was reported for 33 squares in 9 blocks. Most of the data came from observations of broods derived from aerial surveys, and represent the breeding distribution very accurately. There is much suitable habitat unoccupied at present, into which Tundra Swans will probably move in the next few years as the population recovery continues. Productivity of the Ontario stock is likely to be higher than at more northerly latitudes and we should expect a relatively rapid increase.-- *H.G. Lumsden*

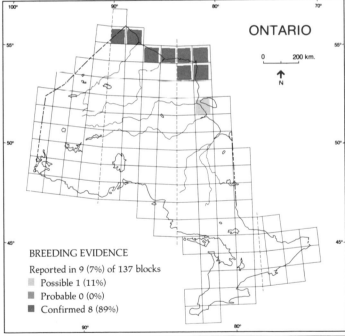

ONTARIO

BREEDING EVIDENCE
Reported in 9 (7%) of 137 blocks
- Possible 1 (11%)
- Probable 0 (0%)
- Confirmed 8 (89%)

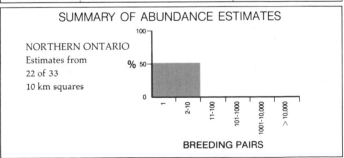

SUMMARY OF ABUNDANCE ESTIMATES

NORTHERN ONTARIO
Estimates from
22 of 33
10 km squares

BREEDING PAIRS

SNOW GOOSE
Oie des neiges
Chen caerulescens

The Snow Goose and the Canada Goose may be the most abundant geese in the world. After a good breeding season as many as 4 million Snow Geese leave the breeding grounds.

The Snow Goose breeds from northwestern Greenland across the islands and mainland coasts of Canada and Alaska to Wrangel Island in Siberia (AOU 1983). It breeds only north of the tree-line on braided river mouths and pond tundra. The Lesser Snow Goose (*C. c. caerulescens*) is the subspecies which breeds in Ontario.

The Snow Geese in the eastern arctic have been increasing in numbers for the past 50 years. The development of agriculture on the spring staging grounds in southern Manitoba and northern North Dakota has provided improved nutrition for the breeding pairs on their way north. The wintering grounds have also been expanded by the development of agriculture on the coastal prairies of Louisiana and Texas.

The first nesting recorded in Ontario was of about 100 pairs west of Cape Henrietta Maria in 1944 (Smith 1944). However, Indians report that nesting occurred well before that. By July 1957, the Cape Henrietta Maria colony had grown to 17,300, and in 1968 to 40,000 birds (Hanson *et al.* 1972). In June 1973, aerial photography yielded an estimate of 29,600 nests (Kerbes 1975), and in 1979 of 54,620 nests (Angehrn 1979). Hundreds and, in some years, thousands of Snow Geese breed away from the main colony on the Ontario coasts of James and Hudson Bays. The blue colour phase of the Snow Goose constitutes 70 to 75% of the breeders at Cape Henrietta Maria.

In the Cape Henrietta Maria colony in 1973, Snow Goose nests were spaced about 30 m apart and were built on islands, or near the shores of lakes and ponds. They were situated in the shelter of stunted willows, near a rock, or

often in the open. After hatch, the broods may walk great distances to graze on the sedge flats near the tideline or move as much as 25 km inland to feed on grassy tundra. Family bonds are tight, and the young accompany their parents throughout the fall and winter. Nesting areas are generally occupied year after year, with females remaining relatively faithful to their home colony, and even their previous nest site. Pairing takes place on the wintering grounds where geese from many colonies mix, so that females may bring back to their natal colony a male that may have hatched thousands of kilometres away.

The atlas map, largely compiled from aerial observations of broods seen in July, gives a good picture of breeding distribution. The great majority of Ontario's Snow Geese breed in the large Cape Henrietta Maria colony, which occupies at least 8 squares in the 3 most northeasterly atlas blocks. The most southerly colony is on Akimiski Island, NWT. Elsewhere, small colonies of 2 to 100 pairs occurred mostly at river mouths from the northwestern shore of James Bay to the Manitoba border. There may, however, be some years with late springs when some of the blocks shown on the map do not have nesting Snow Geese.-- *H.G. Lumsden*

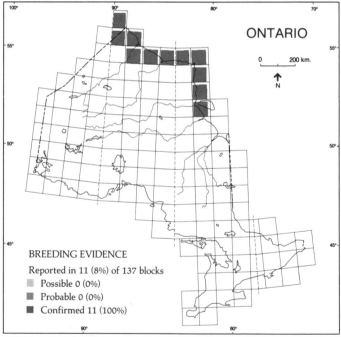

BREEDING EVIDENCE
Reported in 11 (8%) of 137 blocks
☐ Possible 0 (0%)
☐ Probable 0 (0%)
■ Confirmed 11 (100%)

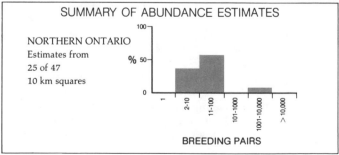

SUMMARY OF ABUNDANCE ESTIMATES

NORTHERN ONTARIO
Estimates from
25 of 47
10 km squares

BREEDING PAIRS

ROSS' GOOSE
Oie de Ross
Chen rossii

Juliana Hawke

This miniature replica of the far more common Snow Goose was first discovered breeding in Ontario in 1975 (Prevett and Johnson 1977). It is strictly a North American species, and, except for occasional extralimital records, was thought to be restricted to the central Canadian Arctic in summer, and California in winter. Then, beginning in 1956, Ross' Geese were found nesting in several Snow Goose breeding colonies along the western periphery of Hudson Bay. Although eastern records date back as far as 1771, it is clear that Ross' Geese have increased in the Hudson Bay area since about 1960, and that this increase is at least partly a result of eastward spread from the main west-central North American population (Prevett and MacInnes 1972).

Migrant Ross' Geese were shot along the southern James Bay coast in Ontario as early as 1953, and many records have accumulated thereafter, from both Hudson and James Bays (Lumsden 1963). However, despite the banding of 20,665 Snow Geese, the first flightless (moulting) adults and prefledging juvenile Ross' Geese in Ontario were not captured until 1975 at the large Snow Goose colony west of Cape Henrietta Maria (Prevett and Johnson 1977). On 13 July 1984, a flightless adult male and 2 young Ross' Geese were caught during Snow Goose banding on Akimiski Island, NWT in James Bay (K. Abraham pers. comm.), providing the only record for the atlas project.

Ross' Geese in eastern North America nest within huge breeding colonies of Snow (including Blue) Geese. Their rarity and close resemblance to Snow Geese make identification extremely difficult except to the most experienced observer. To complicate matters, Ross' and Snow Geese commonly hybridize. In addition, the tundra habitat of the Cape Henrietta Maria region (the most southerly location of true arctic tundra in the world) and Akimiski Island is both remote and inaccessible. These factors seriously reduce chances of encountering and detecting breeding Ross' Geese in Ontario.

Small numbers of Snow Geese nest at other scattered locations along the James and Hudson Bay coasts of Ontario. It is possible that a few Ross' Geese may sometimes also breed among them. However, given their relative overall rarity compared to Snow Geese, these occasions would likely be uncommon and episodic.-- *J.P. Prevett*

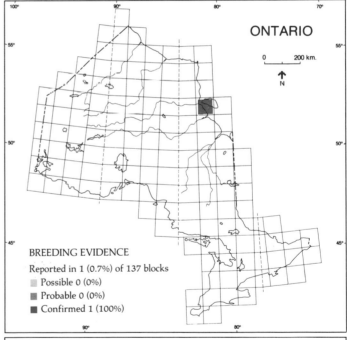

ONTARIO

BREEDING EVIDENCE
Reported in 1 (0.7%) of 137 blocks
▪ Possible 0 (0%)
▪ Probable 0 (0%)
▪ Confirmed 1 (100%)

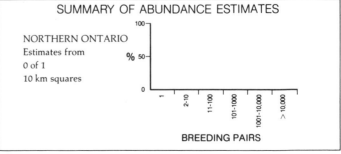

SUMMARY OF ABUNDANCE ESTIMATES

NORTHERN ONTARIO
Estimates from
0 of 1
10 km squares

BREEDING PAIRS

GREATER SCAUP
Grand Morillon
Aythya marila

Three of 4 blocks with confirmed records are in areas of coastal tundra, as expected. Breeding likely also occurs in other blocks on and adjacent to the tundra. The fourth confirmed record was from East Opinnagau Lake, a more inland location. The records from the Sachigo Lake area in northwestern Ontario may be migrant birds, although there is a chance that they would breed there, for they are known to nest on Lake Winnipeg, considerably south of typical tundra habitat. Because of the late nesting season in cold northern areas, birds that are seen even as late as June, away from tundra areas, cannot be assumed to be breeding.

The area of greatest abundance of Greater Scaup in Ontario is probably in the Cape Henrietta Maria area, where suitable habitat is most extensive. There could certainly be 11 to 100 pairs in a number of squares in this area, and possibly more than 100 in certain squares. Numbers are likely to be small along most of the north coast, but probably increase again in the extreme northwest of the province closer to the main breeding range.-- *R.D. James*

The Greater Scaup is a widespread species, whose breeding range extends around the world in subarctic and arctic regions, largely north of 60°N. In North America, it breeds widely from Alaska east to Hudson Bay, with greatest concentrations in coastal western Alaska. There are apparently only smaller and scattered populations summering in northern Quebec and Newfoundland south to Anticosti Island. Greater Scaup winter largely on the Atlantic and Pacific coasts. Smaller, but significant, numbers winter on the Great Lakes and in the Mississippi Valley south to the coast of the Gulf of Mexico.

During the nesting season, it is found on ponds, small lakes, and large rivers in forested areas and on open tundra, with its greatest abundance near the tree-line.

It is probable that the Greater Scaup has nested in Ontario for centuries. It was breeding on the east coast of James Bay in 1890 (Todd 1963), but only in 1940 was the first breeding record obtained in this province, when a female with a shelled egg in its oviduct was collected at Fort Severn; young birds were collected in the Cape Henrietta Maria region in 1948 (Peck and James 1983). The first nests reported in the province were not found until 1984. The author found them relatively easily by walking around appropriate water bodies and flushing the females. However, there are few areas where birds are common enough for this method to be effective, so it is not surprising that they went unnoticed previously. Female scaup with broods are likely to be difficult to flush, making positive identification of the bird difficult. Broods of Greater Scaup are also likely to be late in appearing (late July or August) and few atlassers were still in the field at that time. All these factors combine to limit the number of atlas records available, and hence our understanding of the distribution of the species.

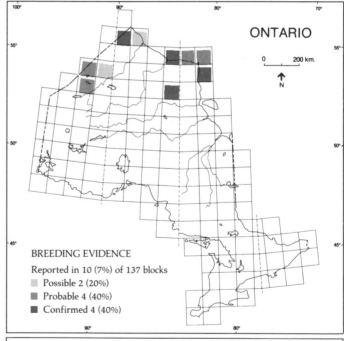

ONTARIO

0   200 km.

↑
N

BREEDING EVIDENCE

Reported in 10 (7%) of 137 blocks

Possible 2 (20%)
Probable 4 (40%)
Confirmed 4 (40%)

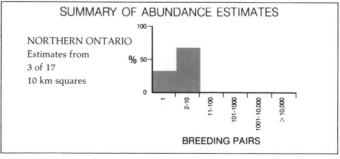

SUMMARY OF ABUNDANCE ESTIMATES

NORTHERN ONTARIO
Estimates from
3 of 17
10 km squares

%

BREEDING PAIRS

COMMON EIDER
Eider à duvet
Somateria mollissima

The Common Eider is a widespread holarctic sea duck, exploited by man in several parts of its range for its warm insulating down feathers, which are collected from the nest. Common Eiders breed from about 45°N on the Atlantic coasts of North America and Europe north to about 80°N in the high arctic. The species has been separated on the basis of plumage and size differences into several subspecies, one of which, the Hudson Bay Eider (*S. m. sedentaria*), nests sparsely in Ontario. The Hudson Bay Eider is the largest subspecies. It winters in the cold waters of Hudson Bay and probably migrates little from its breeding grounds, hence its subspecies name. During the breeding season, it is found only in coastal tundra areas, well away from the tree-line. Nests have been reported only on islands in shallow tundra lakes and off the shore of Hudson Bay.

The first evidence of nesting in Ontario was the sighting of a female with one young at Cape Henrietta Maria on 23 August 1944 (Smith 1944). Subsequently, breeding was documented at East Pen Island in 1960 and 1963 (Baillie 1963b).

Much of our knowledge of the breeding of the Hudson Bay Eider comes from Freeman (1970a) and Schmutz *et al.* (1983). High nesting densities (up to 150/ha) are found on offshore islands or islets in river deltas, but some birds use mainland sites. Unless these islands and deltas are visited, the species can be easily overlooked. The birds are somewhat colonial and usually nest close to water, often in willow or birch shrub cover (Freemark 1977) or beach grass (Freeman 1970a), sometimes in close association with Herring Gulls or Arctic Terns. Males leave soon after incubation begins, and congregate in moulting flocks in coastal regions near the nesting area. Females and their broods also move to the sea, although bodies of fresh water may be used by young broods. Non-breeding or failed breeding females often accompany the broods.

Atlas records conform to the distribution established previously. Specific sites of confirmed breeding during the atlas period were on East Pen Island (near the Manitoba border), near the mouth of the Sutton River, on Little Bear Island, NWT (55°16′N, 82°58′W), on Cape Henrietta Maria and along the James Bay coast about 1 km south of the foot of Cape Henrietta Maria. Seventy-nine nests were counted on Little Bear Island in 1981 and 26 on Cape Henrietta Maria in 1985. Birds were also observed in suitable habitat in two other squares in the Cape Henrietta Maria region.

Abraham and Finney (1986) estimate the Hudson Bay Eider population at 23,000 pairs, but they admit that this is very tentative and may be an underestimate. Most of these nest in Quebec and the islands in Hudson and James Bay. They estimate 300 to 400 pairs in James Bay and 600 to 800 pairs in southern Hudson Bay, most of the latter nesting in Manitoba. From their account and the atlas data, it is evident that the Common Eider is not a common nester in Ontario.-- F. Cooke

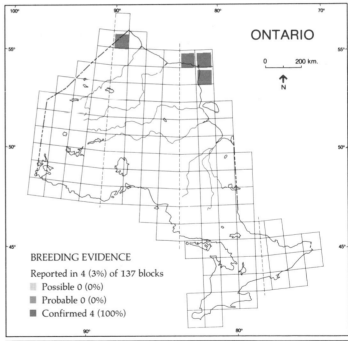

BREEDING EVIDENCE

Reported in 4 (3%) of 137 blocks

Possible 0 (0%)
Probable 0 (0%)
Confirmed 4 (100%)

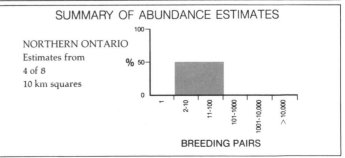

SUMMARY OF ABUNDANCE ESTIMATES

NORTHERN ONTARIO
Estimates from
4 of 8
10 km squares

BREEDING PAIRS

KING EIDER
Eider à tête grise
Somateria spectabilis

The King Eider is a holarctic bird with, on average, a more northerly breeding and wintering range than the Common Eider. Breeding typically in the high arctic islands of northern Canada, it migrates to wintering grounds from Newfoundland south to Maine (Palmer 1976b), and along the coasts of Alaska and the Aleutian Islands. Its more migratory behaviour, in contrast to that of the Hudson Bay Common Eider *(S.m. sedentaria)* may explain why it is seen more frequently in winter on the Great Lakes. Although commoner as a winter visitor to southern parts of Ontario, it is rarer as a breeding bird on our northern sea coast.

In contrast to the Common Eider, the King Eider is a solitary nester, using a variety of habitats. In general, it uses drier, well-drained sites and is often found nesting a considerable distance from water. Not surprisingly, such habitats are common in the high arctic and less frequent on the southern part of Hudson Bay.

The earliest evidence of King Eiders nesting in Ontario was the observation of a female with a brood on Cape Henrietta Maria on 23 August 1944 (Smith 1944). Seven other broods were also thought to have been this species, but may have been Common Eiders. Manning (1952) found 2 broods at the Cape in 1947, one of which hatched from the first nest reported in Ontario (Peck and James 1987). Although summering birds have been reported, no other breeding records were known prior to the atlas period (Peck 1972, Peck and James 1983).

In general, the breeding biology of the King Eider is similar to that of the Common Eider, with clutch sizes of 2 to 6 eggs. The males leave early in the incubation period and females incubate alone. At this stage, the close-sitting and cryptic females could easily be overlooked. Breeding evidence is most easily obtained by observing females with broods of

downy young on fresh water ponds, provided that they can be approached closely enough to be sure that they are not Common Eiders. Broods may be taken to either fresh or salt water for brood rearing, but eventually families move to salt water, where they can often be seen in multi-family assemblages (Parmelee *et al.* 1967).

Atlas data conform to the pattern of previous records, and indicate that the King Eider is a rare breeder on Ontario's tundra. There was only one report of breeding: the single record was of a female with a brood at the mouth of the Sutton River, about 90 km west of Cape Henrietta Maria, in July 1983. Atlassers saw no King Eiders at Cape Henrietta Maria from 20 June to 6 July 1985. This may indicate either a decline of the small population reported in the 1940's or that breeding is irregular.

Abraham and Finney (1986) made no attempt to assess the population of King Eiders in southern Hudson Bay, but it is likely to be very low. It was only in 1977 that the first King Eider nest was reported for Manitoba, where a pair of King Eiders was found within the Common Eider colony at the Mast River delta at La Perouse Bay (Abraham and Cooke 1979). There is no evidence of any change in status over time.-- *F. Cooke and D.J.T. Hussell*

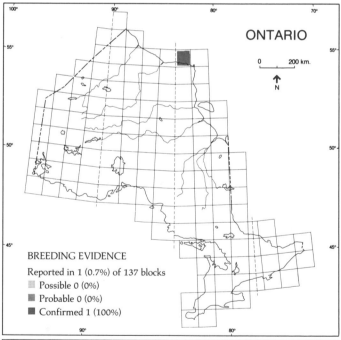

ONTARIO

0   200 km.

N

BREEDING EVIDENCE

Reported in 1 (0.7%) of 137 blocks

Possible 0 (0%)

Probable 0 (0%)

Confirmed 1 (100%)

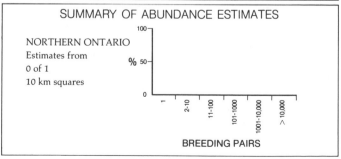

SUMMARY OF ABUNDANCE ESTIMATES

NORTHERN ONTARIO
Estimates from
0 of 1
10 km squares

BREEDING PAIRS

OLDSQUAW
Canard kakawi
Clangula hyemalis

ately to the west of Cape Henrietta Maria (specifically 3 nests with eggs on Little Bear Island, 55⁰16`N, 82⁰58`W, in 1981; and a female performing distraction displays at the mouth of the Sutton River in 1983). In the 2 blocks at, and immediately south of, Cape Henrietta Maria, however, probable breeding pairs were present in 1984 and 1985, but nesting was not confirmed. All Ontario breeding records prior to the atlas (at least 4 nests and 14 broods) came from these same 2 blocks. Departure of atlassers from these 2 tundra blocks in early to mid-July may account for their failure to find broods. Oldsquaws were not reported along the Hudson Bay coast between the Sutton River and East Pen Island, near the Manitoba border. Four broods were observed on East Pen Island, NWT, some 20 km from the Manitoba border, in 1985. These are the first definite breeding records for that part of the coast. Although the observed broods were all within 5 km of each other, by chance they fall into 2 10 km squares representing 2 different 100 km blocks. Thus, the atlas map gives the impression of a wider distribution near the Manitoba border than is warranted by the observations.

The Oldsquaw is an uncommon breeder on Ontario's tundra. There is no evidence of changes in abundance over the last 40 years.-- *D.J.T. Hussell*

The Oldsquaw is a common holarctic duck, which breeds in tundra and alpine habitats south to about 60⁰N in Europe and Asia and to 50⁰N in North America (Cramp and Simmons 1977). It winters wherever there is ice-free water south to about 40⁰N. There is a small breeding population along the Hudson Bay coast, where it is near the southern limit of its arctic breeding range.

Breeding was first reported in Ontario in 1944, when 13 broods were observed between 5 and 22 August in tundra habitat at Cape Henrietta Maria and along the contiguous 50 km of coast extending south into James Bay (Smith 1944). The first nests were found at Cape Henrietta Maria between 19 and 26 July 1947 (Manning 1952). All subsequent breeding records prior to 1981 were also from the Cape region (Peck 1972, Peck and James 1983).

The Oldsquaw is gregarious for most of the year but flocks split up into small groups and pairs which occupy fresh-water tundra ponds prior to nesting (Cramp and Simmons 1977). Nests are usually widely dispersed, often concealed in thick vegetation, but sometimes they are in loose colonies on islands. Males desert females during the first half of the incubation period. Nests are difficult to find except by accidentally flushing the incubating female. The Oldsquaw is a relatively late nester in Ontario: broods not more than 1/3 grown were observed from 5 to 22 August in the Cape Henrietta Maria region (Smith 1944). However, nests with eggs have been found as early as 26 June (atlas data) and downy young as early as 13 July (Peck 1972).

In Ontario, the Oldsquaw is confined to areas strongly influenced by the cold waters of Hudson Bay, well away from the tree-line. Atlas breeding evidence was obtained from 9 squares representing 4 blocks in the Cape Henrietta Maria region. Breeding was confirmed in the block immedi-

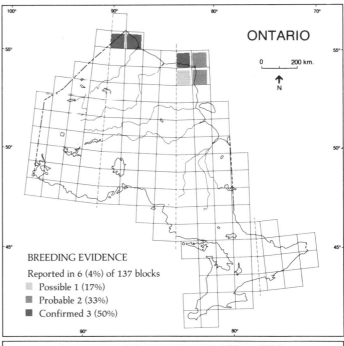

ONTARIO

0   200 km.

N

BREEDING EVIDENCE

Reported in 6 (4%) of 137 blocks

Possible 1 (17%)
Probable 2 (33%)
Confirmed 3 (50%)

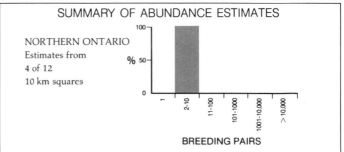

SUMMARY OF ABUNDANCE ESTIMATES

NORTHERN ONTARIO
Estimates from
4 of 12
10 km squares

% 50

100

0

1   2-10   11-100   101-1000   1001-10,000   >10,000

BREEDING PAIRS

SURF SCOTER
Macreuse à front blanc
Melanitta perspicillata

This is the only scoter restricted in its breeding range to North America, where it is found across the continent along the northern limits of the Boreal Forest. It is most common in the zone of forests and barrens at the interface with the tundra and, in Ontario, appears to be restricted to the Hudson Bay Lowland. This scoter usually nests near larger permanent water bodies. In Ontario, birds are most often observed on muskeg ponds and lakes with well-treed shorelines. Most winter off the Atlantic and Pacific coasts.

The Surf Scoter is apparently an extremely rare breeder in Ontario, the first brood having been recorded in 1960 (Simkin 1963) just north of Shagamu Lake (west of Winisk); a second brood was reported in 1980 near Aquatuk Lake east of Winisk (Peck and James 1983). The only confirmed record during the atlas period was of a female with a brood, observed on 4 August 1981 at Kiruna Lake in the Sutton Ridges.

The Surf Scoter is a very difficult species to atlas, given its low breeding densities, its propensity for the largely inaccessible Hudson Bay Lowland, and its secretive nature during the brood rearing period. The nest of this species has rarely been described, and none has ever been reported in Ontario. The best information has been provided by MacFarlane (1891), who noted that the nest resembles that of the White-winged Scoter, but is flimsier and contains less grass and feathers in its structure. The nest is built on the ground, often at considerable distance from open water, and is well hidden beneath tree branches. Concealment beneath clumps of herbaceous vegetation is employed in more open areas. Nest sites are clearly very hard to find because they are not restricted to the shoreline as are those of a number of other waterfowl. Even the positive identification of the species when sighted with a brood is difficult, considering the simi-

larity of the female to the White-winged Scoter, particularly when she is skulking low on the water trying to escape detection. Moreover, atlassing on the Lowland was usually restricted to the vicinity of major rivers, whereas Scoters appear to prefer ponds which are often distant from these rivers. It should also be noted that this is a late breeder, and so pairs seen in suitable habitat, even into the third week of June, may represent migrants, further obscuring any picture of breeding distribution.

Given the above provisos, it is difficult to determine if the atlas map truly represents the extent of the Surf Scoter's breeding distribution in Ontario. The possible breeding record is of 2 males and a female on a muskeg pond, on 22 May 1982, while the 'probable' record is of a pair observed on a 20 ha tundra pond on several occasions between 5 and 16 July 1985. The map allows us to conclude only that the species is restricted to the Hudson Bay Lowland and that it nests in low numbers.

Manning (1952) observed that this was the least abundant of the scoters found along the James and Hudson Bay coast of Ontario during the summer. The one abundance estimate is based on an observation of 14 birds on Kiruna Lake and a nearby lake between 10 July and 4 August 1981.-- *R.K. Ross*

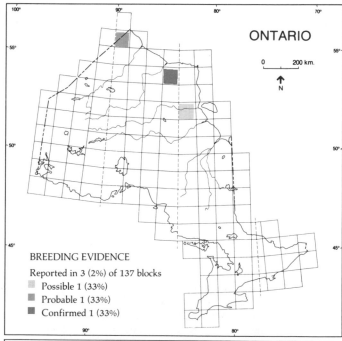

BREEDING EVIDENCE

Reported in 3 (2%) of 137 blocks

Possible 1 (33%)

Probable 1 (33%)

Confirmed 1 (33%)

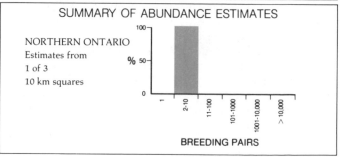

SUMMARY OF ABUNDANCE ESTIMATES

NORTHERN ONTARIO
Estimates from
1 of 3
10 km squares

BREEDING PAIRS

WHITE-WINGED SCOTER
Macreuse à ailes blanches
Melanitta fusca

The White-winged Scoter is the largest of the three scoters. It has a holarctic breeding distribution ranging through northern Eurasia into northwestern North America, east to the James and Hudson Bay coasts, and south from the tree-line to the northern prairies. Although summering birds have been noted east of Ontario, breeding has yet to be documented. In Ontario, the species breeds in low numbers in the Hudson Bay Lowland, and the Boreal Forest of the Canadian Shield. It winters primarily off both the Atlantic and Pacific coasts.

This scoter prefers lakes, muskeg ponds, and slow-moving rivers, frequently nesting on islands. Although often found near the water's edge, nests have been recorded up to 0.8 km from open water (Rawls 1949). The nest is a shallow scoop lined with twigs, grasses, and some feathers, well hidden beneath tree branches or bushes. Nesting density can be locally quite high in the central part of the range, particularly in island situations; 2.3 nests/ha were noted at Jessie Lake, Alberta, by P. Brown (in Bellrose 1978).

Only two confirmed breeding records were available for Ontario prior to the atlas period: a female with young at Ney Lake near the Manitoba border on 8 August 1936 (Baillie 1939), and a nest with eggs at Cool Lake near Hawley Lake, one of which was collected by T.G. Harrison on 31 July 1965 (Peck and James 1983). A further record of young birds in the Hawley Lake area was made during the atlas period.

This is an extremely difficult species to atlas, for many of the same reasons given for the Surf Scoter: it occurs at low breeding density in Ontario, frequents inaccessible areas of the Hudson Bay Lowland, and is very secretive during brood rearing. The distribution shown on the map, while indicative of the general range, should therefore be considered approximate. Clearly, the Hudson Bay Lowland makes up a major portion of its Ontario range, even though it apparently breeds there in small numbers. The most westerly atlas record of the White-winged Scoter was on Bearskin Lake, which is on the Canadian Shield about 100 km southeast of the site of the 1936 brood at Ney Lake. This probably represents a small extension of the breeding habitat used in its better recorded range in western Canada.

Manning (1952) found it in summer along the James and Hudson Bay coasts in higher numbers than the Surf Scoter, which could imply that it also breeds in higher numbers in this area.-- *R.K. Ross*

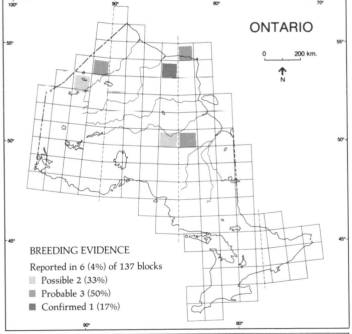

BREEDING EVIDENCE

Reported in 6 (4%) of 137 blocks

Possible 2 (33%)
Probable 3 (50%)
Confirmed 1 (17%)

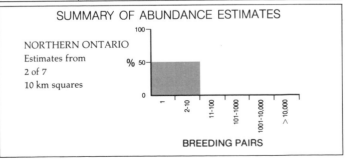

SUMMARY OF ABUNDANCE ESTIMATES

NORTHERN ONTARIO
Estimates from
2 of 7
10 km squares

BREEDING PAIRS

ROUGH-LEGGED HAWK
Buse pattue
Buteo lagopus

The usual clutch is 3 or 4 eggs, and incubation takes 28 to 31 days. Refuse and droppings around the nest in rocky situations form soil which nourishes bright green grasses that become obvious landmarks. Nests are re-used in subsequent years, likely because of the shortage of suitable sites. The female performs the main share of incubation and brooding the young, during which time the male brings all the food, which is almost exclusively small mammals. At the nest, the parents are aggressive in defending the young, which fledge after about 41 days, usually between early July and mid-August.

The distribution shown on the atlas map, with records along the Hudson and northern James Bay coasts, is that which might have been expected based on previous records. The single breeding confirmation was of a pair of birds at a nest at radar site 415. The 3 other records were of single birds on the north coast in tundra habitat. Birds frequently summer along that coast (P. Prevett pers. comm.), but it is not clear whether nesting is ever attempted at locations other than the radar sites near Cape Henrietta Maria. Individuals reported in the breeding season in southern Ontario were considered to be non-breeders, and are not mapped.-- *R.D. Weir*

The Rough-legged Hawk is best known in Ontario as a winter visitor in the south and a migrant during spring and autumn at the hawk lookouts situated along the Great Lakes shorelines. Individuals of the species have either a pale plumage or a dark melanistic plumage, the latter type being in the minority. This large Buteo frequently hovers motionless against the wind as it peers down scanning for prey.

The Rough-legged Hawk is a panboreal species, breeding in the arctic and subarctic regions between 61^0N and 76^0N in Europe, Asia, and North America. In Canada, its breeding range extends farther south to 55^0N around James and Hudson Bays. It is highly migratory, moving south in autumn to escape the arctic winter. Generally, the species avoids forest, preferring to hunt where extensive open ground is available. Its breeding habitat within arctic regions includes rocky outcroppings, ravines, cliffs, and steep river banks, all of which serve as potential sites for its nest.

The species has only recently been found nesting in Ontario. It was thought to have first bred in the province in 1958 on a radar tower at site 416, 40 km south of Cape Henrietta Maria on the Hudson Bay Lowland (Baillie 1963b), but proof was lacking. James *et al.* (1976) defined its summer status as a rare to uncommon resident on the tundra of the Lowland, the first authors to consider this species to be a summer resident anywhere in the province. In 1976, proof of nesting was obtained at radar site 416 (Peck 1976), and subsequently occasional nestings have occurred there or at nearby site 415. All nests were thought to be old Common Ravens' nests, and all were situated on artificial structures at abandoned radar sites, which are the only elevated places in the area (Peck and James 1983).

Normally, nests are placed in elevated situations atop cliffs, on shelves along river banks, or occasionally in trees.

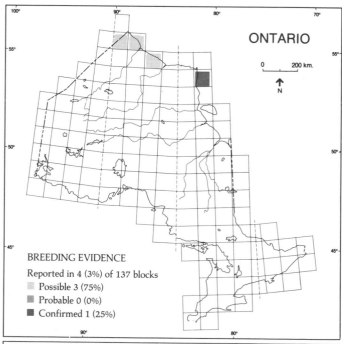

BREEDING EVIDENCE

Reported in 4 (3%) of 137 blocks

Possible 3 (75%)

Probable 0 (0%)

Confirmed 1 (25%)

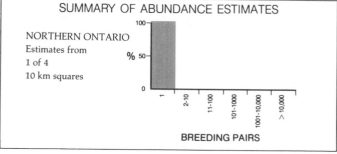

SUMMARY OF ABUNDANCE ESTIMATES

NORTHERN ONTARIO
Estimates from
1 of 4
10 km squares

BREEDING PAIRS

GOLDEN EAGLE
Aigle royal
Aquila chrysaetos

The Golden Eagle is the most widely distributed of the world's large eagles, ranging throughout the northern hemisphere. It nests over much of Canada and in western and northeastern regions of the US, but its distribution is discontinuous, and it is rare in all except western upland regions. Breeding adults are generally associated with wild arid plateaus deeply cut by streams and canyons, or sparsely treed mountain slopes and rock crags. It is also common in western shrub and grassland habitats. Intensely farmed and residential areas are generally avoided, but some territories occur in timbered areas, where they encompass large burns, marshes, or other openings used for hunting.

The Golden Eagle was probably more numerous in Ontario in the past; there are several unsubstantiated historical nesting reports from widely separate locations in both the southern and northern parts of the province (Snyder 1949a, Peck and James 1983). However, it may never have been very common in eastern North America, and, because it was perceived as a threat to domestic livestock, shooting and poisoning campaigns decimated this population. More recent nest records have come mainly from far northern areas near Hudson Bay (Peck and James 1983) and single records from northwestern and northcentral regions of Ontario (De Smet 1987). There have been other summer sightings in northeastern Ontario, and even some in southern Ontario (Holroyd 1968, Sadler 1983).

Atlassing for this species is difficult. In addition to its scarcity, the wariness of the birds (McGahn 1968), and the remoteness of its nest sites, identification of birds is also difficult. The more numerous Bald Eagle, in immature plumage, is easily confused with a Golden Eagle, and several reports of eagles to the atlas were dropped because the identity could not be verified from the descriptions provided. Adult birds are likely to be absent from nests for extended periods when feeding young, so that an observer might have little chance to see a bird even if near a nest.

The few records on the map suggest that this species is a rare breeding bird in Ontario. Even with the atlas project, we still have very inadequate information on its numbers and distribution. The only confirmed record was of a used nest in a region of previously known nesting. The widely spaced records suggest that the bird occurs and possibly nests in some of the more remote areas of the province.

As more has become known about the food habits of Golden Eagles (Olendorff 1976), public attitudes have improved considerably. Bounties and poisoning campaigns have been replaced with strict laws of protection. In Ontario, the Golden Eagle is protected under the Endangered Species Act. Winter trapping and some shooting losses are still serious problems, but in eastern North America, as well as in Ontario, there are encouraging signs of slowly increasing numbers (De Smet 1987). Recent autumn migration totals in Ontario of as many as 60 birds are 3 or 4 times as high as those from the 1950's and 1960's (R.D. Weir pers. comm.), although all these birds may not nest in the province.-- *K. De Smet and R.D. James*

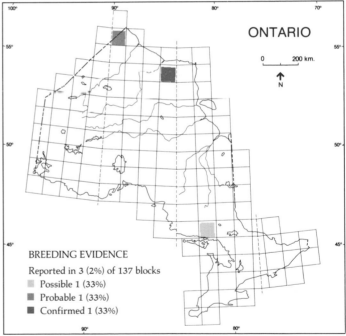

ONTARIO

0   200 km.

N

BREEDING EVIDENCE

Reported in 3 (2%) of 137 blocks

Possible 1 (33%)
Probable 1 (33%)
Confirmed 1 (33%)

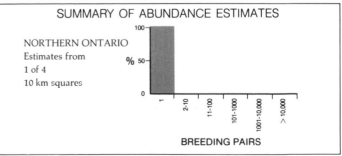

SUMMARY OF ABUNDANCE ESTIMATES

NORTHERN ONTARIO
Estimates from
1 of 4
10 km squares

%

BREEDING PAIRS

WILLOW PTARMIGAN
Lagopède des saules
Lagopus lagopus

Holartic in distribution, the Willow Ptarmigan breeds at or beyond the tree-line in arctic, sub-arctic and northern alpine environments. In North America, 7 races occur across northern Canada and Alaska (AOU 1957). In Ontario it nests in the triangle from just north of the Lakitusaki River mouth (50 km south of Cape Henrietta Maria), northwest to the mouth of the Sutton River. Although there are no confirmed breeding records west of the Winisk River (Peck and James 1983), the Willow Ptarmigan has previously been recorded in July on the 5 to 15 km wide strip of tundra in all blocks on the Hudson Bay coast, and broods have been collected in September near the Pipowitan River mouth, 10 km north of Fort Severn (specimens in ROM). In years of abundance, it has been recorded in summer on open and burnt over areas as much as 60 km inland in June at Hawley Lake (specimens in ROM), and in July near Shagamu and Shellbrook Lakes (H. Lumsden unpubl.).

The Willow Ptarmigan nests in heath-lichen vegetation on the better-drained raised beach ridges, and on permafrost-cored hummocks in sedge fens. Particularly high densities of breeding birds have been reported on the extensive dry elevated beach lines at radar site 415 south of Cape Henrietta Maria. Thomas found 6 pairs in a 30 minute walk in 1980 (30-40 pair/km^2), but in a year of scarcity, R.D. James (pers. comm.) "... saw fewer than 1 bird/km^2 in 1984".

There has probably been little change in the numbers of Willow Ptarmigan since the early days of the fur trade. In the late 1700's, Graham reported "... upwards of 10 thousand caught with nets at Severn River, from the month of November till the beginning of April" (Williams and Glover 1969). In the winter of 1961 to 1962 Angus Miles and Frederick Close (pers. comm.) estimated 1,000 in a day, and 10,000 during the winter, killed by the Indians at Fort Severn.

Breeding Willow Ptarmigan are fairly easy to confirm. Pairs occupy open habitat and the males are very conspicuous in spring when they crow and perform morning and evening song flights. They also respond well to tape recordings. However, nests are not easy to find; they may be in the open with little concealment or be well hidden beneath a stunted willow. The cryptically coloured hen sits tight and only flushes at the closest approach. Both parents attend the brood and often perform vigorous distraction displays when disturbed. Broods hatch late and remain together well into autumn. They are relatively easily found and constitute 18% of atlas records.

Although the general extent of the range is indicated, the Willow Ptarmigan appears to be under-confirmed on the atlas map, probably because most of the atlas field work on the Hudson Bay coast was done in 1984 and 1985 when the birds were relatively scarce. In years of high population levels, Willow Ptarmigan can be expected in all blocks along the Hudson Bay coast and occasionally further inland. All abundance estimates came from the Cape Henrietta Maria region in 1984 and 1985. One square had 1 pair, and 3 had 11 to 100 pairs indicating relatively low population levels.-- *H.G. Lumsden and V.G. Thomas*

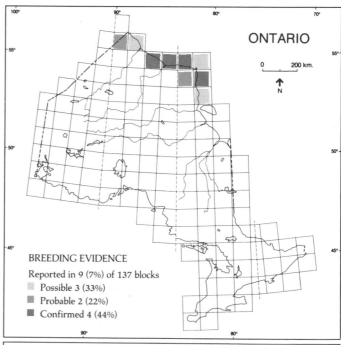

BREEDING EVIDENCE

Reported in 9 (7%) of 137 blocks

Possible 3 (33%)
Probable 2 (22%)
Confirmed 4 (44%)

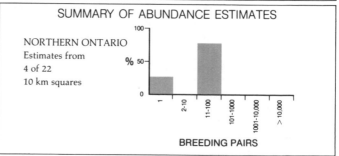

SUMMARY OF ABUNDANCE ESTIMATES

NORTHERN ONTARIO
Estimates from
4 of 22
10 km squares

BREEDING PAIRS

LESSER GOLDEN-PLOVER
Pluvier doré d'Amérique
Pluvialis dominica

S.L. HOUSE

Atlas data were received from 6 squares in 3 blocks in far northern Ontario. Four squares have records of birds in suitable habitat. One of these is from the mainland adjacent to East Pen Island, near the Manitoba border. Another is from the mouth of the Sutton River, where D.J.T. Hussell (pers. comm.) also reported a bird in breeding plumage in 1962. The other 2 'suitable habitat' records are from the block containing radar site 415. Also in this block, 4 nests were found in one square and a bird exhibiting agitated behaviour was reported in another. This evidence suggests that although some breeding may occur locally in the few places where suitable habitat exists along the Hudson Bay coast, the area around site 415 may be the only location where breeding occurs regularly or in any significant numbers.

Abundance estimates help emphasize the importance of the site 415 area. The square in which 4 nests were found was estimated to contain 11 to 100 breeding pairs. The only other estimate came from the square just south of Cape Henrietta Maria, where only one breeding pair was estimated.-- *G.K. Peck*

The Lesser Golden-Plover breeds in tundra regions of the eastern Palearctic and western Nearctic, and, following long migrations, winters well south of its breeding range, as far as the southern hemisphere of both Old and New worlds. It has been found breeding in Ontario only on the narrow strip of coastal tundra bordering Hudson Bay and the northern portion of James Bay. This represents the most southerly known breeding latitude for the species throughout its range. The few Ontario nests have all been found on high, dry, heath-lichen tundra near Cape Henrietta Maria in Kenora District (Peck and James 1983), though broods may move to wet tundra areas. Suitable, dry upland tundra is scarce in Ontario, so the species is restricted in its distribution. Within its limited Ontario habitat, this species appears to be, at best, only moderately common (Peck 1972).

Although probably overlooked previously, the Lesser Golden-Plover was first documented as a breeding bird of Ontario in June 1970, when a nest with 4 eggs was found on the tundra near the abandoned radar site 415, approximately 30 km south of the base of Cape Henrietta Maria (Peck 1972). A second nest was also found near site 415 in 1978, as were 4 others in 1984.

The difficulty of access to the remote, restricted breeding habitat of this species is the most likely reason for there being so few breeding records. It is however, not an exceptionally difficult species to atlas. It can sometimes be seen walking about conspicuously on ridges (D.J.T. Hussell pers. comm.), and performs an obvious display flight early in the season. The nest is inconspicuous and the eggs, like those of most shorebirds, are cryptically coloured. Nests are most easily found by retiring some distance once an alarmed bird or a pair has been located, and watching through binoculars for one to return to the nest.

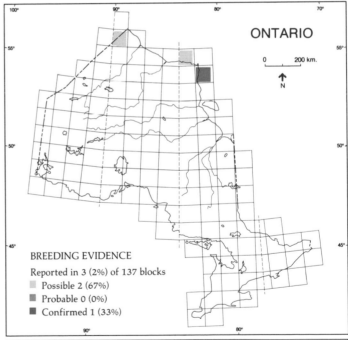

ONTARIO

0   200 km.

N

BREEDING EVIDENCE

Reported in 3 (2%) of 137 blocks

Possible 2 (67%)
Probable 0 (0%)
Confirmed 1 (33%)

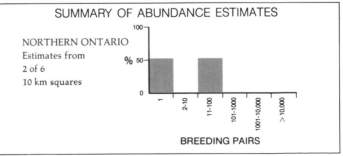

SUMMARY OF ABUNDANCE ESTIMATES

NORTHERN ONTARIO
Estimates from
2 of 6
10 km squares

%   100

50

0

1   2-10   11-100   101-1000   1001-10,000   > 10,000

BREEDING PAIRS

SEMIPALMATED PLOVER
Pluvier semipalmé
Charadrius semipalmatus

S.L.HOUSE

In the breeding season, the Semipalmated Plover inhabits seacoasts and low arctic tundra in Canada and Alaska. In the Old World it is replaced by the closely related Common Ringed Plover (*Charadrius hiaticula*), whose range extends westward into the eastern Canadian arctic, where it overlaps with that of the Semipalmated Plover. The winter range of the latter extends from the southern US to southern Argentina and Chile. In Ontario, its breeding range is confined to the shores of Hudson and James Bays, and inland along some of the major rivers. Breeding habitat consists of sand and gravel bars in rivers and along shorelines, and tundra, including sparsely vegetated raised beach ridges with sand, gravel, and dry peaty ground. In the latter sites, it is often found in association with Horned Larks. The Semipalmated Plover is encountered commonly where such suitable habitat occurs.

The Semipalmated Plover was reported breeding along the James Bay coast by W. Spreadborough as early as 1904 (Macoun and Macoun 1909), but material evidence was not obtained until Manning (1952) collected downy young at Cape Henrietta Maria in 1947. Since then, knowledge of its distribution has become more complete, and there is no indication of change in recent times.

Semipalmated Plovers were readily detected by atlassers in the far north. The birds tend to inhabit the drier tundra ridges and shorelines, which provide good walking conditions, and atlassers canoeing down some of the larger rivers found them easily on sandbars and islands. When it has eggs or young nearby, the Semipalmated Plover behaves conspicuously in an agitated manner, bobbing its head and giving alarm notes. Birds flushed from nests will give elaborate broken-wing distraction displays. Nests and small downy young are easily found by watching the adult bird until it returns to

incubate the eggs or brood the young. Large, unfledged young are more difficult to find as they are well camouflaged and adults attend them less closely. Breeding was confirmed in 26 of 40 squares, indicating that confirmation is indeed relatively easy to obtain. Nests with young were found in 3 squares, nests with eggs in 6, young birds in 10, and distraction displays were reported in 7 squares.

The atlas map confirms that the distribution of the Semipalmated Plover in Ontario is confined to the shores of Hudson and James Bays and adjacent rivers (Peck and James 1983). It may occur at scattered localities farther inland than indicated by the map, as there is an earlier record at Hawley Lake, in a block in which it was not reported during the atlas years (Schueler *et al.* 1974).

The Semipalmated Plover is uncommon to common in suitable habitat within its range. Peck (1972) described it as abundant in the Cape Henrietta Maria region, and atlassers estimated 11 to 100 pairs in one square about 40 km south of Cape Henrietta Maria and in 3 squares at the Cape itself in 1984 and 1985. The highest estimate, of 101 to 1,000 pairs, is from a square at the mouth of the Shagamu River between Winisk and Fort Severn on the Hudson Bay coast.-- *D.J.T. Hussell*

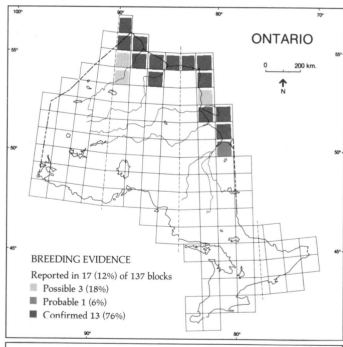

ONTARIO

0   200 km.

BREEDING EVIDENCE

Reported in 17 (12%) of 137 blocks

Possible 3 (18%)
Probable 1 (6%)
Confirmed 13 (76%)

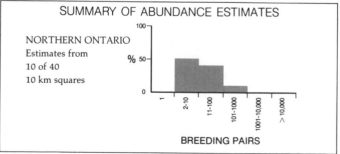

SUMMARY OF ABUNDANCE ESTIMATES

NORTHERN ONTARIO
Estimates from
10 of 40
10 km squares

BREEDING PAIRS

AMERICAN AVOCET
Avocette d'Amérique
Recurvirostra americana

S.L. HOUSE

The American Avocet breeds primarily from the Prairie Provinces southward to southern California, New Mexico, and northern Mexico (Johnsgard 1981), but breeding has also been reported from southeastern British Columbia and northwestern Ontario. It winters from California and southern Texas south through Mexico and Guatemala (Godfrey 1986). Breeding habitat consists of fairly extensive exposed, sparsely vegetated shorelines and mudflats, adjacent to lakes and sloughs. Such habitat is rare in Ontario.

Although the species is seen occasionally during both spring and fall migrations in southern Ontario (Speirs 1985), it was not known to breed in the province until July 1980, when an adult and 1 of 3 flightless young birds were photographed on Sable Island, Lake of the Woods (Peck and James 1983).

The large size and striking plumage of the American Avocet make it unlikely that birds were missed in squares that were well covered. The Rainy River-Lake of the Woods area, which is too large to be properly covered by a few visiting atlassers, has more potential breeding habitat than anywhere else in Ontario, and is near the main breeding range. It is possible that a small number of breeding pairs were overlooked there. Any future nestings in Ontario are likely to occur in that area.

Atlas surveys produced only one record, that being of an apparent pair on 19 May 1981 on Lake of the Woods. The birds were on Sable Island, where the previous breeding record was located. Breeding was not confirmed during the atlas period.-- *M.D. Cadman*

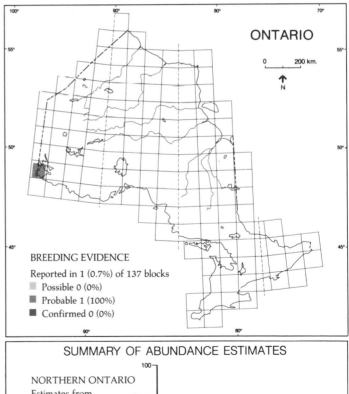

ONTARIO

0   200 km.

N

BREEDING EVIDENCE

Reported in 1 (0.7%) of 137 blocks

Possible 0 (0%)

Probable 1 (100%)

Confirmed 0 (0%)

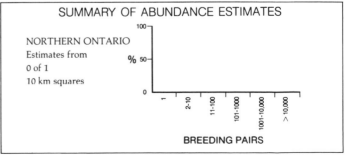

SUMMARY OF ABUNDANCE ESTIMATES

NORTHERN ONTARIO
Estimates from
0 of 1
10 km squares

100

% 50

0

1   2-10   11-100   101-1000   1001-10,000   >10,000

BREEDING PAIRS

523

GREATER YELLOWLEGS
Grand Chevalier
Tringa melanoleuca

S.L.HOUSE

Although it occurs in large numbers in parts of Ontario, we still know almost nothing of the nesting habits of the Greater Yellowlegs (Peck and James 1983), for it is one of the wariest of creatures. A person approaching a nesting area, even when at a considerable distance, is greeted with loud and persistent alarm notes. One can walk for hours through the wetlands of the Hudson Bay Lowland and never have a quiet moment, as one bird after another takes up the call. And, one can walk for days and never discover either a nest or young, both of which are cryptically concealed amongst the wetland vegetation. Therefore, it is perhaps not surprising that the first nest reported in Ontario, at Little Sachigo Lake, was not found until 1977 (McLaren and McLaren 1981b). Subsequently, another nest and flightless young were documented near Aquatuk Lake in 1980 (Nash and Dick 1981).

The Greater Yellowlegs occurs across Canada in summer, in a wide band from the southern Yukon and British Columbia to Newfoundland, in more northern Boreal Forest areas, but south of open tundra. It winters in the coastal and southern US, south to extreme southern South America. Breeding habitat includes fens, bogs, sloughs, and shallow ponds with extensive wet and moss-covered flats, and shorelines surrounded and interspersed with tree and shrub cover. The preferred habitat is generally somewhat wetter than that chosen by the Lesser Yellowlegs. Such wetland habitat is found to some extent in Boreal Forest regions across the country, but is more widespread and abundant in Ontario because of the extensive Hudson Bay Lowland. South of the Lowland, where forest cover is thicker, there are far fewer birds. Some do occur on the tundra in summer, but probably only as non-breeding individuals.

The Greater Yellowlegs is a difficult species to confirm. Atlassers are unlikely to record a range of breeding behav-

iours, as Great Yellowlegs always scold when an observer is near. In fact, 63% of all block records were of agitated behaviour. Young Greater Yellowlegs are likely to hide and not move, as long as an adult is scolding. The nest is nothing but a slight hollow in some moss. Nest-building is rarely seen, and adults do not carry food or faecal sacs; no nests were reported during the atlas period. Breeding was confirmed in 4 atlas squares; all were records of fledged young.

Despite these difficulties, and the few breeding confirmations, the species is readily located, and the atlas map gives the first clear outline of its nesting range in Ontario. As expected, most records are from the Hudson Bay Lowland, with fewer and more scattered records as the southern edge of its breeding range is approached. Three of the most southerly reports, those from Red Lake, Lake Nipigon and Cochrane, are southward extensions of what was previously believed to be its range in the province (Peck and James 1983, Godfrey 1986), but there is every indication that these places are part of the normal range of the species.

Although most abundance estimates are of 10 or fewer pairs per square, more than 100 pairs of Greater Yellowlegs per square are to be expected in extensive areas of prime nesting habitat on the Hudson Bay Lowland. Over 1,000 pairs per square may occur locally.-- *R.D. James*

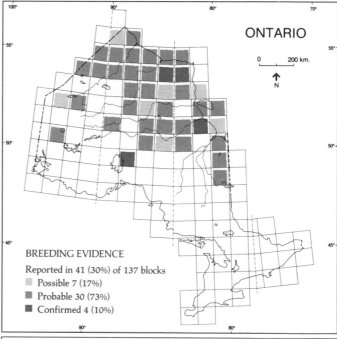

ONTARIO

0   200 km.

↑
N

BREEDING EVIDENCE

Reported in 41 (30%) of 137 blocks
Possible 7 (17%)
Probable 30 (73%)
Confirmed 4 (10%)

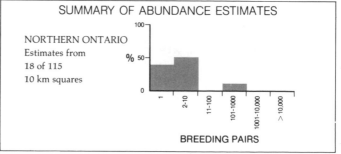

SUMMARY OF ABUNDANCE ESTIMATES

NORTHERN ONTARIO
Estimates from
18 of 115
10 km squares

%

BREEDING PAIRS

LESSER YELLOWLEGS
Petit Chevalier
Tringa flavipes

S. L. HOUSE

The Lesser Yellowlegs nests strictly in North America, in the subarctic muskeg and Boreal Forest, from Alaska to James Bay. In Ontario, it is common in summer in suitable habitat throughout most of the northern third of the province (Godfrey 1986). The Lesser Yellowlegs nests in open woodlands and burnt-over forest interspersed with ponds, lakes, and wetlands, where undergrowth is typically sparse and low. Nest sites are usually on relatively high and dry ground with plenty of deadfalls (Street 1923, Rowan 1929). The Greater Yellowlegs appears to prefer wetter sites in the muskeg proper, as does the Solitary Sandpiper.

Only two previous substantiated breeding records exist for the Lesser Yellowlegs in Ontario. A nest was discovered on 4 June 1938 at Favourable Lake in Kenora District, and downy young were found in July 1940 at Fort Severn on the Hudson Bay coast (Peck and James 1983). In view of its abundance during the nesting season, observations of aggressive and agitated behaviour, and collection of birds with enlarged gonads, the Lesser Yellowlegs is presumed to nest throughout the Hudson Bay Lowland.

The Lesser Yellowlegs is aggressive, noisy, and conspicuous on the breeding grounds, and there is generally no difficulty in detecting this species if it is present. Attempting to confirm breeding by finding nests or young is usually hopeless and exasperating, because of the persistence of the adults' alarm behaviour. Over 40% of all records involved such agitated birds. There is only one atlas record of flightless young, and no nests; the only other confirmed breeding record is of a distraction display.

Given the conspicuousness of this species on the nesting grounds, the atlas data should portray accurately the Ontario breeding distribution of the Lesser Yellowlegs, given 3 assumptions. The first is that the characteristic loud aggres-

sive behaviour is in fact displayed only by nesting birds and not by non-breeders. So little is known about the breeding biology of either species of yellowlegs that this is uncertain, although it is probably true. The second assumption is that there was no confusion by atlassers with the similar Greater Yellowlegs. The last assumption concerns the possibility of confusion of breeding birds with early fall migrants, summer wanderers, and non-breeders. Lesser Yellowlegs regularly occur south, often well south, of their probable nesting range in the 'breeding season', *i.e.*, June and July. Although several records were submitted from south of the records mapped, none were judged to involve acceptable evidence of breeding.

Godfrey (1986) mapped the southern breeding limit of this species in Ontario from approximately 51^{0}N in the east to 52^{0}N in the west. This agrees with atlas results in the east but not in the west, where the southern limit is at about 54^{0}N, indicating that the Lesser Yellowlegs is probably not widely distributed on the Shield. Nevertheless, the atlas data may underrepresent the nesting distribution south of the Hudson Bay Lowland. It is likely that the Lesser Yellowlegs breeds locally in suitable habitat in northern Canadian Shield portions of the Boreal Forest, as indicated by the nest found at Favourable Lake in 1938.-- *R. Harris*

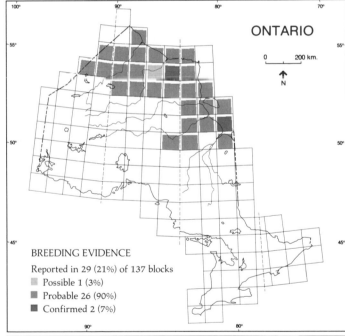

ONTARIO

0   200 km.

N

BREEDING EVIDENCE

Reported in 29 (21%) of 137 blocks

▫ Possible 1 (3%)
▪ Probable 26 (90%)
■ Confirmed 2 (7%)

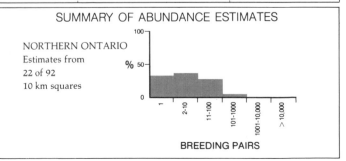

SUMMARY OF ABUNDANCE ESTIMATES

NORTHERN ONTARIO
Estimates from
22 of 92
10 km squares

BREEDING PAIRS

WHIMBREL
Courlis corlieu
Numenius phaeopus

S.L.HOUSE

Partridge Island near Fort Severn. The Whimbrel likely breeds sparsely in the forest-tundra transition zone in most blocks adjacent to the Hudson Bay coast, as reflected on the map, and possibly in some large open bogs farther inland, although confirmation in the latter habitat would be especially difficult. It was not reported on Cape Henrietta Maria during the atlas period.

The Whimbrel is a rare breeder in Ontario, probably because of the relative scarcity of prime breeding habitat, and because the province is at the southernmost extremity of the continental breeding range. Historical records from the northern section of the James Bay coast indicate that the Whimbrel was previously more common there than during the atlas survey. Whether these differences represent a permanent decline in population or irregular breeding at the extremity of the range of the species remains to be determined by future field work.-- *J.P. Prevett*

Four subspecies of the Whimbrel occur non-continuously across the north boreal, subarctic, and low arctic regions of North America and Eurasia (Cramp and Simmons 1983). In North America, the 'Hudsonian Curlew' has two disjunct breeding ranges: west and south of Hudson Bay, and along and inland from the northwestern arctic coast in the NWT, Yukon, and Alaska. The subarctic ecotone between the Boreal Forest and low arctic tundra is perhaps the most characteristic habitat for breeding (Johnsgard 1981). Specific breeding habitats identified at Churchill, Manitoba were variable, including rolling tundra, wet sedge meadows, and (most preferred) hummocky bogs or fens with scattered stunted black spruce and tamarack (Skeel 1976).

In Ontario, the Whimbrel has been found nesting on both wet and dry tundra near the Hudson and James Bay coasts. Shallow ponds and associated fen-bog vegetation between beach ridges, which may support fairly dense spruce forest, appear to be favoured brood-rearing habitat along the coast west of Cape Henrietta Maria.

Most often, Whimbrels fly towards potential predators while still far away (up to 0.7 km) from the nest or flightless young (Bannerman 1961, Cramp and Simmons 1983), then fly about a human intruder raising a loud hue and cry; this makes it relatively easy to record agitated behaviour as evidence of probable breeding, but difficult to find nests. Breeding was confirmed in 5 of 19 squares.

The map shows a breeding range primarily along the Hudson Bay coast from the Sutton River to Fort Severn. No records were obtained from the northern section of the James Bay coast, where most previous nestings were reported (Manning 1952, Peck 1972, Peck and James 1983), even though atlassers visited those areas. A nest with eggs was found further west in Ontario than previously reported: on

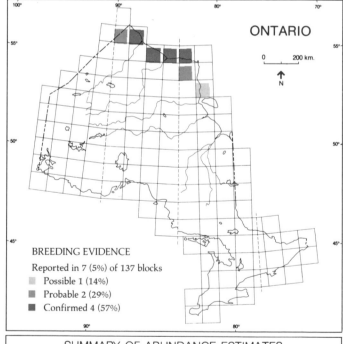

ONTARIO

0   200 km.

N

BREEDING EVIDENCE

Reported in 7 (5%) of 137 blocks

Possible 1 (14%)
Probable 2 (29%)
Confirmed 4 (57%)

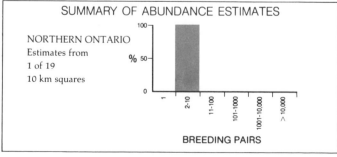

SUMMARY OF ABUNDANCE ESTIMATES

NORTHERN ONTARIO
Estimates from
1 of 19
10 km squares

BREEDING PAIRS

HUDSONIAN GODWIT
Barge hudsonienne
Limosa haemastica

S.L.HOUSE

The breeding range of the Hudsonian Godwit is restricted to North America, although it is probably only incompletely known. In Alaska, the species breeds in the south (Cook Inlet) and possibly in the west (Kotzebue Sound and Norton Bay). In Canada, breeding occurs around the inner deltas of the Mackenzie and Anderson Rivers, in isolated localities on Great Slave Lake and northwestern British Columbia (Chilkat Pass), and on the shores of Hudson Bay from Churchill possibly south to Akimiski Island in James Bay (AOU 1983, Godfrey 1986).

Hagar (1966) described the characteristic nesting habitat of the Hudsonian Godwit at Churchill as extensive sedge marshes and meadows lying on the northern edge of the tree-line, and generally not far from a tidal coast. Such habitat is plentiful along the southwestern coast of Hudson Bay. The principal wintering grounds of the species are in the Tierra del Fuego region of South America, where the birds are found on the intertidal mudflats of large bays and on *restinga* along the coast (Morrison and Ross 1987).

A nest has never been reported in Ontario. Breeding was first confirmed in the province with the collection of a flightless young at the mouth of the Sutton River on 31 July 1962 (Peck and James 1983). The only other confirmed breeding record was of another flightless young bird observed on 16 July 1985, 2.5 km east of the Brant River (D. Evered pers. comm.).

At Churchill, nests are typically located in or under the edge of a low dwarf birch on the dry top of a hummock in a sedge marsh. The surrounding vegetation is thick and tall enough to conceal the sitting bird effectively. Fewer nests are located in more open tundra areas, which may explain the absence of records in the blocks containing Cape Henrietta Maria and the adjacent block along the north shore of James Bay.

Display flights by males range widely over the ground, making the territories or nests themselves difficult to locate, but rendering the birds conspicuous. In the later stages of incubation and after the young have hatched, 'guarding' behaviour of the adults develops, and the birds will alarm vigorously if approached; 8 of 12 probable breeding records are of agitated behaviour. As fledging approaches, the well-grown young are highly mobile and alarms from the guarding parents enable them to keep ahead of approaching danger. The broods probably move from the nesting areas towards the coast during this period, which may explain why the one confirmed atlas record was about 3 km north of the tree-line.

Altogether, observations came from 22 squares in 8 blocks, ranging from the coast of Hudson Bay near the Manitoba border to central James Bay. The Hudsonian Godwit likely breeds in small numbers all the way along the coast of Hudson Bay where suitable habitat exists, and possibly on the north coast of Akimiski Island. These results basically confirm previous information on the species.-- *R.I.G. Morrison*

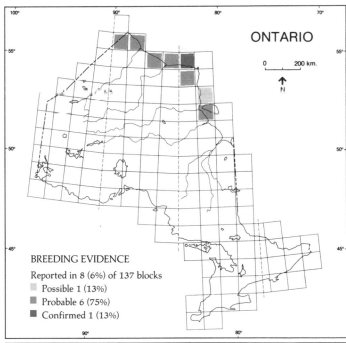

ONTARIO

0   200 km.

N

BREEDING EVIDENCE

Reported in 8 (6%) of 137 blocks

Possible 1 (13%)

Probable 6 (75%)

Confirmed 1 (13%)

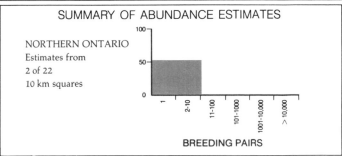

SUMMARY OF ABUNDANCE ESTIMATES

NORTHERN ONTARIO
Estimates from
2 of 22
10 km squares

BREEDING PAIRS

MARBLED GODWIT
Barge marbrée
Limosa fedoa

S.L. HOUSE

The Marbled Godwit is generally thought of as a prairie species. Its breeding range extends from the central Prairie Provinces in Canada southward through Montana, the Dakotas, and western Minnesota. The range has contracted somewhat, having formerly included Nebraska, Iowa and Wisconsin. On the prairies, the Marbled Godwit nests near wetlands, including sloughs and various kinds of ponds and lakes, especially those with extensive grassy borders. Nests are often in short upland grass with relatively sparse cover. The wintering areas extend from California, Texas, and Florida south along both coasts of Mexico and Central America to Guatemala and Belize, and less regularly as far south as northern Chile.

In Ontario, the principal breeding area of the species appears to be along the west coast of James Bay and possibly Hudson Bay. Records of specimens from Hudson Bay date back to 1750 and from James Bay to 1902, and Spreadborough (in Manning 1952) reported it as breeding on both sides of James Bay in 1904. Manning (1952) collected the species near North Point in southern James Bay and on the south coast of Akimiski Island in 1947. The first definite evidence of breeding involved the collection of a downy young near North Point in 1975 (Morrison *et al.* 1976). Further evidence of breeding included the discovery of a nest at North Point in 1978 and another flightless young in 1980, and the observation of a downy young bird on the coastal marshes of southern Akimiski Island in 1978 (Peck and James 1983). The nest on James Bay was located on a raised, grassy ridge between 2 ponds on the open coastal marsh. The available evidence thus suggests that the Marbled Godwit is a long-established breeding bird on the coast of James Bay.

The Marbled Godwit sits tightly on the nest and can on occasion be picked up directly from the eggs. When the nest

or young are threatened, however, the birds alarm noisily, and are generally joined by other birds in the area. Evidence for breeding is thus most likely to be obtained during the latter part of incubation and the early part of the fledging period; agitated behaviour was reported from 6 squares in 4 blocks.

Breeding was not confirmed during the atlas period, but possible breeding was reported from 5 squares, and probable breeding from 11. Six probable records were from northern Akimiski Island. Probable and possible breeding were also reported from 7 squares in the expected range on the James Bay coast between North Point and the southern boundary of Polar Bear Prov. Park. The most northerly record, suggesting a considerable expansion of the known breeding range, was of birds in extensive coastal sedge fen on the Hudson Bay coast southeast of West Pen Island. The probable breeding record from the Rainy River area was of a pair of birds apparently defending a territory from a single displaying bird in a pasture on 18 May 1982. This record suggests that small numbers may breed, at least occasionally, in the agricultural or remnant prairie areas near Rainy River.-- *R.I.G. Morrison*

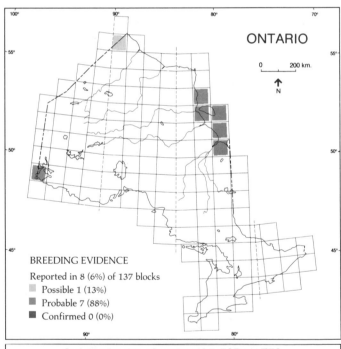

ONTARIO

0   200 km.

↑
N

BREEDING EVIDENCE

Reported in 8 (6%) of 137 blocks

■ Possible 1 (13%)
■ Probable 7 (88%)
■ Confirmed 0 (0%)

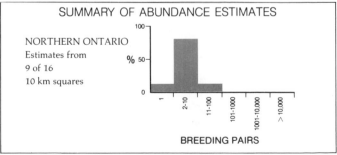

SUMMARY OF ABUNDANCE ESTIMATES

NORTHERN ONTARIO
Estimates from
9 of 16
10 km squares

BREEDING PAIRS

SEMIPALMATED SANDPIPER
Bécasseau semipalmé
Calidris pusilla

S.L.HOUSE

The Semipalmated Sandpiper breeds across much of arctic North America from Baffin Island to Alaska. In Ontario, it is found in a limited region along the Hudson Bay and James Bay coasts where open tundra prevails. Ontario is at the southern edge of the breeding range of the species. It is, in general, a more northerly nester than the similar Least Sandpiper. Where the ranges of the two species overlap, the Semipalmated Sandpiper is found in more open habitats, whereas the Least Sandpiper seems to prefer territories containing a few tall perches such as willow shrubs or small conifers.

The first evidence of breeding in Ontario was obtained by a ROM party in 1948, when a juvenile was collected about 40 km south of Cape Henrietta Maria. Subsequent breeding evidence has come from several sites along the Hudson Bay coast, with most records concentrated in the vicinity of the Cape (Peck and James 1983).

The following account of the breeding biology of the Semipalmated Sandpiper is based on a study carried out on the Hudson Bay coast in northern Manitoba (Gratto *et al.* 1983, 1985). Males arrive in the breeding area in late May or early June and establish territories. At this time, they are conspicuous by their song flights. The females return a little later and spend some time feeding prior to nesting. Nests are established from mid- to late June, depending on the season. The two sexes share incubation, and eggs hatch in early July. Birds attending young become very agitated when an observer approaches and give conspicuous trilling calls; all atlas records of probable breeding are of agitated birds. The precocial young are tended initially by both parents, although the female usually departs well before fledging occurs, leaving parental duties to the male. Families stay close to the nesting territory at first, but may move a considerable distance before the young have fledged.

The atlas results indicate that breeding may occur all along the Hudson Bay coast, but the main concentration of Semipalmated Sandpipers is in the vicinity of Cape Henrietta Maria. Confirmed, probable, and possible breeding records were obtained in 5, 2, and 1 squares, respectively in the 3 most northeasterly blocks, covering Cape Henrietta Maria and adjacent areas. Probable and possible breeding were reported in 1 and 4 other squares, respectively, representing 3 other blocks further west along the Hudson Bay coast. Comparison with the map for Least Sandpiper shows the more widespread and southerly distribution of that species. Such differences also show up locally: atlassers found that Least Sandpipers were absent from the exposed outer ridges of Cape Henrietta Maria, where the Semipalmated Sandpiper was a fairly common breeder.

In preferred habitats, consisting of dry tundra, or hummocks in wetter areas, nesting density may be quite high. A density of 2.3 pairs per hectare was reported at La Perouse Bay, Manitoba (Gratto *et al.* 1983), and comparable densities are to be expected in similar habitats in northern Ontario. Abundance was estimated at 11 to 100 pairs in each of 2 squares, and at 101 to 1,000 pairs in one square, in the Cape Henrietta Maria region.-- *F. Cooke*

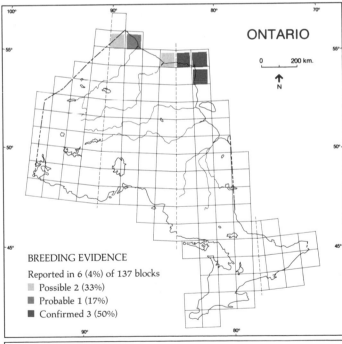

ONTARIO

0   200 km.

BREEDING EVIDENCE

Reported in 6 (4%) of 137 blocks

Possible 2 (33%)
Probable 1 (17%)
Confirmed 3 (50%)

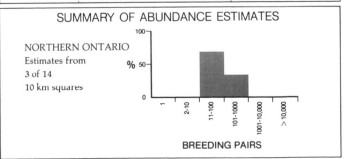

SUMMARY OF ABUNDANCE ESTIMATES

NORTHERN ONTARIO
Estimates from
3 of 14
10 km squares

BREEDING PAIRS

529

LEAST SANDPIPER
Bécasseau minuscule
Calidris minutilla

S.L. HOUSE

The Least Sandpiper nests across subarctic and low arctic North America, and locally in the Maritime Provinces. In Ontario, its breeding range includes the tundra and marshes along the Hudson Bay and James Bay coasts and inland, perhaps as much as 200 km, onto the Hudson Bay Lowland. It prefers wet breeding habitat, nesting in sedgy or grassy bogs, fens, or marshes south of the tree-line, where hummocks or knolls provide dry nest sites. In tundra situations, the Least Sandpiper prefers wet, grassy, tussock tundra and generally avoids the dry tundra areas, where Semipalmated Sandpipers are a more common species (Jehl and Smith 1970).

Few nests have been found since the first Ontario nest was located at Fort Severn in 1940 (Baillie 1961), and most records are of downy young (Peck and James 1983). It was a relatively common breeder in the Cape Henrietta Maria region in 1970 (Peck 1972) and nests commonly on the nearby Twin Islands, NWT, in James Bay (Manning 1981) and at Churchill, Manitoba (Jehl and Smith 1970). All records mapped by Peck and James (1983) were from the Cape Henrietta Maria region and west to Fort Severn along the Hudson Bay coast.

The peak of conspicuous song, display flight, and chase activity by the Least Sandpiper is early in the season, and brief (Miller 1979). Since most northern atlassing was done later, it would be necessary for atlassers to visit preferred habitats during the pre-fledging period to optimize the chances of confirming breeding. Least Sandpipers show agitated distraction behaviour when disturbed with young, whereas they can be quiet and sit tightly on the nest. Of the 19 records of confirmed breeding of this species, 17 were established by observation of distraction display or flightless young, whereas only 2 were by the finding of nests.

Peck and James (1983) postulated that the Least Sandpi-

per's nesting range "extends undoubtedly both for some distance south along the James Bay coast and for a short way inland along the major rivers in the northern coastal regions." The atlas data indicate that this species does not nest only along the Hudson Bay coast. Unfledged young in the block immediately north of Moosonee, together with other breeding evidence from the mouth of the Albany River and Akimiski Island, NWT, indicate that it may breed in suitable habitat elsewhere along the west coast of James Bay. Likewise, observation of a distraction display 100 km south of the Hudson Bay coast, with other evidence of probable breeding along the Winisk, Fawn, Sachigo, and Severn Rivers, indicates a widespread breeding range on the Hudson Bay Lowland. The Least Sandpiper may be presumed to nest, perhaps locally, in more blocks throughout at least the northern portion of the Hudson Bay Lowland and along both the Hudson Bay and James Bay coasts; apparently suitable nesting habitat exists in these areas.

Although the Least Sandpiper is sparsely distributed over much of its range, abundance estimates indicate that it is fairly common in favourable tundra habitat.-- *R. Harris*

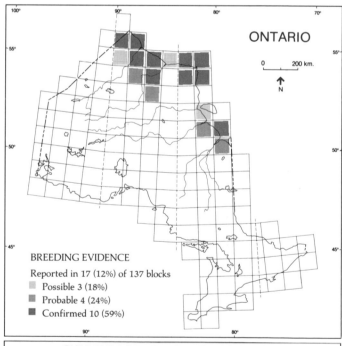

ONTARIO

0   200 km.

N

BREEDING EVIDENCE
Reported in 17 (12%) of 137 blocks
Possible 3 (18%)
Probable 4 (24%)
Confirmed 10 (59%)

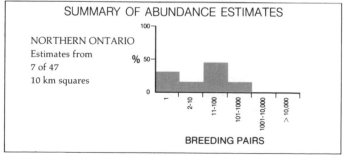

SUMMARY OF ABUNDANCE ESTIMATES

NORTHERN ONTARIO
Estimates from
7 of 47
10 km squares

BREEDING PAIRS

PECTORAL SANDPIPER
Bécasseau à poitrine cendrée
Calidris melanotos

S.L.HOUSE

The Pectoral Sandpiper is known to nest on the coastal tundra of Siberia and east across the northern mainland coast of North America, including some high arctic islands, to the west shores of Hudson Bay (Godfrey 1986). In Ontario, this species is restricted to Ontario's only tundra, along the Hudson Bay coastline and at Cape Henrietta Maria, where it reaches the most southerly point in its breeding range. It occurs over a wide variety of tundra habitats, but is usually associated with flat, poorly drained, and wet tundra with low grass-sedge or shrub cover. Most of its winter range is in southern South America.

The Pectoral Sandpiper was first established as an Ontario breeding species on 5 July 1948, with the collection of 2 flightless young in the Cape Henrietta Maria region (Peck and James 1983). This remains the only confirmed breeding location in the province. Although the Pectoral Sandpiper was believed to nest along most of the southern Hudson Bay coast (Godfrey 1966), there was no evidence that it did so in Ontario (Peck and James 1983) or in the vicinity of Churchill, Manitoba (Jehl and Smith 1970, Cooke *et al.* 1975).

The Pectoral Sandpiper has a promiscuous mating system (Pitelka 1959). Males spend their time seeking matings with females and defending females on their territory from other males. The display of the males is unforgettable. A male flies slowly about his territory, with his streaked bib enlarged into a pendulous 'balloon', while uttering a low hooting call several times per second. Although males characteristically leave the breeding area before the young have hatched, they display throughout their stay and will sit conspicuously atop hummocks or other raised sites. Finding a nest is difficult and one has yet to be discovered in Ontario. Females leave the nest well ahead of a human intruder, often skulking

away, and are well-camouflaged in their grass-sedge habitat.

The only accepted atlas record was of a single bird in suitable breeding habitat about 10 km east of radar site 416 near Cape Henrietta Maria - the same location at which the 2 flightless young were collected in 1948. The bird was in an area of tundra, with both wet and dry areas available, on 10 July 1984. Flying young were reported nearby soon after, but it was considered possible that these birds had been hatched elsewhere and were in fact early migrants.

The species was not uncommon at the same location in 1948, but has been absent or scarce on subsequent visits (Peck 1972, R. James pers. comm.). It has been suggested that shorebirds with promiscuous mating systems tend to vary greatly in density over one region within a season, and also at one locality in different seasons (Pitelka *et al.* 1974). This means that they are likely to breed occasionally in locations that are normally out of range; this may be of some relevance in Ontario, which is at the southern extreme of the range of the Pectoral Sandpiper.-- *R. Harris*

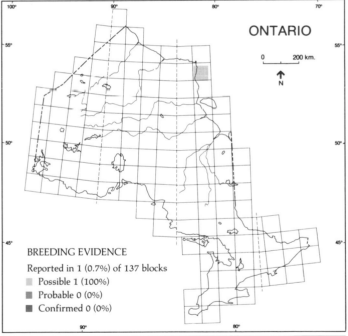

ONTARIO

0   200 km.

↑
N

BREEDING EVIDENCE

Reported in 1 (0.7%) of 137 blocks

Possible 1 (100%)
Probable 0 (0%)
Confirmed 0 (0%)

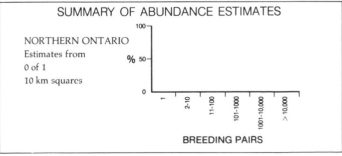

SUMMARY OF ABUNDANCE ESTIMATES

NORTHERN ONTARIO
Estimates from
0 of 1
10 km squares

BREEDING PAIRS

PURPLE SANDPIPER *
Bécasseau violet
Calidris maritima

S.L.HOUSE

in which Purple Sandpipers have been reported nesting elsewhere, the bird was judged to be a possible breeder, but there was no other evidence of breeding. Further exploration of the barren outer ridges of Cape Henrietta Maria, which have rarely been visited by ornithologists, may turn up more conclusive evidence of nesting in the future. Nevertheless, it is unlikely that the Purple Sandpiper will prove to be more than a rare and infrequent breeder in the province.-- *D.J.T. Hussell*

The Purple Sandpiper is one of the most hardy of the shorebirds. It nests in barren and inhospitable habitats in the arctic, and winters farther north than any other arctic-breeding shorebird, on rocky north Atlantic coasts. In winter, it may be encountered as far south as 35^{0}N in North America and 40^{0}N in western Europe, but most of the population is concentrated farther north, wherever there are ice-free rocky shores. It breeds in arctic, subarctic and alpine regions from central Siberia west to the eastern Canadian Arctic, in habitats that include shingly beaches, patches of wet mossy tundra behind beaches, rocky upland tundra, boulder fields near streams on high plateaus, as well as rocky islands and inlets. In the Pacific region, it is replaced by the closely-related Rock Sandpiper *(Calidris ptilocnemis)* (Cramp and Simmons 1983).

In Canada, the Purple Sandpiper breeds from Ellesmere Island in the north southward to the islands in Hudson and James Bays (Godfrey 1986). It has never been reported as a breeding bird in Ontario but it is a potential breeder in far northern Ontario along the coast of Hudson Bay, especially adjacent to the rocky, shingle beaches of Cape Henrietta Maria. The nearest regular nesting area to Ontario is on the Belcher Islands, NWT, 160 km northeast of Cape Henrietta Maria, where it is probably the commonest nesting shorebird (Freeman 1970b, Manning 1976). Breeding has also been reported on North Twin Island in James Bay, which is about 250 km southeast of Cape Henrietta Maria (Manning 1981).

The only atlas record for the Purple Sandpiper was a single bird that was observed briefly along the southeastern shore of Cape Henrietta Maria on 6 July 1985 by the author and Erica Dunn. It was near a mossy tundra pool between shattered limestone beach ridges, a short distance inland from the James Bay shore. As the habitat was similar to that

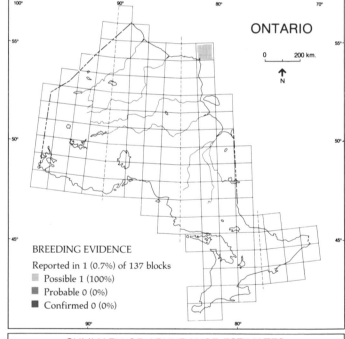

BREEDING EVIDENCE
Reported in 1 (0.7%) of 137 blocks
Possible 1 (100%)
Probable 0 (0%)
Confirmed 0 (0%)

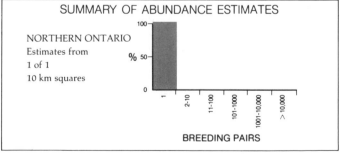

SUMMARY OF ABUNDANCE ESTIMATES

NORTHERN ONTARIO
Estimates from
1 of 1
10 km squares

%

BREEDING PAIRS

DUNLIN
Bécasseau variable
Calidris alpina

S.L. HOUSE

The breeding distribution of the Dunlin is holarctic, ranging from some high arctic islands and mainland tundra, down adjoining oceanic coasts in tundra-like habitats, through subarctic and boreal regions even to temperate areas (Cramp and Simmons 1983). In North America, it nests primarily in the low arctic, along the northern coastal mainland tundra, eastward to Southampton Island and southward along the west coast of Hudson Bay; it is apparently absent from much of the interior Canadian tundra west of Bathurst Inlet (Godfrey 1986). In Ontario, Dunlins have been found nesting only in the northern tundra region. Throughout its range, it breeds primarily in wet tussock tundra and peat-hummock tundra. After hatching occurs, families move to low wet marshes and pond edges, where insect food is more available (Holmes 1966). The winter range in the New World is in the coastal US, British Columbia, and Mexico.

The first reported indication of breeding in Ontario was the account of a Dunlin performing distraction displays on 9 and 10 August 1944, about 30 km south of Cape Henrietta Maria (Smith 1944). A brood of 4 in a nest at Cape Henrietta Maria in 1947 provided the first documented breeding evidence (Manning 1952). Subsequent records indicate that it is a common summer resident in the Cape Henrietta Maria region (Peck 1972), and that it may nest at scattered localities along the entire Hudson Bay coastline of Ontario (Schueler *et al.* 1974, Peck and James 1983). The Dunlin is also known to nest commonly on nearby North Twin Island, NWT, in James Bay (Manning 1981) and at Churchill, Manitoba (Jehl and Smith 1970).

The Dunlin is a relatively conspicuous bird on the tundra and therefore should have been easily located by atlassers. Pair formation usually occurs on the nesting territories shortly after spring arrival. Dunlins are then active and con-

spicuous as they engage in display flights, song, and aerial and ground chases. Although Dunlins can be extremely tight sitters on the nest during incubation, especially as incubation becomes advanced, they often indicate their presence with alarm calls and upright postures in conspicuous locations away from the nest. Confirmation was most often achieved by finding flightless young (62%), or observing distraction displays (31%). A nest was recorded in only one square.

As expected, the atlas data reveal that the Dunlin is restricted to the coastal tundra from the Cape Henrietta Maria region westward. The widest strips of coastal tundra, and therefore the largest areas of Dunlin nesting habitat, are found in the Cape region and in extreme northwestern Ontario adjacent to the Manitoba border. Thirteen of the 22 squares with breeding evidence, and 9 of the 13 confirmations, were in the Cape region. The Dunlin might also be expected to nest at scattered localities in those Hudson Bay coastal blocks where it was not recorded.

In the Cape Henrietta Maria region, atlassers estimated abundances at 101 to 1,000 pairs (1 square) and 11 to 100 pairs (2 squares), indicating that the Dunlin is fairly common in favourable habitat.-- *R. Harris*

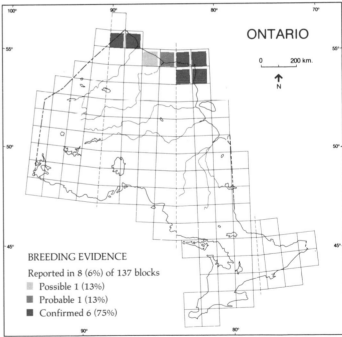

ONTARIO

0   200 km.

BREEDING EVIDENCE
Reported in 8 (6%) of 137 blocks
Possible 1 (13%)
Probable 1 (13%)
Confirmed 6 (75%)

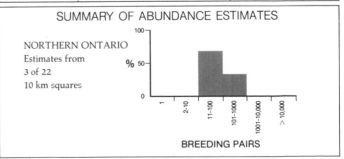

SUMMARY OF ABUNDANCE ESTIMATES

NORTHERN ONTARIO
Estimates from
3 of 22
10 km squares

BREEDING PAIRS

533

STILT SANDPIPER
Bécasseau à échasses
Calidris himantopus

S.L. HOUSE

The breeding range of the Stilt Sandpiper is restricted to the North American tundra, from northern Alaska southeast to the Cape Henrietta Maria region of far northern Ontario, chiefly in the subarctic and low arctic zones (Jehl 1973). It winters primarily in South America from Bolivia and Brazil south to northern Chile and Argentina. In Ontario, it is found breeding in wet tundra meadows, interrupted by raised beach ridges. In this habitat, it is found in association with Lesser Golden-Plovers, Dunlin, Red-necked Phalaropes, Savannah Sparrows, Smith's Longspurs, and Lapland Longspurs. It is uncommon through most of its Ontario breeding range.

The first record of breeding of the Stilt Sandpiper in Ontario was obtained by Manning (1952), when he collected a half-fledged juvenile near the limestone ridges of Cape Henrietta Maria. In 1948, 2 downy young were collected by a ROM party 40 km south of Cape Henrietta Maria (Peck 1972), and in 1962 another downy young was collected by the author from a brood of 4, west of the mouth of the Sutton River on the Hudson Bay coast (Schueler *et al.* 1974). The only nest discovered in Ontario to date was found by R.I.G. Morrison (unpubl. data) near radar site 415 on 20 June 1976.

The Stilt Sandpiper has a wide range of vocalizations, the most spectacular of which is the 'song', which is usually given in flight and ends with a repeated 'ee-haw' note like the braying of a donkey (Jehl 1973). Atlassers on the breeding grounds in late June will have located Stilt Sandpipers easily from these conspicuous displays. Several days after incubation of the eggs has begun, however, 'song' displays are reduced and the birds become so quiet that their presence may go unsuspected. Both sexes incubate, but they can rarely be surprised at the nest; when a human approaches within 75 to 100 m, the incubating bird flies off inconspicu-

ously. Distraction displays are relatively uncommon (Jehl 1973). At hatch time, Stilt Sandpipers fly toward intruders, calling and giving wing-up displays on landing (Jehl 1973), making it easy for atlassers to record probable breeding using the agitated behaviour code. Stilt Sandpipers were reported in only 9 squares in 6 blocks. Probable breeding was reported in 6 squares: agitated behaviour was noted in 5 squares, and a pair of birds in the other. In the remaining 3 squares, birds were observed in suitable habitat. Breeding was not confirmed during the atlas period.

Previous records of juveniles and downy young come from the 2 most easterly blocks of the indicated range and from the Cape Henrietta Maria block, where the species was not seen by atlassers at all in 1985. Nevertheless, the map indicates that scattered populations of Stilt Sandpipers probably occur in appropriate habitat on the tundra strip all along the coast from radar site 415 to the Manitoba border.

The Stilt Sandpiper appears to be scarcer in most localities than several other shorebirds, such as Dunlin, Least Sandpipers, and Semipalmated Sandpipers, but in some places it is fairly common. Atlassers estimated 11 to 100 pairs in a square 40 km south of Cape Henrietta Maria.-- *D.J.T. Hussell*

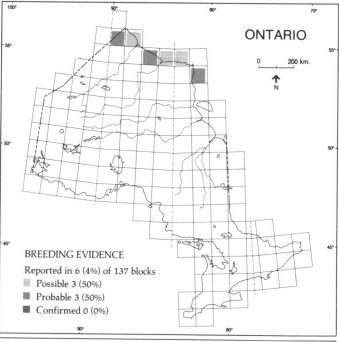

ONTARIO

0   200 km.

N

BREEDING EVIDENCE

Reported in 6 (4%) of 137 blocks

Possible 3 (50%)

Probable 3 (50%)

Confirmed 0 (0%)

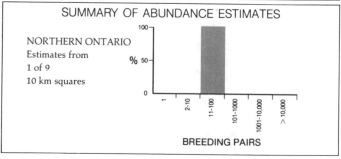

SUMMARY OF ABUNDANCE ESTIMATES

NORTHERN ONTARIO
Estimates from
1 of 9
10 km squares

%

BREEDING PAIRS

SHORT-BILLED DOWITCHER
Bécasseau roux
Limnodromus griseus

S.L. HOUSE

Dowitcher in Ontario. In the limited time available, every effort was made to cover muskeg habitat on foot. Nevertheless, few atlassers were able to walk more than a few kilometres from rivers, so vast areas of the Lowland remained unvisited. The farthest inland record is of a single agitated bird in a fen/pond area on sparsely treed Lowland habitat, about 3 km from the Fawn River. As such habitat is abundant on the Lowland, it seems likely that the species is more widespread on the interior of the Lowland than atlas records would indicate. The single confirmed breeding record is of flightless young, which were eventually photographed in the hand, on Akimiski Island, NWT, on 12 July 1984. This and most other records are from squares adjacent to coastal areas, perhaps indicating that this species is most abundant in Ontario's tree-line habitat.

Only 2 abundance estimates were submitted; one of 11 to 100 pairs in a square near the Hudson Bay coast, and one of one pair in a square on Akimiski Island.-- *R. Harris*

The Short-billed Dowitcher nests across much of boreal and subarctic North America, and winters from the southern US to Brazil and Peru (Godfrey 1986). It is associated with wetlands that are scattered throughout the coniferous forest: muskegs, bogs, hummocky sedge marshes, patterned fens, and other wet areas with low vegetation. In Ontario, its breeding status is very poorly known. It appears to be a rare to uncommon breeder restricted to the Hudson Bay Lowland.

Downy and recently fledged young Short-billed Dowitchers found in 1963 near Winisk by Tuck (1968) established this species as an Ontario breeder. However, because no material evidence was secured, its status remains as 'hypothetical' (Peck and James 1983). It has been found nesting on nearby North Twin Island, NWT (Manning 1981), in James Bay and is a fairly common summer resident at Churchill, Manitoba (Jehl and Smith 1970). To date, despite a considerable amount of atlas field work in suitable habitat, a Short-billed Dowitcher nest has yet to be discovered in Ontario.

This species can very easily be missed by atlassers. Access to its wetland habitat is often difficult, its song is not well known, and for much of the nesting cycle it is quiet and retiring. Rowan (1927) confessed that "The nest of these trustful and confiding birds must be about the hardest of all shorebirds to find". The best time to locate Short-billed Dowitchers is in the 2 to 3 weeks following their spring arrival, before the commencement of incubation, when they are most frequently performing flight songs and aerial chases. However, most northern atlassing was done after this period. Of a total of 8 records, 4 are of single birds, 3 are of agitated behaviour, and one is of young birds.

It is unclear whether the atlas data correctly represent the breeding distribution and abundance of the Short-billed

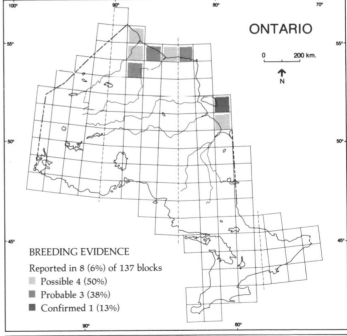

ONTARIO

0   200 km.

BREEDING EVIDENCE
Reported in 8 (6%) of 137 blocks
Possible 4 (50%)
Probable 3 (38%)
Confirmed 1 (13%)

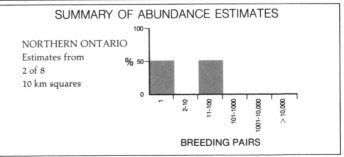

SUMMARY OF ABUNDANCE ESTIMATES

NORTHERN ONTARIO
Estimates from
2 of 8
10 km squares

BREEDING PAIRS

RED-NECKED PHALAROPE
Phalarope hyperboréen
Phalaropus lobatus

S.L.HOUSE

young when they are about 2 weeks old (Hilden and Vuolanto 1972). This behaviour is likely to make it difficult to obtain breeding confirmation late in the season when many atlassers were in the field. Only 3 squares in which the species was reported had confirmed records.

The atlas map and data indicate the possibility of a continuous breeding distribution along the shores of Hudson Bay and south into James Bay. However, 13 of the squares in which Red-necked Phalaropes were reported are within 80 km west or 40 km south of Cape Henrietta Maria. Three of the squares are farther to the west along the Hudson Bay coast, and 2 are farther south on James Bay. The most southerly record is of a pair in the Northbluff Point square about 10 km north of the mouth of the Moose River.

The Red-necked Phalarope is locally common in the vicinity of Cape Henrietta Maria. Abundance estimates for 4 squares at or near the Cape were 11 to 100 pairs in 2 squares and 101 to 1,000 pairs in 2 other squares. No estimates are available elsewhere, but the lack of confirmed breeding and the sparseness of records away from the Cape are probably indicative of less dense populations.-- *D.J.T. Hussell*

The Red-necked Phalarope has a circumpolar breeding distribution in the low arctic and subarctic regions (Cramp and Simmons 1983). In Canada, its range extends from central Baffin Island south to James Bay (Godfrey 1986). In Ontario, it is confined to tundra and tundra-forest transition zones adjacent to Hudson and James Bays. Nesting habitat is typically low-lying wet tundra, including marshes with small ponds and islands in river deltas, but it may also occupy drier ground with sparse vegetation. Nests are usually close to water on moss or among grasses, sedges or other low vegetation (Hilden and Vuolanto 1972, Cramp and Simmons 1983, Peck and James 1983, Reynolds 1987). The Red-necked Phalarope winters at sea in both the Pacific and Atlantic Oceans (Cramp and Simmons 1983; Haney 1985).

The first breeding record for Ontario was obtained in 1947 when downy young were collected at Little Cape, about 80 km west of Cape Henrietta Maria (Manning 1952). In 1948, a ROM party found 2 nests about 40 km south of Cape Henrietta Maria and all subsequent nest records prior to and during the atlas years are from the same region (Peck 1972, atlas data).

Only the male incubates and cares for the young. The Red-necked Phalarope is either monogamous or, less frequently, polygamous or serially polyandrous. If the male loses a clutch he may incubate a second one laid by another female; or a female may lay 2 clutches if a second male is available (Hilden and Voulanto 1972; Reynolds 1987). The adults are not difficult to observe in the breeding areas. Pair bonds last for several days, at least until the start of incubation, and human intruders are often greeted by birds giving mildly agitated 'prek' calls (Cramp and Simmons 1983). Nests and downy young are more difficult to find and the attentiveness of the male decreases rapidly until he abandons the

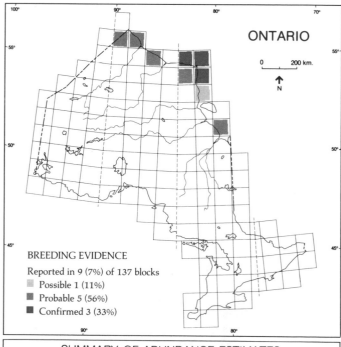

ONTARIO

BREEDING EVIDENCE
Reported in 9 (7%) of 137 blocks
Possible 1 (11%)
Probable 5 (56%)
Confirmed 3 (33%)

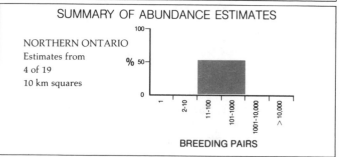

SUMMARY OF ABUNDANCE ESTIMATES

NORTHERN ONTARIO
Estimates from
4 of 19
10 km squares

BREEDING PAIRS

PARASITIC JAEGER
Labbe parasite
Stercorarius parasiticus

S.L. HOUSE

The breeding range of the Parasitic Jaeger is circumpolar, extending from high arctic tundra into subarctic and temperate maritime regions. In North America, it is strictly a tundra nester, occurring in coastal, low and high arctic tundra from coast to coast. Although this species sometimes nests in loose colonies elsewhere, it is known only as a solitary nester in North America (Parmelee *et al.* 1967, Maher 1974), generally nesting on hummocks or otherwise dry sites in low, wet, grassy, or mossy tundra close to ponds or lakes. During winter and in migration, it is mainly pelagic, in both the Atlantic and Pacific Oceans.

The first Ontario breeding evidence for this species, a downy young found dead near a nest, was collected at Cape Henrietta Maria on 22 July 1947 (Manning 1952). On 23 July 1948, a ROM field party collected a half-grown and still flightless young and discovered a recently vacated nest, both in the Cape Henrietta Maria region (Baillie 1962). A nest with 2 eggs was found on 25 June 1970, also in the Cape region, and to date is the only occupied nest reported (Peck 1972).

There are several reasons why more is not known about the Parasitic Jaeger in Ontario, and why there were few breeding confirmations for the atlas. In Ontario, the tundra nesting areas are relatively inaccessible, and inclement weather and wet terrain there make bird surveys difficult. Also, in North America Parasitic Jaegers nest in low densities and maintain relatively large territories, of which only a central core area is defended for much of the season (Parmelee *et al.* 1967, Maher 1974). Consequently, it may be necessary to cover much terrain before encountering a nesting pair defending a territory. Parasitic Jaegers are, however, obviously aggressive when the nest or young are closely approached (within about 100 m) and then usually are easily confirmed as breeding.

No nests or flightless young were located during the atlas survey. However, 3 pairs were observed, in 2 adjacent squares in the same block, performing 'high intensity' distraction displays (Cramp and Simmons 1983) on sedgy, tussock tundra in July 1985. The distraction display noted (D. Shepherd pers. comm.), in which an adult lands near the intruder, keeps its wings elevated, and then hops away dragging its tail and flopping its wings, is characteristic of the late incubation or early chick stage of the breeding cycle (Cramp and Simmons 1983).

The atlas data and apparent habitat suitability suggest that the Parasitic Jaeger nests along the entire coastal tundra from the Cape Henrietta Maria region westward. Four of the 18 squares with data have abundance estimates: one of a single pair, and three of 2 to 10 pairs. This species is apparently uncommon in its tundra breeding habitat, with most of the Ontario breeding population being located in the Cape Henrietta Maria region, where the tundra is most extensive. From within about a 100 km radius of the Cape, there are 13 records, 2 of which are accompanied by abundance estimates in the range of 2 to 10 pairs per square. There has been no apparent change in its numbers in recent times.-- *G.K. Peck*

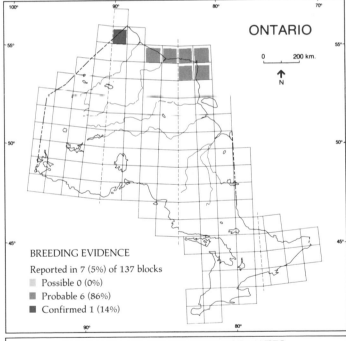

ONTARIO

0   200 km.

N

BREEDING EVIDENCE
Reported in 7 (5%) of 137 blocks
Possible 0 (0%)
Probable 6 (86%)
Confirmed 1 (14%)

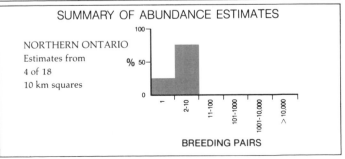

SUMMARY OF ABUNDANCE ESTIMATES

NORTHERN ONTARIO
Estimates from
4 of 18
10 km squares

BREEDING PAIRS

537

BONAPARTE'S GULL
Mouette de Bonaparte
Larus philadelphia

S.L.HOUSE

Unlike most gulls and terns, which normally nest on the ground in fairly open situations, the small and graceful Bonaparte's Gull has the unusual habit of nesting in trees. It is a North American species, nesting in the coniferous belt of western and central Canada as well as in Alaska, and wintering from extreme southern Canada to northern Mexico and the West Indies. In Ontario, the breeding range comprises much of the vast expanse of boreal forest in the northern part of the province (Godfrey 1986).

It is known that the Bonaparte's Gull usually nests in coniferous trees near muskegs, ponds, and lakes (Godfrey 1986), but its breeding biology is virtually unstudied; the first known breeding in the province occurred in 1937 (Peck and James 1983). More than 60 years ago, it was suggested that its habit of nesting singly or in small, widely dispersed colonies in blackfly-infested swamps had deterred most biologists from studying the species on its nesting grounds (Henderson 1926). Nevertheless, it seems a fair assumption that the ecological requirements of the Bonaparte's Gull, unlike those of the Ring-billed Gull, did not change during the last century, and that there have been no major changes in its breeding distribution in recent years.

From the air, nesting Bonaparte's Gulls are usually fairly easy to spot, but on the ground their small colonies are extremely difficult to find unless they happen to be located on bodies of water with good boat access. Thus, the lack of breeding evidence for many blocks in northern Ontario may well be due to inadequate coverage rather than the absence of nesting gulls. Once the presence of gulls has been noticed at a site, it is not difficult to confirm breeding. They will call and circle observers and may dive at them. Also, with a bit of patience, an observer will see conspicuous birds alight on their nests (R. James pers. comm.). Actual nests were found

in only 2 squares, further suggesting that birds were difficult to locate.

In the course of CWS aerial surveys over northern Ontario during the atlas years, it was noted that, on the Hudson Bay Lowland and in the Clay Belt, nesting Bonaparte's Gulls are well distributed and small colonies (10 to 20 birds) probably occur in all blocks. On the northern part of the Canadian Shield, the species is less common and highly localized (R.K. Ross pers. comm.). These overall impressions are in general agreement with the distribution map, which shows a well-defined range that extends from northeastern Ontario to the far northwest but fails to include any of the area south or west of Pickle Lake, part of which is included in Godfrey's (1986) range map. Ferguson (1981) saw small numbers during the 1978 breeding season on the Manitoba side of Lake of the Woods.

Because there are gaps in our knowledge of the breeding distribution and habitat requirements of the Bonaparte's Gull, it is impossible to comment accurately on its abundance as a nesting bird in Ontario. Abundance estimates generally support the observations that numbers on the breeding grounds are usually low, as noted in the preceding paragraph.-- *H. Blokpoel*

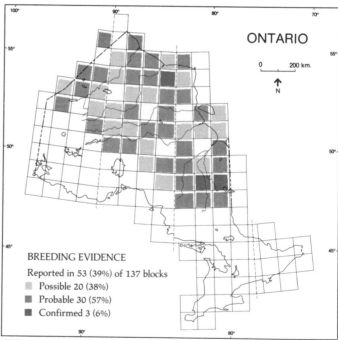

ONTARIO

0   200 km.

N

BREEDING EVIDENCE

Reported in 53 (39%) of 137 blocks

Possible 20 (38%)

Probable 30 (57%)

Confirmed 3 (6%)

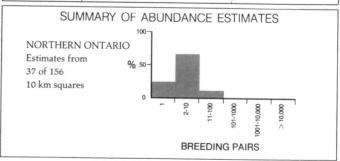

SUMMARY OF ABUNDANCE ESTIMATES

NORTHERN ONTARIO
Estimates from
37 of 156
10 km squares

%

BREEDING PAIRS

538

GLAUCOUS GULL *
Goéland bourgmestre
Larus hyperboreus

S.L. HOUSE

The Glaucous Gull is a holarctic species which, in North America, breeds along the shoreline and on islands of the arctic from the Bering Sea to the north Atlantic. In Ontario, it occurs rarely along the James Bay coast in summer and in small numbers as far south as the lower Great Lakes in winter (James *et al.* 1976). Its preferred nesting habitats are cliffs (usually near the shoreline and often very precipitous and rising to hundreds of metres), where it nests colonially, as well as small islets in tundra pools or at river mouths, where it nests solitarily. However, where the two species occur together, the Herring Gull often competes successfully with the Glaucous Gull for preferred nest-sites such as boulders and islets in lakes (MacPherson 1961).

There are no data to suggest that the distribution of the Glaucous Gull has changed significantly in Ontario over the last 100 years. In the late 1800's and early 1900's, it was recognized as a winter visitor to Lake Ontario (McIlwraith 1894, Taverner 1919, Quilliam 1973); that is still its status today (Godfrey 1986). Its winter occurrence on the northern Great Lakes is better documented today than in past years (Speirs 1985) but it probably always occurred there.

This species is difficult to confirm as breeding in Ontario. It is rare in summer. The most likely habitat is along the relatively inaccessible James Bay and Hudson Bay coasts; the coastal habitat on Hudson Bay is probably marginal at best. On the positive side, the habitat can be flown over with care, and at close range the species can probably be separated from the commoner Herring Gull by wing-tip colouration. As with most gulls, finding the nest should not be the difficult part, but rather finding a bird on territory would be the challenge.

During the atlas field work period, there was only one record of probable breeding for this species. This record

came from the shores of Hudson Bay between Winisk and Cape Henrietta Maria. David Shepherd and George Melvin (pers. comm.) observed an adult pair, which they thought was territorial, on a small islet at the mouth of the Sutton River. They also observed several immature birds. Glaucous Gulls have been seen in the Cape Henrietta Maria area previously. Clifford Hope saw them there in 1948 but Peck did not in 1972 (Peck 1972). Atlassers Erica Dunn, David Hussell and Roy Smith saw two Glaucous Gulls at Cape Henrietta Maria on 26 June and 6 July 1985, but the birds were flying and were not considered to be in suitable breeding habitat. They are known to nest on the Belcher Islands, 160 km northeast in Hudson Bay (Godfrey 1986). Although this species has never been found nesting in Ontario, and is rare in summer, there are large areas of apparently suitable habitat in the coastal tundra along the Hudson and James Bay coasts, where future field work may provide nesting records. However, the distribution of the species in the province may be limited by lack of sea cliffs, the presence of Herring Gulls, and the southerly latitude, which together make it unlikely that the Glaucous Gull will prove to be more than an occasional and irregular breeder in Ontario.-- *D.V. Weseloh*

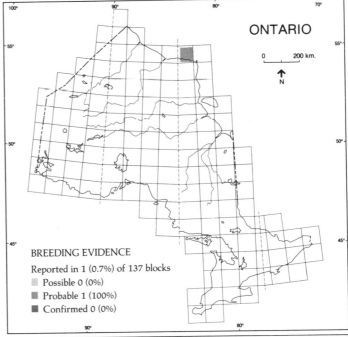

ONTARIO

0   200 km.

N

BREEDING EVIDENCE

Reported in 1 (0.7%) of 137 blocks

Possible 0 (0%)

Probable 1 (100%)

Confirmed 0 (0%)

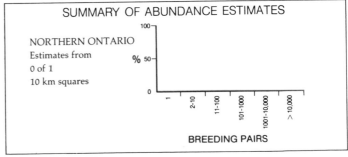

SUMMARY OF ABUNDANCE ESTIMATES

NORTHERN ONTARIO
Estimates from
0 of 1
10 km squares

%

BREEDING PAIRS

ARCTIC TERN
Sterne arctique
Sterna paradisaea

S.L. HOUSE

The Arctic Tern is probably best known for its lengthy annual migrations. After breeding in the northern parts of the Northern Hemisphere, it migrates to the southern sections of the Southern Hemisphere. Having spent the boreal winter near Antarctica, it makes the long flight back, returning on time for the breeding season (Austin 1961).

The breeding range of the Arctic Tern includes the coastlines of northern North America, Greenland, and northern Eurasia (Voous 1960, Cramp 1985). In North America, it nests in Alaska, throughout the Canadian arctic, and along the Atlantic coast south to Massachusetts. In Ontario, nesting is restricted to the coasts of James and Hudson Bays. The Arctic Tern prefers open, uncluttered habitat that provides good visibility, allows easy take-offs and landings, and does not impede its movements on the ground. Nesting habitat includes dunes, sandspits, sand and gravel bars and beaches, rocky shores and islands, and marshy tundra (Peck and James 1983, Godfrey 1986).

The first conclusive evidence of breeding of the Arctic Tern in Ontario was not obtained until 1940, when a female containing an egg was collected at Fort Severn (Baillie 1961). The first nests were found 40 km south of Cape Henrietta Maria in 1948. Between 1948 and the atlas years only a few other nesting sites have been discovered: the Sutton River in 1962, East Pen Island in 1963, the Winisk River in 1965, and Man Island in Akimiski Strait in James Bay in 1978 (Peck and James 1983).

Arctic Terns usually nest in colonies in open areas and vigorously defend their nests against intruders. Thus, it is relatively easy to find nesting sites and to confirm breeding. It is likely that the map properly depicts the main breeding range of Ontario's Arctic Terns.

Specific localities of confirmed breeding include East Pen

Island, near the mouths of the Severn, Shagamu, Winisk, and Sutton Rivers, on Little Bear Island, NWT (about 40 km west of Cape Henrietta Maria), and on the northeast coast of Akimiski Island, NWT. There have been no subsequent nesting records in the block immediately south of Cape Henrietta Maria, where the first Ontario nests were found (Peck 1972), and atlassers found no breeding evidence there in 1984 or on the Cape itself in 1985. A comparison of the maps for the Arctic Tern and the Common Tern shows that there is no overlap in the breeding distribution of the 2 species, with the exception of one block containing the Severn River, just south of the Hudson Bay coast. In that block, there was confirmed nesting by Arctic Terns, and probable nesting by Common Terns in adjacent squares. The Arctic Tern was observed breeding mainly in coastal blocks, whereas the Common Tern breeds farther inland. Elsewhere in North America, the 2 species are sometimes found nesting close together. In Maine and Massachusetts, for example, they share many of their nesting islands along the Atlantic coast (Erwin 1979).

The abundance estimates and distribution data suggest that the Arctic Tern is an uncommon and sparsely distributed breeding bird in Ontario's far north.-- *H. Blokpoel*

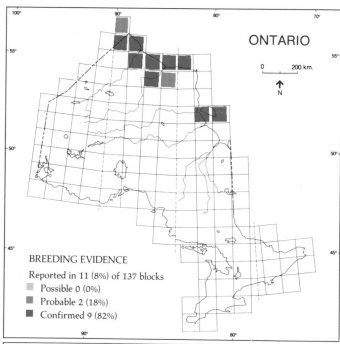

ONTARIO

0   200 km.

BREEDING EVIDENCE
Reported in 11 (8%) of 137 blocks
Possible 0 (0%)
Probable 2 (18%)
Confirmed 9 (82%)

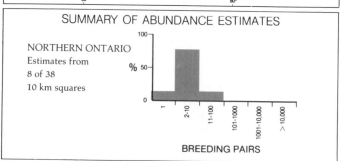

SUMMARY OF ABUNDANCE ESTIMATES

NORTHERN ONTARIO
Estimates from
8 of 38
10 km squares

%

BREEDING PAIRS

SNOWY OWL *
Harfang des neiges
Nyctea scandiaca

The robust Snowy Owl of the arctic is a winter visitor to southern Canada, periodically occurring in greater numbers than usual (Snyder 1949b). It is during these southern incursions that many Canadians have an opportunity to see one of the world's most striking and distinctive owls.

The species is circumpolar, breeding on the arctic tundras of the world. It may winter within its breeding range or irrupt southwards, in North America reaching as far south as the northern US. Major irruptions are correlated with a drop in the population of its main food supply - lemmings.

The early ornithologists of Ontario wrote of the Snowy Owl's winter visits to the southern sections of the province, when many owls were shot. The species was not listed as a breeding bird in Ontario by any of them, including Baillie and Harrington (1936). More recently, Peck and James (1983) did not consider it to be nesting in this province, although Godfrey (1986) noted that it nests on the Belcher Islands (NWT) in Hudson Bay, and at Churchill, Manitoba. Neither location is very far removed from the Ontario coast of Hudson Bay.

The breeding habitat of the Snowy Owl consists of hummocky rolling terrain, and the nest is placed in a slight depression on a raised hummock, outcrop or ridge. Because daylight is virtually continuous in summer within its nesting range, the Snowy Owl has adapted by being active both in daylight and at night. Although well camouflaged against the white snow in winter, the Snowy Owl is very obvious on the brown and green background of the summer tundra, and, if present, is unlikely to be missed by observers. Out of the breeding range, the species frequents open fields or shorelines that resemble the treeless tundra.

The only Ontario breeding season record during the atlas period was on the Hudson Bay coast near Fort Severn from 18 to 21 July 1983. A single adult individual was observed by Bob and Terri Thobaben on the west shore at the mouth of Goose Creek. It was observed hunting in an open sedge meadow containing a few dwarf willows. Other summer sight records are known from the north coast (Peck 1972) and perhaps only a lack of observer coverage has prevented breeding confirmation. However, very few birds occur in Ontario in summer and breeding is likely only during years when populations of small rodents are at peak abundance.-- *R.D. Weir*

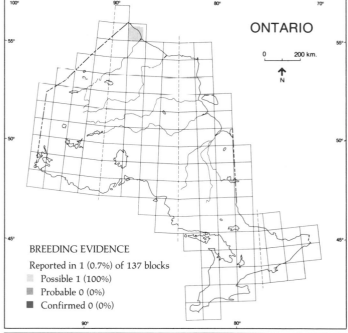

ONTARIO

0   200 km.

N

BREEDING EVIDENCE

Reported in 1 (0.7%) of 137 blocks

Possible 1 (100%)
Probable 0 (0%)
Confirmed 0 (0%)

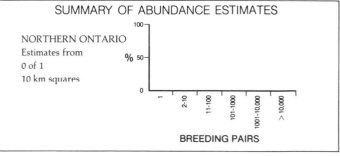

SUMMARY OF ABUNDANCE ESTIMATES

NORTHERN ONTARIO
Estimates from
0 of 1
10 km squares

%

BREEDING PAIRS

NORTHERN HAWK-OWL
Chouette épervière
Surnia ulula

birds, and one was of an adult carrying food for young. The family group remains together for several months.

The map shows all occurrences of this species to be north of the 48th parallel, about the latitude of Kirkland Lake, as expected. However, there is surprisingly little overlap with the previously published breeding records (Peck and James 1983), which were near Longlac, at Thunder Bay, in Rainy River District, near Kenora, and near Ottawa: the only overlap occurred in southern Cochrane District, where there is a lot of suitable habitat. The rarity of the species, combined with a deficiency of previous field work in much of its range, is a more likely explanation for this pattern than any real change in its distribution. The species was found in only 10 blocks in northern Ontario. The relative concentration of records on the Hudson Bay Lowland, where access and travel are most difficult, suggests that that area is an important part of the breeding range of the Northern Hawk-Owl.

All observers who made abundance estimates believed that more than one pair could be found in the square where the observation was made. Nevertheless, the species is nowhere a common bird, even in the core of its range.-- *R.D. Weir*

The Northern Hawk-Owl is most often seen perched atop a tree with its body inclined forward and long tail flicking. Winter irruptions into southern Ontario from its northern breeding areas occur at irregular intervals, probably when food supplies fail. It is only at these times that most Ontario naturalists get the opportunity to view the normally sedentary Northern Hawk-Owl.

This species is a permanent resident of the northern coniferous forests of Eurasia and North America, its range extending north to the tree-line. In Ontario, it breeds in the Boreal Forest zone, with occasional records south to Parry Sound or Ottawa (James *et al.* 1976, Godfrey 1986). Its breeding habitat consists of open coniferous or mixed woods, muskeg, or burned woodlands with standing stubs.

McIlwraith (1894) noted the Northern Hawk-Owl as a rare winter visitor to southern Ontario, but considered it to be more common farther north. MacClement (1915) wrote of its being locally common at irregular intervals in the north. That it nests in northern Ontario from Cochrane west to Lake of the Woods and probably northward through Kenora District to Severn House was first suggested by Baillie and Harrington (1936). Peck and James (1983) confirmed this range and added a 1963 breeding record near Ottawa, thereby showing that nesting may occur farther south in islands of suitable habitat.

Northern Hawk-Owl nests are placed in old woodpecker holes or other tree cavities, especially those at the top of broken tree stubs. This, combined with the remoteness of nesting areas, makes them very difficult to find. A nest with young discovered in 1981 near Kiruna Lake, at the Sutton Ridges on the Hudson Bay Lowland was only the third reported for this species in Ontario (Peck and James 1983). Of the remaining confirmed atlas records, 3 were of young

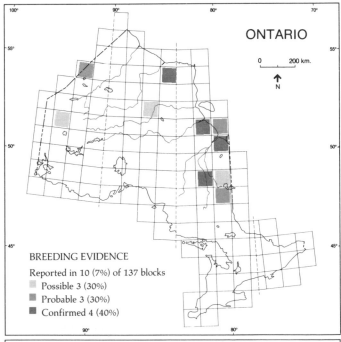

ONTARIO

0   200 km.

N

BREEDING EVIDENCE

Reported in 10 (7%) of 137 blocks

Possible 3 (30%)

Probable 3 (30%)

Confirmed 4 (40%)

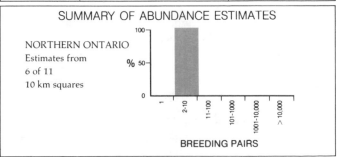

SUMMARY OF ABUNDANCE ESTIMATES

NORTHERN ONTARIO
Estimates from
6 of 11
10 km squares

%

BREEDING PAIRS

WESTERN KINGBIRD *
Tyran de l'Ouest
Tyrannus verticalis

The Western Kingbird breeds across much of western North America, as far north as southern Canada, south to California and Texas, and east to Iowa, Missouri, and rarely Michigan (Bent 1942). In Canada, its breeding range is restricted to southern British Columbia and the Prairie Provinces. It is not known as a species which breeds regularly in Ontario, but this may be because of a lack of coverage in the Lake of the Woods-Rainy River areas of northwestern Ontario. Taverner (1919) and Bent (1942) commented that it appeared to be spreading eastward since the 1890's. It was first recorded in Manitoba in 1907, but by 1919 was rather common throughout the southwestern part of that province. The preferred habitat of open scrubland with scattered small areas of brush and hedgerows is present in the Rainy River District, and this could allow for future expansion and establishment of the species as a regular breeding bird in Ontario. In treeless areas, it will also nest on fence posts, towers, and in nest boxes (Bent 1942).

Sightings of the Western Kingbird in Ontario are primarily restricted to late May, June, or September vagrants. However, a nest was found in Kent Co in southwestern Ontario in 1943 (Speirs 1985). Numerous other June sightings have been reported, including at least two during the atlas period (Speirs 1985), but no evidence of nesting has been found.

Within its range, confirming the nesting of the Western Kingbird is relatively easy. It actively defends its territory, is extremely aggressive and vocal, and its nest is easy to find. Like the Eastern Kingbird, both parent birds feed the young, and the family group will travel together for some time after the young fledge (Bent 1942) - all features that make for relatively easy confirmation of breeding.

Only one record of the Western Kingbird was accepted for the atlas project, that being a sighting of 3 birds in suitable habitat just north of Rainy River. Between 6 and 14 July 1983, B. Jones, L. Fazio, J. Heslop, and D. Elder observed the birds hunting over a grassy field with prominent scattered perches. Two were identified as females and one as a male, based on subtle plumage differences. Interestingly, W. Crins and R. Ridout had observed an adult bird on 26 May 1981 in a square 20 km south of this record. That sighting was not accepted as a breeding record because of the possibility that the bird had been a migrant at such an early date.

The Western Kingbird has never been known to nest near Rainy River, but the knowledge gained during the atlas period suggests that that area is the most likely site for regular breeding in the province.-- *G. Carpentier*

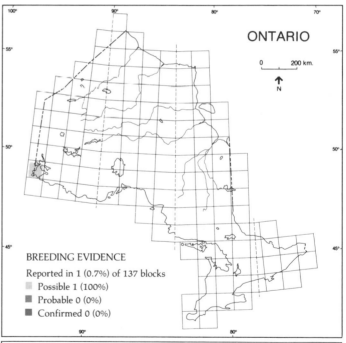

ONTARIO

0   200 km.

BREEDING EVIDENCE

Reported in 1 (0.7%) of 137 blocks
☐ Possible 1 (100%)
☐ Probable 0 (0%)
■ Confirmed 0 (0%)

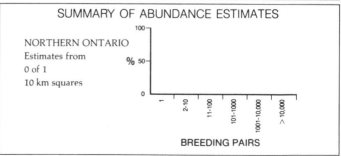

SUMMARY OF ABUNDANCE ESTIMATES

NORTHERN ONTARIO
Estimates from
0 of 1
10 km squares

% 50

BREEDING PAIRS

BLACK-BILLED MAGPIE
Pie bavarde
Pica pica

The Black-billed Magpie is one of the most distinctive of all birds. Its striking black and white plumage and long tail give it an unmistakable appearance. In addition, it is notorious wherever it occurs because of its noisy and aggressive behaviour, which includes raiding the nests of other wild birds and domestic chickens (Taverner 1939).

This species occurs year-round throughout much of Eurasia, as well as northwestern Africa and much of North America (AOU 1983). On this continent, it occupies the open country of the interior plains and surrounding areas to the north and west, almost to the west coast. In Ontario, the Black-billed Magpie nests only in the Rainy River area of northwestern Ontario, where it is a year-round resident in the prairie fringe or 'parkland' agricultural environment, an eastward extension of the range and habitat of the species. The species frequents groves of aspens, willows, or alders from about 2 to 7 m in height. It is frequently seen in the 200 to 300 km^2 of farmland in that corner of Ontario, but is very rare elsewhere in the province.

The Black-billed Magpie, for unknown reasons, has a history of fluctuation in the extent of its range. Macoun and Macoun (1909) reported a decline in numbers in Manitoba, and Speirs (1985) twice used the word "invasion" to describe unprecedented numbers in the Kingston area in 1898, including 2 specimens (Quilliam 1965), and as many as 40 in north-central Ontario in 1972. It is not known how long the Rainy River population of Black-billed Magpies has been in existence, but the first nests (also the first for Ontario) were found there in 1980 (Lamey 1981). Individual birds have recently been found at various places in northwestern Ontario where there is cleared land, including Pickle Lake in 1977 and Dryden in 1983. The species is often kept in captivity, and has frequently been known to escape in areas out-side its normal range, but no such birds have been known to breed in Ontario.

It is doubtful that any nesting Black-billed Magpies could remain undetected for long. The species is extremely vocal, as well as being highly conspicuous, and the voice carries far. Moreover, the domed nest is large, "about 2 feet in height by 1 foot in diameter, or slightly larger" (Bent 1946), although it is often well hidden among the leaves of a grove of trees. There is a fairly close association between Black-billed Magpie habitat and human rural settlements, which further adds to the likelihood of its being found.

The present breeding distribution of the Black-billed Magpie in Ontario is probably almost fully represented on the map. The single block in which a nest with eggs was found (at Rainy River) is undoubtedly the core of the range in this province. It is likely that further effort would confirm that it is breeding in the other blocks with records or in other nearby blocks. This enclave of Black-billed Magpies should be closely monitored in coming years to determine whether they will continue the expansionary trend of the last decade.

At present, there are probably no more than a couple of dozen breeding pairs of Black-billed Magpies in all of Ontario.-- *F.M. Helleiner*

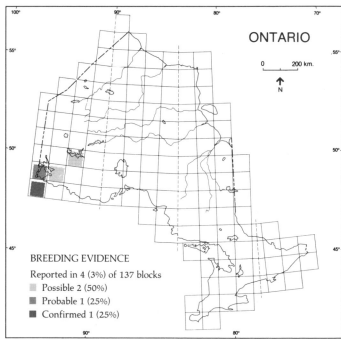

BREEDING EVIDENCE

Reported in 4 (3%) of 137 blocks

Possible 2 (50%)

Probable 1 (25%)

Confirmed 1 (25%)

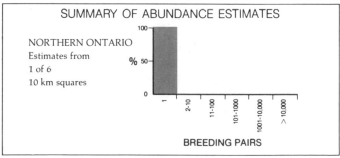

SUMMARY OF ABUNDANCE ESTIMATES

NORTHERN ONTARIO
Estimates from
1 of 6
10 km squares

BREEDING PAIRS

NORTHERN WHEATEAR *
Traquet motteux
Oenanthe oenanthe

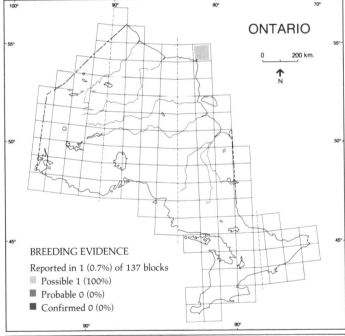

About 18 species of wheatears inhabit the hot deserts and arid open country of the Old World. The Northern Wheatear is the only one whose range extends into the arctic and to North America. In the breeding season, it occurs eastward from Eurasia into Alaska and the Yukon Territory and as far as the Mackenzie River valley, while at the western extremity of its range, a distinctly larger race, the Greenland Wheatear (*O. o. leucohoa*), reaches the eastern Canadian arctic. Both races winter in sub-Saharan Africa (Vaurie 1959, Godfrey 1986). The Greenland Wheatear is the only passerine bird which regularly crosses the North Atlantic Ocean in its migrations.

In Canada, the Greenland race of the Northern Wheatear breeds mainly in northern Labrador, Baffin Island, and Ellesmere Island, but there are isolated records from northern Hudson Bay as far west as Rankin Inlet (Godfrey 1986). It prefers open, rocky terrain, where it nests in crannies in rocks, cliffs, and buildings, and in holes in the ground and under boulders (Sutton and Parmelee 1954a). Breeding has not been reported in Ontario, and there were no known summer records from the coastal tundra of Hudson Bay prior to the atlas survey.

A single male Northern Wheatear was observed for nearly an hour on the barren outer ridges of Cape Henrietta Maria on 2 July 1985 by the author and Erica Dunn. It frequented an area of dry, grassy ground with large, scattered boulders, which was judged to be suitable nesting habitat. The bird did not sing, nor was there any other evidence of breeding behaviour. Two earlier records of female Northern Wheatears on the Hudson and James Bay coasts, one at Winisk on 2 June 1981 and the other at North Point on 4 June 1982 (James 1984c), were not submitted as atlas records, perhaps because they were considered to be migrants. Neverthe-

less, the dates and locations of these records add credence to the suggestion that occasional nesting may occur in Ontario.-- *D.J.T. Hussell*

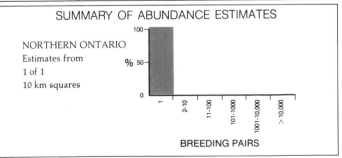

BREEDING EVIDENCE
Reported in 1 (0.7%) of 137 blocks
Possible 1 (100%)
Probable 0 (0%)
Confirmed 0 (0%)

SUMMARY OF ABUNDANCE ESTIMATES

NORTHERN ONTARIO
Estimates from
1 of 1
10 km squares

BREEDING PAIRS

GRAY-CHEEKED THRUSH
Grive à joues grises
Catharus minimus

The breeding range of the Gray-cheeked Thrush extends from Newfoundland across northern Canada to Alaska and even into northeastern Siberia, mainly at or near the limit of trees. Disjunct summering populations have been found in the Appalachian Mountains of the northeastern US and in Quebec, *e.g.*, just north of Ottawa and Quebec City (M. Gosselin pers. comm.). On its northern breeding grounds, it inhabits coniferous woods, denser stands near the tree-line, and areas of dense stunted spruce on coastal islands (Godfrey 1986), but it also frequents dwarf willows and alders (Bent 1949). In most of Ontario, the Gray-cheeked Thrush can be seen only during migration. It winters from Mexico and the West Indies south to South America.

Our lack of knowledge of the distribution of the species in Ontario is reflected by the vagueness of several early accounts. McIlwraith (1894) and Macoun and Macoun (1909) mention the Gray-cheeked Thrush only as a migrant, the former author denoting its home as "in the far north" and not even mentioning its breeding range in Ontario. The only nests reported in Ontario are one with 3 eggs, located in a willow, at Fort Severn in 1940 (Baillie 1961) and another with 3 young at Aquatuk Lake in 1964 (Schueler *et al.* 1974). Godfrey (1986) shows its Ontario range as a narrow strip across the northern Hudson Bay Lowland.

The Gray-cheeked Thrush is notoriously difficult to identify, even when seen clearly in good light. It closely resembles the Swainson's Thrush, whose breeding range overlaps with that of the Gray-cheeked. The song would be a useful identification criterion if it were better known, but few people have heard its full song because it is rarely given during migration. Not only the plumage, but also the behaviour of the species makes it inconspicuous. It is remarkably shy, even for a thrush, singing for only a short period of the sea-

son before the eggs hatch, and frequenting remnant bogs and regenerating forest - environments which are generally difficult for bird-watchers to cover properly. These factors are undoubtedly the reason for the lack of confirmed breeding records. It likely breeds in all possible and probable breeding locations shown on the map, and in several other northern Lowland blocks.

The atlas data suggest that the Gray-cheeked Thrush breeds farther south on the Hudson Bay Lowland than is indicated on the map in Godfrey (1986). The most southerly records are from 2 squares almost at the edge of the Canadian Shield along the Thorne River, a square on the Opinnagau Lakes, and an unspecified square in the vicinity of the Attawapiskat River. The probable breeding record refers to agitated behaviour in a square on the upper reaches of the Black Duck River. The other records are from near the north coast, where Gray-cheeked Thrushes have been reported previously (Baillie 1961, Schueler *et al.* 1974). However, despite intensive atlas work, the species was not found in the vicinity of Hawley Lake, where it was reported as being common in 1964 and 1965 (Schueler *et al.* 1974). The lack of information available on this species makes it impossible to determine whether this is the result of a local population fluctuation or even whether there have been more widespread changes in its population or range.-- *D. Sadler*

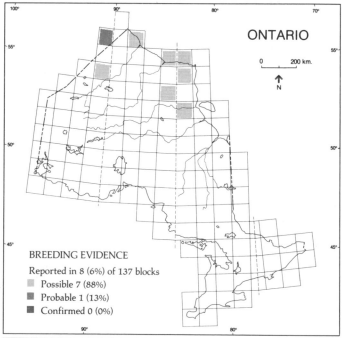

ONTARIO

0   200 km.

↑
N

BREEDING EVIDENCE

Reported in 8 (6%) of 137 blocks

Possible 7 (88%)
Probable 1 (13%)
Confirmed 0 (0%)

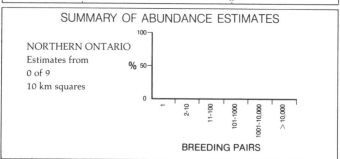

SUMMARY OF ABUNDANCE ESTIMATES

NORTHERN ONTARIO
Estimates from
0 of 9
10 km squares

100

% 50

0

1   2-10   11-100   101-1,000   1001-10,000   > 10,000

BREEDING PAIRS

WATER PIPIT
Pipit spioncelle
Anthus spinoletta

The Water Pipit is a widespread bird of tundra, alpine slopes and meadows, and rocky coasts and islands in the higher latitudes of North America and Eurasia. The taxonomy of the various populations is controversial and the British Ornithologists' Union has recently split the Water Pipit into 3 species - 1 in North America and 2 in Eurasia (BOU 1986). In North America, its breeding range extends from the arctic southward barely into the temperate region at sea level, but as far south as 35^0N in mountain regions in the west (AOU 1983). Except for some coastal populations, the Water Pipit generally winters south of its breeding range in both Eurasia and North America. In Ontario, although it probably breeds on coastal tundra and gravel ridges along most of the Hudson Bay coast, most breeding season reports and all nest records have been from near the Cape Henrietta Maria region (Peck and James 1987).

In 1947, Manning (1952) collected a recently fledged and almost flightless young Water Pipit at Little Cape, in Kenora District, which was the first documented indication of breeding in Ontario. A year later, on 29 June 1948, the first nest (containing 6 eggs) was found by a ROM field party south of Cape Henrietta Maria, near the James Bay coast (Baillie 1962). Three more nests with young were found in 1984, and a fifth nest, with eggs that later hatched, was found in 1985, all in the Cape Henrietta Maria region (Peck and James 1987). These 5 nests remain the only ones thus far reported for the species in the province.

The Water Pipit nests on the ground in tundra and barren ground areas, where its nest is well hidden in recesses in the side of heath, lichen, or mossy hummocks. The eggs, like the adults, are cryptically coloured. These facts, together with the limited and relatively inaccessible breeding habitat, no doubt account in part for the limited number of nests discovered to date. The presence of territorial birds is easily detected by the conspicuous parachuting flight song of the male and by the agitated behaviour of the adults in the vicinity of the nest. Breeding is most easily confirmed by seeing adults carrying food, and such observations represent 44% of the confirmed atlas breeding records. Both parents feed the young, but the male also carries food to the female during incubation (Sutton and Parmelee 1954b, D.J.T. Hussell pers. comm.), so observation of adults with food indicates the presence of eggs or young.

During the atlas survey period, data were received from 16 squares, in 9 of which breeding was confirmed: 4 by nests, 4 by adults with food, and one by a brood of fledged young. All 16 breeding records are from coastal tundra locations, ranging from near the Manitoba border east to the Cape Henrietta Maria region. Except for the latter location, the records are widely scattered but strongly indicate breeding along this entire coastline in suitable habitat.

Three of the 16 records include abundance estimates: 2 of between 11 and 100 pairs, and one of between 101 and 1,000 pairs. These few estimates would suggest an Ontario population far in excess of the numbers indicated by the few reported breeding records.-- *G.K. Peck*

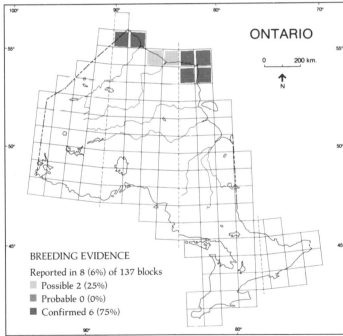

ONTARIO

BREEDING EVIDENCE

Reported in 8 (6%) of 137 blocks

Possible 2 (25%)
Probable 0 (0%)
Confirmed 6 (75%)

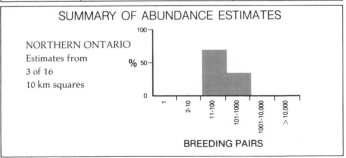

SUMMARY OF ABUNDANCE ESTIMATES

NORTHERN ONTARIO
Estimates from
3 of 16
10 km squares

BREEDING PAIRS

BOHEMIAN WAXWING
Jaseur boréal
Bombycilla garrulus

The Bohemian Waxwing is known to most Ontario naturalists as a sporadic winter visitor, occurring somewhat more regularly in the north and east than in the southwest. It is a beautiful, elegant species, and a welcome addition to the province's breeding bird list: a result of atlas data collection in Ontario's far north.

The species breeds widely in northern Eurasia and northwestern North America, south along the Rocky Mountains to Montana, and east, north of the prairies, to extreme northern Ontario. It has also been recorded in Quebec in summer (Godfrey 1986). In Canada, it has been reported breeding in northern coniferous forest, muskegs, and birch groves (Macoun and Macoun 1909, Godfrey 1986).

Prior to the atlas project, evidence suggested that the Bohemian Waxwing probably bred in the province - it was occasionally reported in the summer months in northern Ontario (Speirs 1985). More convincing evidence was gathered at the junction of the Sutton and Warchesku Rivers, where a female with a brood patch was collected on 19 July 1964 (Schueler *et al.* 1974); they also reported that the birds had been fairly common at the same location during 1962, 1964, and 1965. Although this was the only report of summering birds in suitable habitat, it was east of known nesting locations in Manitoba, thereby suggesting a breeding range across northern Ontario.

Although the call of the Bohemian Waxwing is distinctive, it does not carry far; the species is more often seen than heard. It is an active bird, flying frequently, sitting in prominent locations, and hawking for insects. Its loosely colonial nesting habits mean that it is locally distributed within its breeding range and so is easily missed if field work does not coincide with the small area in which it occurs. No nests were found during atlas field work, but family groups were

discovered in 3 squares, and breeding was confirmed in another when a bird was observed gathering food to feed young.

The Bohemian Waxwing was recorded in 13 atlas squares. All of the records occurred on the Hudson Bay Lowland, despite some intensive field work on the Canadian Shield just south of the Lowland. Eleven records were from squares along major rivers: 5 on the Sachigo, one on the Severn, one on the Fawn, and 4 on the Winisk. While this reflects the fact that coverage of northern blocks was primarily by river, the birds were always found on or near the river banks and not back on the open muskeg. The superior drainage of the river banks creates habitat that the birds seem to prefer. The other 2 records were from squares containing smaller rivers; one was at the mouth of the Shagamu River on the Hudson Bay coast, and the other was on the upper section of the Black Duck River. The lack of records in 3 of the province's most northerly blocks may be due to the fact that coverage in those blocks was concentrated entirely on coastal areas - away from suitable habitat. There was no atlas field work at the Sutton and Warchesku junction, so it is possible that the distribution extends farther east than depicted here, possibly east to James Bay.-- *M.D. Cadman*

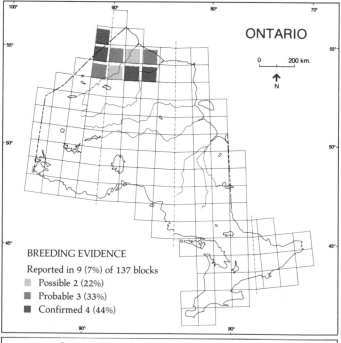

BREEDING EVIDENCE

Reported in 9 (7%) of 137 blocks

▫ Possible 2 (22%)

▪ Probable 3 (33%)

■ Confirmed 4 (44%)

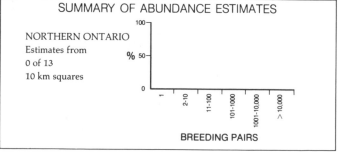

SUMMARY OF ABUNDANCE ESTIMATES

NORTHERN ONTARIO
Estimates from
0 of 13
10 km squares

BREEDING PAIRS

NORTHERN SHRIKE
Pie-grièche grise
Lanius excubitor

The characteristic silhouette of the Northern Shrike perched in a prominent location is a familiar sight in winter in much of Ontario, but it was not until the first year of the atlas project that this bold hunter was added to the province's list of breeding birds.

Known in the Old World as the Great Grey Shrike, this species enjoys a large range in Eurasia, North Africa, and northern North America. In Canada and Alaska, it breeds in the northern Boreal Forest near the tree-line. In Ontario, it uses the sparse coniferous woods of the Hudson Bay Lowland for nesting habitat. In winter, it occurs as far south as the northern and central US.

As early as 1894, McIlwraith suggested that the Northern Shrike "no doubt" bred in the northern portion of the province, likely because it had been reported in similar habitat in Manitoba. Evidence of breeding was lacking in Ontario until 26 May 1972, when a female with enlarged ova was collected at Moosonee (Schueler *et al.* 1974). An undocumented report of a family group at North Point (Manning 1981) provided additional support for breeding in that area. Breeding was finally confirmed for Ontario in the summer of 1981, when an ROM field party located a pair of adults with 4 young in an open tamarack fen northeast of Kiruna Lake (James 1981). Later, 2 more breeding confirmations were reported: an adult flying with food for young near the Fawn River in 1983 and another family group near the Sachigo River in 1984. The relatively large number of recent records is more likely a reflection of the increased coverage of remote areas of the Lowland than an increase in the Northern Shrike population.

The main difficulty in finding this species is gaining access to and travelling in its breeding habitat. Fortunately, the muskeg habitat it favours is fairly open, so visibility is

good, and the bird's habit of perching on the tops of trees make it reasonably easy to detect. Its song is not widely known although it is occasionally given on the southern wintering grounds. The Northern Shrike does not appear to favour river banks, where most atlas work on inland portions of the Hudson Bay Lowland took place. This may have resulted in its not being found in some Lowland blocks where it might have been expected to breed.

Atlas results indicate that the Northen Shrike is widespread on the northern Hudson Bay Lowland, but occurs at very low densities. Low breeding density makes intuitive sense because it is a predatory species in a structurally simple environment with sparse bird populations. The data also support earlier indications that it may breed as far south as Moosonee. It was reported as a possible breeder from 2 squares in one block along the Sachigo River in the western part of far northern Ontario. Otherwise, all block data represent single square records.-- *M.D. Cadman*

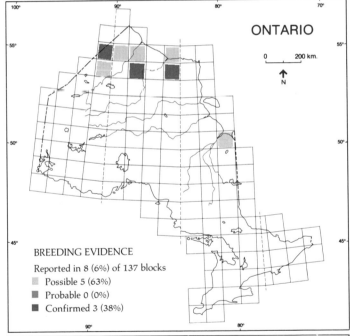

ONTARIO

0   200 km.

BREEDING EVIDENCE

Reported in 8 (6%) of 137 blocks

Possible 5 (63%)

Probable 0 (0%)

Confirmed 3 (38%)

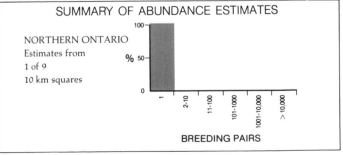

SUMMARY OF ABUNDANCE ESTIMATES

NORTHERN ONTARIO
Estimates from
1 of 9
10 km squares

BREEDING PAIRS

ORANGE-CROWNED WARBLER
Paruline verdâtre
Vermivora celata

The Orange-crowned Warbler is principally a western North American species, breeding from southern California and Texas to northern Yukon and Alaska in the west. Its range also extends eastward from Alaska across the Prairie Provinces and into northern Ontario, Quebec, and Labrador in a band of ever-decreasing width. In Ontario, it is found breeding northward from the northern edge of Lake Superior, with a concentration of records on the Hudson Bay Lowland, particularly in the northwest. Although the species winters as far south as Guatemala (Godfrey 1986), it is a winter resident in the southern US, being found regularly in Georgia, Florida, and Louisiana. It prefers to breed in open woods, particularly of poplar species, with an understory of shrubs such as alder, willow, hazel, and mountain maple. It is one of our most poorly known birds, occurring in southern Ontario only as a migrant.

All of the breeding season records cited by Speirs (1985) for this species are from the present century. They suggest that there has been no significant change in the distribution of the Orange-crowned Warbler in Ontario in that period. Macoun and Macoun (1909), however, describe a nest found in 1906 near Madoc, which is far to the south of its present range. There appears to be some doubt about the reliability of that record. Baillie and Harrington (1937) did not include it as a breeding species. There are very few breeding records in the province. The first nest, which contained 4 eggs, was collected at Favourable Lake in 1938, another nest was subsequently found at Moosonee, and young were collected there and at Lake Attawapiskat (Baillie 1960).

Because the Orange-crowned Warbler is not currently known to breed in southern Ontario, and is common only in the most northerly parts of the province, an atlasser would have to be willing to visit the Boreal Forest or the Hudson Bay Lowland during the black fly season to find it. Regional Coordinators agree that it may easily be overlooked. It is almost always first located by its song, which is a high trill much like that of a Chipping Sparrow, but more musical, and fading as if the bird were tiring. Fully half of the atlas records of the Orange-crowned Warbler are in the 'singing male' breeding category, which is suggestive of a species that is sparsely distributed and is therefore difficult to upgrade to higher categories of breeding evidence. The nest, too, is difficult to find, being located on the ground and often covered with low vegetation. No nests were found during the atlas project. Most of the records of confirmed breeding represent adults carrying faecal sacs or food for young.

The atlas maps show that a concentration of breeding records occurs north of 52^0N latitude. South of that are several possible and probable records, but none of confirmed breeding. Throughout its breeding range in Ontario, the distribution of the Orange-crowned Warbler is spotty, with almost as many gaps as observations. The tendency for records to be concentrated in the western part of far northern Ontario suggests a larger and therefore more detectable population in that area, which is nearer to its principal range. It likely breeds in almost every block from about 50^0N latitude to the tree-line, but is nowhere an abundant species.-- *D.A. Welsh and F.M. Helleiner*

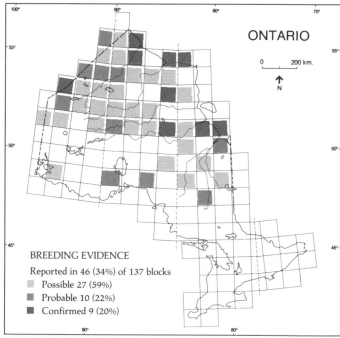

ONTARIO

0   200 km.

BREEDING EVIDENCE

Reported in 46 (34%) of 137 blocks

Possible 27 (59%)

Probable 10 (22%)

Confirmed 9 (20%)

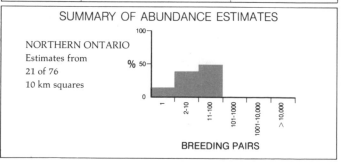

SUMMARY OF ABUNDANCE ESTIMATES

NORTHERN ONTARIO
Estimates from
21 of 76
10 km squares

BREEDING PAIRS

BLACKPOLL WARBLER
Paruline rayée
Dendroica striata

In Ontario, the Blackpoll Warbler is the most exclusively northern breeding warbler, with the southern limit of its breeding range at 51°N. It is one of the most southerly wintering warblers, with its principal winter range extending from Colombia to western Brazil. As a migrant, the most conservative travellers must fly at least 8,000 km, and those summering in the northwestern part of its range must traverse at least 16,000 km twice a year.

The principal breeding range of the Blackpoll Warbler is from Newfoundland and Labrador to Alaska in areas of conifer and mixedwood scrub aptly called "the land of little sticks". It also breeds regularly in the higher mountains of the northeastern US, including the Catskill, Adirondack, Green, and White Mountains. In Ontario, it breeds almost exclusively in areas with open grown conifers, although it can be readily found in alders in northwestern North America north of the Arctic Circle. Old, poorly vegetated beach ridges, backsides of river banks, and dry fens and bogs are its most frequent habitats in our province. The Blackpoll Warbler is a more common breeder both east and west of Ontario, although it is certainly frequent in the subarctic regions of this province.

The original type specimen was described from a pair collected at the Severn River in 1772. Schueler *et al.* (1974) found it to be abundant at Winisk and Hawley Lake. McLaren and McLaren (1981a) reported it to be the most abundant species in spruce scrub, particularly from North Caribou Lake to the Manitoba border, and found it to be common in several other habitats. James *et al.* (1983) estimated a density of 20 pairs/km² on the Sutton Ridges, and it was found in all the plots at Big Trout Lake studied by Lee and Speirs (1977).

The species has an uneven breeding distribution from place to place and from year to year. The song is a thin trill that is among the highest-pitched of all passerine birds. This song is distinctive, but is of sufficiently high frequency that it might be inaudible to many observers. Of the 34 blocks with records, possible breeders were reported in 13, indicating that the species presented considerable difficulty for many atlassers. Overall, 29% of the block records are of confirmed breeding, mostly by the discovery of adults with food or of recently fledged young. One nest was found during the atlas period. The major difficulty in documenting the species is undoubtedly the remoteness and inhospitability of its chosen range.

The atlas survey found the species almost exclusively in far northern Ontario, in its expected range on the Hudson Bay Lowland and adjacent areas of the Shield. The records in the southern portion of northern Ontario seem to reflect very scattered occurrences of territorial males with rare confirmation of breeding. More intensive searching would undoubtedly lead to the discovery of other individuals in increasing density approaching the major breeding population. Welsh and Fillman (1980) found single territorial males to be present throughout the breeding season on several occasions in the southern Boreal Forest zone.

Abundance estimates indicate that the Blackpoll Warbler is a fairly common bird within its northern range.-- *D.A.Welsh*

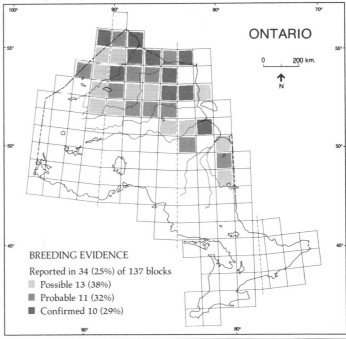

ONTARIO

0   200 km.

BREEDING EVIDENCE

Reported in 34 (25%) of 137 blocks

Possible 13 (38%)

Probable 11 (32%)

Confirmed 10 (29%)

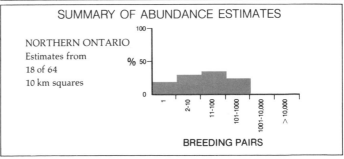

SUMMARY OF ABUNDANCE ESTIMATES

NORTHERN ONTARIO
Estimates from
18 of 64
10 km squares

%

BREEDING PAIRS

AMERICAN TREE SPARROW
Bruant hudsonien
Spizella arborea

The American Tree Sparrow occurs as a breeding bird across the subarctic of North America. In Ontario, it is found along the coasts of Hudson and James Bays along the tree-line in low-lying tundra areas, where there are stands of shrubs and dwarf or stunted trees, especially willow, dwarf birch, and alder. In most of the inhabited parts of the continent, it is known only as a winter bird, since, at that season, it occupies southern Canada and the northern half of the US.

The breeding range of this species in Ontario does not appear to have changed substantially in the past half century. Godfrey (1986) and Peck and James (1987) both give the breeding range as a narrow strip along the Hudson Bay coastline, a situation mirrored in the distribution shown in this atlas.

The species is not particularly difficult to locate on its breeding grounds; a territorial male perches conspicuously on a low tree or shrub and sings frequently. The song is distinctive and quite far-carrying, and the species responds well to tapes and to 'pishing'. It is fairly easy to confirm as a breeding species. Nests are built on the ground or low in a shrub, and are not difficult to locate. However, nests were found in only 2 squares during the atlas period, probably because of the remote location of the breeding range in Ontario. Adults carrying food frequently perch conspicuously before visiting the nest, and, as a result, 40% of the records of confirmed breeding are of adults carrying food or faecal sacs. Not surprisingly, then, breeding was confirmed in 7 of the 10 blocks where the American Tree Sparrow was reported.

American Tree Sparrows winter in southern Ontario, arriving in October and November and leaving by the end of April. Most have left southern portions of northern Ontario by the middle of May. There are, however, several reports of Tree Sparrows in June away from Hudson Bay, south as far

as Frontenac Co, some of which are of males in song (Speirs 1985). During spring migration, individuals frequently sing, especially during warm days. Because song is in a category indicating breeding evidence, a number of reports came from squares south of the known breeding range. These birds were not in known breeding habitat and were removed from the data base by the atlas data review process.

The map shows a distribution of the American Tree Sparrow in Ontario that is similar to that known prior to the atlas project. It extends from the Manitoba border east along the Hudson Bay Lowland to Polar Bear Prov. Park, then south along the west coast of James Bay nearly as far as Attawapiskat. The species is absent from the area south to Moosonee and the Quebec border, presumably because of the absence of suitable habitat. The block containing Cape Henrietta Maria lacks the shrubby patches required for Tree Sparrow breeding.

The 6 abundance estimates provided by atlassers give a wide range of values, presumably dependent upon available habitat. Half of the 6 estimates were of 101 to 1,000 pairs per square, suggesting that some of the coastal squares had substantial populations.-- *R. Knapton*

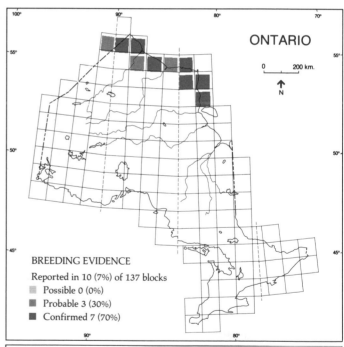

ONTARIO

0   200 km.

N

BREEDING EVIDENCE

Reported in 10 (7%) of 137 blocks

Possible 0 (0%)

Probable 3 (30%)

Confirmed 7 (70%)

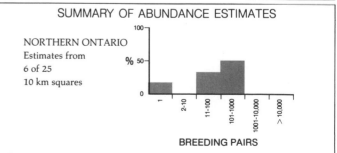

SUMMARY OF ABUNDANCE ESTIMATES

NORTHERN ONTARIO
Estimates from
6 of 25
10 km squares

BREEDING PAIRS

LARK SPARROW
Bruant à joues marron
Chondestes grammacus

The Lark Sparrow is primarily a bird of western and south-western North America, where it reaches its greatest abundance in Texas and Oklahoma (Robbins *et al.* 1986). It breeds from southern Canada to northern Mexico, but is very local east of the Mississippi. The Lark Sparrow prefers open habitat with scattered trees and bushes, including woodland edge, prairies, and farmland. Cemeteries have been reported as nesting habitat in both Texas and Manitoba (Baepler 1968, Walley 1985).

In Ontario, the Lark Sparrow is at the extreme limits of its range and there are very few confirmed breeding records. One or two pairs were reported each summer living in "open sandy fields between pine trees" near London between 1878 and 1891, and a nest was found west of London in 1890 (Saunders and Dale 1933). Two nests were found in High Park, Toronto in 1892 and 1898; there was another at Point Pelee in 1913, and it was also reported to breed in Kent Co in the early years of this century (Baillie and Harrington 1937). The most consistent breeding locality has been on the Norfolk Sand Plain in the vicinity of Walsingham, a few kilometres north of Long Point, in the Haldimand-Norfolk RM. Nests were reported there in 1930, 1935, 1971, and 1976 (ONRS, Peck and James 1987). A nest at Sudbury in 1973 is the most northerly breeding record for the province (Peck and James 1987).

The Lark Sparrow is easily seen and identified in its open habitat; the male has a long and melodious song. Nests are usually placed on the ground, often shaded by an overhanging plant or some other object, such as a tombstone (Walley 1985). The female builds the nest and incubates the 3 to 6 eggs, which take about 11 days to hatch. The young are fed by both parents and leave the nest at about 9 or 10 days of age, when they are barely able to fly. In Ontario, nests with eggs have been found from 24 May to 1 July (ONRS, Peck and James 1987). The main difficulty for atlassers in finding the Lark Sparrow is its sparse and unpredictable distribution.

The only breeding season record during the atlas years was a pair seen by Sarah and Chauncey Wood on open, sandy ground at Elbow Lake, about 48 km east of Thunder Bay on 9 June 1982. The male was heard and seen singing from the tops of mounds of gravel and from fenceposts, while the female fed nearby. As the species breeds regularly in southern Manitoba and northern Wisconsin, it is reasonable to suspect occasional nesting in suitable habitat in the Thunder Bay region. The lack of records during the atlas years probably indicates that the species was absent from its former haunts in Haldimand-Norfolk RM. The 1976 site was an isolated open area among extensive plantations, which has been well-watched in recent years; but Lark Sparrows have not been reported there again. We may expect the species to turn up unpredictably again in the future, and the Norfolk Sand Plain remains the most promising area for establishment of a regular breeding population, both because of the historical records and the continuing presence of pockets of suitable habitat. There would appear to be some potential for managing habitat for this and other species that prefer open and semi-open country.-- *D.J.T. Hussell*

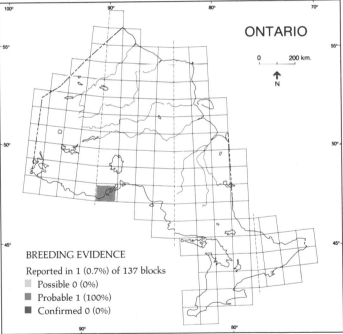

ONTARIO

0   200 km.

N

BREEDING EVIDENCE
Reported in 1 (0.7%) of 137 blocks
Possible 0 (0%)
Probable 1 (100%)
Confirmed 0 (0%)

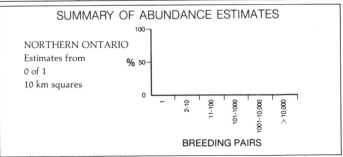

SUMMARY OF ABUNDANCE ESTIMATES

NORTHERN ONTARIO
Estimates from
0 of 1
10 km squares

BREEDING PAIRS

SHARP-TAILED SPARROW
Bruant à queue aiguë
Ammodramus caudacutus

The Sharp-tailed Sparrow is a bird of both fresh and saltwater marshes. It occurs in southern Ontario only during migration, when it is difficult to find.

There are at least 3 disjunct populations of Sharp-tailed Sparrows: one breeds in wet meadows, and the margins of pools, lakes, and marshes, on the northern prairies, possibly east into northwestern Ontario; a second breeds in sedge marshes on the coasts of James and Hudson Bays; and a third breeds in coastal marshes from the south shore of the St. Lawrence River and Nova Scotia south to North Carolina. The Sharp-tailed Sparrow winters in coastal marshes, from New York south to central Florida, and along the coast of the Gulf of Mexico.

Along the Atlantic Coast, where it breeds fairly commonly in coastal marshes, males advertise by singing from an exposed perch on a cattail, tall grass, or small bush. Where uncommon, however, it can be difficult to see, even though it may sing regularly, though not from an exposed perch. Males will sometimes sing all night, and some seem to sing only during the night. Ten of the 24 records supplied by atlassers are of singing males, and 3 more are of 'territorial' (presumably singing) birds. The Ontario birds have not been studied; however, along the Atlantic Coast, the Sharp-tailed Sparrow apparently does not defend a territory, and mating is promiscuous (Woolfenden 1956, Hill 1968). The nest, placed on the ground, or as high as 10 to 15 cm, is well concealed in vegetation. The female alone builds the nest, incubates the eggs, and feeds the young. The nests are cups of coarse grass or sedge, lined with fine grasses, and are placed on the ground on a mat of grasses or sedges. Nests are extremely difficult to locate (Hill 1968, Peck and James 1987). Atlassers found only one nest, and one adult carrying food. In Ontario, the species apparently nests in late June and July

(Schueler *et al.* 1974, Speirs 1985) - the period of peak atlassing activity in the north.

Atlassers found probable or confirmed evidence of nesting only on the coasts of James and Hudson Bays, where the species was previously known to nest. The species is locally common in sedge marshes on James Bay, and less common on Hudson Bay. Over 5 Sharp-tailed Sparrows were observed on 30 June 1979 at Rainy River (Speirs 1985); it seems likely that these birds were nesting, and, if so, were probably representatives of the 'prairie' Sharp-tailed Sparrow population. The atlas report of a singing male from this region may well be of a breeding bird. However, other non-coastal records of birds from Ontario are probably of transient individuals, although the 2 inland records on the northern Hudson Bay Lowland were reported from early July. In general, because of the late nesting of this species, single singing birds recorded in early June do not constitute good evidence of nesting.

All of the highest abundance codes, those of over 10 pairs per square, came from the western shore of James Bay from approximately the border of Polar Bear Prov. Park southward.-- *J.D. Rising*

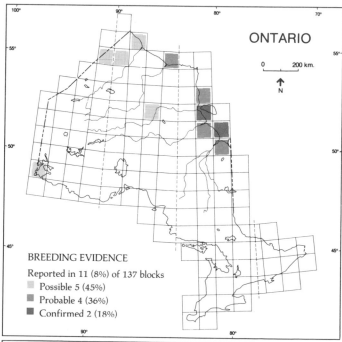

ONTARIO

0   200 km.

BREEDING EVIDENCE

Reported in 11 (8%) of 137 blocks

▢ Possible 5 (45%)
▨ Probable 4 (36%)
■ Confirmed 2 (18%)

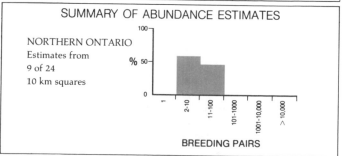

SUMMARY OF ABUNDANCE ESTIMATES

NORTHERN ONTARIO
Estimates from
9 of 24
10 km squares

BREEDING PAIRS

FOX SPARROW
Bruant fauve
Passerella iliaca

Although the Fox Sparrow is the largest North American sparrow, it is not the easiest to find. This is especially true on the breeding grounds, where it inhabits virtually impenetrable thickets. During migration and in winter, Fox Sparrows forage on the ground, and frequently, the sound of their scratching among leaves betrays their presence.

The Fox Sparrow breeds from northern Alaska, east through the central Mackenzie District, northern Ontario, central Quebec, Labrador, and Newfoundland, south on the Pacific coast to Washington, and in the mountains to southern California. In western North America, Fox Sparrows breed in montane brush, chaparral, and large city parks, but in the northeast the species seems restricted to dense, moist brush. In Ontario, the Fox Sparrow breeds in moist, rank growth of black spruce, tamarack, poplar, birch, and especially willow and alder. Although a few Fox Sparrows winter in southern Ontario, the majority move south to the US.

Perhaps in the past, the species was less common in Ontario than today as McIlwraith (1894) ". . . only met with it a few times." Today, it is regular, though never common, on migration.

Although breeding Fox Sparrows are extremely difficult to locate because of the dense habitat that they inhabit, their loud ringing song carries a great distance; 74% of all atlas records are of singing birds. Many of these would be recorded by atlassers canoeing northern rivers where stopping to upgrade records was often not possible. The species responds vigorously to 'pishing' and song playback when on territory. In Ontario, eggs are laid from early June through mid-July; the species is probably double-brooded (Terrill 1968). Nests are placed on the ground in sphagnum, Labrador tea, or laurel, or as high as 2 m (rarely higher) in a spruce or bush (Terrill 1968, Peck and James 1987); 3 to 5

eggs are known to be laid (Terrill 1968), but in Ontario, only clutches of 3 and 4 have been reported (Peck and James 1987). Although no nests were found by atlassers, breeding was confirmed in 3 blocks by observing birds carrying food, and in 2 blocks by the presence of fledged young. Females feign injury to decoy predators near the nest; 11% of all records are of agitated birds.

All confirmed breeding records are on the Hudson Bay Lowland. However, singing males, pairs, and individuals showing agitated behaviour were seen throughout northern Ontario north of the Albany River, and it seems probable that the species breeds throughout this region. From other studies, the species has been known to nest at Attawapiskat, Hawley Lake, Winisk, and Fort Severn (Schueler *et al.* 1974, Speirs 1985). The most southerly atlas record was from the Little Abitibi River approximately 70 km north of Cochrane. This is approximately 100 km south of the previously published range (Godfrey 1986).

One-third of the records from northern Ontario squares have abundance estimates, of which 15 (47%) are of more than 10 pairs per square. These higher estimates are scattered across the Ontario range.-- *J.D. Rising*

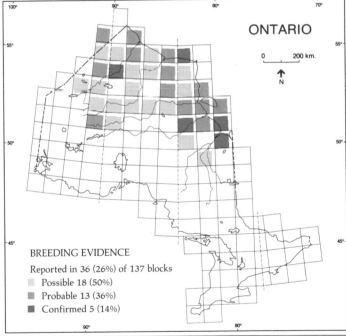

BREEDING EVIDENCE

Reported in 36 (26%) of 137 blocks

Possible 18 (50%)

Probable 13 (36%)

Confirmed 5 (14%)

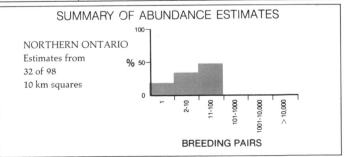

SUMMARY OF ABUNDANCE ESTIMATES

NORTHERN ONTARIO
Estimates from
32 of 98
10 km squares

BREEDING PAIRS

WHITE-CROWNED SPARROW
Bruant à couronne blanche
Zonotrichia leucophrys

This is one of the commonest sparrows in western North America. In the east, however, it nests near the limit of trees, and is poorly known. There have been no detailed studies of the breeding biology of eastern White-crowned Sparrows.

The White-crowned Sparrow breeds from northern Alaska east through the forest and barrens across northern Ontario, central Labrador, and northwestern Newfoundland. In the west, there are somewhat disjunct populations in coastal brush, in the fog zone, and in mountains south to southern California. There are several different subspecies of White-crowned Sparrows, but only one breeds in the east. Although a few White-crowned Sparrows winter in southern Ontario, the great majority winter in the US. The breeding habitat usually consists of shrub growth in an open area, such as woodland edge, forest burns, willow clumps on tundra, and stream edges (Godfrey 1986).

The White-crowned is a large sparrow, is fairly conspicuous, and has a distinctive song. A little over one-third of square records are of possible breeding, and 20% are of probable breeding; 77% of probable breeding records are of agitated behaviour. The adults persistently give their alarm call when an intruder is near the nest. The nest of the White-crowned Sparrow is placed on the ground, generally at the base of a small birch, willow, black spruce, or tamarack, or in vegetation. It is characteristically set into the moss-lichen mat, partially covered, and lined with fine grass (Clement 1968, Peck and James 1987) and is difficult to find. Three to 5 (most commonly 5) eggs are laid, and known egg dates are from 12 June through 24 July (Schueler *et al.* 1974, Peck and James 1987). To distract an intruder approaching her nest, the female may run, mouse-like through the vegetation, and will feign a broken wing if guarding fledglings (Clement 1968). However, most breeding confirmations are of fledged young

or adults carrying food.

In Ontario, the White-crowned Sparrow nests on the Hudson Bay Lowland. The majority of confirmed records come from within 100 km of the Hudson and James Bay coasts and there are few records of any kind more than 200 km inland, indicating that the species occupies primarily the transition zone between the forest and the treeless tundra. Isolated records farther inland indicate the possibility of breeding where there is suitable semi-open habitat south and west of the main range. In general the map agrees well with the previously suspected distribution (*eg.*, Godfrey 1986).

Published records suggest that the species is uncommon along the coast of James Bay and at Kiruna Lake, fairly common at Fort Severn, Winisk, and Cape Henrietta Maria, and common at Hawley Lake (Schueler *et al.* 1974, James *et al.* 1982a, Speirs 1985). Atlas abundance estimates range from 1 pair to 101-1,000 pairs per square. Seven estimates are of more than 10 birds per square and 5 of those are from coastal squares, suggesting higher abundance on the tundra or along the tree-line.-- *J.D. Rising*

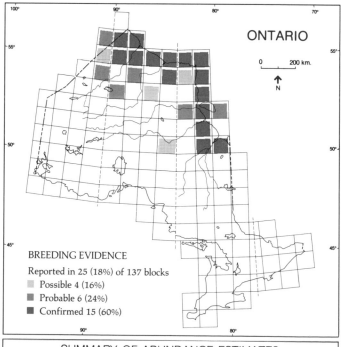

BREEDING EVIDENCE
Reported in 25 (18%) of 137 blocks
Possible 4 (16%)
Probable 6 (24%)
Confirmed 15 (60%)

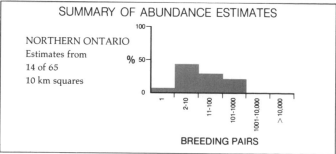

SUMMARY OF ABUNDANCE ESTIMATES

NORTHERN ONTARIO
Estimates from
14 of 65
10 km squares

BREEDING PAIRS

HARRIS' SPARROW
Bruant à face noire
Zonotrichia querula

The Harris' Sparrow has been reported as common at the edge of woodlands near Churchill, Manitoba (Baumgartner 1968), so its occurrence at Fort Severn might have been expected. At Churchill, it is sometimes associated with clearings near railway tracks and with bushy margins in burned-over areas, suggesting that human disturbance of the tundra-muskeg ecotone may favour the species. The Fort Severn nest was in a more disturbed site than those described at Churchill (Baumgartner 1968, Semple and Sutton 1932).

Coverage to the north and west of Fort Severn was primarily on the coastal tundra, so birds on the edge of the muskeg farther inland may have been missed. The town of Winisk and the surrounding area was well covered, and, as no birds were found there, it is unlikely that the Harris' Sparrow breeds that far east. Intensive atlassing was not undertaken at Fort Severn in 1984 or 1985, so it is not possible to say whether Ontario's only known pair of Harris' Sparrows returned to breed after nesting was first discovered.-- *M.D. Cadman*

Except in the Lake of the Woods area, where it can be seen on migration, this robust and attractive sparrow is rarely reported in Ontario. It was therefore a pleasant surprise when it was added to the list of the province's breeding birds during the atlas period.

The breeding range of the Harris' Sparrow is confined to Canada and extends from the Mackenzie River through much of the southern NWT and northern Manitoba to the northern tip of Ontario. It winters in the southern US, primarily in the southwest, north to southern British Columbia (Godfrey 1986). Breeding habitat is the ecotone between spruce forests and tundra, where patches of stunted spruce alternate with open and bushy areas (Baumgartner 1968, Godfrey 1986).

The Harris' Sparrow has a history of being difficult to find. It was not reported in the scientific literature until 1840, and a nest with eggs was not discovered until 1931 (Semple and Sutton 1932). There had been very few breeding season records in the province (Speirs 1985) before a singing male was discovered at Fort Severn in 1983, where the species had also been reported in 1940 (James *et al.* 1976). In 1983, a nest with eggs was eventually found for Ontario's first confirmed breeding record. The birds were on the western edge of the town in an area with a grassy substrate interspersed with the sand and gravel associated with the adjacent airport runway. The nest was located in a patch of willow shrubs and stunted spruces in a predominantly open area about 10 m from the town's radio station. The antenna of the station was a favourite singing post for the male, whose peculiar two-noted, monotone song sounded, appropriately enough, very much like acoustic feedback (pers. obs.). One or 2 fledged young were discovered and photographed in the nearby town dump on 25 July 1983.

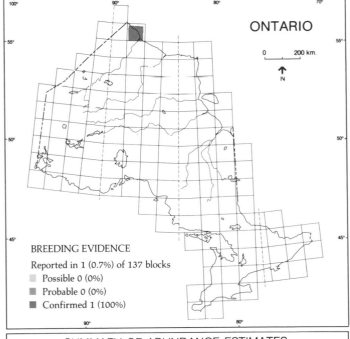

ONTARIO

BREEDING EVIDENCE
Reported in 1 (0.7%) of 137 blocks
☐ Possible 0 (0%)
☐ Probable 0 (0%)
■ Confirmed 1 (100%)

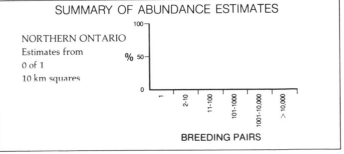

SUMMARY OF ABUNDANCE ESTIMATES

NORTHERN ONTARIO
Estimates from
0 of 1
10 km squares

BREEDING PAIRS

LAPLAND LONGSPUR
Bruant lapon
Calcarius lapponicus

Two species of Longspurs breed in Ontario; the Smith's Longspur has a subarctic distribution while the Lapland Longspur breeds only on the arctic tundra. The Lapland Longspur has a circumpolar breeding range, extending from as far south as 52°N on the Aleutian Islands and Kamchatka Peninsula to at least 80°N on Ellesmere Island. North American populations winter abundantly on the Great Plains of the central US and more sparsely east to the Atlantic seaboard in southern Canada and the northern US. In Ontario, the Lapland Longspur's breeding range is confined to the far north on the tundra strip along the Hudson Bay and James Bay coasts.

It is common in wet fresh-water tundra habitats in most of its range, but it is relatively scarce at high latitudes, where such habitats are poorly represented. Low-lying, hummocky meadows, with a good growth of mosses, grasses, sedges, shrubs, dwarf birches, and dwarf willows are favoured.

Breeding in Ontario was first documented in 1947 at Cape Henrietta Maria (Manning 1952). Subsequent explorations provided additional breeding records from this extreme northeastern part of Ontario (Peck 1972, Peck and James 1987). Atlas field work indicated breeding of the species on the Hudson Bay coast near the Manitoba border, where it had not previously been recorded.

In the early part of the breeding season, male Lapland Longspurs advertise their presence conspicuously with frequent songs given both during aerial displays and from the ground. The inconspicuous females are relatively secretive while nest-building and incubating the eggs. Nests are usually well hidden under overhanging vegetation in the side of a hummock, but occasionally are on level ground or the top of a hummock. The young leave the nest at 8 to 10 days of age, several days before they can fly. Breeding is most easily

confirmed in the nestling and post-fledging stages, when both parents carry food to the young. In 5 of the 8 atlas blocks for which records exist, breeding was confirmed by the finding of nests or of adults carrying food to the young.

Atlas records (19 squares) show a disjunct breeding distribution north of the tree-line along the tundra strip adjacent to Hudson and James Bays. Isolated populations may have gone undetected by atlassers in the intervening section of coast from Winisk to Fort Severn, but there is certainly less suitable habitat there than west of the Severn River and around Cape Henrietta Maria, both of which are strongly influenced by the cold waters of Hudson Bay. The Lapland Longspur's range extends further north and east than that of the Smith's Longspur, especially on Cape Henrietta Maria, where Smith's Longspur is absent. The 2 species broadly overlap at some places along the tree-line.

Lapland Longspurs are locally common in suitable habitat. R.D. James and M.K. Peck estimated 13.3 pairs per ha in wet tussock tundra, but found none on dry heath-lichen plots, at radar site 415, 40 km south of Cape Henrietta Maria, in 1984 (unpubl. data). Abundance estimates of 101 to 1,000 pairs per square were reported in 3 squares in the Cape Henrietta Maria region.-- *D.J.T. Hussell*

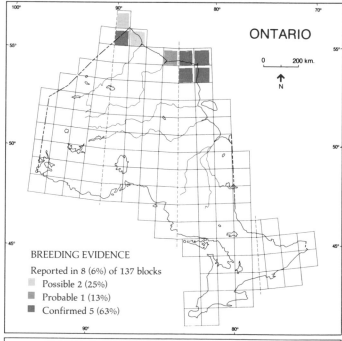

ONTARIO

0   200 km.

BREEDING EVIDENCE

Reported in 8 (6%) of 137 blocks

Possible 2 (25%)
Probable 1 (13%)
Confirmed 5 (63%)

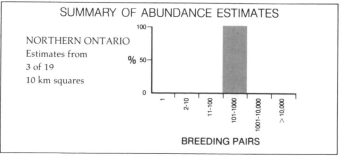

SUMMARY OF ABUNDANCE ESTIMATES

NORTHERN ONTARIO
Estimates from
3 of 19
10 km squares

BREEDING PAIRS

SMITH'S LONGSPUR
Bruant de Smith
Calcarius pictus

The Smith's Longspur is a bird of the forest-tundra, the more northerly part of the transition zone between the Boreal Forest and the treeless tundra (Jehl 1968). Its subarctic breeding range is confined to North America. Although its total population must be smaller than that of the Lapland Longspur, it is locally abundant in the narrow zone of its favoured habitat, stretching from James Bay westward to northern Alaska. Typical breeding habitat consists of drier sedge meadows interrupted by old beach ridges with scattered clumps of black spruce. In this habitat, it is often associated with Stilt Sandpipers, Dunlins, Least Sandpipers, Savannah Sparrows, and sometimes Lapland Longspurs. The Smith's Longspur winters in the central US, from Iowa south to Texas.

Manning (1952) collected a juvenile at Little Cape on the Hudson Bay coast in 1947, but it was not until the following year that the first Ontario nesting was reported by a ROM party, about 40 km south of Cape Henrietta Maria (Peck 1972). Relative numbers appear to vary from year to year. Peck (1972) described it as less abundant than the Lapland Longspur in 1970, whereas R.D. James and M.K. Peck found it more numerous than the Lapland Longspur in the same area in 1984.

Unlike Lapland Longspurs, male Smith's Longspurs do not have a conspicuous flight song, nor apparently do they defend a clearly defined territory (Jehl 1968), but they do sing both from the ground and from trees. Nests are usually in depressions in the tops of dry, flat hummocks and, unlike typical Lapland Longspur nests, are unprotected from above, although they may be at the base of a small shrub or tree. Nest-building and incubation are done by the relatively inconspicuous female. Both sexes feed the nestlings and the fledglings, which vacate the nest before they can fly. The Smith's Longspur was confirmed as a breeding bird in all 7

blocks in which it was reported, either by observing adults carrying food for the young (5 blocks) or by finding nests with young (2 blocks).

The Smith's Longspur was reported from a total of 20 squares. The breeding range parallels and overlaps the strip of tundra along the coast of Hudson Bay, in a zone that is usually much narrower than the width of the 100 km blocks shown on the map. The species is absent from the more exposed outer coast at Cape Henrietta Maria, and was not reported on West Pen Island on the outer coast, near the Manitoba border.

There is little detailed information available on densities on the breeding grounds. In 1984, R.D. James and M.K. Peck estimated 40 and 20 pairs per ha in dry heath-lichen and wet tussock tundra, respectively, at radar site 415, about 40 km south of Cape Henrietta Maria (unpubl. data). Abundance estimates were provided for 3 squares. Two from the mouth of the Shagamu River (50 km east of Fort Severn) were of 11 to 100 pairs, and one from the square containing James' and Peck's study area was of 1,001 to 10,000 pairs.-- *D.J.T. Hussell*

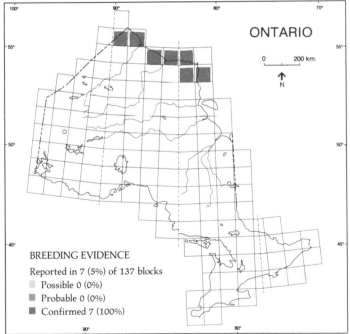

BREEDING EVIDENCE

Reported in 7 (5%) of 137 blocks

Possible 0 (0%)

Probable 0 (0%)

Confirmed 7 (100%)

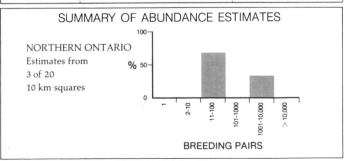

SUMMARY OF ABUNDANCE ESTIMATES

NORTHERN ONTARIO
Estimates from
3 of 20
10 km squares

BREEDING PAIRS

SNOW BUNTING
Bruant des neiges
Plectrophenax nivalis

The circumpolar breeding range of the Snow Bunting extends farther north than that of any other land bird. In Ontario, it is at the southern limit of its range in Canada, and it breeds at a more southerly latitude only in the Aleutian Islands. In summer, the Snow Bunting is a common bird of northern tundras, preferring rough, rocky country, cliffs, and outcrops, which provide it with suitable nesting cavities. Sometimes it nests in artificial sites around human habitation.

The Snow Bunting has been reported in summer at several places along Ontario's Hudson Bay coast. Manning (1952) reported one at Cape Henrietta Maria and another at the Shagamu River in July 1947. Others were seen near Winisk in June 1965 (Schueler *et al.* 1974), at radar site 416, 30 km south of Cape Henrietta Maria, in late June and early July in 1970, and at the forks of the Brant River in 1971 (Peck 1972). It was not until the atlas survey, however, that breeding was confirmed in Ontario.

The Snow Bunting is easily detected during the breeding season. The male is particularly conspicuous in his black and white plumage, as he sings from an exposed rock or in flight. The female builds a rather bulky nest of grass and moss lined with fur, hair, and feathers, usually well-hidden in a cavity formed by a rock fissure, in scree, or under loose rocks on the ground. Clutch size is normally 4 to 7 eggs (Hussell 1972). The male brings food to the female during incubation, which lasts about 12 days, and both sexes feed the nestlings, which remain in the nest cavity until shortly before they can fly at about 12 to 14 days old. The juveniles, which are easily recognized by their gray-streaked body plumage and white in their flight feathers, are fed by their parents for several more days after leaving the nest (Parmelee 1968).

The single confirmed breeding record was an observation of a family party of 2 adults and 2 or 3 newly-fledged young on West Pen Island, near the Manitoba border on 20 July 1985 by David Shepherd and Greg Poole. Although the habitat was not rocky, piles of driftwood along the beaches may have provided a nesting site. Snow Buntings are encountered occasionally elsewhere along the Hudson Bay coast in summer, but a lack of rock cavities renders much of the habitat unsuitable for breeding. At Cape Henrietta Maria, however, the shattered limestone ridges and man-made cairns provide a few marginally suitable sites and occasional nesting there would be expected. Single individuals and flocks of as many as 6 Snow Buntings were seen at Cape Henrietta Maria between 21 June and 6 July 1985 by the author, Roy Smith, and Erica Dunn, but there was no behavioural evidence of breeding.

The nearest regular breeding site of the Snow Bunting is on the Belcher Islands, NWT, which are about 160 km northeast of Cape Henrietta Maria and 550 km east of West Pen Island (Freeman 1970b, Manning 1976). On the west coast of Hudson Bay, the nearest regular breeding area is some 600 km north-northwest of the Pen Islands (Godfrey 1986). The atlas records indicate that breeding in Ontario can probably be regarded as an unusual occurrence.-- *D.J.T. Hussell*

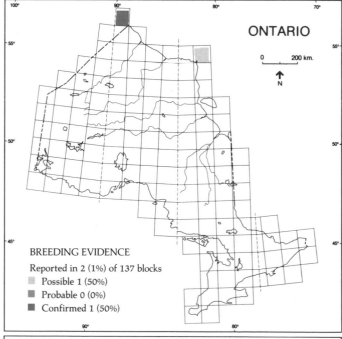

ONTARIO

0   200 km.

N

BREEDING EVIDENCE

Reported in 2 (1%) of 137 blocks

Possible 1 (50%)
Probable 0 (0%)
Confirmed 1 (50%)

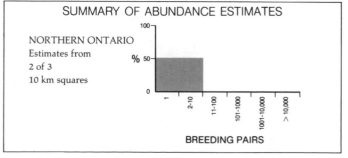

SUMMARY OF ABUNDANCE ESTIMATES

NORTHERN ONTARIO
Estimates from
2 of 3
10 km squares

%

BREEDING PAIRS

COMMON REDPOLL
Sizerin flammé
Carduelis flammea

The Common Redpoll is a truly 'winter finch' as it occurs in the south only during winter. It is superbly adapted to survive in harsh northern environments (Brooks 1968). As with some other members of the genus *Carduelis*, the Common Redpoll is an irruptive species, common in the south in some years and rare in others. Irruptions appear to be closely tied to the appearance of heavy seed crops of trees such as tamarack and alder (Newton 1972).

The Common Redpoll breeds from coast to coast and to the shores of the Arctic Ocean, south to the northern limits of the Boreal Forest. In winter, however, it wanders southward as far as the northern states of the US, irregularly penetrating as far south as Arizona and Texas. In Ontario, the species is a common winter resident throughout the province. As a breeding species, however, it is restricted to the Hudson Bay Lowland north and northwest of the Albany River (James *et al.* 1976, Speirs 1985). Nesting habitat consists of low shrub tundra or barrenlands, often with patches of spruce, alder, or willow thickets.

The Common Redpoll is not well known as a breeding species in Ontario, because of its northern breeding range. Even if observers have access to the barrenlands and tundra, breeding pairs may be difficult to locate because the species shows few territorial displays. When located, the small untidy nest of fine twigs, grass, and plant stems is usually in a dwarf tree. In arctic areas, old nests may be relined and used again in subsequent years (Harrison 1978). Nests are conspicuous because they are in trees, and the Common Redpoll is one of the few tundra species to have its nest above the ground (R. James pers. comm.). The clutch of 4 to 5 pale blue, finely speckled eggs is incubated for about 12 days by the female alone. Young are tended in the nest by both parents and fledge when about 11 to 14 days old. The

breeding season varies with latitude and habitat but usually starts in late May or early June. Double broods have been recorded.

The range shown on the map corresponds well with that indicated by Godfrey (1986). Not surprisingly, the confirmed breeding records for the Common Redpoll occur in the barrens to the west of James Bay, and along the Hudson Bay coast. The open habitat in these coastal areas and the use of small patches of suitable vegetation make for ease of observation of adult birds as they carry food to the young. It is likely that those blocks in this area with possible and probable nesting could have been upgraded to confirmed breeding status, if not for the limited time available. Confirmation of breeding generally decreased away from the coast. Nevertheless, given the appropriate nesting habitat, isolated nesting south of the main range in the north could occur. For example, Godfrey (1986) documents sporadic and local nesting in areas south of the known breeding range.

Abundance estimates suggest that the Common Redpoll is fairly common within its range. The highest abundance estimates, those over 10 pairs per square, come from the James Bay coast and from inland areas west of that coast.-- *A.L.A. Middleton*

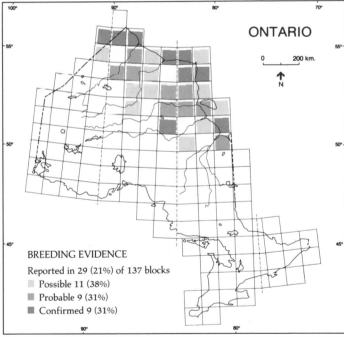

ONTARIO

0   200 km.

N

BREEDING EVIDENCE

Reported in 29 (21%) of 137 blocks

Possible 11 (38%)
Probable 9 (31%)
Confirmed 9 (31%)

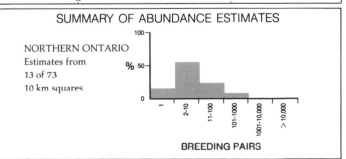

SUMMARY OF ABUNDANCE ESTIMATES

NORTHERN ONTARIO
Estimates from
13 of 73
10 km squares

BREEDING PAIRS

561

HOARY REDPOLL *
Sizerin blanchâtre
Carduelis hornemanni

with brood patches and a male with a cloacal protuberance. According to the atlas definitions of breeding evidence, these observations are categorized as probable breeding. Both records were in extreme northwestern Ontario on coastal tundra with raised sand and shingle beaches with limited ground cover. On 15 July 1985, Duncan Evered and Erik Kiviat observed an adult bird in willow scrub on the tundra 40 km west of Cape Henrietta Maria. The record was accepted as being of an individual in suitable breeding habitat.

The Hoary Redpoll has not been shown conclusively to breed in Ontario, but atlas data suggest that the species may breed in low numbers on the coastal tundra of Hudson Bay.-- *A.L.A. Middleton*

The specific status of the Hoary Redpoll has long been debated by ornithologists. Nevertheless, it was still listed as a separate species in 1983 (AOU 1983). In appearance, the typical Hoary Redpoll closely resembles the Common Redpoll, but its general colouration is decidedly paler and it has an unstreaked white rump which may be tinged with pink in the males (Peterson 1980). A more recent study, however, has demonstrated a continuous range of plumages from Hoary to Common types with no diagnostic characteristics separating one form from the other (Troy 1985).

The breeding distribution of this circumpolar species is restricted to the tundra and areas of the high arctic, with nesting being known as far south as Churchill, Manitoba (Godfrey 1986). In Ontario, it has not been confirmed as a breeding species, and is classed as an uncommon winter resident throughout the province (James *et al.* 1976, Speirs 1985). In winter, it occurs infrequently in flocks of Common Redpolls and may wander southward as far as the US-Canadian border.

Apart from its more northern breeding distribution, which deprives it of substantial shrubs for nesting, the breeding biology is very similar to that of the Common Redpoll. Nesting begins in late May, and the species is characteristically single-brooded. The nest is usually on or near the ground and is a cup of fine grasses lined with feathers, hair, and plant down. The normal clutch is of 4 or 5 pale blue, finely speckled eggs. Incubation, by the female alone, lasts for about 14 days, and young fledge when 10 to 14 days old.

Potential breeding records came from 3 atlas squares in 2 blocks along the Hudson Bay coast. Between 4 and 18 July 1985, David Shepherd and Greg Poole observed a pair at the mouth of Oostaguanako Creek. In the square immediately to the west, in the same time period, they mist-netted 2 females

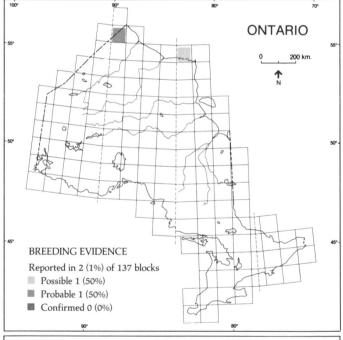

ONTARIO

BREEDING EVIDENCE

Reported in 2 (1%) of 137 blocks

Possible 1 (50%)
Probable 1 (50%)
Confirmed 0 (0%)

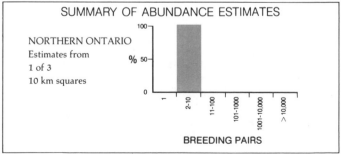

SUMMARY OF ABUNDANCE ESTIMATES

NORTHERN ONTARIO
Estimates from
1 of 3
10 km squares

BREEDING PAIRS

TRUMPETER SWAN *
Cygne trompette
Cygnus buccinator

Although now confined to the west, the Trumpeter Swan probably bred across much of North America prior to the arrival of settlers. Archaeological evidence and reports of early explorers record the bird in southern Ontario and the Hudson Bay Lowland, in circumstances suggesting that it formerly bred in these areas (Lumsden 1984b). It disappeared as a summer resident in southern Ontario sometime in the early 1700's, and the last migrants from the west were recorded crossing Ontario about 100 years ago.

Efforts to restore Trumpeter Swans to Ontario started in 1982. From 1983 to 1986, at Cranberry Marsh near Whitby, Trumpeter Swan eggs were placed in the nests of feral Mute Swans, and the cygnets were foster-raised to flight stage. Eleven cygnets have so far reached independence.-- *H.G. Lumsden*

BRANT *
Bernache cravant
Branta bernicla

The Brant nests on the Eurasian and North American tundra. In Canada, it nests on the coasts from Coats and Southampton Islands to Ellesmere Island, and from the Foxe Basin to Alaska. Brant winter on the coasts of the Atlantic and Pacific Oceans. No evidence of breeding was reported in Ontario during the atlas period.

The Brant nests and raises its brood on low lying terrain closer to the tide line than any other species of goose. While such habitats look barren, they produce unusually nutritious vegetation on which Brant and other geese feed. They may be found in braided river mouths, on désolate coasts among hummocks, and near shallow ponds (Palmer 1976a).

Subadult Brant remain in small numbers throughout the summer and moult on the Hudson Bay coast of Ontario. Small flocks have been seen in July around East Pen Island and on and to the west of Cape Henrietta Maria. There is no evidence that Brant nest now nor that they have ever nested on that coast of Ontario, but there are 2 reports from southern Ontario which appear to be reliable although not substantiated by specimens or photographs.

Baillie (1963b) recorded that 18 Brant summered on Kelly Lake near Copper Cliff in 1954. Frank Fielding kept a Brant brood, containing 5 goslings, under observation from 12 August to 29 September. An adult female, collected on nearby Ramsey Lake on 30 June of that year, by William Morris, is in the Royal Ontario Museum.

Harry McLeod (MNR files 1970) reported that Lloyd Lee, a biologist with the Department of Lands and Forests, saw an adult Brant on 7 August 1970 and caught a gosling on 21 August on small islands near the Robert H. Saunders Dyke

on Lake St. Lawrence. The bird was downy and was just growing feathers. It was released in the goose pen at the Upper Canada Migratory Bird Sanctuary. By 9 September it had "feathered out much more, and could not be mistaken for any other bird". It eventually flew away.-- *H.G. Lumsden*

GREATER PRAIRIE-CHICKEN
Grande Poule-des-prairies
Tympanuchus cupido

The Greater Prairie-Chicken survives in reduced numbers and range in the midwestern US; COSEWIC considers it to be endangered in Canada, with occasional sightings in Saskatchewan not proved to represent recent breeding. It may have been native to the extensive prairies of Kent and Essex Counties prior to settlement. It was certainly there in the early 1800's, and, with forest clearing, it spread to the vicinity of Burlington, Toronto, and Holland Marsh by 1875. It could not survive intensive farming, and, by the early 1900's, had disappeared from most of its acquired range. The last birds were reported on the Indian Reserve on Walpole Island in 1923-24 (Lumsden 1966).

The Rainy River and the southern Kenora Districts were invaded from Minnesota and Manitoba by Greater Prairie-Chickens in the early 1900's. Settlers reported both this species and the Sharp-tailed Grouse near Dryden (Olsen 1959, McGillivray 1963) and in the western Rainy River District (Lumsden unpubl.). A specimen was shot in 1929 about 80 km northwest of Thunder Bay (Baillie 1947b). The last Greater Prairie-Chickens seen in northwestern Ontario were in Sanford Township, Kenora District, on 8 April 1959 (Olsen 1959).

The Sault Ste Marie area was invaded by the Greater Prairie-Chicken from Michigan about 1925. It reached Manitoulin Island about 1928. In the winter of 1932-33, a massive irruption of Northern Sharp-tailed Grouse *(T. p. phasianellus)* took place (Snyder 1935). It spread over 100,000 km² of central Ontario and reached Manitoulin Island (MNR files), where the Sharp-tailed Grouse hybridized with Greater Prairie-Chickens in the spring of 1933. The Northern Sharp-tailed Grouse are adapted to muskeg habitats and cannot survive in grasslands. They disappeared entirely from their acquired range, including Manitoulin Island, by 1934 (Snyder 1935) leaving Greater Prairie-Chickens and hybrids to complete colonization by 1945 (Lumsden unpubl.).

From 1949 to 1951, unsuccessful attempts were made by the Department of Lands and Forests to establish the Manitoulin stock elsewhere by releases at Barriefield (Frontenac Co) and Point Petre (Prince Edward Co). A second invasion of Manitoulin Island by the grassland-adapted Prairie Sharp-tailed Grouse *(T. p. campestris)* took place in the late 1940's. This stock interbred with and finally displaced the hybrid-Prairie-Chicken population.-- *H.G. Lumsden*

CALIFORNIA GULL
Goéland de Californie
Larus californicus

The California Gull is undoubtedly one of the rarest birds in Ontario. The breeding range of this species lies in the western half of North America. In Canada, it nests from the Great Slave Lake area in the Northwest Territories, south through the Boreal Forest of Alberta and Saskatchewan into the parklands and prairies of the 3 Prairie Provinces. The easternmost part of the breeding range in the US is North Dakota (Godfrey 1986). Thus, a nesting California Gull in Ontario is of exceptional rarity. In fact, James *et al.* (1976) do not mention the species, even as an occasional vagrant to Ontario.

California Gulls nest in colonies, often with Ring-billed Gulls and other waterbird species, on islands in lakes, marshes, and rivers (Godfrey 1986). Their breeding biology in Alberta was studied by Vermeer (1970).

During the 1981 to 1985 period, the California Gull has nested in Ontario at least twice. The 2 documented breeding attempts occurred in 1981 and 1982 on the Leslie Street Spit of Toronto's outer harbour. This area was formally named Tommy Thompson Park in 1985. Each year, an apparently unmated adult California Gull attended a nest at the edge of a large Ring-billed Gull colony. The nest in 1982 was less than 10 m away from the location of the 1981 nest. It is, therefore, very likely that the same individual bird was involved in the 2 nesting attempts. The unmated bird was presumably a female because there were eggs in the nest that were larger than those of the Ring-billed Gull.

In 1981, the bird was first seen on 1 May near an empty scrape. On 14 May, it was incubating 2 eggs. Without a mate it would have been unable to provide adequate incubation. Consequently, CWS biologists exchanged its 2 eggs with 3 eggs from a nearby Ring-billed Gull nest, in order to see if the eggs would hatch when incubated by a mated pair of gulls. The California Gull was unable to look after the Ring-billed Gull eggs. On 29 May, only one cold egg was left in the nest, and on 1 June the gull was resting and preening without making further incubation efforts.

The 2 eggs of the California Gull never hatched: on 12 June they were still incubated by the Ring-billed Gull foster parents but one had addled and cracked. On 17 June, the other egg had spoiled as well.

In 1982, the bird was first seen on 6 May. Again, it appeared to be unmated. By 14 May, it had built a normal nest and laid one egg. Between 14 May and 2 June, the bird was often seen incubating its egg but on 6 June, the egg had been depredated and the bird had stopped incubation.

It is extremely unlikely that 2 different, unmated, female California Gulls would nest in the same area in a Ring-billed Gull colony near Toronto in 2 successive years. In 1982, the California Gull did not have a black spot on the lower mandible, whereas that was the case in 1981. The absence of the black spot in 1982 is consistent with Behle and Selander's (1953) account of the plumage cycle of this species, and suggests that the bird was possibly 3 years old when first found, and in its first breeding season.

In the fall, this rare gull possibly may have migrated with the Ring-billed Gulls to the Atlantic Coast, instead of to the usual wintering grounds on the Pacific coast. It is a remarkable coincidence that a single California Gull was seen in coastal New York during the winters 1978-79 through 1981-82 (Weseloh and Blokpoel 1983).

When it became known, during the 1981-82 winter, that a California Gull had nested near Toronto, it was virtually certain that many birders would visit the nesting area in 1982, and it was likely that there would be an undesirable amount of disturbance of the local Caspian Tern colony. As soon as the California Gull was seen near its nest in 1982, the CWS made a bush trail that led to a viewing station from which the gull could be observed without disturbing it or the Caspian Terns (Weseloh and Blokpoel 1983). After a call to Toronto's 'bird hotline' and a write-up in the weekly bird column of the Toronto Globe and Mail, many birders came to see the gull. It may well be years before we get another documented case of this prairie species nesting at the doorstep of Toronto.

Although the only breeding evidence comes from Toronto, a more likely place to look for this species in future is in northwestern Ontario, which is much closer to its prairie breeding range. Its presence could easily go unnoticed unless every individual pair of Herring and Ring-billed Gulls was checked very carefully (D.V. Weseloh pers. comm.).-- *H. Blokpoel*

BLACK GUILLEMOT
Guillemot à miroir
Cepphus grylle

This seabird is widespread in the eastern Canadian arctic, northern Quebec, Labrador, Newfoundland, the Gulf of St. Lawrence, and the Maritime Provinces. It is numerous on the Quebec side of Hudson Bay, is apparently a year-round resident on the Belcher Islands in southern Hudson Bay, and is known to nest as far south as North Twin Island in the centre of James Bay (Todd 1963, Manning 1981), but records on the Ontario side of the bays are very limited. There are no rocky cliffs and practically no areas with even piles of broken rocks where nesting might occur. Despite sporadic sightings dating back to at least the mid-1800's (Manning 1952, Todd 1963), there is only one record of breeding in the province. In 1957, hatched eggshells were found among the rocks on Manchuinagush Island, after a group of about 80 birds was disturbed there (Lumsden 1959). Lumsden described Manchuinagush Island as being about 22 miles west of Cape Henrietta Maria, but recent maps indicate that it is only about 25 km from the Cape at approximately 55°12'N, 82°43'W (H.G. Lumsden pers. comm.). Although it is in Hudson Bay, the island is part of Ontario because it is joined to the mainland at low tide.

Manchuinagush Island was visited by J.P. Prevett about 27 June 1981 but no Black Guillemots or their nests were found there, despite a thorough search (notes in MNR District Office, Moosonee, J.P. Prevett pers. comm.). In view of the relatively thin coverage of the coasts of James and Hudson Bays, it remains possible that Black Guillemots breed at some other site; but most of the habitat is unsuitable and no other potential breeding sites are known.-- *R.D. James*

PASSENGER PIGEON
Tourte
Ectopistes migratorius

This extinct species once occurred in Ontario in numbers that can scarcely be imagined. Flocks of migrants that darkened the sun like clouds passed uninterrupted from sunrise to sunset. Such flocks were estimated to be as much as 500 km long, and, averaging nearly 2 km wide, contained uncountable millions of birds. Nesting colonies were reported to cover areas as large as 350 km^2, with the branches of trees breaking under the weight of birds, and the roar of wings described as like a gale at sea. Colonies were usually established in hardwood bush or swamps, and, as they rapidly expanded outward, trees of any species, coniferous or deciduous, were used to hold nests. The number of colonies is unknown, but was likely limited more and more by available food resources as forest clearing proceeded. They were undoubtedly found throughout southern Ontario, becoming more scattered in the Boreal Forest zone, but occurring at least as far north as the latitude of Moosonee (Mitchell 1935).

They were killed by any means possible without restriction, but the easiest place for market hunters to obtain the millions they shipped for food was at nesting colonies. With the decimation of these colonies and winter roosting sites, the birds rapidly decreased in numbers. The last large flights (with only thousands of birds) in Ontario took place before 1880, and by 1900 the species had all but disappeared. The last nesting reported was in Frontenac Co in 1898 where only about 20 birds attempted to nest, and the last sighting considered to be reliable was made in Simcoe Co in 1902 (Mitchell 1935).-- *R.D. James*

BEWICK'S WREN
Troglodyte de Bewick
Thryomanes bewickii

The Bewick's Wren is the rarest of the wrens that are known to have bred in Ontario. There is no evidence that it still does so, but it appears quite regularly at Point Pelee Nat. Park in the spring (Goodwin 1982), although perhaps less so recently. The Bewick's Wren is found only in North America, where it frequents open woodlands, upland thickets, fence rows, and orchards. It is widespread as a year-round resident across much of the southern US, especially in the west, but has undergone a fairly drastic decline in the eastern and central parts of its range in recent years (Robbins *et al.* 1986). In Canada, it is found in southern British Columbia and has nested in the Carolinian Forest region of southwestern Ontario.

It was first recorded in Ontario in 1898, but the first nest was not found until 1950, when one was located at Point Pelee (Godfrey 1986). Four additional nests were recorded at Pelee in 1956 and 1957 (Kelley *et al.* 1963). Since then, this wren has been sighted during migration along the southern border of the province, but there has been no further confirmation of breeding.-- *M.A. Richard*

MOUNTAIN BLUEBIRD
Merle-bleu azuré
Sialia currucoides

The usual breeding range of the Mountain Bluebird extends from Alaska as far east as Manitoba; however in 1985 one nested in southwestern Ontario for the first time.

The first record of the Mountain Bluebird in Ontario occurred in December 1965. Since then, there have been 10 reports of the species in the province, but only one other Mountain Bluebird has been observed in the summer: a male observed in Carden Township (30 km east of Orillia) from July to September 1980 (Speirs 1985). That bird was seen by many observers, but apparently did not breed.

This species was confirmed as breeding when a male Mountain Bluebird bred in a nest box near Port Stanley, Elgin Co in 1985. It was mated with a female Eastern Bluebird and was seen from 1 May to 25 June. The nest contained 3 eggs at the beginning of May then, after a 3 day gap, 5 new eggs were laid. The final clutch of 8 eggs was larger than normal clutches of either species. Following incubation, only 3 eggs hatched. The 3 nestlings, which appeared to be Eastern Bluebirds, were fed by the Mountain and Eastern Bluebird parents and fledged successfully. The remaining 5 eggs did not hatch and were found to be infertile. It is probable that these 5 eggs were laid by the Mountain-Eastern pairing and were infertile as a result (Field and Hurst unpubl.). Such infertility is known to result from pairings of male hybrids with female Eastern Bluebirds (Rounds and Munro 1982). A Mountain Bluebird, possibly the same male, returned to a nest box in the same area in 1986. It was again mated to an Eastern Bluebird.-- *C.J. Risley*

Applications of atlas data: some preliminary examples

There are many potential applications of the atlas data. The following 'case studies' provide examples of 5 ways in which the data might be used.

The Role of Atlas Data in Determining the Relative Size of Ontario Breeding Bird Populations

Paul F. J. Eagles

Introduction

To know which species deserve priority for conservation, we need to know how many of each species there are, and where they are. These 2 questions underlie many conservation programs, such as those concerned with endangered species, wildlife management, habitat protection and parks planning and management.

Special attention must be given to species that are rare, threatened or endangered. According to the Committee on the Status of Endangered Wildlife in Canada (COSEWIC) there are 3 levels of rarity. A rare species is "any indigenous species of fauna or flora that, because of its biological characteristics or because it occurs at the edge of its range, or for some other reason, exists in low numbers or in very restricted areas of Canada but is not a threatened species." A threatened species is "any indigenous species of fauna or flora that is likely to become endangered in Canada if the factors affecting its vulnerability do not become reversed." An endangered species is "any indigenous species of fauna or flora whose existence in Canada is threatened with immediate extinction through all or a significant portion of its range, owing to the action of man." These definitions are widely accepted across Canada and are now being applied to smaller geographical units such as provinces or counties.

Ontario has an Endangered and Threatened Species Program within the Ministry of Natural Resources. It derives its mandate from the Endangered Species Act which makes it illegal to interfere with or destroy an endangered species or its habitat. The goal of the program in Ontario is "to ensure that no plant or animal species indigenous to Ontario becomes extinct as a result of the action of man" (McKeating and Bowman 1977).

Many regions and counties in Ontario have programs to delineate and protect Environmentally Sensitive Areas (Eagles 1980, 1984). One criterion for the determination of such an area is the presence of a species that is endangered within Canada, within Ontario or within the region or county (Eagles 1981).

At all 3 levels (country, province or county), determining the abundance of a species is a critical component of the various resource management programs. At present, this determination is usually done by soliciting the best professional judgement. This means that the experts on the fauna or flora in question pool their knowledge and experience in an effort to produce a consensual judgement on the status of a species. This method is effective and is the only route available for most species. Eagles and McCauley (1982) provide an example of one such attempt for the breeding birds of Ontario.

There are problems with such a determination. The data on the distribution of the species are often spotty. It is probable that there are no data from some areas of the species' range because of a lack of comprehensive field work. Most species have not been systematically surveyed in the field; some species may have had little or no field attention; some species are secretive and have therefore been recorded by few observers. There may be wide disagreement among experts on the status of a species because of variability in their own field work. For most species, however, the pooled knowledge of many experts is still the most effective method to determine their status.

The atlas project's computer data base provides an unprecedented opportunity for determining a breeding bird's population status. The data were systematically collected over wide geographical areas with objective field criteria and standards of minimum coverage. The number of hours of field observation is very large. This makes the atlas data base the best available source on the present population status of many of Ontario's breeding birds.

The computer data base can be searched in many ways. Searches by block or by square within a wide variety of geographical units are possible. For example, the number of blocks in Ontario in which a species occurs can be determined. The number of squares in which a species occurs in a part of the province, say southeastern Ontario, can also be found. The number of squares in which one breeding species occurred can be compared to the number for another species. Given the flexibility and power of a computerized data base, the potential uses of the data are many.

Relative Frequency of Occurrence at a Variety of Scales

Table 4 is a list of those species that were found in only one block during the 5 year atlas period. Those species that are on the list obviously have limited breeding distributions within Ontario.

TABLE 4: *Species recorded in only one block*

American White Pelican	Glaucous Gull
Cattle Egret	Snowy Owl
Ross' Goose	Western Kingbird
King Eider	Northern Wheatear
Purple Gallinule	Mountain Bluebird
American Avocet	Kirtland's Warbler
Pectoral Sandpiper	Lark Sparrow
Purple Sandpiper	Harris' Sparrow
California Gull	

Table 5 is a list of the 5 species that were found in the highest number of squares in southern Ontario. Those species on this list are widespread and abundant.

TABLE 5: *The 5 species found most widely in southern Ontario*

Species	Number of Squares	Percentage of all squares (out of 1824)
Song Sparrow	1801	98.7
American Robin	1786	97.9
Red-winged Blackbird	1784	97.8
Tree Swallow	1783	97.8
Cedar Waxwing	1779	97.5

Table 6 shows the potential for a search of a small area, such as one atlas region. It is a list of the species that were found in only one of the 53 squares in the atlas Region 7 (Waterloo-Wellington). This region includes parts of 10 counties or regional municipalities.

TABLE 6: *The species found in only one square in atlas region 7 (Waterloo-Wellington)*

Red-necked Grebe	Merlin
Double-crested Cormorant	Wilson's Phalarope
Great Egret	Herring Gull
Northern Shoveler	Carolina Wren
Lesser Scaup	Tennessee Warbler
Canvasback	Prairie Warbler
Ruddy Duck	Le Conte's Sparrow

Tables 4 through 6 show that it is possible to search the computer data base at a variety of scales and in a variety of forms. A valuable list might be that for a single county. Those undertaking environmental assessments might wish to assess the relative frequency of occurrence of a species in a regional area around a proposed development. In that case, a list could be generated for a certain number of squares surrounding the development: for example, the 50 squares around the square of interest.

Abundance Estimates

Estimates of abundance provide important additional information. The estimated numbers of pairs of birds in each square can be used to provide a rough population estimate for a given area. A method for the systematic application of this information has been developed by Ron Weir and Mike Cadman during the preparation of this Atlas. This method is best illustrated by the use of an example: in this case for the Red-shouldered Hawk.

Table 7 shows how abundance estimates can be used to develop ranges of abundance. If the lower limits are totalled and the upper limits are also totalled, the range of estimated numbers becomes 382 to 1,099 pairs. If one assumes that the same proportion of estimates comes from the 205 squares where the atlassers recorded the species but did not submit estimates, then the range becomes 825 to 2,372 pairs. The provincial population of the Red-shouldered Hawk is located almost entirely in southern Ontario. Therefore, the population range estimate of between 825 and 2,372 pairs is also a close approximation of the provincial population.

This estimate assumes that the abundance estimates are accurate, that the number of squares in which the species was found is also accurate, and that there was no difference in abundance between squares with and without abundance estimates. We have no way of knowing the inherent bias due to birds not being recorded in some squares, or to birds moving from one square to another during the atlas period. This method can only be applied to estimate a species' southern

TABLE 7: Red-shouldered Hawk abundance estimates

Abundance estimate	Number of squares	Number range
1 pair	89	89 pairs
2 to 10 pairs	75	150 to 750 pairs
11 to 100 pairs	13	143 to 1,300 pairs
Total	177	382 to 2,139

Since it is not reasonable to expect that there are more than 20 pairs of this species in any one square, the limit of 20 is adopted for the species:

11 to 20 pairs	13	143 to 260 pairs
TOTAL	177	382 to 1,099 pairs

Ontario breeding population because coverage was comprehensive there. In northern Ontario, where only a few squares in each block were covered, we have no way of knowing the population size in the squares which were not visited.

Although the absolute numbers produced through these calculations are not to be taken as anything more than rough approximations of the actual southern Ontario population, the relative abundance is a useful means of ranking species rarity. If this method is applied to 5 species, arbitrarily chosen by the author, one can get an idea of the relative magnitude of the populations, as shown in Table 8. In this table, no arbitrary upper limits were applied.

TABLE 8: Southern Ontario population estimates for 5 species

Selected Species	Population ranges in southern Ontario
Tufted Titmouse	26 to 63 pairs
Carolina Wren	39 pairs (all estimates = 1 pair)
Cerulean Warbler	372 to 2,678 pairs
Rufous-sided Towhee	6,819 to 59,903 pairs
Vesper Sparrow	18,466 to 172,434 pairs

A similar method applied to all species that breed in southern Ontario can provide an objective basis for determining the relative abundance of a species within southern Ontario. This can be very helpful, particularly in conjunction with information on population trends, in deciding whether the species should be considered rare, threatened or endangered as defined by COSEWIC. However, in well known species such as the waterfowl, for which there are population data from other sources, the atlas data would be but one source of information.

It is important to recognize that there are inherent difficulties and limitations with the use of the abundance estimates in the fashion outlined above. The most important will be briefly outlined.

Species have different levels of detectability. Those that are difficult to find will tend to have lower abundance estimates and therefore will appear to be rarer than they are in the province. It might be possible to adjust for such biases by applying a multiplier to those species with low levels of detectability.

Observer effort was not equal across Ontario so those species that inhabit less well-covered areas may tend to have lower abundance estimates. However, this point requires further investigation to see if the bias does occur.

Atlas data are an accumulation of 5 years of year-by year data. The final atlas abundance estimate used for a species in a square is the highest of all individual years' data. This method was adapted because it was felt that there tended to be a general bias amongst atlassers to underestimate the actual field abundance. It is the author's opinion that many of the final abundance estimates are still too low. This is especially the case for quite abundant species, for which atlassers appear to feel that there is validity in estimating conservatively, i.e., low. Many of the rare species might have occurred in an atlas square in only 1 or 2 years of the atlas period. Therefore, for those species, the final abundance estimate is, in a sense, an over-estimation.

Summary

This paper shows the potential of the atlas data for the objective determination of a breeding bird's relative numerical status. A variety of scales of investigation is possible from province-wide down to a local area. The distributional data, combined with abundance estimates, provide a powerful tool for the assignment of conservation priority to Ontario's breeding bird populations.

Uses of Breeding Bird Atlas Data for Environmental Planning

Kenneth W. Dance and Donald M. Fraser

Introduction

Although it is generally recognized that bird atlas data can be used in environmental planning and in the review of projects involving land-use alteration (Laughlin *et al.* 1982), there is a paucity of literature on specific applications of these data.

Dance and Fraser (1985) described 3 case studies in which bird atlas data were used in environmental assessment. This present article outlines the benefits of augmenting conventional sources of information on the breeding birds of Ontario with atlas data. A variety of considerations in using this information for environmental planning and in specific environmental assessments are addressed.

Environmental Assessment

Environmental assessment studies deal with a specific piece of land and a proposal to change the use of that land. Investigations are carried out to determine if significant or sensitive natural features or conditions exist on the piece of land and to determine if impacts associated with proposed changes in land-use are expected to affect those features or conditions. Wildlife populations, including breeding birds, are among the key components of the natural environment that are routinely addressed in environmental assessments.

Often, when a proposal to develop a piece of land is submitted to review agencies, they indicate that concerns exist regarding the biological resources of the area and that they lack sufficient information with which to evaluate the potential consequences of the proposed undertaking on those resources.

The proponent of the undertaking is therefore requested to prepare an Environmental Impact Statement (E.I.S.), a document which provides an assessment of the possible impacts of the project on the environment. The proponent often hires an environmental consultant to prepare the E.I.S.

Under certain circumstances, the E.I.S. is reviewed before a judicial tribunal such as the Ontario Municipal Board. The proponent must therefore commission a study that is defensible, although frequently there are time and cost limitations placed upon the study. In some cases, timing of the assessment is such that detailed observations of bird communities during the breeding season are not possible.

Even if it exists, background information on the breeding birds which inhabit the piece of land in question may be outdated and habitat conditions on the site may have changed. Other sources of information, such as breeding bird atlas data, are therefore required. Even if it is possible to conduct field work on the site, these data are of considerable value.

Atlas Data and Environmental Assessment

One of the first questions regarding breeding birds which must be answered in an environmental assessment is "what species occur where on the site?"

Since atlas data cover 10 km squares, it would appear that these data are of limited value for most assessments, unless the study area is large. However, we have found that by following the series of steps shown in Figure 16, atlas data and the province-wide network of atlassers can make a significant contribution to the assessment. The atlasser network is the collective human resource and has knowledge of what was found where within the square, local habitat conditions, and which breeding species are rare or uncommon.

As a routine part of environmental assessments, one can purchase computer printouts of atlas data for pertinent squares. The printouts indicate species recorded, highest level of breeding status attained, level of effort of atlassing in hours, and abundance estimates.

Printouts are reviewed to determine if any uncommon, rare, or sensitive bird species, collectively referred to as potentially significant species, occur in the square. If, at this stage of the process, it is thought that a significant species might occur on the study site, 2 further types of study ensue:

- confirmation of the species' occurrence on the site, and
- evaluation of the species' status and the significance of such an occurrence.

Thus, if any potentially significant species are identified, the habitat requirements of those species are determined and compared with conditions known to occur on the study site. The knowledge that a potentially significant species might be present assists in designing the site-specific field program to sample for this species, along with the overall characterization of the breeding bird community.

If the project schedule allows, contacts with individual atlassers are initiated through the Atlas Coordinator and Regional Coordinator. The objective of this contact is to talk directly to the atlasser concerning the details of where within the square the potentially significant species was found, whether it occurred in a number of locations in the square and/or in adjacent squares, and whether the species is expected to occur on the site under study. Discussions with the atlasser also aid in evaluating whether the occurrence of a particular species is significant or not.

Significance Determination

Prior to the atlas project, it was often difficult to determine the provincial and regional significance of a particular breeding bird species. In a recent environmental assessment of a proposed development in Durham RM, the presence of a nesting pair of Red-shouldered Hawks, a potentially significant species, emerged as an issue. A review of available literature indicated that various authors adopted different methods and criteria for determining the regional and/or provincial status and abundance of this species.

In evaluating this information, we had to develop a means of distilling a variety of opinions into a defensible summary of status. At the time of the evaluation, data from the first 3 years of the atlas were also available and assisted greatly in reaching the conclusion that the Red-shouldered Hawk was a regionally rare breeding species.

Benefits of Using Atlas Data to Determine Species Status

The proponents of the proposed development in Durham RM mentioned above, the other consultants working on the environmental assessment, and the Ontario Municipal Board all accepted the Ontario Breeding Bird Atlas data as credible. We suspect that the reasons for this acceptance of the atlas data included the following:

- the survey was comprehensive in a geographic sense;
- a standard technique was used to gather, record, and interpret findings;
- results could be expressed numerically, *i.e.*, a species was confirmed in X of Y squares;
- breeding distributions could be mapped, allowing for relatively easy interpretation of the results.

We see atlas results as providing the basis for the preparation of a standard reference on the numerical status and significance of breeding bird species in Ontario. Paul Eagles addresses this issue in his discussion of the role of atlas data in the determination of the status of a species (see page 566). Such a process, producing a ranked list of the rarest species, would eliminate the need for workers across the province to make independent interpretations of the status of a particular bird. The existence of such a list would not, however, eliminate all elements of interpretive decision-making in environmental assessment. The environmental consultant must still review the life history, habitat requirements and sensitivity of a species in light of the type and magnitude of disturbance associated with the proposed undertaking.

Limitations in Use of Atlas Data

As a cautionary note, we recognize that atlas data are not without their inherent limitations. Therefore, we suggest that the following points be considered when using atlas data in environmental assessment:

Level of Effort: the number of hours spent atlassing in a particular square obviously provides some measure of the proportion of total species recorded and how well breeding confirmation levels reflect the actual situation.

Species Expected: Regional Coordinators provided a target species total for every square in southern Ontario. Comparing the actual number of species found in the square to the target number may be a better indicator of effort than the number of hours spent atlassing, which is a productivity criterion varying considerably with atlasser efficiency.

Variety of Habitats in the Square: the square data should be checked to ensure that bird species from all habitats within the square are represented; *e.g.*, were species recorded that are expected to occur on islands or wetlands present, or did inaccessibility prevent coverage of all habitats?

Skill of the Atlasser: we venture to suggest that virtually all atlassers became more skilled as the atlas field work progressed, but that skill level nevertheless varied among atlassers. If a square was covered by more than one person, the coverage disparities between individuals would be partially eliminated. Certain species were undoubtedly overlooked. The point of caution is that atlas data should not be taken to represent a 100% sample of the breeding bird species for a particular square.

Size of Site: if a study site for a planning project is small, *e.g.*, 20 ha in area, atlas data for the surrounding square are primarily of interpretive value; if the study area is 500 ha and contains elements of the major habitats represented in the square, the atlas data could be considered as basic inventory data for the study area. However, as Laughlin *et al.* (1982) caution, atlas data rarely replace site-specific studies.

Abundance Estimates: not all atlassers attempted an estimation of species abundance. The atlas species accounts address the issue of difficulties in finding and confirming certain species and in estimating their numbers. Judicious review of abundance estimates is suggested, if this information is to be used in environmental assessments.

New Initiatives

We are convinced, as routine 'users' of Ontario Breeding Bird Atlas data, that the information gathered to date has many applications in conservation planning, environmental management, and environmental assessment.

Environmental Impact Statements, however, must address more than just birds, and atlas data on other life forms which are sensitive to land-use alteration would also be of value to those preparing environmental assessment reports.

The Ontario Breeding Bird Atlas is a major accomplishment. Using the results of this and other similar initiatives in a way which will influence decision-makers is a challenge which lies ahead.

FIGURE 16: *Use of breeding bird atlas data in environmental assessment*

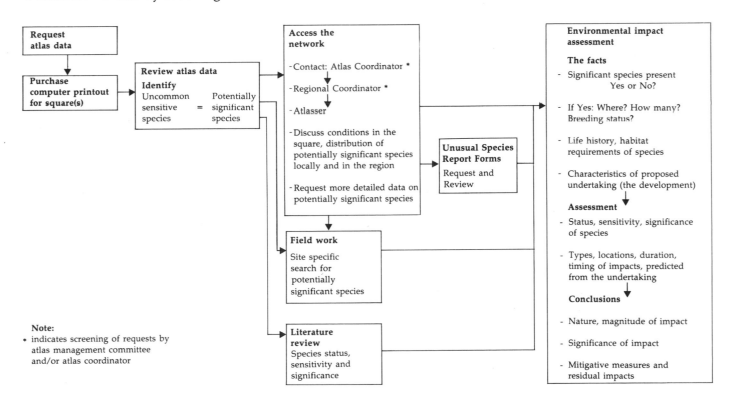

Crossbills, Cone Crops and Irruptions

Harry G. Lumsden and Roy B.H. Smith

Atlas data for most species lend themselves readily to summary on maps. However, irruptive species such as crossbills have a shifting pattern of abundance and breeding distribution because of their dependence on conifer seeds - a food supply which is produced in abundance only at irregular intervals (Tordoff and Dawson 1965). To see these patterns, each year's data must be viewed separately. Using estimates of cone crops made available by MNR, it is possible to compare variations in crossbill distribution with potential food supplies. Cone crop data are not always complete, and are not available for eastern hemlock and balsam fir. We gratefully acknowledge the use of these MNR data.

Crossbills can extract seeds from cones at any stage of ripeness, but most easily from cones which have begun to open naturally (Newton 1972). In Ontario, seeds of most native conifers form in summer and are shed in autumn. Thus, seeds are widely available to crossbills between July and October, but small amounts may be carried over to the succeeding year. Black spruce cones, however, may carry seed for a number of years and may shed at any time. For any given conifer species, heavy crops of seed rarely occur in the same place in successive years (D.A. Skeates pers. comm.). Tamarack may produce a good crop every 3 to 6 years, white spruce every 2 to 8 years and white pine every 3 to 10 years (OMNR 1986). Asynchronous fruiting among different conifer species can lead to local seed availability in successive years, but when this does not happen, crossbills must emigrate in search of food. Hence, crossbill irruptions are believed to be associated with successful conifer fruiting followed by widespread failure, although high population levels are also needed to bring about emigration (Newton 1972).

In 1981, seed crops for most species were poor over much of Ontario. However, some stands of black spruce in the Algonquin Prov. Park and Terrace Bay areas produced heavily. The few White-winged Crossbills recorded were in those areas and on the Sutton Ridges, where confirmed breeding occurred in July. There were only two atlas records of Red Crossbills in 1981.

From North Bay to Lake Simcoe and east to Pembroke, white pine produced a good crop in 1982 as did some stands in the Sault Ste Marie area. White spruce did well from Cornwall to Parry Sound, in the vicinity of Sault Ste Marie, and in the Kirkland Lake-Timmins-Cochrane area. As for tamarack, there was moderate to heavy seed production locally in the Bancroft and Lindsay areas. Both species of crossbill were recorded in many of these regions, but overall numbers were still very low.

In 1983, cone crops of most species were generally poor. However, red pine fruited well south of Algonquin Prov.

Park as did some white spruce stands in the Cornwall area. Although crossbills were more widely reported in the province than in 1982, most records came from the Algonquin region, where both white spruce and white pine cones carried over from 1982. Generally, white spruce cones contain an average of 66 seeds per cone in autumn (D.A. Skeates, pers. comm.). Craig Benkman (1985) found an average of 9 seeds per cone on 1 February when White-winged Crossbills were starting to nest. By 25 March, there were only 0.5 seeds per cone and the birds deserted their nests and left. However, Red Crossbills, feeding mainly on white pine, apparently bred successfully in that area (R. Tozer pers. comm.).

In 1984, cone crops for most species were excellent over wide areas of the province. Between Sault Ste Marie and Cornwall and as far north as Temagami, white pine fruited abundantly, while white spruce had good cone crops almost everywhere. With 12 reports of confirmed breeding and 154 records in total, White-winged Crossbills were more abundant than in any other atlas year, and late summer breeding attempts took place over a large part of northern and south-central Ontario. Red Crossbills, with 48 records, were also more widely distributed in 1984 than in the three previous years and there were 5 confirmed breeding records in late summer.

Breeding by Red Crossbills also occurred in the early part of 1985, sustained by part of the 1984 pine and spruce crop which carried over to spring. Records totalled 171 in 1985 with a notable concentration in areas south of the Shield. In contrast, records of White-winged Crossbill declined by 27% in 1985.

The cone crops of 1985 were generally poor. Red pine fruited well only in the Lindsay and North Bay areas, and white pine near Sudbury. Thus, a general shortage of conifer seed and a high crossbill population set the stage for widespread exodus. For White-winged Crossbills this began in November 1984, when an estimated 17,500 birds were seen moving west along the south shore of Manitoulin Island (Weir 1985a). At that time very few were reported in Michigan and Wisconsin (Tessen 1985). Weir (1985b) noted that thousands stayed in Algonquin Prov. Park until late February 1985, while at Long Point 3,000 were seen heading southwest on 13 March (Shepherd 1985). By spring 1985, White-winged Crossbills, although less common than Red, were widely distributed in northern Michigan and Wisconsin (Powell 1985). In the Little Sachigo Lake area of northern Ontario a major northwesterly movement of White-winged Crossbills was noted on 20 to 23 June 1985 with a maximum of 1,000 on 20 June (Peterson 1985). The Prairie Provinces experienced a major invasion of both crossbill species in July (Harris 1985). The wanderings of Red Crossbills were less apparent in Ontario, but seasonal reports in *American Birds* show that both species were involved in continent-wide irruptive movements in 1984-85.

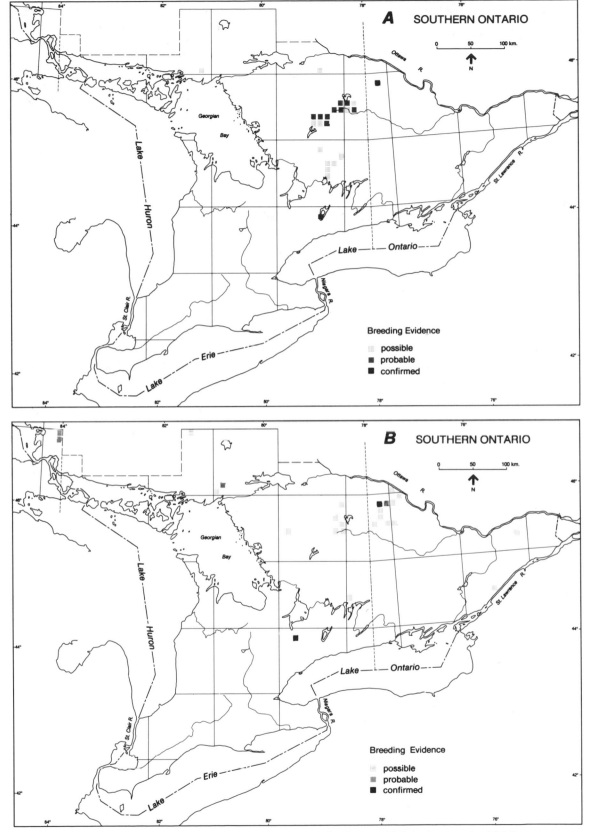

FIGURE 17: *Distribution of crossbill records from 1981 through 1983:*
(A) White-winged Crossbill, and (B) Red Crossbill

Using Atlas Data to Monitor Changes in House Finch Distribution

Erica H. Dunn

One of the most important ways in which atlas data can be used is to document the current range of a species for comparison with past or future distribution. As the Ontario Atlas represents the first time that ranges of many species have been adequately described in this province, particularly those species with distributions extending into northern Ontario, comparisons with the past will have to be made judiciously. Many such comparisons are contained in the individual species accounts in this book. The ideal situation for determining whether species' distributions are changing from what they are now would be to do periodic atlases using the same methods. However, the benchmark data in this atlas will be useful for comparison with other types of surveys when those show presence or absence of a species in a region for which this atlas shows the opposite.

Atlas data can be used not only to demonstrate overall changes in a species' distribution, but also to help document the pattern and, in some cases, the rate, of change. A comparison of new and old distributions can show what habitats are being invaded or abandoned, which in turn can lead to hypotheses on the reasons for the changes. Pace of change can be estimated when consecutive atlases occur at relatively short intervals (5-15 years), and, in certain cases, when a single atlas project covers a period of rapid expansion or decline.

During field work for the Ontario Atlas, House Finches expanded their range rapidly in southern Ontario. A brief examination of data for that species illustrates one way in which atlas data can be used to document changing distribution.

The House Finch was introduced to New York City in 1940 from its native haunts on the North American west coast. Since the late 1940's, it has been expanding rapidly to the west and south and, more slowly, to the north. Documentation of the spread has been primarily through Christmas Bird Counts (Mundinger and Hope 1982). It appeared that House Finches dispersed along river valleys and shorelines, as these were the first parts of newly-invaded regions to be colonized with breeding birds. The first record of a possible House Finch in Ontario was in Kingston in 1970 (Weir 1970).

Field work for the Ontario Breeding Bird Atlas from 1981 to 1985 spanned a period of rapid expansion of the House Finch population in the province. As shown in Table 9, the total number of confirmed records increased each year. However, coverage in the first year was not as intense as later on, and some of the later increases could have been due to better coverage and attention to confirming breeding as time went on. To help overcome this possible bias, the same kind of data are also shown in Table 9 for House Sparrow, a species thought not to have changed in abundance or distribution over the period of atlas field work. Anyone doing a modicum of atlassing in southern Ontario would have confirmed nesting in House Sparrow easily. Because it nests in the same general areas as House Finches (*i.e.*, usually associated with human habitation), an increase in House Sparrow confirmations from year to year probably indicates increases in atlassing effort. As is clear from Table 9, the House Sparrow records did continue to accumulate throughout the atlas period, but at a decelerating rate. Therefore, the more rapidly accumulating House Finch records should be primarily a result of expanding range.

The change in pattern of House Finch distribution is shown in Figure 18. The species appears to have established a foothold in the province first on the Niagara Peninsula. The concentrations of records on the 1981-1984 map suggest that colonists were coming around both ends of Lake Ontario from New York. Thereafter, the birds appear to have spread east and west along the north shores of Lakes Erie and Ontario, and along the St. Lawrence and Ottawa Rivers. It is not clear whether birds breeding in the Windsor-Sarnia area and along the shore of Lake Huron came from Michigan or from the population already breeding in Ontario. In either case, the preponderance of records along shorelines and rivers confirms earlier observations that these features form major routes of dispersal (Mundinger and Hope 1982).

TABLE 9: *Statistics for House Finch and House Sparrow*

Total number of southern Ontario atlas squares with confirmed breeding at the end of each year, and the percentage increase from the previous year: House Finch and House Sparrow.

Year	House Finch		House Sparrow	
	Number	% Increase	Number	% Increase
1981	1	-	196	-
1982	18	*	439	*
1983	35	94	646	47
1984	60	71	839	30
1985	108	80	954	14

* Omitted, since most of the increase from the first to the second year was due to expanding coverage.

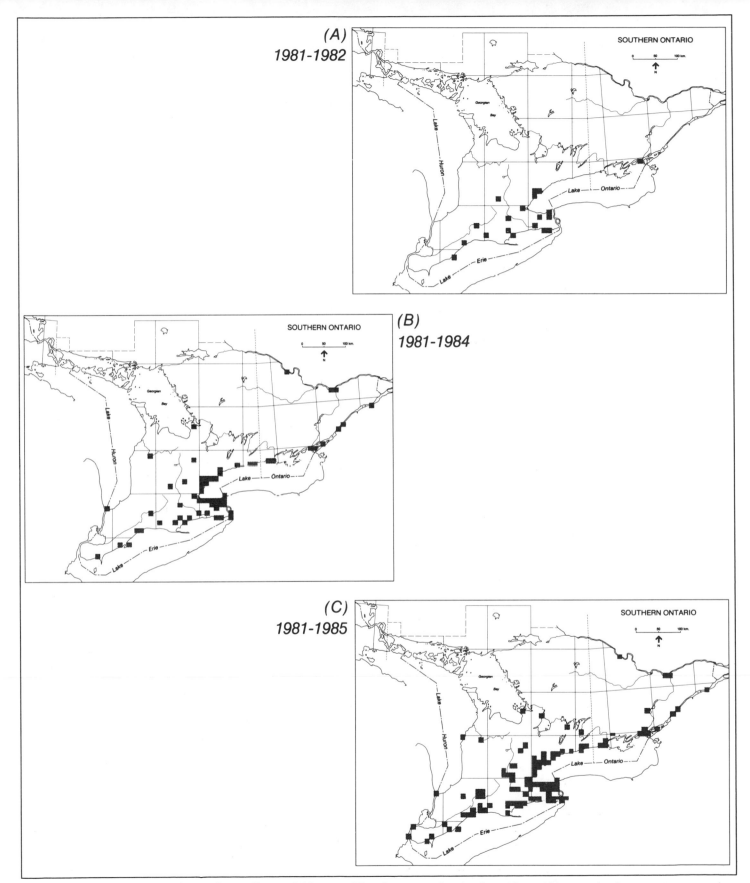

FIGURE 18: *Distribution of confirmed House Finch records during the atlas period*

Multi-species Clusters
of Birds in Southern Ontario

Philip D. Taylor and Stephen M. Smith

Introduction

The primary purpose of generating the atlas data was to produce maps showing the breeding distribution of birds in Ontario. This is an obvious and straightforward use of the masses of data that have been collected. However, in addition to being used for the production of maps, the atlas data can be used to explore the patterns of breeding bird distribution in Ontario. Here we outline an approach to such an analysis and present some introductory findings. We hope that this brief account will illustrate the enormous potential and value of the atlas data in the analysis of both fundamental and applied problems in ornithology and conservation biology.

In a simple sense, the atlas data base is a two-way table (squares by species), recording which species have been found to breed in which squares. Given the large number of squares (1,824 in the south) and species (235 in the south, plus 2 hybrids), it is a daunting exercise to detect general patterns of distribution over a large geographical area such as southern Ontario. Moreover, there may be patterns involving several species jointly that are difficult to detect by examining individual species. A method is needed to condense the vast amount of information in the squares-by-species matrix into a small number of subsets, the characteristics of which might be more readily interpretable than the distributions of the individual species. Several such techniques are available; they fall in the general domain of 'cluster analysis' or 'classification'.

As the name suggests, such techniques form clusters or subsets of objects, based on their similarity. In the case of the atlas data, clusters are groups of squares that are similar in terms of their breeding bird composition. The boundaries between clusters are expected to fall along meaningful biological or environmental lines.

As an analogy, consider the cline in size of the Hairy Woodpecker as we move from south to north - northern individuals are larger than southern individuals. If we were to consider an individual from Tennessee and an individual from the Lake Superior area, we might classify them as subspecies, so different would they be in size. However, the local populations from which they were derived are linked by a continuum of intervening (and interbreeding) populations, so that subspecific designation is arbitrary. Nevertheless, the recognition of a northern and a southern form provides meaningful information on the biology of the species. Similarly, recognition of clusters in a continuum of atlas squares may provide insights into the biology of the groups of species which occupy the various clusters, and may serve as a powerful means of generating hypotheses about the nature of bird communities in Ontario.

By summarizing the information content of the atlas data by a cluster analysis, we hope to increase the accessibility of that information and allow for an easier understanding of the patterns of bird distribution in Ontario. The patterns may be of ecological and/or evolutionary interest in terms of the relationships between species groups and resources, landforms, climate, topography, vegetation, etc. In the future, as additional atlas data on other biota are collected, the relationships between the distribution of birds and plants, for example, will be open to study. Atlas data bases will permit a profound investigation of the distribution and interactions of Ontario's flora and fauna. Without the contributions of atlassers, such detailed, high-resolution studies of communities would be impossible.

Methods

There are many classification techniques available. We have used techniques that are particularly appropriate for a straightforward preliminary investigation of large data sets, *viz.*, the programs COMPCLUS and TWINSPAN. (For an introduction to the techniques of community analysis and a discussion of these programs, interested readers should consult Gauch (1982)). COMPCLUS (COMPosite CLUStering) yields a non-hierarchical clustering of squares and TWINSPAN (TWo-way INdicator SPecies ANalysis) provides a technique for hierarchically clustering objects. COMPCLUS begins at a randomly selected square and progressively clusters squares in an attempt to produce a pre-determined number of clusters. Because of the random starting point, the program is run many times; patterns that repeatedly emerge have a high probability of being real. We used COMPCLUS to produce small (approximately 8) as well as larger numbers (approximately 25) of clusters; the larger numbers of clusters were then analyzed by TWINSPAN to detect hierarchical relationships among the clusters. Agreement between a composite clustering to yield small numbers of clusters and a hierarchical clustering of larger numbers of clusters is evidence of the reality of the patterns detected. The results presented here are preliminary and represent a composite picture of what has emerged from our analyses of the data from southern Ontario.

Results and Discussion

Composite clustering was used to group the 1,824 squares in southern Ontario to yield 30 clusters and outliers. Outliers are squares that were not tightly related to any other squares. We classified 238 squares (approximately 13%) as outliers. Most outlier squares were on the margins of the province or contained mostly water and, as such, comprised incomplete squares. The paucity of species associated with such incomplete squares (a mean of 59.5 species per square) makes it difficult to classify them at the level of analysis presented here. The 30 clusters comprised 1,586 squares (approximately 87%). These 30 clusters were then subjected to a hierarchical clustering by TWINSPAN; the resulting hierarchy is illustrated in Figures 19 to 22. The hierarchy here is typical in that the general patterns illustrated reappeared with repeated clusterings.

Not surprisingly, the first major division (Figure 19) involved a north-south split in which the Canadian Shield played an important role. The split occurred almost exactly along the edge of the Shield; the Shield component actually extended slightly off the Shield onto the Bruce Peninsula and Manitoulin Island, and the southern component contained a few squares at the southern edge of the Shield. The 730 squares of the Shield component had a mean of 95.8 species per square and the 856 squares of the southern component had a mean of 91.7 species per square.

The Shield Component
The Shield component split into a northern-Shield cluster and a southern-Shield cluster at subsequent levels of the hierarchy (Figure 20). There was a relatively depauperate fauna (336 squares: mean of 85.1 species per square) associated with the northern-Shield cluster. The low number of species per square may have been partly attributable to lower effort in this area; southern-Shield squares were atlassed for more than twice as many hours as the northern-Shield squares (a mean of 66 vs. 31 hours per square, respectively).

Within the northern-Shield cluster was a distinctive group of squares (not illustrated) associated with the Algonquin-Haliburton highlands and areas surrounding Lake Nipissing. This group of squares appeared repeatedly in all cluster analyses that we did, showing clearly the characteristic fauna associated with the area. It is not difficult to speculate that the distinctive fauna found there is associated with altitude or climate, or both, as well as with land-use features.

Squares in the southern-Shield cluster (Figure 20: 394 squares) had the highest numbers of species of all clusters (a mean of 104.9 species per square). The large number of species is likely a result of the varied habitats, many wetlands and the influence of both southern and northern species groups found in the area. This cluster represents a transition zone between the north and the south, and as such, is part of a continuum of species change that occurs through this part of Ontario. Several squares that are *geographically* within the northern-Shield cluster emerged within the present group. Relative to surrounding squares, those squares contained a high proportion of altered landscapes, and were located in the area between North Bay, Mattawa and Powassan, along parts of the Highway 60 corridor through Algonquin Prov. Park, and in the intensively utilized areas and abandoned fields surrounding Lake Traverse, Algonquin Prov. Park. These altered landscapes have a wider variety of habitat types, more typical of the areas further south. At least some southern species can find and utilize these areas for breeding. Subsequent partitions of the entire cluster revealed

no obvious *geographical* patterns other than those associated with the Algonquin/Haliburton highlands.

The Southern Component
The southern component was a heterogeneous assemblage of squares that divided into a cluster of squares in southwestern Ontario, and a cluster of squares near the edge of the Shield (off-Shield cluster). The off-Shield cluster (Figure 21: 423 squares: a mean of 99.3 species per square) represents, like the southern-Shield cluster, a transition zone between the southern and northern fauna. The off-Shield cluster made significant inroads into other areas of the province, inroads that are particularly interesting in southwestern Ontario and the Ottawa valley. The southwestern cluster (Figure 21: 433 squares) was distinctive, but had few species (a mean of 84.3 species per square); the depauperate fauna is probably a response of bird communities to the relative biological "sterility" of the intensively cultivated areas of the province. The cluster was comprised of 2 groups: a southern-core cluster (Figure 22: 328 squares: a mean of 86.4 species per square) and a depauperate Kent/Essex counties and Niagara cluster (Figure 22: 105 squares: a mean of 77.7 species per square). At finer levels the latter group separated into two distinct geographical regions, with the intensively drained, cut-over, and farmed areas of Kent and Essex Counties being highly depauperate (a mean of 70.4 species per square), and the Niagara Peninsula being somewhat less so (a mean of 81.1 species per square).

Finer-scale patterns (not illustrated) identified a number of interesting groups of squares in the southern regions. Most distinctive were the small group of squares associated with the pine plantations of the western parts of Haldimand-Norfolk RM and the true Carolinian squares including such sites as Walpole Island, Point Pelee Nat. Park, Rondeau Prov. Park and Long Point. Further analysis may reveal such fine-scale patterns throughout the province.

Indicator Species
Contained within all the species distributions is an enormous quantity of information, much of which is redundant. The purpose of an analysis such as this one is to reduce this mass of information to an understandable subset - the cluster solution. Because each cluster contains information about the distributions of many species, it is not necessarily possible (or desirable) to point to a single species that has a distribution similar to or the same as a given cluster. However, we can examine subsets of species that embody most of the information contained within a cluster. An examination of these 'indicator' species will clarify what makes a cluster distinct. In this section we present such an analysis for the southern-Shield cluster (Figure 20).

All of the clusters presented have clear northern or southern affinities at all levels of the hierarchy; the distributions of most species in the province follow this north-south gradient and many change markedly as one moves north or south through the southern-Shield cluster. The off-Shield cluster (Figure 21) also exhibits these changing distribution patterns.

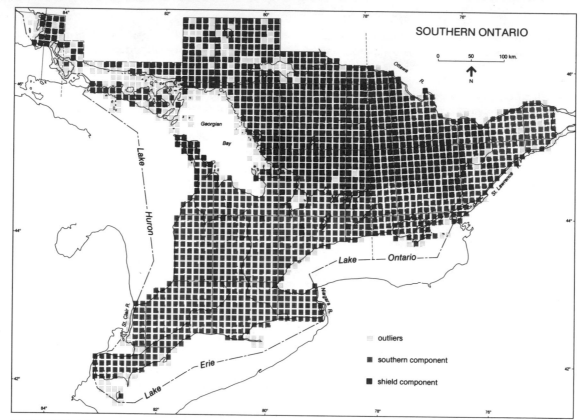

FIGURE 19: *Initial north-south division of province*

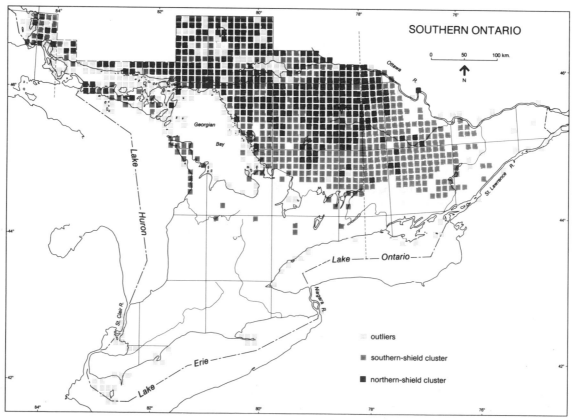

FIGURE 20: *Division of the shield component*

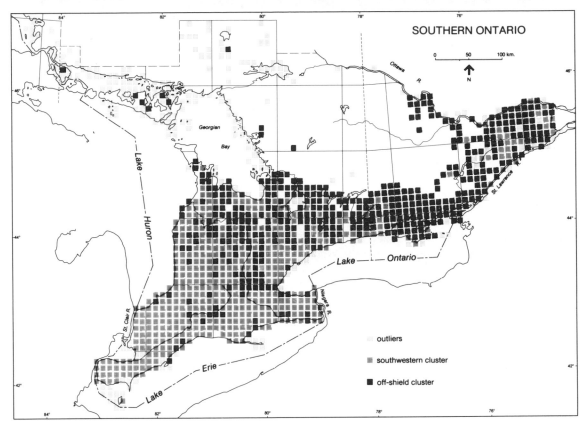

FIGURE 21: *Division of the southern component*

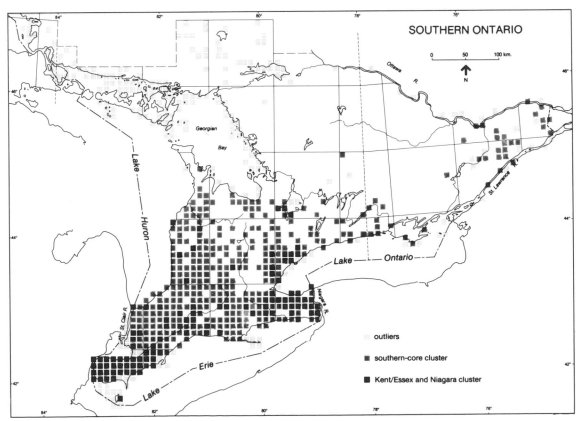

FIGURE 22: *Division of the southwestern cluster*

Both clusters represent transition zones; they differ in that each has a greater similarity to the fauna of their respective parent clusters. The southern-Shield cluster represents a northern transition zone; the off-Shield cluster represents a southern transition zone.

Southern species that decline abruptly (in relation to the number of squares they occupy) at the southern edge of the southern-Shield cluster include Canada Goose, Upland Sandpiper, Eastern Screech Owl, Red-headed Woodpecker, Horned Lark, and Northern Cardinal. Great Horned Owl, Purple Martin, House Wren, Field Sparrow, and House Sparrow all decline within the cluster and Northern Harrier, Mourning Dove and Vesper Sparrow all decline immediately north of the cluster. Northern species that decline abruptly south of the southern-Shield cluster include Olive-sided Flycatcher, Common Raven, Solitary Vireo, Magnolia Warbler, Black-throated Blue Warbler, Blackburnian Warbler, and Evening Grosbeak. Those that decline gradually south of the southern-Shield cluster include Broad-winged Hawk, Red-breasted Nuthatch, Yellow-rumped Warbler, Black-throated Green Warbler, and Canada Warbler. Gray Jay and Cape May Warbler are two species that decline south to the northern edge of the southern-Shield cluster. In addition to these species that show distinct north-south changes in distribution, there are several species that show irregular patterns of distribution and are present more frequently within the southern-Shield cluster than outside it. Two examples are Cooper's Hawk and Golden-winged Warbler.

For most of the above species, simple changes in habitat and land-use parameters can be presented as reasons for the changes in distribution through the Shield border. Taken as a whole, these many-species changes in distribution produce the southern-Shield cluster. The rich and varied fauna within the cluster can be attributed to the meeting of northern and southern species. The clustering solution shows how these changes relate to the overall picture of avian distribution in Ontario. It is noteworthy that, in such a biologically rich and interesting area, there are few major parks or protected areas; this is likely due to the lack of "significant" individual species and immediately shows the tremendous utility of the mass of information contained in the atlas data base.

Patterns of bird-species distribution such as those we have presented here have the potential to be used as creative tools in the management of the intensively used landscape of the province. On a large scale, avifaunal regions could be used to determine how well a system of nature reserves protects the avifauna of a given area, and what suites of species are underrepresented. On a finer scale, nodes of high diversity, or highly distinctive patches of species-groups could be identified from within larger, more homogeneous areas. We will be able to combine data from future atlas projects on other plants and animals with those of the bird atlas to provide a comprehensive data base of the province's biota. Such a data base will prove invaluable to managers, researchers and naturalists alike.

Appendix A

Administrative structure

The Ontario Breeding Bird Atlas was a joint scheme of the Federation of Ontario Naturalists (FON) and the Long Point Bird Observatory (LPBO), but its success was in no small part due to the cooperative assistance of many other groups and agencies. The FON was the administrative home of the project. Most funds went through FON's books and most staff were on its payroll. The project was housed at Locke House, FON Headquarters in Don Mills, from 1981 through 1983, but then moved to the University of Waterloo, the location of the increasingly important computer facilities that were used for the project. Many of the volunteer atlassers were FON members. The LPBO had been running volunteer projects dealing with birds for some time and provided valuable input to the technical aspects of running the project; its membership was also widely involved. The complementary strengths of these two groups melded to give the project a strong scientific and popular base from which to operate.

Initial work on the atlas project began with an organizational meeting of 7 people at the University of Waterloo in March 1979. Over the next two years, a steering committee was formed, a project outline was prepared, funding proposals were submitted, and field methods were tested and refined. In January 1981, a full-time Coordinator was hired to prepare instructional materials and to start developing the Regional Coordinator network, while final word on funding was awaited. The project was officially launched on March 28, 1981 when it became apparent that start-up funding would be forthcoming.

In March 1981, the original steering committee split into the Management and the Technical Advisory Committees, with several members serving on both of the new committees. The Management Committee had overall administrative and policy responsibility for the project, whereas the Technical Advisory Committee was given the vital job of supervising field data collection and data management. Subcommittees of the Management Committee were formed to handle data access and publication matters, and subcommittees of the Technical Advisory Committee were formed to deal with coverage of northern Ontario, data management and data review. There was a conscious attempt to have formal and informal representation on the committees of the major environmental groups, pertinent government agencies, universities and other organizations, including atlassers. These people not only brought to the project their expertise and knowledge, but they also served to keep the cooperating organizations abreast of the progress and requirements of the project. Committee members are listed in the Acknowledgements.

The project budget was initially developed with two full-time employees, a Coordinator and Assistant Coordinator, but after 1983 a computer programmer/data base manager was added to the full-time staff. A number of other people were hired, through government employment programs, to take up clerical, text and data entry, and field work responsibilities. Without the assistance of these programs, through which employees obtained valuable job skills and work experience, much of the success of the atlas project could not have been realized.

The project was dependent upon moderately high levels of funding. Before the project got under way, an initial budget of approximately $55,000 per year was proposed for basic project operation. This became known as the base budget and was the first priority for fund raising. Subsequently, special requirements arose for needs such as computer programming, field work in poorly covered parts of the province, detailed work on endangered species, and data review. In these cases extra funds had to be obtained.

Over the 8 years of the project (1979 through 1987) a total of $670,000 worth of funds, equipment and direct contributions was raised for project operation. In addition to this, several Regional Coordinators obtained funds for the accomplishment of field coverage in their areas. Also, gifts-in-kind, such as the donation of flights for atlassers by the Ontario government and private companies and lodging provided by government agencies and individuals are not included in this amount. A more complete description of project funding is given in Eagles (1982, 1986).

Appendix B

Abundance estimates

David J. T. Hussell

The purpose of estimating abundance was to record information on the general level of abundance of each species throughout the province. The intent was not to determine accurate population levels in each square, but rather to obtain a very approximate estimate that would nevertheless make it possible to compare relative abundances among species and among regions of the province.

It was realized at the outset that asking atlassers to record abundance was to some extent in conflict with the more straightforward primary goal of the atlas of obtaining simple presence or absence records. The latter required observation of only one individual or pair per square and allowed atlassers to ignore a species once it had been confirmed; whereas estimation of abundance required continual awareness by the atlasser of presence and density of a species throughout the square.

Because the main objective of the atlas was to obtain accurate information on the distribution of each species, abundance estimates were given a relatively low priority and were recorded at the option of the atlasser. Atlassers were expected to estimate abundance only if they felt that their experience in a square was adequate to do so. Based on their knowledge of a square, atlassers were asked to estimate the abundance of each breeding species by listing them in one of the following categories: (1) 1 pair, (2) 2 to 10 pairs, (3) 11 to 100 pairs, (4) 101 to 1,000 pairs, (5) 1,001 to 10,000 pairs, and (6) more than 10,000 pairs.

Relatively few atlassers submitted abundance estimates in the first 3 years of the project. Moreover, it was suspected that atlassers were being unrealistically conservative in making estimates. In the last two years of the project, therefore, a special effort was made to encourage atlassers to submit more abundance estimates and to upgrade those previously submitted. The following suggestions for making abundance estimates were published in Newsletter No. 10, May 1984:

"An abundance estimate depends on: (a) your knowledge of the density of pairs of a species in suitable habitat in your square; (b) an estimate of the amount of suitable habitat in your square; (c) your judgement of how representative the density estimate in (a) is of suitable habitat elsewhere in the square. It is then simply a matter of numbers and arithmetic to arrive at a rough estimate of pairs per 10 km square.

A hypothetical example:
You take 1/2 hour to walk 2 km through woodland in your square and you count 12 singing Ovenbirds. If we assume that you can detect birds up to 100 m on each side of your path, then you will have sampled 40 100 m x 100 m units or 40 hectares. Then Ovenbirds have a density of at least 12/40 = 0.3 pairs per hectare.

If your entire 10 km square is made up of woodland suitable for Ovenbirds, then you can estimate that there are 0.3 x 10,000 = 3,000 pairs in the square since there are 10,000 ha in a 10 km square. This would give you an abundance estimate of 5 (1001-10,000). However, if you estimate that only 15% of the square is made up of suitable woodland, then the estimated number of Ovenbird pairs will be reduced to 3,000 x 15/100 = 450, which means that your abundance estimate is now 4 (101-1,000 pairs).

You may perhaps doubt that Ovenbirds are as abundant as 0.3 pairs per hectare in all of the 'suitable' habitat in your square, in which case you may want to reduce your estimate by a fudge factor of, say, 50%. Then your estimate of pairs would be only 225, but your abundance estimate category will still be 4.

The point to be noted here is that what seem to be quite low densities (12 birds observed, 0.3 pairs per hectare) may indicate abundance categories as high as 4 or 5 when extrapolated to the entire square. Obviously, care must be taken and judgement used in making such extrapolations, but abundances of 4, 5, or 6 are not necessarily unrealistic just because thousands of pairs have not been directly recorded.

Conclusion
In the final 2 years of the atlas we would like to obtain new or updated abundance estimates for as many species as possible in as many squares as possible. Be cautious but also be realistic. Use your knowledge of birds and habitat in your square to estimate numbers present. Don't be frightened by the apparently large numbers represented by the higher abundance categories. What we know about bird densities and simple arithmetic tell us the numbers aren't so large after all."

Appendix C

Processing and verification of atlas data

David A. Balser

This Appendix is for those who are curious about the technical side of the project, but it also serves as a record of the extraordinary amount of effort devoted to ensuring that the breeding distributions shown on the atlas maps are accurate. Responsibility for decisions on treatment and acceptability of data described in this Appendix rested with the Technical Advisory and Data Review Committees.

The process by which the information on data cards ultimately became printed maps was long and complex. The Regional Coordinators checked over each card to ensure that all relevant information was included, that the species recorded were reasonable for the area and that erroneous information was not present. The information from each card was manually transferred to a master card, so that over time the Coordinators built up a total record for each square based on all cards submitted. Master cards were updated by the Coordinators each year to indicate the highest breeding code and abundance estimate for each species in every square in their respective regions. The checked atlas cards were then sent to the atlas office. (If Regional Coordinators did not perform these tasks, or others described below, atlas staff or the Data Review Committee did so).

When the field cards arrived at the atlas office, they were checked for legibility and completeness. A self-adhesive label was placed on the front of each card, upon which the 'header' information was clearly written. The header is basically the key information which describes the unique features of a data card. The header for each card consists of the zone-block-square, region, year, atlasser number, dates of first and last visits, and hours of atlassing. Periodically, batches of cards were shipped to a data-entry firm, where they were 'keypunched', that is, electronically encoded on magnetic computer tape. Keypunching a card involved typing the header followed by a series of 'records', one for each species observed. A record is composed of a species name, reduced to a four-letter mnemonic code, *e.g.*, COLO for Common Loon, plus the breeding code and abundance estimate (if given). The species names and codes on the data card proved to be too small to be read accurately and quickly by keypunchers, so the card was placed in a clear template on which were written the 4-letter codes in large block capitals

adjacent to the appropriate card data.

To ensure that the data were keypunched accurately, every data card was punched *twice* and compared by the data entry computer for mismatches. This system of double-entry keypunching resulted in a very low error rate of approximately 1 in 10,000 records. Ultimately, then, the data on the original cards were transferred to the data base with great fidelity.

This system required the minimum possible preparation of data prior to keypunching. This ensured that cards could be handled quickly, and reduced the potential for human error in data transcription. In 1982, a 'casual observation card' was introduced. This small card was to be used to report breeding evidence for fewer than 10 species in a square, the intent being to save funds on the printing of the larger data cards, and to encourage people to report birds from squares other than their own. The same information was required as on the data card, but atlassers were required to write out in longhand the name of the species observed. The card was quite well received by volunteers and was fairly widely used. However, because of the extra time required to prepare this card for keypunching, and because of the potential for errors in transcribing each species name to code, the use of this card was quickly curtailed by the atlas office.

A magnetic tape containing the latest batch of raw data was sent to the atlas office, and the process of integrating it with the existing data base was started. The computerized versions of the data cards (collectively called the raw data), were maintained in 46 separate files: one for each atlas region. These files were called up onto the computer screen and changes made directly to the data when corrections were necessary. Within each regional file, the electronic representation of the data cards was kept in a sequence, according to zone-block-square, and within each square by year, and within each year by a card number (sequence number) assigned prior to keypunching. Any individual card was uniquely identified by the following sequence: region, zone-block-square, year, and card number, as in, for example, 01 17LS25 85 02.

Integrating newly keypunched data into the existing raw data files required the use of a series of computer programs which merged the new data in the proper sequence, after which a new set of 46 regional files replaced the old set. The computer programs used were written using the Statistical Analysis System (SAS (SAS Institute 1985)) and were run on IBM 4341 mainframe computers using the VM/CMS operating system at the University of Waterloo.

The process of converting raw data into the species distributions depicted on the maps involved selecting the *highest* breeding code, and *highest* abundance code, for each species in each square. If 2 or more records for one species were submitted for the same square, the computer program selected the highest breeding code and the highest abundance for the master file for that square. These processed records, then, were often an amalgam of 2 records from the raw data: one contributing the highest breeding code and another the highest abundance estimate. For squares in southern Ontario, the breeding level so determined was mapped on the southern Ontario map and the abundance levels were used for the southern Ontario histograms. For the Ontario block map, the data for all squares in each block were aggregated further to obtain the highest breeding code for each species recorded anywhere in the block, and the corresponding level of breeding evidence was mapped. For example, if only one square out of 100 in the block had a record of a nest with young (coded NY), the computer program would select that record to represent the whole block because NY is the highest breeding code. Thus the original 407,000 raw data records were reduced to 207,000 processed records as redundant or lower-level breeding evidence codes were excluded from the master file.

The job of ensuring that the data in the data base were correct consumed a large proportion of the time and effort expended on the project as a whole. The data were carefully scrutinized by the staff and the Regional Coordinators each year of the project to ensure that the raw data files contained exactly what atlassers recorded on the data cards. Not all errors, however, could be detected. Errors of identification made by atlassers in the field were generally undetectable provided the observation claimed was unremarkable. The type of error to be avoided at all costs was that of allowing an inaccurate or undocumented record, that would have misrepresented the true distribution of the species, to remain in the master file, and eventually to be mapped. Some of these dubious records were the result of simple transcription errors when filling out the data card. A code intended for Chipping Sparrow written in the space above the correct one on the card could result in an American Tree Sparrow record 1,000 km from its true breeding range. The other reasons for errors of this type, and the ones most difficult to handle, were misidentifications and questionable interpretations of breeding evidence.

Despite the definitions provided in Table 2, the breeding categories were interpreted differently by some atlassers. A more thorough explanation of the codes, pointing out particularly difficult species and including specific examples (such as crossbills, which have irregular breeding seasons), would have benefitted the atlassers and saved a considerable amount of repair work 'after the fact.' For example, the code 'P', for "pair of birds in suitable habitat", and 'T' for

"permanent territory presumed through registration of territorial behaviour (song, etc.) on at least two days, a week or more apart, at the same place," were not appropriate for many colonial nesting species. For such species, the presence of a pair bond is most likely to be seen at the nest, and the territory is within at most a few metres of the nest, where a higher breeding category would be justified. Therefore, the following changes were made: 'Pair' and 'Territory' were downgraded to 'Suitable Habitat' for Great Blue Heron, Green-backed Heron, Double-crested Cormorant, and Black-crowned Night-Heron; for Bank Swallow, 'Pair' was downgraded to 'Suitable Habitat.'

To ensure that the raw data file contained exactly what had been submitted on data cards, Regional Coordinators were called upon to compare their records, kept on master cards for each square, with the contents of the data base. When discrepancies occurred, Coordinators reported them on 'Error Report Forms', and submitted them to the atlas office. Discrepancies were checked against the original data cards and resolved, unless, in the opinion of the Regional Coordinator, the data on the field card were incorrect or invalid, in which case both the data card and raw data record were amended. Regional Coordinators, relying on their knowledge of their areas, had almost total discretion to delete records from the data base.

Records which were 'unusual' required special documentation, but deciding which records fell into this category proved to be surprisingly difficult. For the first 3 years of the project, it was largely left to the Regional Coordinators to decide whether an atlasser should document his or her record on an Unusual Species Report Form (USRF), unless he or she had already done so voluntarily. For various reasons, this system did not work as well as expected. Most Coordinators seemed to want explicit instructions in this regard, and this was also desirable from the point of view of ensuring consistency across the province.

Somewhat belatedly, in the winter of 1984-85, a system was developed to identify which species in which regions required additional documentation, and the Data Review Committee began to carry out the job of reviewing records that had been so identified. The first step was the creation of a set of criteria for determining whether a record needed documentation. Six categories were established, in which the species listed or referred to would have to be documented on a USRF. The six categories were:

(1) Species new to the Ontario breeding list, including the following recent additions which were not listed on the atlas data card:

Ross' Goose	Bohemian Waxwing
Cinnamon Teal	Northern Shrike
California Gull	Harris' Sparrow

(2) Species that were officially listed as endangered in Ontario:

American White Pelican	Greater Prairie-Chicken
Golden Eagle	Piping Plover
Bald Eagle	Kirtland's Warbler
Peregrine Falcon	

(3) Species (or hybrids) that were very rare as breeding birds in Ontario:

Horned Grebe	Black Guillemot
Red-necked Grebe	Common Barn-Owl
Cattle Egret	Northern Hawk-Owl
Canvasback	Chuck-will's-widow
Common Eider	Acadian Flycatcher
King Eider	Black-billed Magpie
Surf Scoter	Tufted Titmouse
Wild Turkey	Bewick's Wren
Yellow Rail	White-eyed Vireo
King Rail	Brewster's Warbler
Lesser Golden-Plover	Lawrence's Warbler
Pectoral Sandpiper	Prothonotary Warbler
Stilt Sandpiper	Hooded Warbler
Short-billed Dowitcher	Dickcissel
Little Gull	Lark Sparrow

(4) All records (i.e., those in the possible (PO), probable (P) or confirmed (CO) breeding levels) of species which were outside their previously known breeding ranges. In general, Godfrey (1966) was used as the basic reference.

(5) In addition there were several species that were not necessarily rare in Ontario as breeding birds, but for which there were extremely few breeding records or for which the breeding range was poorly documented. Documentation was required for confirmed breeding records of these species. The 2 owl species on this list were to be documented on USRFs if breeding evidence was probable or confirmed. Documentation of probable or possible records for any of the species on this list might also have been requested, depending on the location and circumstances of the report.

Red-throated Loon	Great Gray Owl
Greater Scaup	Boreal Owl
Oldsquaw	Gray-cheeked Thrush
White-winged Scoter	Orange-crowned Warbler
Bufflehead	Connecticut Warbler
Greater Yellowlegs	Henslow's Sparrow
Lesser Yellowlegs	Le Conte's Sparrow
Solitary Sandpiper	Sharp-tailed Sparrow
Hudsonian Godwit	Pine Grosbeak
Marbled Godwit	Red Crossbill
Parasitic Jaeger	White-winged Crossbill
Bonaparte's Gull	

(6) Species defined by Regional Coordinators as being 'regionally rare.' Identification and/or breeding evidence was required for any other species for which Regional Coordinators considered documentation necessary to substantiate records in their respective regions or in parts of their regions. In addition, some 'difficult' species may have required documentation. For example, in southwestern Ontario the following species were considered to require documentation in some areas within their known ranges:

Great Egret	Willow Flycatcher
Short-eared Owl	Carolina Wren
Red-bellied Woodpecker	Louisiana Waterthrush
Alder Flycatcher	Orchard Oriole

Categories 4 and 6, because they were not explicit, required further elaboration. A simple rule of thumb was established whereby any species found in only one or 2 squares in a region, in the first 4 field seasons (1981 to 1984) had to be documented. In order to select these from the data, a 'USRF table' was devised to allow the computer to check each record as to whether or not it required documentation. The principle adopted here was that any species common enough to be found in 3 or more squares in a region could be excluded from the requirements, unless it fell under one of the other categories in the guidelines. Any species not specifically exempted would automatically require a USRF, which also meant that any species not found in a region by 1984, but which was recorded subsequently, would also automatically require a USRF.

Although simple in concept, this procedure was difficult to formulate precisely, requiring an 'exempt/require' decision for each of 300 species in combination with each of the 46 regions, or 13,800 decisions in all. Once in place however, the system worked well.

Requests for documentation were also made for some Ring-billed Gull and Caspian Tern records. Records of nest with eggs or young on Great Lakes islands or shorelines were not requested, but all other shoreline and island records were requested. All undocumented inland records of these species were downgraded to 'Species Observed' (i.e., were classified as non-breeding) but Regional Coordinators were given the option of documenting what they considered to be legitimate breeding evidence. This procedure was necessary because these species, especially the Ring-billed Gull, travel considerable distances from their nesting colonies, and were frequently reported by atlassers in areas where there was no breeding habitat.

When the raw data were processed, the computer would flag each record in the data base with a special symbol if it required a USRF. Other computer programs made use of these symbols, to extract these records from the data base and print them in various ways.

On the first run through the data base, approximately

4,000 raw data records, or roughly one percent of all records, were identified as requiring additional documentation. In subsequent runs, about 1250 additional records were added to this list, based on a careful perusal of the first complete set of maps produced, in 1986.

Throughout 1985 and 1986, the atlas staff and Data Review Committee sent out thousands of letters to Regional Coordinators and atlassers asking for more precise information on these targeted records. The Coordinators provided their opinion on the acceptability of each record but the ultimate authority lay with the Data Review Committee. Most records were ruled upon by 3 members of the Data Review Committee. If a record was rejected, the Regional Coordinator was informed and could appeal the decision to the whole Data Review Committee. If the appeal was rejected, a further appeal could be launched with the whole Technical Advisory Committee. Through USRFs, letters, phone calls, and many meetings, the process of reviewing each and every one of the targeted 5,250 records was accomplished.

Eventually, the USRF system was developed into a highly automated one. The symbol indicating that a USRF was required for a particular record was replaced by another symbol indicating its status after the USRF had been reviewed by the Data Review Committee and a decision made. In this way, 'USRF Status Reports' were produced every few months in which every 'USRF record' was listed, indicating one of the following:

- Accepted
- Accepted with a change in breeding code
- Accepted as a released or escaped bird
- Breeding code downgraded to 'Species Observed' (no evidence of breeding)
- Rejected
- USRF required (none received).

Rejected records or those downgraded to 'Species Observed' were dropped from the database. Status reports were produced on a region-by-region basis for Coordinators, and on a species-by-species basis to facilitate the work of the Committee and as an aid to the authors of species accounts. Eventually the blank USRFs themselves were computer-generated, with the details of the record and the name and address of the observer printed on them.

Obviously, asking atlassers to provide details of observations made as much as 5 years earlier was not a desirable situation, but it was the only way to determine if they represented legitimate breeding records. Interpretation of breeding evidence was the problem more often than was the identification of the species. The ornithologists on the Data Review Committee spent hundreds of hours reviewing these records. In the end, 349 records were rejected by Regional Coordinators prior to the USRF process; most of these were simple transcription errors. USRFs were received for 3,525 of the 4,901 records for which documentation was required: of the documented records, 2,557 (72%) were accepted for inclusion in the Atlas; the remainder were rejected. Decisions still had

to be taken for records for which a USRF was never secured: 506 (37%) of these were accepted; 870 (63%) were rejected. Thus, the overall rejection rate for the 5,250 targeted records was 42% and it resulted in an average of just over 7 square records per species being deleted which would otherwise have appeared on the map.

Once the Regional Coordinators had carefully examined the data for their respective regions, and the Data Review Committee had reviewed all the unusual and contentious records, a last round of processing was necessary to incorporate the final decisions. The data processing system was designed in such a way that, with the exception of the USRF status reports, all the maps, summaries and statistical reports were produced from master files which ultimately emanated from the raw data. In order to make changes to these files, it was necessary to go back to the raw data and reprocess them. Only the raw data could be edited directly (with any efficiency, at least), and great efforts were made to ensure that the raw data matched the field cards and vice versa. Every change to the raw data was recorded on the appropriate forms, thus leaving a paper trail by which the reasons for corrections to the data base could be determined at any later time.

The task of entering, compiling, correcting and processing the data was enormous. It kept a computer programmer employed full time for 3 years, and involved up to 3 other staff members at various times. In the end, the accuracy of the data base was greatly improved. An unknown number of errors which do not significantly affect the maps (i.e., were not evident to the Data Review Committee) probably still exist. As for errors which would influence the mapped range of a given species, it is our belief that any of these are errors of omission rather than commission; that is, records which may have been correct but were not well enough substantiated to warrant inclusion have not been mapped. In those cases where giving the benefit of the doubt to the atlasser has resulted in putative breeding records which are, in fact (but unknown to us), erroneous, all that can be said is that their documentation was plausible. However, the error rate is certainly very low after the long and involved data processing procedure described above.

The squares and blocks on the species maps in this book were generated entirely by computer. This technology converts a zone-block-square designation into a polygon (usually a square, except along zone lines), at a given location, made up of a pattern of minute dots of 3 different densities. The computer directs a laser in a scanning device to lay down each dot at a precise location. Abundance histograms were produced in a similar way. The maps were produced by the Lands Directorate of Environment Canada.

In many ways, this Atlas is a product of the computer age; the maps were drawn, the type was set, and the statistics were calculated, all by computer. However, this technology can only reproduce, as faithfully as possible, the knowledge that was gained by the small army of human volunteers who worked on the Ontario Breeding Bird Atlas.

Appendix D

Criteria for assigning status to species of wildlife (COSEWIC)

Species: Any species, subspecies or geographically separate population.

Rare: Any indigenous species of fauna or flora that, because of its biological characteristics, or because it occurs at the fringe of its range, or for some other reason exists in low numbers or in very restricted areas in Canada, and so is vulnerable, but is not a threatened species.

Threatened: Any indigenous species of fauna or flora that is likely to become endangered in Canada if the factors affecting its vulnerability do not become reversed.

Endangered: Any indigenous species of fauna or flora whose existence in Canada is threatened with immediate extirpation or extinction throughout all or a significant portion of its range, owing to the actions of man.

Extirpated: Any indigenous species of fauna or flora no longer existing in the wild in Canada but existing elsewhere.

Extinct: Any indigenous species of fauna or flora formerly indigenous to Canada no longer existing elsewhere.

Criteria for assigning status to species of wildlife (MNR)

Species: Any species, subspecies or geographically separate population.

Rare: Any indigenous species of fauna or flora which is represented in Ontario by small but relatively stable populations, and/or which occurs sporadically or in a very restricted area of Ontario, or at the fringe of its range, and which should be monitored periodically for evidence of a possible decline.

Threatened: Any indigenous species of fauna or flora which, on the basis of the best available scientific evidence, is indicated to be experiencing a definite non-cyclical decline throughout all or a major portion of its Ontario range, and which is likely to become an endangered species if the factors responsible for the decline continue unabated.

Endangered: Any indigenous species of fauna or flora which, on the basis of the best available scientific evidence, is indicated to be threatened with immediate extinction throughout all or a significant portion of its Ontario range.

Extirpated: Any indigenous species of fauna or flora no longer existing in the wild in Ontario, but existing elsewhere in that species' range.

Extinct: Any species of fauna or flora formerly indigenous to Ontario which no longer exists anywhere in its former range.

Endangered species as listed by COSEWIC and MNR (as of May 1987)

COSEWIC

Rare
Trumpeter Swan
Cooper's Hawk
Red-shouldered Hawk
Peregrine Falcon
   (subspecies *pealei)*
King Rail
Ross' Gull
Ivory Gull
Caspian Tern
Common Barn-Owl
Great Gray Owl
Eastern Bluebird
Prairie Warbler
Prothonotary Warbler
Ipswich Sparrow

Threatened
Ferruginous Hawk
Peregrine Falcon
   (subspecies *tundrius)*

Roseate Tern
Burrowing Owl
Loggerhead Shrike
Henslow's Sparrow

Endangered
Peregrine Falcon
   (subspecies *anatum)*
Greater Prairie-Chicken
Whooping Crane
Piping Plover
Mountain Plover
Eskimo Curlew
Kirtland's Warbler

Extinct
Great Auk
Labrador Duck
Passenger Pigeon

MNR

Rare
Cooper's Hawk
Red-shouldered Hawk
King Rail
Ivory Gull
Caspian Tern
Great Gray Owl
Eastern Bluebird
Prairie Warbler
Prothonotary Warbler

Threatened
Common Barn-Owl
Loggerhead Shrike
Henslow's Sparrow

Endangered
American White Pelican
Bald Eagle
Golden Eagle
Peregrine Falcon
Piping Plover
Eskimo Curlew

Extirpated
Greater Prairie-Chicken

Extinct
Passenger Pigeon

Literature cited

ABRAHAM, K. F., & F. COOKE. 1979. First record of King Eider nesting in Manitoba. Blue Jay 37: 45-46.

ABRAHAM, K. F., & G. H. FINNEY. 1986. Eiders of the eastern Canadian arctic. Pp. 53-73 in Eider ducks in Canada (A. Reed, Ed.). Can. Wildl. Serv. Rep. Ser. No. 47.

ABRAHAM, K. F., & C. HENDRY. 1985. The nesting of MVP Canada Geese at Winisk, Ontario. Unpublished report, Ont. Min. Nat. Res., Moosonee.

ADKISSON, C. S., & R. N. CONNOR. 1978. Interspecific vocal imitation in White-eyed Vireos. Auk 95: 602-605.

ALDRICH, J. W. 1984. Ecogeographical variation in size and proportions of Song Sparrows (Melospiza melodia). Ornithol. Monogr. No. 35.

ALISON, R. M. 1976. The history of the Wild Turkey in Ontario. Can. Field-Natur. 90: 481-485.

ALLEN, G. 1919. Boston region. In the season. Bird Lore 21: 50-51.

ALLIN, A. E., & L. S. DEAR. 1947. Brewer's Blackbird breeding in Ontario. Wilson Bull. 59: 175-176.

ALLIN, C. C. 1987. Mid-summer Mute Swan survey. Unpublished report to the Atlantic Flyway Council.

ALVO, R. 1981. Marsh nesting of Common Loons (Gavia immer). Can. Field Nat. 95: 357.

ALVO, R., D. J. T. HUSSELL, & M. J. BERRILL. In press. The breeding success of Common Loons in relation to alkalinity and other lake characteristics in Ontario. Can. J. Zool.

AMERICAN ORNITHOLOGISTS' UNION. 1910. Check-list of North American birds, 3rd ed. A.O.U., New York.

AMERICAN ORNITHOLOGISTS' UNION. 1957. Check-list of North American birds, 5th ed. A.O.U., Baltimore.

AMERICAN ORNITHOLOGISTS' UNION. 1973. Thirty-second supplement to the American Ornithologists' Union check-list of North American birds. Auk 90: 411-419.

AMERICAN ORNITHOLOGISTS' UNION. 1983. Check-list of North American birds, 6th ed. A.O.U., Washington.

AMERICAN ORNITHOLOGISTS' UNION. 1985. Thirty-fifth supplement to the American Ornithologists' Union check-list of North American birds. Auk 102: 680-686.

ANDERSON, D. W., J. J. HICKEY, R. W. RISEBROUGH, D. F. HUGHES, & R. E. CHRISTENSEN. 1969. Significance of chlorinated hydrocarbon residues to breeding pelicans and cormorants. Can. Field-Natur. 83: 91-112.

ANDERSON, J. M. 1977. Yellow Rail. Pp. 66-70 in Management of migratory shore and upland game birds in North America (G. C. Sanderson, Ed.). Int. Assoc. Fish Wildl. Agencies, Washington.

ANDRLE, R. F. 1971. Range extension of the Golden-crowned Kinglet in New York. Wilson Bull. 83: 313-316.

ANGEHRN, P. A. M. 1979. A population estimate of the Lesser Snow Goose nesting colony at Cape Henrietta Maria, Ontario, on 14 June 1979. Unpublished report, Can. Wildl. Serv., Ont. Reg.

ANGEHRN, P. A. M., H. BLOKPOEL, & P. COURTNEY. 1979. A review of the status of the Great Black-backed Gull in the Great Lakes area. Ont. Field Biol. 33: 27-33.

APFELBAUM, S., & A. HANEY. 1981. Bird populations before and after wildfire in a Great Lakes pine forest. Condor 83: 347-354.

ARBIB, R. 1972. The Blue List for 1973. Amer. Birds 26: 932-933.

ARBIB, R. 1973. The Blue List for 1974. Amer. Birds 27: 943-945.

ARBIB, R. 1974. The Blue List for 1975. Amer. Birds 28: 971-974.

ARMSTRONG, E. R., & D. L. G. NOAKES. 1981. Food habits of Mourning Doves in southern Ontario. J. Wildl. Manage. 45: 222-227.

ARMSTRONG, E. R., & D. L. G. NOAKES. 1983. Wintering biology of Mourning Doves, Zenaida macroura, in Ontario. Can. Field-Natur. 97: 434-438.

ARMSTRONG, J. T. 1965. Breeding home range in the nighthawk and other birds: its evolutionary and ecological significance. Ecology 46: 619-629.

AUDUBON, J. J. 1840. The birds of America. J.B. Chevalier, Philadelphia.

AUSTIN, O. L., JR. 1961. Birds of the world. Golden Press, New York.

BAEPLER, D. H. 1968. *Chondestes grammacus* (Say). Lark Sparrow. Pp. 886-902 in A.C. Bent. Life histories of North American cardinals, grosbeaks, buntings, towhees, finches, sparrows and allies, part 2 (O.L. Austin, Jr., Ed.). Bull. U.S. Natl. Mus. No. 237.

BAILLIE, J. L., JR. 1925. The Hooded Warbler (*Wilsonia citrina*) in Ontario. Can. Field-Natur. 39: 150-151.

BAILLIE, J. L., JR. 1939. Four additional breeding birds of Ontario. Can. Field-Natur. 53: 130-131.

BAILLIE, J. L., JR. 1940. The summer distribution of the Eastern Evening Grosbeak. Can. Field-Natur. 54: 15-25.

BAILLIE, J. L., JR. 1947a. The Double-crested Cormorant nesting in Ontario. Can. Field-Natur. 61: 119-126.

BAILLIE, J. L., JR. 1947b. Prairie-Chickens in Ontario. Sylva 3: 2-4.

BAILLIE, J. L., JR. 1950a. In birdland. The Toronto Telegram, Sept. 30.

BAILLIE, J. L., JR. 1950b. Spring migration. Ontario - western New York region. Audubon Field Notes 4: 240.

BAILLIE, J. L., JR. 1950c. Nesting season. Ontario - western New York region. Audubon Field Notes 4: 272-274.

BAILLIE, J. L., JR. 1951. Fall migration. Ontario - western New York region. Audubon Field Notes 5: 12-14.

BAILLIE, J. L., JR. 1952. Nesting season. Ontario - western New York region. Audubon Field Notes 6: 279-280.

BAILLIE, J. L., JR. 1953a. Fall migration. Ontario - western New York region. Audubon Field Notes 7: 13-15.

BAILLIE, J. L., JR. 1953b. Spring migration. Ontario - western New York region. Audubon Field Notes 7: 306-307.

BAILLIE, J. L., JR. 1955. Spring migration. Ontario - western New York region. Audubon Field Notes 9: 328-329.

BAILLIE, J. L., JR. 1960. New Ontario breeding birds. Ont. Field Biol. 14: 14-23.

BAILLIE, J. L., JR. 1961. More new Ontario breeding birds. Ont. Field Biol. 15: 1-9.

BAILLIE, J. L., JR. 1962. Fourteen additional Ontario breeding birds. Ont. Field Biol. 16: 1-15.

BAILLIE, J. L., JR. 1963a. Three bird immigrants from the Old World. Trans. R. Can. Inst. 34: 95-105.

BAILLIE, J. L., JR. 1963b. The 13 most recent Ontario nesting birds. Ont. Field Biol. 17: 15-26.

BAILLIE, J. L., JR. 1967. A century of change: birds. Ont. Natur. 5: 14-19.

BAILLIE, J. L., JR. 1968. Birds in Ontario - southern invaders. Ont. Natur. 6: 20-21.

BAILLIE, J. L., JR. 1969. Birds in Ontario - western invaders. Ont. Natur. 7: 28-30.

BAILLIE, J. L., JR., & P. HARRINGTON. 1936. The distribution of breeding birds in Ontario, part 1. Trans. R. Can. Inst. 21: 1-150.

BAILLIE, J. L., JR., & P. HARRINGTON. 1937. The distribution of breeding birds in Ontario, part 2. Trans. R. Can. Inst. 21: 199-283.

BAILLIE, J. L., JR., & C. E. HOPE. 1943. The summer birds of the northeast shore of Lake Superior, Ontario. R. Ont. Mus. Zool. Contrib. 23: 1-27.

BAILLIE, J. L. JR., & C. E. HOPE. 1947. The summer birds of Sudbury District, Ontario. R. Ont. Mus. Zool. Contrib. 28: 1-32.

BAIRD, S. F., T. M. BREWER, & R. RIDGWAY. 1874. A history of North American birds, vol. 1. Little, Brown and Co., Boston.

BANNERMAN, D. A. 1961. The birds of the British Isles, vol. 9. Oliver and Boyd, London.

BARLOW, J. C. 1966. Extralimital occurrences of the House Sparrow in northern Ontario. Ont. Field Biol. 20: 1-3.

BARLOW, J. C., & W. B. MCGILLIVRAY. 1983. Foraging and habitat relationships of the sibling species Willow Flycatcher (*Empidonax traillii*) and Alder Flycatcher (*E. alnorum*) in southern Ontario. Can. J. Zool. 61: 1510-1516.

BARR, J. F. 1986. Population dynamics of the Common Loon (*Gavia immer*) associated with mercury-contaminated waters in northwestern Ontario. Can. Wildl. Serv. Occas. Pap. No. 56.

BARROWS, W. G. 1889. The English Sparrow (*Passer domesticus*) in North America, especially in its relations to agriculture. U.S. Dept. Agric., Div. Econ. Ornith. Mammal. Bull. 1: 1-401.

BARRY, D. 1974. Eastern Bluebird nest-box project. Pp. 347-354 in Birds of the Oshawa-Lake Scugog Region, Ontario (R.G. Tozer and J.M. Richards, Eds.). R.G. Tozer & J.M. Richards, Oshawa.

BART, J., & J. D. SCHOULTZ. 1984. Reliability of singing bird surveys: changes in observer efficiency with avian density. Auk 101: 307-318.

BART, J., R. A. STEHN, J. A. HERRICK, N. A. HEASLIP, T. A. BOOKHOUT, & J. R. STENZEL. 1984. Survey methods for breeding Yellow Rails. J. Wildl. Manage. 48: 1382-1386.

BATEMAN, M. C. 1985. Preliminary report to Canadian Woodcock Singing Ground Cooperators. Unpublished report, Can. Wildl. Serv., Sackville.

BAUMGARTNER, A. M. 1968. *Zonotrichia querula* (Nuttall). Harris' Sparrow. Pp. 1249-1273 in A.C. Bent. Life histories of North American cardinals, grosbeaks, buntings, towhees, finches, sparrows, and allies, part 3 (O.L. Austin, Jr., Ed.). Bull. U.S. Natl. Mus. No. 237.

BAXTER, T. S. H. 1985. The birding handbook: eastern Lake Superior. Superior Lore, Wawa.

BEARDSLEE, C. S., & H. D. MITCHELL. 1965. Birds of the Niagara Frontier region. Bull. Buffalo Soc. Nat. Sci. 22: 1-478.

BEDDALL, B. G. 1963. Range expansion of the Cardinal and other birds in the northeastern states. Wilson Bull. 75: 140-158.

BEHLE, W. H., & R. K. SELANDER. 1953. The plumage cycle of the California Gull (*Larus californicus*) with notes on color changes of soft parts. Auk 70: 239-260.

BELL, F. H. 1978. Status report on Piping Plover (*Charadrius melodus*) in Canada. Committee on the Status of Endangered Wildlife in Canada, Ottawa.

BELLROSE, F. C. 1976. Ducks, geese and swans of North America, 2nd. ed. Stackpole Books, Harrisburg.

BELLROSE, F. C. 1978. Ducks, geese and swans of North America, 2nd ed. (rev.). Stackpole Books, Harrisburg.

BELLROSE, F. C. 1980. Ducks, geese and swans of North America, 3rd ed. Stackpole Books, Harrisburg.

BENDIRE, C. E. 1895. Life histories of North American birds. U.S. Natl. Mus. Spec. Bull. 3.

BENGSTON, S.-A. 1970. Location of nest-sites of ducks in Lake Myvatn area, northeast Iceland. Oikos 21: 218-229.

BENKMAN, C. W. 1985. The foraging ecology of crossbills in eastern North America. Ph.D. dissertation, State Univ. New York, Albany.

BENT, A. C. 1921. Life histories of North American gulls and terns. Bull. U.S. Natl. Mus. No. 113.

BENT, A. C. 1923. Life histories of North American wildfowl, part 1. Bull. U.S. Natl. Mus. No. 126.

BENT, A. C. 1926. Life histories of North American marsh birds. Bull. U.S. Natl. Mus. No. 135.

BENT, A. C. 1929. Life histories of North American shorebirds, part 2. Bull. U.S. Natl. Mus. No. 146.

BENT, A. C. 1937. Life histories of North American birds of prey, part 1. Bull. U.S. Natl. Mus. No. 170.

BENT, A. C. 1939. Life histories of North American woodpeckers. Bull. U.S. Natl. Mus. No. 174.

BENT, A. C. 1940. Life histories of North American cuckoos, goatsuckers, hummingbirds, and allies. Bull. U.S. Natl. Mus. No. 176.

BENT, A. C. 1942. Life histories of North American flycatchers, larks, swallows, and allies. Bull. U.S. Natl. Mus. No. 179.

BENT, A. C. 1946. Life histories of North American jays, crows, and titmice. Bull. U.S. Natl. Mus. No. 191.

BENT, A. C. 1948. Life histories of North American nuthatches, wrens, thrashers, and allies. Bull. U.S. Natl. Mus. No. 195.

BENT, A. C. 1949. Life histories of North American thrushes, kinglets, and allies. Bull. U.S. Natl. Mus. No. 196.

BENT, A. C. 1953. Life histories of North American wood warblers. Bull. U.S. Natl. Mus. No. 203.

BENT, A. C. 1958. Life histories of North American blackbirds, orioles, tanagers, and allies. Bull. U.S. Natl. Mus. No. 211.

BENT, A. C. 1968. Life histories of North American cardinals, grosbeaks, buntings, towhees, finches, sparrows, and allies. (O.L. Austin, Jr., Ed.). Bull. U.S. Natl. Mus. No. 237.

BERGER, A. J. 1958. The Golden-winged - Blue-winged Warbler complex in Michigan and the Great Lakes area. Jack-Pine Warbler 36: 37-73.

BERGER, A. J. 1968. *Pooecetes gramineus gramineus*. Eastern Vesper Sparrow. Pp. 868-882 *in* A.C. Bent. Life histories of North American cardinals, grosbeaks, buntings, towhees,

finches, sparrows, and allies, part 2 (O.L. Austin, Jr., Ed.). Bull. U.S. Natl. Mus. No. 237.

BERGMAN, R. D., P. SWAIN, & M. W. WELLER. 1970. A comparative study of nesting Forster's and Black Terns. Wilson Bull. 82: 435-444.

BERTIN, R. I. 1977. Breeding habitats of the Wood Thrush and Veery. Condor 79: 303-311.

BLACK, C. P. 1975. The ecology and bioenergetics of the Northern Black-throated Blue Warbler (*Dendroica caerulescens caerulescens*). Ph.D. dissertation, Dartmouth Coll., Hanover.

BLAKE, C. H. 1968. *Carpodacus purpureus purpureus* (Gmelin). Eastern Purple Finch. Pp. 264-278 *in* A.C. Bent. Life histories of North American cardinals, grosbeaks, buntings, towhees, finches, sparrows and allies, part 1 (O.L. Austin, Jr., Ed.). Bull. U.S. Natl. Mus. No. 237.

BLEM, C. R. 1973. Geographic variation in the bioenergetics of the House Sparrow. Ornithol. Monogr. 14: 96-120.

BLOKPOEL, H. 1977. Gulls and terns nesting in northern Lake Ontario and the upper St. Lawrence River. Can. Wildl. Serv. Prog. Notes No. 75.

BLOKPOEL, H. 1983. The Canadian Great Lakes. Pp. 86-87 *in* The status of tern populations in northeastern United States and adjacent Canada (S.W. Kress, E.H. Weinstein and I.C.T. Nisbet, Eds.). Colonial Waterbirds 6: 84-106.

BLOKPOEL, H., & A. HARFENIST. 1986. Comparison of 1980 and 1984 inventories of Common Tern, Caspian Tern, and Double-crested Cormorant colonies in the eastern North Channel, Lake Huron, Ontario, Canada. Colonial Waterbirds 9: 61-67.

BLOKPOEL, H., & G. B. MCKEATING. 1978. Fish-eating birds nesting in Canadian Lake Erie and adjacent waters. Can. Wildl. Serv. Prog. Notes No. 87.

BLOKPOEL, H., & G. D. TESSIER. 1986. The Ring-billed Gull in Ontario: a review of a new problem species. Can. Wildl. Serv. Occas. Pap. No. 57.

BOCK, C. E., & L. W. LEPTHIEN. 1972. Winter eruptions of Red-breasted Nuthatches in North America, 1950-1970. Amer. Birds 26: 558-561.

BONDRUP-NIELSEN, S. 1976. First Boreal Owl nest for Ontario, with notes on development of the young. Can. Field-Natur. 90: 477-479.

BOUVIER, J. M. 1974. Breeding biology of the Hooded Merganser in southwestern Quebec, including interactions with Common Goldeneyes and Wood Ducks. Can. Field-Natur. 88: 323-330.

BOYER, G. F., & O. E. DEVITT. 1961. A significant increase in the birds of Luther Marsh, Ontario, following freshwater impoundment. Can. Field-Natur. 75: 225-237.

BRACKBILL, H. 1958. Nesting behavior of the Wood Thrush. Wilson Bull. 70: 70-89.

BRECKENRIDGE, W. J. 1956. Measurements of the habitat niche of the Least Flycatcher. Wilson Bull. 68: 47-51.

BREWER, A. D. 1977. The birds of Wellington County. Guelph Field Natur. Club Spec. Publ., Guelph.

BRITISH ORNITHOLOGISTS' UNION. 1986. Records Committee: twelfth report. Ibis 128: 601-603.

BROLEY, M. J. 1952. Eagle man. Pellegrini and Cudahy, New York.

BROOKS, A., & H. S. SWARTH. 1925. A distributional list of the birds of British Columbia. Pac. Coast Avifauna No. 17.

BROOKS, W. S. 1968. Comparative adaptations of Alaskan redpolls to the arctic environment. Wilson Bull. 80: 253-280.

BROWN, C. R., & M. B. BROWN. 1986. Ectoparasitism as a cost of coloniality in Cliff Swallows (*Hirundo pyrrhonota*). Ecology 67: 1206-1218.

BROWN, D. M., G. A. MCKAY, & L. J. CHAPMAN. 1968. The climate of southern Ontario. Climatolgical Studies No. 5, Dept. Transport (Meteorol. Br.), Toronto.

BROWN, L., & D. AMADON. 1968. Eagles, hawks and falcons of the world, vol. 2. McGraw-Hill, New York.

BRUNN, B. 1968. Migration of Little Gulls *Larus minutus* in the North Atlantic region. Dansk. Ornithol. Foren. Tidsskr. 62: 126-136.

BRUNTON, D. F., & W. J. CRINS. 1975. Status and habitat preference of the Yellow-bellied Flycatcher in Algonquin Park, Ontario. Ont. Field Biol. 29: 25-28.

BRYANT, A. A. 1986. Influence of selective logging on Red-shouldered Hawks, *Buteo lineatus*, in Waterloo Region, Ontario, 1953-1978. Can. Field-Natur. 100: 520-525.

BUECH, R. M. 1982. Nesting ecology and cowbird parasitism of Clay-colored, Chipping, and Field Sparrows in a Christmas tree plantation. J. Field Ornithol. 53: 363-369.

BULL, J. 1974. Birds of New York State. Doubleday, New York.

BULL, W. P. 1936. From hummingbird to eagle: an account of North American birds which appear or have appeared in the County of Peel. Perkins Bull Foundation, Toronto.

BURNS, R. D. 1958. A history of the entry of the Cardinal into Michigan. Jack-Pine Warbler 36: 19-21.

BYELICH, J. W., M. E. DECAPITA, G. W. IRVINE, R. E. RADTKE, N. I. JOHNSON, W. R. JONES, H. MAYFIELD, & W. J. MAHALAK. 1986. Kirtland's Warbler recovery plan. Unpublished report, U.S. Fish Wildl. Serv., Minnesota.

BYSTRAK, D. 1980. The breeding bird survey. Atlantic Natur. 33: 25-28.

BYSTRAK, D. 1981. The North American breeding bird survey. Pp. 34-41 *in* Estimating the numbers of terrestrial birds (C. J. Ralph and J. M. Scott, Eds.). Stud. Avian Biol. 6.

CADE, T. J. 1982. The falcons of the world. Cornell Univ. Press, Ithaca.

CADMAN, M. D. 1986. Status report on the Loggerhead Shrike (*Lanius ludovicianus*) in Canada. Committee on the Status of Endangered Wildlife in Canada, Ottawa.

CAIRNS, W. E., & I. A. McLAREN. 1980. Status of the Piping Plover on the east coast of North America. Amer. Birds 34: 206-208.

CAMPBELL, B., & E. LACK (Eds.). 1985. A dictionary of birds. Buteo Books, Vermillion.

CAMPBELL, C. A. 1975. Distribution and breeding success of the Loggerhead Shrike in southern Ontario. Can. Wildl. Serv. Rep. No. 6065.

CAMPBELL, E. C., & R. W. CAMPBELL. 1984. Status report on the Common Barn Owl (*Tyto alba*) in Canada - 1982. Committee on the Status of Endangered Wildlife in Canada, Ottawa.

CAREY, M., & V. NOLAN, JR. 1979. Population dynamics of Indigo Buntings and the evolution of avian polygyny. Evolution 33: 1180-1192.

CARPENTIER, A. G. 1983. Presumed breeding record of Brewster's x Brewster's Warbler, *Vermivora chrysoptera X pinus*, in Ontario. Can. Field-Natur. 97: 458-459.

CATLING, P. M. 1971. Spring migration of Saw-whet Owls at Toronto, Ontario. Bird-Banding 42: 110-114.

CATLING, P. M. 1972. A study of the Boreal Owl in southern Ontario with particular reference to the irruption of 1968-69. Can. Field-Natur. 86: 223-232.

CHAMBERLAIN, M. 1887. A catalogue of Canadian birds. J. & A. McMillan, St. John.

CHAPMAN, F. M. 1968. The warblers of North America. Dover Publ. Inc., New York.

CHAPMAN, L. J. & M. K. THOMAS. 1968. The climate of northern Ontario. Climatological Studies No. 6, Dept. Transport (Meteorol. Br.), Toronto.

CHRISTY, B. H. 1942. *Empidonax virescens* (Vieillot). Acadian Flycatcher. Pp. 183-197 *in* Life histories of North American flycatchers, larks, swallows, and allies (A.C. Bent, Ed.). Bull. U.S. Natl. Mus. No. 179.

CLAPP, R. B., D. MORGAN-JACOBS, & R. C. BANKS. 1983. Marine birds of the southeastern United States and Gulf of Mexico, part 3. Charadriiformes. U.S. Fish Wildl. Serv. FWS/OBS-83/30.

CLARK, K., D. EULER, & E. ARMSTRONG. 1983. Habitat associations of breeding birds in cottage and natural areas of central Ontario. Wilson Bull. 95: 77-96.

CLARK, R. G., P. J. WEATHERHEAD, H. GREENWOOD, & R. D. TITMAN. 1986. Numerical responses of Red-winged Blackbirds to changes in regional land use patterns. Can. J. Zool. 64: 1944-1950.

CLARK, W. R. 1984. Canada's Capistrano. Nat. Canada 13: 14-15.

CLARKE, C. H. D. 1948. The Wild Turkey in Ontario. Sylva 4: 5-12.

CLARKE, C. H. D. 1954. The Bob-white Quail in Ontario. Fed. Ont. Natur. Bull. 63.

CLARKE, C. H. D., & R. BRAFFETTE. 1947. Ring-necked Pheasant investigations in Ontario, 1946. Unpublished file report, Ont. Dept. Lands For.

CLARKE, C. K. 1890. *Coccyzus erythropthalmus*. Trans. R. Can. Inst. 1: 48-50.

CLEMENT, R. C. 1968. *Zonotrichia leucophrys leucophrys* (Forster). Eastern White-crowned Sparrow. Pp. 1273-1291 *in* A.C. Bent. Life histories of North American cardinals, grosbeaks, buntings, towhees, finches, sparrows, and allies, part 3 (O.L. Austin, Jr., Ed.). Bull. U.S. Natl. Mus. No. 237.

CLEVELAND, N. J., C. W. CUTHBERT, G. D. GRIEEF, G. E. HOLLAND, P. A. HORCH, R. W. KNAPTON, R. F. KOES, N. F. MURDOCH, W. P. NEILY, & I. A. WARD. 1980. Birder's guide to southeastern Manitoba. Man. Natur. Soc. Eco. Ser. No. 1, Winnipeg.

COLLINS, J. M. 1974. The relative abundance of ducks breeding in southern Ontario in 1951 and 1971. Pp. 32-44 *in* Waterfowl studies in eastern Canada: 1969-73 (H. Boyd, Ed.). Can. Wildl. Serv. Rep. Ser. No. 29.

COLLINS, S. L. 1983. Geographic variation in habitat structure of the Black-throated Green Warbler (*Dendroica virens*). Auk 100: 382-389.

COLLINS, S. L., F. C. JAMES, & P. G. RISSER. 1982. Habitat relationships of wood warblers (*Parulidae*) in northern central Minnesota. Oikos 39: 50-58.

CONFER, J. L., & K. KNAPP. 1981. Golden-winged Warblers and Blue-winged Warblers: the relative success of a habitat specialist and a generalist. Auk 98: 108-114.

COOKE, F., R. K. ROSS, R. K. SCHMIDT, & A. J. PAKULAK. 1975. Birds of the tundra biome at Cape Churchill and La Pérouse Bay. Can. Field-Natur. 89: 413-422.

COOPER, W. 1825. Description of a new species of grosbeak inhabiting the northwestern territory of the United States. Ann. Lyc. Nat. Hist. 1: 219-222.

CORE, E. L. 1948. The flora of the Lake Erie Islands. Ohio State Univ., Franz Theadore Stone Lab. Contrib. No. 9.

COURTNEY, P. A., & H. BLOKPOEL. 1983. Distribution and numbers of Common Terns in the Lower Great Lakes during 1900-1980: a review. Colonial Waterbirds 6: 107-120.

COX, G. W. 1960. A life history of the Mourning Warbler. Wilson Bull. 72: 5-28.

CRAMP, S. (Ed.) 1985. Handbook of the birds of Europe, the Middle East and North Africa: the birds of the western Palearctic, vol. 4. Terns - woodpeckers. Oxford Univ. Press, Oxford.

CRAMP, S., & K. E. L. SIMMONS (Eds.). 1977. Handbook of the birds of Europe, the Middle East, and North Africa: the birds of the western Palearctic, vol. 1. Ostriches - ducks. Oxford Univ. Press, Oxford.

CRAMP, S., & K. E. L. SIMMONS (Eds.). 1980. Handbook of the birds of Europe, the Middle East, and North Africa: the birds of the western Palearctic, vol. 2. Hawks - bustards. Oxford Univ. Press, Oxford.

CRAMP, S., & K. E. L. SIMMONS (Eds.). 1983. Handbook of the birds of Europe, the Middle East, and North Africa: the birds of the western Palearctic, vol. 3. Waders - gulls. Oxford Univ. Press, Oxford.

CREIGHTON, W. A. 1965. Lillabelle Lake waterfowl management study. Ont. Fish Wildl. Rev. 4: 19-24.

CRINGAN, A. T. 1957. Notes on the biology of the Red-necked Grebe in western Ontario. Can. Field-Natur. 71: 72-73.

CRINGAN, A. T. 1960. Some changes in the status of the Mallard in southern Ontario. Midwest Fish Wildl. Conf., Toronto.

DANCE, K. W. 1986. Avifauna of an urbanizing environment in southern Ontario, 1921-1982. Ont. Birds 4: 22-29.

DANCE, K. W., & D. M. FRASER. 1985. Use of breeding bird atlas data in environmental planning. Ont. Breed. Bird Atlas Newsletter 14: 1-2.

DAVID, N. 1980. Status and distribution of birds in southern Quebec. Cahiers d'ornithologie Victor-Gaboriault No. 4.

DAVIS, C. M. 1978. A nesting study of the Brown Creeper. Living Bird 17: 237-263.

DAVIS, D. 1983. Breeding biology of Turkey Vultures. Pp. 271-286 *in* Vulture biology and management (S.R. Wilbur and J.A. Jackson, Eds.). Univ. California Press, Los Angeles.

DAVIS, D. E. 1959. Observations on territorial behavior of Least Flycatchers. Wilson Bull. 71: 73-85.

DAVIS, W. J. 1982. Territory size in *Megaceryle alcyon* along a stream habitat. Auk 99: 353-362.

DAWSON, J. B. 1963. Aspects of Hungarian (Gray) Partridge ecology in eastern Ontario. Unpublished report, Ont. Min. Nat. Resour.

DAY, J. H. 1955. An investigation of Luther Marsh, Ontario. Unpublished report, Can. Wildl. Serv.

DEMENT'EV, G. P., & N. A. GLADKOV (Eds.). 1952. Birds of the Soviet Union. Translated by Israel Program for Scientific Translation, Jerusalem.

DE SMET, K. D. 1982. Status report on the Red-necked Grebe (*Podiceps grisegena*) in Canada. Committee on the Status of Endangered Wildlife in Canada, Ottawa.

DE SMET, K. D. 1985. Status report on the Merlin in North America, *Falco columbarius*. Committee on the Status of Endangered Wildlife in Canada, Ottawa.

DE SMET, K. D. 1987. A status report on the Golden Eagle (*Aquila chrysaetos*) in Canada. Committee on the Status of Endangered Wildlife in Canada, Ottawa.

DEAR, L. S. 1940. Breeding birds of the region of Thunder Bay, Lake Superior, Ontario. Trans. R. Can. Inst. 23: 119-143.

DENNIS, D. G. 1974a. Breeding pair surveys of waterfowl in southern Ontario. Pp. 45-52 *in* Waterfowl studies in eastern Canada: 1969-1973 (H. Boyd, Ed.). Can. Wildl. Serv. Rep. Ser. No. 29.

DENNIS, D. G. 1974b. Waterfowl observations during the nesting season in Precambrian and clay belt areas of north-central Ontario. Pp. 53-56 *in* Waterfowl studies in eastern Canada: 1969-73 (H. Boyd, Ed.). Can. Wildl. Serv. Rep. Ser. No. 29.

DENNIS, D. G., & N. R. NORTH. 1984. Waterfowl densities in northwestern Ontario during the 1979 breeding season. Pp. 6-9 in Waterfowl studies in Ontario, 1973-1981 (S. G. Curtis, D. G. Dennis, & H. Boyd, Eds.). Can. Wildl. Serv. Occas. Pap. No. 54.

DEVITT, O. E. 1939. The Yellow Rail breeding in Ontario. Auk 56: 238-243.

DEVITT, O. E. 1964. An extension in the breeding range of Brewer's Blackbird in Ontario. Can. Field-Natur. 78: 42-46.

DEVITT, O. E. 1967. The birds of Simcoe County, Ontario, 2nd ed. Brereton Field Natur. Club, Barrie.

DEVITT, O. E. 1969. First nesting records of Brewer's Blackbird (Euphagus cyanocephalus) for King Township and Simcoe County, Ontario. Ont. Field Biol. 23: 41-42.

DICKERMAN, R. W. 1986a. A review of the Red Crossbill in New York State. Part 1. Historical and nomenclatural background. The Kingbird 36: 73-78.

DICKERMAN, R. W. 1986b. A review of the Red Crossbill in New York State. Part 2. Identification of specimens from New York. The Kingbird 36: 127-137.

DILGER, W. C. 1956a. Adaptive modifications and ecological isolating mechanisms in the thrush genera Catharus and Hylocichla. Wilson Bull. 68: 171-199.

DILGER, W. C. 1956b. Hostile behavior and reproductive isolating mechanisms in the avian genera Catharus and Hylocichla. Auk 73: 313-353.

DOLBEER, R. A. 1980. Blackbirds and corn in Ohio. U.S. Fish Wildl. Serv. Res. Publ. No. 136.

DOLTON, D. L. 1976. 1976 Mourning Dove breeding population status: administrative report. Annual Regulations Conference for Migratory Shore and Upland Game Birds. U.S. Fish Wildl. Serv.

DOW, D. D. 1969a. Home range and habitat of the Cardinal in peripheral and central populations. Can. J. Zool. 47: 103-114.

DOW, D. D. 1969b. Habitat utilization by Cardinals in central and peripheral breeding populations. Can. J. Zool. 47: 409-417.

DOW, D. D., & D. M. SCOTT. 1971. Dispersal and range expansion by the Cardinal: an analysis of banding records. Can. J. Zool. 49: 185-198.

DUNN, E. H. 1979. Nesting biology and development of young in Ontario Black Terns. Can. Field-Natur. 93: 276-281.

DUNN, E. H., H. F. HOWKINS, & R. V. CARTAR. 1975. Red-breasted Nuthatches breeding in nest boxes in pine plantations on the north shore of Lake Erie. Can. Field-Natur. 89: 467-468.

DUNN, E. H., D. J. T. HUSSELL, & J. SIDERIUS. 1985. Status of the Great Blue Heron, Ardea herodias, in Ontario. Can. Field-Natur. 99: 62-70.

EAGLES, P. F. J. 1980. Criteria for the designation of environmentally sensitive areas. Pp. 68-79 in Protection of natural areas in Ontario (S. Barrett and J. Riley, Eds.). Working Paper No. 3, Faculty of Env. Stud., York University, Downsview.

EAGLES, P. F. J. 1981. Environmentally sensitive area planning in Ontario, Canada. A.P.A. Journal 47: 313-323.

EAGLES, P. F. J. 1982. Financing an atlas project. Pp. 103-105 in Proceedings of the northeastern bird atlas conference (S.B. Laughlin, C.S. Robbins and D.P. Kibbe, Eds.). Vermont Inst. Nat. Sci., Woodstock.

EAGLES, P. F. J. 1984. The planning and management of environmentally sensitive areas. Longman, London.

EAGLES, P. F. J. 1986. Funding atlases: learning from projects under way. Pp. 4-6 in Proceedings of the second northeastern breeding bird atlas conference (S.M. Sutcliffe, R.E. Bonney and J.D. Lowe, Eds.). Lab. Ornithol., Cornell Univ., Ithaca.

EAGLES, P. F. J., & M. D. CADMAN. 1981. Ontario Breeding Bird Atlas guide for participants. Atlas Mgmt. Comm., Toronto.

EAGLES, P. F. J., & M. D. CADMAN. 1983. Ontario Breeding Bird Atlas guide for participants, rev. ed. Atlas Mgmt. Comm., Toronto.

EAGLES, P. F. J., & J. D. MCCAULEY. 1982. The rare breeding birds of Ontario. Univ. Waterloo Biol. Ser. No. 24, Waterloo.

EATON, E. H. 1910. Birds of New York, part 1. Univ. New York State, Albany.

EATON, S. W. 1968. Junco hyemalis hyemalis (Linnaeus). Northern Slate-coloured Junco. Pp. 1029-1043 in A.C. Bent. Life histories of North American cardinals, grosbeaks, buntings, towhees, finches, sparrows, and allies, part 3 (O.L. Austin, Jr., Ed.). Bull. U.S. Natl. Mus. No. 237.

ECOLOGISTICS LTD. 1982. Environmentally significant areas study: South Lake Simcoe Conservation Authority.

EDMINSTER, F. C. 1954. American birds of field and forest. Charles Scribner's & Sons, New York.

ELDER, D., D. PRICE, & J. M. SPEIRS. 1971. Thirty-fifth breeding bird census. 57. Leatherleaf bog. Amer. Birds 25: 1005-1006.

ELDER, W. H. 1985. Survivorship in the Tufted Titmouse. Wilson Bull. 97: 517-524.

ELIAS, T. S. 1980. The complete trees of North America. Van Nostrand Reinhold, Toronto.

ELLIOT, D. G. 1895. North American shore birds. Harper, New York.

ELLIOTT, J. J., & R. S. ARBIB, JR. 1953. Origin and status of the House Finch in the eastern United States. Auk 70: 31-37.

ERSKINE, A. J. 1972. Buffleheads. Can. Wildl. Serv. Mono. Ser. No. 4.

ERSKINE, A. J. 1977. Birds in boreal Canada: communities, densities and adaptations. Can. Wildl. Serv. Rep. Ser. No. 41.

ERSKINE, A. J. 1978. The first ten years of the co-operative Breeding Bird Survey in Canada. Can. Wildl. Serv. Rep. No. 42.

ERSKINE, A. J. 1979. Man's influence on potential nesting sites and populations of swallows in Canada. Can. Field-Natur. 93: 371-377.

ERSKINE, A. J. 1980. A preliminary catalogue of bird census studies in Canada, part 4. Can. Wildl. Serv. Prog. Notes No. 112

ERWIN, R. M. 1979. Coastal waterbird colonies: Cape Elizabeth, Maine to Virginia. U.S. Fish Wildl. Serv., Biol. Serv. Prog. FWS/OBS - 79/10.

ESCOTT, N. G. 1986. Thunder Bay's nesting Merlins. Ont. Birds 4: 97-101.

EVANS. C. D., A. S. HAWKINS, & W. H. MARSHALL. 1952. Movements of waterfowl broods in Manitoba. U.S. Fish Wildl. Serv. Spec. Sci. Rep. - Wildl. No. 16.

FAABORG, J. 1976. Habitat selection and territorial behavior of the small grebes of North Dakota. Wilson Bull. 88: 390-399.

FALLS, J. B. 1981. Mapping territories with playback: an accurate census method for songbirds. Pp. 86-91 in Estimating the numbers of terrestrial birds (C.J. Ralph and J.M. Scott, Eds.). Studies in Avian Biology 6.

FARGO, W. G., & M. B. TRAUTMAN. 1930. Late summer birds along the upper Michipicoten River, Ontario. Can. Field-Natur. 44: 30-33.

FEARE, C. J. 1984. The starling. Oxford Univ. Press, Oxford.

FERGUSON, R. S. 1981. Summer birds on the Northwest Angle Provincial Forest and adjacent southeastern Manitoba, Canada. Syllogeus 31. Natl. Mus. Can., Ottawa.

FERGUSON, R. S., & S. G. SEALY. 1983. Breeding ecology of the Horned Grebe, Podiceps auritus, in southwestern Manitoba. Can. Field-Natur. 97: 401-408.

FICKEN, M. S., S. R. WITKIN, & C. M. WEISE. 1981. Associations among members of a Black-capped Chickadee flock. Behav. Ecol. Sociobiol. 8: 245-249.

FLEMING, J. H. 1901. A list of the birds of the Districts of Parry Sound and Muskoka, Ontario. Auk 18: 33-45.

FLEMING, J. H. 1906. Chuck-will's-widow and Mockingbird in Ontario. Auk 23: 343-344.

FLEMING, J. H. 1907. Birds of Toronto, Canada, part 2. Auk 24: 71-89.

FLEMING, J. H. 1930. Ontario bird notes. Auk 47: 64-71.

FLOOD, N. J. 1984. Adaptive significance of delayed plumage maturation in male Northern Orioles. Evolution 38: 267-279.

FOX, G. A., K. S. YOUNG, & S. G. SEALY. 1980. Breeding performance, pollutant burden and eggshell thinning in Common Loon Gavia immer nesting on a boreal forest lake. Ornis Scand. 11: 243-248.

FRANK, R., H. LUMSDEN, J. F. BARR, & H. E. BRAUN. 1983. Residues of organochloride insecticides, industrial chemicals and mercury in eggs and in tissues taken from healthy and emaciated Common Loons, Ontario, Canada, 1968-1980. Arch. Environ. Contam. Toxicol. 12: 641-654.

FRANZREB, K. E. 1984. Foraging habits of Ruby-crowned and Golden-crowned Kinglets in an Arizona montane forest. Condor 86: 139-145.

FREDERICKSON, L. H. 1971. Common Gallinule breeding biology and development. Auk 88: 914-919.

FREEDMAN, B., & J. L. RILEY. 1980. Population trends of various species of birds wintering in southern Ontario. Ont. Field Biol. 34: 49-79.

FREEMAN, M. M. R. 1970a. Observations on the seasonal behavior of the Hudson Bay Eider (Somateria mollissima sedentaria). Can. Field-Natur. 84: 145-153.

FREEMAN, M. M. R. 1970b. The birds of the Belcher Islands, N.W.T., Canada. Can. Field-Natur. 84: 277-290.

FREEMARK, K. E. 1977. Nest site selection by the Hudson Bay Eider (Somateria mollissima sedentaria). B.Sc. thesis, Queen's Univ., Kingston.

FRETWELL, S. 1977. Is the Dickcissel a threatened species? Amer. Birds 31: 923-932.

FRILEY, C. E., JR., L. H. BENNETT, & G. O. HENDRICKSON. 1938. The American Coot in Iowa. Wilson Bull. 50: 81-86.

FYFE, R. W. 1976. Status of Canadian raptor populations. Can. Field-Natur. 90: 370-375.

GAGE, S. H., C. A. MILLER, & L. J. MOOK. 1970. The feeding response of some forest birds to the black-headed budworm. Can. J. Zool. 48: 359-366.

GALLI, A. E., C. F. LECK, & T. T. FORMAN. 1976. Avian distribution patterns in forest islands of different sizes in central New Jersey. Auk 93: 356-364.

GAUCH, H. G., JR. 1982. Multivariate analysis in community ecology. Cambridge Univ. Press, Cambridge.

GERRARD, J. M., P. N. GERRARD, G. R. BORTOLOTTI, & D. W. A. WHITFIELD. 1983. A 14-year study of Bald Eagle reproduction on Besnard Lake, Saskatchewan. Pp. 47-57 in Biology and management of Bald Eagles and Ospreys (D. M. Bird, Ed.). Harpell Press, Ste. Anne de Bellevue.

GILBERTSON, M. E. 1974. Pollutants in breeding Herring Gulls in the lower Great Lakes. Can. Field-Natur. 88: 273-280.

GILL, F. B. 1980. Historical aspects of hybridization between Blue-winged and Golden-winged Warblers. Auk 97: 1-18.

GIRARD, G. L. 1939. Notes on the life history of the shoveler. Trans. N. Amer. Wildl. Nat. Resour. Conf. 4: 364-371.

GLAHN, J. F. 1974. Study of breeding rails with recorded calls in north-central Colorado. Wilson Bull. 86: 206-214.

GLEASON, H. A. 1952. The new Britton and Brown illustrated flora of the northeastern United States and adjacent Canada. Van Nostrand, New York.

GLOVER, F. A. 1953. Nesting ecology of the Pied-billed Grebe in north-eastern Iowa. Wilson Bull. 65: 32-39.

GODFREY, W. E. 1966. The birds of Canada. Natl. Mus. Can. Bull. No. 203.

GODFREY, W. E. 1986. The birds of Canada, rev. ed. Natl. Mus. Can., Ottawa.

GOODWIN, C. E. 1965. The nesting season. Ontario - western New York region. Audubon Field Notes 19: 537-540.

GOODWIN, C. E. 1974. The nesting season. Ontario - western New York region. Amer. Birds 28: 896-900.

GOODWIN, C. E. 1975. The nesting season. Ontario region. Amer. Birds 29: 963-967.

GOODWIN, C. E. 1976. The nesting season. Ontario region. Amer. Birds 30: 948-952.

GOODWIN, C. E. 1977. The nesting season. Ontario region. Amer. Birds 31: 1131-1135.

GOODWIN, C. E. 1978. The nesting season. Ontario region. Amer. Birds 32: 1153-1156.

GOODWIN, C. E. 1979a. The spring migration. Ontario region. Amer. Birds 33: 765-768.

GOODWIN, C. E. 1979b. The nesting season. Ontario region. Amer. Birds 33: 858-860.

GOODWIN, C. E. 1980. The nesting season. Ontario region. Amer. Birds 34: 890-892.

GOODWIN, C. E. 1981. The nesting season. Ontario region. Amer. Birds 35: 934-936.

GOODWIN, C. E. 1982. A bird-finding guide to Ontario. Univ. Toronto Press, Toronto.

GOODWIN, C. E., W. FREEDMAN, & S. M. MCKAY. 1977. Population trends in waterfowl wintering in the Toronto region, 1929-1976. Ont. Field Biol. 31: 1-28.

GOODWIN, C. E., & R. C. ROSCHE. 1970. The nesting season. Ontario - western New York region. Amer. Birds 24: 677-680.

GOODWIN, C. E., & R. C. ROSCHE. 1971. The nesting season. Ontario - western New York region. Amer. Birds 25: 851-856.

GOODWIN, C. E., & R. C. ROSCHE. 1974. The nesting season. Ontario - western New York region. Amer. Birds 28: 896-900.

GRAMZA, A. F. 1967. Responses of breeding nighthawks to a disturbance stimulus. Auk 84: 72-86.

GRATTO, C. L., F. COOKE, & R. I. G. MORRISON. 1983. Nesting success of yearling and older breeders in the Semipalmated Sandpiper, *Calidris pusilla*. Can. J. Zool. 61: 1133-1137.

GRATTO, C. L., R. I. G. MORRISON, & F. COOKE. 1985. Philopatry, site tenacity and mate fidelity in the Semipalmated Sandpiper. Auk 102: 16-24.

GRIER, J. W. 1982. Ban on DDT and subsequent recovery of reproduction in Bald Eagles. Science 218: 1232-1235.

GRIER, J. W. 1985. History and procedures of surveying for Bald Eagles in northwestern Ontario. Pp. 194-200 *in* The Bald Eagle in Canada (J. M. Gerrard and T. M. Ingram, Eds.). White Horse Plains Publ., Headingley.

GRIESE, H. J., R. A. RYDER, & C. E. BRAUN. 1980. Spatial and temporal distribution of rails in Colorado. Wilson Bull. 92: 96-102.

GRISCOM, L. 1938. The birds of Lake Umbagog region of Maine. Bull. Mus. Comp. Zool. 66: 525-560.

GRISCOM, L., & A. SPRUNT, JR. 1957. The warblers of America. Devin-Adair Co., New York.

GRISCOM, L., & A. SPRUNT, JR. 1979. The warblers of America. 2nd ed. Doubleday & Co., Garden City.

GULLION, G. W. 1952. The displays and calls of the American Coot. Wilson Bull. 64: 83-97.

GULLION, G. W. 1954. The reproductive cycle of American Coots in California. Auk 71: 366-412.

GUNN, W. W. H. 1957. Nesting season. Ontario - western New York region. Audubon Field Notes 11: 402-403.

HADFIELD, H. W. 1864. Birds of Canada observed near Kingston during the latter part of summer and autumn of 1857. Zoologist 22: 9297-9310.

HAGAR, J. A. 1966. Nesting of the Hudsonian Godwit at Churchill, Manitoba. Living Bird 5: 5-43.

HAIG, S. 1985. The status of the Piping Plover in Canada. Committee on the Status of Endangered Wildlife in Canada, Ottawa.

HAIG, S., & L. W. ORING. 1985. Distribution and status of the Piping Plover throughout the annual cycle. J. Field Ornithol. 56: 334-345.

HALL, G. A. 1984. A long-term bird population study in an Appalachian spruce forest. Wilson Bull. 96: 228-240.

HALL, H. M. 1955. The Greater Yellow-Legs. Audubon 57: 154-155.

HALL-ARMSTRONG, J., & E. R. ARMSTRONG. 1982. The status of Sandhill Cranes in the Cochrane District of northeastern Ontario. Unpublished report, Ont. Min. Nat. Resour.

HANEY, C. J. 1985. Wintering phalaropes off the southeastern United States: application of remote sensing imagery to seabird habitat analysis at oceanic fronts. J. Field Ornithol. 56: 321-333.

HANRAHAN, C., & B. DILABIO. 1984. Recent bird sightings. Trail and Landscape 18: 233-238.

HANSON, H. C. 1950. Canada Geese of the Mississippi Flyway. Ill. Nat. Hist. Surv. 25: 67-210.

HANSON, H. C., H. G. LUMSDEN, J. J. LYNCH, & H. W. NORTON. 1972. Population characteristics of three mainland colonies of Blue and Lesser Snow Geese nesting in the southern Hudson Bay region. Ont. Min. Natur. Resour. Rep. No. 93.

HARDING, D. C. 1931. Nesting habits of the Black-throated Blue Warbler. Auk 48: 512-522.

HARMESON, J. P. 1974. Breeding ecology of the Dickcissel. Auk 91: 348-359.

HARRIS, W. C. 1985. The nesting season. Prairie Provinces region. Amer. Birds. 39: 927-928.

HARRISON, C. 1978. A field guide to the nests, eggs and nestlings of North American birds. Collins, Cleveland.

HARRISON, H. H. 1975. A field guide to birds' nests. Houghton Mifflin, Boston.

HARRISON, H. H. 1984. Wood warblers' world. Simon & Schuster, New York.

HATCH, D. R. M. 1972. Breeding status of the Forster's Tern in Manitoba. Blue Jay 30: 102-104.

HAYS, H. 1972. Polyandry in the Spotted Sandpiper. Living Bird 11: 43-57.

HEADSTROM, R. 1970. A complete field guide to nests in the United States. Ives Washburn, New York.

HEIMBERGER, M., D. EULER, & J. BARR. 1983. The impact of cottage development on Common Loon reporductive success in central Ontario. Wilson Bull. 95: 431-439.

HENDERSON, A. D. 1926. Bonaparte's Gull nesting in northern Alberta. Auk 43: 288-294.

HENRY, A. 1809. Travels and adventures in Canada and the Indian territories between the years 1760 and 1776, part 1. Riley, New York.

HICKEY, J. J. (Ed.). 1969. Peregrine Falcon populations: their biology and decline. Univ. Wisconsin Press, Madison.

HILDEN, O., & S. VUOLANTO. 1972. Breeding biology of the Red-necked Phalarope (Phalaropus lobatus) in Finland. Ornis Fenn. 49: 57-85.

HILL, N. P. 1968. Ammospiza caudacuta caudacuta (Gmelin). Eastern Sharp-tailed Sparrow. Pp. 795-812 in A.C. Bent. Life histories of North American cardinals, grosbeaks, buntings, towhees, finches, sparrows, and allies, part 2 (O.L. Austin, Jr., Ed.). Bull. U.S. Natl. Mus. No. 237.

HOFSLUND, P. B. 1954. Incubation period of the Mourning Warbler. Wilson Bull. 26: 198-205.

HOHN, E. O. 1967. Observations of the breeding biology of Wilson's Phalarope (Steganopus tricolor) in central Alberta. Auk 84: 220-244.

HOLDSWORTH, C. 1973. Winter study of Bobwhite Quail in southwestern Ontario, 1972-1973. Unpublished report, Ont. Min. Nat. Resour.

HOLMES, R. T. 1966. Breeding ecology and annual cycle adaptations in the Red-backed Sandpiper in northern Alaska. Condor 68: 3-46.

HOLROYD, G. L. 1968. Recent Golden Eagle sightings in eastern Ontario. Ont. Field Biol. 22: 30-33.

HOLROYD, G. L. 1975. Nest site availability as a factor limiting population size of swallows. Can. Field-Natur. 89: 60-64.

HOPE, C. E. 1938. Birds of Favourable Lake Mine, Patricia portion of Kenora District, Ontario. Unpublished report, Dept. Ornith., R. Ont. Mus., Toronto.

HOPE, C. E. 1940. Journals 13 and 14. Log of the Royal Ontario Museum field trip to Fort Severn. Unpublished diary, R. Ont. Mus., Toronto.

HORTON, K. W., & G. H. D. BEDELL. 1960. White and red pine: ecology, silviculture and management. Dept. North Aff. Nat. Res. (For. Br.) Bull. 124, Ottawa.

HOSIE, R. C. 1969. Native trees of Canada. Can. For. Serv., Ottawa.

HOWE, H. F. 1976. Egg size, hatching asynchrony, sex, and brood reduction in the Common Grackle. Ecology 57: 1195-1207.

HOWE, M. A. 1975a. Social interactions in flocks of courting Wilson's Phalaropes (Phalaropus tricolor). Condor 77: 24-33.

HOWE, M. A. 1975b. Behavioral aspects of the pair bond in Wilson's Phalarope. Wilson Bull. 87: 248-270.

HOWES-JONES, D. 1985. Relationships among song activity, context, and social behavior in the Warbling Vireo. Wilson Bull. 97: 4-20.

HOYT, S. F. 1957. The ecology of the Pileated Woodpecker. Ecology 38: 246-256.

HUBER, K. R. 1982. The Kirtland's Warbler (Dendroica kirtlandii) - an annotated bibliography. Ann Arbor Mus. Zool., Univ. Michigan.

HUSSELL, D. J. T. 1972. Factors affecting clutch size in arctic passerines. Ecol. Monogr. 42: 317-364.

HUSSELL, D. J. T. 1981. Migrations of the Least Flycatcher in southern Ontario. J. Field Ornithol. 52: 97-111.

HUSSELL, D. J. T. 1982. Migrations of the Yellow-bellied Flycatcher in southern Ontario. J. Field Ornithol. 53: 223-234.

HUTCHINSON, C. D., & B. NEATH. 1978. Little Gulls in Britain and Ireland. Brit. Birds 71: 563-582.

INGRAM, C. 1974. The migration of the swallow. Witherby, London.

JACKSON, J. A. 1983. Nesting phenology, nest site selection, and reproductive success of Black and Turkey Vultures. Pp. 245-270 in Vulture biology and management (S.R. Wilbur and J.A. Jackson, Eds.). Univ. California Press, Los Angeles.

JAMES, E. (Ed.) 1956. A narrative of the captivity and adventures of John Tanner during thirty years residence among the Indians in the interior of North America. Ross and Haines Inc., Minneapolis.

JAMES, R. D. 1976. Foraging behavior and habitat selection of three species of vireos in southern Ontario. Wilson Bull. 88: 62-75.

JAMES, R. D. 1978a. Pairing and nest site selection in Solitary and Yellow-throated Vireos with a description of a ritualized nest building display. Can. J. Zool. 56: 1163-1169.

JAMES, R. D. 1978b. Nesting of the House Finch (Carpodacus mexicanus) in Ontario. Ont. Field Biol. 32: 30-32.

JAMES, R. D. 1979. The comparative foraging behavior of Yellow-throated and Solitary Vireos: the effect of habitat and sympatry. Pp. 137-163 in The role of insectivorous birds in forest ecosystems (J. G. Dickson, R. N. Connor, R. R. Fleet, J. A. Jackson and J. C. Kroll, Eds.). Academic Press, New York.

JAMES, R. D. 1980. Notes on the summer birds of Pickle Lake, Ontario. Ont. Field Biol. 34: 80-92.

JAMES, R. D. 1981. Northern Shrike confirmed as a breeding species in Ontario. Ont. Field Biol. 35: 93-94.

JAMES, R. D. 1984a. Structure, frequency of usage, and apparent learning in the primary song of the Yellow-throated Vireo, with comparative notes on Solitary Vireos. Can. J. Zool. 62: 468-472.

JAMES, R. D. 1984b. The breeding bird list for Ontario: additions and comments. Ont. Birds. 2: 24-29.

JAMES, R. D. 1984c. Ontario Bird Records Committee report for 1983. Ont. Birds 2: 53-65.

JAMES, R. D., J. A. DICK, S. V. NASH, M. K. PECK, & B. E. TOMLINSON. 1983. Avian breeding and occurrence notes from the Sutton Ridges area of northeastern Ontario. Can. Field-Natur. 97: 187-193.

JAMES, R. D., P. L. McLAREN, & J. C. BARLOW. 1976. Annotated checklist of the birds of Ontario. Life Sci. Misc. Publ., R. Ont. Mus., Toronto.

JAMES, R. D., S. V. NASH, & M. K. PECK. 1982a. Distribution, abundance and natural history of birds at Kiruna Lake - 1981. Unpublished report, Dept. Ornith., R. Ont. Mus., Toronto.

JAMES, R. D., S. V. NASH, & M. K. PECK. 1982b. A survey of birdlife near the mouth of the Harricanaw River - 1982. Pp. 43-78 in A faunal study of the Hudson Bay Lowland. Field report 1982: Harricanaw River study area. Unpublished report, Dept. Ornith., R. Ont. Mus., Toronto.

JEHL, J. R., JR. 1968. The breeding biology of Smith's Longspur. Wilson Bull. 80: 123-149.

JEHL, J. R., JR. 1973. Breeding biology and systematic relationships of the Stilt Sandpiper. Wilson Bull. 85: 115-147.

JEHL, J. R., JR., & B. A. SMITH. 1970. Birds of the Churchill region, Manitoba. Man. Mus. Man Nat. Spec. Publ. 1: 1-87.

JENKINS, D. 1961. Social behaviour in the partridge Perdix perdix. Ibis 103: 157-188.

JOHNS, J. E. 1969. Field studies of Wilson's Phalarope. Auk 86: 660-670.

JOHNSGARD, P. A. 1973. Grouse and quails of North America. Univ. Nebraska Press, Lincoln.

JOHNSGARD, P. A. 1981. The plovers, sandpipers, and snipes of the world. Univ. Nebraska Press, Lincoln.

JOHNSON, R. R., & J. L. DINSMORE. 1985. Brood-rearing and postbreeding habitat use by Virginia Rails and Soras. Wilson Bull. 97: 551-554.

JOHNSON, S. R., & W. J. ADAMS. 1977. The Little Gull (Larus minutus) in arctic North America. Can. Field-Natur. 91: 294-296.

JONES, I. 1982. Nesting of the Yellow Rail at Richmond Fen 1982. The Shrike 7: 3.

KADLEC, J. A., & W. H. DRURY. 1968. Structure of the New England Herring Gull population. Ecology 49: 644-676.

KANE, R., & P. A. BUCKLEY. 1975. The spring migration. Hudson - St. Lawrence region. Amer. Birds 29: 827-832.

KAUFMANN, G. W. 1983. Displays and vocalizations of the Sora and the Virginia Rail. Wilson Bull. 95: 42-59.

KELLEY, A. H. 1978. Birds of southeastern Michigan and southwestern Ontario. Cranbrook Inst. Sci., Bloomfield Hills.

KELLEY, A. H., D. S. MIDDLETON, & W. P. NICKELL. 1963. Birds of the Detroit-Windsor Area. Cranbrook Inst. Sci., Bloomfield Hills.

KEMSIES, E. 1968. Calcarius pictus (Swainson). Smith's Longspur. Pp. 1628-1635 in A.C. Bent. Life histories of North American cardinals, grosbeaks, buntings, towhees, finches, sparrows, and allies, part 3 (O.L. Austin, Jr., Ed.). Bull. U.S. Natl. Mus. No. 237.

KENDEIGH, S. C. 1947. Bird population studies in the coniferous forest biome during a spruce budworm outbreak. Ont. Dept. Lands For. Biol. Bull. No. 1.

KERBES, R. H. 1975. The nesting population of Lesser Snow Geese in the eastern Canadian arctic. Can. Wildl. Serv. Rep. Ser. No. 35.

KIBBE, D. P., & C. M. BOISE. 1984. The nesting season. Niagara-Champlain region. Amer. Birds 38: 1017-1019.

KIBBE, D. P., & C. M. BOISE. 1985. The Winter Season. Niagara-Champlain region. Amer. Birds 39: 164-166.

KILHAM, L. 1964. The relation of breeding Yellow-bellied Sapsuckers to wounded birches and other trees. Auk 81: 520-527.

KILHAM, L. 1971. Reproductive behavior of Yellow-bellied Sapsuckers. I. Preference for nesting in Fomes-infected aspens, and nest hole interrelations with flying squirrels, raccoons, and other animals. Wilson Bull. 83: 159-171.

KILHAM, L. 1983. Life history studies of woodpeckers of eastern North America. Publ. Nuttall Ornithol. Club No. 20.

KLIMSTRA, W. D. 1950. Notes on Bob-whites nesting behavior. Iowa Bird Life 20: 2-7.

KLUGH, A. B. 1905. The birds of Wellington County, Ontario. Ont. Nat. Sci. Bull. 1: 1-10.

KNAPTON, R. W. 1982. The Henslow's Sparrow (Ammodramus henslowii) in Canada: a status report. Committee on the Status of Endangered Wildlife in Canada, Ottawa.

KNAPTON, R. W. 1984. The Henslow's Sparrow in Ontario: a historical perspective. Ont. Birds 2: 70-74.

KNAPTON, R. W., & J. B. FALLS. 1982. Polymorphism in the White-throated Sparrow: habitat occupancy and nest-site selection. Can. J. Zool. 60: 452-459.

KNIGHT, O. W. 1904. Contributions to the life history of the Yellow Palm Warbler. J. Maine Ornithol. Soc. 6: 36-41.

KOONZ, W. H. 1987. Status update, the American White Pelican Pelicanus erythrorhynchos in Canada. Committee on the Status of Endangered Wildlife in Canada, Ottawa.

KOONZ, W. H., & P. W. RAKOWSKI. 1985. Status of colonial waterbirds nesting in southern Manitoba. Can. Field-Natur. 99: 19-29.

KUERZI, R. G. 1941. Life history studies of the Tree Swallow. Proc. Linn. Soc. N.Y. 52, 53: 1-52.

KUSHLAN, J. A. 1973. Least Bittern nesting colonially. Auk 90: 685-686.

LACK, D. 1968. Ecological adaptations for breeding in birds. Methuen & Co., London.

LAJEUNESSE, E. J. 1960. The Windsor border region. Univ. Toronto Press, Toronto.

LAMBERT, A. B., & R. B. H. SMITH. 1984a. The status of the Prairie Warbler (Dendroica discolor) in Canada. Committee on the Status of Endangered Wildlife in Canada, Ottawa.

LAMBERT, A. B., & R. B. H. SMITH. 1984b. The status and distribution of the Prairie Warbler in Ontario. Ont. Birds 2: 99-115.

LAMEY, J. 1981. Unusual records of birds for Ontario's Rainy River District. Ontario Bird Banding 14: 38-42.

LANYON, W. E. 1979. Hybrid sterility in meadowlarks. Nature 279: 557-558.

LAUGHLIN, S. B., & D. P. KIBBE. 1985. The atlas of breeding birds of Vermont. Univ. Press of New England, Hanover.

LAUGHLIN, S. B., D. P. KIBBE, & P. F. J. EAGLES. 1982. Atlassing the distribution of the breeding birds of North America. Amer. Birds 36: 6-19.

LAURENZI, A. W., B. W. ANDERSON, & R. D. OHMART. 1982. Wintering biology of Ruby-crowned Kinglets in the lower Colorado River valley. Condor 84: 385-398.

LAWRENCE, L. DE K. 1949. The Red Crossbill at Pimisi Bay, Ontario. Can. Field-Natur. 63: 1470-160.

LAWRENCE, L. DE K. 1953a. Notes on the nesting behavior of the Blackburnian Warbler. Wilson Bull. 65: 135-144.

LAWRENCE, L. DE K. 1953b. Nesting life and behavior of the Red-eyed Vireo. Can. Field-Natur. 67: 47-77.

LAYCOCK, G. 1984. The Belted Kingfisher. Bird Watcher's Digest 6: 17-23.

LECK, C. F., & F. L. CANTOR 1979. Seasonality, clutch size, and hatching success in the Cedar Waxwing. Auk 96: 196-198.

LEE, D. 1978. An annotated list of the birds of the Big Trout Lake area, Kenora District. Ont. Field Biol. 32: 17-36.

LEE, D., & J. M. SPEIRS. 1977. Breeding bird census at Big Trout Lake, 1975. Ont. Field Biol. 31: 48-54.

LEPTHIEN, L. W., & C. E. BOCK. 1976. Winter abundance patterns of North American kinglets. Wilson Bull. 88: 483-485.

LEWIS, H. F. 1928. Recent records of the European Starling. Can. Field-Natur. 42: 48.

LLOYD, A. C. 1934. A northern record for the Starling, Sturnus vulgaris. Can. Field-Natur. 48: 82.

LLOYD, H. 1923. The birds of Ottawa, 1923. Can. Field-Natur. 37: 101-105.

LOKEMOEN, J. T., & H. F. DUEBBERT. 1973. An upland nest of the Redhead far from water. Wilson Bull. 85: 468.

LONG, J. L. 1981. Introduced birds of the world. Universe Books, New York.

LORD, D. 1955. Occurrence of the Prairie Warbler at Georgian Bay, Ontario. Ont. Field Biol. 9: 23-24.

LOWTHER, J. K. 1961. Polymorphism in the White-throated Sparrow, Zonotrichia albicollis (Gmelin). Can. J. Zool. 39: 281-292.

LOWTHER, J. K. 1977. Nesting biology of the Sora at Vermilion, Alberta. Can. Field-Natur. 91: 63-67.

LOWTHER, J. K., & J. B. FALLS. 1968. Zonotrichia albicollis (Gmelin). White-throated Sparrow. Pp. 1364-1392 in A.C. Bent. Life histories of North American cardinals, grosbeaks, buntings, towhees, finches, sparrows, and allies, part 3 (O.L. Austin, Jr., Ed.). Bull. U.S. Natl. Mus. No. 237.

LUDWIG, J. P. 1965. Biology and structure of the Caspian Tern (Hydroprogne caspia) population of the Great Lakes from 1896-1964. Bird-Banding 36: 217-233.

LUDWIG, J. P. 1974. Recent changes in the Ring-billed Gull population and biology in the Laurentian Great Lakes. Auk 91: 575-594.

LUDWIG, J. P. 1985. Decline, resurgence and population dynamics of Michigan and Great Lakes Double-crested Cormorants. Jack-Pine Warbler 62: 91-102.

LUMSDEN, H. G. 1951. Breeding diving ducks on Lake St. Clair, Ontario. Can. Field-Natur. 65: 31-32

LUMSDEN, H. G. 1959. Mandt's Black Guillemot breeding on the Hudson Bay coast of Ontario. Can. Field-Natur. 73: 54-55.

LUMSDEN, H. G. 1963. Further records of the Ross' Goose in Ontario. Can. Field-Natur. 77: 174-175.

LUMSDEN, H. G. 1966. The Prairie-Chicken in southwestern Ontario. Can Field Natur. 80: 33-45.

LUMSDEN, H. G. 1971. The status of the Sandhill Crane in northern Ontario. Can. Field-Natur. 85: 285-293.

LUMSDEN, H. G. 1976. The Whistling Swan in James Bay and the southern region of Hudson Bay. Arctic 28: 194-200.

LUMSDEN, H. G. 1980. Starling nest sites and cleared land. J. Field Ornithol. 51: 178-179.

LUMSDEN, H. G. 1981. History of breeding Canada Geese (Branta canadensis) in southwestern Ontario. Ont. Field Biol. 35: 49-56

LUMSDEN, H. G. 1984a. The breeding status of Tundra Swans (Cygnus columbianus) in northern Ontario. Ont. Field Biol. 38: 1-4.

LUMSDEN, H. G. 1984b. The pre-settlement breeding distribution of Trumpeter, (Cygnus buccinator), and Tundra Swans (C. columbianus) in eastern Canada. Can. Field-Natur. 98: 415-424.

LUMSDEN, H. G. 1986. Choice of nest boxes by Tree Swallows, Tachycineta bicolor, House Wrens, Troglodytes aedon, Eastern Bluebirds, Sialia sialis, and European Starlings, Sturnus vulgaris. Can. Field-Natur. 100: 343-349.

LUNK, W. A. 1962. The Rough-winged Swallow, Stelgidopteryx ruficollis (Vieillot), a study based on its breeding biology in Michigan. Publ. Nuttall Ornithol. Club No. 4.

LYNCH, J. F., E. S. MORTON, & M. E. VAN DER VOORT. 1985. Habitat segregation between the sexes of wintering Hooded Warblers (Wilsonia citrina). Auk 102: 714-721.

MACARTHUR, R. H. 1958. Population ecology of some warblers of northeastern coniferous forests. Ecology 39: 599-619.

MACCLEMENT, W. T. 1915. The new Canadian bird book. Dominion, Toronto.

MACCRIMMON, H. R. 1977. Animals, man and change: alien and extinct wildlife of Ontario. McClelland and Stewart, Toronto.

MACFARLANE, F. 1891. Notes on and list of birds and eggs collected in arctic America, 1861-1866. Proc. U.S. Natl. Mus. 14: 413-446.

MACINS, V. 1977. White Pelican position paper. Unpublished report, Ont. Migratory Birds Tech. Comm.

MACLULICH, D. A. 1938. Birds of Algonquin Provincial Park, Ontario. R. Ont. Mus. Zool. Contrib. 13: 1-47.

MACOUN, J. 1900. Catalogue of Canadian birds, part 1. Geol. Surv. Can., Dept. Mines.

MACOUN, J., & J. M. MACOUN. 1909. Catalogue of Canadian birds. Geol. Surv. Can., Dept. Mines.

MACPHERSON, A. H. 1961. Observations on Canadian arctic *Larus* gulls, and on the taxonomy of *L. thayeri* Brooks. Arctic Inst. N. Amer. Tech. Paper No. 7.

MACQUEEN, P. M. 1950. Territory and song in the Least Flycatcher. Wilson Bull. 62: 194-205.

MAHER, W. J. 1974. Ecology of Pomarine, Parasitic, and Long-tailed Jaegers in northern Alaska. Pac. Coast Avifauna No. 37.

MANNING, T. H. 1952. Birds of the west James Bay and southern Hudson Bay coasts. Natl. Mus. Can. Bull. 125.

MANNING, T. H. 1976. Birds and mammals of the Belcher, Sleeper, Ottawa and King George Islands, Northwest Territories. Can. Wildl. Serv. Occas. Pap. 28: 1-40.

MANNING, T. H. 1981. Birds of the Twin Islands, James Bay, N.W.T., Canada. Syllogeus 30. Natl. Mus. Can., Ottawa.

MANSELL, W. 1983. The Forester Huntsville. Reprinted in Bird Watcher's Digest 6: 42.

MARKHAM, B. J. 1978. Status report on White Pelican *Pelicanus erythrorhynchos* in Canada. Committee on the Status of Endangered Wildlife in Canada, Ottawa.

MARTIN, N. D. 1960. An analysis of bird populations in relation to forest succession in Algonquin Provincial Park, Ontario. Ecology 41: 126-140.

MARTIN, S. G. 1974. Adaptations for polygynous breeding in the Bobolink, *Dolichonyx oryzivorus*. Amer. Zool. 14: 109-119.

MAYFIELD, H. 1960. The Kirtland's Warbler. Cranbrook Inst. Sci., Bloomfield Hills.

MCCRACKEN, J. D. 1981. Status report on the Prothonotary Warbler (*Protonotaria citrea*) in Canada. Committee on the Status of Endangered Wildlife in Canada, Ottawa.

MCCRACKEN, J. D. 1987. Annotated checklist to the birds of Haldimand-Norfolk. *In* The natural areas inventory of the Regional Municipality of Haldimand-Norfolk, vol. 2. Annotated checklists. Norfolk Field Natur., Simcoe.

MCCRACKEN, J. D., M. S. W. BRADSTREET, & G. L. HOLROYD. 1981. Breeding birds of Long Point, Lake Erie: a study in community succession. Can. Wildl. Serv. Rep. Ser. No. 44.

MCGAHAN, J. 1968. Ecology of the Golden Eagle. Auk 85: 1-12.

MCGILLIVRAY, R. W. 1963. The present status of the Sharp-tailed Grouse in the Kenora District. Unpublished report, Ont. Min. Nat. Resour.

MCILWRAITH, T. 1886. The birds of Ontario. Lawson, Hamilton.

MCILWRAITH, T. 1894. The birds of Ontario, 2nd ed. William Briggs, Toronto.

MCKEATING, G. 1985. Charles Broley: eagles then and now in southern Ontario. Pp. 25-34 *in* The Bald Eagle in Canada (J. M. Gerrard and T. M. Ingram, Eds.). White Horse Plains Publ., Headingley.

MCKEATING, G., & I. BOWMAN. 1977. The Ontario endangered and threatened species program. Ont. Fish Wildl. Rev. 16: 3-5.

MCLAREN, M. A. 1975. Breeding biology of the Boreal Chickadee. Wilson Bull. 87: 344-354.

MCLAREN, M. A. 1976. Vocalizations of the Boreal Chickadee. Auk 93: 451-463.

MCLAREN, M. A., & P. L. MCLAREN. 1981a. Relative abundances of birds in the boreal and subarctic habitats of northwestern Ontario and northeastern Manitoba. Can. Field-Natur. 95: 418-427.

MCLAREN, P. L. 1975. Habitat selection and resource utilization in four species of wood warblers. Ph.D. dissertation, Univ. Toronto, Toronto.

MCLAREN, P. L., & M. A. MCLAREN. 1981b. Bird observations in northwestern Ontario, 1976-77. Ont. Field Biol. 35: 1-6.

MCNICHOLL, M. K. 1969. The courtship behavior of Forster's Tern (*Sterna forsteri*) in Delta marsh. Univ. Man. Field Stn. (Delta Marsh) Ann. Rep. 4: 47-53.

MCNICHOLL, M. K. 1971. The breeding biology and ecology of Forster's Tern (*Sterna forsteri*) at Delta, Manitoba. M.Sc. thesis, Univ. Manitoba, Winnipeg.

MCNICHOLL, M. K. 1975. Larid site tenacity and group adherence in relation to habitat. Auk 92: 98-104.

MCNICOL, D. K., R. J. ROBERTSON, & P. J. WEATHERHEAD. 1982. Seasonal, habitat, and sex-specific food habits of Red-winged Blackbirds: implications for agriculture. Can. J. Zool. 60: 3282-3289.

MCRAE, R. D. 1984. First nesting of the Little Gull in Manitoba. Amer. Birds 38: 368-369.

MEANLEY, B. 1957. Notes on the courtship behaviour of the King Rail. Auk 74: 433-440.

MEANLEY, B. 1969. Natural history of the King Rail. U.S. Dept. Inter., Fish Wildl. Serv., N. Amer. Fauna No. 67.

MENDALL, H. L. 1958. The Ring-necked Duck in the northeast. Univ. Maine Press, Orono.

MESSINEO, D. J. 1985. The 1985 nesting of Pine Siskin, Red Crossbill, and White-winged Crossbill in Chenago County, N.Y. The Kingbird 35: 233-237.

METROPOLITAN TORONTO AND REGION CONSERVATION AUTHORITY. 1982. Environmentally significant areas study.

MIDDLETON, A. L. A. 1977a. The molt of the American Gold-finch. Condor 79: 440-444.

MIDDLETON, A. L. A. 1977b. Increase in overwintering by the American Goldfinch, *Carduelis tristis*, in Ontario. Can. Field-Natur. 91: 165-172.

MIDDLETON, A. L. A. 1979. Influence of age and habitat on reproduction by the American Goldfinch. Ecology 60: 418-432.

MILLER, A. H. 1931. Systematic revision and natural history of the American shrikes (*Lanius*). Univ. Calif. Publ. Zool. 38: 11-242.

MILLER, E. H. 1979. Functions of display flights by males of the Least Sandpiper, *Calidris minutilla* (Vieill.), on Sable Island, Nova Scotia. Can. J. Zool. 57: 876-893.

MILLER, R. S., & R. W. NERO. 1983. Hummingbird (*Archilochus colubris*) and sapsucker (*Sphyrapicus varius*) associations in northern climates. Can. J. Zool. 61: 1540-1546.

MILLS, A. 1981. A cottager's guide to the birds of Muskoka and Parry Sound. Alex Mills, Guelph.

MITCHELL, M. H. 1935. The Passenger Pigeon in Ontario. R. Ont. Mus. Zool. Contrib. 7: 1-181.

MOORE, G. A. 1954. Cerulean Warbler and Blue-gray Gnatcatch-er nesting in Bruce County, Ontario. Can Field-Natur. 68: 44.

MORDEN, J. A., & W. E. SAUNDERS. 1882. List of the birds of western Ontario. Can. Sportsman Natur. 2: 187, 192-194.

MORRIS, R. F., W. F. CHESHIRE, C. A. MILLER, & D. G. MOTT. 1958. The numerical response of avian and mammalian predators during a gradation of the spruce budworm. Ecology 39: 487-494.

MORRISON, R. I. G., & T. H. MANNING. 1976. First breeding records of Wilson's Phalarope for James Bay, Ontario. Auk 93: 656-657.

MORRISON, R. I. G., T. H. MANNING, & J. A. HAGAR. 1976. Breeding of the Marbled Godwit, *Limosa fedoa*, in James Bay. Can. Field-Natur. 90: 487-490.

MORRISON, R. I. G., & R. K. ROSS. 1987. Atlas of distribution of nearctic shorebirds on the coast of South America. Can. Wildl. Serv. Spec. Publ., Ottawa. In press.

MORSE, D. H. 1971. Effects of the arrival of a new species upon habitat utilization of two forest thrushes in Maine. Wilson Bull. 83: 57-65.

MORSE, D. H. 1976. Hostile encounters among spruce-woods warblers (*Dendroica*: Parulidae). Anim. Behav. 24: 764-771.

MORSE, T. E., J. L. JAKABOSKY, & V. P. MCCROW. 1969. Some aspects of the breeding biology of the Hooded Merganser. J. Wildl. Manage. 33: 596-604.

MOUSLEY, H. 1916. Five years personal notes and observations on the birds of Hatley, Stanstead County, Quebec, 1911-1915. Auk 33: 57-73, 168-186.

MOUSLEY, H. 1937. A study of a Virginia Rail and Sora Rail at their nests. Wilson Bull. 49: 80-84.

MOUSSEAU, P. 1984. Etablissement du Goéland à bec cerclé, *Larus delawarensis*, au Québec. Can. Field- Natur. 98: 29-37.

MUNDINGER, P. C. & S. HOPE. 1982. Expansion of the winter range of the House Finch: 1947-1979. Amer. Birds 36: 347-353.

MURRAY, B. G., JR. 1969. A comparative study of Le Conte's and Sharp-tailed Sparrows. Auk 86: 199-231.

MURRAY, B. G., JR. 1983. Notes on the breeding biology of Wilson's Phalarope. Wilson Bull. 95: 472-475.

NASH, C. W. 1908. Checklist of the birds of Ontario. Pp. 7-82 *in* Manual of the vertebrates of Ontario. Dept. Education, Toronto.

NASH, C. W. 1913. Birds of Ontario in relation to agriculture, 2nd ed. Ont. Dept. Agric. Bull. No. 218.

NASH, S. V., & J. A. DICK. 1981. First documentation of Greater Yellowlegs breeding in Ontario. Ont. Field Biol. 35: 48.

NATIONAL GEOGRAPHIC SOCIETY. 1983. Field guide to the birds of North America. National Geographic Society, Washington.

NERO, R. W. 1979. Status report on the Great Gray Owl, (*Strix nebulosa*), in Canada, 1979. Committee on the Status of Endangered Wildlife in Canada, Ottawa.

NERO, R. W. 1980. The Great Gray Owl, phantom of the northern forest. Smithsonian Inst. Press, Washington.

NESTLER, R. B. 1946. Vitamin A, vital factor in the survival of Bobwhites. Trans. N. Amer. Wildl. Conf. 11: 176-195.

NEWTON, I. 1972. Finches. Collins, London.

NICE, M. M. 1930a. A study of a nesting of Black-throated Blue Warblers. Auk 47: 338-345.

NICE, M. M. 1930b. Observations at a nest of Myrtle Warblers. Wilson Bull. 42: 60-61.

NICE, M. M. 1932. Habits of the Blackburnian Warbler in Pelham, Massachusetts. Auk 49: 92-93.

NICE, M. M. 1968. *Melospiza melodia euphonia* (Wetmore). Mississippi Song Sparrow. Pp. 1513-1523 *in* A.C. Bent. Life histories of North American cardinals, grosbeaks, buntings, towhees, finches, sparrows, and allies, part 3 (O.L. Austin, Jr., Ed.). Bull. U.S. Natl. Mus. No. 237.

NICHOLLS, T. H., & D. W. WARNER. 1972. Barred Owl habitat use as determined by radio-telemetry. J. Wildl. Manage. 36: 213-224.

NICHOLSON, J. C. 1972. The birds of Manitoulin Island - a species accounts summary. J. C. Nicholson, Sudbury.

NICHOLSON, J. C. 1974. The birds of the Sudbury district - a species accounts summary. J. C. Nicholson, Sudbury.

NICHOLSON, J. C. 1981. The birds of Manitoulin Island and adjacent islands within Manitoulin District, 2nd ed. J. C. Nicholson, Sudbury.

NICKELL, W. P. 1951. Studies of habitats, territory and nests of the Eastern Goldfinch. Auk 68: 447-470.

NIEMI, G. J., & J. M. HANOWSKI. 1984. Effects of a transmission line on bird populations in the Red Lake peatland, northern Minnesota. Auk 101: 487-498.

NOL, E., & A. LAMBERT. 1984. Comparison of Killdeers (Charadrius vociferus) breeding in mainland and peninsular sites in southern Ontario. Can. Field-Natur. 98: 7-11.

NOLAN, V., JR. 1968. Melospiza melodia melodia (Wilson). Eastern Song Sparrow. Pp. 1492-1512. in A.C. Bent. Life histories of North American cardinals, grosbeaks, buntings, towhees, finches, sparrows, and allies, part 3 (O.L. Austin, Jr., Ed.). Bull. U.S. Natl. Mus. No. 237.

NOLAN, V., JR. 1978. The ecology and behavior of the Prairie Warbler Dendroica discolor. Ornithol. Monogr. No. 26.

NORTH, C. A. 1973. Movement patterns of House Sparrows in Oklahoma. Ornithol. Monogr. 14: 79-81.

NOVAK, M. 1976. The beaver in Ontario. Ont. Min. Nat. Resour.

NUDDS, T. D. 1982. Ecological separation of grebes and coots: interference competition or microhabitat selection? Wilson Bull. 94: 505-514.

ODUM, E. P. 1941. Annual cycle of the Black-capped Chickadee - 1. Auk 58: 314-333.

OLENDORFF, R. R. 1976. The food habits of North American Golden Eagles. Amer. Midl. Natur. 95: 231-236.

OLSEN, A. R. 1959. Report on the status of Sharp-tailed Grouse, Kenora District. Unpublished report, Ont. Min. Nat. Resour.

OMAND, D. N. 1947. The cormorant in Ontario. Sylva 3: 18-23.

ONTARIO MINISTRY OF NATURAL RESOURCES. 1986. Guidelines for tree seed crop forecasting. Ont. Min. Nat. Resour., Toronto.

ONTARIO MINISTRY OF NATURAL RESOURCES, & CANADIAN WILDLIFE SERVICE. 1983. An evaluation system for wetlands of Ontario south of the Precambrian Shield. Ont. Min. Nat. Resour. (Wildl. Br.), and Can. Wildl. Serv.

ORING, L. W. 1964. Behavior and ecology of certain ducks during the postbreeding period. J. Wildl. Manage. 28: 223-233.

ORING, L. W. 1973. Solitary Sandpiper early reproductive behavior. Auk 90: 652-663.

ORING, L. W., & M. L. KNUDSON. 1972. Monogamy and polyandry in the Spotted Sandpiper. Living Bird 11: 59-73.

ORING, L. W., D. B. LANK, & S. J. MAXSON. 1983. Population studies of the polyandrous Spotted Sandpiper. Auk 100: 272-285.

OUELLET, H., & M. GOSSELIN. 1983. Les noms français des oiseaux d'Amérique du Nord. Syllogeus 43. Natl. Mus. Can., Ottawa.

OUELLET, H., S. J. O'DONNELL, & R. S. FOXALL. 1976. Gray Jay nesting in the Mer Bleue Bog, Ottawa, Ontario. Can. Field-Natur. 90: 5-10.

PALMER, R. S. (Ed.) 1962. Handbook of North American birds, vol. 1. Loons through flamingos. Yale Univ. Press, New Haven.

PALMER, R. S. 1968. Spinus pinus (Wilson). Pine Siskin. Pp. 424-447 in A.C. Bent. Life histories of North American cardinals, grosbeaks, buntings, towhees, finches, sparrows, and allies, part 1 (O.L. Austin, Jr., Ed.). Bull. U.S. Natl. Mus. No. 237.

PALMER, R. S. (Ed.) 1976a. Handbook of North American birds, vol. 2. Waterfowl, part 1. Whistling geese, sheld-ducks, dabbling ducks. Yale Univ. Press, New Haven.

PALMER, R. S. (Ed.) 1976b. Handbook of North American birds, vol. 3. Waterfowl, part 2. Eiders, wood ducks, diving ducks, mergansers, stifftails. Yale Univ. Press, New Haven.

PARKES, K. C. 1951. The genetics of the Golden-winged x Blue-winged Warbler complex. Wilson Bull. 63: 5-15.

PARMELEE, D. F. 1968. Plectrophenax nivalis nivalis (Linnaeus). Snow Bunting. Pp. 1652-1675 in A.C. Bent. Life histories of North American cardinals, grosbeaks, buntings, towhees, finches, sparrows, and allies, part 3 (O.L. Austin, Jr., Ed.). Bull. U.S. Natl. Mus. No. 237.

PARMELEE, D. F., H. A. STEPHENS, & R. H. SCHMIDT. 1967. The birds of southeastern Victoria Island and adjacent small islands. Natl. Mus. Can. Bull. No. 222.

PAYNE, R. B. 1982. Ecological consequences of song matching: breeding success and intraspecific song mimicry in Indigo Buntings. Ecology 63: 401-411.

PEAKALL, D. B. 1970. Pesticides and the reproduction of birds. Sci. Amer. 224: 72-78.

PEARSON, T. G. 1936. Birds of America. Garden City Books, Garden City.

PECK, G. K. 1966. First published breeding record of Mute Swan for Ontario. Ont. Field. Biol. 20: 43.

PECK, G. K. 1970. First Ontario nest records of Arctic Loon (Gavia arctica) and Snow Goose (Chen hyperborea). Ont. Field. Biol. 24: 25-28.

PECK, G. K. 1972. Birds of the Cape Henrietta Maria region, Ontario. Can. Field-Natur. 86: 333-348.

PECK, G. K. 1973. Nest of the Evening Grosbeak (Hesperiphona vespertina) in Ontario. Ont. Field Biol. 27: 38-40.

PECK, G. K. 1976. Recent revisions to the list of Ontario's breeding birds. Ont. Field. Biol. 30: 9-16.

PECK, G. K., & R. D. JAMES. 1983. Breeding birds of Ontario: nidiology and distribution, vol. 1. Non-passerines. Life Sci. Misc. Publ., R. Ont. Mus., Toronto.

PECK, G. K., & R. D. JAMES. 1987. Breeding birds of Ontario: nidiology and distribution, vol. 2. Passerines. Life Sci. Misc. Publ., R. Ont. Mus., Toronto.

PENAK, B. L. 1983. The status of the Cooper's Hawk (Accipiter cooperii) in Ontario, with an overview of the status in Canada. Committee on the Status of Endangered Wildlife in Canada, Ottawa.

PETERSON, J. M. C. 1985. Birds of Little Sachigo Lake and Thorne-Sachigo Rivers, Ontario. Ont. Birds 3: 87-99.

PETERSON, R. T. 1980. A field guide to the birds east of the Rockies, 4th ed. Houghton Mifflin, Boston.

PETTINGILL, O. S., JR. 1936. The American Woodcock *Philohela minor* (Gmelin). Mem. Boston Soc. Nat. Hist. 9: 167-391.

PHILIPP, P. B., & B. S. BOWDISH. 1917. Some summer birds of northern New Brunswick. Auk 34: 265-275.

PICKWELL, G. B. 1931. The Prairie Horned Lark. Trans. Acad. Sci. St. Louis 27: 1-153.

PICKWELL, G. B. 1942. *Otocoris alpestris praticola* Henshaw. Prairie Horned Lark. Pp. 342-356 *in* Life histories of North American flycatchers, larks, swallows, and allies (A. C. Bent, Ed.). Bull. U.S. Natl. Mus. No. 179.

PINKOWSKI, B. C. 1978. Michigan bird survey, summer 1978. Jack-Pine Warbler 56: 200-206.

PITELKA, F. A. 1940. Breeding behavior of the Black-throated Green Warbler. Wilson Bull. 52: 3-18.

PITELKA, F. A. 1959. Numbers, breeding schedule, and territoriality in Pectoral Sandpipers of northern Alaska. Condor 61: 233-264.

PITELKA, F. A., R. T. HOLMES, & S. F. MACLEAN, JR. 1974. Ecology and evolution of social organization in arctic sandpipers. Amer. Zool. 14: 185-204.

PIZZEY, G. 1980. A field guide to the birds of Australia. Princeton Univ. Press, Princeton.

POWELL, D. J. 1985. The spring migration. Western Great Lakes region. Amer. Birds 39: 302-305.

PREECE, W. H. A. 1923. *Cardinalis cardinalis cardinalis*, the Cardinal, taken near Sault Ste. Marie, Ontario. Can. Field-Natur. 37: 169.

PRESCOTT, D. R. C. In press. Territorial responses to song playback in allopatric and sympatric populations of Alder (*Empidonax alnorum*) and Willow (*E. traillii*) Flycatchers. Wilson Bull.

PREVETT, J. P., & F. C. JOHNSON. 1977. Continued eastern expansion of breeding range of Ross' Goose. Condor 79: 121-123.

PREVETT, J. P., & C. D. MACINNES. 1972. The number of Ross' Geese in central North America. Condor 74: 431-438.

PRICE I. M., & D. V. WESELOH. 1986. Increased numbers and productivity of Double-crested Cormorants, *Phalacrocorax auritus*, on Lake Ontario. Can. Field-Natur. 100: 474-482.

PUTNAM, L. S. 1949. The life history of the Cedar Waxwing. Wilson Bull. 61: 141-182.

QUILLIAM, H. R. 1965. History of the birds of Kingston, Ontario. Kingston Field Natur., Kingston.

QUILLIAM, H. R. 1973. History of the birds of Kingston, Ontario, 2nd ed. Kingston Field Natur., Kingston.

QUILLIAM, H. R. 1986. History of birds in the Kingston area relative to European settlement and habitat changes. Blue Bill 33: 57-62.

QUINNEY, T. E. 1986. Male and female parental care in Tree Swallows. Wilson Bull. 98: 147-150.

RAND, A. L. 1948. Variation in the Spruce Grouse in Canada. Auk 65: 33-40.

RAVELING, D. G., & H. G. LUMSDEN. 1977. Nesting ecology of Canada Geese in the Hudson Bay Lowlands of Ontario: evolution and population regulation. Ont. Min. Nat. Resour. Fish Wildl. Res. Rep. No. 98.

RAWLS, C. K., JR. 1949. An investigation of the life history of the White-winged Scoter (*Melanitta fusca deglandi*). M.Sc. thesis, Univ. Minnesota, Minneapolis.

REYNARD, G. B. 1974. Some vocalizations of the Black, Yellow, and Virginia Rails. Auk 91: 747-756.

REYNOLDS, J. D. 1987. Mating system and nesting biology of the Red-necked Phalarope: what constrains polyandry? Ibis 129: In press.

REYNOLDS, J. D., & R. W. KNAPTON. 1984. Nest-site selection and breeding biology of the Chipping Sparrow. Wilson Bull. 96: 488-493.

REYNOLDS, J. W. 1977. The earthworms (*Lumbricidae* and *Sparganophilidae*) of Ontario. Life Sci. Misc. Publ., R. Ont. Mus.

RICE, D. W. 1956. Dynamics of range expansion of Cattle Egrets in Florida. Auk 73: 259-266.

RICE, J. C. 1978a. Behavioral interactions of interspecifically territorial vireos. I. Song discrimination and natural interactions. Anim. Behav. 26: 527-549.

RICE, J. C. 1978b. Ecological relationships of two interspecifically territorial vireos. Ecology 59: 526-538.

RICHARDS, J. M. 1977. A summary of nesting records for Ruddy Duck (*Oxyura jamaicensis*) in Ontario, with particular reference to the Regional Municipality of Durham. Ont. Field. Biol. 31: 45-47.

RICHARDS, J. M., & G. K. PECK. 1968. Nesting of Brewer's Blackbird (*Euphagus cyanocephalus*) in Ontario and Durham Counties. Ont. Field. Biol. 22: 25-27.

RIGDEN, E., & H. LANG-RUNTZ. 1984. Peregrine Falcons nesting in Arnprior. Trail and Landscape 18: 22-24.

RILEY, J. L. 1982. Habitats of Sandhill Cranes in the southern Hudson Bay Lowland, Ontario. Can. Field-Natur. 96: 51-55.

RIPLEY, S. D. 1977. Rails of the world: a monograph of the family *Rallidae*. M. F. Feheley Publ. Ltd., Toronto.

RISLEY, C. J. 1981. The status of the Eastern Bluebird (*Sialia sialis*) in Canada with particular reference to Ontario. Committee on the Status of Endangered Wildlife in Canada, Ottawa.

RISLEY, C. J. 1982. The status of the Red-shouldered Hawk (*Buteo lineatus*) in Ontario, with an overview of the status in Canada. Committee on the Status of Endangered Wildlife in Canada, Ottawa.

ROBBINS, C. S., B. BRUUN, & H. S. ZIM. 1966. Birds of North America. Golden Press, New York.

ROBBINS, C. S., D. BYSTRAK, & P. H. GEISSLER. 1986. The breeding bird survey: its first fifteen years, 1965 - 1979. U.S. Dept. Int., Fish Wildl. Serv. Res. Publ. No. 157.

ROBBINS, C. S., & W. T. VAN VELZEN. 1974. Progress report on the North American Breeding Bird Survey. Acta Ornithol. 14: 170-191.

ROBERTS, T. S. 1936. The birds of Minnesota, vol. 1. Univ. Minnesota Press, Minneapolis.

ROSS, R. K., D. G. DENNIS, & G. BUTLER. 1984. Population trends of the five most common duck species breeding in southern Ontario, 1971-76. Pp. 22-26 in Waterfowl studies in Ontario, 1973-81 (S. G. Curtis, D. G. Dennis and H. Boyd, Eds.). Can. Wildl. Serv. Occas. Pap. No. 54.

ROUNDS, R. C., & H. L. MUNRO. 1982. A review of hybridization between *Sialia sialis* and *S. currucoides* Wilson Bull. 94: 219-223.

ROWAN, W. 1927. Notes on Alberta waders included on the British list. Part VI: Dowitcher and Spotted Sandpiper. Brit. Birds 20: 210-221.

ROWAN, W. 1929. Notes on Alberta waders included on the British list. Part VII. *Tringa flavipes* Yellowshank. Brit. Birds 23: 2-17.

ROWE, J. S. 1972. Forest regions of Canada. Can. For. Serv. Publ. 1300, Ottawa.

ROWNTREE, A. 1979. Lowdown on wetlands. Ont. Fish. Wildl. Rev. 18: 11-18.

RUPERT, D. L. 1974. The seventy-fourth Christmas Bird Count. Ontario region. Amer. Birds 28: 150-151.

RUPERT, D. L. 1984. The eighty-fourth Christmas Bird Count. Ontario region. Amer. Birds 38: 401-407.

RUSSELL, R. P., JR. 1983. The Piping Plover in the Great Lakes region. Amer. Birds 37: 951-955.

RYDER, J. P., P. L. RYDER, & B. TERMAAT. 1983. Newly discovered Ring-billed Gull colonies in Lake-of-the-Woods. Loon 55: 156-157.

SADLER, D. 1983. Our heritage of birds: Peterborough County in the Kawarthas. Orchid Press, Peterborough.

SALT, G. W. 1957. An analysis of avifaunas in the Teton Mountains and Jackson Hole, Wyoming. Condor 59: 373-393.

SAMUELS, E. A. 1880. Birds of New England and adjacent states. Lockwood, Brooks & Co., Boston.

SANDERS, C. J. 1970. Populations of breeding birds in the spruce-fir forests of northwestern Ontario. Can. Field-Natur. 84: 131-135.

SANDILANDS, A. P. 1984. Annotated checklist of the vascular plants and vertebrates of Luther Marsh, Ontario. Ont. Field Biol. Spec. Publ. No. 2.

SAS INSTITUTE. 1985. SAS User's Guide. SAS Inst., Cary.

SAUNDERS, W. E., & E. M. S. DALE. 1933. History and list of birds of Middlesex County, Ontario. Trans. R. Can. Inst. 19: 161-251.

SAWYER, M., & M. I. DYER. 1968. Yellow-headed Blackbird nesting in southern Ontario. Wilson Bull. 80: 236-237.

SCHARF, W. C., & G. W. SHUGART. 1984. Distribution and phenology of nesting Forster's Terns in eastern Lake Huron and Lake St. Clair. Wilson Bull. 96: 306-309.

SCHMUTZ, J. K., R. J. ROBERTSON, & F. COOKE. 1983. Colonial nesting of the Hudson Bay Eider Duck. Can. J. Zool. 61: 2424-2433.

SCHUELER, F. W., D. H. BALDWIN, & J. D. RISING. 1974. The status of birds at selected sites in northern Ontario. Can. Field-Natur. 88: 141-150.

SCOTT, G. A. 1963. First nesting of the Little Gull (*Larus minutus*) in Ontario and the New World. Auk 80: 548-549.

SCOTT, P. 1972. The swans. Michael Joseph, London.

SEALY, S. G. 1980. Breeding biology of Orchard Orioles in a new population in Manitoba. Can. Field-Natur. 94: 154-158.

SEALY, S. G., D. A. SEXTON, & K. M. COLLINS. 1980. Observations of a White-winged Crossbill invasion of southeastern Manitoba. Wilson Bull. 92: 114-116.

SEMPLE, J. B., & G. M. SUTTON. 1932. Nesting of Harris's Sparrow (*Zonotrichia querula*) at Churchill, Manitoba. Auk 49: 166-168.

SETON, E. E. T. 1884. Nest and habits of the Connecticut Warbler (*Oporornis agilis*) Auk 1: 192-193.

SHARP, M. J., & C. A. CAMPBELL. 1982. Breeding ecology and status of Red-shouldered Hawks (*Buteo lineatus*) in Waterloo Region. Ont. Field Biol. 36: 1-10.

SHARROCK, J. T. R. 1976. The atlas of the breeding birds in Britain and Ireland. T. and A. D. Poyser, Hertfordshire, England.

SHEPHERD, D. 1985. Whew - that was the spring that was! LPBO Newsletter 17: 8-16.

SHIGO, A. L., & L. KILHAM. 1968. Sapsuckers and *Fomes igniarius var. populinus*. U.S. For. Serv., N.E. For. Expt. Sta. Res. Note NE-84: 1-2.

SHIPMAN, C. N. 1927. Nesting of the Herring Gull and some other birds on Lake Erie islands. Auk 44: 242-243.

SILIEFF, E., & G. H. FINNEY. 1981. The co-operative breeding bird survey in Canada, 1980. Can. Wildl. Serv. Prog. Notes No. 122.

SIMKIN, D. W. 1963. A Surf Scoter nesting record for northwestern Ontario. Can. Field-Natur. 77: 60.

SIMKIN, D. W. 1968. Red-throated Loon nesting in northern Ontario. Can. Field-Natur. 82: 49.

SINCLAIR, V. L. 1978. Breeding biology of Wilson's Phalaropes (*Phalaropus tricolor* Vieillot) at South Marsh, James Bay, Ontario. B.Sc. thesis, Acadia Univ., Wolfville.

SKALES, H. 1906. Migration report of the Wellington Field Naturalists Club. Ont. Nat. Sci. Bull. 2: 20-23.

SKEEL, M. A. 1976. Nesting strategies and other aspects of the breeding biology of the Whimbrel (*Numenius phaeopus*) at Churchill, Manitoba. M.Sc. thesis, Univ. Toronto, Toronto.

SMITH, J. L. 1966. The Gray Jay in Saskatchewan. Blue Jay 24: 65-70.

SMITH, R. H. 1944. An investigation of the waterfowl resources of the west coast of James Bay. Unpublished report, Can. Wildl. Serv., Ottawa.

SMITH, R. L. 1968. *Ammodramus savannarum* (Gmelin). Grasshopper Sparrow. Pp. 725-745 *in* A.C. Bent. Life histories of North American cardinals, grosbeaks, buntings, towhees, finches, sparrows, and allies, part 2 (O.L. Austin, Jr., Ed.). Bull. U.S. Natl. Mus. No. 237.

SMITH, S. M. 1967. Seasonal changes in survival of the Black-capped Chickadee. Condor 69: 344-359.

SMITH, S. M. 1976. Ecological aspects of dominance hierarchies in Black-capped Chickadees. Auk 93: 95-107.

SMITH, W. J. 1957. Birds of the Clay Belt region of northern Ontario and Quebec. Can. Field-Natur. 71: 163-181.

SNYDER, L. L. 1928. The summer birds of Lake Nipigon. Trans. R. Can. Inst. 16: 251-277.

SNYDER, L. L. 1935. A study of the Sharp-tailed Grouse. Univ. Toronto Biol. Ser. No. 40.

SNYDER, L. L. 1938. The summer birds of western Rainy River District, Ontario, part 1. Trans. R. Can. Inst. 22: 121-153.

SNYDER, L. L. 1941. The birds of Prince Edward County, Ontario. Univ. Toronto Biol. Ser. 48: 25-92.

SNYDER, L. L. 1949a. On the distribution of the Golden Eagle in eastern Canada. Can. Field-Natur. 63: 39-41.

SNYDER, L. L. 1949b. The Snowy Owl migration of 1946-47. Wilson Bull. 61: 99-102.

SNYDER, L. L. 1951. Ontario birds. Clarke, Irwin & Co., Toronto.

SNYDER, L. L. 1953. Summer birds of western Ontario. Trans. R. Can. Inst. 30: 47-95.

SNYDER, L. L. 1957a. Arctic birds of Canada. Univ. Toronto Press, Toronto.

SNYDER, L. L. 1957b. Changes in the avifauna of Ontario. Pp. 26-42 *in* Changes in the fauna of Ontario (F. A. Urquhart, Ed.). Univ. Toronto Press, Toronto.

SNYDER, L. L., & E. B. S. LOGIER (Eds.). 1931. A faunal investigation of Long Point and vicinity, Norfolk County, Ontario. Trans. R. Can. Inst. 18: 117-236.

SNYDER, L. L., E. B. S. LOGIER, & T. B. KURATA. 1942. A faunal investigation of the Sault Ste. Marie region, Ontario. Trans. R. Can. Inst. 24: 99-165.

SNYDER, L. L., E. B. S. LOGIER, T. B. KURATA, F. A. URQUHART, & J. F. BRIMLEY. 1941. A faunal investigation of Prince Edward County, Ontario. R. Ont. Mus. Zool. Contrib. No. 19.

SOPER, J. D. 1923. The birds of Wellington and Waterloo Counties, Ontario. Auk 40: 489-513.

SOUTHERN, W. E. 1974. Seasonal distribution of Great Lakes region Ring-billed Gulls. Jack-Pine Warbler 52: 155-179.

SOUTHWICK, E. E., & A. K. SOUTHWICK. 1980. Energetics of feeding on tree sap by Ruby-throated Hummingbirds in Michigan. Amer. Midl. Natur. 104: 328-334.

SOWLS, L. K. 1955. Prairie ducks: a study of their behavior, ecology and management. Stackpole Books, Harrisburg.

SPEIRS, D. H. 1984. The first breeding record of Kirtland's Warbler in Ontario. Ont. Birds 2: 80-84.

SPEIRS, J. M. 1949. The relation of DDT spraying to the vertebrate life of the forest. Pp. 141-158 *in* Forest spraying and some effects of DDT. Ont. Dept. Lands For., Div. Res. Biol. Bull. 2: 141-158.

SPEIRS, J. M. 1954. Brewer's Blackbird nesting at Sault Ste. Marie, Ontario. Bull. Fed. Ont. Natur. 65: 29.

SPEIRS, J. M. 1959. Worth noting. Bull. Fed. Ont. Natur. 85: 22-31.

SPEIRS, J. M. 1977. Birds of Ontario County, part 5: Turkey Vulture to Northern Phalarope. Fed. Ont. Natur., Don Mills.

SPEIRS, J. M. 1985. Birds of Ontario, vol. 2. Natural Heritage, Toronto.

SPEIRS, J. M., & J. FRANK. 1970. Thirty-fourth breeding bird census. 4. Beech forest. Audubon Field Notes 24: 741-742.

SPEIRS, J. M., & D. H. SPEIRS. 1968. *Melospiza lincolnii lincolnii* (Audubon). Lincoln's Sparrow. Pp. 1434-1463 *in* A.C. Bent. Life Histories of North American cardinals, grosbeaks, buntings, towhees, finches, sparrows, and allies, part 3 (O.L. Austin, Jr., Ed.). Bull. U.S. Natl. Mus. No. 237.

SPRAGUE, R. T., & R. D. WEIR. 1984. The birds of Prince Edward County. Kingston Field Natur., Kingston.

STEIN, R. C. 1963. Isolating mechanisms between populations of Traill's Flycatchers. Proc. Amer. Phil. Soc. 107: 21-50.

STEPNEY, P. H. R. 1975. Wintering distribution of Brewer's Blackbird: historical aspect, recent changes, and fluctuations. Bird Banding 16: 106-125.

STEWART, R. E. 1953. A life history study of the Yellow-throat. Wilson Bull. 65: 99-115.

STIRRETT, G. M. 1945. The Kentucky Warbler in Canada. Can. Field-Natur. 59: 70.

STIRRETT, G. M. 1973. The springbirds of Point Pelee National Park of Ontario. Can. Dept. North. Affairs, Ottawa.

STOKES, A. W. 1950. Breeding behavior of the goldfinch. Wilson Bull. 62: 107-127.

STOKES, A. W. 1954. Population studies of the Ring-necked Pheasant on Pelee Island, Ontario. Ont. Dept. Lands For. Tech. Bull., Wildl. Ser. No. 4.

STOKES, D. W. 1979. A guide to the behavior of common birds. Little, Brown & Co., Boston.

STREET, J. F. 1923. On the nesting grounds of the Solitary Sandpiper and the Lesser Yellow-legs. Auk 40: 577-583.

STRICKLAND, D., R. TOZER., & R. J. RUTTER. 1980. Birds of Algonquin Park. Ont. Min. Nat. Resour.

STRICKLAND, R. D. 1968. Ecologie, nidification et comportement social du Geai Gris (*Perisoreus canadensis*) M.Sc. thesis, Univ. Montréal, Montréal.

STRONG, R. M. 1912. Some observations on the life history of the Red-breasted Merganser, *Mergus merganser* Linn. Auk 29: 479-488.

STULL, W. D. 1968. *Spizella passerina* (Bechstein). Eastern and Canadian Chipping Sparrows. Pp. 1166-1184. *in* A.C. Bent. Life histories of North American cardinals, grosbeaks, buntings, towhees, finches, sparrows, and allies, part 2 (O.L. Austin, Jr., Ed.). Bull. U.S. Natl. Mus. No. 237.

SUGDEN, L. G. 1977. Horned Grebe breeding habitat in Saskatchewan parklands. Can. Field-Natur. 91: 372-376.

SUGDEN, L. G. 1979. Habitat use by nesting American Coots in Saskatchewan parklands. Wilson Bull. 91: 599-607.

SUTHERLAND, C. A. 1963. Notes on the behavior of Common Nighthawks in Florida. Living Bird 2: 31-39.

SUTHERLAND, D. 1986. Magnolia Warbler breeding in the Regional Municipality of Halton, Ontario. Ont. Birds 4: 69-71.

SUTTON, G. M., & D. F. PARMELEE. 1954a. Nesting of the Greenland Wheatear on Baffin Island. Condor 56: 295-306.

SUTTON, G. M., & D. F. PARMELEE. 1954b. Survival problems of the Water-Pipit in Baffin Island. Arctic 7: 81-92.

SZUBA, K. J., & J. F. BENDELL. 1983. Population densities and habitats of Spruce Grouse in Ontario. Pp. 199-213 *in* Resources and dynamics of the boreal zone (R. W. Wein, R. R. Riewe and R. Methven, Eds.). Assoc. Can. Univ. North. Stud., Ottawa.

TATE, J., JR. 1981. The Blue List for 1981. Amer. Birds 35: 3-10.

TATE, J., JR. 1986. The Blue List for 1986. Amer. Birds 40: 227-236.

TATE, J., JR., & D. J. TATE. 1982. The Blue List for 1982. Amer. Birds 36: 126-135.

TAVERNER, P. A. 1919. Birds of eastern Canada. Geol. Surv. Can., Dept. Mines Mem. No. 104.

TAVERNER, P. A. 1921. The Evening Grosbeak in Canada. Can. Field-Natur. 35: 41-45.

TAVERNER, P. A. 1922a. Birds of eastern Canada, 2nd ed. Geol. Surv. Can., Dept. Mines Mem. 104., Geol. Surv. Can. Biol. Ser. 3.

TAVERNER, P. A. 1922b. The disappearance and recovery of the Eastern Bluebird. Can. Field-Natur. 36: 71-72.

TAVERNER, P. A. 1934. Birds of Canada. Can. Dept. Mines Misc. Bull. No. 72.

TAVERNER, P. A. 1939. Canadian land birds: a pocket field guide. Musson, Toronto.

TAVERNER, P. A. 1953. Birds of Canada. Musson Book Co., Toronto.

TAYLOR, P. 1985. Wings along the Winnipeg: the birds of the Pinawa-Lac du Bonnet region, Manitoba, rev. ed. Man. Nat. Soc. Eco. Ser. No. 2.

TEBBEL, P. D., & C. D. ANKNEY. 1982. Status of Sandhill Cranes (*Grus canadensis*) in central Ontario. Can. Field-Natur. 96: 163-166.

TEIXEIRA, R. M. 1979. Atlas van de Nederlandse Broedvogels. Stichting Ornithologisch Veldonderzoek Nederland, Amsterdam.

TERBORGH, J. W. 1980. The conservation status of Neotropical migrants: present and future. Pp. 21-30 *in* Migrant birds in the neotropics (A. Keast and E. S. Morton, Eds.). Smithsonian Inst. Press, Washington.

TERRES, J. K. 1980. The Audubon encyclopedia of North American birds. Alfred A. Knopf, New York.

TERRILL, L. M. 1968. *Passerella iliaca iliaca* (Merrem). Eastern Fox Sparrow. Pp. 1395-1415 *in* A.C. Bent. Life histories of North American cardinals, grosbeaks, buntings, towhees, finches, sparrows and allies, part 3 (O.L. Austin, Jr., Ed.). Bull. U.S. Natl. Mus. No. 237.

TESSEN, D. D. 1985. The autumn migration. Western Great Lakes region. Amer. Birds 39: 55-58.

THOMPSON, S. L. 1922. The birds of North Bay, Ontario, and vicinity in 1904. Can. Field-Natur. 36: 161-168.

THOMPSON, T. 1983. Birding in Ohio. Indiana Univ. Press, Bloomington.

TODD, W. E. C. 1963. Birds of the Labrador Peninsula. Univ. Toronto Press, Toronto.

TOMLINSON, R. E., & R. L. TODD. 1973. Distribution of two western Clapper Rail races as determined by responses to taped calls. Condor 75: 177-183.

TONER, G. C., W. E. EDWARDS, & M. W. CURTIS. 1942. Birds of Leeds County, Ontario. Can. Field-Natur. 56: 50-56.

TORDOFF, H. B., & W. R. DAWSON. 1965. Influence of day-length on reproductive timing in the Red Crossbill. Condor 67: 416-422.

TOZER, R. G., & J. M. RICHARDS. 1974. Birds of the Oshawa - Lake Scugog region, Ontario. R.G. Tozer & J.M. Richards, Oshawa.

TROY, D. M. 1985. A phenetic analysis of the redpolls *Carduelis flammea flammea* and *C. hornemanni exilipes* Auk 102: 82-96.

TUCK, L. M. 1968. Dowitcher breeding in Ontario. Ont. Field Biol. 21: 39.

TUCK, L. M. 1972. The snipes: a study of the genus *Capella*. Can. Wildl. Serv. Monogr. Ser. No. 5.

TUCKER, J. A. 1972. ABA's 5 most wanted birds. Birding 4: 189-190.

TUFTS, R. W. 1961. The birds of Nova Scotia. Prov. Mus. Nova Scotia, Halifax.

TUFTS, R. W. 1973. The birds of Nova Scotia, 2nd ed. Prov. Mus. Nova Scotia, Halifax.

TWOMEY, A. C. 1936. Climographic studies of certain introduced and migratory birds. Ecology 17: 122-132.

U.S. FEDERAL REGISTER. 1985. Part III, 11 December 1985.

U.S. FISH AND WILDLIFE SERVICE, & CANADIAN WILDLIFE SERVICE. 1985. Status of waterfowl and fall flight forecast. Special joint report of U.S. Fish Wildl. Serv. and Can. Wildl. Serv. No. 32.

U.S. FISH AND WILDLIFE SERVICE, & CANADIAN WILDLIFE SERVICE. 1986. 1986 status of waterfowl and fall flight forecasts. Special joint report of U.S. Fish Wildl. Serv. and Can. Wildl. Serv.

VAN BUSKIRK, J. 1985. Species-area relationship of birds on small islands at Isle Royale, Michigan. Wilson Bull. 97: 566-569.

VAURIE, C. 1959. The birds of the Palearctic fauna: a systematic reference. Passeriformes. Witherby, London.

VAURIE, C. 1965. The birds of the Palearctic fauna: a systematic reference. Non-Passeriformes. Witherby, London.

VERMEER, K. 1970. Breeding biology of California and Ring-billed Gulls: a study of ecological adaptation to the inland habitat. Can. Wildl. Serv. Rep. Ser. No. 12.

VERMEER, K., & L. RANKIN. 1984. Population trends in nesting Double-crested and Pelagic Cormorants in British Columbia. Murrelet 65: 1-9.

VERNER, J. 1965. Breeding biology of the Long-billed Marsh Wren. Condor 67: 6-30.

VERNER, J., & G. H. ENGELSEN. 1970. Territories, multiple nest-building, and polygyny in the Long-billed Marsh Wren. Auk 87: 557-567.

VICKERY, P. D. 1983. Sedge Wren. Pp. 351-352 in The Audubon Society master guide to birding, vol. 2. Gulls to dippers (J. Farrand, Ed.). Alfred A. Knopf, New York.

VOOUS, K. H. 1960. Atlas of European birds. T. Nelson & Sons Ltd., New York.

WALKINSHAW, L. H. 1937. The Virginia Rail in Michigan. Auk 54: 464-475.

WALKINSHAW, L. H. 1941. The Prothonotary Warbler, a comparison of nesting conditions in Tennessee and Michigan. Wilson Bull. 53: 3-21.

WALKINSHAW, L. H. 1966. Summer biology of Traill's Flycatcher. Wilson Bull. 78: 31-46.

WALKINSHAW, L. H. 1968. Passerherbulus caudacutus (Latham). Le Conte's Sparrow. Pp. 765-776 in A.C. Bent. Life histories of North American cardinals, grosbeaks, buntings, towhees, finches, sparrows, and allies, part 2. (O.L. Austin, Jr., Ed.). Bull. U.S. Natl. Mus. No. 237.

WALKINSHAW, L. H. 1983. Kirtland's Warbler - the natural history of an endangered species. Cranbrook Inst. Sci., Bloomfield Hills.

WALKINSHAW. L. H., & C. J. HENRY. 1957. Yellow-bellied Flycatcher nesting in Michigan. Auk 74: 293-304.

WALKINSHAW, L. H., & D. A. ZIMMERMAN. 1961. Range expansion of the Brewer's Blackbird in eastern North America. Condor 63: 162-177.

WALLEY, W. J. 1985. Breeding range extension of the Lark Sparrow into west-central Manitoba. Blue Jay 43: 18-24.

WALTON, C. 1986. Bluebird bungalow. Harrowsmith 11: 86-91.

WEATHERHEAD, P. J., S. TINKER, & H. GREENWOOD. 1982. Indirect assessment of avian damage to agriculture. J. Appl. Ecol. 19: 773-782.

WEEKES, F. 1974. A survey of Bald Eagle nesting attempts in southern Ontario, 1969-1973. Can. Field-Natur. 88: 415-416.

WEIR, R. D. 1970. Possible House Finch. Blue Bill 17: 60-61.

WEIR, R. D. 1982. The nesting season. Ontario region. Amer. Birds 36: 970-974.

WEIR, R. D. 1983. The nesting season. Ontario region. Amer. Birds 37: 982-985.

WEIR, R. D. 1984. The autumn migration. Ontario region. Amer. Birds 38: 195-199.

WEIR, R. D. 1985a. The autumn migration. Ontario region. Amer. Birds 39: 46-50.

WEIR, R. D. 1985b. The winter season. Ontario region. Amer. Birds 39: 161-164.

WEIR, R. D. 1985c. The nesting season. Ontario region. Amer. Birds 39: 905-909.

WEIR, R. D., & H. R. QUILLIAM. 1980. Supplement to history of the birds of Kingston, Ontario. Kingston Field Natur., Kingston.

WELLER, M. W., & C. S. SPATCHER. 1965. Role of habitat in the distribution and abundance of marsh birds. Iowa State Univ. Dept. Zool. Entomol. Spec. Rep. No. 43.

WELSH, D. A. 1971. Breeding and territoriality of the Palm Warbler in a Nova Scotia bog. Can. Field-Natur. 85: 31-37.

WELSH, D. A., & D. R. FILLMAN. 1980. The impact of forest cutting on boreal bird populations. Amer. Birds 34: 84-94.

WELTER, W. A. 1935. The natural history of the Long-billed Marsh Wren. Wilson Bull. 47: 3-34.

WESELOH, C., & H. BLOKPOEL. 1983. Precautions for a rare visitor. Seasons 23: 42-43.

WESELOH, D. V. 1984. Characteristics of a Greater Black-backed Gull colony on Lake Ontario, New York, 1981-1983. The Kingbird 34: 91-95.

WESELOH, D. V., P. MINEAU, S. M. TEEPLE, H. BLOKPOEL, & B. RATCLIFF. 1986. Colonial waterbirds nesting in Canadian Lake Huron in 1980. Can. Wildl. Serv. Prog. Notes No. 165.

WESELOH, D. V., S. M. TEEPLE, & H. BLOKPOEL. In press. The distribution and status of colonial waterbirds nesting in western Lake Erie. In Biogeography of the island region of western Lake Erie (J. F. Downhower, Ed.). Ninth Biosciences Colloquim of the College of Biological Science, Ohio State Univ., Columbus.

WESELOH, D. V., S. M. TEEPLE, & M. GILBERTSON. 1983. Double-crested Cormorants of the Great Lakes: egg laying parameters, reproductive failure and contaminant residues in eggs, Lake Huron 1972-73. Can. J. Zool. 61: 427-436.

WEST, J. D., & J. M. SPEIRS. 1959. The 1956-1957 invasion of Three-toed Woodpeckers. Wilson Bull. 71: 348-363.

WETHERBEE, D. K. 1968. Melospiza georgiana georgiana (Latham). Southern Swamp Sparrow. Pp. 1475-1490 in A.C. Bent. Life histories of North American cardinals, grosbeaks, buntings, towhees, finches, sparrows, and allies, part 3 (O.L. Austin, Jr., Ed.). Bull. U.S. Natl. Mus. No. 237.

WIENS, T. P., & F. J. CUTHBERT. 1984. Status and reproductive success of the Piping Plover in Lake-of-the-Woods. Loon 56: 106-109.

WILBUR, S. R. 1983. The status of vultures in the western hemisphere. Pp. 113-123 in Vulture biology and management (S. R. Wilbur and J. A. Jackson, Eds.). Univ. California Press, Los Angeles.

WILL, R. L. 1973. Breeding success, numbers, and movements of House Sparrows at McLeansboro, Illinois. Ornithol. Monogr. 14: 60-78.

WILLIAMS, G., & R. GLOVER. 1969. Andrew Graham's observations on Hudson Bay 1767-1791. Hudson Bay Record Soc. 27.

WILLIAMS, L. E., JR. 1981. The book of the Wild Turkey. Winchester Press, Oklahoma.

WILLIAMS, M. Y. 1942. Notes on the fauna of the Bruce Peninsula, Manitoulin, and adjacent islands. Can. Field-Natur. 56: 70-81.

WILLSON, M. F. 1966. Polygamy among Swamp Sparrows. Auk 83: 666.

WILSON, A. 1814. American ornithology; or, the natural history of the birds of the United States, vol. 9. Bradford and Inskeep, Philadelphia.

WOOD, D. M. 1955. Nesting of Brewer's Blackbird at Sault Ste. Marie, Ontario. Ont. Field Biol. 9: 23.

WOODFORD, J. 1962. The Tufted Titmouse "invades" Ontario. Fed. Ont. Natur. Bull. 95: 18-20.

WOODFORD, J., & J. LUNN. 1961. The nesting season. Ontario - western New York region. Audubon Field Notes 15: 464-466.

WOOLFENDEN, G. E. 1956. Comparative breeding behavior of Ammospiza caudacuta and A. maritima. Univ. Kansas Mus. Nat. Hist. Publ. 10: 45-75.

WOODS, R. S. 1968. Carpodacus mexicanus frontalis (Say). House Finch. Pp. 290-314 in A.C. Bent. Life histories of North American cardinals, grosbeaks, buntings, towhees, finches, sparrows, and allies, part 1 (O.L. Austin, Jr., Ed.). Bull. U.S. Natl. Mus. No. 237.

WORLD WILDLIFE FUND. 1986. Endangered species in Canada, 1986. World Wildlife Fund (Canada), Toronto.

WORMINGTON, A., & R. D. JAMES. 1984. Ontario Bird Records Committee, checklist of the birds of Ontario. Ont. Birds 2: 13-23.

ZELENY, L. 1976. The bluebird. Univ. Indiana Press, Bloomington.

ZELENY, L. 1977. Nesting box programs for bluebirds and other passerines. Pp. 55-60 in Endangered birds - management techniques for preserving threatened species (S. A. Temple, Ed.). Univ. Wisconsin Press, Madison.

ZIMMERMAN, J. L. 1966. Polygyny in the Dickcissel. Auk 83: 534-546.

ZIMMERMAN, J. L. 1971. The territory and its density dependent effect in Spiza americana. Auk 88: 591-612.

ZIMMERMAN, J. L. 1984. Nest predation and its relationship to habitat and nest density in Dickcissels. Condor 86: 68-72.

ZINK, R. M., & B. A. FALL. 1981. Breeding distribution, song and habitat of the Alder Flycatcher and Willow Flycatcher in Minnesota. Loon 53: 208-214.

Index of English, French and scientific names

W

X

Y

Z